SOLIDIFICATION

Engineering Sciences *Materials*

SOLIDIFICATION

J. A. Dantzig and M. Rappaz

EPFL Press
A Swiss academic publisher distributed by CRC Press

Taylor and Francis Group, LLC
6000 Broken Sound Parkway, NW, Suite 300,
Boca Raton, FL 33487

Distribution and Customer Service
orders@crcpress.com

www.crcpress.com

Library of Congress Cataloging-in-Publication Data
A catalog record for this book is available from the Library of Congress.

The authors and publisher express their thanks to the Ecole polytechnique fédérale de Lausanne (EPFL) for its generous support towards the publication of this book.

The series *Materials* is published under the editorial direction of Professor Michel Rappaz (EPFL).

Previous volumes:
Corrosion and Surface Chemistry of Metals
Dieter Landolt

Cover design by Sébastien Rappaz

is an imprint owned by Presses polytechniques et universitaires romandes, a Swiss academic publishing company whose main purpose is to publish the teaching and research works of the Ecole polytechnique fédérale de Lausanne.

Presses polytechniques et universitaires romandes, EPFL – Centre Midi,
Post office box 119, CH-1015 Lausanne, Switzerland
E-Mail: ppur@epfl.ch, Phone: 021 / 693 21 30, Fax: 021 / 693 40 27

www.epflpress.org

ISBN 978-2-940222-17-9 (EPFL Press)
ISBN 978-0-8493-8238-3 (CRC Press)

Printed in Italy

PREFACE

The modern science of solidification began in the 1940's, when engineers began to use analytical methods and models to describe solidification processes. In 1940, Chvorinov applied the analysis of heat flow to predict solidification patterns and defect in sand castings. In the 1950's Chalmers and co-workers analyzed the heat and solute balance at the moving solid-liquid interface to understand why planar interfaces become unstable during unidirectional growth. This body of work culminated in a seminal text, *Principles of Solidification*, written by Chalmers in 1964.

In the 1960's, Mullins and Sekerka put Chalmers' analysis on a firmer mathematical footing by performing a formal stability analysis. Later in the 1960's and in the 1970's, Flemings and co-workers developed models for segregation and other microstructural features by applying heat and solute balances at the scale of the microstructural features themselves. Flemings followed Chalmers' text with *Solidification Processing* in 1974, presenting this next generation of achievements.

The next decade saw a great deal of activity in the study of microstructure as a pattern selection problem through the competition between the transport of heat and solute and inherent length scales in the material owing to surface energy. This body of work was summarized in 1984 in *Fundamentals of Solidification*, by Kurz and Fisher.

Kurz and Fisher's book appeared just at the beginning of a revolution in modeling of solidification, when low-cost powerful computers became available. Computational approaches allowed more accurate and detailed models to be constructed, shedding light on many important phenomena. Today, industrial users regularly model the solidification of geometrically complex parts ranging from directionally solidified turbine blades to automotive engines. At the microscopic scale, computational models have been used to great effect to understand the pattern selection process in ways that were only hinted at using the analytical techniques available earlier. In the 1990's, methods were developed to combine these microscopic and macroscopic views of solidification processes.

Although there have been a few specialty texts written in the intervening time between Kurz and Fisher's book and the present, none of them provides a comprehensive presentation of the fundamentals, analytical models, and computational approaches. Also in the 1990's and 2000's, a short course on solidification was developed in collaboration with

Ecole des Mines de Nancy, EPFL and Calcom, that incorporated both the fundamental aspects described earlier and the developing computational techniques. Through teaching this course, and at our respective universities, we felt that there was a need for a new text, which led us to write the book you now hold in your hands. The subject is presented in three parts: *Fundamentals*, which provides the basics of thermodynamics, phase diagrams, and modeling techniques. *Microstructure* then uses these techniques to describe the evolution of the solid at the microscopic scale, from nucleation to dendrites, eutectics and peritectics to microsegregation. This section concludes with a chapter on coupling macro- and micro-models of solidification. The final part, *Defects*, uses the same principles to describe porosity, hot tearing and macrosegregation. We have striven to present this wide range of topics in a comprehensive way, and in particular to use consistent notation throughout.

Acknowledgment

This work represents the culmination of our education, training and practice over the last 25 years. We have had many mentors, colleagues and friends who have helped us along the way, too numerous to name them all. We would particularly like to express our gratitude to our mentors, Wilfried Kurz, Stephen Davis and Robert Pond, Sr. In addition to the authors mentioned above, we would like to acknowledge the very fruitful discussions we have had over the years with many esteemed colleagues: William Boettinger, Martin Glicksman, John Hunt, Alain Karma and Rohit Trivedi, to name just a few. Much of the structure of this text derived from the short course described above, and we would like to thank our colleagues and fellow teachers in that course, past and present, in particular Philippe Thévoz and Marco Gremaud who organize the course. Special thanks are due to Christoph Beckermann, Hervé Combeau, Arne Dahle and Mathis Plapp for their helpful comments and contributions to the manuscript, and to Sébastien Rappaz for the design of the cover. We also owe a special debt of gratitude to our students, postdocs and coworkers, whose contributions made this book possible. We would also like to thank our many colleagues and friends who graciously allowed us to use figures and movie sequences that appear throughout the book.

Most importantly, we would like to thank our families for their love and support.

Jonathan A. Dantzig
Michel Rappaz
Lausanne, 2009

CONTENTS

NOMENCLATURE AND DIMENSIONLESS GROUPS

Principal Nomenclature

A book that covers as many topics as this one does is bound to encounter some problems with nomenclature. We have tried to use standard notation wherever possible, and to be consistent in usage throughout. In order to help distinguish between dimensional quantities and their dimensionless counterparts after scaling, we use Roman alphabet symbols for dimensional variables and corresponding Greek letters for the dimensionless ones. For example, the dimensional coordinates $(x, y, z) \rightarrow (\xi, \eta, \zeta)$. For some variables, such as velocity $\boldsymbol{v}, v_i$, this scheme is not possible because there is no Greek counterpart. Further, we need symbols for both the vector and its components. We handle this by using italic symbols for dimensional quantities and Roman symbols for the dimensionless ones, e.g., $(\boldsymbol{v}, v_i) \rightarrow (\mathbf{v}, \mathrm{v}_i)$.

Subscripts and superscripts can be complicated as well. We use upper case Roman letters to designate components, lower case Greek to represent phases, s or ℓ to designate solid and liquid, respectively, and a superscript '$*$' to designate quantities evaluated at the solid-liquid interface. Whereas the '$*$' will always appear as a superscript, the other indices may appear as subscripts or superscripts, depending on what form provides the clearest description in the current context. As an example, the most complicated symbol used in the text is $C_{J\ell}^{*\alpha}$, which means the mass fraction of component J in the liquid, evaluated at the solid-liquid interface ahead of phase α. This symbol appears in the discussion of the solidification of eutectic and peritectic alloys. The most important symbols are given below.

Roman alphabet

A, B, ...	species (component) A
A, $A_{s\ell}$	area, solid-liquid interfacial area
$A_{f\ell}$, A_{fs}	surface area between foreign substrate and liquid, or foreign substrate and solid
A_C, A_R	growth constants for eutectics
$A(\boldsymbol{n})$	surface energy anisotropy function

$a_1, a_2, a_3, \ldots$	surface energy anisotropy coefficients
$a_{A\alpha}$	chemical activity of species A in phase α
B	ratio of solutal and thermal expansion coefficients ($= \beta_C/(m_\ell \beta_T)$)
$[\boldsymbol{B}^e]$	spatial derivatives of element shape functions
$\boldsymbol{b}$	vector of body force per unit mass; design vector
$[\boldsymbol{C}^e], [\boldsymbol{C}]$	element and global capacitance matrices in FEM
C	mass fraction of solute in a binary alloy
C_J	mass fraction of species J in a mixture
C_s^*, C_ℓ^*	mass fractions of solute in the solid and liquid phases of a binary alloy at the solid-liquid interface
$c_0, c_1, \ldots$	constants of integration
c_p, c_V	specific heat at constant pressure; at constant volume
D	chemical diffusivity of solute; diameter
d	diameter of a sphere
$\boldsymbol{D}$	rate-of-deformation tensor ($= (\nabla \boldsymbol{v} + (\nabla \boldsymbol{v})^T)/2$)
d_0, d_0^C	thermal capillary length; chemical capillary length
E	Young's modulus
E, E^m, e	total, molar and specific internal energy
$\dot{E}$	cumulative average deformation rate
$\boldsymbol{\mathcal{E}}, \mathcal{E}_{ijkl}$	elasticity tensor, indecial form
e_J^I	second-order interaction coefficient between solute element I and gas element J
$\mathcal{F}, F$	total and volumetric free energy in phase-field model
$\boldsymbol{F}$	deformation gradient tensor, $F_{ij} = \partial x_i / \partial X_j$
f_A, f_V	geometric factors for nucleation in a conical pit
$f_{J\ell}, f_{J\ell}^0$	activity coefficient for species J in an alloy and in a pure material
f_α	mass fraction of phase α
G	temperature gradient
G, G^m, g	total, molar and specific Gibbs free energy
G_C	composition gradient
g	acceleration due to gravity, 9.82 m s^{-2}
g_α	volume fraction of phase α
g_d, g_e, g_g	volume fraction of interdendritic liquid, extradendritic liquid and grain
g_s	volume fraction of solid
g_{se}	extended volume fraction of the solid phase
g_{si}	internal volume fraction of the solid phase in a grain
H, H^m, h	total, molar and specific enthalpy
h_T	heat transfer coefficient
$\boldsymbol{I}$	unit tensor (identity tensor); the ij component is δ_{ij}
I^{homo}, I^{heter}	homogeneous or heterogeneous nucleation rate
I_0^{homo}, I_0^{heter}	prefactors for homogeneous or heterogeneous nucleation rate
$\mathrm{Iv}_{2D}, \mathrm{Iv}_{3D}$	Ivantsov function in 2-D or 3-D
i	$\sqrt{-1}$
j_A	mass fraction flux for species A
J	Jacobian (det $\boldsymbol{F}$)
$[\boldsymbol{J}]$	element Jacobian for isoparametric FEM

$\boldsymbol{K}, K$	permeability tensor; value of isotropic permeability
$[\boldsymbol{K}^e], [\boldsymbol{K}]$	element and global conductance matrices in FEM
$\boldsymbol{k}, k$	thermal conductivity tensor; value of isotropic conductivity
k_T	thermal conductivity ratio ($= k_s/k_\ell$)
k_0, k_0^m	partition coefficient (mass); partition coefficient (molar)
k_B	Boltzmann's constant, 1.38×10^{-23} J K^{-1}
L, L_c	characteristic length
L_f	latent heat of fusion per unit mass
L_v	latent heat of vaporization per unit mass
M	mass of a system; morphological number
$\mathcal{M}_J$	molecular weight of species J
m_ℓ, m_s	slopes of the liquidus and solidus curve (mass fractions)
$[\boldsymbol{N}^e]$	element shape functions in FEM
N_0	Avagadro's number, 6.02×10^{23} atoms mol^{-1}
N_C	number of components
N_F	number of degrees of freedom for phase equilibria
N_I	total number of atoms of species I in a mixture
N_b	bond coordination number
N_g	number of grid points in a computational domain
N_ϕ	number of phases
n	number of moles
$\boldsymbol{n}, (n_x, n_y, n_z)$	unit vector normal to a surface and its Cartesian components
n, n_g	density of grains
n_{max}	maximum density of particles available
n_p	density of potent nucleant particles; density of pores
$\mathcal{O}(\cdot)$	order of magnitude
P	power input to a system; penalty parameter
$P(r; R_{tip})$	surface of a paraboloid of revolution (dimensional)
$\mathcal{P}(\varrho)$	surface of a paraboloid of revolution (dimensionless)
p, p_a	pressure, atmospheric pressure
p_c	probability of capture
$p(\phi)$	orientation distribution function
$\tilde{p}, p'$	intermediate pressure and pressure correction in SIMPLE algorithm
$\hat{p}$	modified pressure ($= p + \rho_0 g h$)
Q	heat input to a system
q_b	boundary heat flux
$\boldsymbol{q}$	heat flux vector
R	radius
$R_g; R_{g0}$	radius of grain; final grain radius
R_1, R_2	principal radii of curvature
$\mathcal{R}$	gas constant, 8.31 J mol^{-1} K^{-1}
$\mathfrak{R}$	dimensionless radius for a spherical solid particle
$\dot{R}_q$	specific heat generation rate
R_c	radius of a critical nucleus
R_p	pore radius
R_{tip}	tip radius of a paraboloid
$\{\boldsymbol{R}\}$	residual vector
r, θ, z	cylindrical coordinates

r, θ, ϕ	spherical coordinates
r_J^I	second-order interaction coefficient between solute element I and gas element J
S	bounding surface
S, S^m, s	total, molar and specific entropy
S_{mix}^m	molar entropy of mixing
$\mathcal{S}_V, \mathcal{S}_V^{s\ell}$	solid-liquid interfacial area per unit volume
$\mathcal{S}_V^{de}$	interfacial area per unit volume for inter-extradendritic liquid
$\mathcal{S}_V^{sd}$	interfacial area per unit volume for solid-interdendritic liquid
T	temperature
$\dot{T}$	cooling rate
T_0	boundary temperature; temperature where $G_s^m = G_\ell^m$
T^*	solid-liquid interface temperature
T_b	boundary temperature
T_{col}	temperature of a columnar front
T_{eut}	eutectic temperature
T_f	equilibrium melting temperature of pure material
T_{liq}	liquidus temperature
T_{per}	peritectic temperature
T_{ref}, T_0	reference temperature
T_{sol}	solidus temperature
T_v	vaporization temperature at atmospheric pressure
$\{\boldsymbol{T}^e\}, \{\boldsymbol{T}\}$	local and global vector of nodal temperatures
t, t_c	time, characteristic time
t_f	local solidification time
t_n	time of nucleation
$\boldsymbol{t}$	surface traction vector
$\boldsymbol{u}$	displacement vector
V, V^m, v	total, molar and specific volume
V_R	volume of representative volume element
V_s, V_ℓ	volume of solid and liquid phases in representative volume element
v	scalar velocity
v_g	velocity normal to the surface of a grain
$\boldsymbol{v}$, v_i	(dimensional) velocity vector and its ith component
$\mathbf{v}$, v_i	(dimensionless) velocity vector and its ith component
$\mathbf{v}_K$	velocity vector for species K
v_n	normal component of velocity of the solid-liquid interface
v_{sound}	speed of sound
v_T	isotherm velocity
W	width; total work done by external forces
W_0	phase-field interface width
X_I	molar composition of species I
$\boldsymbol{X}$	material coordinate vector
$\{\mathbf{X}\}, \{\mathbf{Y}\}, \{\mathbf{Z}\}$	element vectors of nodal coordinates in FEM
$\boldsymbol{x}$	position vector
x^*	interface position in 1-D problems
x, y, z	Cartesian coordinates; also x_1, x_2, x_3
$\hat{\mathbf{x}}, \hat{\mathbf{y}}, \hat{\mathbf{z}}$	unit vectors in Cartesian coordinates

Greek alphabet

α	thermal diffusivity ($= k/(\rho c_p)$)
α, β, γ	generic phases
α_T	linear thermal expansion coefficient ($= \beta_T/3$)
β	Solidification shrinkage ($= \rho_s/\rho_\ell - 1$)
β_T	volumetric thermal expansion coefficient ($= 3\alpha_T$)
β_C	volumetric solutal expansion coefficient
β_p	coefficient of compressibility
$\Gamma_{s\ell}$	Gibbs-Thomson coefficient ($= \gamma_{s\ell} T_f/(\rho_s L_f)$)
$\Gamma^{m*}_{s,\ell}$, $\mathbf{\Gamma}^{\sigma*}_{s,\ell}$, $\Gamma^{h*}_{s,\ell}$, $\Gamma^{C*}_{s,\ell}$	interfacial mass, momentum, energy or species term for solid or liquid
$\gamma_{f\ell}$	surface energy between foreign substrate and liquid
γ_{fs}	surface energy between foreign substrate and solid
γ_{gb}	grain boundary energy
$\gamma_{s\ell}, \gamma^0_{s\ell}$	surface energy between solid and liquid; value of isotropic surface energy
Δ	dimensionless undercooling $c_p \Delta T/L_f$ (Stefan number)
ΔC_0	difference in compositions across eutectic plateau
$\Delta G_n^{homo}, \Delta G_n^{hetero}$	free energy barrier for homogeneous or heterogeneous nucleation
ΔH^m_{mix}	molar enthalpy of mixing
ΔS^m_f	molar entropy difference between solid and liquid
Δs^J_f	specific entropy of fusion of species J ($= L^J_f/T^J_f$)
ΔT	total undercooling
ΔT_b	undercooling for bridging or coalescence
ΔT_c	characteristic temperature difference
ΔT_0	Equilibrium freezing range ($= T_{liq} - T_{sol}$)
ΔT_k	kinetic undercooling
ΔT_n	nucleation undercooling
ΔT_R	curvature undercooling
ΔT_C	solutal undercooling
ΔT_T	thermal undercooling
$\Delta x, \Delta y, \Delta z$	grid spacing in various coordinate directions
δ	dimensionless solidified layer thickness; boundary layer thickness
ε_{JK}	bond energy between atoms of J and K
ε	strain tensor
ε_{eq}	equivalent strain
$\varepsilon_4, \varepsilon_n$	4-fold, n-fold coefficient for the planar anisotropy of $\gamma_{s\ell}$
η	dimensionless $y-$coordinate; paraboloidal coordinate
ζ	dimensionless $z-$coordinate; fractional time step
θ	dimensionless temperature; angular coordinate; wetting angle
$\bar{\kappa}, \kappa_G$	mean and Gaussian curvature of a surface
Λ	ratio of eutectic spacing to extremum value ($= \lambda/\lambda_{ext}$)
λ	wavelength of instability; eutectic spacing
λ_1, λ_2	primary, secondary dendrite arm spacing
μ_ℓ	shear viscosity of a Newtonian fluid
$\mu_{J\alpha}$	chemical potential of species J in phase α

μ_k	kinetic attachment coefficient
ν_ℓ	kinematic viscosity ($= \mu_\ell/\rho_\ell$)
ν_0	atomic vibration frequency
ν_e	Poisson's ratio
ξ	dimensionless $x-$coordinate; parabolic coordinate
$\boldsymbol{\xi}$	Cahn-Hoffmann vector ($= \nabla(r\gamma_{s\ell}(\boldsymbol{n}))$)
π	$3.14159\ldots$
Π	dimensionless scaled pressure
ρ	density
ρ_0	density at reference temperature and pressure
ϱ	dimensionless radial coordinate
$\boldsymbol{\sigma}$	total stress tensor
$\hat{\boldsymbol{\sigma}}$	effective stress tensor ($= \boldsymbol{\sigma} + p\boldsymbol{I}$)
σ_{eq}	equivalent stress
σ^*	dendrite tip selection constant
σ_n	instability growth rate exponent for mode n
σ_y	yield stress
$\boldsymbol{\tau}$	extra stress tensor
τ	dimensionless time
τ_0	time scale factor in phase-field model
Υ	noise in phase-field equation
ϕ	constant used to describe interface position
ϕ_s, ϕ_ℓ	existence function for solid, liquid phase
χ_α	mole fraction of phase α
ψ	phase-field order parameter
Ψ	surface stiffness
Ω^m	regular solution parameter
Ω	supersaturation
$\boldsymbol{\omega}$	vorticity vector ($= \nabla \times \boldsymbol{v}$)

Subscripts, superscripts and indices

A^*	evaluated on the solid-liquid interface
A_C	composition
A_c	characteristic value
A_{col}	columnar zone
A^{el}	elastic deformation
A_{eut}	eutectic
A_ℓ	liquid phase
A_g	gas phase
A_k	attachment kinetics
A_I, A_J	species I, species J
A_{liq}	liquidus
A^m	amount per mole
A_n	component of vector A normal to the interface
A_p	pores
A^R	surface with radius of curvature R
A_s	solid phase
A_{sol}	solidus
$A_{s\ell}$	solid-liquid interface

A^{th}	thermal deformation
A^{tr}	transformational deformation
A^{vp}	viscoplastic deformation
$A_\alpha, A_\beta, \ldots$	quantity in phase α, β, ...
A_x, A_y, A_z	x, y, z components of a vector
A_0	nominal or reference value
A^∞	flat surface ($R \to \infty$)

Mathematical functions

Symbol	Meaning	Representation
$\mathrm{E}_1(u)$	exponential integral	$\int_u^\infty \frac{e^{-s}}{s} ds$
$\mathrm{erf}\,(u)$	error function	$\frac{2}{\sqrt{\pi}} \int_0^u e^{-s^2} ds$
$\mathrm{erfc}\,(u)$	complementary error function	$1 - \mathrm{erf}\,(u)$
$f(\theta)$	nucleation geometric factor	$\frac{(2+\cos\theta)(1-\cos\theta)^2}{4}$
$\mathcal{L}_n(x)$	Laguerre polynomial	$\frac{e^x}{n!}\frac{d^n}{dx^n}\left(e^{-x}x^n\right)$
$P_{nm}(x)$	associated Legendre polynomial	$\frac{(-1)^m}{2^n n!}\left(1-x^2\right)^{m/2}\frac{d^{n+m}}{dx^{n+m}}\left(x^2-1\right)^n$
Q_4	first cubic harmonic function	$n_x^4 + n_y^4 + n_z^4$
S_4	second cubic harmonic function	$n_x^2 n_y^2 n_z^2$
$Y_{nm}(\theta,\phi)$	spherical harmonic function	$\sqrt{\frac{(2n+1)(n-m)!}{4\pi(n+m)!}} e^{-im\phi} P_{nm}(\cos\theta)$
$\delta(x)$	Dirac δ-function	$\delta(x) = \begin{cases} +\infty & x = 0 \\ 0 & x \neq 0 \end{cases}$ $\int_{-\infty}^{\infty} \delta(x)dx = 1$
δ_{ij}	unit tensor (Kronecker delta)	$\delta_{ij} = \begin{cases} 1 & i = j \\ 0 & i \neq j \end{cases}$
ε_{ijk}	permutation tensor	$\begin{cases} 1 & i,j,k \text{ even permutations} \\ -1 & i,j,k \text{ odd permutations} \\ 0 & \text{otherwise} \end{cases}$

Mathematical operators

Symbol	Meaning	Representation
$\boldsymbol{A} \cdot \boldsymbol{B}$	dot product of two vectors	$a_i b_i$
$\boldsymbol{A}^T$	transpose of a second rank tensor	a_{ji}
$\boldsymbol{A} : \boldsymbol{B}$	scalar product of second rank tensors	$a_{ij} b_{ji}$

$D\psi/Dt$	material derivative of ψ	$\frac{\partial \psi}{\partial t} + (\boldsymbol{v} \cdot \nabla)\psi$
$\text{tr}\boldsymbol{A}$	trace of a second rank tensor	a_{ii}
∇A	gradient of a scalar	$\frac{\partial A}{\partial x_i}$
$\nabla \cdot \boldsymbol{A}$	divergence of $\boldsymbol{A}$	$\frac{\partial a_i}{\partial x_i}$
$\nabla \times \boldsymbol{A}$	curl of a vector	$\varepsilon_{ijk} \frac{\partial a_j}{\partial x_k}$
$\nabla^2 A$	Laplacian of A	$\frac{\partial^2 A}{\partial x_i \partial x_i}$
$\Vert\boldsymbol{A}\Vert$	L_2 norm of a vector	$\sqrt{a_i a_i}$
$\langle A \rangle$	volume average of A	$\frac{1}{V_R} \int_{V_R} A \, dV$
$\langle A_{s,\ell} \rangle$	phase average of A_s or A_ℓ	$\frac{1}{V_R} \int_{V_R} \phi_{s,\ell} A \, dV$
$\langle A \rangle_{s,\ell}$	intrinsic average of A_s or A_ℓ	$\frac{1}{V_{s,\ell}} \int_{V_R} \phi_{s,\ell} A \, dV$
$\langle A^*_{s,\ell} \boldsymbol{n} \rangle^*$	interfacial average of A_s or A_ℓ	$\frac{1}{A_{s\ell}} \int_{A_{s\ell}} A^*_{s,\ell} \boldsymbol{n} dA$
$\langle C \rangle_M$	mass average composition	$\int_0^{f_s} C_s df_s + \int_0^{f_\ell} C_\ell df_\ell$

Classical dimensionless numbers

Name	Expression	Physical Meaning
Biot	$\text{Bi} = \frac{h_T L_c}{k}$	ratio of heat advection from a surface to heat conduction inside
Boussinesq	$\text{Bo} = \frac{g \beta_T \Delta T_c L_c^3}{\alpha_0^2}$	ratio of heat advected by buoyancy to conducted heat
Fourier	$\text{Fo} = \frac{\alpha t_c}{L_c^2}$	ratio of characteristic time t_c to the time for conduction L_c^2/α
Grashof	$\text{Gr} = \frac{g \beta_T \Delta T_c L_c^3}{(\mu_\ell / \rho_{\ell 0})^2}$	ratio of buoyant advective flow to viscosity
Lewis	$\text{Le} = \frac{\alpha}{D}$	ratio of thermal diffusion to mass diffusion
Péclet	$\text{Pe} = \frac{v_c L_c}{\alpha}$	ratio of heat advection to heat conduction
Péclet (solutal)	$\text{Pe}_C = \frac{v_c L_c}{D}$	ratio of solute advection to solute diffusion
Prandtl	$\text{Pr} = \frac{c_p \mu_\ell}{k} = \frac{\nu_\ell}{\alpha}$	ratio of momentum and thermal diffusivities in a fluid

Rayleigh	$\mathrm{Ra} = \dfrac{\rho_0 g \beta_T \Delta T_c L_c^3}{\mu_\ell \alpha_0}$	ratio of buoyant advection to the product of viscosity and heat conduction
Reynolds	$\mathrm{Re} = \dfrac{\rho v_c L_c}{\mu_\ell} = \dfrac{v_c L_c}{\nu_\ell}$	ratio of inertia to viscosity
Schmidt	$\mathrm{Sc} = \dfrac{\mu_\ell}{\rho_\ell D_\ell} = \dfrac{\nu_\ell}{D_\ell}$	ratio of momentum diffusivity to mass diffusivity
Stefan	$\mathrm{Ste} = \dfrac{c_p \Delta T}{L_f}$	ratio of sensible heat to latent heat

CHAPTER 1

OVERVIEW

1.1 INTRODUCTION

Solidification processes are familiar to all of us, whether they concern the formation of frost on windows or ice in trays, the freezing of solders in electronic circuits, or the casting of aluminum and steel in industrial practice. Solidification has long represented a major force in human development, and some of the "Ages" of man have even been classified by the alloys that the inhabitants were able to melt and cast. During the Bronze Age, ca. 4000 BC - 1200 BC, copper-based weapons and other artifacts of daily life were common throughout Europe and Asia. Examples are shown in Fig. 1.1(a). However, once it became possible to melt and alloy iron, ca. 1200 BC, this metal quickly replaced bronze for weapons and other applications because of its superior properties. Figure 1.1(b) shows an Iron Age axe. Several variants of steel, the most famous of which is the legendary Damascus steel, were produced in antiquity by mechanical means.

The invention of the Bessemer process in 1858 led to the mass production of steel in liquid form, which was then cast into shapes and ingots for wrought processing. This was one of the key inventions of the industrial revolution, and provided the foundation for transportation by rail, and later by automobile. Similarly, the Hall-Héroult process for producing aluminum, invented in 1886, enabled the mass production of aluminum cast products, which in turn gave rise the aircraft industry in the following century.

The ability to produce these metals in liquid form made it possible to easily manufacture alloys of controlled composition, which could then be cast into either final products or into ingots that, in turn, would be deformed in the solid state into plates, sheets, billets, and other wrought products. The solidification process marked the stage of production where the composition and structure were set for all future processing. Through the first half of the 20th century, metallurgists developed an understanding of how the properties of cast products were related to the conditions extant during the solidification process.

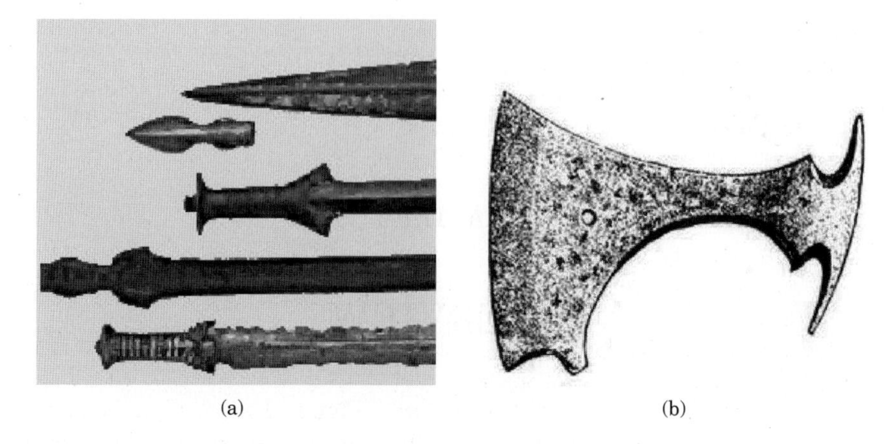

(a) (b)

Fig. 1.1 (a) Bronze age weapons (Reproduced with permission from images. encarta.msn.com); (b) Iron age axe (Photograph taken by Glenn McKechnie, April 2005).

One could argue that the art and practice of solidification entered the realms of engineering and science with the publication of Chalmers' landmark text *Principles of Solidification* in 1964 [3], which presented some of the basic models for solute partitioning during the freezing of alloys, and helped to explain how microstructural patterns such as dendrites evolve during planar or spherical growth. Ten years later, Flemings' *Solidification Processing* [4] extended this modeling approach to develop models for the evolution of measurable microstructural features, such as dendrite arm spacing and segregation patterns. These models began to quantify the effect of processing parameters such as the cooling rate and the temperature gradient, as well as their interaction with alloy properties such as the freezing range and the underlying phase diagram on the final structure. Over the next decade, many important advances were made in the understanding of pattern formation in solidification microstructures, in particular regarding length scales in dendritic growth. Largely as a result of these advances, Kurz and Fisher published *Fundamentals of Solidification*, which focused in greater detail on the evolution of microstructure [6].

The present book is intended to be the next entry in this line. The time since the publication of Kurz and Fisher's text has seen the advent of large scale computation as a tool for studying solidification. This has allowed significant advances to be made in both theory and application. The development of phase-field methods has permitted a further understanding of the evolution of complex microstructures, and the availability of inexpensive large-scale computers and commercial software packages now allows process engineers to perform realistic simulations of macroscopic heat transfer, solute transport and fluid flow in realistic geometries. The development of volume averaging methods and the statistical representation of microstructures provide a bridge between the microscopic and macroscopic scales. Our objective in this book is to place the models described in these earlier texts, as well as more recently developed ones, in a context that begins with fundamental concepts and culminates in analytical and/or numerical implementations for practical applications.

The unifying theme for our approach is the classification of various phenomena over a range of length and time scales, as illustrated in Fig. 1.2. At the macroscopic length scale, geometry and processing conditions determine the progress of solidification at various locations. The figure shows as an example a six-cylinder engine block, roughly one meter in length, that freezes over a time period as long as 30 minutes. We also show the temperature distribution at a particular time, computed with a finite element simulation of the heat transfer process.

By focusing on smaller length scales, roughly between 1 μm and 1 mm, we are able to observe the *microstructure*. Figure 1.2 shows an array of dendrites, having grown from an initially planar interface into the liquid phase. The dendrites are visible because of a chemical segregation that occurs on the microscopic scale. When solidification takes place in the presence of fluid flow, solute can be advected to distances much larger than the local microscopic scale, thereby causing *macrosegregation*.

Certain processes occurring at the atomic to nanometer length scale are also important for solidification. The properties of the solid-liquid interface, and the manner in which atoms attach to it, can affect the growth patterns. For example, the anisotropy of the surface energy determines to a large extent the morphology of the dendrite patterns observed at the microscopic scale.

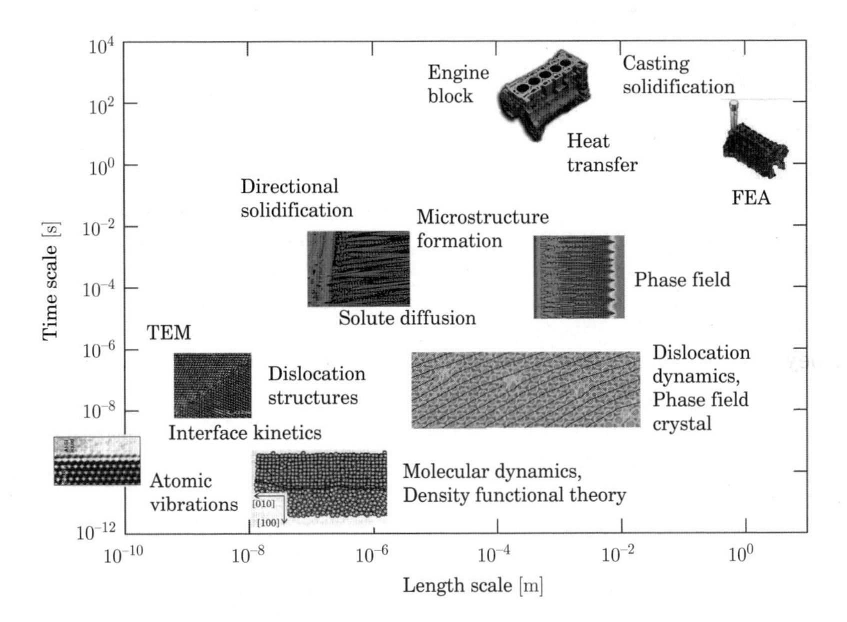

Fig. 1.2 A schematic illustration of the various phenomena associated with solidification at various length and time scales.

1.1.1 Organization of the text

We begin our book with the present overview chapter, which concludes with a section describing various solidification processes. These are presented in sufficient detail for the reader to understand the important aspects of each process, and in order to motivate the remainder of the book. A deeper treatment is left to others.

The rest of the book is divided into three parts: *Fundamentals*, *Microstructure* and *Defects*. Fundamentals, as its name implies, provides the basic principles needed to study solidification. We begin with Chap. 2, *Thermodynamics*, in which the concepts of equilibria for condensed phases, including models for the free energy and chemical potential, are developed. We also introduce departures from equilibrium, including the effect of curvature, and kinetics. These concepts are then used in Chap. 3, *Phase Diagrams*, to motivate and study equilibrium phase diagrams in binary and ternary systems as the result of chemical equilibria between phases.

Equilibrium implies that the processes take place over sufficient time for there to be no spatial variations in temperature, composition, pressure, etc. Thus, there exists no effective time or length scale in equilibrium. The finite time associated with real processes, however, leads to spatial variations over length scales that affect the properties of the solidified part. The microstructure shown in the previous section are prime examples. The governing equations for transport of mass, momentum, energy and species are developed in Chap. 4 for this purpose. In particular, we derive governing equations for single phases, and for the interface between phases. We also develop volume-averaged forms appropriate for control volumes that contain a mixture of solid and liquid phases. This latter formulation is extremely useful in subsequent chapters for developing mesoscale models that bridge the microscopic and macroscopic length scales.

The focus of this book is on solidification processes, which are characterized by a moving boundary between the solid and liquid phases. This represents a modeling challenge, as one must apply boundary conditions on a phase boundary whose position is *a priori* unknown. Chapter 5 is devoted to the study of the class of such problems that have an analytical solution. These problems are very useful in identifying the important physical phenomena that control solidification processes. In order to go beyond these model problems, which are mostly one-dimensional and also require constant material properties and simple boundary conditions, numerical methods are required. These are developed in Chap. 6.

With these fundamentals in hand, we proceed to Part II, the study of microstructure. Continuing the theme of organizing by length scale, we begin with Chap. 7, *Nucleation*, in which we explore how the first solid forms from the melt as it cools. Thermodynamics play a crucial role, and we see that the sub-microscopic length scale for nucleation is set by the balance between surface energy associated with the solid-liquid interface and free energy associated with the bulk phases. As such nuclei grow to microscopic size, they begin to express their underlying crystallography.

The most common morphology is that of the dendrite, studied in detail in Chap. 8. The length scale associated with dendrites is very important for the use of cast products since it strongly influences their properties, as well as the formation of defects. We therefore expend some effort to investigate how the length scales are affected by processing. We present the analyses in some detail so that the interested reader can appreciate the underlying theory, but we also emphasize the key results that are needed in subsequent chapters. In Chap. 9, we study eutectic and peritectic alloy systems, in which the solidification involves the melt and two distinct solid phases. We focus in particular on the evolution of length scales in the microstructure of these alloys as it is affected by processing conditions.

These chapters identify processes that lead to chemical segregation in the final product. The amount and extent of this segregation affects the material properties, and is important to understand for subsequent heat treatment (e.g., homogenization and precipitation). Models for this phenomenon are presented in Chap. 10, and constitute direct applications of the principles developed in Chaps. 5 and 6 at the microscopic scale. We conclude this part of the book with Chap. 11, Macro-micro modeling, which shows how to integrate the models for microstructure developed in Chaps. 7-10 with the macroscale modeling methods developed in particular in Chap. 6. This provides a powerful tool for the analysis of real casting processes.

Finally, in Part III we present a detailed analysis of some of the most common defects found in solidified parts. Chapter 12, *Porosity*, describes how solidification shrinkage and the evolution of dissolved gases lead to porosity in the final product. Almost all materials increase their density upon freezing, with water and semiconductors such as silicon and germanium being well-known exceptions. As solidification proceeds, liquid flow is needed to compensate for the volume change upon freezing. This flow is opposed by viscous forces exerted by the microstructure, and if the impediment is large enough, pores can develop. If, in addition to the viscous effects, there are superimposed strains, e.g., from thermal contraction of the solid, then hot tears can form; a phenomenon explored in Chap. 13. Finally, in Chap. 14, *Macrosegregation*, we demonstrate how relative movement of the solid and liquid can lead to segregation on the macroscopic scale.

1.2 SOLIDIFICATION PROCESSES

1.2.1 Shape casting

Casting is the most cost-effective way for manufacturing parts of complex shape. Applications range from mass-produced automotive parts (blocks, cylinder heads, suspension and brake components, etc.), to individual products such as jewelry and statuary. All of these processes have in common a mold with a cavity corresponding to a “negative” of the final product,

Table 1.1 Process characteristics for shape casting.

Process	Mold material	Cavity	Core materials
Foundry casting	Bonded sand	Wood, metal pattern	Baked sand
Investment casting	Fired ceramic	Wax, polymer	Leachable ceramic
Permanent mold and die casting	Tool steel, Copper, Graphite	Machined cavity	Metal

which is initially filled by liquid, after which solidification takes place by heat extraction through the mold. The processes differ mainly in the mold material and how the cavity is formed, as listed in Table 1.1. Each of these processes is described in more detail in the following sections.

Foundry casting

Foundry casting, sometimes called sand casting, is one of the most common processes for mass production of parts with complex shapes. A re-usable pattern is made from wood, metal, or other suitable materials. The pattern has the shape of the intended part, augmented in several ways to accommodate the solidification process. For example, the part dimensions are increased by a "shrink factor" that compensates for the volume change (typically a few percent) associated with thermal contraction during cooling from the solidification temperature to room temperature. The pattern may also have to differ from the desired final product in order to allow it to be easily removed from the mold before casting, as well as the addition of risers to compensate for solidification shrinkage and gating to conduct the liquid metal into the mold.

The mold, illustrated in Fig. 1.3, is formed in a *flask*, which generally consists of two parts: the lower *drag* and the upper *cope*. In a typical hand-molding operation, the drag part of the pattern is placed on a flat surface, and the drag case is inverted over it. A sand mixture is then poured over the pattern and compacted until it has sufficient strength so as to hold together after the pattern is removed. A typical composition (by weight) for the molding sand is 96 parts silica sand, 4 parts bentonite (clay) and 4 parts water. The particle size is controlled by sieving the mixture. After formation of the drag, it is inverted and the cope is fitted to it. The cope part of the pattern is then connected to the drag part. A parting compound, e.g., ground bone, is sprinkled onto the surface and the cope is filled with sand and compacted. The two mold halves are subsequently separated and the pattern is removed, thus forming the mold cavity. The cope may be formed separately after which the two halves are assembled. This is the more typical approach in automated processes. After the pattern has been removed, *cores* can be placed in the mold cavity to produce passages inside the final cast product. One can usually discern the *parting line*,

(a)

(b)

Fig. 1.3 (a) A sand mold for the manufacturing of simple cast parts. Cores have been placed in the drag to create internal cavities in the cast product. Notice the hole in the center of the cope corresponding to the down-sprue, and the runners from the central sprue to the individual castings. (b) Bronze and Al castings obtained with this mold. (Source: en.wikipedia.org/wiki/Sand_casting [14].)

corresponding to the location of the joint between the cope and drag on the surface of the casting.

In addition to the part pattern, the mold must also have the "plumbing" to allow metal to fill the mold cavity. As illustrated in Fig. 1.3, this consists of a basin to absorb the impact of pouring, a down-sprue to conduct metal to the level of the mold cavity, and a set of runners and in-gates to feed the metal into the cavity. The main objectives of this *feeding system* is to introduce the liquid into the cavity with as little turbulence as possible, so as to avoid incorporating surface oxides and other undesirable impurities into the casting. The sand mold is porous, allowing air to escape readily during filling. The runner system may also incorporate a ceramic filter in order to capture undesirable materials, such as oxide films and stray sand particles before they enter the mold. In addition to the runners and in-gates, most castings also have *risers* that are strategically placed on the casting. These provide additional metal to feed solidification shrinkage, which would otherwise lead to internal cavities in the solidified part. For a more detailed description of the practical requirements of the feeding system, the reader is referred to the texts by Heine, et al. [5] and by Campbell [1]. After the mold cavity is filled and the part solidifies by heat extraction into the mold, the sand mold is broken away, leaving the final casting.

Heat is extracted from the metal by the cold mold, thus effecting solidification. Although methods and results for the analysis of this heat transfer problem are discussed in Chaps. 5 and 6, a brief, more general discussion is useful at this point. The heat removed from the casting includes both the *sensible heat*, as measured by the specific heat c_p, and the latent heat of fusion L_f. In most materials, the latent heat contribution is much larger than the sensible heat. The latent heat flows through the solidified region of the casting, across the solid-mold interface and into the mold. It is demonstrated in Chap. 5 that, for sand casting, the overall heat transfer is dominated by heat conduction in the mold. It is moreover found that the solidified thickness x^* as a function of time t for a pure metal can be approximated by

$$x^* \approx \frac{2(T_f - T_0)}{\rho_s L_f \sqrt{\pi}} \sqrt{k_m \rho_m c_{pm}} \sqrt{t} \tag{1.1}$$

where T_f is the equilibrium freezing temperature of the metal, T_0 is the initial temperature of the mold, ρ_s is the density of the solid metal, k_m is the thermal conductivity of the mold, ρ_m is the density of the mold material, and c_{pm} is its specific heat. In the case where heat transfer is dominated by the mold, this result can be generalized by replacing x^* by the ratio of the volume V of the part to its surface area A. When inverted to obtain an expression for the solidification time t_f, the result is called *Chvorinov's Rule*:

$$t_f = \frac{\pi}{4 k_m \rho_m c_{pm}} \left(\frac{\rho_s L_f}{T_f - T_0} \right)^2 \left(\frac{V}{A} \right)^2 \tag{1.2}$$

Chvorinov's rule is used in daily foundry practice to estimate the solidification time for parts. It can be further extended to assess the solidification pattern in parts with varying section sizes, and thus predict the overall progress of solidification through the part. Casting designers can then place risers appropriately to ensure that the last regions to freeze, where shrinkage porosity is likely (see Chap. 12), are situated outside the final product. In many cases, this simple calculation is sufficient for designing an acceptable casting. However, many castings are either too complicated, or require too much precision, for this approximate approach to be adequate, and in these cases computer simulations of the filling and heat transfer are carried out, using the methods described in Chap. 6.

Before leaving this topic, we should mention an interesting variant of the foundry casting process known as *lost foam casting*. In this procedure, the pattern is constituted of foamed polystyrene, and is left in the mold during casting. The heat of the incoming metal vaporizes the polystyrene, with the gases escaping through the mold. This process offers several advantages. Since the pattern does not need to be removed from the mold, there is more flexibility in the pattern design. Similarly, since the mold does not have to be manipulated to remove the pattern, unbonded sand can be used to form it, thus reducing the cost of reprocessing the sand between uses.

Investment casting
Sand casting is an excellent technique for mass production of complex shapes. There are some applications, however, where the need for a superior surface finish or more controlled solidification conditions leads to the use of the *investment* or *lost wax* casting process. This process is more expensive than sand casting, and is for this reason used most often in applications where the material itself is costly, or where the final shape is very difficult to obtain by machining. Examples include gold and silver jewelry, bronze sculptures, titanium parts for medical or sporting applications, and superalloy components for high temperature engine applications.

The pattern is made from a material with a low melting temperature such as plastic or wax, which is then coated with ceramic to form a mold. The ceramic can be simply cast around the pattern, e.g., from plaster of Paris, or built up as a series of layers. The first layer, i.e., the one closest to the part, is usually fabricated from a low viscosity slurry with fine particles so that the ceramic can penetrate into intricate designs on the pattern, and provide a good surface finish to the cast component. This layer is then strengthened by having sand sprinkled over it. After partial drying, the "invested" pattern/mold combination is dipped into a low viscosity slurry and then coated again with sand. The process is repeated until a thick, strong mold has been created. The mold/pattern is subsequently heated to melt the wax pattern material, which is then poured out of the mold, leaving a cavity of the desired shape. The ceramic mold is ultimately fired to achieve final strength. Ceramic cores can be incorporated by forming the wax pattern around them before investing the pattern into the mold.

The casting is obtained by pouring the melt into the ceramic mold. The mold is usually pre-heated to reduce thermal stresses during filling, and to ensure that the melt flows into intricate details of the mold before freezing. In some cases, thermal insulation is applied at certain locations on the mold to ensure a desirable heat flow pattern during solidification. The filling and solidification is sometimes done under vacuum to ensure that all of the air is removed from the mold, and to prevent contamination of the melt during filling. In that case, the heat transfer is controlled by radiation from the mold assembly to the environment. If a hollow casting is desired, the mold may be inverted a short time after pouring so as to decant the remaining liquid. After completion of solidification, the ceramic mold is broken off the casting. In some cases, internal cores are removed by dipping the casting in a caustic solution to dissolve the core. Figure 1.4 shows an example of a turbocharger rotor made by investment casting.

Fig. 1.4 Left to right: A ceramic mold after removal of the pattern; a superalloy turbocharger rotor (etched to reveal grain structure); and a half-section of a ceramic mold. Notice the smooth ceramic at the casting surface, giving way to a coarser consistency.

Another example of an industrial application is the production of superalloy blades for turbine applications. When in service, the blades are subjected to the high combustion temperature in the engine as they rotate at high speed. This produces a tensile stress directed along the axis of the blade, and the predominant failure mechanism occurs by creep deformation along grain boundaries. For this reason, the best solidification microstructure is one where all grain boundaries are aligned with the blade axis, or even better, where there are no grain boundaries at all, i.e., a single crystal. To achieve this microstructure, the casting is solidified by placing the mold assembly in a furnace that maintains a large temperature gradient, and then slowly withdrawing the mold assembly through the furnace to obtain *directional solidification*. Figure 1.5(a) shows an example of a cast turbine blade. Notice that the blade has a “pigtail” at the bottom serving to select only one grain from the many that form at the lowermost chill plate, thus producing a single crystal in the blade. The process by which the grains are eliminated is described in detail in Chap. 11, where we present methods for simulating the microstructure. An example of a result from such a simulation is given in Fig. 1.5(b). In this particular case, a stray grain has formed at the bottom blade platform above the pigtail and grown into the blade, producing a bicrystal, which would be cause for rejection during inspection.

Permanent mold and die casting

In the processes described in the previous sections, each casting had its own mold, which was destroyed after solidification in order to recover the

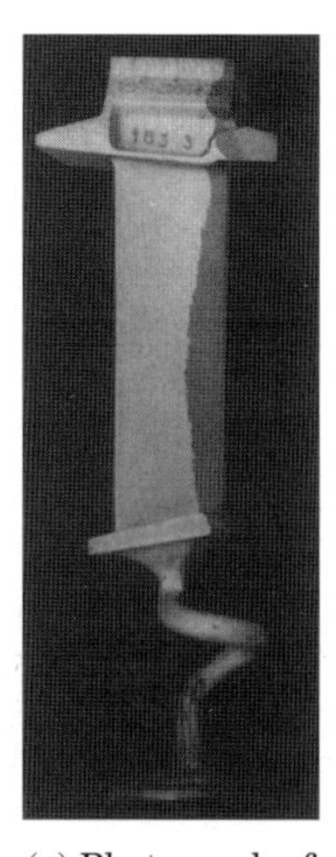

(a) Photograph of a turbine blade

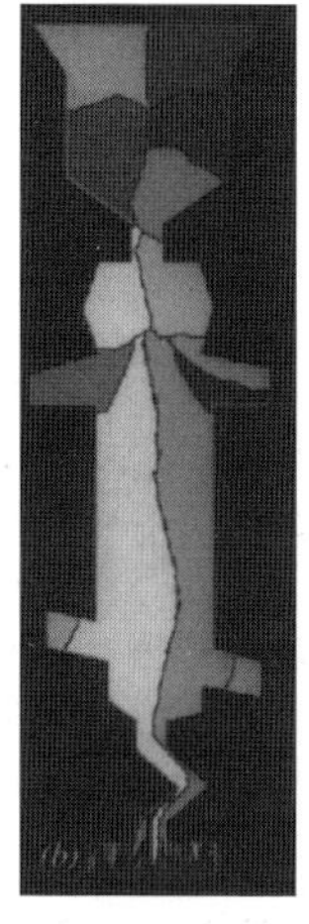

(b) Simulated microstructure for this part

Fig. 1.5 A defective single crystal superalloy turbine blade, and a simulation of the microstructure formed during the casting process.

cast part. Some parts lend themselves to production in a re-usable mold, made of tool steel, copper, or graphite, depending on the application. Such parts tend to be somewhat less complex, especially with respect to any internal passages. The mold normally consists of several parts that are coated with a slurry (e.g., a suspension of graphite in alcohol), preventing the metal from reacting with the mold, as well as assisting in releasing the part after solidification. The mold parts are then assembled prior to casting. Since the mold is re-used, this process is called *permanent mold casting*. The mold may have internal heating and cooling passages to control the solidification pattern. The casting is carried out by filling the mold, waiting long enough to allow the solidification to take place, and then disassembling the mold in order to retrieve the casting. The need to disassemble the mold places some obvious limitations on the complexity of the shapes that can be cast with this process. Note that in this case, the mold is not permeable to air, and some allowance must be made for venting the air from the mold during filling.

The heat transfer characteristics in permanent mold casting are somewhat different from those in sand casting. In the latter two processes, heat transfer is determined almost exclusively by conduction through the mold. This allows the use of Chvorinov's rule to estimate the solidification time. In permanent mold casting, however, the conductivity of the mold and casting are of similar magnitude, and the dominant resistance to heat transfer is the interface between the mold and metal part. In this case, it can be assumed that the mold is maintained at temperature T_0 and that the casting is approximately isothermal at the freezing temperature T_f.

One can estimate the rate of solidification by equating the heat flux across the metal-mold interface to the rate at which latent heat is evolved by solidification. In such a case, the thickness of the solidified layer becomes

$$x^* = \frac{h_T(T_f - T_0)}{\rho_s L_f} t \tag{1.3}$$

where h_T is the heat transfer coefficient at the metal-mold interface, and the other terms have the same meaning as in Eq. (1.1). Notice that, in this case, the thickness solidified is proportional to time, rather than proportional to the square root of time as in Eq. (1.1). Chvorinov's rule *cannot be used* to analyze casting processes where the solidification rate is controlled by heat transfer across the mold-metal interface. We should note that h_T depends strongly on thermal strains, which can affect the size of the air gap between the metal and mold. This is detailed further in Chaps. 4 and 13.

Another casting process that uses a permanent mold is *die casting*. The distinction between the two processes is that, in permanent mold casting, any pressure that is applied tends to be low, whereas in die casting, liquid metal is injected at high speed and high pressure into a die, usually made of tool steel. The process is very similar to injection molding of polymers. The reader may have seen such a process at a museum, for the manufacturing of souvenirs. Die casting is of course done at higher temperature for most metals, and at higher pressure, as well. In a typical process, the dies are coated with a release agent, and then closed. A sufficient charge of liquid is poured into a receptacle outside the die, known as a "shot sleeve," after which the charge is injected under high pressure into the die cavity. The dies are typically cooled via internal air and/or water passages. After solidification, they are opened and the part is ejected.

Die casting is used for relatively thin-walled parts, and the production rate can be quite high, making it very cost-effective for high volume production. Liquid metals tend to have a high surface tension and low viscosity in comparison to polymers. As a result, the incoming stream has a tendency to separate, or even atomize during filling, which leads to defects such as porosity and entrained oxides in the final part. This limits the applications where die casting can be used to those where mechanical properties such as fatigue resistance are not too demanding. Figure 1.6 shows sample parts for automotive use: an aluminum alloy engine cover and a magnesium alloy cross beam that serves as a lightweight support for an automobile dashboard. These parts illustrate the remarkable complexity of products achievable with this process. A variant of the process, called *low-pressure die casting*, is used to produce automotive wheels. The mold is assembled above a holding furnace, and the furnace is then sealed and pressurized to force liquid up into the die. The filling tends to produce much less agitation than high-pressure die casting.

In another variant of the process, known as *thixocasting*, the charge is semi-solid, rather than liquid. The name derives from the term *thixotropic*, which signifies that the viscosity of the material decreases with

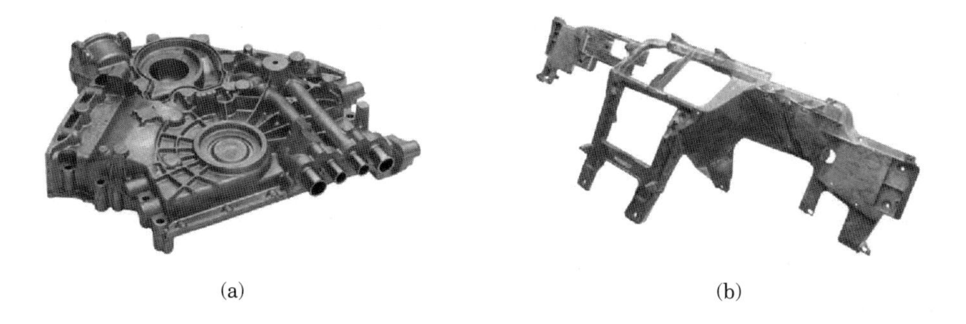

(a) (b)

Fig. 1.6 Examples of die cast parts. (a) An aluminum alloy engine cover. (b) A magnesium alloy automotive dashboard support. (Source: diecasting.org/castings.)

time while undergoing shear. Many metal alloy systems demonstrate this property since the structure can evolve to accommodate the deformation. The process has several advantages, the primary one being that the metal shows much less tendency to atomize during injection as a result of the viscosity of the semi-solid being much higher than that of the corresponding liquid. Further, the fact that the charge is only partially liquid reduces the amount of heat that has to be extracted from the dies. The principal problem in thixocasting is that the cost of preparing semi-solid materials with an appropriate microstructure can offset other advantages.

1.2.2 Continuous and semi-continuous casting

The casting processes described in the previous section are used for fabricating parts that are intended to be used essentially in their as-cast condition, perhaps after some machining and heat treatment. Other applications, such as sheets, rails, tubes, etc. are more practically produced by first casting a relatively simple form (slabs and billets) and then mechanically deforming it into the desired shape, using processes such as rolling and extrusion. We note that the alloys used for these applications, called *wrought alloys*, tend to have much lower solute contents than their counterparts used for shape casting. These products were once made by casting large quantities of metal into molds that consisted of holes dug in the ground, then removing them and performing the necessary mechanical treatments. The drawbacks of this process include very long solidification times (a matter of days for some very large ingots), and a tendency to produce separation of solute elements over very large distances in the ingot, a process called *macrosegregation*, described in Chap. 14.

Large ingot casting for subsequent mechanical processing has largely been supplanted by *continuous* or *semi-continuous casting*. In such processes, a metal is introduced into a water-cooled mold with an open exit. The process begins with a starter block in the mold exit. The metal is then introduced into the mold through a nozzle, where it proceeds to freeze against the mold and the starter block. The block is subsequently withdrawn at a controlled rate and metal is continuously fed to maintain a constant level in the mold. This process is used to produce steel, aluminum, copper and other alloys for wrought applications, and the particular form of the continuous casting process depends to a certain extent on the material being cast.

More than 90% of the world's steel production is manufactured by continuous casting. A typical apparatus for making slabs is illustrated in Fig. 1.7(a). Liquid steel is fed from the melting units into a *tundish*, from which it is conducted into the water-cooled copper mold by a submerged entry nozzle (SEN). Typical mold dimensions are 5-25 cm by 0.5-2 m, and the casting speed for conventional slabs is 1-8 m min^{-1}. The steel is cast vertically, and as the slab exits the mold, a series of rolls gradually turns it to the horizontal direction over a distance of about 10 m. This is possible because of the low thermal conductivity and high casting speed typical of

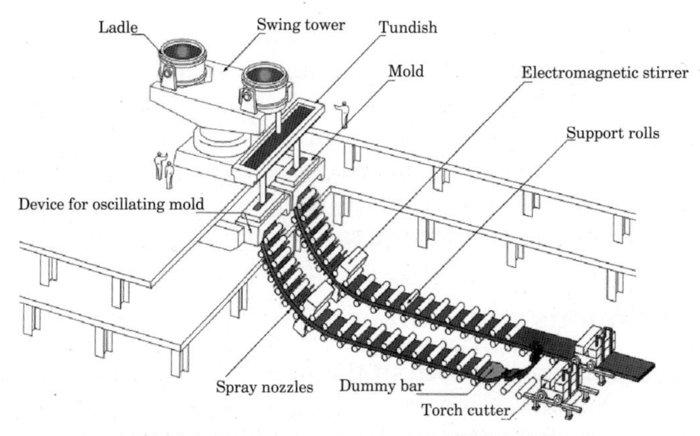

(a) A steel slab casting facility (Reproduced with permission from JFE 21st Century Foundation).

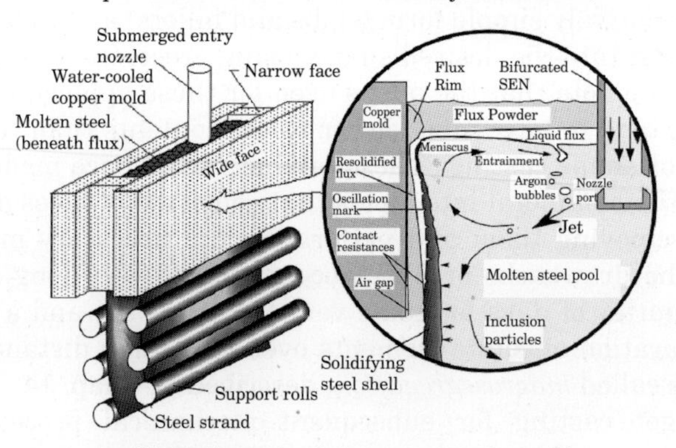

(b) A schematic of phenomena in the mold (Reproduced with permission from B. G. Thomas).

Fig. 1.7 A schematic view of a facility for continuous casting of steel slabs, and the phenomena that occur in the casting mold.

steels, which also have sufficient ductility for the bending operation. Once the steel is entirely solid, it can be cut on the fly into convenient lengths for subsequent processing. Many of the important phenomena that occur in the mold are illustrated in the close-up view shown in Fig. 1.7(b). A slag layer covers the liquid steel in the mold to protect it from oxidation. The mold is oscillated as the slab is withdrawn to help prevent sticking of the slab to the mold wall. This draws the slag into the region between the slab and the mold. It is clear from the illustration that only a small portion of the slab solidifies before the slab leaves the mold. The remaining solidification takes places below the mold by the application of water sprays between the guide rolls.

The heat transfer process in the mold is the result of several competing phenomena. We show in Chap. 4 that one can estimate the relative importance of conduction and advection (heat carried by the moving material) by computing the Péclet number, Pe

$$\mathrm{Pe} = \frac{V_c L_c}{\alpha} \tag{1.4}$$

where V_c is the casting speed, L_c is a characteristic length, e.g., the slab half-width, and $\alpha = k/(\rho c_p)$ is the thermal diffusivity of the steel. When employing values typical for steel continuous casting, one finds that $\mathrm{Pe} \approx 200$, which implies that conduction is less important than advection in the axial direction. For this reason, models of the process where the slab is treated as a 2D "traveling slice" moving with the casting speed are often encountered.

At the top of the slab, where solidification begins, there is a liquid flux layer between the slab and the mold, which makes for very effective heat transfer. This locally heats the mold, causing it to expand, and the cooling slab tends to shrink, ultimately resulting in an air gap that significantly reduces the heat transfer. We note that certain molds are designed with an inward taper in order to compensate for the thermal distortion. Accounting for these phenomena in an accurate process model can be complicated, due to the thermal distortion depending on the mold design, details of the roll positions below the mold, etc.

In steel casting, thermal stresses also play an important role below the mold. The solidification zone typically extends as much as 10 m below the mold. The solidified shell remains quite hot and ductile in this region, and the ferro-static pressure of the liquid sump deforms the shell outward between the containment/turning rolls, as illustrated in Fig. 1.8. The alternating bulging and compression of the shell can lead to centerline segregation in the final slab, as the composition that is drawn in and then pushed back out can change. Deformation of the solidifying solid is addressed in Chap. 13, and the associated segregation is considered in detail in Chap. 14.

The continuous casting process for aluminum alloys is performed somewhat differently than it is for steel. Aluminum is a much better conductor than steel, and aluminum alloys tend to have much lower high-temperature strength. This renders it impractical to bend aluminum

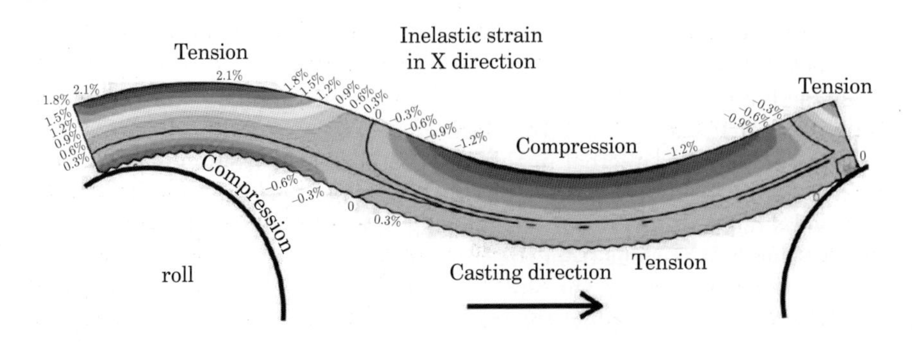

Fig. 1.8 A computer simulation of the thermal stresses that develop in a steel slab between rolls. (Courtesy B. G. Thomas).

alloys slabs, and they are therefore cast vertically into deep pits, typically 8-10 m. Such a procedure would more properly be called semi-continuous casting, but is often referred to as *direct chill* or DC casting. The molds are generally made from aluminum, and there is no mold oscillation. A mold release grease is sometimes applied before casting, or, alternatively, a continuous film of oil is fed to the meniscus that forms between the aluminum and the mold surface. Water sprays are applied below the mold to finish the solidification. DC cast aluminum slabs are typically as thick as 400-500 mm, and are cast at 60-70 mm min^{-1}. If one computes the Péclet number using these parameters, one finds that $\mathrm{Pe} \approx 1$, which implies that the solidification process is truly three-dimensional. After the entire length of the ingot has been cast, the process is brought to a halt, the molds are removed, and the ingots are taken away for further processing. Macrosegregation and thermal stresses are significant issues in aluminum casting, but tend to be manifested in different ways than they are in the casting of steel. The geometry of the process often produces significant macrosegregation in the slabs, which can be a particularly vexing problem in some alloys. The same processes of shell solidification and air gap formation as those described for steels also occur in aluminum alloys. Thermal stresses tend to appear as distortions at the base of the ingot, and in certain severe cases, this can lead to fracture and scrapping of the entire ingot. We address this particular problem in detail in Chap. 13.

One would prefer to cast both aluminum and steel horizontally, as opposed to vertically, since this makes the plant layout more efficient. Such processes tend to be limited to rather small cross-sections, up to 100 mm or so. In a relatively recent variant of the continuous casting process, metal is poured into a fairly small gap between to rotating rolls. This process is useful for making sheets thin enough to be processed by cold rolling, which offers a significant reduction in production cost.

1.2.3 Crystal growth processes

The invention of the transistor, and the subsequent miniaturization of electronic circuits has led to a complete revolution in technology, which needs no further description here. We note, though, that virtually all of this technology relies on the ability to produce Si crystals of extremely high quality, with very low, but at the same time well controlled, levels of impurities, and extremely low levels of mechanical defects such as dislocations. Other important applications include the fabrication of quartz and sapphire crystals for the watch industry, as well as GaAs crystals for diodes. Such crystals are grown from the melt using a variety of methods, all of which feature a high temperature gradient and small growth velocity. It is demonstrated in Chap. 8 that these conditions are required in order to maintain a stable planar interface between the solid and liquid when solutal impurities are present.

Two common crystal growth processes are illustrated in Fig. 1.9. In Czochralski growth, typically used for Si, a single crystal is produced by first preparing a large bath of a high-purity melt. A seed crystal is usually used to start the solidification process in the desired crystallographic orientation. The seed is dipped into the surface of the melt and then withdrawn upward at a controlled rate, and solidification occurs as heat is lost to the environment by radiation. The crystal and/or the crucible may be rotated during growth to minimize segregation, and to ensure that the shape is cylindrical. The principal means for heat extraction is radiation, which tends to limit the growth rate. For a 200 mm diameter crystal, the maximum growth rate is about 0.5 mm min^{-1}, which translates to the production of a 3 m long crystal in about 5 days. This is the basic process used for the preparation of a majority of the Si crystals employed in electronic applications.

In the Bridgman growth process, shown schematically in Fig. 1.9(b), a baffle separates the hot and cold zones, establishing a large temperature

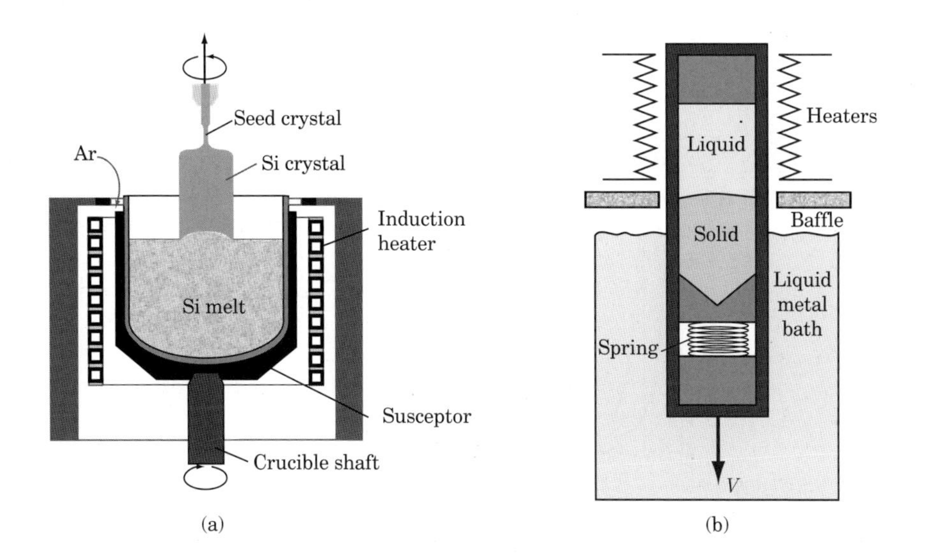

Fig. 1.9 A schematic view of (a) the Czochralski and (b) the Bridgman crystal growth processes.

gradient. The raw material for the crystal is first sealed inside a container called an ampoule, which is then placed in the hot zone of the furnace. The ampoule has a sharp point at the bottom to ensure the nucleation of only a single grain (see Chap. 7). Alternatively, there may be a seed crystal placed at the bottom of the ampoule, in which case the tip is not melted. After the material is melted, the ampoule is withdrawn at constant velocity. Once again, the objective is to produce a crystal grown under stringently controlled thermal conditions. In certain applications, the lower zone of the apparatus consists of a liquid metal bath (usually Ga) in order to give rise to the highest possible heat extraction rate. A similar process is used for directional solidification of single crystal turbine blades, as discussed previously in the section on investment casting.

Composition control is one of the most important aspects of crystal growth. Let us consider a simple application that motivates the more detailed studies of chemical segregation appearing in Chaps. 5 and 10. Suppose that there is a small amount of impurity present in the raw material, and that, upon melting, the composition of the liquid is uniform. Most materials for which this process is used demonstrate very low solubilities for impurities. As the material solidifies, the solute is rejected ahead of the interface into the liquid, gradually increasing its concentration. This problem is presented and solved in detail in Chap. 5, and we therefore limit ourselves here to a more descriptive approach to the behavior. Figure 1.10 shows the development of the composition in the solid behind, and the liquid ahead of the moving interface as it moves at constant velocity. One can see an initial solute-depleted region at the start of solidification, which eventually gives way to a steady growth region where the composition in the solid is constant. Eventually, there is also a final transient at the opposite end of the sample, where the additional solute in the boundary layer ahead of the interface is deposited. See Chap. 5 for further details.

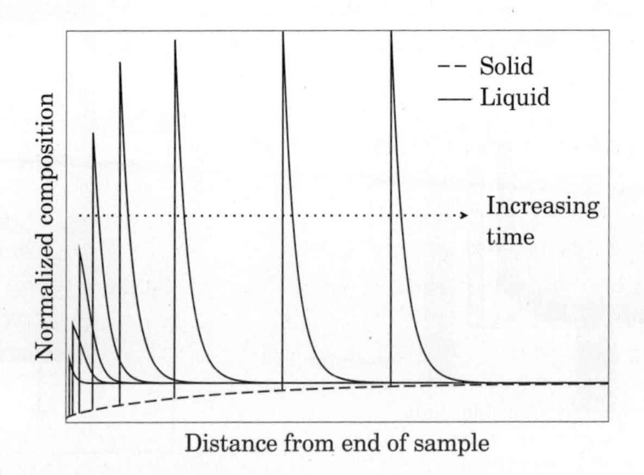

Fig. 1.10 Development of solute layer ahead of the solid-liquid interface, and the corresponding composition in the solid.

This is an example of macrosegregation; i.e., separation by large distances between regions of significantly different composition.

Macrosegregation is not always a bad thing. There is a variant of the Bridgman process called *zone refining* which takes advantage of the segregation in order to create materials of high purity. Instead of melting the entire sample, zone melting consists in melting a small portion of the sample, e.g., using an induction coil, and then passing the heater over the container, moving from one end to the other. In doing so, the segregation process described above "pushes" the solute toward one end. If the process is repeated, the sample is continuously refined. This is the most common process used for preparing high purity samples of many materials.

1.2.4 Welding

One of the most common methods for joining materials is welding, in which two materials are melted by the application of some sort of directed energy, after which they re-solidify, creating a solid joint. A filler rod may also be used to help fill any gaps between the materials, or to control the chemistry of the weldment. Welding processes are often characterized by their means for melting. The most versatile method is to use an oxygen-acetylene torch; a method that is somewhat difficult to control. In arc welding, a current passing between a welding rod and the workpiece provides the necessary heat. A shielding gas or flux is used to prevent oxidation. In certain applications, the gas is ionized and forms a plasma that performs the actual energy transfer to the sample. Heat for melting can also be generated by a laser or electron beam. The energy transfer processes in welding are far more complicated than those associated with solidification, and we leave their description to other texts.

In most applications, the weld beam passes over the sample, melting the material along its path, which then re-solidifies as the weld beam moves away. Thermal stresses develop due to the large thermal gradients. One can develop a fairly simple analytical model of the process by considering the weld beam to be a point source of heat Q, traveling at constant velocity V in the $x-$direction along the surface of an infinite medium. It is convenient for a solution to transform from the Cartesian coordinate system (x, y, z) fixed on the workpiece to a reference frame (ξ, y, z) traveling with the weld beam, placed at the origin in this frame. The coordinate y measures the transverse distance from the weld centerline, and z measures the depth from the surface. Latent heat can be neglected to a first approximation, in which case the solution for the temperature field in the material is given by the so-called "Rosenthal solution"

$$T = T_0 + \frac{Q}{2\pi k r} \exp\left(-\frac{V(r+\xi)}{2\alpha}\right) \tag{1.5}$$

where T_0 is the base temperature of the material before welding, Q is the power of the heat source, k is the thermal conductivity, $\alpha = k/(\rho c_p)$ is the

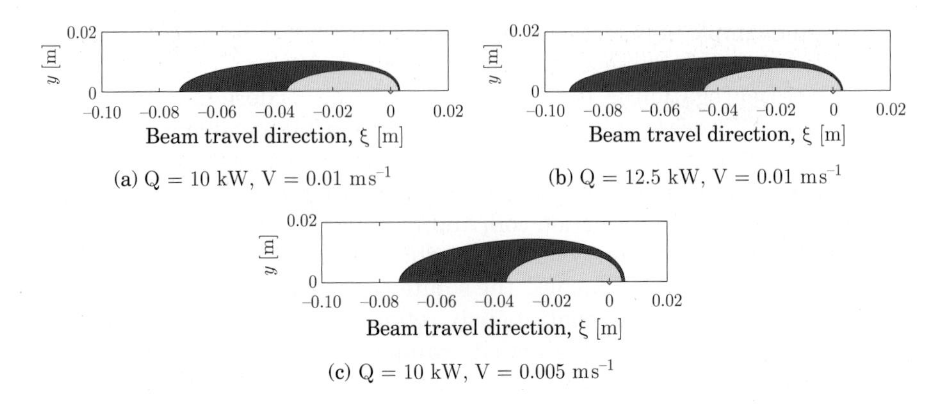

Fig. 1.11 Surface temperatures computed with Eq. (1.5) for various values of the welding parameters.

thermal diffusivity, and $r = (\xi^2 + y^2 + z^2)$ is the radial distance from the weld beam. Notice that the solution is symmetric with respect to y and z, signifying that it is sufficient to consider the solution on the surface.

Figure 1.11 shows several cases obtained by varying the parameters in Eq. (1.5). In the base case, shown in Fig. 1.11(a), we set $Q = 10$ kW and $V = 0.01$ m s^{-1}. These values are appropriate for the welding of steel. The molten pool ($T > 1500°$C) is shown in light gray, and in this frame, the material moves from right to left in the figure (i.e., the weld power source moves from left to right in the laboratory frame). The weld beam is located at the origin, and one can see that the pool extends ahead of the beam, and that there is a "tail" of heated material behind it. Material that has been melted is said to be in the *fusion zone.* We also show in darker gray a portion of the material whose temperature has been raised above 800°C, roughly corresponding to the transition to austenite. The portion of the material that undergoes the solid state transformation, but is not melted, is called the *heat-affected zone* or HAZ. This region is usually weaker than the base metal or the weldment, and is thus a source of mechanical failure. The analytical expression for the temperature renders it possible to explore the importance of the process parameters on width. Figure 1.11(b) shows the same contours as the base case, after the input heat has been increased by 25%, and Fig. 1.11(c) portrays the effect of decreasing the welding speed by 50%. Both changes increase the width (and depth) of both the fusion zone and the HAZ.

The Rosenthal solution provides guidance and helps us understand the role of welding speed and power input. Nevertheless, a number of very important aspects of the welding process are naturally omitted. The most obvious is that real weld beam would not be a point source. In many cases, the power can be well represented by a Gaussian distribution, whose total energy is equal to Q as above, and whose width depends on processing variables, such as beam shape, plasma flow, stand-off distance,

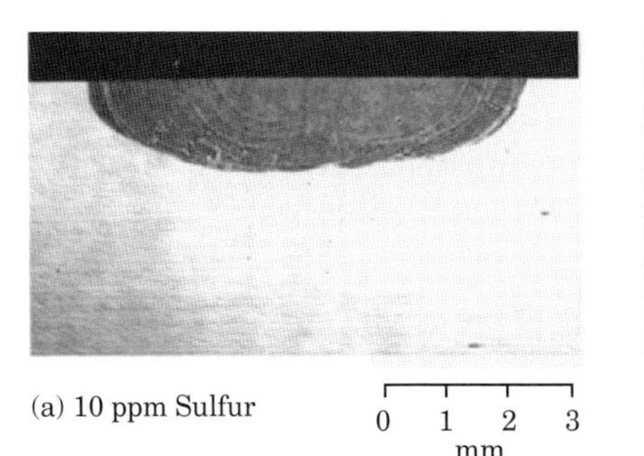

(a) 10 ppm Sulfur

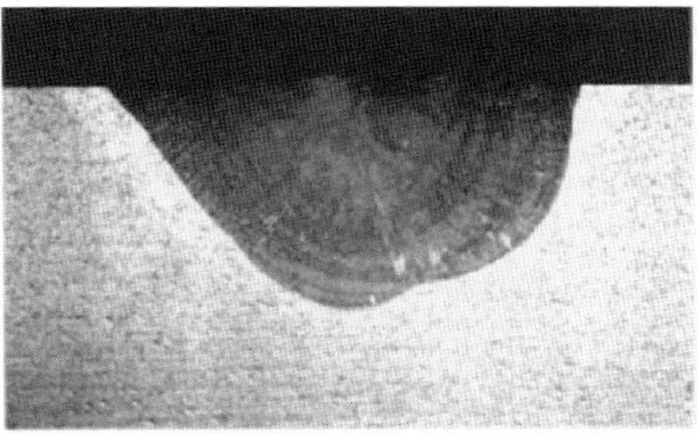

(b) 150 ppm Sulfur

Fig. 1.12 Cross-sections of type 304 stainless steels after GTA welding. (a) 10 ppm sulfur content. (b) 150 ppm sulfur content. Reproduced with permission from Pierce, et al. [7].

etc., relative to the particular type of welding. Far away from the source, however, the Rosenthal solution might still provide a reasonable estimate.

Another very important phenomenon that is not included is the effect of fluid flow within the weld pool. The flow may be driven by electromagnetic effects in arc welding, but perhaps a more interesting source is flow driven by gradients in surface tension on the weld pool surface, a phenomenon sometimes referred to as the *Marangoni effect*. Because of the strongly concentrated energy source, a large temperature gradient is often induced on the surface.[1] If the surface tension $\gamma_{\ell a}$ depends on temperature, then the temperature gradient induces a surface flow that can have a considerable influence on the melting pattern. If the surface tension increases with temperature ($d\gamma_{\ell a}/dT > 0$), the liquid moves toward the weld beam on the surface, inducing a narrower and deeper pool. On the other hand, if the surface tension decreases with temperature ($d\gamma_{\ell a}/dT < 0$), the surface flow moves away from the weld beam, leading to a wider and shallower pool. This phenomenon is very important in the welding of steels, where the sign of $d\gamma_{\ell a}/dT$ is actually reversed as the sulfur content in the steel increases. Low-sulfur steels are often specified for welding to avoid cracking problems associated with MnS "stringers" in the HAZ. However, $d\gamma_{\ell a}/dT < 0$ for these compositions, which renders it difficult to ensure sufficient penetration for welding of thick plates. Welders thus sometimes add sulfur to the weld shielding gas to increase the concentration in the weld pool enough for $d\gamma_{\ell a}/dT > 0$, thereby allowing deep penetration welds to be made.

This phenomenon is illustrated in Fig. 1.12, which shows cross-sections of gas-tungsten arc (GTA) welds made on type 304 stainless steel [7]. The melted region is clearly delineated by etching. When the

[1]In the Rosenthal solution, where there is a point source, the temperature gradient at the weld beam is infinite. This is clearly not the case in reality, but it nevertheless indicates that the temperature gradient can be very large.

sulfur content is low (10 ppm, Fig. 1.12a), the weld pool is relatively shallow, because at this sulfur content, $d\gamma_{\ell a}/dT < 0$, leading to a surface flow on the weld pool away from the beam. At higher sulfur content (150 ppm, Fig. 1.12b), the sign of $d\gamma_{\ell a}/dT$ reverses, and the surface flow is directed toward the weld beam, leading to a deeper penetration weld.

1.3 SUMMARY

This chapter provides an overview of the content and coverage of this book. In the three parts that follow, we first describe the fundamental aspects of thermodynamics, phase diagrams and analytical techniques that are employed in the subsequent parts. These concepts are then used in Part II to describe the evolution of microstructures, beginning with nucleation, continuing with dendritic growth and structures in eutectic and peritectic alloys, and concluding with a chapter on microsegregation. The final part combines the fundamentals from Part I with the understanding of microstructure to examine defects that may appear during solidification. This chapter also describes several important solidification processes, focusing on those aspects that motivate the further studies addressed in the remainder of the text.

1.4 REFERENCES

[1] J. Campbell. *Castings: the new metallurgy of cast metals.* Butterworth-Heinemann, 2003.

[2] B. Cantor and K. O'Reilly, editors. *Solidification and casting: an Oxford-Kobe materials text.* CRC Press, 2003.

[3] B. Chalmers. *Principles of solidification.* Wiley, New York, 1964.

[4] M. C. Flemings. *Solidification processing.* McGraw-Hill, New York, 1974.

[5] R. W. Heine, C. R. Loper, and P. C. Rosenthal. *Principles of metal casting.* McGraw-Hill, New York, 1967.

[6] W. Kurz and D. J. Fisher. *Fundamentals of solidification.* Trans. Tech. Publ., Aedermansdorf, Switzerland, 4th edition, 1998.

[7] S. W. Pierce, P. Burgardt, and D. L. Olson. Thermocapillary and arc phenmoena in stainless steel welding. *Welding Journal*, 78:45S – 52S, 1999.

[8] D. M. Stefanescu. *Science and engineering of casting solidification.* Kluwer Academic/Plenum, New York, 2nd edition, 2002.

[9] R. Trivedi. *Materials in art and technology.* Taylor Knowlton, 1998.

[10] K.-O. Yu. *Modeling for casting and solidification processing.* CRC Press, 2001.

[11] *ASM Handbook - Volume 6, Welding.* ASM International, Metals Park, OH, 1993.

[12] *ASM Handbook - Volume 15, Casting.* ASM International, Metals Park, OH, 2008.

[13] North American Die Casting Association. URL, 2008. http://diecasting.org/castings.

[14] *Sand Casting.* Wikipedia, 2008. http://en.wikipedia.org/wiki/Sand_casting.

Part I

Fundamentals and Macroscale Phenomena

This part of the book treats the most classical aspects of solidification, i.e., those that can be viewed at the macroscopic scale. It also introduces several fundamental concepts that form the foundation for study. In Chap. 2, the thermodynamics of solidification processes is considered, introducing the important concepts of free energy, chemical potential and equilibrium. We also discuss departure from equilibrium induced by capillarity and kinetics. These concepts, which are then applied throughout the book, are generalized and put into a compact form in Chap. 3, where we describe equilibrium phase diagrams for binary and ternary systems. However, thermodynamics tells only what *should* happen, i.e., it identifies the state of the system that has the lowest energy at equilibrium. It does not guarantee that this *will* happen, nor does it include any information about how the system evolves from one state to another in time. For this, we need to include transport phenomena and departures from equilibrium associated with curved surfaces and moving interfaces.

Chapter 4 presents the balance equations for mass, momentum, energy and solute transport. We also introduce the constitutive models for heat and solute transport most often used to devise a complete set of governing equations. These conservation equations are integrated over a representative volume element of the microstructure in order to obtain two sets of equations: average equations applicable at the macroscopic scale, and interfacial balances at moving interfaces that can be used at the microstructure level. Finally, this chapter introduces the concepts and methods for scaling the governing equations to identify the important (and negligible) terms. The scaling process is essential for the study of solidification microstructures because it allows for the separation of time and length scales for heat and solute transport. These are the basic concepts behind microstructure pattern formation, mediated by surface tension forces.

In Chap. 5, several fundamental solidification problems having analytical solutions are presented. Some of the most important dynamic features of solidification are identified in these problems: the separation of length scales for heat and solute transport; the role of surface tension; the rate of growth of solidified layers in molds; and the curious inability of transport alone to predict growth patterns. These features become the basis for studying microstructure formation in Part II.

Finally, Chap. 6 describes some of the important aspects of numerical methods for analyzing solidification. First, fixed grid methods based on average balance equations are presented in order to describe solidification processes at the macroscopic scale. Then, we describe front tracking methods that follow the moving phase boundary on a discrete grid. Some of the problems presented in Chap. 5 are used as test cases for the numerical schemes in order to illustrate the techniques used to attack the moving boundary problem.

CHAPTER 2

THERMODYNAMICS

2.1 INTRODUCTION

Thermodynamics is the study of the behavior of systems of matter under the action of external fields such as temperature and pressure. It is used in particular to describe equilibrium states, as well as transitions from one equilibrium state to another. In this chapter, we introduce the concepts pertinent to solidification. It is assumed that the reader has had some exposure to classical thermodynamics; thus, the presentation is more in the form of definitions and reviews than what might be found in a standard thermodynamics text. We begin with a brief review of basic concepts, and then proceed to develop the relationships needed in future chapters, with a focus on condensed phases, i.e., liquids and solids.

The *system* consists of all of the matter that can interact within a space of defined boundary. A system may contain several *components*, chemically distinct entities such as elements or molecules. There may also be several *phases*, which are defined as portions of a system that are physically distinct in terms of their state (solid, liquid, vapor), crystal structure or composition. For example, pure water is considered as a single component, H_2O, that may exist as several different phases, such as ice, liquid or steam. In binary alloys, such as mixtures of Al and Cu, various phases can be present, including a liquid in which both elements are completely miscible; limited solid solutions of Cu in Al and Al in Cu; and intermetallic compounds such as Al_2Cu, etc.

For a single component, or *unary* system, there are three thermodynamic variables: the temperature T, the pressure p and the volume V. These variables define the state of the system. However, only two are independent, and the third is obtained through an *equation of state*, e.g., $V = V(p,T)$. The ideal gas law is one of the most familiar equations of state, but such an equation of state exists for all materials. Note that T and p are *intensive* variables, meaning that they do not depend on the amount of material present, whereas V is an *extensive* variable, which means that its value does depend on the quantity of material present in the system. A *state variable* is one which can be written as a function of the thermodynamic variables. For example, the *internal energy* of the system E is a state variable which is also extensive. We will use upper case characters to designate extensive state variables such as E, G, H, S, V (these symbols will be defined as they appear below). An intensive form of any extensive state variable can be formed by dividing it by the total mass M, or by the total number of moles n in the system. The former is called the *specific* form and is represented by the corresponding lower case character, e, g, h, s, v. The latter is called the *molar* form, denoted by the superscript m, i.e., E^m, G^m, H^m, S^m, V^m. For example, the specific internal energy of the system is $e(T, V) = E/M$, whereas the molar internal energy is given by $E^m(T, V) = E/n$.

In the following sections, we first introduce the concepts of classical thermodynamics for systems having one component (unary systems), then we treat the case of binary alloys, introducing the entropy and enthalpy of mixing as well as ideal and regular solutions. In particular, the regular solution model, despite its relative simplicity, is able to produce a rich variety of solidification reactions, e.g., eutectics, peritectics or monotectics, and solid state transformations (e.g., spinodal decomposition, ordering, etc.). This makes the model a very useful tool for describing the relationship between the molar Gibbs free energy and phase transformations. Differentiation of the free energy leads to the *chemical potential*, which is an indispensable concept for understanding phase transformations.

The subject then turns to the thermodynamics of multi-component systems. It will be seen that, as the composition space is enriched by adding more components, more phases can co-exist. The treatment above is then generalized to define the conditions of equilibrium for a multi-component, multi-phase system. This takes the form of *Gibbs' phase rule*, which gives the number of degrees of freedom of a multi-component system as a function of the number of co-existing phases. The concepts developed in this chapter are later used and enlarged in Chap. 3, which is devoted to equilibrium phase diagrams for binary and ternary systems.

Finally, we discuss the phenomena that lead to modification or departure from equilibrium. The effect of surface energy and curvature on thermodynamic equilibrium is first derived for a static interface. This is extremely important for the understanding of microstructure development in later chapters. We also introduce the concept of disequilibrium at a moving solid-liquid interface due to kinetic effects; this effect is related to the ability of atoms and species in the liquid to rearrange themselves into a crystalline phase.

2.2 THERMODYNAMICS OF UNARY SYSTEMS

2.2.1 Single phase systems

In thermodynamics, one compares different states and the corresponding relative changes in the various thermodynamic quantities. Often, the states being compared are "close" to each other, meaning that the kinds of change under scrutiny are incrementally small. Consider the internal energy of the system E, which is a state variable for a whole system of volume V. The difference dE between two states is an exact differential. Choosing T and V as the independent state variables, we may write the following:

$$dE(T,V) = \left(\frac{\partial E}{\partial T}\right)_V dT + \left(\frac{\partial E}{\partial V}\right)_T dV \tag{2.1}$$

One changes variables frequently in this subject, so it is necessary to designate which variables are held constant by use of the subscript on the parenthesis. The partial derivatives that arise from such forms are often given special names and symbols. For example, $(\partial e/\partial T)_V = c_V$ is called the specific heat at constant volume (units $\mathrm{J\,kg^{-1}\,K^{-1}}$).

We shall see that T and V are not the natural variables to express the energy for condensed phases. The change in internal energy in the system is the sum of any heat added to the system, designated δQ, plus work done by external forces, designated δW. The *First Law of Thermodynamics* is a translation of this statement into the equation

$$dE = \delta Q + \delta W \tag{2.2}$$

For a gas or a liquid, the work arises only by compression, and thus $\delta W = -pdV$. However, for a solid, there may be other contributions from the other components of the stress tensor, as will be seen in Chap. 4. Electromagnetic contributions can also enter into δW. The variations δQ and δW are written with the symbol δ to indicate that their value is path-dependent, and thus both Q and W are not state variables. In this chapter, we consider only the hydrostatic compression/expansion term, $\delta W = -pdV$.

It is useful to separate δQ into a *reversible* part δQ_{rev} and an *irreversible* part δQ_{irr}. When following a reversible path, i.e, a succession of infinitesimally close equilibrium states, $\delta Q_{irr} = 0$. This concept of irreversibility leads to definition of the *entropy* $S(T,V)$, a state variable, defined such that for a reversible path one has

$$dS = \frac{\delta Q_{rev}}{T} \tag{2.3}$$

Notice that even though δQ_{rev} is not a state variable, the entropy is. The *Second Law of Thermodynamics* is the statement that, for any process, $\delta Q_{irr} \geq 0$. We will be concerned almost entirely with equilibrium processes, where $\delta Q_{irr} = 0$. This also implies that at equilibrium, the system has attained its minimum energy and maximum entropy. The first law, Eq. (2.2), can now be written in terms of the entropy

$$dE = \delta Q - pdV = TdS - pdV \tag{2.4}$$

In this form, the internal energy seems to be naturally expressed in terms of the two extensive variables, S and V. Unfortunately, this form is not very convenient for condensed phases. For example, how would you measure the specific heat at constant volume, c_V, i.e., the increment of internal energy of the system while keeping its volume constant? You would of course use a calorimeter. But constraining the volume to remain constant during the test would be very difficult. Therefore, one defines another state variable, the enthalpy H:

$$H = E + pV \tag{2.5}$$

The enthalpy will be used extensively in the analysis of solidification. Using Eq. (2.4), the differential of H is given by

$$dH = dE + pdV + Vdp = TdS + Vdp \tag{2.6}$$

Considering $H = H(S, p)$, we have

$$dH = \left(\frac{\partial H}{\partial S}\right)_p dS + \left(\frac{\partial H}{\partial p}\right)_S dp \tag{2.7}$$

By comparison with Eq. (2.6) we may identify

$$\left(\frac{\partial H}{\partial S}\right)_p = T; \quad \left(\frac{\partial H}{\partial p}\right)_S = V \tag{2.8}$$

Dividing by the constant mass of the system, M, and using Eq. (2.8) gives

$$dh = Tds + vdp \tag{2.9}$$

At constant pressure, the second term vanishes, and thus only the heat brought to the system contributes to dh. For this reason, the enthalpy is also sometimes called the *heat content*. By writing $h = h(T, p)$, and applying the chain rule for differentiation, we can define c_p, the *specific heat at constant pressure* as:

$$c_p = (\partial h/\partial T)_p \tag{2.10}$$

Thus, at constant pressure $dh = c_p dT$. It follows from Eq. (2.10) that c_p can be measured in an adiabatic calorimeter maintained at constant pressure, rather than at constant volume, which is certainly much more convenient!

In practice, the enthalpy is computed by integrating measured values of c_p over temperature, with the constant of integration adjusted in such a way that the enthalpy is zero at 298 K:

$$h(T) = \int_{298\ \mathrm{K}}^{T} c_p(\theta) d\theta \tag{2.11}$$

It also follows from Eq. (2.9) that the specific entropy, $s(T)$, is given at constant pressure by

$$ds = \left(\frac{dh}{T}\right)_p = \frac{c_p dT}{T} \Rightarrow s(T) = \int_{0\ K}^{T} \frac{c_p(\theta)}{\theta} d\theta \tag{2.12}$$

Note that the convention is to choose the reference temperature for the entropy to be 0 K, rather than ambient temperature, so that the entropy is zero at absolute zero. It is left as an exercise to show that

$$c_p - c_V = \left(\left(\frac{\partial e}{\partial v} \right)_T + p \right) \left(\frac{\partial v}{\partial T} \right)_p \tag{2.13}$$

Perhaps the most important thermodynamic state variable for solidification is the *Gibbs free energy*, G, which is used extensively throughout this book. It is defined as

$$G = H - TS \tag{2.14}$$

Computing the differential dG and using Eq. (2.6), we obtain

$$dG = dH - SdT - TdS = Vdp - SdT \tag{2.15}$$

from which we see that the Gibbs free energy is naturally expressed in terms of the two intensive variables, p and T. In other words, if the pressure and the temperature are known for a unary system, its Gibbs free energy per unit mass or per mole is unambiguously specified. From Eq. (2.15), one sees that

$$S = -\left(\frac{\partial G}{\partial T} \right)_p ; \qquad V = \left(\frac{\partial G}{\partial p} \right)_T \tag{2.16}$$

Figure 2.1 shows schematically the variation of the enthalpy of a single phase with temperature. Note that the value of c_p may be approximately constant at room temperature (i.e., linear $h(T)$), but as $T \rightarrow 0$ K, the specific heat must tend towards zero faster than the temperature. Figure 2.1 also shows the specific Gibbs free energy.

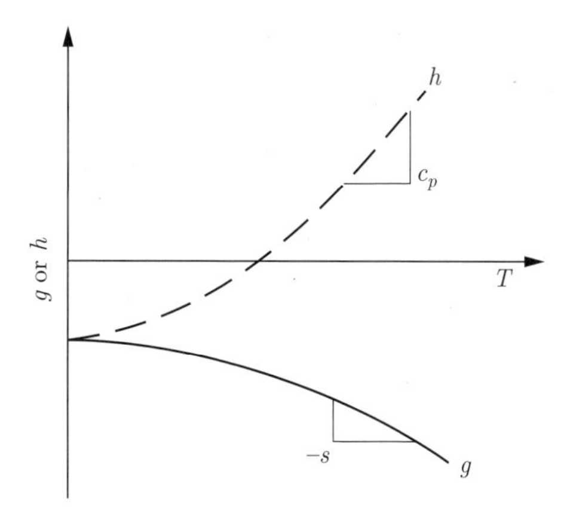

Fig. 2.1 Free energy and enthalpy as a function of temperature.

2.2.2 Equilibrium of phases

The second law of thermodynamics stated that the free energy of a closed system at equilibrium must be minimum (or its entropy maximum). For a system consisting of a single, homogeneous phase, Eq. (2.15) implies that both the temperature and pressure are uniform.

Consider now a closed system consisting of two phases α and β (e.g., solid and liquid, liquid and vapor, two crystalline phases, etc.). Assume that mechanical and thermal equilibrium have been established, i.e., that the temperature and pressure are fixed. The Gibbs free energy of the system can be written as the following sum:

$$G = n_\alpha G_\alpha^m + n_\beta G_\beta^m \tag{2.17}$$

where G_α^m and G_β^m are the molar Gibbs free energies of the two phases, and n_α and n_β are the number of moles of each phase. Since the system is closed, the constraint $(n_\alpha + n_\beta) = n = \text{constant}$ must be satisfied. It is convenient to choose one mole for the whole system, in which case one has $G^m = \chi_\alpha G_\alpha^m + \chi_\beta G_\beta^m$, where we have introduced the mole fractions of the phases χ_i, which have the property $\chi_\alpha + \chi_\beta = 1$. Since equilibrium corresponds to a minimum energy, matter must be repartitioned between the two phases until at equilibrium we have

$$\frac{\partial G^m}{\partial \chi_\alpha} = 0 = G_\alpha^m - G_\beta^m \tag{2.18}$$

We obtain the important property that at equilibrium, a unary system consisting of two phases must satisfy the following conditions:

- $T_\alpha = T_\beta = T$ (thermal equilibrium)
- $p_\alpha = p_\beta = p$ (mechanical equilibrium)
- $G_\alpha^m = G_\beta^m = G^m$ (phase equilibrium)

These relations can be thought of as constraints on the system. Gibbs defined the *degrees of freedom* of a system, designated N_F, as the number of independent variables (T, p and, as we will see later, composition) which can be changed and still maintain the same number of phases in equilibrium. When two phases are present, their pressure and temperature must be equal, so from four potential variables $(T_\alpha, T_\beta, p_\alpha, p_\beta)$ the number of independent variables is reduced to two (T and p) by the constraint of equilibrium. The equality of the Gibbs free energies of the two phases further reduces the number of degrees of freedom by one, e.g., fixing the pressure implies that the temperature is also given. This is familiar for water: at atmospheric pressure at sea level, ice melts at $0°$C (co-existence of ice and water) and water boils at $100°$C (co-existence of water and vapor).

Figure 2.2 shows schematically the variation with temperature of the molar enthalpy and Gibbs free energy of the solid and liquid phases of a

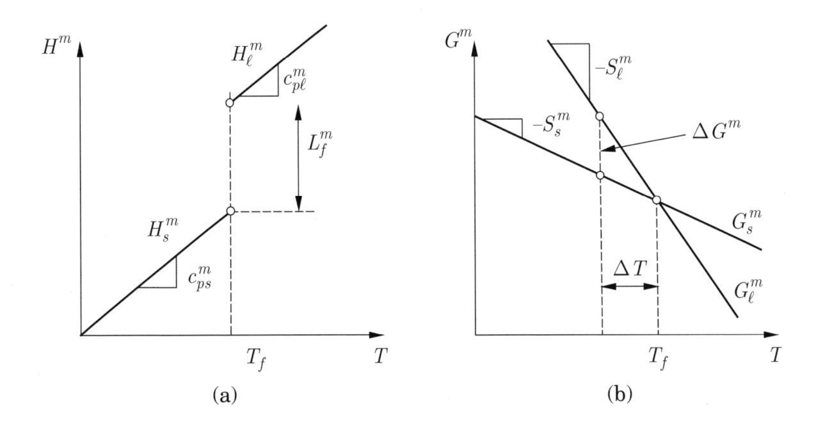

Fig. 2.2 Molar enthalpy (a) and molar Gibbs free energy (b) for the liquid and solid phases of a pure substance as a function of temperature.

pure substance at atmospheric pressure. The point where the free energies of the two phases are equal defines the equilibrium melting point T_f. Note that the slope of G_ℓ^m is greater (in absolute value) than that of G_s^m, since the liquid is more disordered than the solid. The difference in slopes, i.e., $\Delta S_f^m = S_\ell^m - S_s^m$, is called the *entropy of fusion*, and it may be computed using the fact that the molar Gibbs free energy of the liquid and solid phases are equal at T_f.

$$\begin{aligned} G_\ell^m(T_f) - G_s^m(T_f) = 0 &= [H_\ell^m(T_f) - T_f S_\ell^m(T_f)] - [H_s^m(T_f) - T_f S_s^m(T_f)] \\ &= [H_\ell^m(T_f) - H_s^m(T_f)] - T_f\,[S_\ell^m(T_f) - S_s^m(T_f)] \\ 0 &= L_f^m - T_f \Delta S_f^m \\ \text{or } \Delta S_f^m &= \frac{L_f^m}{T_f} \end{aligned} \tag{2.19}$$

where L_f^m is the molar enthalpy of fusion, more often called the *latent heat of fusion* per mole. It corresponds to the increase of enthalpy in the liquid phase at the melting point, as illustrated in Fig. 2.2. It will be useful in later chapters to consider the free energy change upon solidification at some temperature T below the melting temperature $T = T_f - \Delta T$, where ΔT is called the *undercooling* (see Fig. 2.2). The Gibbs free energy change upon melting at temperature T is

$$G_\ell^m(T) - G_s^m(T) = +\Delta G^m(T) = (H_\ell^m(T) - H_s^m(T)) - T(S_\ell^m(T) - S_s^m(T)) \tag{2.20}$$

If the undercooling is small, then it can be shown (see Exercise 2.5) that the enthalpy change at temperature T, $(H_\ell^m(T) - H_s^m(T))$ is the same as at temperature T_f, i.e., $L_f(T) = L_f(T_f)$, and similarly that $\Delta S_f^m(T) = \Delta S_f^m(T_f)$. Substituting these relations into Eq. (2.20) and using Eq. (2.19) yields the result:

$$\Delta G^m(T) = +\Delta S_f^m \Delta T \tag{2.21}$$

What happens to the equilibrium melting point if the pressure changes by an incremental amount dp? The incremental variation in G^m for each phase must be the same for equilibrium to be re-established. Applying Eq. (2.15) in molar form to each phase, we have

$$dG_\ell^m = V_\ell^m dp - S_\ell^m dT \tag{2.22}$$

$$dG_s^m = V_s^m dp - S_s^m dT \tag{2.23}$$

Setting $dG_\ell^m = dG_s^m$ and subtracting Eq. (2.23) from Eq. (2.22) gives the *Clausius-Clapeyron* equation

$$\left.\frac{dp}{dT}\right|_{s\ell} = \frac{\Delta S_f^m}{\Delta V_f^m} = \frac{L_f^m}{T_f \Delta V_f^m} \tag{2.24}$$

where ΔV_f^m is the molar volume change upon melting. For most materials, $\Delta V_f^m > 0$. Water and bismuth are two rare exceptions. The subscript $s\ell$ indicates that the slope corresponds to the solid-liquid phase boundary. One can derive a similar expression for the vaporization reaction, replacing the volumes, enthalpies and entropies by the appropriate values.

When three phases are present (e.g., solid, liquid and vapor), one obtains, by similar arguments, the result that equilibrium is defined by: $T_\alpha = T_\beta = T_\gamma = T; p_\alpha = p_\beta = p_\gamma = p; G_\alpha^m = G_\beta^m = G_\gamma^m = G^m$. This imposes another constraint on the system, reducing N_F to zero. Such points in phase diagrams, where $N_F = 0$, are called *invariant points*. This means that the three phases can coexist only at a unique pressure and temperature (called the *triple point* in this case).

The information we have just described can be summarized in an equilibrium phase diagram, which shows the phases present at various values of T and p. Figure 2.3 shows an example for water. The solid curves are phase boundaries, also called curves of two-phase coexistence. The intersections of two-phase coexistence curves where all three phases coexist are called triple points. Notice that there are two such points in the diagram, one for water, vapor and Ice I at 0.01°C and 612 Pa, and a second involving water, Ice I and Ice III at about −20°C and about 5×10^8 Pa. The slope of the two-phase coexistence curves is given by the Clausius-Clapeyron equation, Eq. (2.24). Water is an unusual material in that its molar volume decreases upon melting, which results in a liquid-solid coexistence curve of slightly negative slope. For most materials, the slope is positive, like the liquid-vapor and solid-vapor curves.

Let us conduct some "thought" experiments at atmospheric pressure p_a, as indicated in Fig. 2.3. Beginning at high temperature, where $T > T_v(p_a)$, only the vapor phase exists. Now, the system is cooled with a fixed heat extraction rate per unit mass, $\dot{h} < 0$. The cooling rate $\dot{T}$ of the vapor phase is given by $\dot{h}/c_{pg}$, as indicated in Fig. 2.4. As the temperature decreases, the liquid-vapor phase boundary is encountered at temperature T_v, commonly called the boiling point. The two phases coexist at T_v until all of the vapor has transformed to liquid. Note the very large isothermal

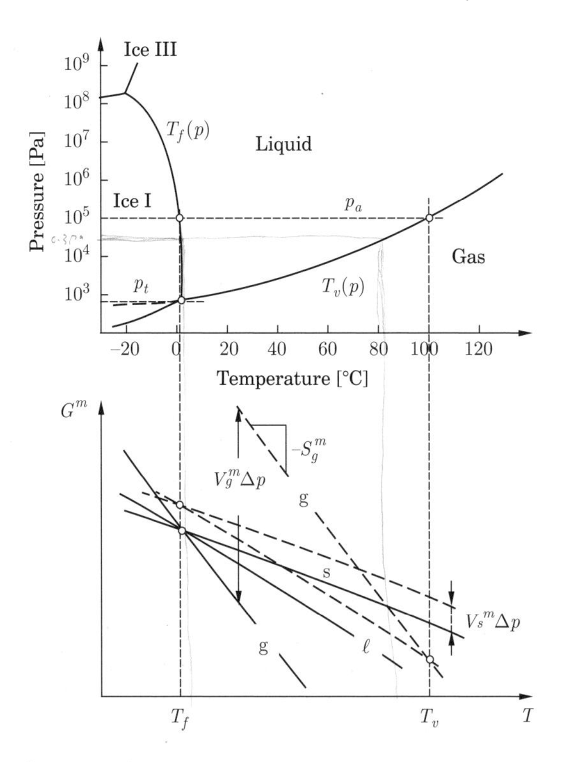

Fig. 2.3 Equilibrium phase diagram for water and schematic Gibbs free energies of the solid, liquid and gas phases at atmospheric pressure p_a (dashed curves) and at the triple point pressure p_t (solid lines). This figure is only schematic as $V_g^m \approx 1250\, V_\ell^m$.

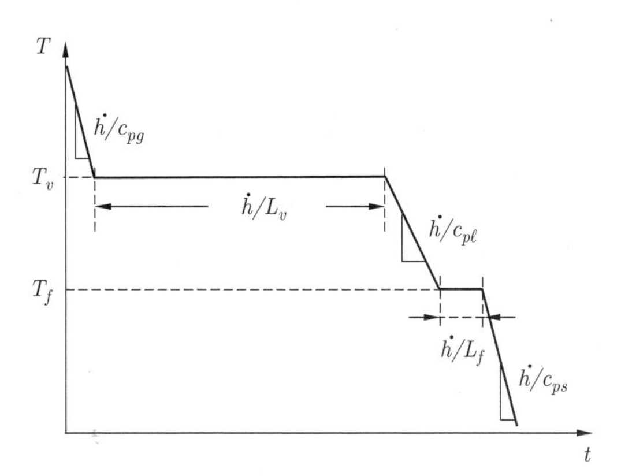

Fig. 2.4 Thermal history of a water sample cooled with a constant heat extraction rate, $\dot{h} < 0$, at atmospheric pressure. As $c_{p\ell} \approx 2c_{pg} \approx 2c_{ps} = 4.18$ J g^{-1}, note that the cooling rate of the liquid is half that of the two other phases. Also, since $L_V = 2250$ J g^{-1} and $L_f = 333$ J g^{-1}, the time for condensation is much longer than the time for solidification.

plateau associated with condensation of water: it is a direct measure of the very large enthalpy of vaporization (or condensation). If the cooling curve had instead continued with the same slope, this would correspond to a temperature at the end of the plateau more than 1000°C below T_v. After all the vapor has condensed, the temperature can once again decrease. Because the specific heat of water is about twice that of the vapor, the cooling rate ($\dot{T}$) is about half of the value before condensation if we maintain a constant rate of heat extraction $\dot{h}$. When reaching the melting point T_f, the temperature again remains at the plateau until all of the liquid has solidified. Notice the difference in the duration of the arrest at T_f compared to the arrest at T_v, reflecting the difference in magnitude of the latent heats of fusion and vaporization.

Suppose that we were to repeat the preceding thought experiment at slightly lower pressure, e.g., on the top of Mount Everest where the pressure is only about $0.3\,p_a$. What would change? We would traverse the phase diagram along an isobar parallel to p_a at $p = 30$ kPa. The results would be qualitatively similar, but the values we would observe for T_f and T_v would change. Note, however, that because the slopes of the lines of two-phase separation are different, the change in boiling point would be much larger than the change in melting point (water boils at around 70°C at the top of Mount Everest). It also should be pointed out that the experiment of boiling water when performed in a kitchen is different from the one proposed here: indeed, the kitchen's atmosphere is not a unary system as we have a mixture of vapor and air, itself made of many species (nitrogen, oxygen, carbon dioxide, etc.). On the other hand, if water vapor is initially the only species in a closed vessel, the equation of state for the vapor phase ($pV = nRT$ if it is assumed to be ideal), imposes that the pressure cannot remain constant upon cooling and later during condensation (see Exercise 2.2).

Figure 2.3 also includes Gibbs free energy curves for the three phases at atmospheric pressure p_a (dashed lines) and at the triple-point pressure, $p_t = 612$ Pa (solid lines). Note that these curves are only schematic, as the molar volume of the vapor phase is about 1250 times that of the condensed phases. Free energy is a very powerful tool for understanding the relationships between the phases. At p_t, the three curves intersect at the triple point (0.01°C), where the free energies of all three phases are equal. Recall from Eq. (2.16) that

$$S^m = -\left(\frac{\partial G^m}{\partial T}\right)_p ; \qquad V^m = \left(\frac{\partial G^m}{\partial p}\right)_T \tag{2.25}$$

Thus, the slope of each curve is negative, and the magnitudes of the slopes, i.e., the entropy of the phases, are in the order $S_s^m < S_\ell^m < S_g^m$. When the pressure is increased to p_a, each curve shifts upward, resulting in a new curve approximately parallel to the original. The magnitude of the shift is proportional to its specific volume (note that the specific volume will also vary between p_t and p_a). Because the specific volume of liquid water is smaller than that of the solid near the melting point, the liquid free energy curve shifts slightly less than the corresponding curve for the solid. This decreases the solid-liquid equilibrium from $T_t = 0.01°C$ to $T_f = 0°C$. By the same reasoning, we see that the free energy curve for the gas shifts more than those of the other two phases. The intersection points of the phase curves are projected back to the phase diagram, as shown. Fig. 2.3 also shows (as a heavy dashed line) the continuation of the liquid-gas two phase equilibrium curve past the triple point, extending into the "Ice I" region. If the solid Ice I were to somehow fail to appear, then the system would remain liquid, i.e., it would follow this projection for this range of temperatures and pressures. This is called *metastable equilibrium*, because there is a different transformation between solid and gas that has lower energy. The concept of metastable extensions to the lines in the phase diagrams will prove to be very useful in later chapters for understanding microstructure development.

2.3 BINARY ALLOYS

2.3.1 Thermodynamics of a single phase solution

We now extend the treatment developed in the previous section to the thermodynamics of binary mixtures. For the moment, the mixture will be assumed to consist of a single phase, i.e., a *solution* of the two components. Consider the volume illustrated in Fig. 2.5, composed of a homogeneous phase made up of two components, labeled A and B. The components A and B may be atoms or molecules. B is assumed to be the minor component and is called the *solute*, whereas the major component A is the *solvent*. In addition to the thermodynamic variables introduced in the previous section, we now need additional variables to completely specify the state of the system; specifically we need to describe the *composition*. The system contains n_A moles of component A, and n_B moles of species B. The total

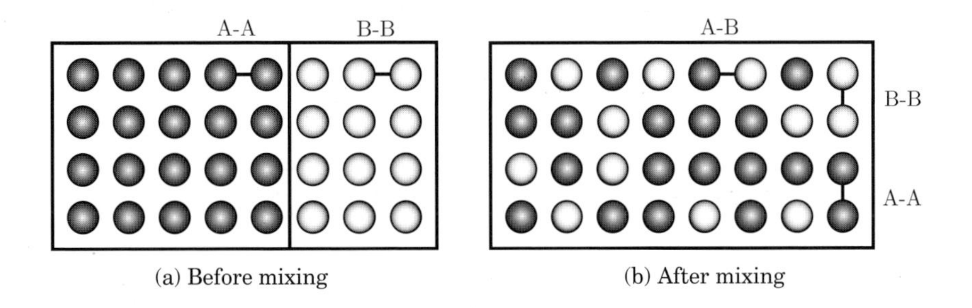

Fig. 2.5 Schematic of the mixing of n_A moles of A atoms/molecules with n_B moles of B atoms/molecules. Before the components are mixed, there are only A-A and B-B bonds; upon forming a solution, some of these bonds are replaced by A-B bonds. This becomes important when we discuss the free energy of the solution.

number of moles $n = n_A + n_B$. We can also specify the composition using the mole fractions of each component, X_A and X_B, defined as

$$X_A = \frac{n_A}{n_A + n_B}; \qquad X_B = \frac{n_B}{n_A + n_B} \tag{2.26}$$

Other possible representations of the composition are discussed in Chap. 3. By their definitions, it follows that $X_A + X_B = 1$. One can choose to use the n_J or the X_J to represent the composition, as long as the constraints on the X_J are kept in mind.

Because the molecules or atoms interact when they are mixed, the Gibbs free energy of the solution varies with composition, as well as with temperature and pressure. Writing $G = G(T, p, n_A, n_B)$, the differential dG becomes

$$\begin{aligned} dG(p, T, n_A, n_B) = {} & V(p, T, n_A, n_B)dp - S(p, T, n_A, n_B)dT \\ & + \mu_A(p, T, X_B)dn_A + \mu_B(p, T, X_B)dn_B \end{aligned} \tag{2.27}$$

where we have introduced the *chemical potential* of each species, μ_A and μ_B, defined as

$$\mu_A = \left(\frac{\partial G}{\partial n_A}\right)_{p,T,n_B}; \qquad \mu_B = \left(\frac{\partial G}{\partial n_B}\right)_{p,T,n_A} \tag{2.28}$$

Note that the chemical potentials are intensive variables, which is why in Eq. (2.26) they are not written as functions of n_A and n_B, but rather as functions of the mole fraction X_B. We could have chosen X_A instead of X_B, since we know that $X_A + X_B = 1$. Dividing Eq. (2.27) by n allows us to write the expression in terms of molar quantities:

$$\begin{aligned} dG^m(p, T, X_B) = {} & V^m(p, T, X_B)dp - S^m(p, T, X_B)dT \\ & + (\mu_B(p, T, X_B) - \mu_A(p, T, X_B))dX_B \end{aligned} \tag{2.29}$$

where we have invoked the relation $dX_A = -dX_B$.

One might then wonder how the molar Gibbs free energy of the system varies with composition at constant temperature and pressure. G is called a *homogeneous function* of the n_J, meaning that its magnitude is directly proportional to the amount of its constituents, so at constant temperature and pressure we have

$$G = n_A\mu_A + n_B\mu_B \quad (\text{Fixed } T, p) \tag{2.30}$$

This expression is shown graphically in Fig. 2.6. Computing the differential of Eq. (2.30) gives:

$$dG = n_A d\mu_A + n_B d\mu_B + \mu_A dn_A + \mu_B dn_B \tag{2.31}$$

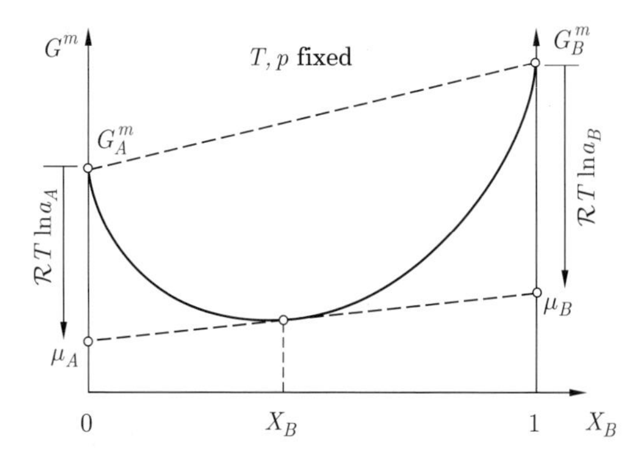

Fig. 2.6 Molar free energy of a binary solution as a function of the mole fraction of element B, showing the tangent construction to compute the chemical potentials μ_A and μ_B.

Taking T and p constant in Eq. (2.27), and equating dG with Eq. (2.31) yields the *Gibbs-Duhem* equation:

$$n_A d\mu_A + n_B d\mu_B = 0 \tag{2.32}$$

Dividing Eq. (2.32) by n, we obtain the following expression:

$$X_A d\mu_A + X_B d\mu_B = 0 \tag{2.33}$$

Next, we divide Eq. (2.30) by n, yielding

$$G^m = X_A \mu_A + X_B \mu_B \tag{2.34}$$

and then differentiate with respect to X_B:

$$\frac{\partial G^m}{\partial X_B} = \mu_B - \mu_A \tag{2.35}$$

Finally, combining Eqs. (2.34) and (2.35) yields the following two equations:

$$\begin{aligned} \mu_A(T,p,X_B) &= G^m(T,p,X_B) - X_B \left(\frac{\partial G^m}{\partial X_B}\right)_{T,p} \\ \mu_B(T,p,X_B) &= G^m(T,p,X_B) + (1 - X_B) \left(\frac{\partial G^m}{\partial X_B}\right)_{T,p} \end{aligned} \tag{2.36}$$

These equations define the *tangent rule construction* giving the chemical potential of components A and B for a solution having a composition X_B (see Fig. 2.6). The molar free energy G^m at that composition (and pressure and temperature) being known, the tangent to G^m at that point intersects the vertical axes, $X_A = 1$ and $X_B = 1$, at the values μ_A and μ_B, respectively.

Some thermodynamics texts find it useful to define the *chemical activity*, designated as $a_A(T, p, X_B)$ and $a_B(T, p, X_B)$ of components A and B in the solution of composition X_B, defined as

$$a_A(T, p, X_B) = \exp\left(-\frac{G_A^m - \mu_A}{\mathcal{R}T}\right)$$
$$a_B(T, p, X_B) = \exp\left(-\frac{G_B^m - \mu_B}{\mathcal{R}T}\right) \tag{2.37}$$

where $\mathcal{R}$ is the universal gas constant ($\mathcal{R} = 8.3144\,\mathrm{J\,mole^{-1}\,K^{-1}}$) The chemical activity of an element thus measures, in the form of a thermal activation energy, the difference between the molar free energy of the component (G_J^m) when it is pure and when it is diluted in the solution (μ_J). Using these expressions, the chemical activity of both components can be read off directly from a figure such as Fig. 2.6, where $\mathcal{R}T \ln a_A$ and $\mathcal{R}T \ln a_B$ correspond to the differences ($G_A^m - \mu_A$) and ($G_B^m - \mu_B$), respectively.

2.3.2 Ideal and regular solutions

We will need to understand the effect of thermodynamic variables on the free energy of the solution. This is most conveniently done using a model, and in this section, we introduce two of the most common: the ideal and regular solution models. For a totally disordered solution such as that illustrated in Fig. 2.5, the formation of one mole of solution with X_A moles of A and X_B moles of B will create on average

$$\begin{aligned} &0.5(X_A N_0)(X_A N_b) \quad \text{bonds} \quad (A - A),\\ &0.5\left[(X_A N_0)(X_B N_b) + (X_B N_0)(X_A N_b)\right] \quad \text{bonds} \quad (A - B)\\ &0.5(X_B N_0)(X_B N_b) \quad \text{bonds} \quad (B - B) \end{aligned} \tag{2.38}$$

Here, N_0 is Avagadro's number and N_b is the coordination number or the number of atomic bonds per atom (e.g., 6 in a 3-dimensional simple cubic arrangement). There are $X_J N_0$ atoms/molecules of species J and each one is surrounded by N_b neighbors, whose types A/B are precisely in the ratio X_A/X_B if the solution is random. The internal energy associated with the bonds is thus given by

$$E_i^m = 0.5 N_0 N_b \left(X_A^2 \varepsilon_{AA} + X_B^2 \varepsilon_{BB} + 2 X_A X_B \varepsilon_{AB}\right) \tag{2.39}$$

where ε_{IJ} is the bond energy between components I and J. Note that $\varepsilon_{IJ} < 0$. This relationship can also be written as

$$\begin{aligned} E_i^m &= 0.5 N_0 N_b \left(X_A^2 \varepsilon_{AA} + X_B^2 \varepsilon_{BB} + X_A X_B \varepsilon_{AA} + X_A X_B \varepsilon_{BB}\right)\\ &\quad + 0.5 N_0 N_b X_A X_B \left(2\varepsilon_{AB} - \varepsilon_{AA} - \varepsilon_{BB}\right) \end{aligned} \tag{2.40}$$

Taking into account that $X_A + X_B = 1$, we find

$$E_i^m = X_A E_{iA}^m + X_B E_{iB}^m + \Omega^m X_A X_B \tag{2.41}$$

where the last term in this equation, $\Delta H_{mix} = \Omega^m X_A X_B$ is called the *enthalpy of mixing* and

$$E_{iA}^m = 0.5 N_0 N_b \varepsilon_{AA} \tag{2.42}$$

$$E_{iB}^m = 0.5 N_0 N_b \varepsilon_{BB} \tag{2.43}$$

$$\Omega^m = N_0 N_b \bar{\varepsilon}_{AB} = N_0 N_b \left(\varepsilon_{AB} - \frac{\varepsilon_{AA} + \varepsilon_{BB}}{2} \right) \tag{2.44}$$

The bonding energy $\bar{\varepsilon}_{AB} = \varepsilon_{AB} - 0.5(\varepsilon_{AA} + \varepsilon_{BB})$ measures the relative affinity of atoms A and B. When $\bar{\varepsilon}_{AB} > 0$, the magnitude of the bond energy between A and B atoms is smaller than the average A-A and B-B bonding energy. The tendency in this case is to have a *demixing* between A and B species. On the other hand, when $\bar{\varepsilon}_{AB} < 0$, the system can lower its energy by forming more A-B bonds, and thus will have a tendency to form an *ordered phase* in which A atoms have a B-environment and vice-versa.

We now determine the free energy of the solution. The contribution of bonding to the internal energy described above is only one part of the total. We must also consider the vibrational energy at finite temperature associated with the simple mixing of various species. A helpful analogy is to compare the separation of mixed inert objects, such as coins on a table, to the separation of animated ones, such as boys and girls in a nursery school! In the first case, corresponding to the absolute zero temperature in our atomic analogy, it does not cost energy to separate mixed objects. However, at a temperature $T > 0$ K, atomic vibration induces an energy cost of separation even in the case where $\varepsilon_{AA} = \varepsilon_{BB} = \varepsilon_{AB}$. The simplest model consistent with the internal energy computed above is one with only two contributions to the free energy beyond the simple rule of mixtures: the excess energy $\Omega^m X_A X_B$ and the *entropy of mixing* S_{mix}^m (described below). The free energy in this case is derived as follows:

$$\begin{aligned}
G^m &= E^m + pV^m - TS^m \\
&= (X_A E_A^m + X_B E_B^m) + p(X_A V_A^m + X_B V_B^m) - T(X_A S_A^m + X_B S_B^m) \\
&\quad + \Omega^m X_A X_B - TS_{mix}^m \\
&= X_A(E_A^m + pV_A^m - TS_A^m) + X_B(E_B^m + pV_B^m - TS_B^m) + \Omega^m X_A X_B - TS_{mix}^m \\
G^m &= X_A G_A^m + X_B G_B^m + \Omega^m X_A X_B - TS_{mix}^m
\end{aligned} \tag{2.45}$$

where S_A^m and S_B^m are the molar entropies of the pure components before mixing, and G_A^m and G_B^m are the molar free energies of the pure components A and B. The first two terms in Eq. (2.45) represent a simple linear combination of the $A-A$ and $B-B$ bonds, often referred to as a "rule of mixtures." The entropy of mixing is modeled by enumerating the number of distinguishable configurations of a collection containing N_A A-atoms and N_B B-atoms ($N_A + N_B = N_0$). Following Boltzmann, the molar entropy of mixing is then given by

$$S_{mix}^m = k_B \ln \frac{(N_A + N_B)!}{(N_A)!(N_B)!} \tag{2.46}$$

where k_B is Boltzmann's constant ($k_B = 1.3807 \times 10^{-23}\,\mathrm{J\,K^{-1}}$). Using Stirling's formula for the factorial operation ($x!$):

$$\ln x! \approx x \ln x - x \qquad \text{for } x \to \infty \tag{2.47}$$

the entropy of mixing S^m_{mix} can be written as:

$$S^m_{mix} = -\mathcal{R}\,(X_A \ln X_A + X_B \ln X_B) \tag{2.48}$$

since $\mathcal{R} = k_B N_0$. Inserting this expression into Eq. (2.45), one finally gets the general expression for the free energy of a *regular solution*:

$$G^m = X_A G^m_A + X_B G^m_B + \Omega^m X_A X_B + \mathcal{R}T\,(X_A \ln X_A + X_B \ln X_B) \tag{2.49}$$

When $\Omega^m = 0$, i.e., enthalpy of mixing is zero (the bonds between A and B atoms are equivalent to those between atoms of the pure substance), this expression reduces to the so-called *ideal solution*. Let us analyze in slightly more detail the various contributions to the molar Gibbs free energy of such an ideal solution, as illustrated in Fig. 2.7. The first two terms of Eq. (2.49) correspond to a simple linear interpolation between G^m_A and G^m_B. Note that as the temperature increases, the free energies of both pure components decrease as a result of the entropy term ($-TS^m_J$), as shown in Fig. 2.1. Since $\Omega^m = 0$, the free energy of the ideal solution deviates from this linear interpolation only via the entropy of mixing. This term is negative and increases as the temperature increases, but is of course zero for pure A or B. The maximum difference is reached when $X_A = X_B = 0.5$, and the slope of G^m is equal to $-\infty$ for $X_A = 1$ and $+\infty$ for $X_B = 1$.

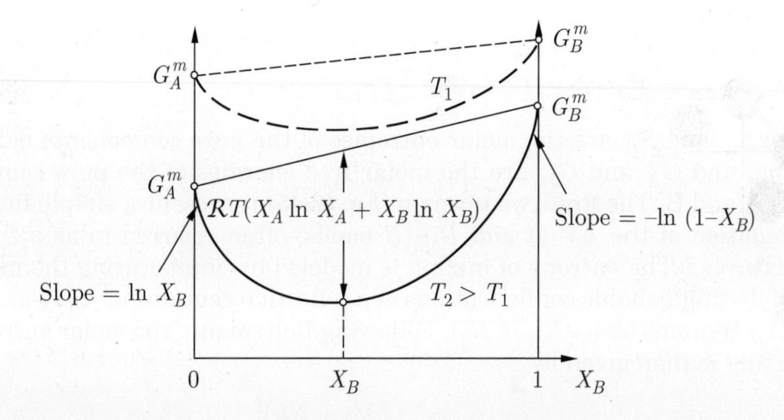

Fig. 2.7 Molar Gibbs free energy for an ideal solution at two temperatures $T_1 < T_2$.

The chemical potentials for the ideal solution are obtained at any given T and p by the tangent construction (see Fig. 2.6) and Eq. (2.36)

$$\mu_A(X_B) = G^m(X_B) - X_B \frac{\partial G^m}{\partial X_B}$$
$$\mu_B(X_B) = G^m(X_B) + (1 - X_B) \frac{\partial G^m}{\partial X_B} \tag{2.50}$$

Now, let us consider solutions where $\Omega^m \neq 0$. When $\Omega^m < 0$, the enthalpy of mixing adds to the entropy of mixing contribution, i.e., makes the curve G^m even deeper (see Fig. 2.8). As noted earlier, this corresponds to stronger attraction of A and B atoms, possibly leading to the formation of ordered phases, such as those illustrated in Fig. 2.9 for the Cu-Au system.

At high temperature, despite the attraction of A and B atoms, the locations of the atoms might be totally random, at least at long range. As the temperature decreases, ordering might occur over fairly long distances.

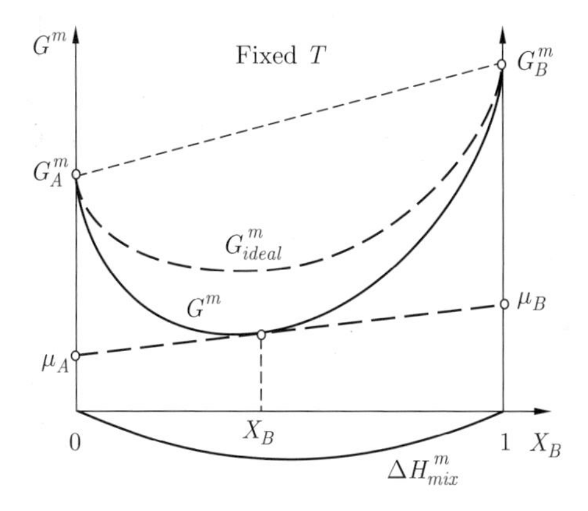

Fig. 2.8 Molar Gibbs free energy for a regular solution with $\Omega^m < 0$ at a fixed temperature. The chemical potentials are also indicated in this figure for a given concentration.

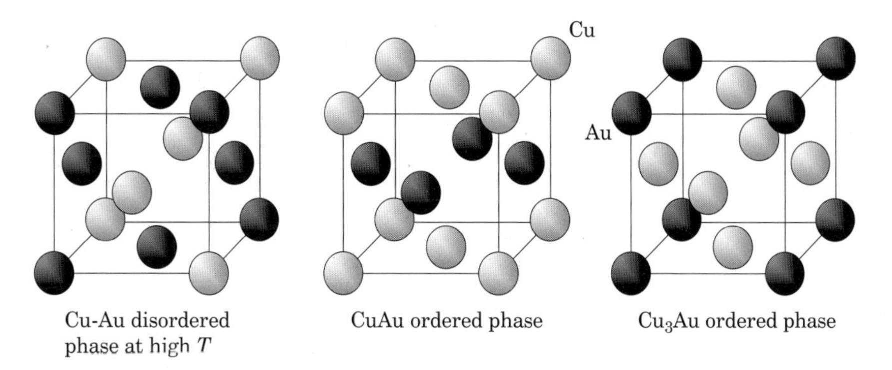

Fig. 2.9 Ordered phases in the Cu-Au system.

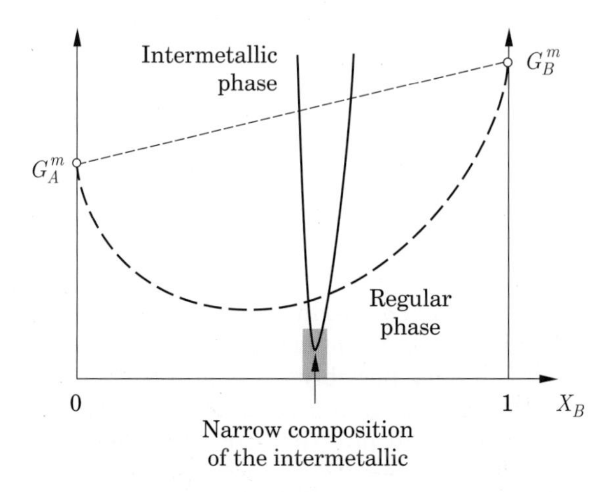

Fig. 2.10 Schematic of the Gibbs free energy of an intermetallic phase and intersecting a regular phase.

This is shown schematically for the face centered cubic (fcc) phase of the Cu-Au alloy (see Fig. 2.9). For a composition of about 0.5, lowering the temperature causes the Cu and Au atoms to occupy alternating layers of the fcc structure (center). Note that there are 4 Au atoms on the vertical faces of the unit cell (2 intrinsic atoms), 2 Cu atoms on the upper and bottom faces (1 intrinsic atom) and 8 at the corners (1 intrinsic atom). At lower mole fraction of gold (around $X_{Au} = 0.25$), the Au atoms occupy the corners of the unit cell (1 intrinsic atom), while the Cu atoms are on the 6 faces (3 intrinsic atoms).

In such cases, the attraction of components might lead to the formation of *intermediate phases*, i.e., a new solid having its own crystal structure. When this occurs for a very narrow composition range, it is usually called an *intermetallic phase*. Such phases are usually characterized by a very sharp Gibbs free energy curve, as illustrated in Fig. 2.10. The significance of the shape of the curve will be made more clear in Chap. 3.

For $\Omega^m > 0$, the enthalpy of mixing has an effect that opposes that of the entropy of mixing. At high temperature, the entropy contribution may still dominate and the G^m curve would be overall concave (see Fig. 2.11). At lower temperature, the contribution from the entropy of mixing decreases, while the enthalpy of mixing remains unchanged. As a result, it is possible for curves to become convex, leading to phase separation. This will be discussed in more detail in the next chapter.

2.3.3 Equilibrium of two phases

The conditions for equilibrium of several phases in a unary system (Sect. 2.2.2) is now extended to binary alloys containing species A and B (atoms or molecules). By the same argument as before, if several phases

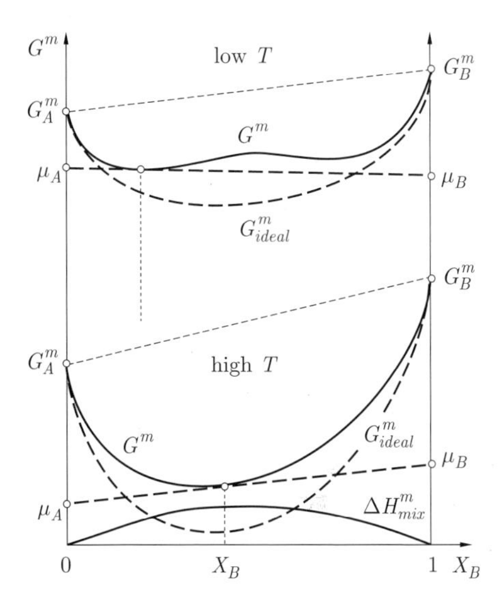

Fig. 2.11 Molar Gibbs free energy for a regular solution with $\Omega^m > 0$ at low and high temperatures.

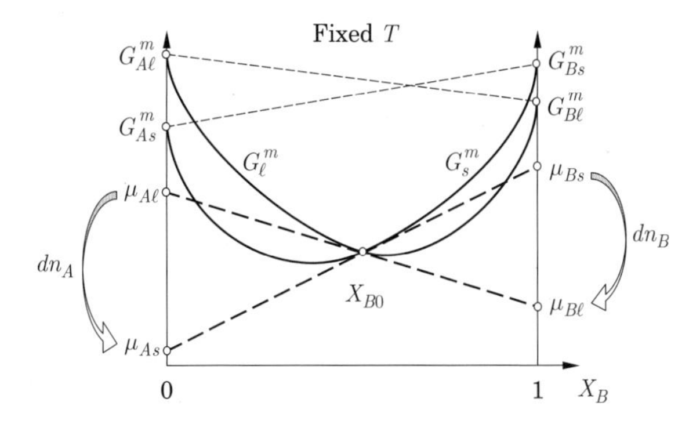

Fig. 2.12 Free energies of two phases, solid and liquid, as a function of the composition X_B and chemical potentials at the intersection of these two curves.

exist in equilibrium in a binary alloy, the temperature and pressure within all the phases must be equal. What can we say about the compositions in the various phases? For the sake of simplicity, consider just a solid and liquid phase: the compositions in these two phases are written now X_{Bs} and $X_{B\ell}$, where we have added a second subscript to identify the phase. Note that the composition of each phase must be uniform, otherwise the molar Gibbs free energy, G_s^m or G_ℓ^m, will not be constant within each phase. We now deduce the relationship between X_{Bs} and $X_{B\ell}$.

Consider, at fixed pressure and temperature, the molar free energies of the two phases, G_s^m and G_ℓ^m (Fig. 2.12). If the two free energy curves do not cross, the minimum energy of the system would always be described by the lowest energy curve over the whole composition field, and thus only one phase would be present in the system. If the two curves intersect, as illustrated in Fig. 2.12, one might be tempted to say that the phase that exists in equilibrium for a particular composition is the one with the lowest free energy. This implies that one would have solid for $X_B < X_{B0}$ and liquid for $X_B > X_{B0}$ at this temperature. Two-phase equilibrium would exist only when $G_s^m = G_\ell^m$, i.e., at point X_{B0}. Let's now understand why this analysis is wrong. At the point of intersection, called the *T_0 point* for reasons we will see later, the chemical potentials of the two components in the solid and liquid phases are as shown in Fig. 2.12. Since $\mu_{As} \neq \mu_{A\ell}$ and $\mu_{Bs} \neq \mu_{B\ell}$, the system could decrease its energy by transferring solute atoms B from the solid to the liquid (gain of energy $\mu_{B\ell} - \mu_{Bs} < 0$) and solvent atoms A from the liquid to the solid (gain of energy $\mu_{As} - \mu_{A\ell} < 0$). Indeed, the variation of free energy of the system during this operation would be given by

$$dG = (\mu_{As} - \mu_{A\ell})\, dn_A + (\mu_{B\ell} - \mu_{Bs})\, dn_B < 0 \tag{2.51}$$

For the alloy, the equilibrium condition for two co-existing phases therefore corresponds to (Fig. 2.13)

$$\mu_{As} = \mu_{A\ell}; \qquad \text{and} \qquad \mu_{Bs} = \mu_{B\ell} \tag{2.52}$$

This equilibrium condition is illustrated in Fig. 2.13. The conditions that $\mu_{As} = \mu_{A\ell}$ and $\mu_{Bs} = \mu_{B\ell}$ require that the compositions of the solid and liquid lie on the *common tangent*, as shown. At this temperature and pressure, the solid phase has the lowest free energy below X_{Bs}, whereas above $X_{B\ell}$, the phase with the lowest free energy is the liquid. For any

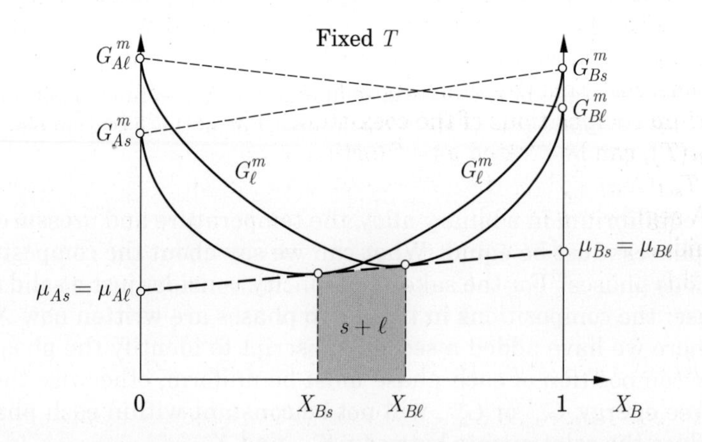

Fig. 2.13 Free energies of two phases, solid and liquid, as a function of the composition X_B showing the condition of equilibrium when they co-exist.

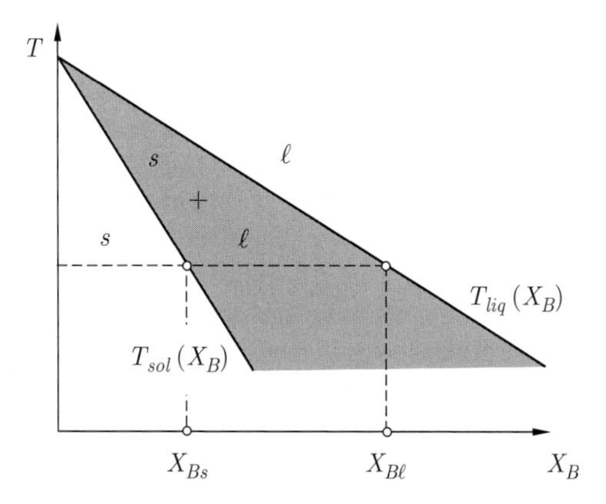

Fig. 2.14 Simple phase diagram showing the region where the two phases, solid and liquid, co-exist.

alloy whose composition lies in the interval $X_{Bs} < X_B < X_{B\ell}$, the system will minimize its molar free energy by having both phases, with a composition X_{Bs} for the solid and $X_{B\ell}$ for the liquid. In order to ensure solute conservation, the fraction of phases will change from fully solid for an alloy of composition $X_B = X_{Bs}$ to fully liquid for a composition $X_B = X_{B\ell}$. For any composition between X_{Bs} and $X_{B\ell}$, solute conservation gives the mole fractions χ_s and χ_ℓ as

$$\chi_s X_{Bs} + \chi_\ell X_{B\ell} = X_B \tag{2.53}$$

The ratio $X_{Bs}/X_{B\ell}$ of the solute element B in the solvent A is called the *partition coefficient*, k_0^m. As there is only one solute element (B) in a binary alloy, no index "B" will be used in this case. For a multicomponent alloy, one will specify k_{0B}^m, k_{0C}^m, etc, for each solute element (see Chap. 3). In the particular case shown in Fig. 2.13, k_0^m is smaller than unity.

If one performs the same analysis over a range of temperatures, the equilibrium compositions of the coexistent solid and liquid phases, $X_{Bs}(T)$ and $X_{B\ell}(T)$, can be plotted as a function of temperature. This defines the *solidus* $T_{sol}(X_B)$ and *liquidus* curves $T_{liq}(X_B)$ on an equilibrium phase diagram. A segment of such a construction is shown in Fig. 2.14. Notice that the liquidus and solidus intersect at the melting point of pure A. We will pursue this subject in much greater depth in the next chapter.

2.3.4 Multi-component alloys and Gibbs' phase rule

There is a similar procedure for dealing with a multi-component alloy consisting of N_c chemical species (A, B, C,...). Assuming that element A is the solvent, the molar Gibbs free energy is a function of the $(N_c - 1)$ solute compositions, i.e., $G^m(p, T, X_B, X_C, \ldots, X_{N_c})$. Consider a regular solution model with only two-element interactions. The molar Gibbs free energy is given by:

$$G^m = \sum_{I=1}^{N_c} X_I G_I^m + \sum_{I=1}^{N_c} \sum_{J>I}^{N_c} \Omega_{IJ}^m X_I X_J + \mathcal{R}T \sum_{I=1}^{N_c} X_I \ln X_I \tag{2.54}$$

where G_I^m is the molar Gibbs free energy of the pure form of component I, and Ω_{IJ}^m represents the interaction parameter between chemical components I and J. The chemical potentials are still given by Eq. 2.28 for any of the components of the system. At fixed T and p, the free energy curves we found for binary alloys become hypersurfaces in $(N_c - 1)$-space, i.e., surfaces in a ternary alloy, volumes in a quaternary alloy, etc. The conditions for having N_ϕ phases in equilibrium is to have the temperature, pressure and chemical potentials of all the components equal in all of the phases, i.e.:

- $T_\alpha = T_\beta = T_\gamma = \ldots = T_{N_\phi} = T$
- $p_\alpha = p_\beta = p_\gamma = \ldots = p_{N_\phi} = p$
- $\mu_{A\alpha} = \mu_{A\beta} = \mu_{A\gamma} = \ldots = \mu_{AN_\phi} = \mu_A$
- $\mu_{B\alpha} = \mu_{B\beta} = \mu_{B\gamma} = \ldots = \mu_{BN_\phi} = \mu_B$
- $\mu_{C\alpha} = \mu_{C\beta} = \mu_{C\gamma} = \ldots = \mu_{CN_\phi} = \mu_C$
- $\vdots$
- $\mu_{N_c\alpha} = \mu_{N_c\beta} = \mu_{N_c\gamma} = \ldots = \mu_{N_cN_\phi} = \mu_{N_c}$

The first subscript of the chemical potentials again indicates the component while the second identifies the phase. At fixed p and T, the condition of equilibrium given by the chemical potentials can be represented as a common *hyperplane* tangent to the molar Gibbs free energy hypersurfaces of all the co-existing phases (i.e., a line tangent to the G^m curve in a binary alloy, a plane tangent to a surface for a ternary alloy, etc.). More details will be given in the next chapter.

The question now arises as to how many phases can co-exist in multicomponent alloys? We have seen in a unary system that a maximum of three phases can co-exist if the pressure and temperature can vary. Considering now a multicomponent alloy in which N_ϕ phases are present, the free variables are as follows:

- temperature, one variable;
- pressure, one variable;
- $N_c - 1$ "free" compositions in each phase, i.e., $N_\phi(N_c - 1)$ variables.

The number of constraints imposed by the equality of the chemical potentials is given by $(N_\phi - 1)N_c$. Therefore, the number of free variables, including the pressure and temperature of the system, is equal to $N_c + 2 - N_\phi$. This relationship is known as *Gibbs' phase rule*, usually written as

$$N_F = N_c - N_\phi + 2 \tag{2.55}$$

Let us take a few examples.

Binary alloy: $N_c = 2$
When there is only one phase ($N_\phi = 1$), there are three independent variables: T, p, X_B. When two phases are present, this number is reduced to two: if p and T are fixed, we have seen that the compositions of the two phases (say solid and liquid) are fixed (given by X_{Bs} and $X_{B\ell}$). For three co-existing phases, only one variable remains, the pressure. Assuming that this is fixed (e.g., $p_a = 10^5$ Pa), no variable is left and the co-existence of three phases occurs at a specific temperature, called an *invariant point*, corresponding to the common tangent to the three G^m curves of the phases considered. The co-existence of four phases can be achieved only if the pressure is allowed to vary until the chemical potentials of the two components in the four phases are equal.

Ternary alloy: $N_c = 3$
In this case, there are four independent variables when only one phase is considered: T, p, X_B and X_C. At constant pressure, the maximum number of co-existing phases is thus four for ternary alloys, at the invariant temperature. For the case where three phases are present and p is fixed, there is still one degree of freedom: fixing the temperature, the compositions of B and C in the three phases are given (say $(X_{B\alpha}, X_{C\alpha}), (X_{B\beta}, X_{C\beta})$, $(X_{B\gamma}, X_{C\gamma})$). The case where there is one active degree of freedom is sometimes called *monovariant*.

The practical application of the thermodynamic rules presented in this chapter will be seen in the next chapter for both binary and ternary alloys. It will be shown that various types of phase diagrams can be drawn using these conditions of equilibrium.

2.4 DEPARTURE FROM EQUILIBRIUM

The systems encountered in materials processing in general, and for solidification in particular, are almost never in a state of equilibrium. The reason is simple: most processes are run too fast to allow enough time for complete diffusion of heat or solute, leading to non-equilibrium conditions. This is not necessarily a bad thing. Indeed, out-of-equilibrium phenomena enable engineers to retain *metastable* phases, such as diamond in pure carbon, or martensite in steel, that would normally not form under equilibrium conditions. Faster processing also tends to refine the microstructure, as we shall see later, leading to improved mechanical properties. This raises the question: since equilibrium conditions will not be met, is the thermodynamic approach presented in this and in the following chapter of any use? In this section we describe why the answer is a definite "yes".

2.4.1 Interfacial equilibrium

The first departure from equilibrium relevant to our study of solidification arises from *gradients*, in temperature and/or composition of the various phases, that form during processing. For a unary system, this means, for example, that a thermocouple placed in one of the phases (e.g., the liquid) measures a local temperature which is not equal to that of the solid or of the solid-liquid interface (Fig. 2.15, left). Thus, an experimental measurement of the temperature would record the melting point of the substance only when the interface intersects the thermocouple position. Note that the temperature is continuous across the solid-liquid interface, whereas the enthalpy is discontinuous as seen earlier. In alloys, composition gradients may exist in all of the phases, as solute diffusion is even slower than thermal diffusion. In that case, the compositions in the solid and in the liquid might be lower than the interfacial compositions (X^*_{Bs} and $X^*_{B\ell}$, respectively) predicted from the equilibrium condition (Fig. 2.13). Such a situation is represented on the right of Fig. 2.15. Note that the composition is discontinuous across the interface, whereas the chemical potentials are continuous from the tangent construction.

In the situations where the departure from equilibrium is due only to gradients in the various phases, the equilibrium conditions derived in

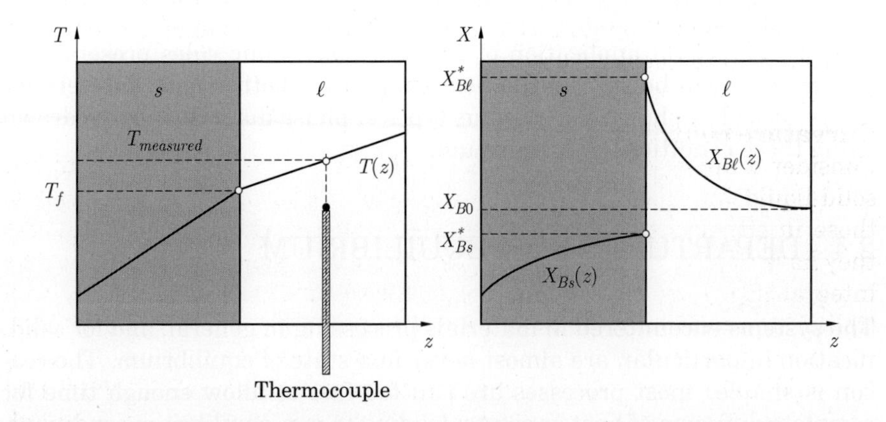

Fig. 2.15 Schematic of the temperature gradients and solute gradients that might exist in a unary or binary system.

the preceding sections can still be applied *at the interface*. This is sometimes called the assumption of *local equilibrium*. We will use an asterisk to denote values of the various variables at the interface. For example, T^* will denote the temperature of the interface, and we will have $T^* = T_f$ when the interface itself is at equilibrium. For the solute field, which makes a jump at the interface, X^*_{Bs} and $X^*_{B\ell}$ will be used for the interfacial compositions in the solid and liquid, respectively. The notation $X_B^{s\ell}$ and $X_B^{\ell s}$ is also quite often encountered in the literature to denote compositions at the liquid-solid interface, the first index identifying the phase (i.e., $X_B^{s\ell} = X^*_{Bs}$ and $X_B^{\ell s} = X^*_{B\ell}$).

2.4.2 True departure from equilibrium

Thermal and solute gradients are of course departures from equilibrium, since the overall system is not at the lowest Gibbs free energy. For example, after solidification, an alloy usually contains solute gradients which could evolve as a function of time if it is heated (e.g., homogenization treatment). However, as we have seen, the interface might still be at equilibrium during the process, in which case the phase diagram can still be used, but only to determine the interfacial values. In other situations, the solid-liquid interface itself might be “truly” out of equilibrium, i.e., the temperature or compositions at the interface are no longer those given by the equilibrium phase diagram. The main sources of deviations from equilibrium are:

- surface energy of a curved interface;
- attachment kinetics of the atoms/molecules;
- trapping of solute elements.

The departure from the equilibrium phase diagram due to curvature occurs even for a static interface, whereas the latter two contributions appear only when the interface is moving. These phenomena are described separately in the next subsections.

Curvature contribution

Consider a unary system at equilibrium. The atoms or molecules at the solid-liquid interface do not have the same molar Gibbs free energy as those in either the solid or liquid: they have an excess energy because they have to accommodate the slight structural changes on both sides. The integral of this excess free energy over the thickness of the interface, multiplied by some form of molar volume, is the *solid-liquid interfacial energy*, $\gamma_{s\ell}$, having units of $\mathrm{J\,m^{-2}}$ (Fig. 2.16). (Note that the molar volumes of the solid and liquid are slightly different, so the molar volume used here should be an average over the atoms within the interface). In general, since the solid phase is crystalline, the surface energy will be anisotropic. This becomes important later when we consider microstructures, but for

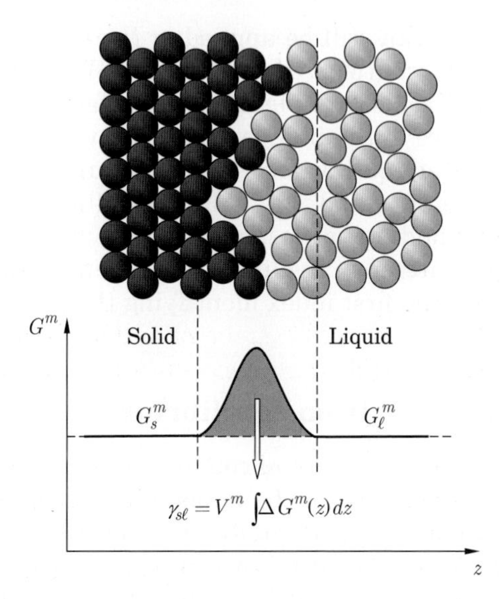

Fig. 2.16 Excess Gibbs free energy of atoms located within a diffuse solid-liquid interface and interfacial energy.

the moment, let us assume that $\gamma_{s\ell}$ is isotropic. The total Gibbs free energy of the system may be expressed as

$$G = G_s^m n_s + G_\ell^m n_\ell + A_{s\ell}\gamma_{s\ell} \tag{2.56}$$

where n_s and n_ℓ are the number of moles in the solid and liquid, respectively, and $A_{s\ell}$ is the interfacial area. Taking a fixed number of moles n, so that $n_\ell = n - n_s$, the minimum Gibbs free energy configuration satisfies

$$\frac{\partial G}{\partial n_s} = 0 = G_s^m - G_\ell^m + \frac{\partial A_{s\ell}}{\partial n_s}\gamma_{s\ell} \tag{2.57}$$

Note that we have further assumed that the interfacial energy is independent of n_s, i.e., it is not only isotropic but also independent of the size and shape of the volume element. Since the volume of the solid V_s is given by $V_s^m n_s$, where V_s^m is the molar volume of the solid, the partial derivative in the last term in Eq. (2.57) can be written as

$$\frac{\partial A_{s\ell}}{\partial n_s} = V_s^m \frac{\partial A_{s\ell}}{\partial V_s} = V_s^m 2\bar{\kappa} \tag{2.58}$$

where $\bar{\kappa}$ is the *mean curvature of the solid*:

$$\bar{\kappa} = \frac{1}{2}\frac{\partial A_{s\ell}}{\partial V_s} = \frac{1}{2}\left(\frac{1}{R_1} + \frac{1}{R_2}\right) \tag{2.59}$$

R_1 and R_2 are the radii of curvature measured for any orthogonal pair of directions in the surface. There is always one particular choice of direction where R_1 and R_2 attain their maximum and minimum values simultaneously, and these values are called the *principal curvatures*. The mean curvature $\bar{\kappa}$ and the Gaussian curvature, defined as $1/(R_1R_2)$, are invariants. For a sphere, $A_{s\ell} = 4\pi R^2$ and $V_s = 4\pi R^3/3$, therefore $\bar{\kappa} = 1/R$. For a cylinder of radius R and infinite length, $\bar{\kappa} = 1/(2R)$. For a saddle point (Fig. 2.17), one principal radius of curvature is positive while the other one is negative. Therefore, the mean curvature might be zero despite the fact the interface looks curved! This would be indicated by a negative value of the Gaussian curvature.

In summary, for a unary system, the condition of equilibrium between the liquid and a solid particle of mean curvature $\bar{\kappa}$ is given by Eq. (2.57):

$$G_\ell^m = G_s^m + 2V^m\bar{\kappa}\gamma_{s\ell} \tag{2.60}$$

where the index s for the molar volume has been dropped. This term can also be found by considering the pressure difference between the liquid and the solid (Fig. 2.18, left). The surface tension, equal to the surface

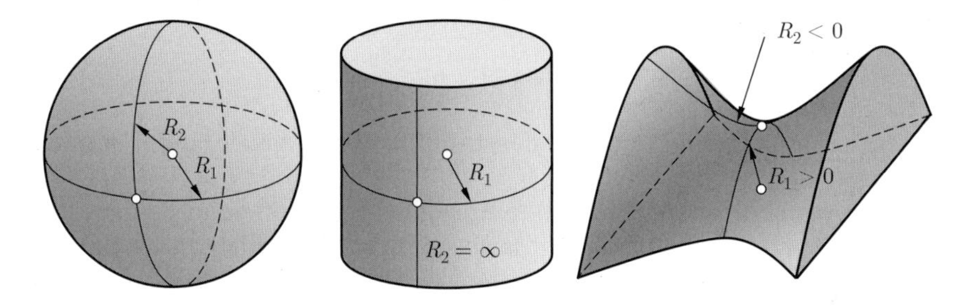

Fig. 2.17 Principal radii of curvature for a sphere, cylinder and saddle point.

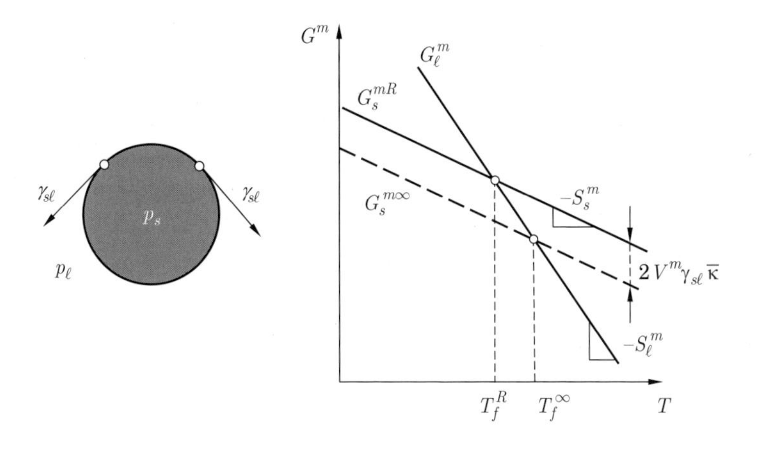

Fig. 2.18 Gibbs free energy for a solid sphere of radius R, and the condition of equilibrium with the liquid.

energy with the above assumptions, increases the pressure of the positively curved solid particle in the same way as encountered when inflating a balloon. It is straightforward to show that the difference in pressure is given by: $\Delta p = p_s - p_\ell = 2\bar{\kappa}\gamma_{s\ell}$. Multiplying this value by the molar volume gives the difference in the molar Gibbs free energy.

For a unary system, one can see in Fig. 2.18 that the melting point of a positively curved (convex) solid, T_f^R, is lower than that of the planar solid (labeled T_f^∞). The decrease of the melting point associated with curvature, called the *curvature undercooling*, can be calculated using Eq. (2.21):

$$\Delta T_R = T_f^\infty - T_f^R = 2\bar{\kappa}\frac{\gamma_{s\ell}V^m}{\Delta S_f^m} = 2\Gamma_{s\ell}\bar{\kappa} \tag{2.61}$$

where we have introduced the *Gibbs-Thomson coefficient*, $\Gamma_{s\ell}$, given by

$$\Gamma_{s\ell} = \frac{\gamma_{s\ell}V^m}{\Delta S_f^m} = \frac{\gamma_{s\ell}}{\rho_s \Delta s_f} = \frac{\gamma_{s\ell}T_f}{\rho_s L_f} \tag{2.62}$$

The expression that gives the melting point for the pure material with a curved surface is called the *Gibbs-Thomson equation*:

$$T_f^R = T_f^\infty - 2\Gamma_{s\ell}\bar{\kappa} \tag{2.63}$$

This equation is crucial to the understanding of the development of microstructure.

The value of $\Gamma_{s\ell}$ is about 10^{-7} Km. Thus, a small solid particle with radius $10\,\mu$m has its melting point lowered by 0.02 K, which, at first glance, does not seem very important. However, such small differences are responsible for many features of microstructure formation (see Chap. 8). Indeed, if this solid particle is close to a flat solid-liquid interface, say within a distance equal to its radius, this induces a heat flux given by $k(T_f^\infty - T_f^R)/R$, where k is the thermal conductivity of the liquid separating the two solids. Taking $k = 100$ W m^{-1} K^{-1}, the small particle will be "heated" by the flat interface with a heat flux of 200 kW m^{-2}, thus inducing rapid melting of the particle. As we shall see in Chap. 8, a similar mechanism driven by curvature-induced solute diffusion, (rather than heat diffusion) is responsible for the *ripening* or *coarsening* of secondary dendrite arms. Note that for nanoparticles, the curvature contribution increases dramatically, as shown in Fig. 2.19: gold particles of about 2-nm (20 Å) radius have been observed to melt around 250 K below the tabulated melting temperature (1337.6 K).

In alloys, even if the situation is more complicated, it can be treated in a very similar way. Taking the molar Gibbs free energy of both the solid and liquid phases (Fig. 2.13), the contribution of the curvature can be added to that of the solid phase. As shown in Fig. 2.20, for a fixed temperature and an alloy that has a partition coefficient $k_0 < 1$, this shifts the concentrations of a positively curved solid-liquid interface, X_{Bs}^R and $X_{B\ell}^R$, to lower values as compared with those of a flat interface, X_{Bs}^∞ and $X_{B\ell}^\infty$.

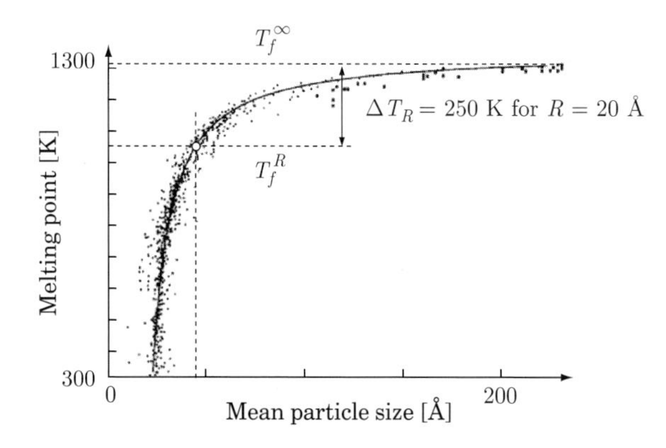

Fig. 2.19 Melting point of small gold particles as measured by electron diffraction techniques (courtesy of Ph. Buffat, EPFL [2]).

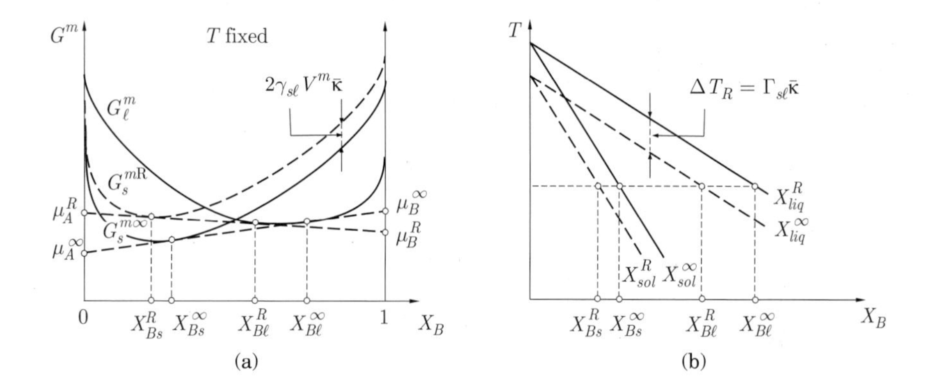

Fig. 2.20 Equilibrium of solid and liquid phases for a binary alloy including the curvature contribution. (a) shift in free energy curves, and (b) the shift in the liquidus and solidus curves on the equilibrium phase diagram.

For dilute solutions, it can be shown that it is a good approximation to simply shift the liquidus and solidus curves vertically by ΔT_R from Eq. (2.61), as illustrated in Fig. 2.20.

A few final remarks can be made concerning the effect of the surface energy on the solid-liquid equilibrium. Droplets of liquid (i.e., concave solid-liquid interface viewed from the solid side) have a melting point that is slightly greater than the equilibrium value. As we will see in Chap. 7, a solid nucleating in the groove of a pit can have a negative curvature. In this case, it can remain solid above the equilibrium melting point! A saddle-shaped solid (Fig. 2.17) has zero undercooling if $R_1 = -R_2$. However, it can be shown that such a shape is unstable: any perturbation to the saddle will make it go to a plane (of zero final curvature), through intermediate states that do not have a zero mean curvature.

The equilibrium shape of a pure crystal in contact with the liquid (or the gas) phase is such that the molar Gibbs free energy is constant everywhere. For an alloy, this condition applies to the chemical potentials of all the elements. The crystal is therefore spherical if $\gamma_{s\ell}$ is isotropic. However, when the interfacial energy is weakly anisotropic, equilibrium of the interface implies that the curvature undercooling is constant at every point of the interface. As a consequence, the curvature will adjust, with a mean curvature that becomes a function of the normal $\boldsymbol{n}$ to the interface. In a 2-D situation with one principal radius of curvature $R_1(\theta)$, the equilibrium condition for a crystal is given by Herring's relation:

$$\Delta T_R(\theta) = \frac{1}{R_1(\theta)} V^m \frac{\gamma_{s\ell} + (d^2\gamma_{s\ell}/d\theta^2)}{\Delta S_f^m} = \text{constant} \tag{2.64}$$

where θ is the angle between the local surface normal vector and the direction of maximum surface energy. The expression in the numerator is called the *surface stiffness*. For example, cubic symmetry is often represented by writing

$$\gamma_{s\ell} = \gamma_{s\ell}^0[1 + \epsilon_4 \cos(4\theta)] \tag{2.65}$$

where ϵ_4 is a measure of the strength of the anisotropy. The surface stiffness is then

$$\gamma_{s\ell} + \frac{d^2\gamma_{s\ell}}{d\theta^2} = \gamma_{s\ell}^0[1 - 15\epsilon_4 \cos(4\theta)] \tag{2.66}$$

Notice that this implies that the surface stiffness is positive for all values of θ so long as $\epsilon_4 < 1/15$. This is what we mean by "weak anisotropy." For larger values of ϵ_4, i.e., "strong anisotropy," some values of θ give rise to negative values, implying that these orientations cannot appear. This leads to *faceted* crystals, for which some orientations are missing. These crystals are discussed further in the next section. For a 3-D unary system, Herring's relation, written in terms of the molar Gibbs free energy becomes

$$\Delta G_s^{mR} = V^m \left[\frac{1}{R_1}\left(\gamma_{s\ell} + \frac{\partial^2\gamma_{s\ell}}{\partial\theta_1^2}\right) + \frac{1}{R_2}\left(\gamma_{s\ell} + \frac{\partial^2\gamma_{s\ell}}{\partial\theta_2^2}\right)\right] \tag{2.67}$$

where θ_1 and θ_2 are angles measuring deviations from the normal direction along the two principal curvature directions $\hat{t}_1$ and $\hat{t}_2$ associated with R_1 and R_2.

Attachment kinetics

The curvature contribution is the only "static" out-of-equilibrium effect of the interface for both pure systems and alloys. However, if the interface of a unary system is in motion, as in the case of a propagating solid-liquid phase boundary during solidification, another effect can contribute to deviations from equilibrium: the *kinetics of attachment* of the atoms/molecules to the interface. Before discussing this phenomenon, it is first necessary to describe the nature of the solid-liquid interface itself. We start by distinguishing two types of interfaces, as illustrated schematically in Fig. 2.21.

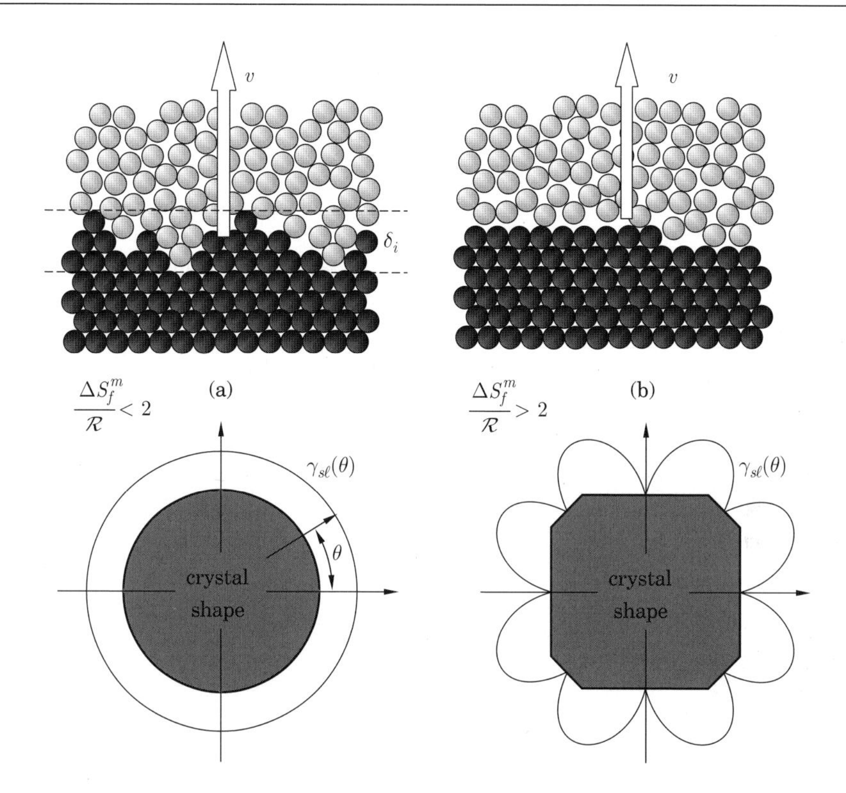

Fig. 2.21 Types of solid-liquid interfaces encountered in solidification (top): (a) diffuse interfaces typical of metals; b) smooth interfaces typical of faceted materials. The associated γ-plots (thin line) and equilibrium crystal shapes (gray shapes) are indicated at the bottom.

Atomically rough interfaces:
The structure of most liquid metals does not differ very much from that of the solid. The average spacing between the atoms might be a few percent larger than that characterizing the solid, i.e., the density is lower, and the atomic arrangement is simply disordered, at least at long distance. Some local order can exist even in the liquid, e.g., local arrangement of atoms into an icosahedral phase. Consequently, the difference of molar entropy between the two phases is fairly small, and one typically works with the following expression:

$$\Delta S_f^m = S_\ell^m - S_s^m < 2\mathcal{R} \qquad (2.68)$$

where $\mathcal{R}$ is again the ideal gas constant. Under such circumstances, the transition from liquid to solid occurs over some distance, and the interface is *diffuse* or *atomically rough*. The thickness δ_i over which the transition occurs is called the *interface thickness*. Molecular dynamics calculations indicate that δ_i is on the order of 1-3 nm, i.e, a few interatomic spacings. As a result, it is reasonable to assume that the interfacial energy $\gamma_{s\ell}$ is nearly *isotropic*, with typically $\delta\gamma_{s\ell}(\boldsymbol{n})/\gamma_{s\ell}$ on the order of 1-5%, where $\boldsymbol{n}$ defines the orientation of the surface being considered. Since the interfacial energy is fairly isotropic, the equilibrium shape of a crystal is close to spherical and does not exhibit facets, i.e., its shape is macroscopically smooth. At the bottom half of Fig. 2.21(a), we have drawn the γ-plot (thin line) as well as the resulting equilibrium shape corresponding to such a case.

It should be noted that it is possible for metals that solidify from the melt with a macroscopically smooth interface to form faceted crystals when they grow from the vapor phase. This can be understood again by considering the entropy of sublimation, which will be much larger than $2\mathcal{R}$ in this case.

Atomically smooth or faceted interfaces:
In semiconductors, oxides, carbides, compounds, polymers and other materials consisting of complex molecules, the liquid structure differs significantly from that of the solid. For these materials, it is typically the case that

$$\Delta S_f^m = S_\ell^m - S_s^m > 2\mathcal{R} \tag{2.69}$$

When Eq. (2.69) holds, the interface follows well-defined crystallographic planes and is *smooth* at the atomic scale. The interfacial energy itself can be very anisotropic and may exhibit cusps as indicated at the bottom of Fig. 2.21(b). The equilibrium shape of the crystal, given by the so-called Wulff construction, does not look like the γ-plot. It is obtained by drawing all the normals to and passing by the extremities of the vectors $\boldsymbol{r} = \gamma_{s\ell}(\boldsymbol{n})\boldsymbol{n}$, and then by taking the inner envelope to all these normals. In the simplest case, the crystal has facets defined by the inner cusps, but this is not always observed (see Chap. 8). In particular, a general formalism based on the construction of a so-called ξ-vector has been given by Cahn and Hofmann and captures both Herring relations, i.e., weak anisotropy, and the Wulff construction usually used for very anisotropic γ-plots.

For such interfaces, molecules or atoms usually attach to the solid at steps. Therefore, solidification occurs by completion of rows in the direction perpendicular to the apparent growth direction. When a row is completed, it is necessary to start a new row. This can occur by what is called *secondary nucleation*, i.e., by the attachment of one molecule to a flat surface (Fig. 2.22a). The Gibbs free energy of a molecule or atom positioned at the solid surface but otherwise completely surrounded by liquid is fairly high. These events are therefore very unlikely, and nature prefers another mechanism: the use of *defects* to generate easy attachment sites. Typically, screw dislocations or double twins play this role. Figure 2.22 illustrates these two mechanisms, showing a fcc lattice for the double twin and a simple cubic for the screw dislocation. In the first case, a set of vertical (111)-planes are shown with the $\langle 110 \rangle$ directions indicated by arrows. Two (111) twin planes have been introduced: a twin can occur during the growth of an fcc lattice when the natural stacking sequence of ABCABCA... planes

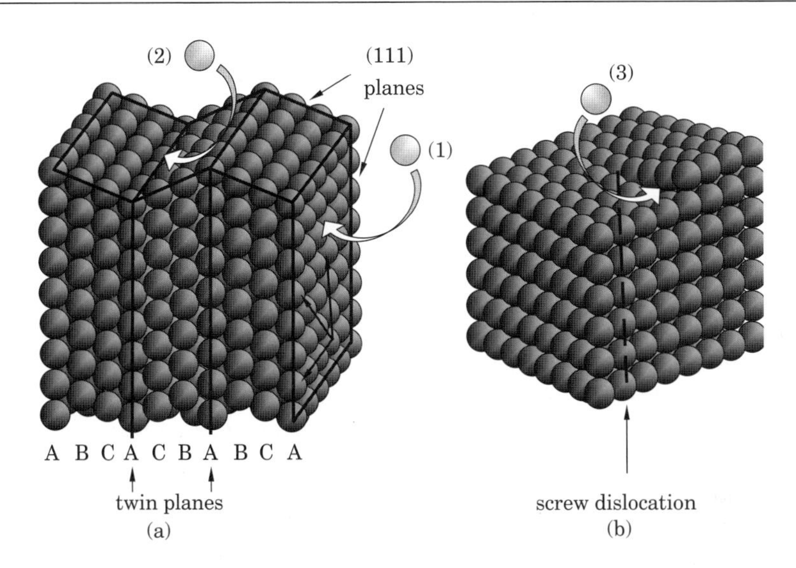

Fig. 2.22 (a) Illustration of secondary nucleation of an atom (1) on a (111) plane in a fcc lattice. In the same diagram, the double-twin mechanism of crystal growth is indicated for an atom (2). (b) Illustration of the screw dislocation growth mechanism in a simple cubic lattice with the attachment of atom (3) on the step.

is broken by a misplaced (111) plane, i.e., a plane C replacing a plane B in the first stacking fault and a plane B replacing a plane C in the second one in Fig. 2.22. In both occurrences, this creates a twin symmetry of the crystal around a plane A, i.e., sequence ABC**A**CB**A**BCA.... Locally, around the twin planes A, the packing of atoms follows a hexagonal configuration, i.e., CAC and BAB respectively, thus creating grooves or edges at the other sets of (111) planes shown in Fig. 2.22. Liquid atoms can start filling the grooves with a much lower attachment energy. The screw dislocation mechanism is illustrated for a simple cubic lattice in which a [100] screw dislocation (dashed line) naturally propagates a step on a (100) plane, onto which atoms can grow as a spiral of Archimedes.

For pure metals, the attachment kinetics, i.e., the relationship between the velocity of the solid-liquid interface v_n and the kinetic undercooling ΔT_k, is given by

$$v_n = \mu_k \Delta T_k = \mu_k (T_f - T^*) \tag{2.70}$$

where μ_k is the attachment-kinetics coefficient. Turnbull estimated this coefficient for metallic systems to be

$$\mu_k = \frac{v_{sound} L_f^m}{\mathcal{R} T_f^2} \tag{2.71}$$

where v_{sound} is the velocity of sound (on the order of a few thousands of m s^{-1}). For nickel, this gives an attachment-kinetics coefficient of

$\mu_k \approx 2 \times 10^4$ m s^{-1}K^{-1}. However, recent estimates based on molecular dynamics calculations and high undercooling experiments give values nearly one order of magnitude lower. Furthermore, the attachment kinetics depend on the crystallographic plane considered, i.e., μ_k is anisotropic and $\mu_k^{100} > \mu_k^{110} > \mu_k^{111}$. Unless solidification occurs at very high speed, ΔT_k in pure metals is small (typically 2-3 K at 1 m s^{-1}) and may be neglected in comparison to the other contributions under most conditions. However, under very special circumstances, such as atomization or planar flow casting, it may be necessary to consider this contribution. When the velocity is very large, typically 100-1000 m s^{-1}, the undercooling of the front is so large that the atoms cannot move fast enough to avoid being captured by the moving isotherms. In this case, the structure of the liquid is frozen without rearrangement, giving what is called a *metallic glass*. Let us call the velocity limit for crystallization to occur v_o. Complex metallic alloys in which a eutectic reaction occurs at very low temperature are also characterized by a much lower v_o. They can be produced in large size specimens called *bulk metallic glasses* having quite unique mechanical properties. In complex molecular systems, such as polymers and silicates, v_o is considerably smaller than for metals, and normal solidification conditions naturally produce glasses or amorphous structures.

When more complex systems solidify with a faceted morphology, the attachment kinetics can become the dominant contribution. In this case, the attachment-kinetics law will depend on the growth mechanism (steps, secondary nucleation, screw dislocations, twins, ...) and the relationship between the velocity and the undercooling is often written as a power law:

$$v_n = \mu_k \Delta T^n \tag{2.72}$$

where n is a coefficient that is determined from experiments.

Solute trapping

For metallic alloys, another possible kinetic departure from equilibrium is called "solute trapping," which can be understood with the help of Fig. 2.23. Under conditions of low propagation speeds (left), the interface is at local equilibrium. Atoms and molecules have enough time to diffuse locally in a region close to the interface. If the liquid just in front of the moving interface has a concentration X_ℓ^*, then in equilibrium the solid just after the interface has passed has composition $X_s^* = k_0^m X_\ell^*$. For $k_0 < 1$, this means that solute is being rejected into the liquid, as illustrated schematically in Fig. 2.23(left). As we have seen in this chapter, this ensures equality of the chemical potentials of both the solvent and solute across the interface. As the speed of the interface propagation increases, there is less time available to accommodate solute diffusion. We will see in Chap. 5 that the *solute boundary layer* ahead of such a growing interface is given by D_ℓ / v_n, where D_ℓ is the diffusion coefficient in the liquid. As the velocity increases, a point is reached where D_ℓ / v_n decreases to a value on the same order as the thickness of the diffuse interface, which we denote as δ_i. At

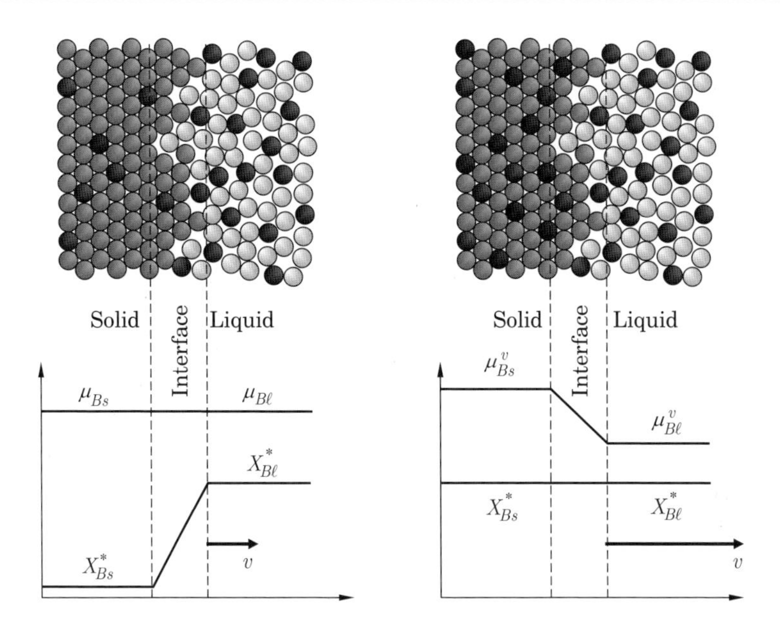

Fig. 2.23 Illustration of solute trapping for a diffuse solid-liquid interface. Notice that at higher velocity (right hand side) more solute atoms are "trapped" in the solid.

higher velocity, atoms do not have enough time to "escape" from the moving interface and are instead *trapped*, at least partially. This means that $X_s^* > k_0^m X_\ell^*$. At very high speed, all of the solute in the liquid goes into the solid, so that $X_s^* = X_\ell^*$. Of course, under these conditions the chemical potentials of the solute and solvent no longer satisfy the equilibrium condition, as shown in Fig. 2.23(right).

When solute trapping occurs, the temperature of the interface also changes. We will see in Chap. 5 that at low speed, a steady-state planar solid-liquid interface grows at the solidus temperature of the alloy. Suppose that the alloy composition is X_{B0}. If the interface is in local equilibrium, then it follows that the concentration in the solid is equal to X_{B0} while that in the liquid is given by $X_\ell^* = X_{B0}/k_0^m$. This situation is illustrated in the Gibbs free energy curves shown in Fig. 2.24 for the temperature $T_{sol}(X_{B0})$. Now suppose that we gradually accelerate the interface so that at any time we are still at steady state. For complete solute trapping, $X_\ell^* = X_{B0}$, and therefore the Gibbs free energies of both phases should be such that $G_s^m(X_{B0}) = G_\ell^m(X_{B0})$. This must occur at the T_0-point of this alloy defined in Fig. 2.12. Therefore, the interface temperature under these conditions is given by $T_0(X_{B0})$. The Gibbs-free energy diagram, sketched in Fig. 2.24, shows that it is necessary to *increase* the temperature from $T_{sol}(X_{B0})$ because the liquid phase has a higher entropy than the solid, i.e., the curve $G_\ell^m(T, X)$ moves downward faster than $G_s^m(T, X)$ as the temperature increases.

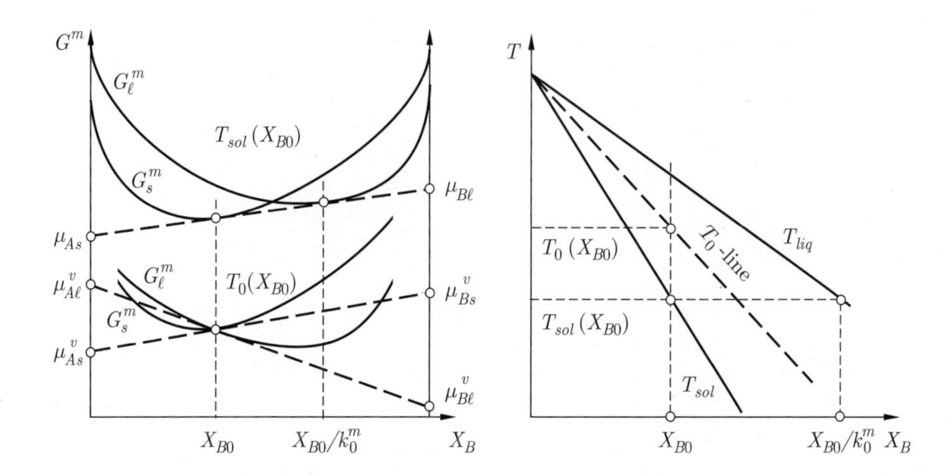

Fig. 2.24 Gibbs free energies for an alloy of concentration X_{B0} solidifying under equilibrium conditions at the solidus temperature $T_{sol}(X_{B0})$ and at high speed when complete solute trapping is occurring, i.e., solidification at $T_0(X_{B0})$ for which $G_\ell^m(X_{B0}, T_0) = G_s^m(X_{B0}, T_0)$. The crossing points of G_s^m and G_ℓ^m when X_{B0} is varied defines the T_0-line (right).

Complete solute trapping occurs when $D_\ell/v_n \ll \delta_i$, and this case corresponds to an *effective partition coefficient* k_{0v}^m equal to one. At low speed, i.e., $D_\ell/v_n \gg \delta_i$, k_{0v}^m is simply that given by the equilibrium phase diagram k_0^m. Aziz [1] proposed a model for intermediate values, in which the effective partition coefficient varies continuously between the two extrema according to the ratio of the two length scales:

$$k_{0v}^m = \frac{k_0^m + (\delta_i v_n / D_\ell)}{1 + (\delta_i v_n / D_\ell)} \tag{2.73}$$

This function is plotted in Fig. 2.25 for $k_0^m = 0.5$. In the original work, instead of D_ℓ, Aziz used a diffusion coefficient of elements within the interface. In the case of solidification, this is roughly equivalent to D_ℓ.

It can be shown, for dilute solutions, that the temperature of the interface consistent with Eq. (2.73) is given by

$$T^* = T_f + m_v X_\ell^* - \frac{\mathcal{R} T_f}{\Delta S_f^m} \frac{v}{v_0} \tag{2.74}$$

where v_0 is the limit at which crystallization still occurs (see previous section) and m_v is an apparent liquidus slope given by:

$$m_v = m_\ell \left(1 + \frac{k_0^m - k_{0v}^m (1 + \ln(k_0^m / k_{0v}^m))}{1 - k_0^m} \right) \tag{2.75}$$

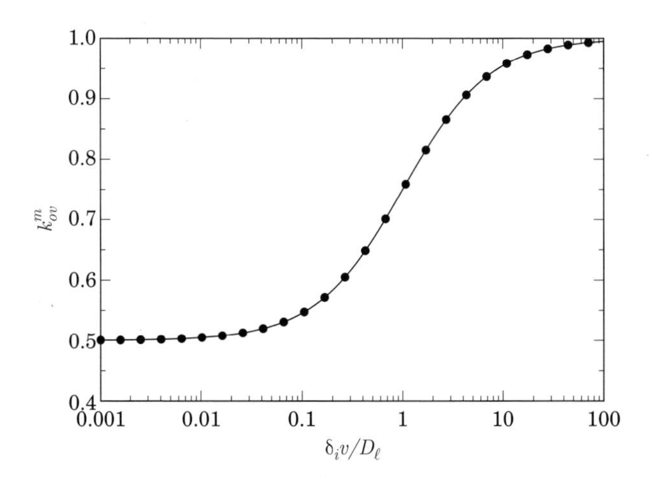

Fig. 2.25 Effective partition coefficient k_{0v} as a function of the dimensionless interface speed $\delta_i V/D_\ell$. For purposes of illustration, we have taken $k_0^m = 0.5$.

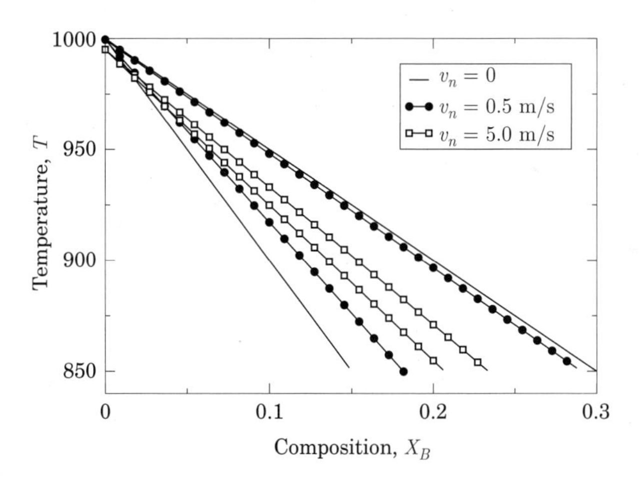

Fig. 2.26 Velocity-dependence of a phase diagram in which both attachment kinetics and solute trapping have been accounted for. $\Delta S_f^m = \mathcal{R}$, $m_\ell = -500$°C, $k_0^m = 0.5$, $\delta_i = 2$ nm, $D_\ell = 3 \times 10^{-9}$ m^2 s^{-1}, $v_0 = 1000$ m s^{-1}.

Setting $X_s^* = k_{0v} X_\ell^*$, Fig. 2.26 shows the "apparent" phase diagram giving the interfacial concentrations, $X_s^*(T, v)$ and $X_\ell^*(T, v)$ for two velocities, $v = 0.5$ m s^{-1} (filled circles) and $v = 5$ m s^{-1} (open squares), as calculated with these relationships and the values listed in the figure caption. One can see that the solidus line moves faster towards the T_0-line than the liquidus. At the highest velocity, one also starts to observe the effects of the attachment kinetics of the pure substance, i.e., the depression of the effective melting point T_f.

2.5 SUMMARY

In this chapter, we have introduced the concepts of thermodynamic equilibrium, laying an important foundation for the study of general phase transformations. The thermodynamic state is defined by any pair of the three variables temperature, pressure and volume, along with the composition in multicomponent systems. Because our focus in this text is on solidification from the liquid state, we choose temperature and pressure as the most convenient pair. The Gibbs free energy was defined, and equilibrium is identified as equality of the free energy between two possible states. We first discussed unary systems, and introduced the concept of invariant points, where the equality of the thermodynamic variables constrains the number of phases that can coexist in equilibrium.

We applied these same concepts to alloys containing multiple components. When the composition is also variable, the number of phases can increase. The constraints on the system were summarized in Gibbs' phase rule. The regular solution model for the free energy of alloys was derived from consideration of nearest-neighbor interactions among atoms or molecules. We introduced the concept of chemical potential to define equilibrium between phases in alloys, and introduced graphical constructions that will prove very useful in the next chapter in the development and understanding of equilibrium phase diagrams.

Finally, several important departures from equilibrium were discussed. The simplest of these is when a given phase is inhomogeneous. True departures from equilibrium owing to interface curvature and finite attachment kinetics were demonstrated. These phenomena are crucial to the understanding of the development of solidification microstructures.

2.6 EXERCISES

Exercise 2.1. Relating c_p and c_V.
Show that for any system one has the following relationship:

$$c_p - c_V = \left(\left(\frac{\partial e}{\partial v} \right)_T + p \right) \left(\frac{\partial v}{\partial T} \right)_p$$

Exercise 2.2. Liquid-vapor equilibrium.
Consider a closed container of fixed volume V and initial temperature T_0 containing only a vapor phase at initial pressure p_0. The vapor phase may be modeled as a perfect gas. Assuming that the system is always at equilibrium and considering the relationship $T_v(p)$ describing the liquid-vapor

co-existence (Fig. 2.3), describe how the pressure and the number of moles in the vapor and liquid evolve as the system is cooled from T_0 to 0°C.

Exercise 2.3. Chemical potential in regular solutions.
For a regular solution of a binary alloy, show that the chemical potential of each species is given by:

$$\mu_A = G_A^m + \mathcal{R}T \ln X_A + \Omega^m(1 - X_A^2)$$
$$\mu_B = G_B^m + \mathcal{R}T \ln X_B + \Omega^m(1 - X_B^2) \tag{2.76}$$

(a) Develop an expression for the chemical activities a_A and a_B when $\Omega^m = 0$. How is it modified when $\Omega^m \neq 0$?

(b) Draw the chemical activity $a_A(X_B)$ and $a_B(X_B)$ for the three cases:

(i) $\Omega^m = 0$
(ii) $\Omega^m < 0$
(iii) $\Omega^m > 0$

What can you deduce about both activities near $X_B = 0$ and $X_B = 1$ for all three cases?

Exercise 2.4. Free energy of a spherical particle.
Consider a spherical solid particle of radius R surrounded by liquid in a unary system. Demonstrate that the pressure difference $p_s - p_\ell$ is given by $2\gamma_{s\ell}/R$. From this, show that

$$\Delta G_s^{mR} = G_s^{mR} - G_s^{m\infty} = 2V^m \gamma_{s\ell}/R$$

Exercise 2.5. Temperature and heat of fusion.
Find the latent heat of fusion and entropy of fusion at a temperature T below the equilibrium freezing temperature T_f by following the procedure outlined below.

(a) Start with Eq. (2.11) and write expressions for the enthalpy of the solid and liquid phases at temperature T and at T_f, and then manipulate them to find an expression for their difference, the latent heat at T. Call this $L_f(T)$, and use the symbol $L_f = L_f(T_f)$. Note that the expressions normally given for the specific heat tend to be valid only in a limited range, so you should try to express all of your results in terms of known values at the melting point, and corrections to those values.

(b) One often finds specific heat data in the form given below. Using this form, derive an expression for $L_f(T)$ in terms of the coefficients in the representation.

$$c_{ps} = a_s + b_s T + c_s T^{-2}$$
$$c_{p\ell} = a_\ell + b_\ell T + c_\ell T^{-2}$$

(c) The following data for pure Al were obtained from Kubaschewski et al. [9]. Compute $L_f(650°C)$. Is the change significant?

	T_f [K]	L_f [J/g]	$c_{p\ell}$ [J/gK]	c_{ps} [J/gK]
Al	933	387.4	1.176	$0.7657 + 4.586 \times 10^{-4}T$

(d) Repeat this procedure to find an expression for the entropy of fusion, and compute the value of Δs_f for Al at 650°C.

Exercise 2.6. Clausius-Clapeyron equation.
Compute the slope of the liquid-solid phase boundary, $(dp/dT)|_{s\ell}$ for pure Cu and pure Fe using Eq. (2.24). Consider the search for the necessary material property data to be part of the exercise.

Exercise 2.7. Latent heat of fusion for a binary eutectic alloy.
Al and Si form a binary eutectic at 13 wt % Si that freezes at 577°C. Thermodynamic data for the solidification of the pure phases are given in the table below. The solubility in both phases is very limited in the solid state, so you may assume that the two phases are simply pure Al and pure Si.

	T_f [K]	L_f [J/g]	$c_{p\ell}$ [J/gK]	c_{ps} [J/gK]
Al	933	387.4	1.176	$0.7657 + 4.586 \times 10^{-4}T$
Si	1683	1802	0.912	$0.8535 + 8.786 \times 10^{-5}T - 1.473{\times}10^4/T^2$

(a) Use the results from Exercise 2.5 to compute the latent heat of fusion of the pure phases at the eutectic temperature.

(b) Use a simple rule of mixtures to estimate the latent heat of fusion of the eutectic, and compare the result to the latent heat of fusion for pure Al. Is the difference significant?

2.7 REFERENCES

[1] M. J. Aziz. Model for solute redistribution during rapid solidification. *J. Appl. Phys.*, 53: 1158-1168, 1982.
[2] P. Buffat and J.-P. Borel. Size effect on melting temperature of gold particles. *Phys. Rev. A*, 13:2287-2298, 1976.
[3] L. S. Darken and R. W. Gurry. *Physical chemistry of metals*. McGraw-Hill Co., New York, 1953.
[4] D. R. Gaskell. *Introduction to metallurgical thermodynamics*. McGraw-Hill, 1973.
[5] C. Herring. Surface tension as a motivation for sintering. In W. E. Kingston, editor, *The physics of powder metallurgy*, Chap. 8, pages 143-177. McGraw-Hill, 1951.
[6] C. Herring. The use of classical macroscopic concepts in surface energy problems. In R. Gomer and C.S. Smith, editors, *Structure and properties of solid surfaces*, pages 5-72. University of Chicago Press, 1953.

[7] D. W. Hoffman and J. W. Cahn. A vector thermodynamics for anisotropic surfaces. I. fundamentals and application to plane surface junctions. *Surf. Sci.*, 31:368-388, 1972.
[8] K. A. Jackson. On the theory of crystal growth: Growth of small crystals using periodic boundary conditions. J. Cryst. *Growth 3/4* 507, 1968.
[9] O. Kubaschewski and C. B. Alcock. *Metallurgical thermochemistry*. Pergamon Press: Oxford, 1979.
[10] W. Kurz and D. J. Fisher. *Fundamentals of solidification*. Trans. Tech. Publ., Aedermansdorf, Switzerland, 4th edition, 1998.
[11] C. H. P. Lupis. *Chemical thermodynamics of materials*. Elsevier Science Publishing Company Inc. New York, 1983.
[12] D. A. Porter and K. Easterling. *Phase transformations in metals and alloys*. Chapman and Hall, London, 2nd edition, 1992.
[13] G. Wulff. Zur Frage der Geschwindigkeit des Wachsthums und der Aufloesung der Krystallflachen. *Z. Kristallogr. Mineral*, 34:449-530, 1901.

CHAPTER 3

PHASE DIAGRAMS

3.1 MOTIVATION

One of the most powerful tools for studying the development of microstructure is the equilibrium phase diagram. A phase diagram embodies information derived from the thermodynamic principles described in Chap. 2, specialized for a particular range of compositions and presented in a form that makes the data readily accessible. The diagram shows the phases present in equilibrium, as well as the composition of the phases over a range of temperatures and pressures. In this text, we will not normally be concerned with the effect of pressure on thermodynamic equilibrium as we will work almost entirely with condensed phases (solids and liquids) in systems open to the atmosphere. We will thus for the most part assume that the pressure is constant.

Before we begin, it should be noted that thermodynamics tells us what *should* happen, i.e., it identifies the lowest energy configuration. However, as already stated in Chap. 2, it does not guarantee that that is what *will* happen, as the behavior of the real system may be constrained by kinetic processes, such as solute or heat transport. Nevertheless, we will frequently make the assumption that there is sufficient time for these processes to establish local thermodynamic equilibrium at the solid-liquid interface, if not in the entire system. This will allow us to use the information in equilibrium phase diagrams, even in cases where transport is incomplete.

The present chapter provides some fundamental concepts and tools for analyzing phase diagrams building upon the basics of thermodynamics and deviations from equilibrium that were presented in Chap. 2. We then apply these tools to binary and ternary systems. Although examples related to real systems are occasionally given, the emphasis in this chapter is on general features and classifications, rather than on particular systems. We will also describe how phase diagrams can be derived from thermodynamic data.

The two main ingredients derived in Chap. 2 used extensively in this chapter are the tangent rule construction to find equilibrium conditions, and Gibbs' phase rule. The first rule ensures that the chemical potentials of all chemical elements are equal in the various phases present. The second, derived from this equilibrium principle, states that:

$$N_F = N_c - N_\phi + 2 \tag{3.1}$$

where N_F is the number of degrees of freedom of a system made of N_c components when N_ϕ phases are present. The integer 2 comes from the two independent thermodynamic variables, i.e., the pressure p and the temperature T. If we restrict ourselves to condensed phases open to the atmosphere, i.e., to a constant pressure, Eq. (3.1) becomes

$$N_F = N_c - N_\phi + 1 \tag{3.2}$$

The utility of Gibbs' phase rule will become clearer through the following examples and applications.

3.2 BINARY SYSTEMS

Unary systems have been discussed in Chap. 2, and we here turn our attention to more interesting phenomena presented by *alloys*, i.e., mixtures of two or more components. Confining our attention to systems at constant pressure, where Eq. (3.2) applies, one can immediately see that things become more interesting when there are two components. Two phases may coexist over a range of compositions and temperatures, whereas invariant points now involve three phases, as opposed to two in unary systems. Since this book deals with solidification, we will focus on binary systems in which the liquid phase coexists with one or possibly two solid phases (invariant points). As a matter of fact, the various types of binary phase diagrams are classified according to the types of transformations and invariant reactions present.

As we shall see, the equilibrium phase diagram of a system is a direct result of the common tangent construction rule applied to the Gibbs free energy curves of the various phases that may be present. The liquid phase will be approximated by a regular solution model. For pedagogical purposes, the same approximation will be made for the solid phase although it certainly does not represent an accurate description of all systems. For example, an intermetallic phase cannot be well represented by such an approximation. On the other hand, invariant points, where the liquid phase is in equilibrium with two distinct solid phases, necessitate the introduction of two Gibbs free energy curves for these solid phases. We will see how such invariant situations can be modeled in a simple way, using only one regular solution curve for the description of two solid phases. However, we will first describe the most simple case in which a liquid is in equilibrium with only one solid phase, i.e., the *isomorphous system*. This will allow us to see how solidification proceeds under equilibrium conditions and to define more precisely the evolutions of the composition and phase fractions in a binary system.

3.2.1 Isomorphous systems: preliminary concepts

The simplest possible binary alloy phase diagram is the *isomorphous* system, in which the liquid and solid phases are completely miscible over the entire composition range. In such systems, the individual components must have identical crystal structures, the atomic sizes of the components must be within about 15% of each other, and the components must have similar electronegativities. Example systems that possess these attributes are Cu-Ni and Fe-Mn, but much more complicated molecular systems such as the pagioclase mixture $NaAlSi_3O_8$(albite)-$CaAl_2Si_2O_8$ (anorthite) are also known to exhibit isomorphous phase diagrams.

A schematic isomorphous phase diagram is shown in Fig. 3.1, displaying the phases present at various temperatures and compositions. The diagram consists of three regions: at high temperature, there is a liquid solution, designated ℓ; at low temperature there is a solid solution designated α, and within the gray region, there is a mixture of solid and liquid phases. The *liquidus* is defined as the curve $T_{liq}(X_B)$ above which the system is fully liquid, and the *solidus* is defined as the curve $T_{sol}(X_B)$ below which it is fully solid. The liquidus and solidus curves must meet at the two ends of the diagram, corresponding to the pure components A or B, since these are actually unary systems. At each temperature T and alloy composition X_{B0}, we will want to keep track of the following quantities: the phases present (α, ℓ), the composition of each phase $(X_{B\alpha}, X_{B\ell})$, and the fraction of the total amount of material corresponding to each individual phase (χ_α, χ_ℓ). The χ_i are expressed as fractions of the total, so that $\chi_\alpha + \chi_\ell = 1$.

Consider the alloy with composition X_{B0} as shown in Fig. 3.1. Above the liquidus, or below the solidus, we will have at equilibrium a homogeneous liquid or solid, respectively, with composition X_{B0}. At temperature T_0, the liquid and solid phases coexist for alloy X_{B0}. Notice that this would be true for any alloy with a composition between $X_{B\alpha}$ and $X_{B\ell}$,

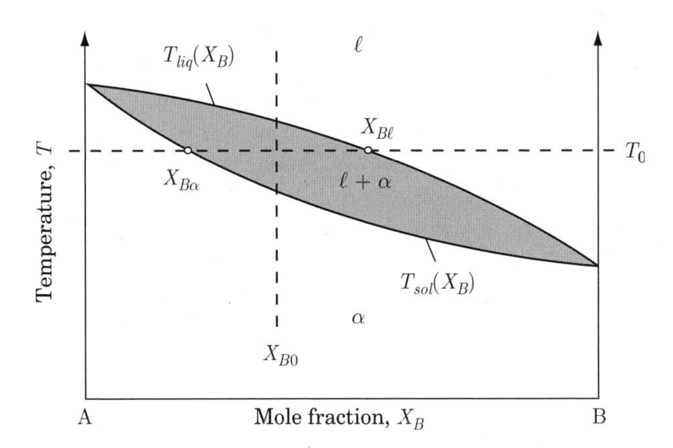

Fig. 3.1 An isomorphous phase diagram.

indicated on the diagram. Indeed, the temperature being fixed, the number of degrees of freedom is reduced by one and Gibbs' phase rule becomes $N_F = N_c - N_\phi$. For a system with two phases and two components, there is no degree of freedom left and the compositions of the solid and liquid phases remain constant for any alloy with an overall composition in the interval $X_{B\alpha} \leq X_{B0} \leq X_{B\ell}$ at temperature T_0. Since the composition of the liquid is fixed at $X_{B\ell}(T_0)$ and that of the solid at $X_{B\alpha}(T_0)$ when these two phases coexist, the amount of each phase must vary when the composition X_{B0} is changed. As was demonstrated in Chap. 2, these two compositions are given by the intersection of the common tangent to the two Gibbs free energy curves $G_\alpha^m(T_0)$ and $G_\ell^m(T_0)$. The segment connecting the two compositions located at the phase boundaries (liquidus and solidus lines in this case) is called a *tie-line* and is a general feature of not only binary but also multi-component systems.

Since the compositions of the individual phases are known in the two-phase region, we may readily compute the fraction of each phase present, χ_α and χ_ℓ. The total amount of solute must be equal to X_{B0}, thus giving

$$\chi_\alpha X_{B\alpha} + \chi_\ell X_{B\ell} = X_{B0} \tag{3.3}$$

Substituting for $\chi_\ell = 1 - \chi_\alpha$ and solving Eq. (3.3) for χ_α yields the *inverse lever rule*

$$\chi_\alpha = \frac{X_{B\ell} - X_{B0}}{X_{B\ell} - X_{B\alpha}}; \qquad \chi_\ell = \frac{X_{B0} - X_{B\alpha}}{X_{B\ell} - X_{B\alpha}} \tag{3.4}$$

Note that Eq. (3.4) applies for any two-phase region in a binary alloy system, and is not restricted to isomorphous systems. It may also sometimes be known simply as the lever rule (omitting "inverse"), and these names will be used interchangeably.

Up to this point, we have described everything related to alloys in terms of the mole fraction. This is convenient for thermodynamic development, and is the most appropriate choice for computing phase diagrams as will be described later in this chapter. However, materials scientists often work with other measures of composition. For example, when preparing an alloy, one normally begins by weighing the components before mixing, and, in this case, it would therefore be more natural to work with mass fractions, rather than mole fractions. When we consider a microstructure in a metallographic section, however, phases appear in proportion to their surface area or equivalently to their volume fraction. Depending on the particular components and phases, the three measures may be quite different. The various measures for the compositions and phase fractions are defined in Table 3.1. Notice that we have not included a measure for volume fraction of components in individual phases, as this is generally not a meaningful quantity. Instead, one uses the *concentration*, measured as ρ_J, the mass per unit volume of component J. Note that the composition can be directly obtained from the product ρC_J, where ρ is the density of the phase and C_J the mass fraction of component J in this phase. It can

Table 3.1 Composition measures and symbols.

Base	Phase composition	Definition
moles	X_J	moles of component J / total moles
mass	C_J	mass of component J / total mass
Base	**Relative amount**	**Definition**
moles	χ_ν	moles of phase ν / total moles
mass	f_ν	mass of phase ν / total mass
volume	g_ν	volume of phase ν / total volume

naturally be very confusing with all these different definitions and we thus take the time to define everything carefully.

One can readily convert the different measures of composition using the molecular weights of the components and the densities of the phases. Continuing with the example introduced in Fig. 3.1, the composition of the α phase at temperature T_0 can be expressed as a mass fraction $C_{B\alpha}$, given by

$$C_{B\alpha} = \frac{\text{Mass B}}{\text{Total mass}} = \frac{\mathcal{M}_B X_{B\alpha}}{\mathcal{M}_B X_{B\alpha} + \mathcal{M}_A(1 - X_{B\alpha})} = \frac{\mathcal{M}_B X_{B\alpha}}{\mathcal{M}_A + (\mathcal{M}_B - \mathcal{M}_A)X_{B\alpha}} \tag{3.5}$$

where $\mathcal{M}_A$ and $\mathcal{M}_B$ are the molecular weights of A and B, respectively. One can write similar expressions for $C_{B\ell}$ and C_{B0} by replacing $X_{B\alpha}$ with the appropriate mole-based composition, viz.

$$C_{B\ell} = \frac{\mathcal{M}_B X_{B\ell}}{\mathcal{M}_A + (\mathcal{M}_B - \mathcal{M}_A)X_{B\ell}}; \quad C_{B0} = \frac{\mathcal{M}_B X_{B0}}{\mathcal{M}_A + (\mathcal{M}_B - \mathcal{M}_A)X_{B0}} \tag{3.6}$$

Most phase diagram reference books provide diagrams drawn with respect to both mole and mass fraction.

Following the same lever rule construction gives f_α and f_ℓ, the mass fractions of α and liquid, respectively

$$f_\alpha = \frac{C_{B\ell} - C_{B0}}{C_{B\ell} - C_{B\alpha}}; \qquad f_\ell = \frac{C_{B0} - C_{B\alpha}}{C_{B\ell} - C_{B\alpha}} \tag{3.7}$$

Finally, we can use the densities of the phases to also compute volume fractions.

$$g_\alpha = \frac{f_\alpha/\rho_\alpha}{f_\alpha/\rho_\alpha + f_\ell/\rho_\ell} = \frac{\rho_\ell f_\alpha}{\rho_\ell f_\alpha + \rho_\alpha f_\ell}; \qquad g_\ell = \frac{\rho_\alpha f_\ell}{\rho_\ell f_\alpha + \rho_\alpha f_\ell} \tag{3.8}$$

Note that if $\mathcal{M}_A = \mathcal{M}_B$, then $C_{B\nu} = X_{B\nu}$ and $\chi_\nu = f_\nu$. Also, if $\rho_\ell = \rho_\alpha$, then $f_\nu = g_\nu$. Obviously, this is never exactly the case, but for some alloys, such as Al-Si, whose densities are very similar, it is a reasonable approximation.

Phase fraction calculations are most often carried out using mass fractions, and we thus proceed using this form. In such calculations, it is often convenient to idealize the phase diagram, representing the solidus and liquidus curves as straight lines

$$T_{liq} = T_f + m_\ell C_\ell; \qquad T_{sol} = T_f + m_s C_s \tag{3.9}$$

where T_f is the melting temperature of the pure component, and m_ℓ and m_s are the slopes of the liquidus and solidus lines, respectively. Note that the index B has been omitted in the composition and that index α has been replaced by s for the sake of clarity. Solving Eqs. (3.9) for the compositions, gives us

$$C_\ell = \frac{T_{liq} - T_f}{m_\ell}; \qquad C_s = \frac{T_{sol} - T_f}{m_s} \tag{3.10}$$

A measure of the difference in composition between the solid and liquid is the *segregation coefficient* k_0, defined as the ratio C_s/C_ℓ. When the liquidus and solidus are linearized, then $k_0 = m_\ell/m_s$ is constant. However, this is usually applicable only over small ranges of temperature and composition. When the linear form is used, the inverse lever rule, Eq. (3.4), can be written in terms of the temperature and k_0. For example, for an alloy of composition C_0 at a temperature T in the two phase region, the solid fraction f_s is given by

$$f_s = \frac{1}{1-k_0} \frac{T - T_{liq}(C_0)}{T - T_f} \tag{3.11}$$

where $T_{liq}(C_0)$ is the liquidus temperature for this alloy. One can also track the *solidification path*, i.e., the sequence of compositions taken by the solid and liquid phases during solidification, as illustrated in Fig. 3.2. The solid phase composition follows the solidus curve, while the liquid phase follows that of the liquidus. Note that the composition of each phase is uniform at any instant since equilibrium is assumed, i.e., the transformation

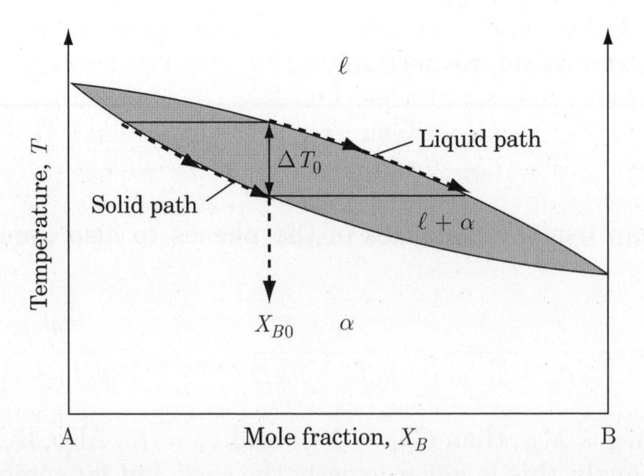

Fig. 3.2 The solidification path for alloy X_{B0} in a simple isomorphous system.

takes place slowly enough that diffusion in all phases reaches steady state. The *solidification interval* or *freezing range* of an alloy, written $\Delta T_0(C_0)$, is simply given by the temperature interval over which the solid and liquid phases co-exist, i.e., $\Delta T_0(C_0) = T_{liq}(C_0) - T_{sol}(C_0)$ (Fig. 3.2).

3.2.2 Isomorphous systems: construction from Gibbs free energy curves

As already stated in the introduction and in Chap. 2, the phase diagrams result directly from the equilibrium conditions, i.e., the equality of the chemical potentials, and the shape of the free energy curves or surfaces. The regular solution model for binary mixtures, introduced in Chap. 2, will be used for the liquid phase as well as for the unique solid phase assumed in this section. Consider binary alloys made from components A and B, with entropies of fusion ΔS^m_{fA} and ΔS^m_{fB}, and equilibrium melting points T_f^A and T_f^B, respectively. The regular solution model for the molar free energy of a binary alloy was given in Eq. (2.49). Taking the liquid as the reference state (where G^m_A and G^m_B are zero), the molar free energies of the liquid and solid are given by

$$G^m_\ell(X_B, T) = RT(X_B \ln X_B + (1 - X_B)\ln(1 - X_B)) + \Omega^m_\ell X_B(1 - X_B) \tag{3.12}$$

$$\begin{aligned} G^m_s(X_B, T) = \; & (1 - X_B)\Delta S^m_{fA}(T - T_f^A) + X_B \Delta S^m_{fB}(T - T_f^B) \\ & + RT(X_B \ln X_B + (1 - X_B)\ln(1 - X_B)) + \Omega^m_s X_B(1 - X_B) \end{aligned} \tag{3.13}$$

where Ω^m_ℓ and Ω^m_s are phenomenological interaction parameters for the regular solution model, described in Chap. 2. As we shall see in the following sections, various types of phase diagrams can be obtained by systematically varying the parameters Ω^m_ℓ and Ω^m_s. In the example calculations that follow, we choose the melting points and entropies of fusion for the pure components given in Table 3.2.

As a first example to illustrate the isomorphous system, we consider an ideal solution for both phases, i.e., $\Omega^m_s = \Omega^m_\ell = 0$. The two curves $G^m_\ell(X_B)$ and $G^m_s(X_B)$ calculated with these parameters at four temperatures are shown in Fig. 3.3. When the temperature is above T_f^A, Fig. 3.3(a), the Gibbs free energy curve for the liquid is below that of the solid at all points: The minimum energy principle states that, at this temperature, an

Table 3.2 Material properties for the pure components in example phase diagrams.

Component	T_f [K]	ΔS^m_f [J mol^{-1} K^{-1}]
A	1200	10
B	800	10

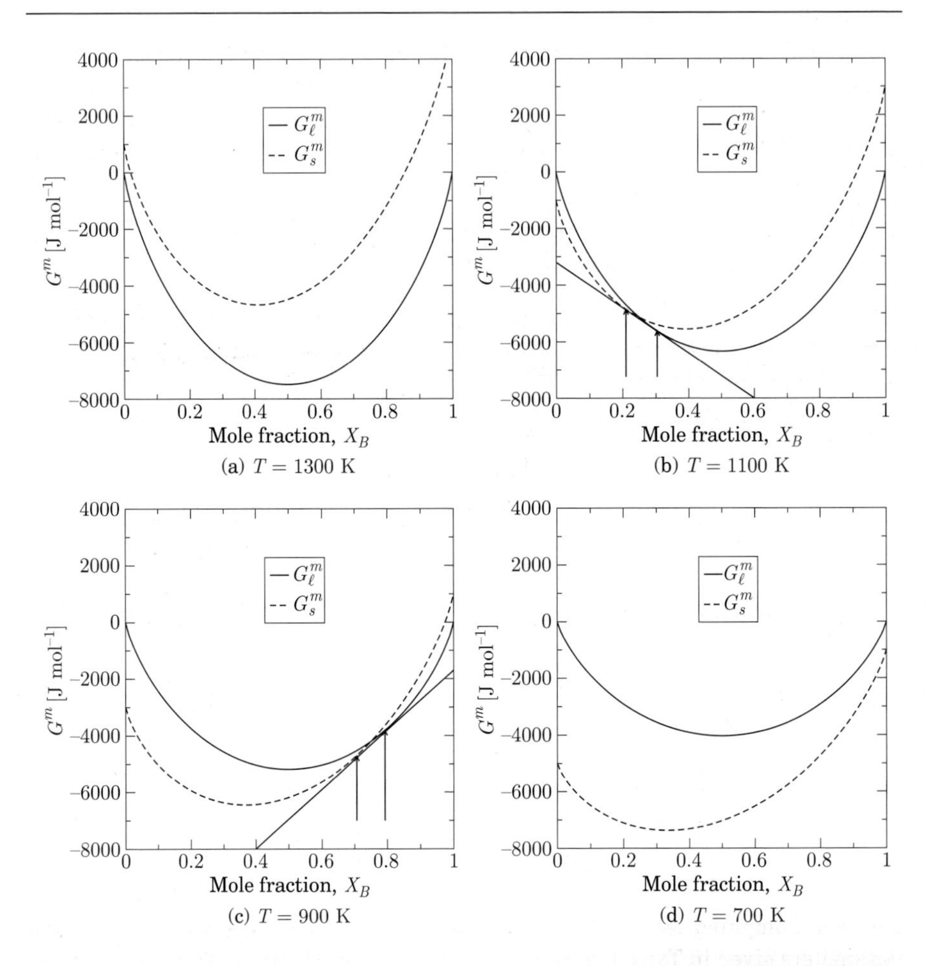

Fig. 3.3 Free energy curves at four different temperatures for an ideal solution with properties as given in Table 3.2, and with $\Omega_s^m = \Omega_\ell^m = 0$.

alloy of any composition X_B is entirely liquid. Similarly, when the temperature is below T_f^B, Fig. 3.3(d), $G_s^m(X_B)$ is lower than $G_\ell^m(X_B)$ for any X_B, corresponding to the solid solution. At intermediate temperatures, as in Fig. 3.3(b) and (c), the curves intersect, indicating a co-existence of the two phases. The phase boundaries at this temperature are determined by constructing the common tangent to the free energy curves, as seen in Chap. 2. The intersections of the tangent with $G_s^m(X_B)$ and $G_\ell^m(X_B)$, indicated by the arrows in Figs. 3.3(b) and (c), give the compositions, $X_{Bs}(T)$ and $X_{B\ell}(T)$, of the solidus and liquidus lines at this temperature, respectively, i.e., $T = T_{sol}(X_{Bs})$ and $T = T_{liq}(X_{B\ell})$. The minimum energy criterion states in this case that: for $X_B < X_{Bs}$, only the solid is present; for $X_B > X_{B\ell}$, only the liquid phase is present; for $X_{Bs} \leq X_B \leq X_{B\ell}$, the two phases co-exist with phase fractions given by the lever rule and compositions given by the ends of the tie-line, i.e., by X_{Bs} and $X_{B\ell}$.

Notice that G_ℓ^m has fixed extremities since $G_{A\ell}^m$ and $G_{B\ell}^m$ have been set to zero, but its curvature decreases with temperature due to the entropy of the mixing term. The same occurs for G_s^m but, additionally, this curve rises as the temperature increases. This seems to contradict what has been seen in Chap. 2, i.e., $dG_\nu^m = -S_\nu^m dT$ is negative when $dT > 0$. However, it should be kept in mind that $S_{\ell J}^m > S_{sJ}^m$ and that the curve G_ℓ^m has been fixed, i.e., $\Delta S_{fJ}^m = S_{\ell J}^m - S_{sJ}^m$ has been added to the solid. In other words, relative to the liquid curve, the solid curve G_s^m increases with temperature. As a result of the relative movements of the two Gibbs free energy curves, the common tangent points move with temperature, eventually sweeping out the liquidus and solidus curves shown in Fig. 3.4.

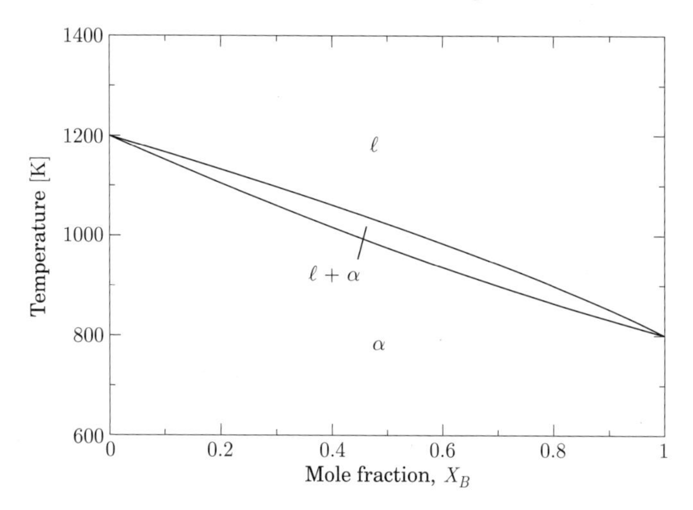

Fig. 3.4 Computed isomorphous phase diagrams for an ideal solution with the parameters given in Table 3.2.

Isomorphous systems do not have to be ideal solutions. Figure 3.5(a) and (b) show possible isomorphous phase diagrams, corresponding to $\Omega_s^m = 0$ and either large positive or large negative values of Ω_ℓ^m. Both cases produce an isomorphous phase diagram, one with a liquidus maximum, and the other with a minimum. Each of these phase diagrams has two separate two-phase regions, but both regions contain exactly the same phases, differing only in their compositions. The free energy curves corresponding to "interesting" temperatures for the two phase diagrams are given in Fig. 3.5(c) and (d), in which one can see that the pair of two-phase regions comes about as a result of two sets of common tangent pairs. The particular composition X_{Ba} and temperature T_a for which the two phase domain has a solidification interval ΔT_0 reduced to zero (apart from the pure elements) is called an *azeotrope*. This azeotrope corresponds to the temperature at which the two curves G_s^m and G_ℓ^m have only one tangent pair. Later, in Sect. 3.2.6, we will discuss the computation of phase diagrams from solution models in more detail.

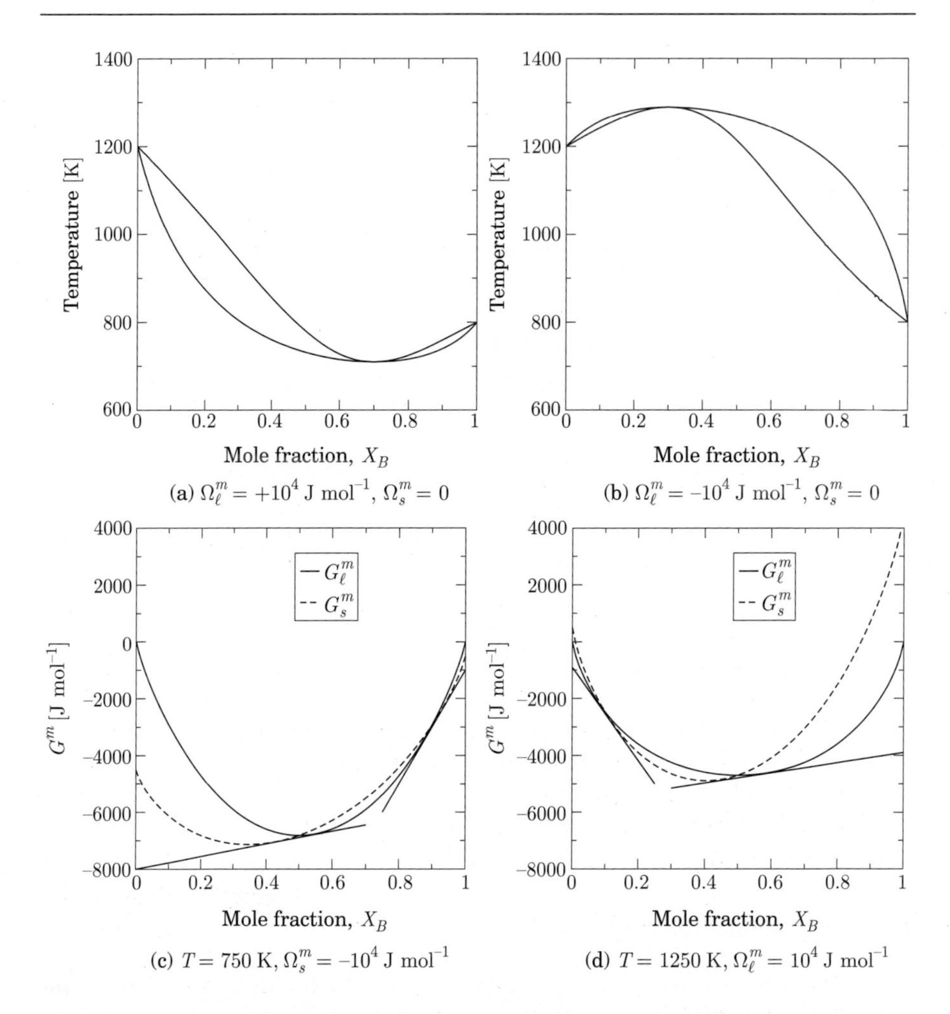

Fig. 3.5 Computed isomorphous phase diagrams for regular solutions, corresponding to the parameters given in Table 3.2, and (a) $\Omega_\ell^m = -10^4$ and $\Omega_s^m = 0\,\mathrm{J\,mol^{-1}}$. (b) $\Omega_\ell^m = 10^4$ and $\Omega_s^m = 0\,\mathrm{J\,mol^{-1}}$. (c) and (d) show free energy curves at "interesting" temperatures.

3.2.3 Eutectic systems

According to Gibbs' phase rule, a maximum of 3 phases can co-exist at a fixed pressure in a binary system (invariant point), e.g., two solids co-existing with the liquid phase. A second phase can appear for several reasons. When the two solute elements are not sufficiently similar, e.g., if their atomic sizes are very dissimilar, or their crystal structures are different, they cannot form a continuous solid solution over the entire composition range. When the two elements are particularly attracted by each other, they can form intermediate or even intermetallic phases. In such cases, it is necessary to consider a separate Gibbs free energy curve $G_\nu^m(T, X_B)$ for

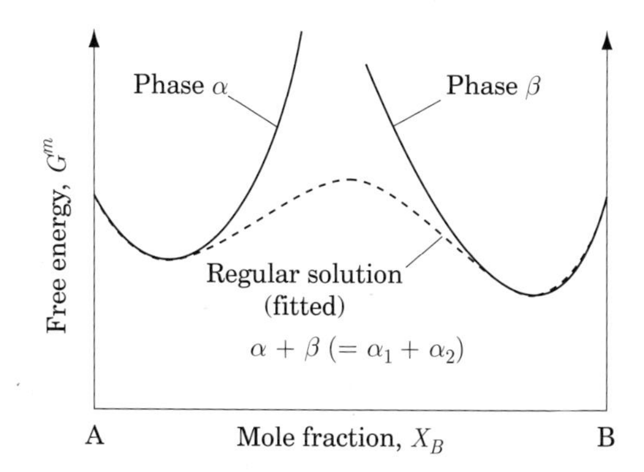

Fig. 3.6 An illustration of the fitting of a regular solution model to the free energy curves of two different solids. The regular solution model implies that the two phases are essentially the same, so the two solid phases α and β are instead written as α_1 and α_2 to emphasize this point.

each phase ν as shown in Fig. 3.6. This is the approach used in commercial thermodynamic software such ThermoCalc®, FactSage® or Pandat®. In the following, we use instead a single regular solution expression with $\Omega_s^m > 0$ for the description of two phases. As shown in Fig. 3.6, the regular solution model can produce a "double-well" curve G_s^m with minima close to those of the curves G_α^m and G_β^m. This procedure replaces two distinct phases α and β by a single solid α exhibiting two minima which will from hereon be denoted α_1 and α_2. As we shall see, such a simple approach is capable of reproducing, at least qualitatively, most of the features observed in actual phase diagrams.

Two solids that prefer not to bond to each other correspond to a case where Ω_s^m is large. Free energy curves at several temperatures for such a system are shown in Fig. 3.7, and the corresponding equilibrium phase diagram is shown in Fig. 3.8. It is instructive to discuss the free energy curves and phase diagrams together, in order to emphasize the relationship between them.

The large positive value of Ω_s^m results in a convex portion of the free energy curve for the solid solution at low temperatures, Fig. 3.7(a). The corresponding common tangent indicates the compositions of two solids in equilibrium. This leads to the region marked "$\alpha_1 + \alpha_2$" in the phase diagram shown in Fig. 3.8. Notice that the solid free energy is lower at all points as compared to the liquid at this temperature. When the temperature increases, as in Fig. 3.7(c), the free energy curves for the solid and liquid pass through each other while the convexity of G_s^m decreases due to the increasing entropy of mixing partially overcoming Ω_s^m. At this temperature, there are two sets of common tangents between the liquid and solid, as shown. These correspond to the regions marked "$\ell + \alpha_1$" and "$\ell + \alpha_2$" in Fig. 3.8. As the temperature increases further, as seen

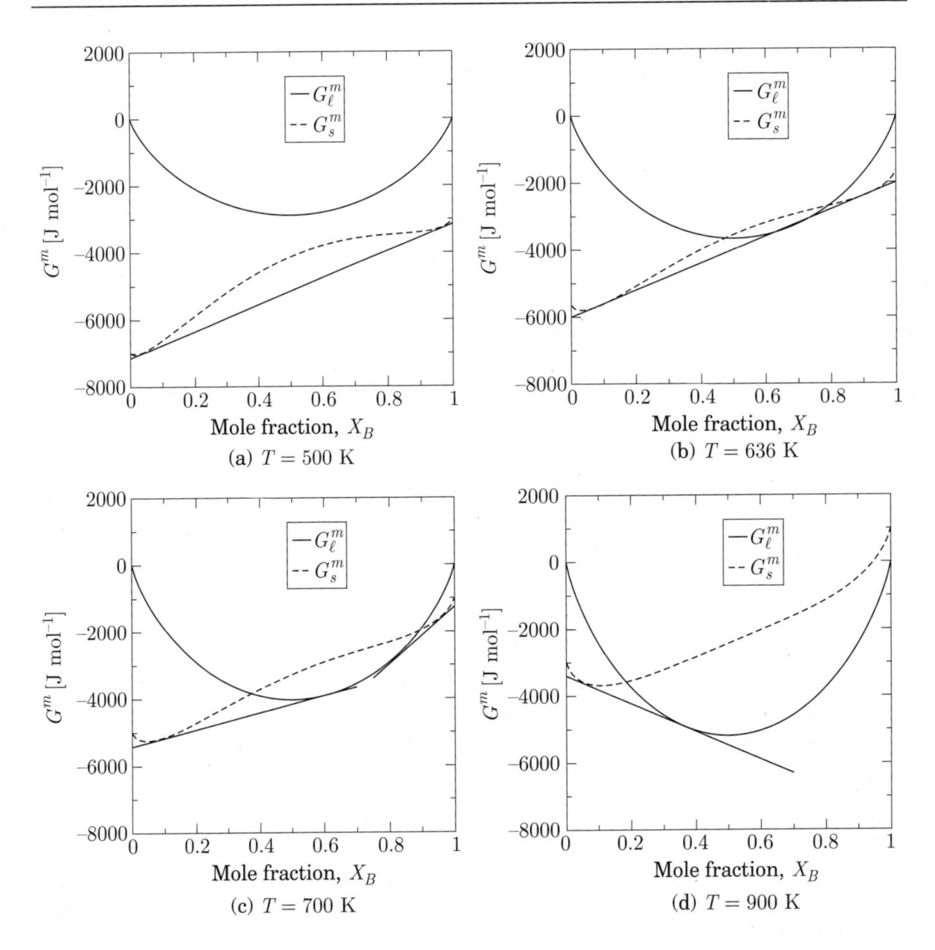

Fig. 3.7 Free energy curves for four different temperatures for a regular solution with properties as given in Table 3.2, and with $\Omega_s^m = 1.5 \times 10^4$ J mol^{-1} and $\Omega_\ell^m = 0$.

in Fig. 3.7(d), the solid has a free energy lower than the liquid only near $X_B = 0$. Moreover, there is only one region of two-phase equilibrium, $\ell + \alpha_1$. At one unique temperature, Fig. 3.7(b), the common tangent connects three compositions: two solids and one liquid. These three phases are in equilibrium, and according to the Gibbs phase rule, this gives us an invariant point ($N_F = 2 - 3 + 1 = 0$). Such an invariant point is called a *eutectic*, and the temperature at which it occurs is labeled T_{eut}. Under equilibrium conditions, an alloy with a composition X_{eut} equal to that of the liquid coexisting with the two solid phases at T_{eut} (in this case 0.68) will transform according to the eutectic reaction: $\ell \rightarrow \alpha_1 + \alpha_2$, i.e., it will solidify with a solidification interval $\Delta T_0 = 0$. Examples of simple binary systems with eutectic reactions are Al-Si, Pb-Sn and Bi-Cd.

The eutectic and isomorphous systems are actually more similar than they first appear. We can illustrate this using the regular solution model by

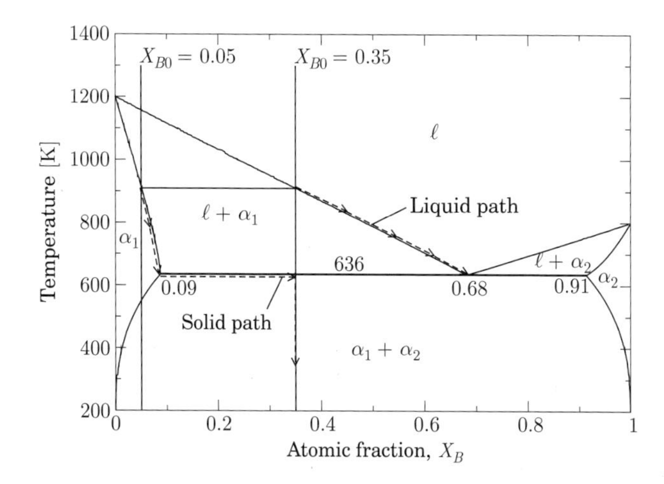

Fig. 3.8 Computed eutectic phase diagram for a regular solution, corresponding to the parameters given in Table 3.2, and with $\Omega_\ell^m = 0$ and $\Omega_s^m = 1.5 \times 10^4$ J mol^{-1}.

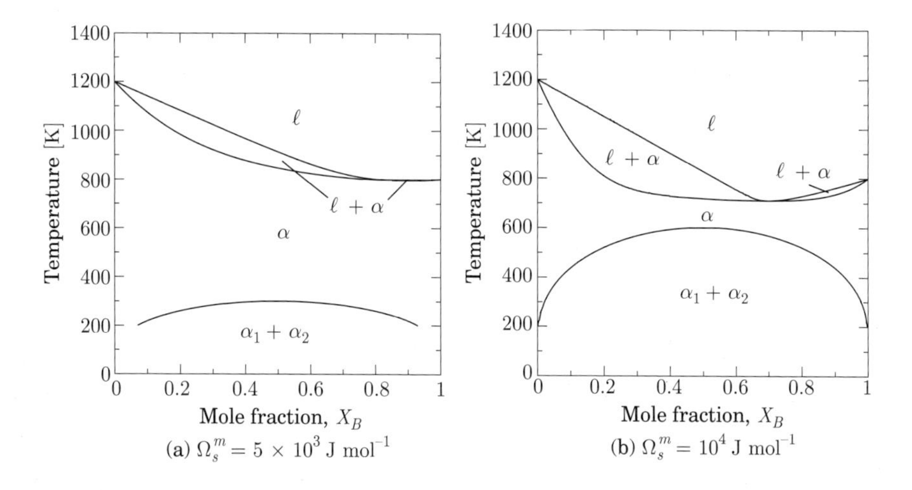

Fig. 3.9 Computed isomorphous phase diagrams for regular solutions, corresponding to the parameters given in Table 3.2, and with $\Omega_\ell^m = 0$ and $\Omega_s^m = 5 \times 10^3$ or 10^4 J mol^{-1}.

fixing $\Omega_\ell^m = 0$ and gradually increasing the value of Ω_s^m while observing the effect on the phase diagram. Figure 3.9 shows such a progression. At modest values of Ω_s^m, the solid solution gives way to two solids at low temperature. (In fact, since we used only one curve to represent the solid phase, the transformation of a homogeneous solid phase at high temperature into the same phase but with different compositions at lower temperature is called a *spinodal decomposition*). As Ω_s^m increases, the phase diagram is topologically similar, and the two-phase region extends to higher temperatures.

Eventually, when Ω_s^m is large enough, the $\alpha_1 + \alpha_2$ region overlaps with the $\ell + \alpha$ region, leading to the eutectic reaction. Notice that, in Fig. 3.7(c), one could also construct a common tangent between two solid compositions at this temperature. This *metastable* equilibrium has a higher free energy than the $\ell + \alpha_1$ equilibrium. This will be further discussed in Sect. 3.2.6.

Solid fraction calculations for eutectic systems are performed similarly to those for isomorphous systems. To simplify the discussion, let us assume that the molecular weights of the two components are identical. The mass and mole fractions are thus the same, and it is possible to use the same phase diagram that we just computed. Consider first the alloy marked as $X_{B0} = 0.05$ in Fig. 3.8. When studying the phase diagram, we immediately deduce that the liquid will transform into the α_1 phase whereas an alloy with a composition close to pure B, say $X_{B0} = 0.95$, would solidify as α_2. On cooling from the liquid under equilibrium conditions, solidification begins at the liquidus temperature of T_{liq}, and ends at the solidus temperature T_{sol}. Solid fractions are computed at intermediate values using the lever rule, as described in Sect. 3.2.1. In order to perform an actual calculation, the liquidus and solidus curves will be taken to be straight lines. Using the data in Fig. 3.8, we can make the following estimates:

$$\begin{aligned} m_{\alpha_1}^m &= \frac{636 - 1200}{0.09} = -6267\,\mathrm{K}(\mathrm{mol\,frac.})^{-1}; \\ m_\ell^m &= \frac{636 - 1200}{0.68} = -829\,\mathrm{K}(\mathrm{mol\,frac.})^{-1}; \\ k_{0\alpha_1}^m &= \frac{m_{\alpha_1}}{m_\ell} = 0.13 \end{aligned} \tag{3.14}$$

Note that we have used the superscript m to indicate that these are molar quantities.

From these data, one can readily compute for alloy $X_{B0} = 0.05$ that $T_{liq} = 1159$ K, and $T_{sol} = 887$ K. We can also calculate the solid fraction as a function of temperature, using Eq. (3.11). The result is shown graphically in Fig. 3.10. Note that the solidified fraction is decidedly nonlinear over the freezing range. Figure 3.10 shows the results of a similar calculation for the second alloy in Fig. 3.8, for which $X_{B0} = 0.35$. This calculation proceeds similarly to the one for $X_{B0} = 0.05$ until the eutectic temperature is reached. At that point, $\chi_{\alpha_1} = 0.559$ and the remaining liquid of composition X_{eut} will freeze isothermally with the eutectic structure. Just below T_{eut}, the amount of α_1 and α_2 phases is still given by the lever rule, i.e., the nominal composition of the alloy ($X_{B0} = 0.35$ is such that $0.35 = 0.09\chi_{\alpha_1} + 0.91(1 - \chi_{\alpha_1}) = 0.09(1 - \chi_{\alpha_2}) + 0.91\chi_{\alpha_2}$). The reader can verify that the same amount of phases is obtained if one considers the fraction of α_1 phase, $\chi_{\alpha_1} = 0.559$, formed just above the eutectic temperature and the α_1 and α_2 phases that form in the remaining liquid ($\chi_\ell = 0.441$) of a eutectic composition 0.68. The solidification path for the alloy $X_{B0} = 0.35$ is also shown in Fig. 3.8. Note that after the eutectic reaction, the average solid composition jumps to $X_{B0} = 0.35$, although the solute is distributed

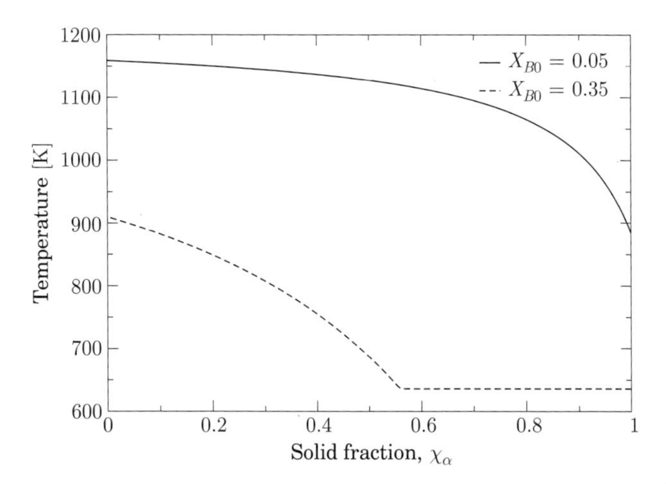

Fig. 3.10 The computed equilibrium solid fraction as a function of temperature for the two alloys indicated in Fig. 3.8.

among the α_1 and α_2 phases. The formation of eutectic microstructures will be discussed extensively in Chap. 9.

3.2.4 Peritectic systems

When both the solid and liquid phases prefer not to mix (corresponding to both parameters Ω_ℓ^m and Ω_s^m in the regular solution model being large and positive), a *peritectic reaction* may result. To illustrate this behavior, we begin with a large value of Ω_ℓ^m, and then gradually increase the value of Ω_s^m as we did for Figs. 3.8 and 3.9. The resulting sequence of phase diagrams is shown in Fig. 3.11. As Ω_s^m increases, the two-phase domain $(\alpha_1 + \alpha_2)$ intersects the $(\ell + \alpha_1)$ and $(\ell + \alpha_2)$ domains, leading again to an invariant point where the three phases co-exist and which is characterized by a temperature T_{per} (Fig. 3.11(c), $T_{per} = 868\,\mathrm{K}$ in this case). However, the main difference with respect to the eutectic case seen before is that the liquid composition $X_\ell(T_{per})$ does not lie in between $X_{\alpha_1}(T_{per})$ and $X_{\alpha_2}(T_{per})$ but outside (in this case on the right). The shape of this phase diagram is such that, above the invariant temperature, the only solid phase that is able to form is the α_1 phase. The other phase that is able to form below T_{per} is called the peritectic phase for reasons that will be discussed in more detail in Chap. 9. At the peritectic temperature, an alloy with a composition of the peritectic phase α_2 ($X_{per} = X_{\alpha_2}(T_{per}) = 0.66$ in this case) will solidify according to the reaction: $\ell + \alpha_1 \rightarrow \alpha_2$. Of course, if the system is cooled further below T_{per}, the α_2 phase will partially transform into α_1 as we enter the two-phase region $(\alpha_1 + \alpha_2)$ due to the slope of the solvus curve. When the solubility decreases with temperature, the solvus curve is called "retrograde". It should be noted that not all peritectics present such strong retrogrades, and the peritectic phase may be perfectly stable below

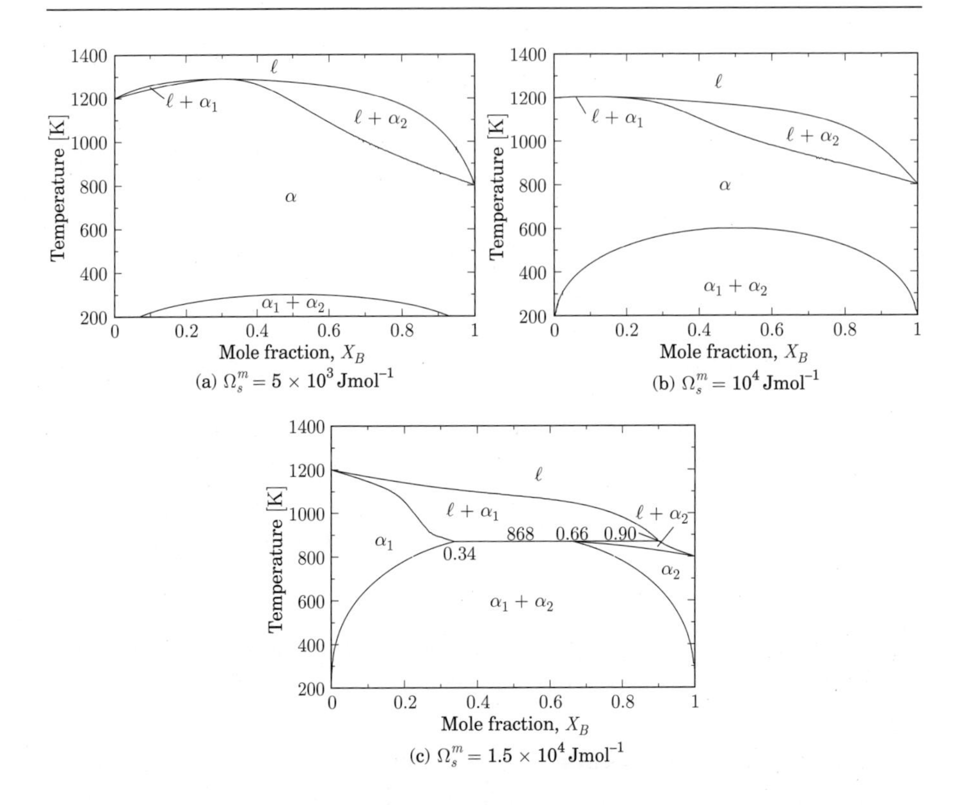

Fig. 3.11 Computed isomorphous phase diagrams for regular solutions, corresponding to the parameters given in Table 3.2, and with $\Omega_\ell^m = 1.5 \times 10^4$ and $\Omega_s^m = 5 \times 10^3 - 1.5 \times 10^4\,\mathrm{J\,mol^{-1}}$.

T_{per}. Examples of simple binary peritectic systems include Ag-Pt, Fe-Ni or Ni-Ru.

Free energy curves at several temperatures for the peritectic system of Fig. 3.11(c) are shown in Fig. 3.12. The sequence is similar to that observed for the eutectic system, beginning with two solids at low temperatures, followed by an intersection of the solid and liquid free energy curves producing a three-phase equilibrium at an invariant point (868 K, Fig. 3.12(c)). Note that although the free energy curve of the solid appears to be flat in this case it still exhibits two minima. As stated before, the difference between the eutectic and peritectic cases is the sequential order of the phases along the common tangent line. For the eutectic system, the sequence was (from left to right) $\alpha_1 - \ell - \alpha_2$, whereas the peritectic system displayed the sequence $\alpha_1 - \alpha_2 - \ell$.

3.2.5 Other binary systems

There exist many other possible reactions, and corresponding phase diagrams, and we refer to the reader to the references at the end of this

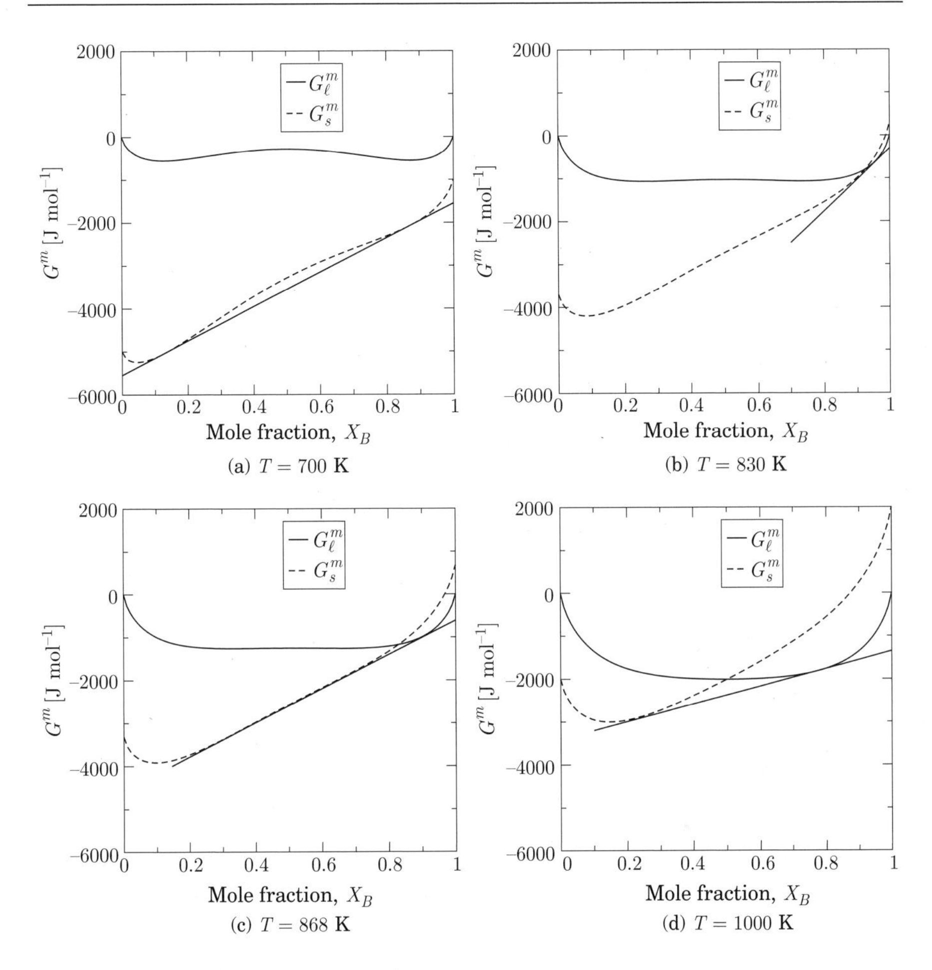

Fig. 3.12 Free energy curves for four different temperatures for a regular solution with properties as given in Table 3.2, and with $\Omega_s^m = 1.5 \times 10^4\,\mathrm{J\,mol^{-1}}$, and $\Omega_\ell^m = 1.5 \times 10^4$.

chapter for further information. However, there are a few types of diagrams that are worth mentioning at this point. Figure 3.13 shows a *monotectic* system, which has a miscibility gap in the liquid at high temperature. At the invariant point, 1160 K, the monotectic reaction occurs, $\ell_1 \rightarrow \ell_2 + \alpha$. The Cu-Pb system provides a real example of this reaction. There is also a peritectic reaction at 810 K and $X_B = 0.97$.

When intermediate compounds form, e.g., A_xB_y, the equilibrium diagram becomes divided, as illustrated in Fig. 3.14. Although the diagram appears complicated, it is really less so than it may seem. There are actually two separate phase diagrams here, one between A and A_xB_y, and the other between A_xB_y and B. The former is a simple eutectic system, and the latter a simple peritectic system. Each segment follows the description given in the preceding section for the corresponding simple system.

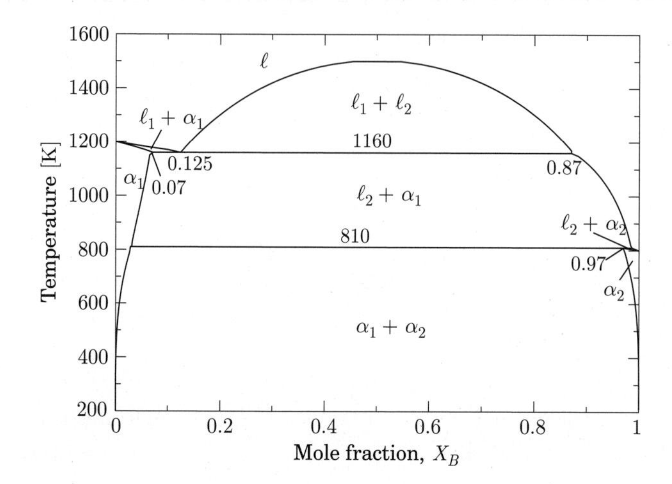

Fig. 3.13 An example of a monotectic system computed with the properties in Table 3.2, and with $\Omega_s^m = 2.5 \times 10^4\,\mathrm{J\,mol^{-1}}$ and $\Omega_\ell^m = 2.5 \times 10^4\,\mathrm{J\,mol^{-1}}$.

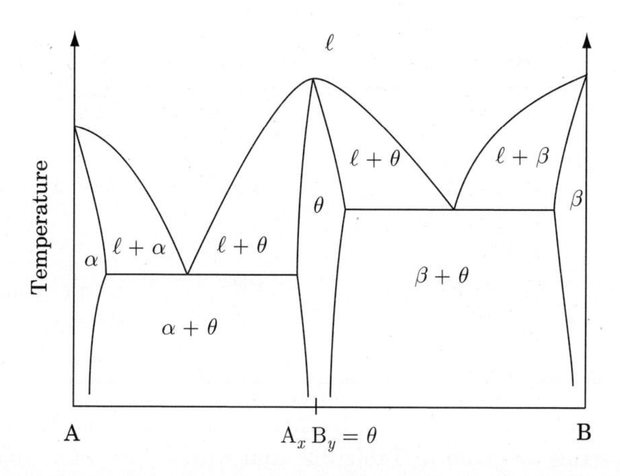

Fig. 3.14 A hypothetical diagram for a compound-containing system.

Intermediate phases may also appear at different temperatures, corresponding to different allotropic forms of a particular composition. Figure 3.15, adapted from Rhines [5], shows a collection of the possibilities, which are also tabulated in Table 3.3. There are several reactions shown in the diagram that have not been discussed, including several that occur in the solid state.

3.2.6 Calculation of binary alloy phase diagrams

This section describes an approach to computing binary alloy phase diagrams using thermodynamic data. The general idea is to find the compositions that minimize the free energy of the mixture inside a two-phase

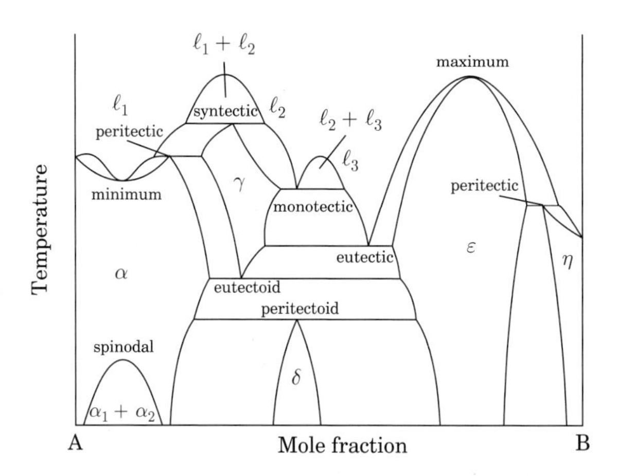

Fig. 3.15 A hypothetical diagram displaying an array of possible invariant reactions in a binary alloy system. (After Rhines [5]).

Table 3.3 Table of invariant binary reactions.

Name	Invariant reaction
Eutectic	$\ell \rightarrow \alpha + \beta$
Peritectic	$\ell + \alpha \rightarrow \beta$
Monotectic	$\ell_1 \rightarrow \ell_2 + \alpha$
Syntectic	$\ell_1 + \ell_2 \rightarrow \alpha$
Congruent	$\ell \rightarrow \alpha$
Spinodal	$\alpha \rightarrow \alpha_1 + \alpha_2$
Eutectoid	$\gamma \rightarrow \alpha + \beta$
Peritectoid	$\gamma + \alpha \rightarrow \beta$

region. The key issue is to locate the two-phase regions, as well as the compositions that form their boundaries. Referring back to Fig. 3.3, note that the temperatures corresponding to the two-phase region are those at which the free energy curves intersect. Thus, to compute the phase diagram, one may scan through a predetermined range of temperatures, and determine for each temperature whether the curves intersect. If the two curves do not intersect, it is a simple matter of determining which curve has the lower free energy in order to decide which phase is present. When the curves do intersect, one can compute the compositions of the respective phases minimizing the free energy to determine the compositions on the two-phase boundary. The intersection point in the two-phase region is itself of interest, as, at this point, the solid and liquid phases have the same free energy, and it is possible to obtain a direct transformation from liquid to solid with no separation of phases. This is called the "T_0 point," and it is important for understanding metastable phase formation during rapid solidification. (Recall that this point was discussed in the context of solute trapping in Chap. 2.) Once the T_0 point is determined, it remains to identify the common tangent points for the phase boundaries.

We describe briefly one possible approach to this problem. A more detailed treatment of the calculation of phase diagrams can be found in Kaufman and Bernstein [2]. The algorithm begins with the identification of the T_0 point, with a composition designated X_{B0}. After determining which phase lies to the left and which lies to the right of the T_0 point, putative values are chosen for the compositions, say X_B^α and X_B^ℓ. The lever rule then gives the fraction of each phase present as

$$\chi_\alpha = \frac{X_{B\ell} - X_{B0}}{X_{B\ell} - X_{B\alpha}}; \qquad \chi_\ell = \frac{X_{B0} - X_{B\alpha}}{X_{B\ell} - X_{B\alpha}} \tag{3.15}$$

The total free energy of the mixture corresponding to these compositions is given by

$$G_{total}^m(X_{B0}, T; [X_{B\alpha}, X_{B\ell}]) = \chi_\alpha G_s^m(X_{B\alpha}, T) + \chi_\ell G_\ell^m(X_{B\ell}, T) \tag{3.16}$$

where the parametric dependence of G_{total}^m on the phase compositions χ_S and χ_ℓ has been explicitly indicated.

The equilibrium phase compositions are those that minimize the total free energy G_{total}^m. This can be posed as an optimization problem: Identify those values of $X_{B\alpha}$ and $X_{B\ell}$ that minimize G_{total}^m in Eq. (3.16). The possible compositions are constrained to lie between X_{B0} and either 0 or 1, depending on which side of X_{B0} that corresponds to α and ℓ. For the example shown in Fig. 3.3, we have $0 \leq X_{B\alpha} < X_{B0}$ and $X_{B0} < X_{B\ell} \leq 1$. Note that if the two curves cross in the opposite sense, i.e, $G_\ell^m > G_s^m$ for $X_B < X_{B0}$, then the bounds on $X_{B\alpha}$ and $X_{B\ell}$ are reversed. There are many optimization algorithms suitable for this problem.

If the free energy curve for either the liquid or solid is convex at some composition, a common tangent must then exist between two compositions on the free energy curve. When this occurs, the possible solid-solid and/or liquid-liquid equilibria must also be computed, along with the free energy of this potential equilibrium point. Of the multiple possible equilibrium points thus determined, the one with the lowest energy is the one that appears in the phase diagram.

The metastable equilibrium diagram, consisting of the higher energy potential equilibrium points just determined, can be useful for understanding solidification structures that appear under non-equilibrium conditions, such as in rapid solidification. An example, using the same parameters as for the eutectic phase diagram in Fig. 3.8, is shown in Fig. 3.16. The T_0 curve is included, as is the metastable extension of the solid-solid equilibrium curve.

As an aid for understanding the relationship between free energy and phase diagrams, a program called PDtool that implements the algorithm described above for binary alloys assuming that both the solid and liquid

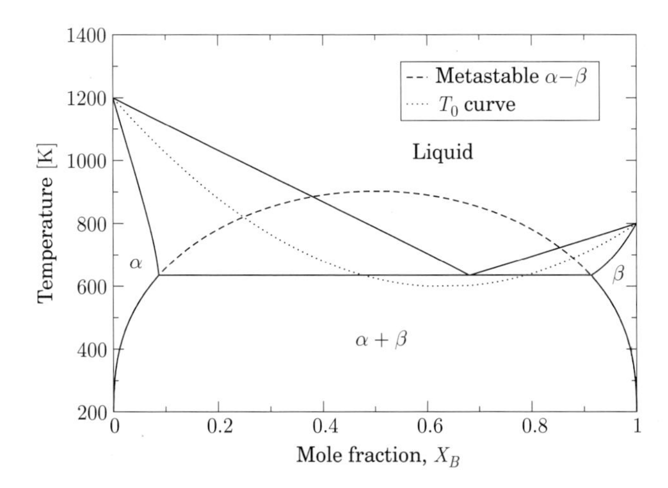

Fig. 3.16 A computed eutectic phase diagram for a regular solution, corresponding to the parameters given in Table 3.2, and with $\Omega_\ell^m = 0$ and $\Omega_s^m = 1.5 \times 10^4\,\mathrm{J\,mol^{-1}}$. The T_0 and metastable extension of the solid-solid equilibrium curves are also shown.

Table 3.4 Some examples of phase diagram types for various values of the regular solution parameters. The pure material properties are fixed at $T_f^A = 800\,\mathrm{K}$, $T_f^B = 1200\,\mathrm{K}$, $\Delta S_{fA}^m = 10\,[\mathrm{J\,mol^{-1}\,K^{-1}}]$ and $\Delta S_{fB}^m = 10\,[\mathrm{J\,mol^{-1}\,K^{-1}}]$.

Ω_s^m [J mol^{-1}]	Ω_ℓ^m [J mol^{-1}]	Phase Diagram	Example
0	0	Isomorphous	Cu-Ni
0	-1.5×10^4	Liquidus minimum	Nb-Mo
1.5×10^4	-1.5×10^4	Eutectic	Ag-Cu
1.0×10^4	1.0×10^4	Isomorphous with spinodal	Au-Pt
1.5×10^4	1.0×10^4	Peritectic	Ag-Pt
1.0×10^4	3.0×10^4	Syntectic with spinodal	Na-Zn
3.0×10^4	2.0×10^4	Eutectic and Monotectic	Al-In

are regular solutions is distributed together with this text.[1] The main difference between PDtool and commercial software for computing phase diagrams, such as ThermoCalc®, PANDAT®, CALPHAD®, etc., is that the latter employ validated thermodynamic databases to obtain the free energy for the liquid and solid phases. This access to more complete thermodynamic databases enables them to compute phase diagrams for systems with intermetallic phases, as well as for systems with more than two components. The basic ideas, however, are the same. To encourage experimentation, Table 3.4 lists several types of phase diagrams corresponding to various values of regular solution model parameters.

[1]PDtool for Mac OS/X, Linux and Windows can be downloaded from `http://www.solidification.org`.

3.3 TERNARY SYSTEMS

Most commercial alloys consist of more than two components, sometimes up to a dozen, as in Ni-base superalloys. In this section, we examine systems with three components, called ternary systems. Some of the most common commercial aluminum alloys are the 6000-series (Al-Mg-Si) and 7000-series (Al-Mg-Zn). Note that some of these alloys also contain significant amounts of other elements, such as Cu or Mn, and a more complete study of these alloy systems would have to include them as well. This section describes in general terms how to read and construct ternary phase diagrams, without reference to specific systems. We will present in some detail three types of diagrams: isomorphous systems, of which one has a two-phase equilibrium, as well as example systems with three-phase and four-phase equilibria. A detailed description of all of the different types of phase diagrams for ternary systems is beyond the scope of this book. Rather, the reader is referred to the excellent texts devoted to the subject listed in the chapter bibliography.

Since $X_A + X_B + X_C = 1$ in a ternary alloy, where X_J is again the molar fraction of component J, the composition is represented using a triangle, as shown in Fig. 3.17. The individual components are placed at the vertices of this triangle. A line emanating from a vertex, such as line Bb in the figure, has a fixed ratio A : C, whereas a line parallel to one edge, such as line yz, has a constant fraction B. A little arithmetic will show that, based on these properties, each point in the triangle represents a unique fraction of the three components. Note that this property applies to all triangles, a fact that we will find useful when computing phase fractions.

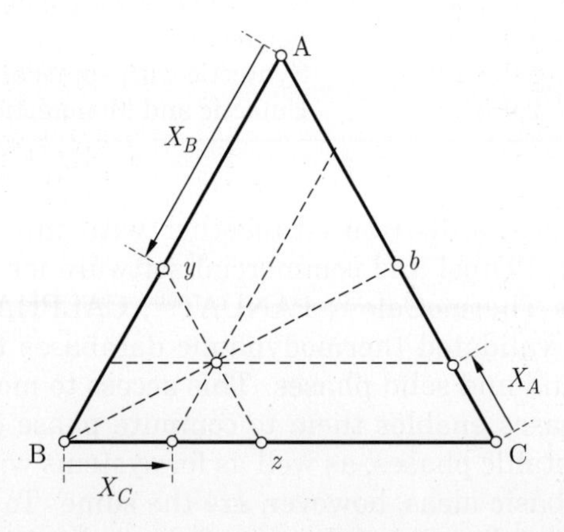

Fig. 3.17 A triangle representing the compositions in a ternary system. The compositions along the line Bb have a constant ratio $X_A : X_C$, and those on the line yz have a constant fraction X_B.

3.3.1 Ternary isomorphous systems

In a ternary phase diagram, the temperature is plotted on an axis perpendicular to the page above the composition triangle shown in Fig. 3.17, thus giving rise to a three-dimensional figure. This somewhat complicates the reading and understanding of ternary diagrams, in comparison to binary phase diagrams. Figure 3.18 shows a ternary isomorphous system with complete solubility in the liquid and solid states. Each vertical face of the diagram corresponds to the binary phase diagram for two of the components. The liquidus and solidus curves in the binary system thus become surfaces in the ternary system as do the Gibbs free energies, as we shall see later. Since the two-phase region is reduced to one point along the vertical axes passing through the corners of the triangle, it appears as a double-sided cap cut by 3 vertical planes crossing each other at the edge of the cap.

Figure 3.18(b) shows an isothermal section through the ternary diagram, indicating the phases present at that temperature by a shading of the various phase regions. The intersection of this isotherm with the liquidus and solidus surfaces now produces lines rather than points as for binary alloys. Recall that for the binary isomorphous system, once the temperature was fixed, we had $N_F = N_c - N_\phi = 2 - 2 = 0$ in the two-phase region. This implied that the compositions of the liquid and solid phases in

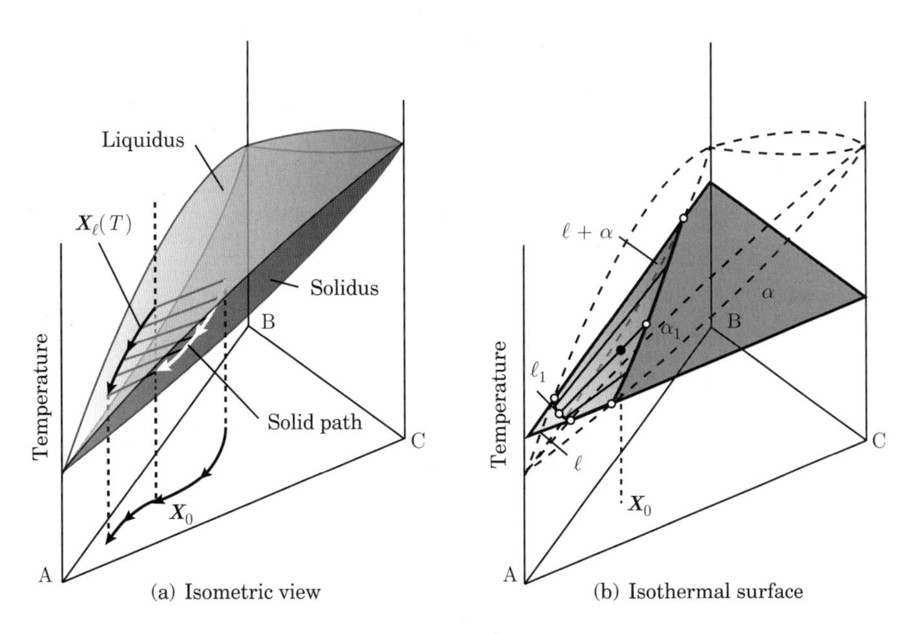

Fig. 3.18 (a) An isometric view of an isomorphous system, displaying the liquidus and solidus surfaces, as well as the solidification path for the alloy marked x. (b) An isothermal section, intersecting both surfaces. The tie-lines through the two-phase region are also included.

the two-phase region were fixed once the temperature was specified. With a third component present, however, $N_F = 1$, and the compositions of the phases are no longer unique. The diagram must specify this additional information in the form of *tie-lines* that indicate the compositions that are in equilibrium. In other words, each tie-line specifies which set of compositions in the liquid $(X_{A\ell}(T), X_{B\ell}(T), X_{C\ell}(T))$ that is in equilibrium with that in the solid $(X_{As}(T), X_{Bs}(T), X_{Cs}(T))$. The tie-lines are indicated on the isothermal section given in Fig. 3.18(b). Note that the tie-line on each face of the diagram, i.e., corresponding to each binary system, is known. However, we do not have any *a priori* knowledge of the placement of the tie-lines within the diagram. As we shall see, the tie-lines are given by the construction of common tangent planes to the Gibbs free energies.

Ternary phase diagrams are most easily understood through the use of horizontal (isothermal) and vertical (pseudo-binary) sections. Figure 3.19 shows a series of isothermal sections of the isomorphous ternary system, taken at a variety of temperatures, and viewed from above. The melting points were chosen in the order $T_f^A < T_f^B < T_f^C$. Thus, as the temperature increases, the two-phase region is displaced toward the vertices B and C. The solution is also close enough to ideal that no internal maxima or minima exist in the liquidus surface. Such extrema are nonetheless possible, and the reader is referred to Prince [4] (p. 136 ff) or Rhines [5] (p. 123 ff) for further discussion of such cases.

The phase diagram can be used to track the freezing path of a ternary alloy. Consider the alloy of composition $\boldsymbol{X}_0 = (X_{A0}, X_{B0}, X_{C0})$ in Fig. 3.18. The ends of the tie-lines passing through point $\boldsymbol{X}_0$ indicate the compositions of α and ℓ that are in equilibrium for any temperature where $\boldsymbol{X}_0(T)$ lies within the two phase region. The sequence of tie-lines over the solidification range is indicated as dashed lines in Fig. 3.18(a). At the start of solidification, when $T = T_{liq}(\boldsymbol{X}_0)$, the liquid composition is simply $\boldsymbol{X}_0$. The liquid composition moves along the liquidus surface toward the A vertex as the solidification proceeds, following the curve on the liquidus surface shown in the figure. At the same time, the solid composition moves along the solidus surface, following the other end to the respective tie-lines, until

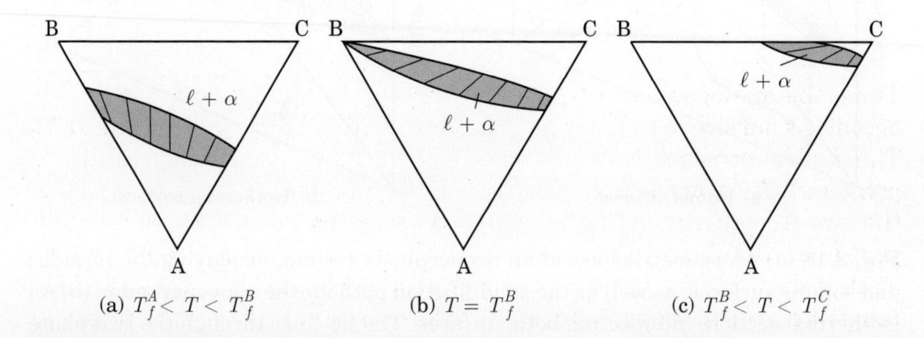

Fig. 3.19 A series of isothermal sections through a ternary isomorphous phase diagram, demonstrating the displacement of the two-phase field with temperature.

the solidification comes to an end, at $T = T_{sol}(\boldsymbol{X}_0)$ where the solid composition is $\boldsymbol{X}_0$. The projection of the solid and liquid paths for alloy $\boldsymbol{X}_0$ on the ternary composition triangle is also indicated. The lines are purposely drawn curved to emphasize that not all tie-lines line up in a vertical plane: the only restriction is that they all have to pass through the nominal composition $\boldsymbol{X}_0$.

We can also employ the ternary phase diagram to compute phase fractions using the inverse lever rule as we did for binary systems. At the temperature shown in Fig. 3.18(b),

$$\chi_\alpha = \frac{\boldsymbol{X}_0\ell_1}{\alpha_1\ell_1}; \qquad \chi_\ell = \frac{\alpha_1\boldsymbol{X}_0}{\alpha_1\ell_1} \tag{3.17}$$

where $\boldsymbol{X}_0\ell_1$, $\alpha_1\ell_1$ and $\alpha_1\boldsymbol{X}_0$ are the lengths of the three segments connecting the compositions $\boldsymbol{X}_0$, $\boldsymbol{X}_\ell(T_1) = \ell_1$ and $\boldsymbol{X}_\alpha(T_1) = \alpha_1$ along the tie-line corresponding to temperature T_1. Note that all the tie-lines pass through $\boldsymbol{X}_0$. Clearly, the liquidus and solidus temperatures are degenerate cases of Eq. (3.17), with χ_α equal to zero or one, respectively. Simple geometry considerations show that Eq. (3.17) can also be written for any of the element compositions, i.e.,

$$\chi_\alpha = \frac{X_{J\ell_1} - X_{J0}}{X_{J\ell_1} - X_{J\alpha_1}}; \qquad \chi_\ell = \frac{X_{J0} - X_{J\alpha_1}}{X_{J\ell_1} - X_{J\alpha_1}}; \qquad J = \mathrm{A, B, or\ C} \tag{3.18}$$

As for binary alloys, the phase diagram of multi-component systems, including the tie-line positions, can be determined from the Gibbs free energies of the various phases. Let us continue to use the regular solution model, extended to three components, but in which only two-element interactions are included

$$\begin{aligned} G_\ell^m(X_A, X_B, X_C, T) =& RT(X_A \ln X_A + X_B \ln X_B + X_C \ln X_C) \\ &+ \Omega_{\ell BC}^m X_B X_C + \Omega_{\ell AC}^m X_A X_C + \Omega_{\ell AB}^m X_A X_B \end{aligned} \tag{3.19}$$

$$\begin{aligned} G_s^m(X_A, X_B, X_C, T) =& X_A \Delta S_{fA}^m (T - T_f^A) + X_B \Delta S_{fB}^m (T - T_f^B) \\ &+ X_C \Delta S_{fC}^m (T - T_f^C) \\ &+ RT(X_A \ln X_A + X_B \ln X_B + X_C \ln X_C) \\ &+ \Omega_{sBC}^m X_B X_C + \Omega_{sAC}^m X_A X_C + \Omega_{sAB}^m X_A X_B \end{aligned} \tag{3.20}$$

Thus, the analogous form of the free energy curves for two components becomes a surface over the composition triangle, as illustrated in Fig. 3.20. This figure corresponds to a temperature in the two-phase region. The T_0 point in 2-D becomes a space curve corresponding to the intersection of the two free energy surfaces. There is also a collection of common tangent lines, that become the tie-lines, as illustrated in Figs. 3.20(a) and (b). Each of the tie-lines is given by the two contact points of one of the common tangent planes to these two surfaces. The common tangent plane construction ensures that the chemical potential of the three elements μ_J, given by the intersection of this plane with the vertical axes passing through the corners of the triangle, are equal in the two phases. The tie-lines can be

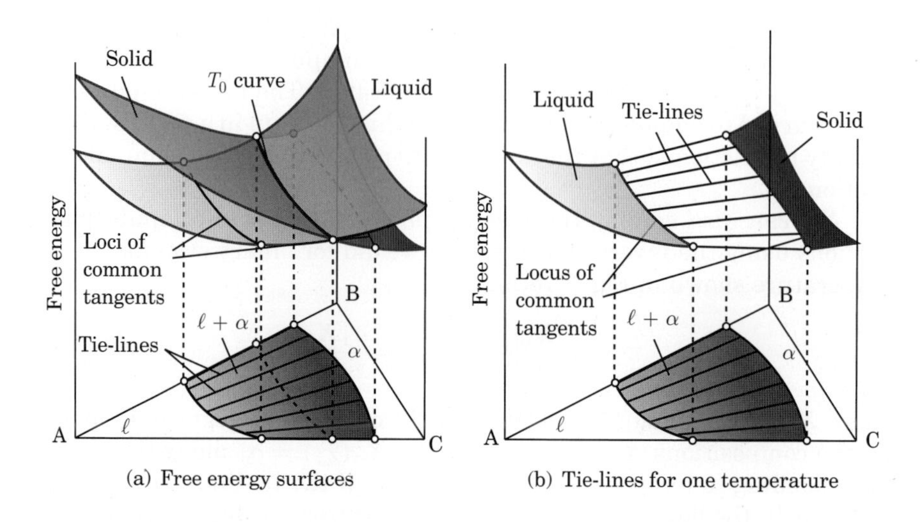

Fig. 3.20 Free energy surfaces for the liquid and solid phases at a temperature inside the two-phase region for an isomorphous system. (a) The free energy surfaces, and the intersection T_0 curve, as well as the projection of the tie-lines onto an isothermal surface. (b) The same figure, with parts of the free energy surfaces stripped away. Note that the tie-lines do not all occur at the same free energy, even though they occur at the same temperature.

determined by an approach similar to the one described in the previous section for binary alloys, where one traverses the T_0 curve, and performs a series of optimizations to find the compositions of the two phases that minimize the free energy of the mixture. Note that this represents much more work for ternary systems as opposed to binary systems, since, at each temperature in the two-phase region, tie-lines must be found across the ternary composition diagram. Further, each optimization problem is somewhat more complicated as a result of the search domain for the compositions being larger. The reader is referred to the chapter bibliography for further details.

Ternary phase diagrams for real systems are only rarely rendered in three-dimensions, as in Fig. 3.18, due to difficulties in extracting the necessary data. Instead, one normally sees isothermal sections with indications of the tie-lines, as in Fig. 3.19, as well as surface contour plots, e.g., for the liquidus and solidus surfaces, as shown in Fig. 3.21. In this latter representation, lines obtained by cutting the corresponding surface with isotherms are simply drawn in the composition triangle. In Fig. 3.21, certain isotherms on the liquidus and solidus surfaces have been drawn with dashed and solid lines, respectively. It is also informative to construct vertical *pseudobinary* sections (sometimes called "cuts") through the phase diagrams, which can be done readily with the aid of the surface contours and isothermal sections. The prefix "pseudo" indicates that while the section may resemble a binary phase diagram, the tie-lines do not generally lie entirely within the plane of the section, and thus a horizontal line in

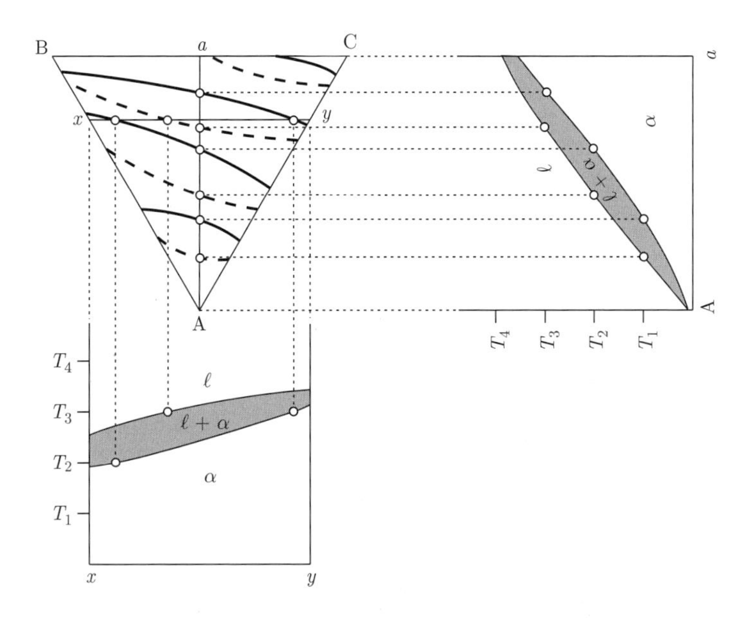

Fig. 3.21 A schematic drawing of the construction of vertical sections from isothermal sections. Isotherms on the liquidus surface are shown as solid curves, and isotherms on the solidus are shown as dashed curves inside the ternary triangle.

a two-phase region does not necessarily connect compositions that are in equilibrium with each other. Figure 3.21 illustrates the construction of pseudobinary sections using contour data for the liquidus and solidus surfaces. The vertical section is obtained by locating the intersections of contour lines from the liquidus and solidus surfaces with the vertical section, as shown.

3.3.2 Ternary three-phase equilibrium

Let us now consider a system in which two of the binary systems are eutectics, say A-B and C-B, and the third is isomorphous (A-C). The new feature of this system is that there is an extended region of three-phase equilibrium, corresponding to the reaction $\ell \to \alpha + \beta$. In the binary systems, this reaction is invariant, because, with two components, $N_F = N_c - N_\phi + 1 = 2 - 3 + 1 = 0$. However, when a third component is added, N_F evaluates to one. This means that the three-phase equilibrium can extend over a range of temperatures and compositions.

In order to understand the solidification path, it is helpful to consider a series of isothermal sections, as illustrated in Fig. 3.22. Assume that the melting points of the pure components T_f^A, T_f^B and T_f^C, and the eutectic temperatures in the binary systems $\mathrm{A-B}$ and $\mathrm{B-C}$ are such that $T_f^A > T_f^B > T_f^C > T_{eut}^{AB} > T_{eut}^{BC}$. Figure 3.22(a) shows an isothermal section at temperature T_1, where $T_f^C > T_1 > T_{eut}^{AB}$. We find two two-phase regions, corresponding to $\ell + \alpha$ and $\ell + \beta$. The interpretation of this diagram is

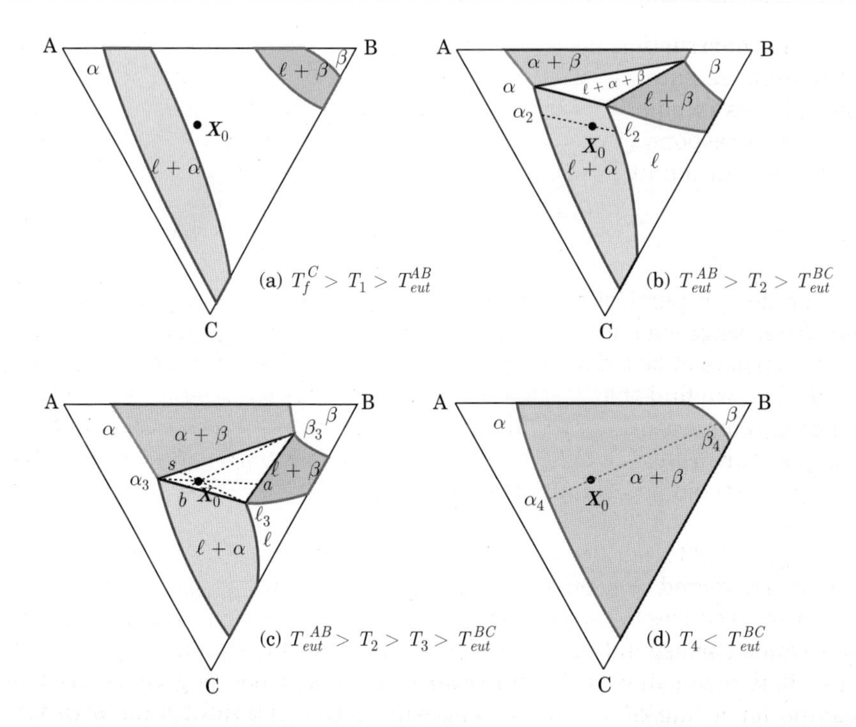

Fig. 3.22 A series of isothermal sections through a ternary phase diagram showing three-phase equilibria.

a straightforward extension of the observations we made for the isomorphous system. At temperature T_{eut}^{AB} (not shown in the figure), the liquid phase boundaries on each of the two phase regions intersect at the eutectic composition in the A − B binary system. This point corresponds to a three-phase equilibrium of ℓ, α and β. In the binary system, this is an invariant point ($N_F = 0$), but in the ternary system, $N_F = 1$. Three-phase equilibrium may thus occur over a range of temperatures and compositions, as indicated by the triangle $\ell + \alpha + \beta$ in Fig. 3.22(b).

One such temperature is T_2, where $T_{eut}^{AB} > T_2 > T_{eut}^{BC}$, as shown in Fig. 3.22(b). For the alloy $\boldsymbol{X}_0$, which lies within this triangle, the compositions of the phases in equilibrium correspond to those at the vertices of the triangle. Figures 3.22(b) and (c) illustrate the movement of the region of three phase equilibrium as the temperature decreases. The diagram also shows a particular alloy composition $\boldsymbol{X}_0$, whose solidification path we now analyze. At temperature T_1, the alloy is entirely liquid, with composition $\boldsymbol{X}_0$. At temperature T_2, the alloy lies within the two-phase $\ell + \alpha$ region, and the phase fractions and compositions are determined by using the lever rule on the tie-line (represented as a dashed line), i.e., by the lengths of the segments connecting the three compositions on the tie-line, $\boldsymbol{X}_0$, $\boldsymbol{X}_\ell$ and $\boldsymbol{X}_s$ (as in Eq. (3.7)):

$$\chi_\alpha = \frac{\ell_2 \boldsymbol{X}_0}{\ell_2 \alpha_2}; \qquad \chi_\ell = \frac{\alpha_2 \boldsymbol{X}_0}{\ell_2 \alpha_2} \tag{3.21}$$

The construction becomes somewhat more interesting in the region of three-phase equilibrium, at temperature T_3. The compositions of the three phases in equilibrium are at the vertices of the triangle, the edges of which correspond to the tie-lines, and the phase fractions are computed, using the notation in Fig. 3.22(c), as (see Exercise 3.7):

$$\chi_\alpha = \frac{a\boldsymbol{X}_0}{\alpha_3 a}; \quad \chi_\ell = \frac{s\boldsymbol{X}_0}{\ell_3 s}; \quad \chi_\beta = \frac{b\boldsymbol{X}_0}{\beta_3 b} \tag{3.22}$$

The simplex property of the triangle ensures that the χ_i sum to one. Note that the *average* composition of the solid at T_3 is given by point s, although it now consists of two distinct phases α and β. Following the logic of this calculation, we find that the solidification for alloy $\boldsymbol{X}_0$ begins at the corresponding temperature on the liquidus surface of α, $T_{liq\alpha}(\boldsymbol{X}_0)$, and ends at a temperature T_{sol} such that $\boldsymbol{X}_0$ lies on the corresponding $\alpha - \beta$ edge of the three-phase triangle. Finally, at temperature T_4, alloy $\boldsymbol{X}_0$ consists entirely of $\alpha + \beta$.

Figure 3.23 shows an isometric view of the ternary system. The locus of points traversed by the liquid composition of the triangles of three-phase equilibria produces a space curve, called a *binary crystallization curve*, whose ends correspond to the eutectic points in the respective binary systems. It is also called a *monovariant line* as only one degree of freedom remains for a liquid with this composition and in equilibrium with two solid phases. On the B-rich side of the binary crystallization curve, the first solid to form (the *primary crystallization product*) is β, and on the other side of the curve the primary crystallization product is α. The α and

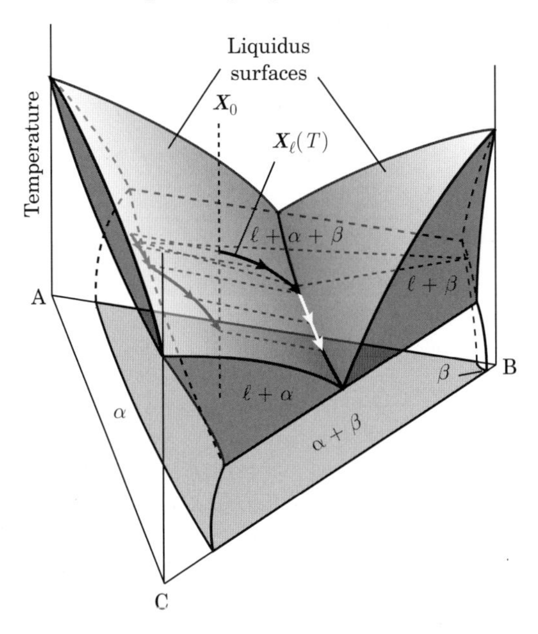

Fig. 3.23 A ternary phase diagram showing a three-phase equilibrium, as well as the solidification path for alloy $\boldsymbol{X}_0$.

β compositions that participate in the three-phase equilibrium trace out curves of maximum solid solubility, shown as dashed lines in the figure. Note that it is not always the case that these compositions correspond to maximum solubility, but more often than not the solubility decreases with temperature, and the slopes of the solvus curves are as shown. The figure also displays the solidification path for alloy $\boldsymbol{X}_0$. It is clear that during solidification of the primary phase α, the composition of the liquid moves along the liquidus surface of α toward the binary crystallization curve, while the composition of the solid moves along the solidus surface of α. When the liquid composition encounters the binary crystallization curve, the liquid composition begins to follow this curve, whereas the solid composition (now $\alpha + \beta$) tracks along the side of the three-phase triangle that is opposite to the liquid composition. This corresponds to composition s in Fig. 3.22(c). The solidification ends when the solid composition s reaches the nominal composition $\boldsymbol{X}_0$, which also corresponds to the temperature where $\boldsymbol{X}_0$ lies on the side of the three-phase equilibrium triangle opposite the liquid composition.

Figure 3.24 shows the phase diagram viewed from above, displaying the trace of the binary crystallization curve as a solid line. The arrows indicate the direction of decreasing temperature, and the dashed lines represent the curves of the maximum solid solubility. There are also two pseudobinary sections. Why does the $\ell + \alpha + \beta$ region terminate in a point on the right-hand side of section $y - z$?

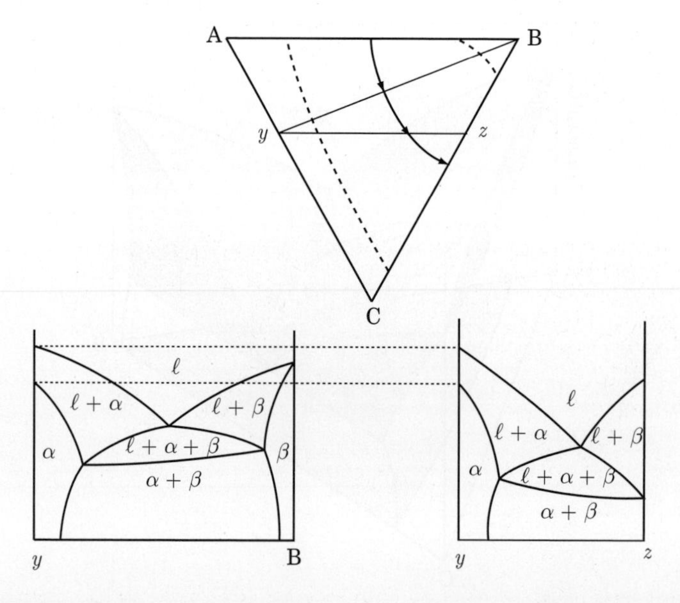

Fig. 3.24 Vertical sections through the ternary three-phase equilibrium diagram.

3.3.3 Ternary four-phase equilibrium: ternary eutectic

The final ternary phase diagram presented in detail is a ternary eutectic system that contains a reaction of four-phase equilibrium $\ell \to \alpha + \beta + \gamma$. In a three component system at fixed pressure, this reaction is an invariant. Assuming that each of the binary systems is a simple eutectic system, with $T_f^A > T_f^B > T_f^C > T_{eut}^{AB} > T_{eut}^{AC} > T_{eut}^{BC} > T_{eut}^{ABC}$, where T_{eut}^{ABC} is the temperature of the ternary eutectic reaction, the series of isothermal sections shown in Fig. 3.25 follows the progress of solidification.

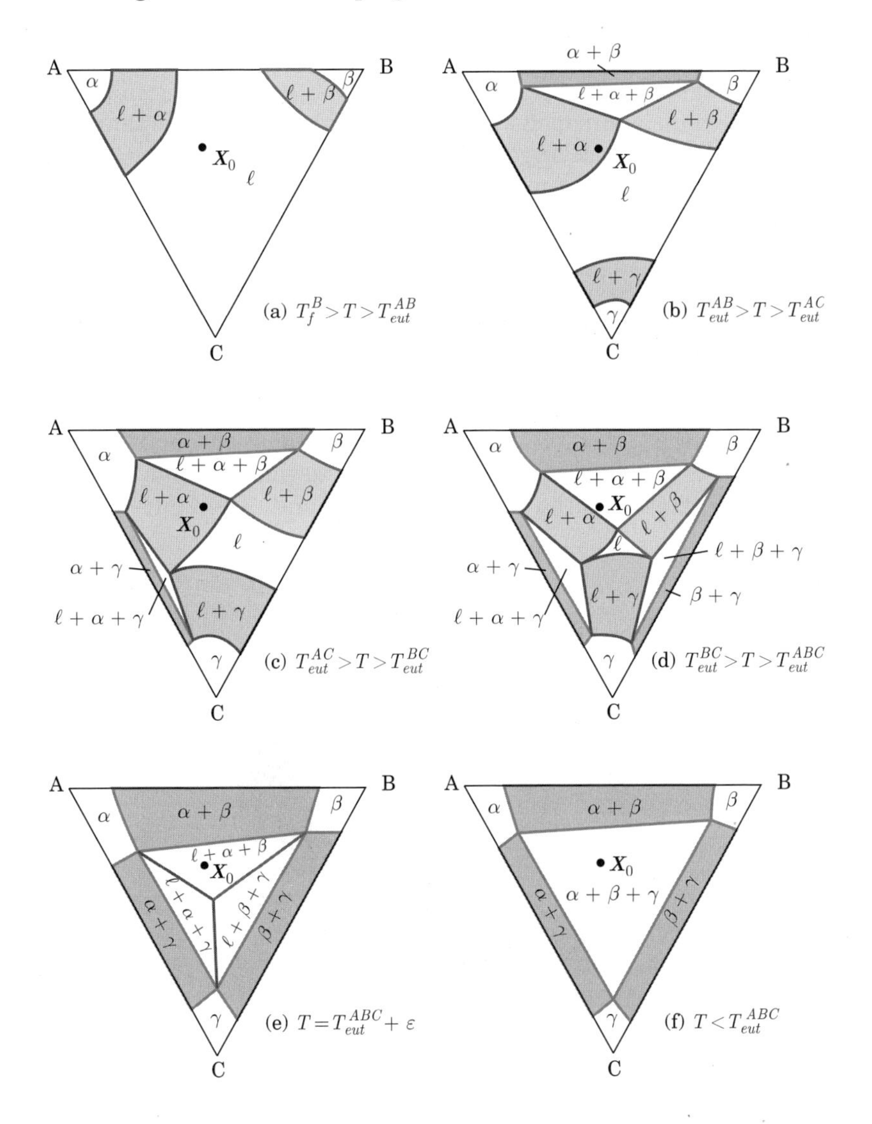

Fig. 3.25 A series of isothermal sections through a ternary phase diagram with a ternary eutectic reaction.

Figure 3.25(a) shows an isothermal section at a temperature below the melting points of the pure components A and B, but above any of the binary eutectic temperatures or the melting point of pure C. Notice the solid solutions α and β in their respective corners of the phase diagram, as well as two regions of two-phase equilibrium. The only difference between this isothermal section and that corresponding to a three-phase equilibrium, Fig. 3.22(a), is that the $\ell + \alpha$ region in this case does not extend up to the B – C binary system. As in the previous case, the two-phase regions approach each other when the a temperature decreases: they meet at $T = T_{eut}^{AB}$, the eutectic temperature in the A – B binary system. At a temperature below T_{eut}^{AB}, but still above the eutectic temperatures in the other binary systems, Fig. 3.25(b), the region corresponding to the three-phase equilibrium again opens up into a triangle. We also see that a region of two-phase equilibrium appears near the C-rich corner. Figures 3.25(c) and (d) illustrate the movement of the phase regions as the temperature decreases below the remaining two binary eutectic temperatures, leading to the formation of two additional three-phase equilibrium triangles $\ell + \alpha + \gamma$ and $\ell + \beta + \gamma$. The regions converge at the ternary eutectic temperature T_{eut}^{ABC}, where a four-phase equilibrium exists between the liquid, α, β and γ. This point is invariant, as $N_F = N_c - N_\phi + 1 = 3 - 4 + 1 = 0$. Below this temperature, one of the phases must disappear (in this case the liquid) and the ternary eutectic reaction thus becomes $\ell \rightarrow \alpha + \beta + \gamma$.

An isometric view of the ternary eutectic diagram is shown in Fig. 3.26. The three binary crystallization curves or monovariant lines form "valleys" leading to the ternary eutectic point. The figure also shows

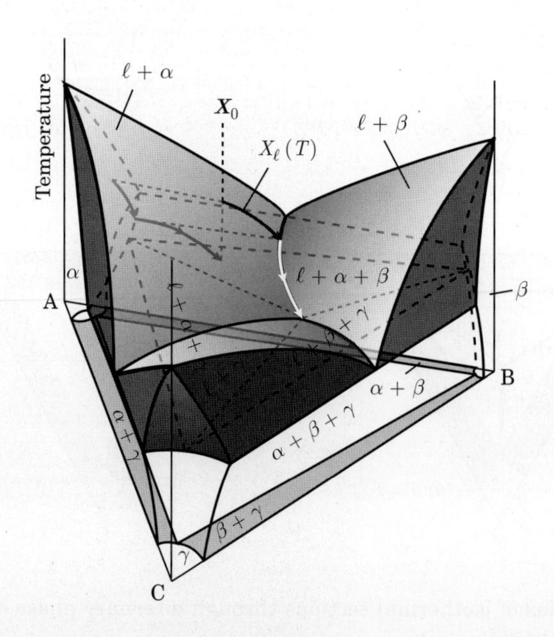

Fig. 3.26 A ternary phase diagram with a ternary eutectic reaction.

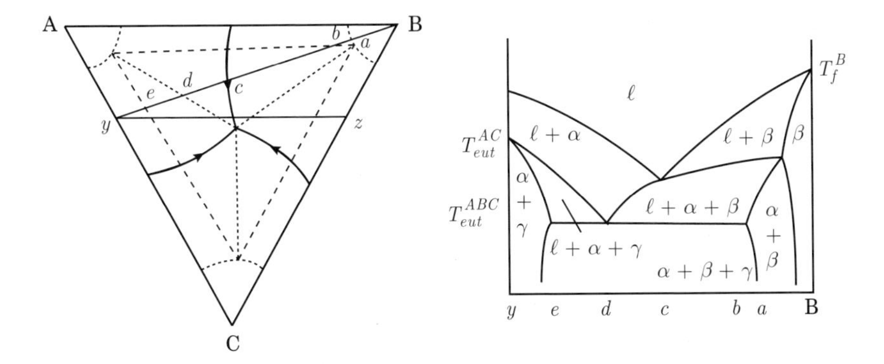

Fig. 3.27 The projection of the ternary phase diagram onto the composition triangle, and the vertical section $B - y$.

the regions corresponding to simple solid solutions in the corners, two-phase equilibria near the edges, and a three phase-equilibrium for the compositions in the center of the triangle. The solidification path for an alloy experiencing the ternary eutectic reaction is also shown in Fig. 3.26. The analysis of the solidification path for this alloy differs only slightly from the one we performed in the previous section for the three-phase equilibrium. The only exception is that the solidification terminates when the liquid reaches the ternary eutectic point. The final solid to freeze thus also has the eutectic composition, which causes the average solid composition to jump to the average value for the alloy. Analysis of the solidification path for other alloys is left as an exercise.

It is also informative to construct pseudobinary sections through this phase diagram. Since these sections become more difficult as the diagrams become more complex, we describe a procedure for their construction. Consider the pseudobinary section B $- y$ in Fig. 3.27(a). We have designated by the points a through e the places where the pseudobinary section intersects a phase boundary. The temperatures and significance of these points are collected in Table 3.5, in order to assist in producing the pseudobinary

Table 3.5 Phase boundaries in the vertical section B $- y$.

Point	Temperature	Significance
B	T_f^B	Melting point of B
a	$T_f^B > T > T_{eut}^{ABC}$	Separates β from $\alpha + \beta$
b	T_{eut}^{ABC}	Separates $\alpha + \beta$ from $\alpha + \beta + \gamma$
c	$T_{eut}^{AB} > T > T_{eut}^{ABC}$	Binary crystallization curve $\ell \rightarrow \alpha + \beta$
d	T_{eut}^{ABC}	Separates $\ell + \alpha + \beta$ from $\ell + \alpha + \gamma$
e	T_{eut}^{ABC}	Separates $\alpha + \gamma$ from $\alpha + \beta + \gamma$
y	$T_f^A > T > T_{eut}^{AC}$	Liquidus curve in A $-$ C system
	T_{eut}^{AC}	Binary eutectic; convergence of $\ell + \alpha + \gamma$

section, shown in Fig. 3.27(b). The pseudobinary section $y - z$ is left as an exercise.

3.4 SUMMARY

In this chapter, we have discussed unary, binary and ternary phase diagrams. We have extensively used the Gibbs phase rule, which tells us how many of the variables specifying the system (temperature, pressure, and the number and compositions of phases) can be changed independently. The Gibbs phase rule was written in terms of the number N_C of components, as well as the number N_ϕ of phases present, i.e. $N_F = N_c - N_\phi + 2$. When the pressure remains fixed, as we assumed in the development of binary and ternary phase diagrams, we obtained $N_F = N_c - N_\phi + 1$.

For binary systems, we examined in detail the relationship between the phase diagrams and the free energy of the phases. In particular, we showed that the phase diagram can be constructed entirely once the Gibbs free energy is known for all the phases. For that purpose, we used a minimum energy principle that includes the tangent construction rule seen in the previous chapter. We used the regular solution model as an example, and demonstrated that one could generate several kinds of diagrams just by varying the parameters of the regular solution model. We then moved on to ternary systems, where it became necessary to represent phase diagrams in three dimensions. The composition was plotted in the plane, and the temperature was plotted perpendicularly to that plane. We examined a relatively small number of the possible ternary systems, and explained how to read and interpret these diagrams.

3.5 EXERCISES

Exercise 3.1. Degeneracy of Gibbs phase rule for binary systems.
Explain, using the Gibbs phase rule, why the liquidus and solidus curves must meet at the sides of the phase diagram corresponding to the pure components.

Exercise 3.2. Lever rule for Cu-Ni alloy.
Consider an alloy of 50% Ni in Cu. What phases are present, what are their compositions, and what are their relative amounts at the temperatures corresponding to the fully liquid, fully solid, and $\ell + \alpha$ regions?

Exercise 3.3. Analytical expression for $f_s - T$ curve.
Derive Eq. (3.11) from Eqs. (3.4) and (3.10).

Exercise 3.4. Constructing free energy curves for liquidus extrema.
Sketch the free energy curves at the temperatures corresponding to the liquidus extrema for the two phase diagrams shown in Fig. 3.5.

Exercise 3.5. Computation of phase binary phase diagrams.
Use PDTOOL to determine the regular solution parameters for the following binary phase diagrams. You will find a materials data handbook useful. You should be able to obtain a good fit by varying the remaining parameters. Note that the computed phase diagram is placed in the file "PhaseDiagram.dat."

Fill in the table below. "System" refers to isomorphous, eutectic, etc.

A	T_f [K]	ΔS_f [J mol^{-1} K]	B	T_f [K]	ΔS_f [J mol^{-1} K]	System	Ω_s^m	Ω_ℓ^m
Cu			Ni					
Bi			Cd					
Al			Bi					
Ag			Pt					
Cu			Pb					

Exercise 3.6. Metasatable extension of phase boundaries.
Explain why the dashed line extension of the solvus phase diagram shown to the right below is not possible. Hint: Construct (sketch) free energy curves corresponding to the temperatures just below and just above the eutectic diagram for the "possible" and "impossible diagrams."

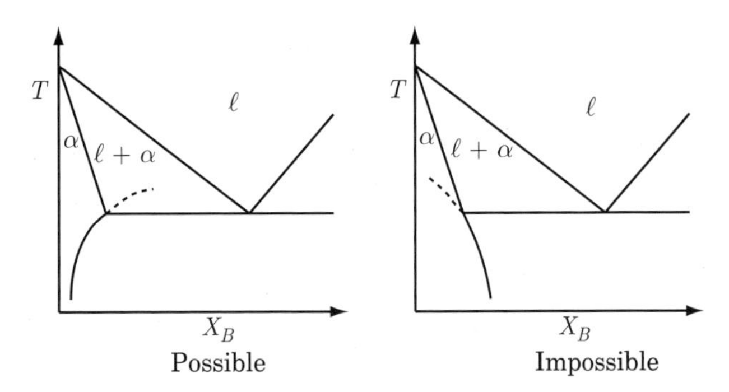

Exercise 3.7. Ternary eutectic phase diagram calculations.
Consider the ternary eutectic system shown below. Answer the following questions.

(a) Consider the alloys marked as x, y and z above. What is the solidification path for each alloy (e.g., $\ell \rightarrow \ell + \alpha \ldots$)

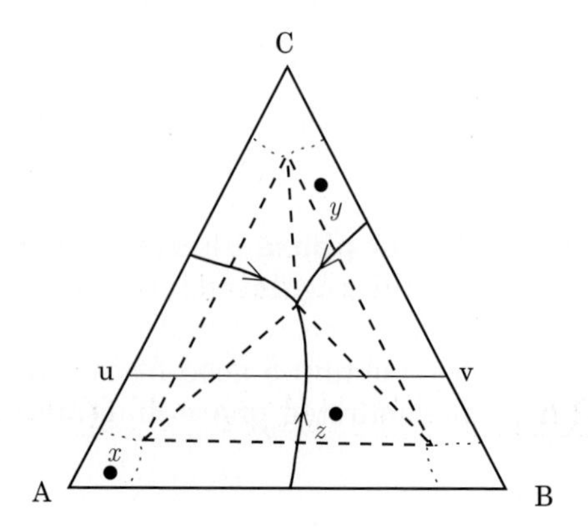

(b) Compute (symbolically) the fractions of the primary, binary and ternary solidification products.

(c) Sketch a microstructure for each alloy, clearly indicating the primary, binary and ternary constituents.

(d) Construct the vertical section u-v.

Exercise 3.8. Al-Mg_2Si quasibinary.
Locate the ternary Al-Mg-Si phase diagram.

(a) Construct an *accurate* Al-Mg_2Si quasibinary section.

(b) The nominal composition of alloy AA6061 is Al-1.0% Mg-0.6% Si. Show this alloy on your quasibinary section.

(c) Determine the expected equilibrium liquidus and solidus temperatures.

3.6 REFERENCES

[1] J. W. Gibbs. *The scientific papers of J. Willard Gibbs*. Dover, 1961.
[2] L. Kaufman and H. Bernstein. *Computer calculation of phase diagrams*. Academic Press, 1970.
[3] Th. B. Massalski, editor. *Binary alloy phase diagrams*. ASM International, 1986.
[4] A. Prince. *Alloy phase equilibria*. Elsevier, 1966.
[5] F. Rhines. *Phase diagrams in metallurgy - their development and application*. McGraw-Hill, 1956.
[6] G. Masing (translated by B. A. Rogers). *Ternary systems*. Reinhold, 1944.

CHAPTER 4

BALANCE EQUATIONS

4.1 INTRODUCTION

Chapter 3 described the equilibrium between phases of different composition at various temperatures. The term equilibrium implies, in particular, that there is no variation of the quantities in either space or time. It is obvious, however, that solidification processes and specifically microstructure development must include spatial and temporal evolution, i.e., transport phenomena. In this chapter, we introduce the governing equations for the transport of mass, solute, momentum and energy. These are often referred to as *balance equations* since they are obtained by applying the basic balance, i.e., that the rate of accumulation of any quantity within a volume is equal to the sum of inputs through the surface and direct production within the volume. This is a subject that has been thoroughly treated (see the chapter references for some examples), and we therefore include only as much as is needed to provide a basis for the material in later chapters.

The balance equations by themselves are correct, but incomplete, since there are more unknowns than equations. The balance equations are supplemented by *constitutive relations* that describe the materials' responses to applied fields. Some common examples are Fourier's law for heat conduction and Fick's law for diffusion of species. In many cases, we find terms in the equations that represent phenomena that are unimportant for the solidification problems that we deal with in this book. We identify these terms, and then usually neglect them in order to concentrate on the most important physical phenomena for our purposes. Our focus on the special forms needed for solidification leads us to examine balances at the solid-liquid interface, as well as in control volumes consisting of a mixture of both phases.

4.1.1 Reference frames and definitions

Before considering balances, we introduce several definitions and results that are needed for the remainder of the chapter. This is done in the

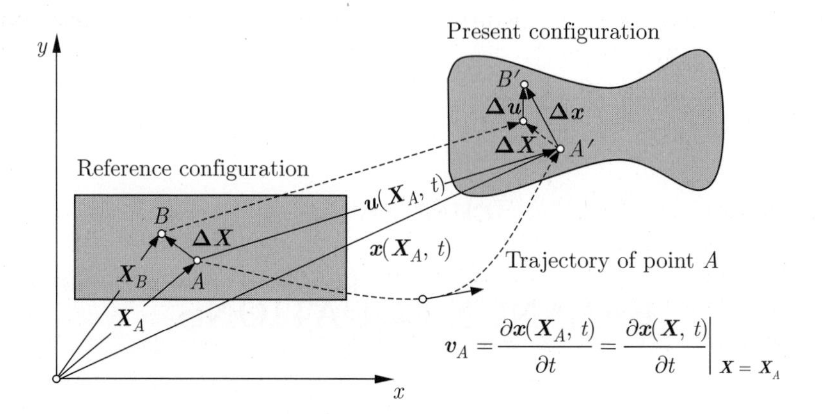

Fig. 4.1 Reference and current configurations during deformation.

context of a reference frame. Our primary interest is in the behavior of materials and their responses to external forces. There are two reference frames of particular importance: the *material* or *Lagrangian* frame, in which the trajectory of each individual particle of the material is followed, together with any of its time-dependent fields (temperature T, velocity $\boldsymbol{v} = (v_x, v_y, v_z)$, etc). We start with a body in some initial reference configuration, attaching a label $\boldsymbol{X}$ to each material point, corresponding to its position (See Fig. 4.1). The temporal evolution of a certain physical quantity ψ at a material point is written as $\psi^{\mathcal{L}}(\boldsymbol{X}, t)$, where the superscript $\mathcal{L}$ indicates the Lagrangian frame.

It is also natural to consider measurements made at a point $\boldsymbol{x}$, measured relative to a fixed location in space. Various authors refer to this as the *Eulerian*, *laboratory* or *spatial* frame. In this frame, we write the field variable $\psi(\boldsymbol{x}, t)$, without any index. To visualize the difference between the Eulerian and Lagrangian frames, suppose that you step outside on a cold day. After a short distance, you notice that the material points on your face have decreased in temperature. A temperature monitoring device focused on the same location that your face now occupies would register an *increase* in temperature as compared to the air occupying that space before your arrival. Clearly, at the moment when your face occupies the monitored point, $\psi^{\mathcal{L}} = \psi$. Nevertheless, the histories of $\psi^{\mathcal{L}}$ and ψ differ since the variables measure different things most of the time.

Although the majority of this book uses a Eulerian description for the various fields we encounter, constitutive equations that describe how the material behaves in response to these fields are naturally expressed in a Lagrangian representation. In order to relate the fields measured in the two frames, we formally write the relations between $\boldsymbol{x}$ and $\boldsymbol{X}$ as

$$\boldsymbol{x} = \boldsymbol{x}(\boldsymbol{X}, t)\ ; \qquad \boldsymbol{X} = \boldsymbol{X}(\boldsymbol{x}, t) \tag{4.1}$$

where it is assumed that the functions are invertible, which implies that the body does not fragment. Next, we define the *velocity* of a material point as

$$\boldsymbol{v} = \left.\frac{\partial \boldsymbol{x}(\boldsymbol{X}, t)}{\partial t}\right|_{\boldsymbol{X}} \tag{4.2}$$

The components of the velocity field will be labeled $\boldsymbol{v} = (v_x, v_y, v_z) = (v_1, v_2, v_3)$ depending on the context. The time derivative of any quantity at a fixed material point is called the *material derivative*, denoted D/Dt.

The material derivative of any generic physical quantity $\psi(\boldsymbol{x}, t)$ in the Eulerian frame is computed using the chain rule as

$$\frac{D\psi}{Dt} = \left.\frac{\partial \psi}{\partial t}\right|_{\boldsymbol{x}} + \sum_{i=1}^{3} \left.\frac{\partial \psi}{\partial x_i}\frac{\partial x_i}{\partial t}\right|_{\boldsymbol{X}} = \frac{\partial \psi}{\partial t} + \sum_{i=1}^{3} v_i \frac{\partial \psi}{\partial x_i} \tag{4.3}$$

We often use shorthand notation to write Eq. (4.3) in more compact form, using either the summation convention for repeated indices, or a vectorial form

$$\frac{D\psi}{Dt} = \frac{\partial \psi}{\partial t} + v_i \frac{\partial \psi}{\partial x_i} = \frac{\partial \psi}{\partial t} + \boldsymbol{v} \cdot \nabla \psi \tag{4.4}$$

Finally, we are interested in the deformation of the material over time. The function $x(\boldsymbol{X}, t)$ describes a mapping of the reference configuration onto the current configuration, as shown in Fig. 4.1. Relative displacements of nearby material points are described by the *deformation gradient* tensor $\boldsymbol{F}$

$$F_{ij} = \frac{\partial x_i}{\partial X_j} \tag{4.5}$$

The determinant $J = \det \boldsymbol{F}$, also called the *Jacobian* of the mapping from the reference to the deformed configuration, is a measure of the volume change during deformation, i.e., it maps an incremental reference volume dV_r to the corresponding volume increment in the current configuration dV

$$dV = J d\mathrm{V_r} \tag{4.6}$$

The *displacement* $\boldsymbol{u}(\boldsymbol{X}, t)$ of a material point $\boldsymbol{X}$ is defined as the difference between its current position $\boldsymbol{x}(\boldsymbol{X}, t)$ and its original position in the reference frame, i.e.,

$$\boldsymbol{x}(\boldsymbol{X}, t) = \boldsymbol{X} + \boldsymbol{u}(\boldsymbol{X}, t); \qquad x_i(\boldsymbol{X}, t) = X_i + u_i \tag{4.7}$$

Differentiating the indecial form of Eq. (4.7) with respect to X_j gives

$$F_{ij} = \frac{\partial x_i}{\partial X_j} = \delta_{ij} + \frac{\partial u_i}{\partial X_j} \tag{4.8}$$

Now consider a vector $d\boldsymbol{X}$ in the reference configuration (a finite $\Delta\boldsymbol{X}$ separating points A and B is shown in Fig. 4.1). The differential of Eq. (4.7) is

$$dx_i = dX_i + \frac{\partial u_i}{\partial X_j} dX_j = \left(\delta_{ij} + \frac{\partial u_i}{\partial X_j} \right) dX_j \tag{4.9}$$

Combining Eqs. (4.8) and (4.9) yields

$$dx_i = F_{ij} dX_j \tag{4.10}$$

If we were to consider another vector $d\boldsymbol{X}'$, we would find by analogy to Eq. (4.10) that

$$dx'_i = F_{ik} dX'_k \tag{4.11}$$

The scalar that measures the variations of length and/or angle between the original and deformed configurations is the dot product of the two vectors, such that

$$d\boldsymbol{x}' \cdot d\boldsymbol{x} = dx'_i dx_i = F_{ik} F_{ij} dX'_k dX_j \tag{4.12}$$

Expanding the product $F_{ik}F_{ij}$ using Eq. (4.8) yields

$$C^G_{jk} = F_{ik} F_{ij} = \delta_{jk} + \frac{\partial u_j}{\partial X_k} + \frac{\partial u_k}{\partial X_j} + \frac{\partial u_i}{\partial X_k} \frac{\partial u_i}{\partial X_j} \tag{4.13}$$

$\boldsymbol{C}^G$ is called the *Cauchy-Green deformation tensor* or the *Cauchy dilatation tensor*. The *Green-Lagrange strain tensor* ε^G_{jk}, defined as $\frac{1}{2}(C^G_{jk} - \delta_{jk})$, measures the distortion during deformation:

$$\varepsilon^G_{jk} = \frac{1}{2}(C^G_{jk} - \delta_{jk}) = \frac{1}{2} \left(\frac{\partial u_j}{\partial X_k} + \frac{\partial u_k}{\partial X_j} + \frac{\partial u_i}{\partial X_k} \frac{\partial u_i}{\partial X_j} \right) \tag{4.14}$$

Note that the diagonal terms ($k = j$) correspond to changes in length, whereas the off-diagonal terms ($k \neq j$) correspond to changes in angle between orthogonal vectors, or *shear*. When the deformations are small, the nonlinear terms become negligible in comparison with the linear terms, and we thus obtain the *linearized strain tensor*

$$\varepsilon_{jk} = \frac{1}{2} \left(\frac{\partial u_j}{\partial X_k} + \frac{\partial u_k}{\partial X_j} \right) \tag{4.15}$$

In a condensed notation, one can introduce the gradient of the displacement vector $\boldsymbol{u}$ as:

$$(\nabla \boldsymbol{u})_{jk} = \frac{\partial u_j}{\partial X_k} \tag{4.16}$$

Using this notation, the Green-Lagrange tensor becomes

$$\boldsymbol{\varepsilon}^G = \frac{1}{2} \left(\nabla \boldsymbol{u} + (\nabla \boldsymbol{u})^T + (\nabla \boldsymbol{u})^T \cdot (\nabla \boldsymbol{u}) \right) \tag{4.17}$$

The last term on the right hand side is neglected in the linearized form. In a similar way, the *rate of deformation* tensor $\boldsymbol{D}$ is defined as

$$\boldsymbol{D} = \frac{1}{2} \left(\nabla \boldsymbol{v} + (\nabla \boldsymbol{v})^T \right); \qquad D_{jk} = \frac{1}{2} \left(\frac{\partial v_j}{\partial x_k} + \frac{\partial v_k}{\partial x_j} \right) \tag{4.18}$$

The rate of deformation tensor involves derivatives of the velocity field, and is taken with respect to the current, and not the reference configuration (small x versus capital $\boldsymbol{X}$). Note also that $\mathrm{tr}(\boldsymbol{D}) = \nabla \cdot \boldsymbol{v}$.

4.1.2 Control volumes

Consider the volume element shown in Fig. 4.2, which is located in the mushy zone, including thus both the solid and liquid phases. It is easier to derive the balance equations for such a volume using an Eulerian representation, i.e., the volume element is fixed in the laboratory. The volume element is assumed to be sufficiently large and diverse to accurately represent the local structure at the mesoscopic length scale, i.e, at a scale that includes multiple microstructural features, yet small enough that important variations in the temperature, enthalpy and volume fraction of the solid are resolved for the problem of interest. This latter length scale might, depending on the context, be as small as individual dendrite arms, or as large as clusters of multiple grains. We refer to this volume as a "Representative Volume Element" or RVE. We also consider smaller control volume elements, as illustrated in the figure. These include control volumes that contain only one phase, marked "Solid" or "Liquid"; control volumes that shrink to zero thickness at an interface between liquid and solid, such as the volume marked "Interface"; and volumes that include finite amounts of both solid and liquid, as well as multiple interfaces, such as the RVE itself.

In the sections that follow, balance equations are first developed for the single phase control volumes, and the interface conditions are subsequently derived using the control volume that contains the interface.

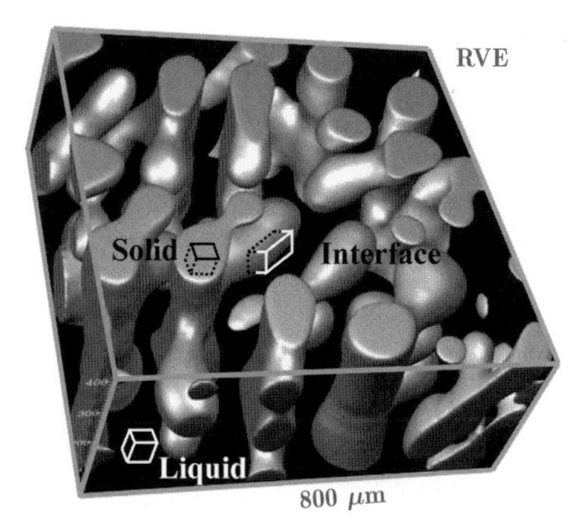

Fig. 4.2 A representative volume element (RVE) containing multiple, disconnected solid and liquid regions. This figure was obtained by reconstruction of a sequence of micrographs obtained by serial sectioning of a Sn-Pb alloy. (Reproduced with permission of A. Genau and P. Voorhees.)

When we consider the RVE with extended segments of solid and liquid, average forms of the balance equations are constructed. We follow the approach of Sun and Beckermann [9], and refer the reader to the original paper for a more in-depth treatment. Let us begin by defining some terms. The RVE has a volume V_R, consisting of solid with the total volume V_s and liquid with the total volume V_ℓ. Clearly, $V_R = V_s + V_\ell$. The volume fractions of the two phases are defined as

$$g_s = \frac{V_s}{V_R}; \qquad g_\ell = \frac{V_\ell}{V_R} \tag{4.19}$$

The *volume average* of any quantity ψ is defined as

$$\langle \psi \rangle = \frac{1}{V_R} \int\limits_{V_R} \psi \, dV \tag{4.20}$$

Next, we introduce "indicator functions" for the solid and liquid phases, ϕ_s and ϕ_ℓ, defined as

$$\phi_s = \begin{cases} 1 & \text{solid} \\ 0 & \text{liquid} \end{cases}; \qquad \phi_\ell = 1 - \phi_s = \begin{cases} 0 & \text{solid} \\ 1 & \text{liquid} \end{cases} \tag{4.21}$$

The function ϕ_s is illustrated for a 2-D domain in Fig. 4.3. We can then compute the *phase average* of any quantity ψ as

$$\begin{aligned} \langle \psi_s \rangle &= \langle \phi_s \psi \rangle = \frac{1}{V_R} \int\limits_{V_R} \phi_s \psi dV = \frac{1}{V_R} \int\limits_{V_s} \phi_s \psi dV \\ \langle \psi_\ell \rangle &= \langle \phi_s \psi \rangle = \frac{1}{V_R} \int\limits_{V_R} \phi_\ell \psi dV = \frac{1}{V_R} \int\limits_{V_\ell} \phi_\ell \psi dV \end{aligned} \tag{4.22}$$

Notice that the integrals over V_R have been restricted to integrals over either V_s or V_ℓ, taking advantage of the fact that $\phi_s = 1$ inside V_s and

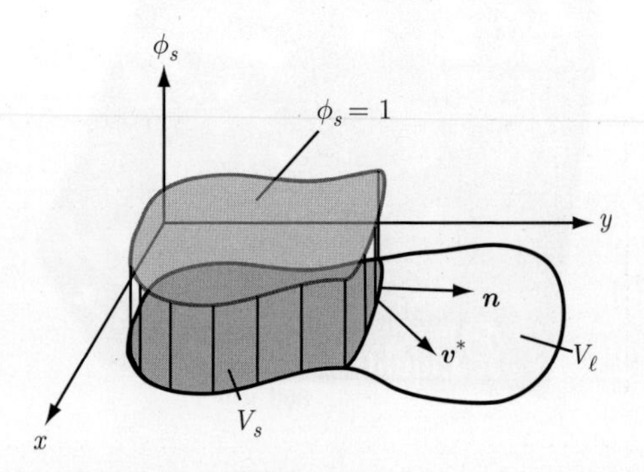

Fig. 4.3 A definition of the solid phase distribution $\phi_s(x, t)$ within a solidifying 2D domain, divided into two domains V_s and V_ℓ. The interface has a normal vector n and moves with velocity v^*.

$\phi_s = 0$ in V_ℓ. The opposite applies for ϕ_ℓ. This property becomes useful later. As an example, let $\psi = \rho$, the density. In that case, $\langle \rho_s \rangle$ represents the total mass of solid in V_R divided by V_R thus giving the average solid density in V_R. We also sometimes use the *intrinsic average* of a quantity, defined as the average value over a single phase.

$$\langle \psi \rangle_s = \frac{1}{V_s} \int_{V_R} \phi_s \psi \, dV_R = \frac{\langle \psi_s \rangle}{g_s} = \frac{\langle \phi_s \psi \rangle}{g_s}; \qquad \langle \psi \rangle_\ell = \frac{\langle \psi_\ell \rangle}{g_\ell} = \frac{\langle \phi_\ell \psi \rangle}{g_\ell} \tag{4.23}$$

Continuing with density as an example, the integrals again produce the total mass in the RVE, but we now divide by the volume of just the solid or liquid phase. Thus, $\langle \rho \rangle_s$ represents the average density of the solid phase alone. If local variations of the density in each phase are small at the scale of the RVE, then $\langle \rho \rangle_s = \rho_s$ while $\langle \rho_s \rangle = g_s \rho_s$.

When we write the balance equations in terms of average quantities, we will need certain intermediate results for the derivatives of average quantities. Consider the volume element V_R illustrated in Fig. 4.3. It contains an interface between the solid and liquid, with a local normal unit vector $\boldsymbol{n}$ defined as positive pointing out of the solid (this convention is followed throughout this book). As the interface moves at a velocity $\boldsymbol{v}^*$, liquid is converted into solid, so that

$$\frac{\partial \phi_s}{\partial t} + \boldsymbol{v}^* \cdot \nabla \phi_s = 0; \qquad \frac{\partial \phi_\ell}{\partial t} + \boldsymbol{v}^* \cdot \nabla \phi_\ell = 0 \tag{4.24}$$

The most important general results needed when we develop the average form of the balance equations relate to the average of quantities including $\nabla \phi_s$ and $\nabla \phi_\ell$. Since $\phi_s = 1 - \phi_\ell$, $\nabla \phi_s = -\nabla \phi_\ell$ and it is sufficient to examine ϕ_s. Consider first the one-dimensional case, where $\phi_s(x)$ is a step function, as illustrated in Fig. 4.4. Its derivative, $d\phi_s/dx$, has the

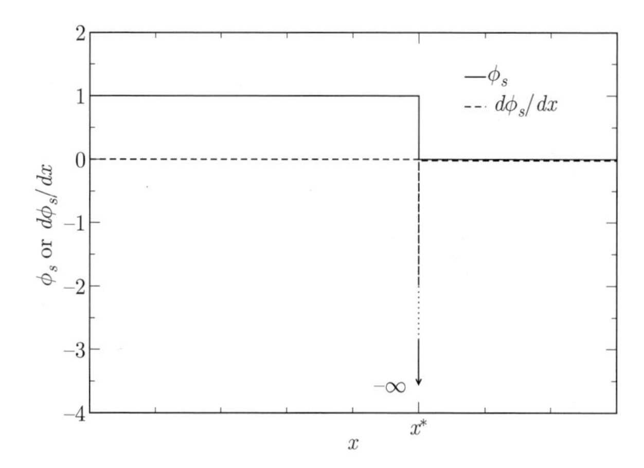

Fig. 4.4 A graph of ϕ_s and its derivative in one dimension. The derivative for $x > x^*$ has been moved to just below zero for the sake of clarity.

properties of a Dirac δ–function at the interface, i.e., infinite height, zero width and integral one. We can thus write

$$\frac{d\phi_s}{dx} = -\delta(x - x^*) \tag{4.25}$$

The negative sign appears since the step goes from one to zero at x^*. Further, it is easy to show that the Dirac δ–function has the following property

$$\int_{-\infty}^{\infty} \psi(x)\delta(x - x^*)\, dx = \psi(x^*) \tag{4.26}$$

Generalizing this concept to three dimensions, $\nabla\phi_s$ has the properties of a δ–function *normal* to the interface, and all other components are zero.

$$\nabla\phi_s = -|\nabla\phi_s|\boldsymbol{n} = -\delta(\boldsymbol{x} - \boldsymbol{x}^*)\boldsymbol{n} \tag{4.27}$$

Here, the minus sign comes from the definition that the normal vector points outward from the solid. The collection of all points on the interface $\boldsymbol{x}^*$ is the solid-liquid interface $A_{s\ell}$, shown in Fig. 4.5. The corresponding form in 3-D to Eq. (4.26) is

$$\int_{V_R} \psi(\boldsymbol{x})\delta(\boldsymbol{x} - \boldsymbol{x}^*)\, dV = \int_{A_{s\ell}} \psi(\boldsymbol{x}^*)\, dA \tag{4.28}$$

When the average forms of the balance equations are developed, the basic procedure is to take the general form of the balance equation, multiply it by ϕ_s (or ϕ_ℓ) and then compute the average. This leads to terms such as $\langle\phi_s\nabla\psi\rangle$, and "averaging theorems" can now be derived for such quantities. We write integrals over V_s and V_ℓ, and it should be understood from Fig. 4.2 that V_s and V_ℓ are generally multiply connected domains. Beginning with the definition of the volume average, we have

$$\langle\phi_s\nabla\psi\rangle = \frac{1}{V_R}\int_{V_R} \phi_s\nabla\psi\, dV \tag{4.29}$$

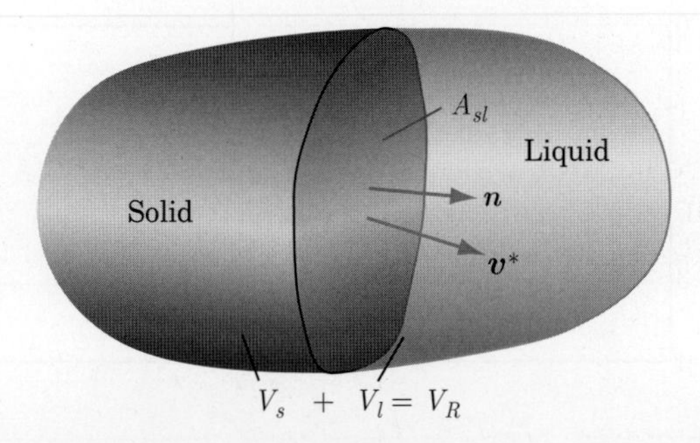

Fig. 4.5 A sketch of a control volume including an interface with area $A_{s\ell}$ with normal vector $\boldsymbol{n}$ and moving at velocity $\boldsymbol{v}^*$.

Since $\phi_s = 0$ in V_ℓ, the integral may be restricted to the volume V_s. We also rewrite the integrand in an equivalent form to obtain

$$\langle \phi_s \nabla \psi \rangle = \frac{1}{V_R} \int\limits_{V_s} \nabla(\phi_s \psi)\, dV - \frac{1}{V_R} \int\limits_{V_s} \psi \nabla \phi_s \, dV \tag{4.30}$$

The first integral remains unchanged if it is converted back to an integral over V_R. We also substitute Eq. (4.27) for $\nabla \phi_s$, which leads to

$$\langle \phi_s \nabla \psi \rangle = \frac{1}{V_R} \int\limits_{V_R} \nabla(\phi_s \psi)\, dV + \frac{1}{V_R} \int\limits_{V_s} \psi \boldsymbol{n} \delta(\boldsymbol{x} - \boldsymbol{x}^*)\, dV \tag{4.31}$$

Because V_R is fixed, the gradient can be taken outside the first integral (this is called Gauss' rule). Referring to Fig. 4.5, the δ–function reduces the integral over the volume V_R to an integral over the solid-liquid interface $A_{s\ell}$, so that we may write

$$\langle \phi_s \nabla \psi \rangle = \nabla \left(\frac{1}{V_R} \int\limits_{V_R} \phi_s \psi \, dV \right) + \frac{1}{V_R} \int\limits_{A_{s\ell}} \psi_s^* \boldsymbol{n}\, dA \tag{4.32}$$

We recognize the term in the parentheses as $\langle \phi_s \psi \rangle = \langle \psi_s \rangle$. The second term is interpreted more easily in a slightly different form,

$$\langle \phi_s \nabla \psi \rangle = \nabla \langle \phi_s \psi \rangle + \underbrace{\frac{A_{s\ell}}{V_R}}_{\mathcal{S}_V} \underbrace{\left(\frac{1}{A_{s\ell}} \int\limits_{A_{s\ell}} \psi_s^* \boldsymbol{n} dA \right)}_{\langle \psi_s^* \boldsymbol{n} \rangle^*} \tag{4.33}$$

where we have defined the *specific surface area* $\mathcal{S}_V = A_{s\ell}/V_R$, and the *interfacial average* $\langle \cdot \rangle^*$

$$\langle \psi_s^* \boldsymbol{n} \rangle^* = \frac{1}{A_{s\ell}} \int\limits_{A_{s\ell}} \psi_s^* \boldsymbol{n} dA \tag{4.34}$$

Thus, we finally obtain

$$\langle \phi_s \nabla \psi \rangle = \nabla \langle \phi_s \psi \rangle + \mathcal{S}_V \langle \psi_s^* \boldsymbol{n} \rangle^* = \nabla \langle \psi_s \rangle + \mathcal{S}_V \langle \psi_s^* \boldsymbol{n} \rangle^* \tag{4.35}$$

Applying Eq. (4.23) gives the equivalent form

$$\langle \phi_s \nabla \psi \rangle = \nabla(g_s \langle \psi \rangle_s) + \mathcal{S}_V \langle \psi_s^* \boldsymbol{n} \rangle^* \tag{4.36}$$

Averaging theorems for quantities involving divergences of vector quantities $\boldsymbol{\psi}$, as well as corresponding forms for liquids, are readily obtained using an equivalent procedure (see Exercise 4.1).

$$\langle \phi_\ell \nabla \psi \rangle = \nabla(g_\ell \langle \psi \rangle_\ell) - \mathcal{S}_V \langle \psi_\ell^* \boldsymbol{n} \rangle^* \tag{4.37}$$

$$\langle \phi_s \nabla \cdot \boldsymbol{\psi} \rangle = \nabla \cdot (g_s \langle \boldsymbol{\psi} \rangle_s) + \mathcal{S}_V \langle \boldsymbol{\psi}_s^* \cdot \boldsymbol{n} \rangle^* \tag{4.38}$$

$$\langle \phi_\ell \nabla \cdot \boldsymbol{\psi} \rangle = \nabla \cdot (g_\ell \langle \boldsymbol{\psi} \rangle_\ell) - \mathcal{S}_V \langle \boldsymbol{\psi}_\ell^* \cdot \boldsymbol{n} \rangle^* \tag{4.39}$$

Note that the change of sign for the surface terms in the theorems involving the liquid phase arise from the choice of the direction of $\boldsymbol{n}$.

The final averaging theorems required for the balance equations involve time derivatives. The procedure is similar to the one given above for the gradient, and we therefore skip a few steps in the derivation.

$$\left\langle \phi_s \frac{\partial \psi}{\partial t} \right\rangle = \frac{1}{V_R} \int\limits_{V_R} \phi_s \frac{\partial \psi}{\partial t} \, dV$$

$$= \frac{1}{V_R} \int\limits_{V_R} \frac{\partial (\phi_s \psi)}{\partial t} \, dV - \frac{1}{V_R} \int\limits_{V_s} \psi \frac{\partial \phi_s}{\partial t} \, dV \tag{4.40}$$

The time derivative can be taken outside of the first integral since V_R is fixed (this is called Leibniz' rule), while Eqs. (4.24) and (4.27) are used to replace the integrand in the second integral, resulting in

$$\left\langle \phi_s \frac{\partial \psi}{\partial t} \right\rangle = \frac{\partial}{\partial t} \left(\frac{1}{V_R} \int\limits_{V_R} \phi_s \psi \, dV \right) - \frac{1}{V_R} \int\limits_{V_s} \psi \boldsymbol{v}^* \cdot \boldsymbol{n} \, \delta(\boldsymbol{x} - \boldsymbol{x}^*) \, dV$$

$$\left\langle \phi_s \frac{\partial \psi}{\partial t} \right\rangle = \frac{\partial \left\langle \phi_s \psi \right\rangle}{\partial t} - \mathcal{S}_V \left\langle \psi_s^* \boldsymbol{v}^* \cdot \boldsymbol{n} \right\rangle^* = \frac{\partial (g_s \left\langle \psi \right\rangle_s)}{\partial t} - \mathcal{S}_V \left\langle \psi_s^* \boldsymbol{v}^* \cdot \boldsymbol{n} \right\rangle^* \tag{4.41}$$

The analogous form for ϕ_ℓ is

$$\left\langle \phi_\ell \frac{\partial \psi}{\partial t} \right\rangle = \frac{\partial \left\langle \phi_\ell \psi \right\rangle}{\partial t} + \mathcal{S}_V \left\langle \psi_\ell^* \boldsymbol{v}^* \cdot \boldsymbol{n} \right\rangle^* = \frac{\partial (g_\ell \left\langle \psi \right\rangle_\ell)}{\partial t} + \mathcal{S}_V \left\langle \psi_\ell^* \boldsymbol{v}^* \cdot \boldsymbol{n} \right\rangle^* \tag{4.42}$$

One further useful result can be obtained by considering Eq. (4.41) with $\psi = 1$. Clearly, the left hand side is zero, so we have

$$0 = \frac{\partial \left\langle \phi_s \right\rangle}{\partial t} - \mathcal{S}_V \left\langle \boldsymbol{v}^* \cdot \boldsymbol{n} \right\rangle^* \tag{4.43}$$

Since $\langle \phi_s \rangle = g_s$, we then obtain

$$\frac{\partial g_s}{\partial t} = \mathcal{S}_V \left\langle \boldsymbol{v}^* \cdot \boldsymbol{n} \right\rangle^* \tag{4.44}$$

Explicitly, Eq. (4.44) states that the increase in volume fraction of the solid phase is equal to the average interface normal velocity multiplied by the amount of interfacial area in the volume. This turns out to be useful for modeling microsegregation.

This completes the preliminaries, and we now proceed to derive the balance equations for mass, momentum, energy and solute.

4.2 MASS BALANCE

The mass balance in the x direction for a single phase control volume is constructed with the aid of Fig. 4.6. Mass is neither created nor destroyed

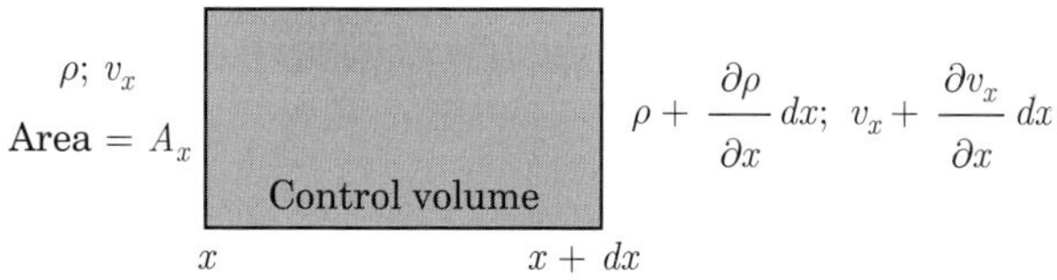

Fig. 4.6 One-dimensional control volume for mass balance.

within the control volume. Therefore, the rate of mass accumulation within the fixed volume element $[x, x + dx]$ balances the net influx of matter through the external surfaces. In equation form, this is written as

$$\frac{\partial \rho}{\partial t} A_x dx = \rho v_x A_x - \left[\left(\rho + \frac{\partial \rho}{\partial x} dx \right) \left(v_x + \frac{\partial v_x}{\partial x} dx \right) \right] A_x \tag{4.45}$$

Canceling terms of opposing sign and neglecting terms of order dx^2 leaves

$$\frac{\partial \rho}{\partial t} + v_x \frac{\partial \rho}{\partial x} + \rho \frac{\partial v_x}{\partial x} = 0 \tag{4.46}$$

Considering the y and z directions gives additional terms of equivalent form as Eq. (4.46), so that in three dimensions we have

$$\frac{\partial \rho}{\partial t} + \boldsymbol{v} \cdot \nabla \rho + \rho \nabla \cdot \boldsymbol{v} = \frac{D\rho}{Dt} + \rho \nabla \cdot \boldsymbol{v} = 0 \tag{4.47}$$

The first two terms in Eq. (4.47) comprise the material derivative of ρ, $D\rho/Dt$, as defined in Eq. (4.4). Note that when ρ is constant, Eq. (4.47) reduces to

$$\nabla \cdot \boldsymbol{v} = 0 \tag{4.48}$$

When considering integral forms, it is also useful to group the second and third terms of Eq. (4.47), thus providing the *conservative form* of the mass balance equation

$$\frac{\partial \rho}{\partial t} + \nabla \cdot (\rho \boldsymbol{v}) = 0 \tag{4.49}$$

This conservative form provides a convenient interpretation when it is integrated over a control volume V

$$\int_V \frac{\partial \rho}{\partial t} \, dV + \int_V \nabla \cdot (\rho \boldsymbol{v}) \, dV = 0$$

$$\int_V \frac{\partial \rho}{\partial t} \, dV + \int_A \rho \boldsymbol{v} \cdot \boldsymbol{n} \, dA = 0 \tag{4.50}$$

The divergence theorem has been applied to convert the second term from a volume to a surface integral. As for the one-dimensional case, this form simply states that the rate of change of mass inside the volume is due to the mass flux through the surface.

Example 4.1 Volume element inside a continuously cast ingot

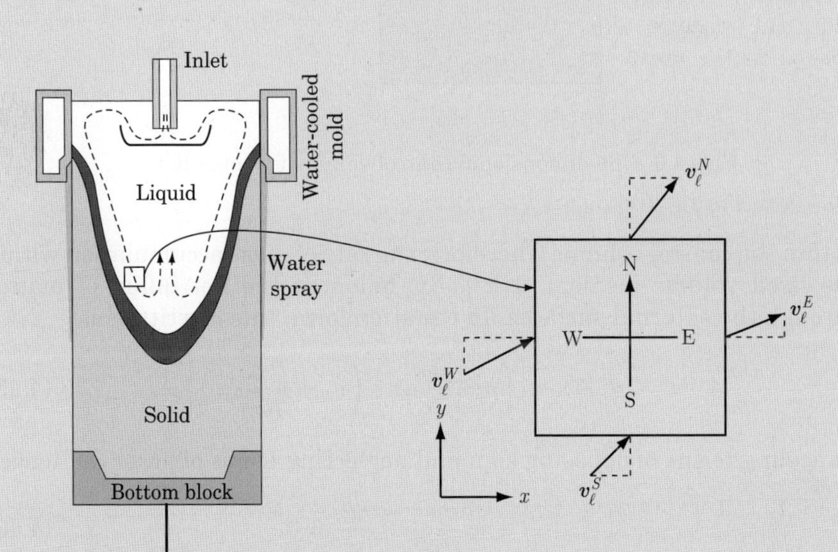

Consider the continuous casting process for aluminum alloys, illustrated schematically in the accompanying figure. A fixed rectangular open mold has a bottom block initially inserted that moves down as the melt is supplied from the upper nozzle.

Consider for the sake of simplicity the fixed small volume element V located in the liquid pool in two dimensions. The mass balance over V is obtained by applying Eq. (4.50).

$$\frac{d}{dt}\int_V \rho dV = \rho_\ell^W v_{\ell x}^W A^W - \rho_\ell^E v_{\ell x}^E A^E + \rho_\ell^S v_{\ell y}^S A^S - \rho_\ell^N v_{\ell y}^N A^N$$

Let us consider, as an example, the liquid phase of a multi-component alloy. Recall from Chap. 2 that the density can be written in terms of the composition and thermodynamic variables, i.e.,

$$\rho_\ell = \rho_\ell(T, p, C_1, \ldots, C_{N_c-1}) \tag{4.51}$$

where C_J is the mass fraction of component J, and N_c is the number of chemical components present in the liquid. The density is often expressed in the form of a truncated Taylor series, expanded around a reference density $\rho_{\ell 0} = \rho_\ell(T_0, p_0, C_{J0})$, such that

$$\rho_\ell = \rho_{\ell 0}\left(1 - \beta_T(T - T_0) - \beta_p(p - p_0) - \sum_{J=1}^{N_c-1} \beta_J(C_J - C_{J0})\right) \tag{4.52}$$

Here, the various β_J are *volumetric expansion coefficients* with respect to component J. Note that $-1/\beta_p$ is called the *compressibility*. We can then write the material derivative of ρ_ℓ in terms of the thermodynamic variables, and the mass balance equation becomes

$$-\rho_{\ell 0}\left(\beta_T \frac{DT}{Dt} + \beta_p \frac{Dp}{Dt} + \sum_{J=1}^{N_c-1} \beta_J \frac{DC_J}{Dt}\right) + \rho_\ell \nabla \cdot \boldsymbol{v}_\ell = 0 \tag{4.53}$$

This equation is very useful when studying solute inhomogeneities induced by density changes and natural convection in Chap. 14.

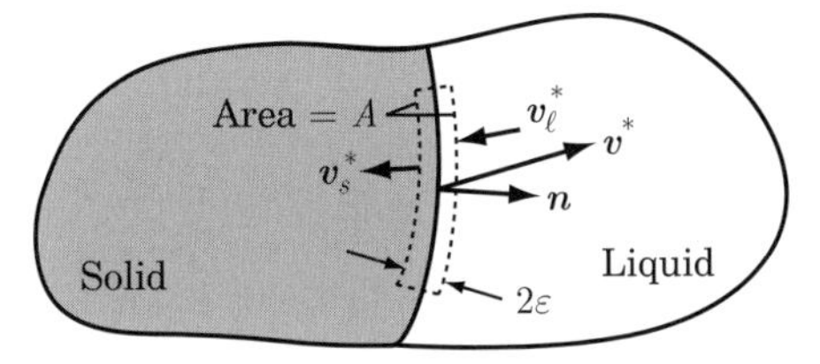

Fig. 4.7 The control volume encapsulating part of the liquid-solid interface.

Next, we derive the mass balance at a solid-liquid interface moving at velocity $\boldsymbol{v}^*$. To this end, we apply Eq. (4.50) to a control volume that surrounds the interface as it moves with it, as shown in Fig. 4.7. To develop an interface balance, we allow the thickness of the control volume $\varepsilon \to 0$. In that case, the volume integral in Eq. (4.50) also approaches zero. Similarly, the integral over the boundary has contributions only from the surfaces marked with area A in Fig. 4.7. In the limit of $\varepsilon \to 0$, the velocities in the liquid and solid are denoted by a superscript "$*$" to indicate that they are evaluated at the interface. Note that, since the control volume is itself moving with velocity $\boldsymbol{v}^*$, it is the *differences* in velocity that contribute to the integral, i.e., $(\boldsymbol{v}_s^* - \boldsymbol{v}^*)$ and $(\boldsymbol{v}_\ell^* - \boldsymbol{v}^*)$. Thus, Eq. (4.50) reduces as follows:

$$0 = \lim_{\varepsilon \to 0}\left(\int_V \frac{\partial \rho}{\partial t}\, dV + \int_A \rho \boldsymbol{v} \cdot \boldsymbol{n}\, dA\right) \tag{4.54}$$

$$= \rho_\ell A(\boldsymbol{v}_\ell^* - \boldsymbol{v}^*) \cdot \boldsymbol{n} - \rho_s A(\boldsymbol{v}_s^* - \boldsymbol{v}^*) \cdot \boldsymbol{n} \tag{4.55}$$

A rearrangement of the terms gives the final form for the interfacial mass balance

$$(\rho_\ell - \rho_s)\boldsymbol{v}^* \cdot \boldsymbol{n} = \rho_\ell \boldsymbol{v}_\ell^* \cdot \boldsymbol{n} - \rho_s \boldsymbol{v}_s^* \cdot \boldsymbol{n} \tag{4.56}$$

The easiest way to provide a physical interpretation of this equation is to consider the solid to be fixed ($\boldsymbol{v}_s^* = 0$), i.e., the solid does not undergo any deformation or contraction once formed behind the interface. Solving for the velocity in the liquid normal to the interface gives

$$\boldsymbol{v}_\ell^* \cdot \boldsymbol{n} = \frac{\rho_\ell - \rho_s}{\rho_\ell} \boldsymbol{v}^* \cdot \boldsymbol{n} = -\beta \boldsymbol{v}^* \cdot \boldsymbol{n} \tag{4.57}$$

where $\beta = (\rho_s - \rho_\ell)/\rho_\ell$ is called the *solidification shrinkage factor*. For almost all materials, water and semiconductors such as Si, Ge or Bi being notable exceptions, $\beta > 0$. In that case, there is a net flow of the liquid toward the interface in order to satisfy the volume shrinkage upon solidification. This equation turns out to be crucial to the understanding of porosity and hot tearing (Chaps. 12 and 13).

Finally, we derive the average form of the mass balance. Beginning with the solid phase, Eq. (4.49) is first multiplied by ϕ_s and then the average is taken to obtain

$$\left\langle \phi_s \frac{\partial \rho}{\partial t} \right\rangle + \langle \phi_s \nabla \cdot (\rho \boldsymbol{v}) \rangle = 0 \tag{4.58}$$

Applying the averaging theorems in Eqs. (4.38) and (4.41) gives

$$\frac{\partial \langle \phi_s \rho \rangle}{\partial t} + \nabla \cdot \langle \phi_s \rho \boldsymbol{v} \rangle = -\mathcal{S}_V \langle \rho_s^* \boldsymbol{v}_s^* \cdot \boldsymbol{n} \rangle^* + \mathcal{S}_V \langle \rho_s^* \boldsymbol{v}^* \cdot \boldsymbol{n} \rangle^* \tag{4.59}$$

The volume averages on the left hand side can be expanded with the help of Eq. (4.23), while the right-hand-side terms can be consolidated, thus giving the average mass balance equation for the solid phase

$$\frac{\partial (g_s \langle \rho \rangle_s)}{\partial t} + \nabla \cdot (g_s \langle \rho \boldsymbol{v} \rangle_s) = -\mathcal{S}_V \langle \rho_s (\boldsymbol{v}_s^* - \boldsymbol{v}^*) \cdot \boldsymbol{n} \rangle^* = -\Gamma_s^{m*} \tag{4.60}$$

where Γ_s^{m*} represents the average rate of production of mass by the solid phase at the moving interface. The average mass balance for the liquid is computed using an identical procedure to obtain

$$\frac{\partial (g_\ell \langle \rho \rangle_\ell)}{\partial t} + \nabla \cdot (g_\ell \langle \rho \boldsymbol{v} \rangle_\ell) = \mathcal{S}_V \langle \rho_\ell (\boldsymbol{v}_\ell^* - \boldsymbol{v}^*) \cdot \boldsymbol{n} \rangle^* = \Gamma_\ell^{m*} \tag{4.61}$$

where Γ_ℓ^{m*} is defined, analogously to Γ_s^{m*}, as the average rate of production of mass by the liquid phase at the moving interface. Recall that the change in sign on the right hand side arises from the definition of the direction of the normal.

Summing Eq. (4.60) and (4.61) gives

$$\frac{\partial (g_s \langle \rho \rangle_s + g_\ell \langle \rho \rangle_\ell)}{\partial t} + \nabla \cdot (g_s \langle \rho \boldsymbol{v} \rangle_s + g_\ell \langle \rho \boldsymbol{v} \rangle_\ell) =$$
$$-\Gamma_s^{m*} + \Gamma_\ell^{m*} = -\mathcal{S}_V \langle [\rho_s (\boldsymbol{v}_s^* - \boldsymbol{v}^*) - \rho_\ell (\boldsymbol{v}_\ell^* - \boldsymbol{v}^*)] \cdot \boldsymbol{n} \rangle^* = 0 \tag{4.62}$$

The right hand side is zero by virtue of the interface balance in Eq. (4.56). This turns out to be the case for all of the balances: the sum of the interfacial contributions satisfies the continuum interface balance condition. The average mass balance for the mixture is then written as

$$\frac{\partial \langle \rho \rangle}{\partial t} + \nabla \cdot \langle \rho \boldsymbol{v} \rangle = 0 \tag{4.63}$$

where the average density and the average mass flux are defined as

$$\langle \rho \rangle = g_s \langle \rho \rangle_s + g_\ell \langle \rho \rangle_\ell ; \qquad \langle \rho \boldsymbol{v} \rangle = g_s \langle \rho \boldsymbol{v} \rangle_s + g_\ell \langle \rho \boldsymbol{v} \rangle_\ell \tag{4.64}$$

We often neglect local variations in the density, for which we have $\langle \rho \rangle_s \approx \rho_s \neq \langle \rho \rangle_\ell \approx \rho_\ell$.

4.3 MOMENTUM BALANCE

We now focus our attention on the motion and deformation induced in materials by the action of external forces. Forces that are applied by contact at a surface are called *surface tractions*: these forces are represented by the vector $\boldsymbol{t}(\boldsymbol{x}, t)$ in units of force per unit area (Nm^{-2}=Pa). The internal forces induced by such tractions are illustrated in Fig. 4.8. The components of the force per unit area on a material surface normal to the x axis are written as σ_{xx}, σ_{xy} and σ_{xz}. The various components can be arranged in the *Cauchy stress tensor*

$$\boldsymbol{\sigma} = \begin{bmatrix} \sigma_{xx} & \sigma_{xy} & \sigma_{xz} \\ \sigma_{yx} & \sigma_{yy} & \sigma_{yz} \\ \sigma_{zx} & \sigma_{zy} & \sigma_{zz} \end{bmatrix} \tag{4.65}$$

For materials without internal couples, the condition that the material does not rotate in static equilibrium requires that $\boldsymbol{\sigma}$ be symmetric, i.e., $\sigma_{ij} = \sigma_{ji}$. The Cauchy stress tensor has the useful property that, for any arbitrary surface with normal $\boldsymbol{n}$, the force density on that surface is given by

$$\boldsymbol{t} = \boldsymbol{\sigma} \cdot \boldsymbol{n} \tag{4.66}$$

The eigenvalues of the Cauchy stress tensor $(\sigma_1, \sigma_2, \sigma_3)$ are called the *principal stresses*. When the coordinate axes are oriented along the eigenvectors associated with the principal stresses, called the *principal stress axes*, the Cauchy stress tensor is diagonal, with entries σ_1, σ_2 and σ_3 in that frame. The stress tensor is often decomposed into the sum of a hydrostatic part and an "extra stress" tensor $\boldsymbol{\tau}$

$$\boldsymbol{\sigma} = -p\boldsymbol{I} + \boldsymbol{\tau} \tag{4.67}$$

where p is the pressure, introduced in Chap. 2, and $\boldsymbol{I}$ is the identity tensor. This decomposition is clearly not unique. For the special choice

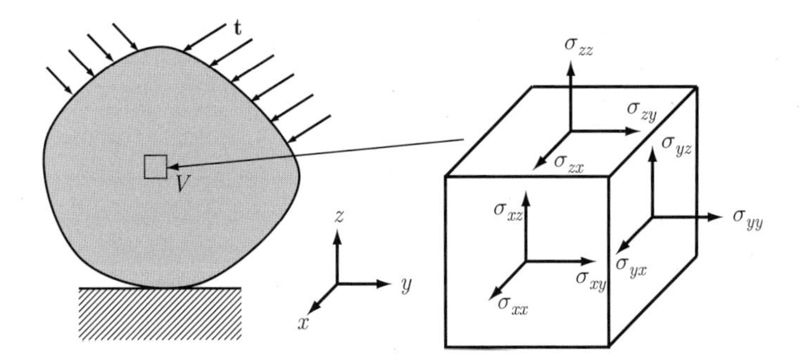

Fig. 4.8 External forces applied on the surface of a domain (left) and internal contact forces exerted on the surfaces of a small volume element V within this domain.

$p = -\text{tr}(\boldsymbol{\sigma})/3$, we have $\text{tr}(\boldsymbol{\tau}) = 0$, and the extra stress is then called the *deviatoric stress*. This form is especially useful in solid mechanics, where the visco-plastic behavior of many materials can be taken to be independent of the hydrostatic stress.

Forces can also be applied to a material without contact, and we call these *body forces* $\boldsymbol{b}(\boldsymbol{x}, t)$, with units of force per unit mass. Examples of body forces include gravity and electromagnetic forces.

Newton's second law of motion states that the time rate of change of linear momentum balances the sum of the external forces acting on a body. Consider a deformable body whose volume is given by $V(t)$ and whose surface is given by $A(t)$. In equation form, Newton's law is written as

$$\frac{d}{dt} \int\limits_{V(t)} \rho \boldsymbol{v} dV = \int\limits_{A(t)} \boldsymbol{t} dS + \int\limits_{V(t)} \rho \boldsymbol{b} dV \tag{4.68}$$

Equation (4.66) and the divergence theorem allow conversion of the first integral on the right hand side to a volume integral. Next, the time derivative on the left hand side is brought inside the integral, bearing in mind that the volume $V(t)$ changes with time. Using the mass conservation equation (Eq. (4.47)), it is left as an exercise to show that one obtains

$$\frac{d}{dt} \int\limits_{V(t)} \rho \boldsymbol{v} dV = \int\limits_{V(t)} \rho \frac{D\boldsymbol{v}}{Dt} dV \tag{4.69}$$

This operation has introduced the material derivative of the velocity. Since Eq. (4.68) is valid for any volume $V(t)$ attached to the material, the momentum conservation equation can be written in local form as

$$\rho \frac{D\boldsymbol{v}}{Dt} = \nabla \cdot \boldsymbol{\sigma} + \rho \boldsymbol{b} \tag{4.70}$$

The conservative form of Eq. (4.70) is obtained by multiplying Eq. (4.49) by $\boldsymbol{v}$ and adding it to Eq. (4.70), with the result

$$\frac{\partial(\rho \boldsymbol{v})}{\partial t} + \nabla \cdot (\rho \boldsymbol{v}\boldsymbol{v}) = \nabla \cdot \boldsymbol{\sigma} + \rho \boldsymbol{b} \tag{4.71}$$

The product $(\boldsymbol{v}\boldsymbol{v})$ is called the *tensor product* of two vectors, written in component form as $(\boldsymbol{v}\boldsymbol{v})_{ij} = v_i v_j$. With this definition, the i component of Eq. (4.71) becomes

$$\frac{\partial(\rho v_i)}{\partial t} + \frac{\partial(\rho v_i v_j)}{\partial x_j} = \frac{\partial \sigma_{ij}}{\partial x_j} + \rho b_i \tag{4.72}$$

Separating the total stress as in Eq. (4.67) gives

$$\frac{\partial(\rho \boldsymbol{v})}{\partial t} + \nabla \cdot (\rho \boldsymbol{v}\boldsymbol{v}) = -\nabla p + \nabla \cdot \boldsymbol{\tau} + \rho \boldsymbol{b} \tag{4.73}$$

We may now derive the boundary condition at a moving interface starting from Eq. (4.71). Following a procedure similar to that used in the derivation of the interfacial mass balance, we integrate over the control volume shown in Fig. 4.7. As the volume shrinks to zero width at the interface, the first and last terms in Eq. (4.71) integrate to zero. Converting the remaining terms to surface integrals using the divergence theorem, and invoking Eq. (4.56) gives the surface force contributions $\mathbf{\Gamma}_\ell^{\sigma *}$ and $\mathbf{\Gamma}_s^{\sigma *}$ corresponding to the liquid and solid phases, respectively

$$[\rho_\ell \boldsymbol{v}_\ell^*(\boldsymbol{v}_\ell^* - \boldsymbol{v}^*) - \boldsymbol{\sigma}_\ell^*] \cdot \boldsymbol{n} - [\rho_s \boldsymbol{v}_s^*(\boldsymbol{v}_s^* - \boldsymbol{v}^*) - \boldsymbol{\sigma}_s^*] \cdot \boldsymbol{n} = \mathbf{\Gamma}_\ell^{\sigma *} - \mathbf{\Gamma}_s^{\sigma *} \tag{4.74}$$

In contrast to the interfacial mass balance, the sum of these two contributions is not zero, as a result of surface forces (e.g., surface tension) associated with the interface (see Chap. 2). Figure 4.9 illustrates the forces acting on a two-dimensional segment of the interface, assuming that the interfacial surface tension is an isotropic tensor (i.e., equal to the surface energy) but potentially varying with position. The momentum balance at a moving interface becomes

$$\rho_s \boldsymbol{v}_s^*(\boldsymbol{v}_s^* - \boldsymbol{v}^*) \cdot \boldsymbol{n} - \rho_\ell \boldsymbol{v}_\ell^*(\boldsymbol{v}_\ell^* - \boldsymbol{v}^*) \cdot \boldsymbol{n} - \boldsymbol{\sigma}_s^* \cdot \boldsymbol{n} + \boldsymbol{\sigma}_\ell^* \cdot \boldsymbol{n} = 2\gamma_{s\ell}\bar{\kappa}\boldsymbol{n} + \nabla_{surf}(\gamma_{s\ell}) \tag{4.75}$$

where the notation ∇_{surf} indicates the gradient in the surface. Two components of the surface forces appear on the right hand side of this equation. The first one, involving the surface tension $\gamma_{s\ell}$ and the mean curvature $\bar{\kappa}$, has already been introduced in Chap. 2. If the solid, liquid and interface are immobile and the stresses are only hydrostatic, then the projection of Eq. (4.75) on the normal to the interface becomes

$$(\boldsymbol{\sigma}_\ell^* \cdot \boldsymbol{n} - \boldsymbol{\sigma}_s^* \cdot \boldsymbol{n}) \cdot \boldsymbol{n} = p_s^* - p_\ell^* = 2\gamma_{s\ell}\bar{\kappa} \tag{4.76}$$

This demonstrates that the pressure inside a convex solid surrounded by the liquid is higher than the external pressure, which is consistent with Fig. 2.18. This equation is particularly useful when treating porosity, cf. Chap. 12. The second contribution that appears in Eq. (4.75) is associated with gradients in the surface tension, e.g., due to temperature or compositional effects. These gradients produce differences in the shear stress across the interface. Such a phenomenon, called the *Marangoni effect*, can be particularly important in welding, as discussed in Chap. 1.

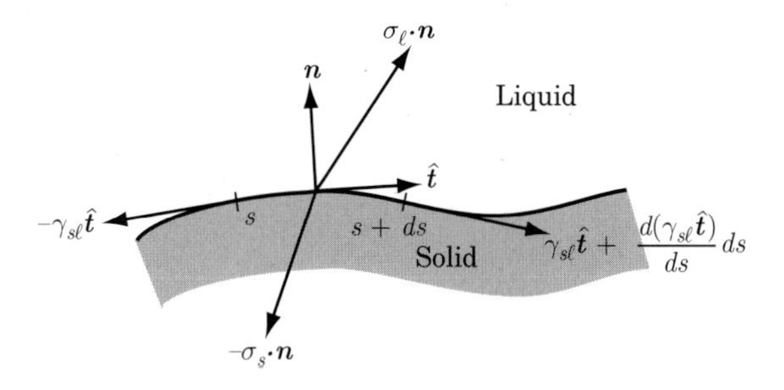

Fig. 4.9 A segment of an interface displaying the forces on a small segment of length ds due to surface tension.

Equations (4.49) and (4.73) apply to any material. However, they are not sufficient for solving most problems since there are more unknowns (three components of velocity, pressure, and six independent components of τ) than equations (one scalar mass balance and three vector components of momentum balance). A *constitutive relation* is required to relate the stress components to the velocities in order to bring the number of unknowns and equations into agreement. In the next two subsections we introduce certain constitutive relations for solids as well as for a Newtonian fluid.

4.3.1 Linear elastic solids

When dealing with the deformation of solids, one can often make the assumption that the accelerations or inertia, represented by the left hand side of Eq. (4.73), are negligible with respect to the terms on the right hand side. In this *quasi-static* approximation, the momentum balance is reduced to a balance of external forces, namely

$$0 = \nabla \cdot \boldsymbol{\sigma} + \rho \boldsymbol{b} \tag{4.77}$$

The unknowns (six components of σ) still outnumber the three scalar equations obtained for each direction of the coordinate system. Hooke's Law posits a linear relationship between the stress and strain tensors. It is consistent with this form to also assume that the deformations are small, and we thus write the constitutive relation in terms of the linearized form of the strain tensor (Eq. (4.15)) as

$$\sigma_{ij} = \mathcal{E}_{ijkl}\varepsilon_{kl} \tag{4.78}$$

The *elasticity tensor* $\mathcal{E}$ contains 81 entries (3^4), but the application of symmetry rules significantly reduces that number. The symmetry of σ and ε implies that there are at most 36 independent constants since $\mathcal{E}_{ijkl} = \mathcal{E}_{jikl} = \mathcal{E}_{ijlk}$. When one can define an energy associated with the deformation, it can be shown that $\mathcal{E}_{ijkl} = \mathcal{E}_{klij}$ and the number of parameters reduces to 21. Finally, crystalline symmetry reduces the number of independent constants further, e.g., materials with cubic crystal structures have at most three independent constants.

We most often deal with materials that are *isotropic*, i.e., where all directions are equivalent, in which case there are merely two independent constants. Hooke's law for isotropic materials reduces to

$$\sigma_{ij} = \frac{E}{1+\nu_e}\varepsilon_{ij} + \frac{\nu_e E}{(1+\nu_e)(1-2\nu_e)}\varepsilon_{kk}\delta_{ij} \tag{4.79}$$

where E is Young's modulus and ν_e is Poisson's ratio. Note the implicit summation on k to indicate the trace of the strain tensor. When the stress is entirely axial, e.g., only $\sigma_{xx} \neq 0$, it can be shown (see Exercise 4.2) that we have

$$\sigma_{xx} = E\varepsilon_{xx}; \qquad \varepsilon_{yy} = \varepsilon_{zz} = -\nu_e\varepsilon_{xx} \tag{4.80}$$

4.3.2 Plastic deformation

Elasticity is a good approximation of the deformation of solids at low stress, where a material returns to its original configuration when the load is removed. At higher stresses, some of the deformation remains after unloading. This is called *plastic deformation*. In the uniaxial tension test, plastic deformation occurs when the stress exceeds the *yield stress* or *flow stress* σ_Y. It can be difficult to locate σ_Y exactly, so it is usually estimated from the intersection between the stress-strain curve and a line of slope E passing through the abscissa at 0.2% deformation (see Fig. 4.10). The residual strain that remains after unloading to $\sigma_{xx} = 0$ is called the plastic strain and is labeled ε_{xx}^{pl}. In such cases, the total strain ε_{ij} is decomposed as follows

$$\varepsilon_{ij} = \frac{1}{2}\left(\frac{\partial u_i}{\partial x_j} + \frac{\partial u_j}{\partial x_i}\right) = \varepsilon_{ij}^{el} + \varepsilon_{ij}^{pl} \tag{4.81}$$

The elastic contribution ε_{ij}^{el} is still given by Eq. (4.79), but we now need an additional constitutive relation for the plastic strain. The simplest such form is called the "elastic-perfectly plastic" solid, as shown in Fig. 4.10(a). In 1-D, the behavior of the material is given by

$$\begin{aligned} \varepsilon_{xx} &= \varepsilon_{xx}^{el} = \frac{\sigma_{xx}}{E} && \text{If } \sigma_{xx} < \sigma_Y \\ \varepsilon_{xx} &= \varepsilon_{xx}^{el} + \varepsilon_{xx}^{pl} = \frac{\sigma_Y}{E} + \varepsilon_{xx}^{pl} && \text{If } \sigma_{xx} = \sigma_Y \end{aligned} \tag{4.82}$$

A more realistic model includes *strain hardening*, where the elastic limit σ_Y becomes a function of the local strain history of the material (Fig. 4.10(a)). In this case, we define the local slope of the stress strain curve in the plastic regime as E_T. Thus

$$d\sigma_{xx} = E_T d\varepsilon_{xx} \tag{4.83}$$

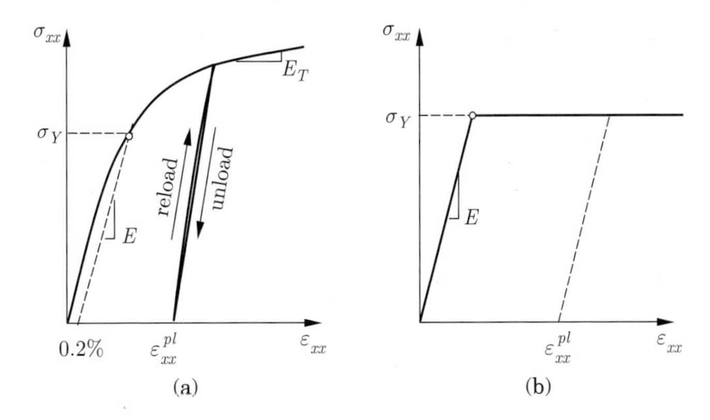

Fig. 4.10 The real (a) and ideal (b) visco-plastic behavior of a solid in a uniaxial tensile test.

However, using Eqs. (4.80) and (4.81), one can also write

$$d\varepsilon_{xx} = d\varepsilon_{xx}^{pl} + d\varepsilon_{xx}^{el} = d\varepsilon_{xx}^{pl} + \frac{d\sigma_{xx}}{E} \tag{4.84}$$

Combining Eqs. (4.83) and (4.84) makes it possible to relate the stress increment to the plastic strain increment:

$$d\sigma_{xx} = E_p d\varepsilon_{xx}^{pl}; \text{ where } E_p = \frac{EE_T}{E - E_T} \tag{4.85}$$

In order to be useful, the constitutive relations established in the uniaxial test should be extended to situations involving multi-axial stresses. There are three commonly used models for this purpose, each most conveniently expressed in terms of the principal stresses.

1. *Maximum principal stress:* The material fails when the maximum principal stress exceeds σ_Y. This criterion is useful for brittle materials that typically fracture with little or no plastic deformation. In such cases, σ_Y would more properly be called the fracture stress. The yield surface forms a cube of edge $2\sigma_Y$ in principal stress space, and its intersection with the plane $\sigma_3 = 0$ is a square, as illustrated in Fig. 4.11.

2. *Tresca condition:* In this model, the material is assumed to yield when the magnitude of the maximum shear stress exceeds the shear stress at yield in the one-dimensional tension test. Written in terms of the principal stresses, the magnitudes of the three shear stresses are $|\sigma_i - \sigma_j|/2$ for $i \neq j$, each running from 1 to 3. It is thus easy to see that the maximum shear stress at yield in the one-dimensional

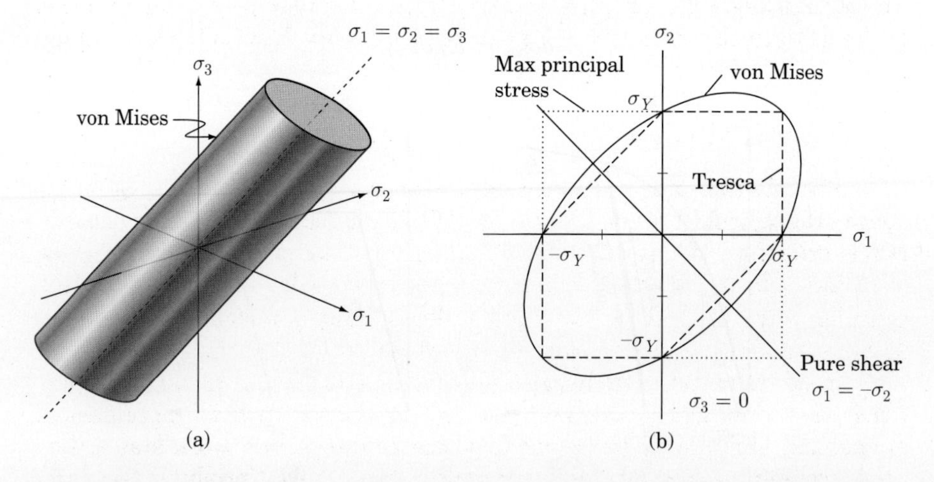

Fig. 4.11 (a) The von Mises yield surface in principal stress space. (b) The intersection of the various yield surfaces with the plane $\sigma_3 = 0$. Notice the difference in behavior for the three models in pure shear ($\sigma_1 = -\sigma_2$).

tension test (where $\sigma_1 = \sigma_Y$ and $\sigma_2 = \sigma_3 = 0$) is simply $\sigma_Y/2$. The Tresca condition can be expressed in equation form as

$$\max(|\sigma_1 - \sigma_2|, |\sigma_1 - \sigma_3|, |\sigma_2 - \sigma_3|) = \sigma_Y \tag{4.86}$$

Equation (4.86) defines a hexagonal prism in principal stress space, oriented along the axis $\sigma_1 = \sigma_2 = \sigma_3$. The intersection of this surface with the plane $\sigma_3 = 0$ is illustrated in Fig. 4.11. The Tresca condition can be inconvenient to implement due to the sharp edges in principal stress space, where the normal vector is undefined. The following von Mises model is thus seen more frequently.

3. *Equivalent stress or von Mises model:* In this model, described further below, the material is assumed to yield when the strain energy in the 3-D stress state is equal to the strain energy at yield in the uniaxial test.

The von Mises condition can also be expressed as the criterion that yield occurs when the equivalent stress $\sigma_{eq} = \sigma_Y$, where σ_{eq} is given by

$$\sigma_{eq} = \frac{1}{\sqrt{2}} \left[(\sigma_1 - \sigma_2)^2 + (\sigma_2 - \sigma_3)^2 + (\sigma_3 - \sigma_1)^2 \right]^{1/2} \tag{4.87}$$

It is also useful to have the equivalent stress in an arbitrary coordinate system, where we have

$$\sigma_{eq} = \frac{1}{\sqrt{2}} \left[(\sigma_{xx} - \sigma_{yy})^2 + (\sigma_{yy} - \sigma_{zz})^2 + (\sigma_{zz} - \sigma_{xx})^2 + 6\sigma_{xy}^2 + 6\sigma_{yz}^2 + 6\sigma_{zx}^2 \right]^{1/2} \tag{4.88}$$

Finally, it is left as an exercise to show that σ_{eq} also can be written in terms of the deviatoric stress $\boldsymbol{\tau} = \boldsymbol{\sigma} - \frac{1}{3}\mathrm{tr}(\boldsymbol{\sigma})\boldsymbol{I}$, defined in Eq. (4.67), as

$$\sigma_{eq} = \sqrt{\frac{3}{2}} \left[\tau_{xx}^2 + \tau_{yy}^2 + \tau_{zz}^2 + 2\tau_{xy}^2 + 2\tau_{xz}^2 + 2\tau_{yz}^2 \right]^{1/2} \tag{4.89}$$

The *yield surface* is defined as the locus of points in principal stress space where yield occurs. From Eq. (4.87) one can see that for the von Mises criterion, the yield surface is a cylinder in principal stress space of radius $\sqrt{2/3}\sigma_Y$ about the axis $\sigma_1 = \sigma_2 = \sigma_3$. The intersection of this cylinder with the plane $\sigma_3 = 0$ is an ellipse with major axes $2\sigma_Y\sqrt{2}$ and $2\sigma_Y\sqrt{2/3}$ at $+45°$ and $-45°$, respectively (see Fig. 4.11).

In addition to the yield criterion, one also needs a *flow rule* that provides the increment of plastic strain $d\varepsilon^{pl}$ associated with an increment of stress past yield $d\boldsymbol{\sigma}$. The principle of maximum plastic work states that the increment of plastic strain past yield is the one that produces the maximum work. It can be shown (cf. [7]) that this principle is satisfied if the increment of plastic strain is normal to the yield surface. This criterion is difficult to enforce for the Tresca condition, since the sharp edges on the yield surface make the normal vector ill-defined at this position. We now develop the flow rule for the von Mises criterion.

To aid in the development, we define an *equivalent plastic strain increment* $d\varepsilon_{eq}^{pl}$ given by

$$d\varepsilon_{eq}^{pl} = \frac{\sqrt{2}}{3}\left[(d\varepsilon_{xx}^{pl} - d\varepsilon_{yy}^{pl})^2 + (d\varepsilon_{yy}^{pl} - d\varepsilon_{zz}^{pl})^2 + (d\varepsilon_{zz}^{pl} - d\varepsilon_{xx}^{pl})^2 + 6d\varepsilon_{xy}^{pl\,2} + 6d\varepsilon_{yz}^{pl\,2} + 6d\varepsilon_{zx}^{pl\,2}\right]^{1/2} \tag{4.90}$$

It is also convenient to define several new row vectors:

$$\begin{aligned}
d\boldsymbol{\sigma} &= \{d\sigma_{xx}\ d\sigma_{yy}\ d\sigma_{zz}\ d\sigma_{xy}\ d\sigma_{yz}\ d\sigma_{zx}\} \\
d\boldsymbol{\tau} &= \{d\tau_{xx}\ d\tau_{yy}\ d\tau_{zz}\ d\tau_{xy}\ d\tau_{yz}\ d\tau_{zx}\} \\
d\boldsymbol{\varepsilon}^{pl} &= \{d\varepsilon_{xx}^{pl}\ d\varepsilon_{yy}^{pl}\ d\varepsilon_{zz}^{pl}\ d\varepsilon_{xy}^{pl}\ d\varepsilon_{yz}^{pl}\ d\varepsilon_{zx}^{pl}\} \\
d\boldsymbol{\varepsilon}^{el} &= \{d\varepsilon_{xx}^{el}\ d\varepsilon_{yy}^{el}\ d\varepsilon_{zz}^{el}\ d\varepsilon_{xy}^{el}\ d\varepsilon_{yz}^{el}\ d\varepsilon_{zx}^{el}\} \\
\boldsymbol{Q} &= \frac{3}{2\sigma_{eq}}\{\tau_{xx}\ \tau_{yy}\ \tau_{zz}\ 2\tau_{xy}\ 2\tau_{yz}\ 2\tau_{zx}\}
\end{aligned} \tag{4.91}$$

Differentiating Eq. (4.89) gives

$$d\sigma_{eq} = \boldsymbol{Q}^T d\boldsymbol{\tau} \tag{4.92}$$

The flow rule relates the plastic strain increment to the deviatoric stress via

$$d\boldsymbol{\varepsilon}^{pl} = \boldsymbol{Q} d\varepsilon_{eq}^{pl} \tag{4.93}$$

The final relation needed is that of elasticity,

$$d\boldsymbol{\sigma} = [\boldsymbol{E}]d\boldsymbol{\varepsilon}^{el} \tag{4.94}$$

where the symmetric elasticity matrix $[\boldsymbol{E}]$ for an isotropic material is defined by Eq. (4.79). Combining all of these relations provides the following set of equations for determining the elastic and plastic strain increments:

$$d\boldsymbol{\sigma} = [\boldsymbol{E}_{ep}]d\boldsymbol{\varepsilon}^{pl};\ \text{where}\ [\boldsymbol{E}_{ep}] = [\boldsymbol{E}] - [\boldsymbol{E}]\frac{\boldsymbol{Q}\boldsymbol{Q}^T[\boldsymbol{E}]}{E_p + \boldsymbol{Q}^T[\boldsymbol{E}]\boldsymbol{Q}} \tag{4.95}$$

where E_p is defined in Eq. (4.85).

The difficulty in implementing Eq. (4.95) as a procedure lies in its nonlinearity. Suppose that we have a particular stress state, and that an increment of stress is imposed. If one knew the value of E_p, it would be possible to start with Eq. (4.95) and work backward to find the elastic and plastic strain components. However, the result might not give a consistent value of E_p in Eq. (4.85). In that case, the estimate for E_p is updated and the process repeated until convergence is achieved. This algorithm, or some variant of it, is implemented in many finite element codes. Finally, we note that there exist several other theories of plasticity that consider, among other phenomena, a difference in yield behavior under tension as opposed to compression, as well as hardening rules that change the shape of the yield surface as deformation proceeds. These are beyond the scope of this text.

4.3.3 Newtonian fluids

Most liquid metals are well represented as Newtonian fluids, where the extra stress $\boldsymbol{\tau}$ is proportional to the velocity gradient $\nabla \boldsymbol{v}$. Since $\boldsymbol{\tau}$ is symmetric, a proper constitutive relation may involve only the symmetric part of the velocity gradient, i.e., the deformation rate tensor $\boldsymbol{D}$ defined in Eq. (4.18). The most general linear form for an isotropic material is

$$\boldsymbol{\tau} = 2\mu \boldsymbol{D} + \lambda(\mathrm{tr}\boldsymbol{D})\boldsymbol{I} \tag{4.96}$$

where μ is the shear viscosity and λ is the dilatational viscosity. Note that this relation is formally equivalent to that introduced in Eq. (4.79) for linear elasticity, provided that the strain is replaced by the deformation rate and the Cauchy stress by the deviatoric stress. Using the definition of $\boldsymbol{\tau}$ (Eq. (4.67)) and substituting Eq. (4.96) into Eq. (4.73) yields

$$\frac{\partial(\rho \boldsymbol{v})}{\partial t} + \nabla \cdot (\rho \boldsymbol{v}\boldsymbol{v}) = -\nabla p + \nabla \cdot (2\mu \boldsymbol{D} + \lambda(\mathrm{tr}\boldsymbol{D})\boldsymbol{I}) + \rho \boldsymbol{b} \tag{4.97}$$

Equations (4.97) and (4.49) provide a complete set of equations that may be used to solve fluid flow problems. There are now four equations (the mass balance and three components of the momentum balance) and four unknowns (pressure and three components of velocity).

This set of equations is further reduced for a constant density fluid, for which $\mathrm{tr}(\boldsymbol{D}) = 0$, and thus λ has no effect on the flow. Equation (4.97) then becomes the *Navier-Stokes* equation,

$$\frac{\partial(\rho \boldsymbol{v})}{\partial t} + \nabla \cdot (\rho \boldsymbol{v}\boldsymbol{v}) = -\nabla p + \nabla \cdot (2\mu \boldsymbol{D}) + \rho \boldsymbol{b} \tag{4.98}$$

Finally, if the viscosity is constant, we have

$$\frac{\partial(\rho \boldsymbol{v})}{\partial t} + \nabla \cdot (\rho \boldsymbol{v}\boldsymbol{v}) = -\nabla p + \mu \nabla^2 \boldsymbol{v} + \rho \boldsymbol{b} \tag{4.99}$$

Using the mass conservation equation for the fluid (Eq. (4.49)), we obtain the non-conservative form

$$\rho \frac{\partial \boldsymbol{v}}{\partial t} + (\rho \boldsymbol{v} \cdot \nabla)\boldsymbol{v} = -\nabla p + \mu \nabla^2 \boldsymbol{v} + \rho \boldsymbol{b} \tag{4.100}$$

Note that the Laplacian of the velocity field, $\nabla^2 \boldsymbol{v}$, applies to each component of this vector.

4.3.4 Average form of the momentum balance

The average form of the momentum balance equations for the solid and liquid phases are now obtained, following Sun and Beckermann [9]. This is by far the most complicated of the averaging procedures, and we refer the reader to their paper for further details. In particular, there are numerous phenomena of importance in high speed, two phase flows that do not concern us in this text. We identify (and neglect) some of them as we go along, but the reader should be aware that the equations developed here are simplified versions appropriate for our application.

We begin with a straightforward application of the averaging procedure to the momentum balance for the solid. As for the mass balance, we multiply the basic form of Eq. (4.73) by ϕ_s and average. Applying the averaging theorems in Eqs. (4.36), (4.38) and (4.41), and again assuming that each of the densities of the solid and liquid phases are constant, gives

$$\frac{\partial}{\partial t}(g_s\rho_s\langle \boldsymbol{v}\rangle_s)+\nabla\cdot(g_s\rho_s\langle \boldsymbol{vv}\rangle_s)+\nabla(g_s\langle p\rangle_s)-\nabla\cdot(g_s\langle \boldsymbol{\tau}\rangle_s)-g_s\rho_s\langle \boldsymbol{b}\rangle_s = -\mathcal{S}_V\langle\rho_s\boldsymbol{v}_s^*(\boldsymbol{v}_s^*-\boldsymbol{v}^*)\cdot\boldsymbol{n}\rangle^* + \mathcal{S}_V\langle(-p_s^*\boldsymbol{I}+\boldsymbol{\tau}_s^*)\cdot\boldsymbol{n}\rangle^* \quad (4.101)$$

The term $\langle \boldsymbol{vv}\rangle_s$ can be expanded by writing $\boldsymbol{v}_s$ as the sum of an average and "fluctuating" part: $\boldsymbol{v}_s = \langle\boldsymbol{v}\rangle_s + \hat{\boldsymbol{v}}_s$. We then have

$$\nabla\cdot(g_s\rho_s\langle \boldsymbol{vv}\rangle_s) = \nabla\cdot(g_s\rho_s\langle \boldsymbol{v}\rangle_s\langle \boldsymbol{v}\rangle_s)+\nabla\cdot(\hat{\boldsymbol{v}}_s\hat{\boldsymbol{v}}_s) \quad (4.102)$$

The term involving the fluctuations may be more familiar as "Reynolds stresses" in the context of turbulent flow. As we are mostly concerned with flows at relatively high volume fraction inside the mushy region, we neglect these terms, which leads to the form

$$\frac{\partial}{\partial t}(g_s\rho_s\langle \boldsymbol{v}\rangle_s)+\nabla\cdot(g_s\rho_s\langle \boldsymbol{v}\rangle_s\langle \boldsymbol{v}\rangle_s)+\nabla(g_s\langle p\rangle_s)-\nabla\cdot(g_s\langle \boldsymbol{\tau}\rangle_s)-g_s\rho_s\langle \boldsymbol{b}\rangle_s = -\boldsymbol{\Gamma}_s^{\sigma *} \quad (4.103)$$

where the interfacial force $\boldsymbol{\Gamma}_s^{\sigma *}$ is given by

$$\boldsymbol{\Gamma}_s^{\sigma *} = \mathcal{S}_V\langle\rho_s\boldsymbol{v}_s^*(\boldsymbol{v}_s^*-\boldsymbol{v}^*)\cdot\boldsymbol{n}\rangle^* + \mathcal{S}_V\langle(p_s^*\boldsymbol{I}-\boldsymbol{\tau}_s^*)\cdot\boldsymbol{n}\rangle^* \quad (4.104)$$

There is of course a corresponding form for the liquid phase (also written neglecting the Reynolds stresses) given by

$$\frac{\partial}{\partial t}(g_\ell\rho_\ell\langle \boldsymbol{v}\rangle_\ell)+\nabla\cdot(g_\ell\rho_\ell\langle \boldsymbol{v}\rangle_\ell\langle \boldsymbol{v}\rangle_\ell)+\nabla(g_\ell\langle p\rangle_\ell)-\nabla\cdot(g_\ell\langle \boldsymbol{\tau}\rangle_\ell)-g_\ell\rho_\ell\langle \boldsymbol{b}\rangle_\ell = \boldsymbol{\Gamma}_\ell^{\sigma *} \quad (4.105)$$

and

$$\boldsymbol{\Gamma}_\ell^{\sigma *} = \mathcal{S}_V\langle\rho_\ell\boldsymbol{v}_\ell^*(\boldsymbol{v}_\ell^*-\boldsymbol{v}^*)\cdot\boldsymbol{n}\rangle^* + \mathcal{S}_V\langle(p_\ell^*\boldsymbol{I}-\boldsymbol{\tau}_\ell^*)\cdot\boldsymbol{n}\rangle^* \quad (4.106)$$

Summing Eqs. (4.103) and (4.105) and comparing the result to Eq. (4.75), we see that the sum of the interfacial contributions to the average momentum equation corresponds to

$$-\boldsymbol{\Gamma}_s^{\sigma *}+\boldsymbol{\Gamma}_\ell^{\sigma *} = -\mathcal{S}_V\langle 2\gamma_{s\ell}\bar{\kappa}\boldsymbol{n}\rangle^* - \mathcal{S}_V\langle\nabla_{surf}(\gamma_{s\ell})\rangle^* \quad (4.107)$$

It is important to understand the physical interpretation of the interfacial contributions in Eqs. (4.104) and (4.106). The first term on the right hand side represents momentum transport between phases due to phase change. The pressure term is often separated into two parts, such that

$$\mathcal{S}_V\langle p_s^*\boldsymbol{n}\rangle^* = p_s'\nabla g_s + \mathcal{S}_V\langle(p_s^*-p_s')\boldsymbol{n}\rangle^*;\quad \mathcal{S}_V\langle p_\ell^*\boldsymbol{n}\rangle^* = p_\ell'\nabla g_\ell - \mathcal{S}_V\langle(p_\ell^*-p_\ell')\boldsymbol{n}\rangle^* \quad (4.108)$$

The choice of p'_s and p'_ℓ is clearly not unique. A simple model that is sufficiently accurate for our purposes is to take $p'_\alpha = \langle p \rangle_\alpha$, i.e., equal to the internal pressure in phase α. The remaining term represents an apparent buoyant force on the solid phase due to the average pressure.

Finally, the dissipative part, $\langle \boldsymbol{\tau}_\ell \cdot \boldsymbol{n} \rangle^*$, represents the effect of viscous drag caused by the relative motion of the two phases. Modeling this term can be quite complicated, because it depends on the details of the motion inside the control volume, i.e., the physical phenomena that we are trying to avoid modeling by using the averaging approach. When $\boldsymbol{v}_s \neq \boldsymbol{v}_\ell$, e.g., for equiaxed crystals settling in the melt, the interfacial stress can be modeled as viscous drag acting over the specific solid-liquid interfacial area

$$\mathcal{S}_V \langle \boldsymbol{\tau}_\ell \cdot \boldsymbol{n} \rangle^* = \rho_\ell C_D \frac{\mathcal{S}_V}{2} | \langle \boldsymbol{v} \rangle_\ell - \langle \boldsymbol{v} \rangle_s | (\langle \boldsymbol{v} \rangle_\ell - \langle \boldsymbol{v} \rangle_s) \tag{4.109}$$

The drag coefficient C_D should depend on the microstructure of the solid phase, and could thus be considered to be a sort of collection point for the physical phenomena eliminated by averaging.

There are two cases of interest for us in later chapters where the interfacial stresses take on simpler forms. If the solid and liquid phases move together without separation, Eq. (4.109) indicates that the interfacial shear stresses are zero. This might be a reasonable assumption for small equiaxed crystals at low g_s, i.e., in the early stages of solidification. The other case occurs when the solid fraction is high enough for the solid to form a continuous structure. One can then write the interfacial stress in the form of *Darcy's law*

$$\mathcal{S}_V \langle \boldsymbol{\tau}^*_\ell \cdot \boldsymbol{n} \rangle = \mu_\ell g_\ell^2 \boldsymbol{K}^{-1} \left(\langle \boldsymbol{v} \rangle^*_\ell - \langle \boldsymbol{v} \rangle^*_s \right) \tag{4.110}$$

where $\boldsymbol{K}$ is the permeability tensor, with units of length^2, that represents the ease with which fluid can flow through, or permeate the structure. This equation is further simplified when one assumes that $\boldsymbol{v}_s = 0$, corresponding to columnar growth, or to late-stage equiaxed solidification, when the crystals are sufficiently packed so as to become immobile. This case, which is important for the formation of porosity as discussed in Chap. 12, is therefore developed in more detail.

When $\boldsymbol{v}_s = 0$, it is necessary to solve only for the velocity in the liquid phase starting with Eq. (4.105). The flow field for this case is shown schematically in Fig. 4.12. At the length scale of the individual particles, the velocity profile might look like that shown on the right hand side of the figure, indicating that there is no slip at the solid-liquid interface between the particles. When viewed at the lower resolution of the averaging volume in Fig. 4.12(a), the average flow is usually written in terms of the *superficial velocity* $\langle \boldsymbol{v}_\ell \rangle = g_\ell \langle \boldsymbol{v} \rangle_\ell$, which represents the flow that is apparent to an observer. At this point, it is also appropriate to introduce a Newtonian constitutive model for the liquid. Since we have assumed that

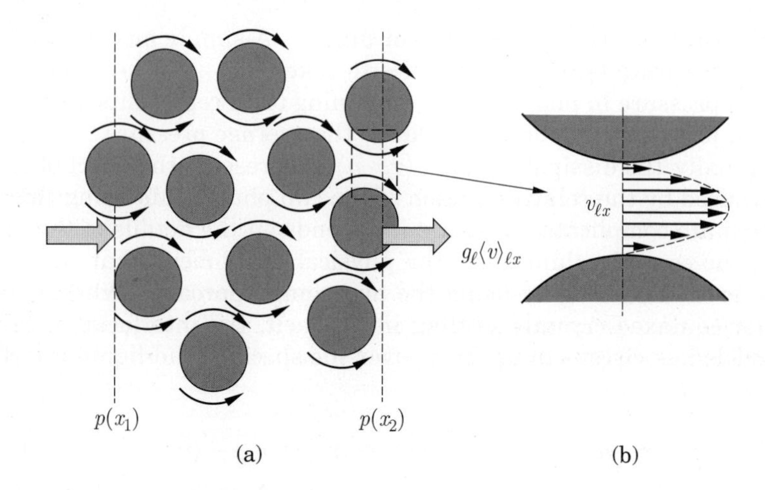

Fig. 4.12 The flow through a packed-bed. (a) average flow over a control volume; (b) local flow between particles.

ρ_ℓ is constant, the dilatational viscosity gives no contribution. Averaging Eq. (4.96) and writing it in terms of the intrinsic averages gives

$$\langle \boldsymbol{\tau} \rangle_\ell = \mu_\ell^{eff} \left(\nabla (g_\ell \langle \boldsymbol{v} \rangle_\ell) + (\nabla (g_\ell \langle \boldsymbol{v} \rangle_\ell))^T \right) \tag{4.111}$$

Often, μ_ℓ^{eff} is taken to be equal to μ_ℓ, but in general, it should depend on the temperature. Notice that the right hand side of Eq. (4.111) can be written in terms of $\langle \boldsymbol{v}_\ell \rangle$, thus finally giving us

$$\langle \boldsymbol{\tau} \rangle_\ell = \mu_\ell^{eff} \left(\nabla \langle \boldsymbol{v}_\ell \rangle + (\nabla \langle \boldsymbol{v}_\ell \rangle)^T \right) \tag{4.112}$$

Substituting Eqs. (4.108), (4.110) and (4.112) into Eq. (4.105) yields

$$\frac{\partial}{\partial t}(\rho_\ell g_\ell \langle \boldsymbol{v} \rangle_\ell) + \nabla \cdot (\rho_\ell g_\ell \langle \boldsymbol{v} \rangle_\ell \langle \boldsymbol{v} \rangle_\ell) + \nabla (g_\ell \langle p \rangle_\ell) - \nabla \cdot \left(\mu_\ell^{eff} \left[\nabla \langle \boldsymbol{v}_\ell \rangle + \nabla \langle \boldsymbol{v}_\ell \rangle^T \right] \right)$$
$$- \rho_\ell g_\ell \langle \boldsymbol{b} \rangle_\ell = \mathcal{S}_V \langle \rho_\ell \boldsymbol{v}_\ell^* (\boldsymbol{v}_\ell^* - \boldsymbol{v}^*) \cdot \boldsymbol{n} \rangle^* - \mu_\ell g_\ell^2 \boldsymbol{K}^{-1} (\langle \boldsymbol{v} \rangle_\ell - \langle \boldsymbol{v} \rangle_s) \tag{4.113}$$

The right hand side of Eq. (4.113) represents the forces exerted by the fixed solid on the moving liquid. When $g_\ell = 1$, the first term on the right hand side of Eq. (4.113) goes to zero since there is no true interface, and thus $\boldsymbol{v}_\ell = \boldsymbol{v}^*$. The second term on the right hand side also goes to zero as a result of the permeability becoming infinite in the fluid. Thus, Eq. (4.113) is reduced to the standard form derived in Sect. 4.3.3 for a Newtonian fluid when $g_\ell = 1$.

The permeability is strongly dependent on the volume fraction of solid. For values of g_s higher than about 0.2-0.3, the permeability is so low that it dominates all of those terms in which $\langle \boldsymbol{v}_\ell \rangle$ appears. Equation (4.113) reduces to a more familiar form of Darcy's Law

$$\langle \boldsymbol{v}_\ell \rangle = -\frac{\boldsymbol{K}}{\mu_\ell}(\nabla \langle p \rangle_\ell - \rho_\ell \boldsymbol{b}_\ell) \tag{4.114}$$

In this case, the superficial velocity is proportional to the gradient of the pressure in the liquid as well as to the body forces (see Fig. 4.12). If the structure is assumed to be isotropic, the permeability tensor reduces to $K\boldsymbol{I}$ where K is now a scalar function of g_ℓ and of the morphology of the permeable solid. The most widely used correlation for the permeability in solidifying systems is called the Carman-Kozeny relation

$$K(g_\ell) = \frac{g_\ell^3}{c\mathcal{S}_V^2 \bar{\tau}^2} \tag{4.115}$$

Here, c is a constant, $\mathcal{S}_V = A_{s\ell}/V_R$ and $\bar{\tau}$ is called the *tortuosity*, since it measures the average effective path taken by the liquid when going from one point to another (e.g., from x_1 to x_2 in Fig. 4.12) divided by the shortest path without the solid (e.g, $x_2 - x_1$). For packed spheres, $\bar{\tau} \approx \sqrt{2}$ and $c \approx 2.5$. Finally, Eq. (4.115) becomes

$$K(g_s) = \frac{\bar{d}^2}{180} \frac{(1 - g_s)^3}{g_s^2} \tag{4.116}$$

where $\bar{d}$ is the average diameter of the spheres. The permeability is plotted as a function of g_s in Fig. 4.13. At high g_s, the permeability is very low and Darcy's Law is a good approximation of the flow. Chapter 12 demonstrates how Eq. (4.114) is used to calculate the pressure drop in the mushy zone due to solidification shrinkage. At lower values of g_s, the velocity can be fairly high and some correction to Darcy's Law becomes necessary.

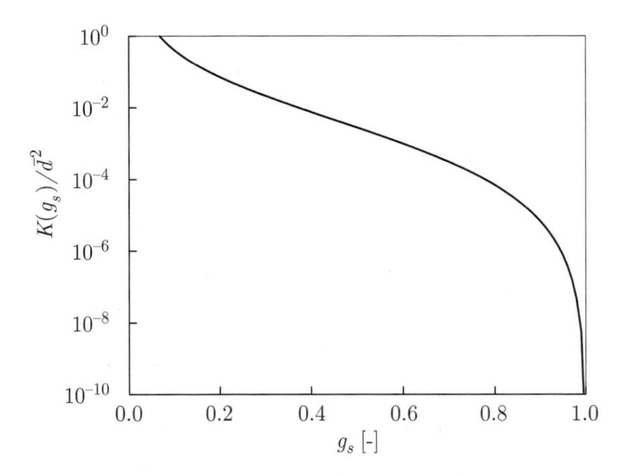

Fig. 4.13 The normalized permeability as a function of the volume fraction of solid, as calculated with the Carman-Kozeny relationship.

Finally, we can also develop an average momentum balance for the mixture by summing the individual equations for the solid and liquid phases. The solid is modeled as a Newtonian fluid, but it is now essential to include the dependence of μ_s^{eff} on g_s. The effective viscosity for the solid should become infinite at some value of the solid fraction where the solid forms a rigid skeleton, typically at $g_s \approx 0.3 - 0.4$. A power-law model is frequently used to implement this behavior. If the contribution of the interfacial energy is neglected, the average momentum balance for the mixture is given by

$$\begin{aligned}\frac{\partial}{\partial t}(g_\ell \rho_\ell \langle \boldsymbol{v} \rangle_\ell + g_s \rho_s \langle \boldsymbol{v} \rangle_s) + \nabla \cdot (g_\ell \rho_\ell \langle \boldsymbol{v} \rangle_\ell \langle \boldsymbol{v} \rangle_\ell + g_s \rho_s \langle \boldsymbol{v} \rangle_s \langle \boldsymbol{v} \rangle_s) = \\ - g_\ell \nabla \langle p \rangle_\ell - g_s \nabla \langle p \rangle_s + \nabla \cdot \left(g_\ell \mu_\ell^{eff} \left[\nabla \langle \boldsymbol{v} \rangle_\ell + (\nabla \langle \boldsymbol{v} \rangle)_\ell^T \right] \right) \\ + \nabla \cdot \left(g_s \mu_s^{eff} \left[\nabla \langle \boldsymbol{v} \rangle_s + (\nabla \langle \boldsymbol{v} \rangle)_s^T \right] \right) + g_\ell \rho_\ell \langle \boldsymbol{b} \rangle_\ell + g_s \rho_s \langle \boldsymbol{b} \rangle_s \end{aligned} \tag{4.117}$$

4.4 ENERGY BALANCE

Solidification processes necessarily involve the transport of heat, and in this section we derive the appropriate balance equations. In fact, we have already encountered the energy balance in Chap. 2, where it was called the First Law of Thermodynamics. It is really no more than a budget of the total energy in the system. In the current context, we rewrite the First Law including a variation in time. The total energy in the system is the sum of the kinetic energy K, and the internal energy E. Their time rate of change balances the rate of heat added to the system $\dot{Q}$ as well as the rate of work (i.e., the mechanical power) performed by mechanical forces $\dot{W}$. Writing a budget for the energy gives us

$$\frac{d}{dt}(K + E) = \dot{Q} + \dot{W} \tag{4.118}$$

Here, K, E and $\dot{W}$ can be written in terms of the velocity $\boldsymbol{v}$, the specific internal energy e, the pressure p and the extra stress $\boldsymbol{\tau}$. The only term that introduces new quantities is $\dot{Q}$, which is given by

$$\dot{Q} = -\int_{S(t)} \boldsymbol{q} \cdot \boldsymbol{n}\, dS + \int_{V(t)} \rho \dot{R}_q dV \tag{4.119}$$

Heat added or lost at the surface of the domain is represented by the *heat flux* vector $\boldsymbol{q}$ with units W m^{-2}. Heat may also be generated within the volume (e.g. by chemical reactions or electrical currents), and is symbolized by the heat generation per unit mass $\dot{R}_q$. The minus sign is a convention: when $q_n = \boldsymbol{q} \cdot \boldsymbol{n} > 0$, heat is lost by the domain. Developing the various terms (see for example Dantzig and Tucker [2], or Rappaz et al. [7] for a detailed derivation), one obtains the local energy balance equation

$$\rho\frac{De}{Dt} = -\nabla\cdot\boldsymbol{q} - p\nabla\cdot\boldsymbol{v} + \boldsymbol{\tau} : \boldsymbol{D} + \rho\dot{R}_q \qquad (4.120)$$

The third term on the right hand side is equal to $\tau_{ij}D_{ij}$, where the summation is carried out over both indices. The physical interpretation of Eq. (4.120) is that the rate of change in internal energy can be attributed to four contributions: heat flux associated with conduction, mechanical power associated with the expansion of a compressible system (i.e., $\nabla\cdot\boldsymbol{v}$), mechanical power of deformation and internal heat generation. Equation (4.120) takes a more convenient form when we introduce the specific enthalpy, h, defined in Chap. 2 as

$$h(T,p) = e + \frac{p}{\rho} \qquad (4.121)$$

Applying Eq. (4.47) and some manipulation of terms results in

$$\rho\frac{Dh}{Dt} - \frac{Dp}{Dt} = -\nabla\cdot\boldsymbol{q} + \boldsymbol{\tau} : \boldsymbol{D} + \rho\dot{R}_q \qquad (4.122)$$

The pressure term is almost always negligible, and since we are concerned with processes where the pressure is nearly constant, this term is neglected. The viscous dissipation is also usually negligible for solidification processes, except for processes such as semi-solid forging, where the forces and deformation rates can be quite large. We omit both terms from here on.

The heat flux vector is written in terms of Fourier's Law

$$\boldsymbol{q} = -\boldsymbol{k}\cdot\nabla T \qquad (4.123)$$

where $\boldsymbol{k}$ is the thermal conductivity tensor. Substituting Eq. (4.123) into Eq. (4.122) and dropping the negligible terms, gives us

$$\rho\frac{Dh}{Dt} = \nabla\cdot(\boldsymbol{k}\cdot\nabla T) + \rho\dot{R}_q \qquad (4.124)$$

If the material is isotropic, then $\boldsymbol{k} = k\boldsymbol{I}$, and Eq. (4.124) becomes

$$\rho\frac{Dh}{Dt} = \rho\frac{\partial h}{\partial t} + \rho\boldsymbol{v}\cdot\nabla h = \nabla\cdot(k\nabla T) + \rho\dot{R}_q \qquad (4.125)$$

If the thermal conductivity is also constant, then k can be taken outside the divergence. The conservative form can be obtained, as for the momentum balance, by multiplying Eq. (4.49) by h and adding it to Eq. (4.124), thus giving

$$\frac{\partial(\rho h)}{\partial t} + \nabla\cdot(\rho h\boldsymbol{v}) = \nabla\cdot(k\nabla T) + \rho\dot{R}_q \qquad (4.126)$$

Equation (4.126) is the basic equation for heat flow that is used in the remainder of the text. Note that if there is no phase change, Eq. (4.126) can be written in terms of the specific heat,

$$\rho c_p \frac{\partial T}{\partial t} + \rho c_p \boldsymbol{v} \cdot T = \nabla \cdot (k \nabla T) + \rho \dot{R}_q \tag{4.127}$$

The interfacial energy balance boundary condition can be derived in a similar way to the interface condition for the mass balance. We begin with Eq. (4.126) in its integral form, and evaluate the integrals in the same manner as we did for the interfacial mass balance over a small volume element surrounding and moving with the interface. As the thickness of this volume element is reduced to zero, both the first and last terms of Eq. (4.126) tend toward zero, leaving only the two terms in which a divergence appears. After invoking the divergence theorem, we have

$$\rho_\ell h_\ell^* (\boldsymbol{v}_\ell^* - \boldsymbol{v}^*) \cdot \boldsymbol{n} - \rho_s h_s^* (\boldsymbol{v}_s^* - \boldsymbol{v}^*) \cdot \boldsymbol{n} + \left(k_s \frac{\partial T_s}{\partial n} - k_\ell \frac{\partial T_\ell}{\partial n} \right)^* = -\Gamma_s^{h*} + \Gamma_\ell^{h*} \tag{4.128}$$

As for the mass and momentum balances, we have introduced the interfacial contribution $\Gamma_\alpha^{h*} = [\rho_\alpha h_\alpha^* (\boldsymbol{v}_\alpha^* - \boldsymbol{v}^*) \cdot \boldsymbol{n} - (k_\alpha \partial T / \partial n)^*]$ of each phase α to the heat balance. It is possible to show that the sum of the interfacial contributions balances the rate of surface energy generation associated with the moving solid-liquid interface

$$-\left(\Gamma_\ell^{h*} - \Gamma_s^{h*}\right) = 2\gamma_{s\ell} \bar{\kappa} \boldsymbol{v}^* \cdot \boldsymbol{n} \tag{4.129}$$

where $\bar{\kappa}$ is the local mean curvature of the interface. However, this contribution is usually negligible with respect to the other terms. As an example, consider a spherical particle of radius $100\,\mu\text{m}$ ($\bar{\kappa} = 10^4$ m^{-1}) that grows at $1\,\text{mm s}^{-1}$. The right hand side of Eq. (4.129) evaluates to 2 W m^{-2} for $\gamma_{s\ell} = 0.1$ J m^{-2}. On the other hand, the energy associated with the enthalpy jump at the solid-liquid interface (the first two terms in Eq. (4.128)) is approximately 10^6 W m^{-2} for a volumetric latent heat of fusion of 10^9 J m^{-3}. We therefore omit this term from this point forward. Rearranging terms on the left hand side gives

$$\rho_\ell h_\ell^* (\boldsymbol{v}^* - \boldsymbol{v}_\ell^*) \cdot \boldsymbol{n} - \rho_s h_s^* (\boldsymbol{v}^* - \boldsymbol{v}_s^*) \cdot \boldsymbol{n} = \left(k_s \frac{\partial T_s}{\partial n} \right)^* - \left(k_\ell \frac{\partial T_\ell}{\partial n} \right)^* \tag{4.130}$$

Now, multiplying Eq. (4.56) by h_ℓ^* and using the result to simplify the first term provides

$$\rho_s (h_\ell^* - h_s^*)(\boldsymbol{v}^* - \boldsymbol{v}_s^*) \cdot \boldsymbol{n} = \left(k_s \frac{\partial T_s}{\partial n} \right)^* - \left(k_\ell \frac{\partial T_\ell}{\partial n} \right)^* \tag{4.131}$$

Finally, we recognize from Chap. 2 that $h_\ell^* - h_s^*$ is simply the latent heat, L_f, leaving the *Stefan condition*

$$\rho_s L_f (\boldsymbol{v}^* - \boldsymbol{v}_s^*) \cdot \boldsymbol{n} = \left(k_s \frac{\partial T_s}{\partial n} \right)^* - \left(k_\ell \frac{\partial T_\ell}{\partial n} \right)^* \tag{4.132}$$

which is usually seen in the form where $\boldsymbol{v}_s^* = 0$,

$$\rho_s L_f \boldsymbol{v}^* \cdot \boldsymbol{n} = \left(k_s \frac{\partial T_s}{\partial n} \right)^* - \left(k_\ell \frac{\partial T_\ell}{\partial n} \right)^* \tag{4.133}$$

The physical meaning of the Stefan condition is that the latent heat generated by the moving interface is balanced by the net conduction away from the interface in the solid and liquid phases. The boundary condition is illustrated graphically in Fig. 4.14. It may seem somewhat odd that only ρ_s appears in Eq. (4.133). The underlying reason is that the latent heat is defined per unit mass, and therefore subtracting the mass balance from Eq. (4.130) exactly eliminates terms involving ρ_ℓ.

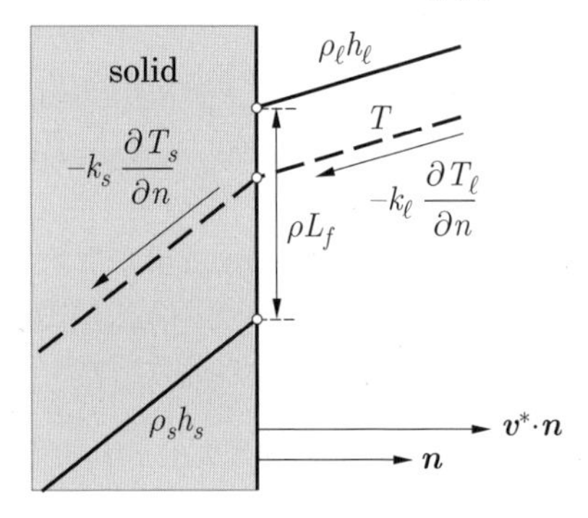

Fig. 4.14 A schematic representation of the enthalpy and heat flow for the Stefan condition.

We now turn to averaging Eq. (4.126). The process is equivalent to the one used for averaging the mass and momentum balance equations, so most of the derivation is omitted for the sake of brevity. The averaged equations for the individual phases are given by

$$\frac{\partial(g_s \rho_s \langle h \rangle_s)}{\partial t} + \nabla \cdot (g_s \rho_s \langle h \rangle_s \langle \boldsymbol{v} \rangle_s) - \nabla \cdot \langle k_s \nabla T_s \rangle - g_s \rho_s \left\langle \dot{R}_q \right\rangle_s = -\Gamma_s^{h*} \tag{4.134}$$

where

$$\Gamma_s^{h*} = \mathcal{S}_V \langle \rho_s h_s^* (\boldsymbol{v}_s^* - \boldsymbol{v}^*) \cdot \boldsymbol{n} \rangle^* - \mathcal{S}_V \langle k_s \nabla T_s \cdot \boldsymbol{n} \rangle^*$$

$$\frac{\partial(g_\ell \rho_\ell \langle h \rangle_\ell)}{\partial t} + \nabla \cdot (g_\ell \rho_\ell \langle h \rangle_\ell \langle \boldsymbol{v} \rangle_\ell) - \nabla \cdot \langle k_\ell \nabla T_\ell \rangle - g_\ell \rho_\ell \left\langle \dot{R}_q \right\rangle_\ell = \Gamma_\ell^{h*} \tag{4.135}$$

where

$$\Gamma_\ell^{h*} = \mathcal{S}_V \langle \rho_\ell h_\ell^* (\boldsymbol{v}_\ell^* - \boldsymbol{v}^*) \cdot \boldsymbol{n} \rangle^* - \mathcal{S}_V \langle k_\ell \nabla T_\ell \cdot \boldsymbol{n} \rangle^*$$

To obtain this form of the advective terms, we have again introduced the average and fluctuating parts of h_s, h_ℓ, $\boldsymbol{v}_s$ and $\boldsymbol{v}_\ell$, and neglected products of the fluctuating quantities. The terms $\nabla \cdot \langle k_s \nabla T_s \rangle$ and $\nabla \cdot \langle k_\ell \nabla T_\ell \rangle$ can be quite complicated to evaluate, since the average flux in a given phase depends strongly on the microstructure. To proceed, a choice must be made. We thus make the simplest possible one and write

$$\langle k_s \nabla T_s \rangle = g_s k_s \nabla \langle T \rangle_s \, ; \quad \langle k_\ell \nabla T_\ell \rangle = g_\ell k_\ell \nabla \langle T \rangle_\ell \tag{4.136}$$

Note that this choice omits many phenomena relating to the microstructure within the RVE. Additional reading of the chapter references is highly recommended for further details. We also take k_s and k_ℓ to be constant, but not necessarily equal.

It is clear from taking the average of Eq. (4.130) that the right hand sides of Eqs. (4.134) and (4.135) sum to zero. The average heat flow equation for the mixture is then

$$\frac{\partial \langle \rho h \rangle}{\partial t} + \nabla \cdot \langle \rho h \boldsymbol{v} \rangle = \nabla \cdot \langle k \nabla T \rangle + \langle \rho \dot{R}_q \rangle \tag{4.137}$$

where:

$$\begin{aligned}
\langle \rho h \rangle &= g_s \rho_s \langle h \rangle_s + g_\ell \rho_\ell \langle h \rangle_\ell \\
\langle \rho h \boldsymbol{v} \rangle &= g_s \rho_s \langle h \rangle_s \langle \boldsymbol{v} \rangle_s + g_\ell \rho_\ell \langle h \rangle_\ell \langle \boldsymbol{v} \rangle_\ell \\
\langle k \nabla T \rangle &= g_s k_s \nabla \langle T \rangle_s + g_\ell k_\ell \nabla \langle T \rangle_\ell \\
\langle \rho \dot{R}_q \rangle &= g_s \rho_s \left\langle \dot{R}_q \right\rangle_s + g_\ell \rho_\ell \left\langle \dot{R}_q \right\rangle_\ell
\end{aligned} \tag{4.138}$$

It is important to understand that even if k_s and k_ℓ are constant, the average heat flux $\langle k \nabla T \rangle$ can still depend strongly on microstructure, and is in general anisotropic. This matter is explored further in Exercise 4.5. Equation (4.137) is the form implemented in most casting solidification codes. To obtain closure, a relationship between enthalpy and temperature is required. In Chap. 10, we introduce microsegregation models that provide a relationship between g_ℓ and T, thus also giving $\langle \rho h(T) \rangle$. Examples of such relationships for a pure material and a binary alloy with primary and two-phase solidification are presented in Fig. 4.15.

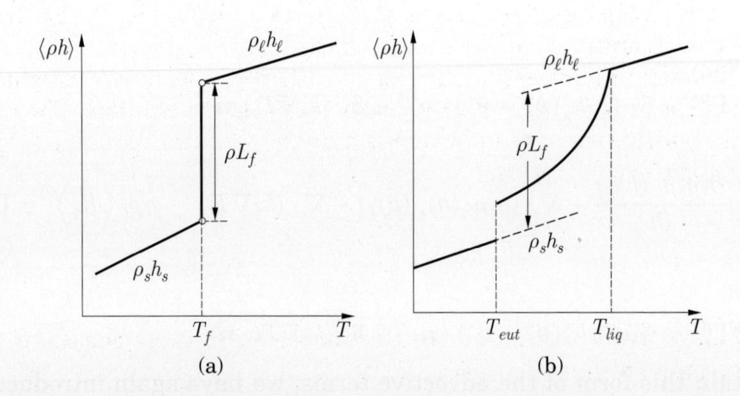

Fig. 4.15 The enthalpy as a function of temperature for (a) a pure material, and (b) a binary alloy having both primary and binary solidification.

The contribution of advection to Eq. (4.137) is discussed in more detail in Chap. 11, when we study grain movement. However, the same two simplified cases that were considered for the momentum balance are worthy of examination. When the solid is stationary, i.e., $\boldsymbol{v}_s = 0$, the advective term in Eq. (4.137) reduces to

$$\nabla \cdot \langle \rho h \boldsymbol{v} \rangle = g_\ell \rho_\ell c_{p\ell} \boldsymbol{v}_\ell \cdot \nabla T \qquad (\boldsymbol{v}_s = 0) \tag{4.139}$$

Thus, for a stationary solid, only specific heat is transported by convection. On the other hand, when the solid and liquid move together without phase separation, we have $\boldsymbol{v}_s = \boldsymbol{v}_\ell = \boldsymbol{v}$. Assuming $\langle \rho \rangle = cst$, i.e. $\nabla \cdot \boldsymbol{v} = 0$, the advective term in this case becomes

$$\nabla \cdot \langle \rho h \boldsymbol{v} \rangle = \boldsymbol{v} \cdot \nabla \langle \rho h \rangle \qquad (\boldsymbol{v}_s = \boldsymbol{v}_\ell) \tag{4.140}$$

indicating that both latent and specific heat are advected.

4.5 SOLUTE BALANCE IN MULTICOMPONENT SYSTEMS

The last balance equations to develop involve the transport of species within the various phases. We are still concerned with solid and liquid phases, but now consider the possibility that each phase may be chemically inhomogeneous. The treatment is entirely analogous to heat transport, and we therefore make use in various places of the development in the preceding section to shorten the discussion.

We compute averages over control volumes for multicomponent systems, and it is therefore most convenient to use the mass concentration ρC_K to represent the mass of species K per unit volume in a phase of density ρ. (Recall that C_K is defined as the mass fraction of species K in the mixture.) We assume that C_K is a continuous function in each phase, which implies that all species are present at all locations in the domain. Note that this already implies a certain level of averaging above the atomic scale. We also introduce the *species velocity* $\boldsymbol{v}_K$, and ignore all chemical reactions that might convert one phase or species to another. The sum of the mass fractions is one, and we may therefore define the mixture density (total mass per unit volume) as

$$\sum_{K=1}^{N_c} \rho C_K = \rho \sum_{K=1}^{N_c} C_K = \rho \tag{4.141}$$

Taking the now familiar control volume for a single phase, and computing a mass balance for species K gives

$$\frac{\partial (\rho C_K)}{\partial t} + \nabla \cdot (\rho C_k \boldsymbol{v}_K) = 0 \tag{4.142}$$

Summing over all N_c species provides a mass balance equation for the mixture

$$\frac{\partial \rho}{\partial t} + \nabla \cdot \left(\rho \sum_{K=1}^{N_c} C_K \boldsymbol{v}_K \right) = 0 \tag{4.143}$$

By analogy to the mass balance equation developed earlier, Eq. (4.49), we can identify a velocity for the mixture $\boldsymbol{v}$ given by

$$\boldsymbol{v} = \sum_{K=1}^{N_c} C_K \boldsymbol{v}_K \tag{4.144}$$

This velocity is called the *barycentric* or *mass-averaged* velocity.

The *diffusive flux* measures the relative separation of the different species from a single velocity common to all species. Choosing $\boldsymbol{v}$ as this common velocity, the diffusive mass flux for species K is defined as

$$\boldsymbol{j}_K = \rho C_K (\boldsymbol{v}_K - \boldsymbol{v}) \tag{4.145}$$

We should note that other average velocities can be computed by using various representations for the composition, such as volume fraction or mole fraction, and corresponding volume or mole fluxes can then be defined. The barycentric form used here is particularly convenient in that the equations that follow are consistent with those used for the other balance equations. Inserting the definition of $\boldsymbol{j}_K$ into Eq. (4.142) gives

$$\frac{\partial (\rho C_K)}{\partial t} + \nabla \cdot (\rho C_K \boldsymbol{v}) = -\nabla \cdot \boldsymbol{j}_K \tag{4.146}$$

A constitutive model is required to relate $\boldsymbol{j}_K$ to the composition.

The theory of diffusion, that provides such a relation, is the subject of numerous books. We will restrict ourselves to simple cases pertinent to solidification processes. As discussed in Chaps. 2 and 3, a variation of chemical potential provides a driving force for the minimization of Gibbs free energy in the system. Figure 4.16 illustrates a binary solid solution of A and B atoms with an inhomogeneous composition along the x-direction. Two positions x_1 and x_2 located on opposite sides of the centerline differ with respect to their chemical potentials owing to their different compositions. As shown in Fig. 4.16(b), $\mu_A(x_1) > \mu_A(x_2)$ and $\mu_B(x_2) > \mu_B(x_1)$. Therefore, as shown in Exercise 4.3, the system can lower its total energy by exchanging A atoms at x_1 with B atoms at x_2.

Fick's first law posits that the rate at which this exchange occurs, i.e., the flux of species K, is proportional to the gradient of its chemical potential, which can be written in the general form

$$\boldsymbol{j}_K = -M_K C_K \nabla (\mu_K / T) \tag{4.147}$$

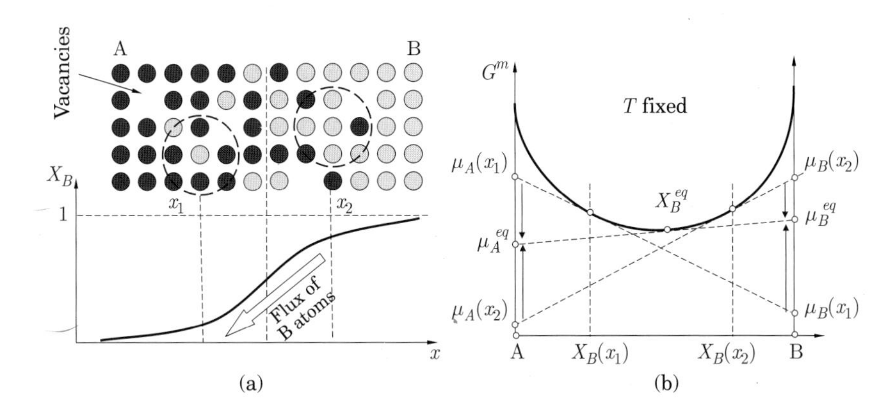

Fig. 4.16 (a) A schematic of a solid solution of A and B atoms inhomogeneous along one dimension. The overall composition profile averaged over vertical lines is shown at the bottom. (b) The Gibbs free energy for this phase at this particular temperature, with indication of the composition and chemical potentials for the positions x_1 and x_2.

where M_K is called the *mobility* of species K. Since the chemical potential depends on composition and temperature, i.e., $\mu_K = \mu_K(T, C_1, C_2, \ldots, C_{N_C-1})$ we can expand the gradient in Eq. (4.147) to obtain

$$\begin{aligned} j_K &= -\rho \sum_{J=1}^{N_c} \left(\frac{M_K C_K}{T} \frac{\partial \mu_K}{\partial C_J} \right) \nabla C_J + \frac{M_K C_K \mu_K}{T^2} \nabla T \\ &= -\rho \sum_{J=1}^{N_c} D_{KJ} \nabla C_J + \frac{M_K C_K \mu_K}{T^2} \nabla T \end{aligned} \tag{4.148}$$

The second term on the right hand side represents the *Soret effect*, where a temperature gradient can cause species diffusion. This effect, and its reciprocal, the *Dufour effect* (where composition gradients lead to heat diffusion), can be important in gases, but are not significant in most solidification problems, and are thus neglected beyond this point. The elements D_{KJ} of the diffusivity matrix, representing the diffusion of species K due to gradients in the composition of species J, are called *intrinsic diffusivities*. The off-diagonal terms can be important in solid state diffusion in some multicomponent alloys, leading in certain cases to "uphill diffusion," where the flux of certain species K actually goes from lower to higher composition C_K due to the influence of composition gradients in other species. We consider this phenomenon in Chap. 10. For simplicity here, we will assume that the diffusivity matrix is diagonal. In that case, it is unambiguous to simply write D_K instead of D_{KK}, and so we adopt this notation. The simplified version of Fick's first law is then

$$j_K = -\rho D_K \nabla C_K \tag{4.149}$$

Substituting this form of Fick's first law into the solute balance of Eq. (4.146) gives what is often called Fick's second law

$$\frac{\partial(\rho C_K)}{\partial t} + \nabla \cdot (\rho C_K \boldsymbol{v}) = \nabla \cdot (\rho D_K \nabla C_K) \tag{4.150}$$

Notice the similarity between Eq. (4.150) and the heat balance in Eq. (4.126). If chemical reactions had been retained in the treatment, a term analogous to $\rho \dot{R}_q$ would have appeared in Eq. (4.150).

The interfacial solute balance is obtained in the same way that we obtained the Stefan condition. Integrating Eq. (4.150) over a small control volume containing and following the interface gives

$$\begin{aligned} &\rho_\ell C^*_{K\ell}(\boldsymbol{v}^*_\ell - \boldsymbol{v}^*) \cdot \boldsymbol{n} - \rho_s C^*_{Ks}(\boldsymbol{v}^*_s - \boldsymbol{v}^*) \cdot \boldsymbol{n} \\ &\qquad - \rho_\ell D_{K\ell} \left(\frac{\partial C_{K\ell}}{\partial n}\right)^* + \rho_s D_{Ks} \left(\frac{\partial C_{Ks}}{\partial n}\right)^* = \Gamma^{C*}_\ell - \Gamma^{C*}_s \end{aligned} \tag{4.151}$$

The interfacial contribution $\Gamma^{C*}_\nu = [\rho_\nu C^*_{K\nu}(\boldsymbol{v}^*_\nu - \boldsymbol{v}^*) \cdot \boldsymbol{n} - \rho_\nu D_{K\nu} \partial C_{K\nu}/\partial n]$ of each phase ν to the solute balance has been introduced. Multiplying Eq. (4.49) by $C^*_{K\ell}$ and using the result to reduce Eq. (4.151), followed by subsequently invoking thermodynamic equilibrium at the interface in the form $C^*_{Ks} = k_{0K} C^*_{K\ell}$ gives

$$\rho_s(1 - k_{0K}) C^*_{K\ell}(\boldsymbol{v}^* - \boldsymbol{v}^*_s) \cdot \boldsymbol{n} = \rho_s D_{Ks} \left(\frac{\partial C_{Ks}}{\partial n}\right)^* - \rho_\ell D_{K\ell} \left(\frac{\partial C_{K\ell}}{\partial n}\right)^* \tag{4.152}$$

Notice that ρ_s appears on the right hand side of Eq. (4.152), as we saw for the Stefan condition. A sketch of the composition distribution near the interface is shown in Fig. 4.17. If the velocity of the solid is zero, the jump of solute concentration at the interface given by $(1 - k_{0K}) C^*_{K\ell}$ is directly proportional to the solute flux away from the interface.

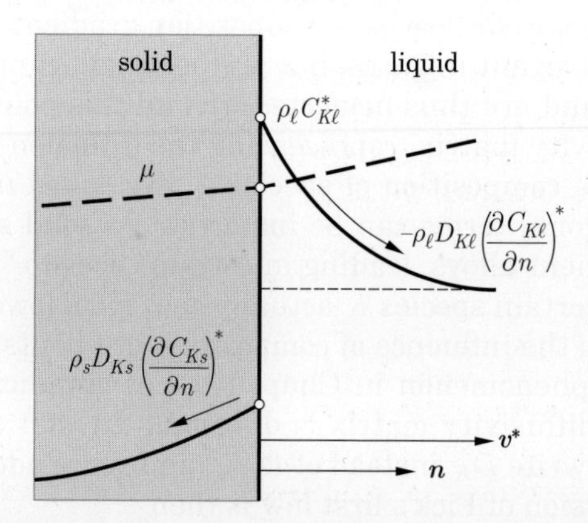

Fig. 4.17 A schematic of the composition and solute mass fluxes near the solid-liquid interface.

Averaging of the solute balance equations also proceeds in exactly the same way as for heat conduction, so we simply state the results

$$\frac{\partial(g_s\rho_s \langle C_K\rangle_s)}{\partial t} + \nabla\cdot(g_s\rho_s\langle C_K\rangle_s\langle \boldsymbol{v}\rangle_s) - \nabla\cdot\langle\rho_s D_{Ks}\nabla C_{Ks}\rangle$$
$$= -\mathcal{S}_V\langle\rho_s C^*_{Ks}(\boldsymbol{v}^*_s - \boldsymbol{v}^*)\cdot\boldsymbol{n}\rangle^* + \mathcal{S}_V\langle\rho_s D_{Ks}(\nabla C_{Ks})^*\cdot\boldsymbol{n}\rangle^* = -\Gamma^{C*}_s \tag{4.153}$$

$$\frac{\partial(g_\ell\rho_\ell \langle C_K\rangle_\ell)}{\partial t} + \nabla\cdot(g_\ell\rho_\ell\langle C_K\rangle_\ell\langle \boldsymbol{v}\rangle_\ell) - \nabla\cdot\langle\rho_\ell D_{K\ell}\nabla C_{K\ell}\rangle$$
$$= \mathcal{S}_V\langle\rho_\ell C^*_{K\ell}(\boldsymbol{v}^*_\ell - \boldsymbol{v}^*)\cdot\boldsymbol{n}\rangle^* - \mathcal{S}_V\langle\rho_\ell D_{K\ell}(\nabla C_{K\ell})^*\cdot\boldsymbol{n}\rangle^* = \Gamma^{C*}_\ell \tag{4.154}$$

As we found for the energy equation, the interfacial contributions appearing in these volume averages cancel since $\Gamma^{C*}_\ell - \Gamma^{C*}_s = 0$ (Eq. (4.151)). The same caveats apply here concerning the importance of microstructure in the gradient terms. In fact, since in general $D_{Ks} \ll D_{K\ell} \ll \alpha = k(\rho c_p)^{-1}$, a simple average is likely to be a far worse estimate for the average solute diffusivity than it was for the thermal diffusivity. Nevertheless, we simply perform the same sum as we did for the conduction equation and write for the mixture

$$\frac{\partial\langle\rho C_K\rangle}{\partial t} + \nabla\cdot\langle\rho C_K\boldsymbol{v}\rangle = \nabla\cdot\langle\rho D_K\nabla C_K\rangle \tag{4.155}$$

where

$$\begin{aligned}
\langle\rho C_K\rangle &= g_s\rho_s\langle C_K\rangle_s + g_\ell\rho_\ell\langle C_K\rangle_\ell \\
\langle\rho C_K\boldsymbol{v}\rangle &= g_s\rho_s\langle C_K\rangle_s\langle\boldsymbol{v}\rangle_s + g_\ell\rho_\ell\langle C_K\rangle_\ell\langle\boldsymbol{v}\rangle_\ell \\
\langle\rho D_K\nabla C_K\rangle &= g_s\rho_s D_{Ks}\nabla\langle C_K\rangle_s + g_\ell\rho_\ell D_{K\ell}\nabla\langle C_K\rangle_\ell
\end{aligned} \tag{4.156}$$

Since the diffusion in the liquid is quite fast at the level of the RVE, the composition of the liquid $C_{K\ell}$ can be assumed to be uniform. If the same assumption is made for the solid phase, i.e., that C_{Ks} is uniform, the lever rule already encountered in Chap. 2 results. In other cases, the average composition $\langle C_K\rangle_s$ needs to be calculated using a microsegregation model. As we shall see in Chap. 10, this problem can be made tractable by approximating the complex microstructure in the RVE as a one-dimensional geometry in which the axis is normal to the interface.

4.6 SCALING

In developing the governing equations, we have tried to be as complete as possible. Nevertheless, certain terms have been neglected since they were deemed to be of secondary importance (e.g., thermomigration coefficient in the diffusion equation, or deformation heating in the heat flow equation). When we consider a particular problem, though, it is frequently the case that at least some of the remaining terms are less important than others.

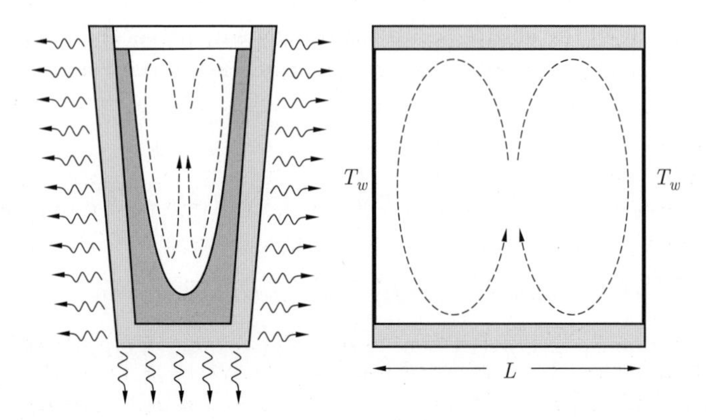

Fig. 4.18 Macrosegregation in large ingots, and a simplification of the process for understanding scaling of the governing equations.

Scaling analysis is a systematic means for determining which terms can be safely neglected, and which must be included. Scaling can also provide us with additional information: approximate values for unknown quantities such as pressure and velocity, as well as the way that the process is affected by changes in the physical parameters. In this section, we introduce scaling through an example.

To illustrate the procedure, consider a simplified version of the buoyancy-driven flow that leads to macrosegregation in large ingots. The process and its idealization are shown in Fig. 4.18. In the real process, a large ingot is cooled from the sides and bottom, and as the ingot solidifies, a buoyant flow occurs due to the temperature gradient, and also due to solute gradients that derive from the segregation inherent in the solidification of alloys. We simplify the problem by considering the convection of a liquid under a horizontal temperature gradient, in a cavity of similar size to the ingot. Thus, there is no solidification or solute to worry about. This may seem like a lot of assumptions, which is in fact true, but the goal here is to demonstrate the scaling procedure, and there is enough left to make the problem interesting. The macrosegregation problem is treated in a more rigorous manner in Chap. 14.

The scaling analysis is somewhat complicated, so we begin with a roadmap for the process. We start by collecting the material properties and important process parameters, then adopt the following procedure:

1. Utilize the process data to find characteristic values, and define scaled variables that are dimensionless and of order one. Some variables, such as pressure and time, do not always have obvious characteristic values. We will just define one symbolically, e.g., Δp_c for the pressure scale, and then let the process guide us in determining the others.
2. Substitute these new variables into the governing equations. Then render the governing equations dimensionless by dividing through

by one of the coefficients, chosen such that all of the dimensionless coefficients are of order one or smaller. This allows us to systematically eliminate terms based on their value relative to one.

3. Coefficients containing an unknown characteristic value are then set to one in order to determine the actual value of the characteristic. Some of the dimensionless groups appear with sufficient frequency that they are given names, such as the Reynolds number.

As shown in Fig. 4.18, the cavity has dimension L in each direction. The liquid is initially at temperature T_0, and at the start of the experiment ($t = 0$), the vertical faces are suddenly cooled to temperature $T_w < T_0$, and subsequently maintained at that temperature. All other faces are insulated. We assume that the fluid is Newtonian, and that the density varies linearly with pressure and temperature, as given in Eq. (4.52). Typical values for the material properties of steel are given in Table 4.1. These properties are used to determine the relative magnitude of the terms in the scaled equations below.

First, a note on notation. Insofar as possible, we use letters from the English alphabet for dimensional variables, and the closest Greek letter for their dimensionless counterparts (e.g., $x \rightarrow \xi, t \rightarrow \tau, T \rightarrow \theta$, etc.). Certain variables, such as velocity, have no Greek counterparts, and also have both vector and scalar forms. In this case, we use italic font for the dimensional quantity ($\boldsymbol{v}, v_x$), and Roman font for the corresponding dimensionless quantity ($\mathrm{v}, \mathrm{v}_\xi$). Notice also the change in subscript for the dimensionless component.

Begin by collecting *characteristic values* for the experiment, and use them to define scaled variables. We expect the temperature to be in the range $T_w < T < T_0$. We therefore define a scaled temperature $\theta \in [0, 1]$

$$\theta = \frac{T - T_w}{T_0 - T_w} = \frac{T - T_w}{\Delta T_c} \tag{4.157}$$

Table 4.1 Approximate material properties and process variables for buoyancy-driven flow in liquid steel.

Symbol	Property	Units	Value
ρ_0	Density	$\mathrm{kg\,m^{-3}}$	7×10^3
c_p	Specific heat	$\mathrm{J\,kg^{-1}K^{-1}}$	800
k	Thermal conductivity	$\mathrm{W\,m^{-1}K^{-1}}$	25
μ	Shear viscosity	Pa s	0.37×10^{-3}
β_T	Volumetric thermal expansion	$\mathrm{K^{-1}}$	1.3×10^{-5}
β_p	Volumetric pressure expansion	$\mathrm{Pa^{-1}}$	5.8×10^{-12}
g	Acceleration of gravity	$\mathrm{m\,s^{-2}}$	9.8
L	Characteristic length	m	0.5
T	Absolute temperature	K	1800
ΔT_c	Characteristic temperature difference	K	10

where $\Delta T = T_0 - T_w$ corresponds to the characteristic temperature difference. Scaled variables are always dimensionless and of $\mathcal{O}(1)$, and we are thus able to maintain or neglect terms based on their magnitude relative to unity. Following the same prescription for the coordinate directions, we define the scaled dimensions (ξ, η, ζ) as

$$\boldsymbol{\xi} = \{\xi, \eta, \zeta\} = \frac{1}{L}\{x, y, z\} = \frac{1}{L}\boldsymbol{x} \tag{4.158}$$

To develop the scaled equations, consider first the mass balance equation in the form of Eq. (4.53). Under the assumption that the density varies linearly with temperature, Eq. (4.53) reduces to

$$-\rho_0\beta_T\frac{DT}{Dt} - \rho_0\beta_p\frac{Dp}{Dt} + \rho\left(\nabla \cdot \boldsymbol{v}\right) = 0 \tag{4.159}$$

We need characteristic values for the velocity v_c and pressure Δp_c in order to scale the equation. (v_c and Δp_c come out of the scaling of the momentum equation, below. For now, we just use the symbols.) We employ the same velocity scale for all directions, and define the scaled variables as

$$\mathrm{v} = \frac{\boldsymbol{v}}{v_c}; \quad \Pi = \frac{p - p_0}{\Delta p_c}; \quad \tau = \frac{L}{v_c} \tag{4.160}$$

Substituting the various scales into Eq. (4.159) and dividing by $\rho_0 v_c/L$ yields the dimensionless form

$$-\beta_T\Delta T_c\frac{D\theta}{D\tau} - \beta_p\Delta p_c\frac{D\Pi}{D\tau} + \left(1 - \beta_T\Delta T_c\theta\right)\nabla \cdot \mathrm{v} = 0 \tag{4.161}$$

It is to be understood that the material derivative and gradient operator for scaled variables also uses scaled variables, i.e.,

$$\frac{D}{D\tau} = \frac{\partial}{\partial\tau} + \mathrm{v}_j\frac{\partial}{\partial\xi_j} \tag{4.162}$$

Now, incorporating the parameters listed in Table 4.1, one can see that for a temperature difference of 10 K, which is typical for steel castings, $\beta_T\Delta T_c \sim 1.3 \times 10^{-4} \ll 1$. In the scaling of the momentum balance equation below, we find Δp_c and v_c, and it turns out that $\beta_p\Delta p_c \ll 1$, as well. Let us simply assert for now that this is the case, remembering to verify it later. Thus, the mass balance equation is simplified by scaling to

$$\nabla \cdot \mathrm{v} = 0 \tag{4.163}$$

The relative importance of the other terms in the equation is also maintained in the dimensional form of the equation, thus giving

$$\nabla \cdot \boldsymbol{v} = 0 \tag{4.164}$$

This *looks* like the mass balance equation for a constant density fluid, but it is in fact an approximation, valid only when $\beta\Delta T_c \ll 1$ and $\beta\Delta p_c \ll 1$.

Next, we consider the momentum balance equation, written for the Newtonian fluid in Eq. (4.97). In view of Eq. (4.164), we can neglect the term $\lambda\nabla \cdot \mathbf{v}$, and use the form for the constant viscosity fluid given in Eq. (4.99). There is a body force due to gravity of magnitude ρg, acting in the $-\hat{\mathbf{z}}$ direction. The steady form of the momentum balance equation then becomes

$$\rho_0 \left[1 - \beta_T(T - T_0) - \beta_p(p - p_0)\right] \frac{D\boldsymbol{v}}{Dt} = -\nabla p + \mu\nabla^2\boldsymbol{v} - \rho_0 g\hat{\mathbf{z}} + \rho_0\beta(T - T_0)g\hat{\mathbf{z}} \tag{4.165}$$

It is common to lump the static portion of the pressure into the pressure gradient term by defining a so-called *modified pressure* $\hat{p} \equiv p + \rho_0 g z$. Equation (4.165) thus becomes

$$\rho_0 \left[1 - \beta_T(T - T_0) - \beta_p(p - p_0)\right] \frac{D\boldsymbol{v}}{Dt} = -\nabla\hat{p} + \mu\nabla^2\boldsymbol{v} + \rho_0\beta_T(T - T_0)g\hat{\mathbf{z}} \tag{4.166}$$

Substitution of the dimensionless variables into this equation gives

$$\left[\frac{\rho_0 v_c^2}{L}\right](1 - \beta_T\Delta T_c\theta - \beta_p\Delta p_c\Pi)\frac{D\mathbf{v}}{D\tau} = -\left[\frac{\Delta p_c}{L}\right]\nabla\hat{\Pi} + \left[\frac{\mu v_c}{L^2}\right]\nabla^2\mathbf{v} + \left[\rho_0 g\beta\Delta T_c\right]\theta\,\hat{\mathbf{z}} \tag{4.167}$$

As we did for the mass balance equation, we simplify the left hand side by neglecting $\beta_T\Delta T_c$ and $\beta_T\Delta p_c$ in comparison to one.

There is no source of pressure gradient to drive the flow in this problem, indeed in most natural convection problems. Thus, the buoyant forces, represented by $[\rho_0 g\beta\Delta T_c]\,\theta\,\hat{\mathbf{z}}$ must be balanced by either the inertial or viscous forces. When scaling the equations, one sometimes has to guess which terms are most important. The equation is then divided by the coefficient that multiplies that term, and the remaining groups of parameters are evaluated. If the dimensionless groups so formed are large in comparison with one, then the guess was wrong. One should thus divide through by that dimensionless group to find terms that are small in comparison to one, and that can be safely neglected. For the purpose of our illustration, let us suppose that the buoyancy forces are balanced by the inertial forces. We then divide Eq. (4.167) by $\rho_0 v_c^2/L$, and obtain

$$\frac{D\mathbf{v}}{D\tau} = -\left\{\frac{\Delta p_c}{\rho_0 v_c^2}\right\}\nabla\hat{\Pi} + \left\{\frac{\mu}{\rho_0 v_c L}\right\}\nabla^2\mathbf{v} + \left\{\frac{g\beta_T\Delta T_c L}{v_c^2}\right\}\theta\,\hat{\mathbf{z}} \tag{4.168}$$

From Eq. (4.168) we deduce that $\Delta p_c = \rho_0 v_c^2$. Since the inertial terms, now order one, were assumed to balance the buoyancy forces, we can evaluate the characteristic velocity v_c by setting the coefficient multiplying the buoyancy to one, thereby yielding $v_c = \sqrt{g\beta_T\Delta TL}$. By using this result, and substituting for v_c in the coefficient multiplying the viscous term, we identify

$$\frac{\mu}{\rho_0 v_c L} = \frac{\mu}{\rho_0\sqrt{g\beta_T \Delta T_c L^3}} = \frac{1}{\sqrt{\mathrm{Gr}}}; \quad \mathrm{Gr} = \frac{g\beta_T \Delta T_c L^3}{(\mu/\rho_0)^2} \tag{4.169}$$

where Gr is called the *Grashof number*, a measure of the relative importance of inertial and viscous forces in natural convection problems. Note that to get to Eq. (4.168), we assumed that $\mathrm{Gr} \gg 1$. Dimensionless groups, such as the Grashof number, appear sufficiently often that they are given names, as a sort of shorthand to facilitate communication. They always represent the ratio of two phenomena in the governing equations, and arise naturally from the scaling of the governing equations. In general, the ratio of inertial to viscous forces $\rho_0 v_c L/\mu$, is called the *Reynolds number*, Re. In this particular case, where we have an estimate of v_c in terms of the other parameters in the problem, $\mathrm{Re} = \sqrt{\mathrm{Gr}}$. As we continue, we identify several other named dimensionless groups. There is nothing magic or mysterious about the names, or their origin in the scaling of the equations; it is just a matter of convenience.

Before going further, the property and process data can be used to determine certain numerical values. First, we compute $v_c = \sqrt{g\beta_T \Delta T L} = 0.025$ m/s. Next, $\mathrm{Re} = \sqrt{\mathrm{Gr}} = 2.36 \times 10^5 \gg 1$, which is consistent with our expectations. Note that the Reynolds number is high enough that the flow is likely to be turbulent, and a more complete model would be needed to get everything right. We can now also compute $\Delta p_c = \rho_0 v_c^2 = 4.4$ Pa, which leads to

$$\beta_p \Delta p_c \sim 2.5 \times 10^{-11} \ll 1 \tag{4.170}$$

This confirms that it was reasonable to neglect the pressure contribution to the density.

To complete the description of the governing equations, consider the energy balance, starting with the form Eq. (4.127). There is no source of external heating, so we set $\dot{R}_q = 0$. After substituting the scaled variables already defined, we have

$$\frac{\rho_0 c_p \Delta T_c}{L/v_c}\frac{D\theta}{D\tau} - \frac{\Delta p_c}{L/v_c}\frac{D\Pi}{D\tau} = \frac{k\Delta T_c}{L^2}\nabla^2\theta + \frac{2\mu v_c^2}{L^2}\mathrm{tr}\mathbf{D}^2 \tag{4.171}$$

Note that we have put back the terms involving the contribution of the pressure change to the energy, as well as the viscous dissipation, in order to verify that they are small. Also, by writing ρ_0 on the left hand side, we have already invoked the fact that $\beta_T \Delta T$ and $\beta_T \Delta p_c$ are both small in comparison with one. Conduction is expected to be important, and so we divide the entire equation by $k\Delta T_c/L^2$ to obtain a dimensionless equation

$$\frac{v_c L}{\alpha_0}\frac{D\theta}{D\tau} - \frac{\rho_0 v_c^3 L}{k\Delta T_c}\frac{D\Pi}{D\tau} = \nabla^2\theta + \frac{2\mu v_c^2}{k\Delta T_c}\mathrm{tr}\mathbf{D}^2 \tag{4.172}$$

where $\alpha_0 = k/(\rho_0 c_p)$ is the thermal diffusivity. The dimensionless group $v_c L/\alpha_0$, measuring the relative importance of advection to conduction of heat, is called the *Péclet number*. The group $2\mu v_c^2/k\Delta T$ represents the ratio of the rate of viscous generation of heat to conduction, and is called the *Brinkman number*. The dimensionless group multiplying the pressure time derivative, which is almost always negligible, does not have a specific name.

Now substitute $v_c = \sqrt{g\beta_T \Delta T_c L}$ in Eq. (4.172) to find

$$\frac{\sqrt{g\beta_T \Delta T_c L^3}}{\alpha_0}\frac{D\theta}{D\tau} - \frac{\rho_0}{k}\sqrt{g^3\beta_T^3\Delta T_c L^5}\frac{D\Pi}{D\tau} = \nabla^2\theta + \frac{2\mu g\beta_T L}{k}\mathrm{tr}\mathbf{D}^2 \tag{4.173}$$

The form of the Péclet number for this problem, i.e., the coefficient of the $D\theta/D\tau$ term, is the square root of the *Boussinesq number*, Bo, i.e.,

$$\mathrm{Bo} = \frac{g\beta_T \Delta T_c L^3}{\alpha_0^2} \tag{4.174}$$

Two more dimensionless groups often appear in buoyancy problems: the *Prandtl number*, a material property defined as the ratio of the kinematic viscosity μ/ρ_0 to the thermal diffusivity α_0

$$\mathrm{Pr} = \frac{\mu/\rho_0}{\alpha_0} \tag{4.175}$$

and the *Rayleigh number*, defined as

$$\mathrm{Ra} = \frac{\mathrm{g}\beta_\mathrm{T}\Delta \mathrm{T}\mathrm{L}^3}{\alpha_0(\mu/\rho_0)} \tag{4.176}$$

The various dimensionless groups are related according to

$$\mathrm{Bo} = \mathrm{Pr}^2\mathrm{Gr}; \quad \mathrm{Ra} = \mathrm{Pr}\mathrm{Gr} \tag{4.177}$$

Using the properties in Table 4.1 to evaluate these dimensionless groups gives: $\mathrm{Pr} = 0.012$; $\mathrm{Pe} = 2800$; $\mathrm{Bo} = 7.9 \times 10^6$; $\mathrm{Ra} = 6.7 \times 10^8$; and $\mathrm{Br} = 1.8 \times 10^{-7}$. Finally, the coefficient multiplying $D\Pi/D\tau$ is evaluated to approximately 2×10^{-4}. Neglecting terms whose coefficients are much smaller than one, Eq. (4.173) becomes

$$\mathrm{Pe}\frac{D\theta}{D\tau} = \nabla^2\theta \tag{4.178}$$

It took a fair amount of work to get to this point, but the reduction in complexity is clearly worthwhile.

Let us summarize what we learned from scaling the governing equations for this problem. First, we found that although the density varies with temperature, the only place where this variation is significant is as a body force in the momentum balance equation. The mass balance, energy balance, and inertial terms in the momentum balance equations use only the constant reference density ρ_0. This is sometimes called the *Boussinesq approximation*, and it is valid when $\beta_T\Delta T \ll 1$ and $\beta_T \Delta p_c \ll 1$, which is frequently the case for liquids, but not always for gases. For the particular case we considered, i.e., natural convection in a large steel ingot, the inertial terms in the momentum equation were much more important than the viscous terms in the momentum balance ($\mathrm{Re} \gg 1$), and the advection terms were much more important than conduction in the energy equation ($\mathrm{Pe} \gg 1$). Viscous dissipation was also negligible.

We identified several dimensionless groups during the scaling analysis, and saw how they arose from the scaling of the governing equations. Different physical processes lead to the apparition of different dimensionless groups in the scaling process. A listing of some of the more common dimensionless groups, adapted from [2] is given in the frontmatter of this book.

4.7 SUMMARY

In this chapter, we developed the balance equations for mass, momentum, energy and solute. We found that these equations, while correct, were not sufficient to solve problems, since there were more unknown variables than there were equations. The description was completed by introducing constitutive equations to relate the unknown variables, using material properties. We also computed averages of these equations over a small representative volume element and obtained two types of equations: volumetric average equations and interfacial balances. These equations serve as the basis for the analysis that is presented in the remainder of this book. Finally, we described a process for scaling the governing equations for a particular problem in order to identify the important terms (and corresponding phenomena).

4.8 EXERCISES

Exercise 4.1. Averaging theorems.
Following the procedure used to derive Eq. (4.36), derive Eqs. (4.37) and (4.38).

Exercise 4.2. Unidirectional deformation.
Begin with Eq. (4.79), and derive Eq. (4.80) using only the assumption that the only non-zero component of $\boldsymbol{\sigma}$ is σ_{xx}.

(a) Show first that the shear strains ε_{ij}, $i \neq j$, are zero.

(b) Next, write the three remaining equations for σ_{xx}, σ_{yy} and σ_{zz}. Show first that $\varepsilon_{yy} = \varepsilon_{zz}$.

(c) Finally, develop by subtraction of the appropriate equations, expressions for ε_{yy} and ε_{zz} in terms of σ_{xx} and ε_{xx}. Substitute these results into the equation for σ_{xx} and then derive Eq. (4.80).

Exercise 4.3. Chemical potential in diffusion.
Show that the total free energy of a system with composition gradients can be lowered by transferring atoms in such a way as to eliminate the composition gradients. You may find the discussion surrounding Fig. 4.16 a useful guide.

Exercise 4.4. Measuring $\mathcal{S}_V$.
The surface area per unit volume $\mathcal{S}_V$ can be measured in a surprisingly easy way. Draw a straight line crossing the image, and then count the number of intersections on the line corresponding to the surface of interest and divide that number by the length of the line. Call this number N_L. An estimate for $\mathcal{S}_V$ can be obtained by randomly placing several such lines, recording N_L for each, and then setting

$$\mathcal{S}_V = 2\bar{N}_L$$

where the overbar indicates the average of all readings. One can also use a fixed grid of lines to make the measurement, as long as it is sufficiently dense.

(a) Verify this formula by constructing a "microstructure" consisting of a white square containing one, two, or four gray circles of known diameter, then apply the recipe described above. Experiment with the number of grid lines to visualize the effect on the accuracy.

(b) Consider Fig. 9.19(a). Make a photocopy of the image, and then use the same technique to estimate $\mathcal{S}_V$ for the graphite nodules.

Exercise 4.5. Average properties.
Consider the representative volume element shown, where the "microstructure" consists of unequal amounts of solid and liquid, arrayed as two parallel strips.

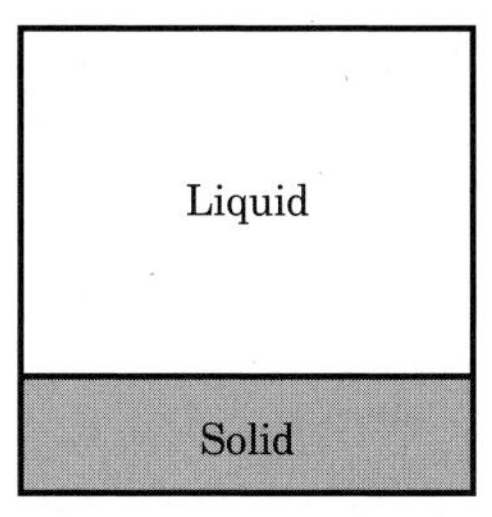

(a) Compute the average density using a rule of mixtures. Write your answer in terms of g_s, g_ℓ, ρ_s, etc.

(b) Suppose that the left and right hand faces are maintained at temperatures T_1 and T_2, while the top and bottom faces are insulated.

Compute the total heat flux, and manipulate the result in such a way that you can determine the effective thermal conductivity $k_{\parallel}^{eff}$ for this orientation, i.e., in the form $q_{\parallel} = k_{\parallel}^{eff}\Delta T$.

(c) Repeat the calculation to find the effective thermal conductivity $k_{\perp}^{eff}$ for the case where the top and bottom faces are maintained at different temperatures and the left and right hand sides are insulated.

(d) Finally, compute k_{avg}^{eff} using a simple rule of mixtures, and compare the results for the three effective conductivities when $g_s = 0.25$ and $k_s = 2k_\ell$.

(e) It would be completely analogous to define effective diffusion coefficient in the same way as the conductivity. In this case, however, one might find more typically that $D_s = D_\ell/1000$. Compare the effective computed diffusivities using the same three methods, again for $g_s = 0.25$.

4.9 REFERENCES

[1] R. B. Bird, W. E. Stewart, and E. N. Lightfoot. *Transport phenomena*. Wiley, New York, 1960.

[2] J. A. Dantzig and C. L. Tucker III. *Modeling in materials processing*. Cambridge University Press, New York, 2001.

[3] S. H. Davis. *Theory of solidification*. Cambridge University Press, Cambridge, 1999.

[4] Y. C. Fung. *First course in continuum mechanics*. Prentice-Hall, New York, 1994.

[5] J. Ni and C. Beckermann. A volume-averaged two-phase model for transport phenomena during solidification. *Metall. Trans.*, 22B:349–361, 1991.

[6] J. Philibert. *Atom movements diffusion and mass transport in solids*. Les éditions de physique, 1991. translated by S. J. Rothman.

[7] M. Rappaz, M. Bellet, and M. Deville. *Numerical modeling in materials science and engineering*. Springer-Verlag, Berlin, 2002.

[8] C. J. Smithells, editor. *Metals reference book*. Butterworths, London, fifth edition, 1976.

[9] Y. Sun and C. Beckermann. Diffuse interface modeling of two-phase flows based on averaging: Mass and momentum equations. *Physica D*, 198:281–308, 2004.

CHAPTER 5

ANALYTICAL SOLUTIONS FOR SOLIDIFICATION

5.1 INTRODUCTION

This chapter describes how the governing equations developed in Chap. 4 and the thermodynamic principles introduced in Chaps. 2 and 3 are applied to model the solidification process. For the time being, we consider only problems that have analytical solutions, reserving the discussion concerning computational approaches to a subsequent chapter. These problems provide valuable insight into the behavior of solidifying systems, as well as into the roles of various parameters. In particular, the analytical solutions demonstrate how the solidification front moves over time, and how solute boundary layers build up ahead of the interface.

For example, we will find that the thickness of a solidified layer growing from a fixed-temperature wall increases in proportion to $\sqrt{t}$ and that consequently the characteristic growth velocity of the interface v^* is proportional to $1/\sqrt{t}$. In alloys, there is a solute boundary layer in the liquid ahead of the interface with a thickness of order D_ℓ/v^*, where D_ℓ is the diffusion coefficient. These characteristics will be very important later on when we consider the development of microstructure. Our study begins with solidification of a superheated melt from a cold wall. Subsequently, we consider solidification in an undercooled melt, developing solutions for planar, paraboloidal and spherical front growth.

5.2 SOLIDIFICATION IN A SUPERHEATED MELT

5.2.1 Pure materials

Consider the one-dimensional solidification of a pure material in a mold, as depicted in Fig. 5.1. The initial condition is a pure liquid at uniform temperature T_∞ greater than the melting point T_f, and filling the semi-infinite domain $x \geq 0$. A semi-infinite mold, initially at uniform temperature

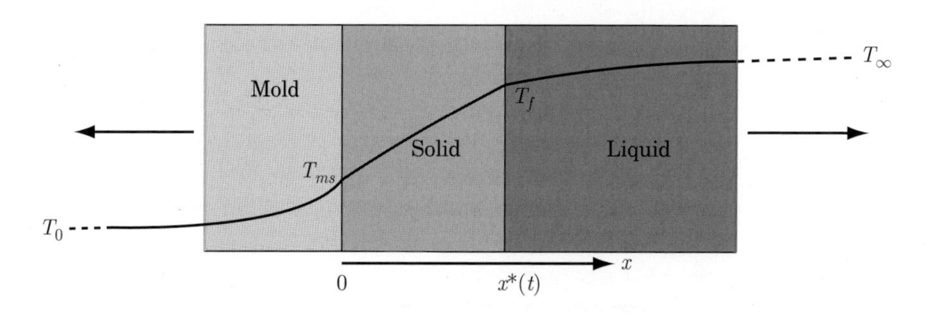

Fig. 5.1 A schematic drawing of the temperature distribution in the mold, solid and liquid at a certain time after solidification begins.

$T_0 < T_f$, occupies the domain $x < 0$. The solidification begins at time $t = 0$ by conduction of heat into the mold. At some point in time, the temperature distributions in the mold, solid and liquid will correspond to the schematic depicted in Fig. 5.1. The position of the solid-liquid interface is designated $x^*(t)$, and part of the problem will be to find an analytical expression for x^*. We have assumed perfect contact between the solid and the mold, and the interface temperature T_{ms} is thus identical in both materials. (This assumption is necessary in order to obtain an analytical solution for the movement of the interface.)

The goal in this section is to compute the temperature distributions in all regions at all times, as well as the movement of the solid-liquid interface. To accomplish this task, we will solve the equation for heat conduction in each material, subject to appropriate boundary conditions. Assume that all material properties are constant, that there is no liquid flow ($\boldsymbol{v}_\ell = 0$), which also implies that there is no viscous dissipation, and that there is no internal heat generation. The energy balance (Eq. (4.127)) for each material takes the reduced form

$$\frac{\partial T_\nu}{\partial t} = \alpha_\nu \frac{\partial^2 T_\nu}{\partial x^2} \tag{5.1}$$

where $\nu = m, s, \ell$ refers to the mold, solid or liquid, respectively, and $\alpha_\nu = k_\nu/\rho_\nu c_{p\nu}$ is the thermal diffusivity of the material or phase ν. The thermal diffusivities of the solid and liquid are normally within approximately 10% of each other. However, α_m may be much smaller than α_s and α_ℓ, for example when casting metals in sand molds. In other cases, α_m may be much larger than α_s and α_ℓ, such as for polymers freezing in metal molds. We will first solve the problem in its general form, and consider these special cases afterward. Since we intend to keep all the terms without making any judgments regarding their magnitude, there is no particular advantage to be gained from scaling the equations in the general form. Accordingly, the solution is presented in dimensional form. Nevertheless, the solution is naturally expressed in terms of dimensionless groups, which will appear in due course.

The governing equation, boundary and initial conditions for the mold domain are

$$\frac{\partial T_m}{\partial t} = \alpha_m \frac{\partial^2 T_m}{\partial x^2} \qquad -\infty < x \le 0 \tag{5.2}$$

$$T_m = T_0 \qquad x \to \infty \tag{5.3}$$

$$T_m = T_{ms} \qquad x = 0 \tag{5.4}$$

$$k_m \frac{\partial T_m}{\partial x} = k_s \frac{\partial T_s}{\partial x} \qquad x = 0 \tag{5.5}$$

$$T_m = T_0 \qquad t = 0 \tag{5.6}$$

Note that there are two boundary conditions at $x = 0$ as a result of the mold-solid interface temperature T_{ms} not actually being known at this point. The heat flux must be continuous across the interface, as stated in Eq. (5.5). The governing equations for the solid are

$$\frac{\partial T_s}{\partial t} = \alpha_s \frac{\partial^2 T_s}{\partial x^2} \qquad 0 \le x \le x^*(t) \tag{5.7}$$

$$T_s = T_{ms} \qquad x = 0 \tag{5.8}$$

$$k_m \frac{\partial T_m}{\partial x} = k_s \frac{\partial T_s}{\partial x} \qquad x = 0 \tag{5.9}$$

$$T_s = T_f \qquad x = x^*(t) \tag{5.10}$$

$$\rho_s L_f \frac{dx^*}{dt} = k_s \frac{\partial T_s}{\partial x} - k_\ell \frac{\partial T_\ell}{\partial x} \qquad x = x^*(t) \tag{5.11}$$

There is no initial condition for the solid, because the solidifying material is initially entirely liquid. Two boundary conditions are required at the solid-liquid interface because, although the temperature T_f is known, the position of the interface $x^*(t)$ is not. The second condition is the Stefan condition, first introduced as Eq. (4.133). Finally, for the liquid, we have

$$\frac{\partial T_\ell}{\partial t} = \alpha_\ell \frac{\partial^2 T_\ell}{\partial x^2} \qquad x^* \le x < \infty \tag{5.12}$$

$$T_\ell = T_\infty \qquad x \to \infty \tag{5.13}$$

$$T_\ell = T_f \qquad x = x^*(t) \tag{5.14}$$

$$\rho_s L_f \frac{dx^*}{dt} = k_s \frac{\partial T_s}{\partial x} - k_\ell \frac{\partial T_\ell}{\partial x} \qquad x = x^*(t) \tag{5.15}$$

$$T_\ell = T_\infty \qquad t = 0 \tag{5.16}$$

We have repeated the Stefan condition for the sake of clarity, because it applies to both the liquid and the solid.

Equation 5.1 admits the possibility of a *similarity solution*, where the partial differential equation reduces to an ordinary differential equation in terms of a new variable that combines both x and t. One can show by substitution into Eq. (5.1) that the following expression is a solution to the equation:

$$T_\nu(x,t) = A_\nu + B_\nu \text{erf}\left(\frac{x}{2\sqrt{\alpha_\nu t}}\right) \tag{5.17}$$

where the *error function* erf (u) is defined as

$$\operatorname{erf}(u) = \frac{2}{\sqrt{\pi}} \int_0^u e^{-r^2} dr; \qquad \frac{\partial(\operatorname{erf}(u))}{\partial x} = \frac{2}{\sqrt{\pi}} e^{-u^2} \frac{\partial u}{\partial x} \tag{5.18}$$

Note that $\operatorname{erf}(0) = 0$ and $\operatorname{erf}(\pm\infty) = \pm 1$. We will also occasionally use the *complementary error function*, $\operatorname{erfc}(u) = 1 - \operatorname{erf}(u)$. If coefficients A_ν and B_ν can be determined such that all of the boundary and initial conditions are satisfied, then Eq. (5.17) is a valid solution to the problem at hand. Such is the case for all three media (mold, solid and liquid) in this problem.

It is easy to demonstrate that the solution for the mold temperature that satisfies the boundary and initial conditions is

$$T_m(x,t) = T_{ms} + (T_{ms} - T_0)\operatorname{erf}\left(\frac{x}{2\sqrt{\alpha_m t}}\right) \qquad -\infty < x \leq 0 \tag{5.19}$$

Recall that, at this point, T_{ms} is still unknown. Proceeding in a similar way to construct the solution for the temperature in the solid, substitution of the boundary conditions at $x = 0$ and $x = x^*(t)$ into Eq. (5.17) yields

$$T_f = T_{ms} + B_s \operatorname{erf}\left(\frac{x^*(t)}{2\sqrt{\alpha_s t}}\right) \tag{5.20}$$

Since the left hand side of Eq. (5.20) is constant, and T_{ms} has been assumed to be constant also, then $x^*(t)$ must be proportional to $\sqrt{t}$ for a solution to exist, i.e.,

$$x^*(t) = 2\phi\sqrt{\alpha_s t} \tag{5.21}$$

Here, ϕ is an as yet unknown constant. Note that the interface speed is given by

$$\frac{dx^*}{dt} = \frac{\phi\sqrt{\alpha_s}}{\sqrt{t}} \tag{5.22}$$

The infinite speed at time zero arises from the discontinuity of the boundary and initial conditions, and is obviously not physical.

We can now determine B_s in terms of ϕ, with the result

$$T_s(x,t) = T_{ms} + \frac{T_f - T_{ms}}{\operatorname{erf}(\phi)} \operatorname{erf}\left(\frac{x}{2\sqrt{\alpha_s t}}\right) \qquad 0 \leq x \leq x^*(t) \tag{5.23}$$

By equating the heat fluxes in the solid and the mold at $x = 0$, Eq. (5.23) can be solved for T_{ms}

$$T_{ms} = \frac{T_0\sqrt{k_m \rho_m c_{pm}} \operatorname{erf}(\phi) + T_f\sqrt{k_s \rho_s c_{ps}}}{\sqrt{k_m \rho_m c_{pm}} \operatorname{erf}(\phi) + \sqrt{k_s \rho_s c_{ps}}} \tag{5.24}$$

The product $\sqrt{k_\nu \rho_\nu c_{p\nu}}$ is called the *effusivity* of material ν. Note that T_{ms} is indeed constant, which is consistent with the assumptions made at the start of the problem. The constant ϕ is still unknown, and will be determined next.

The solution for the temperature in the liquid is determined after substituting Eq. (5.21) into Eq. (5.15) for the interface position $x^*(t)$, and then using the boundary and initial conditions to obtain

$$T_\ell = T_\infty - \frac{T_\infty - T_f}{\text{erfc}\left(\phi\sqrt{\alpha_s/\alpha_\ell}\right)} \text{erfc}\left(\frac{x}{2\sqrt{\alpha_\ell t}}\right) \qquad x^*(t) < x < \infty \tag{5.25}$$

Next, the substitution of the expressions for T_s from Eq. (5.23) and T_ℓ from Eq. (5.25) into the Stefan condition, Eq. (5.11), produces the following transcendental equation for ϕ

$$\left\{\phi \exp\left(\phi^2\right) - \frac{c_{ps}(T_\infty - T_f)}{L_f\sqrt{\pi}} \frac{\exp\left([1 - \alpha_s/\alpha_\ell]\,\phi^2\right)}{\text{erfc}\left(\phi\sqrt{\alpha_s/\alpha_\ell}\right)} \sqrt{\frac{k_\ell \rho_\ell c_{p\ell}}{k_s \rho_s c_{ps}}}\right\} \times \left\{\text{erf}\,(\phi) + \sqrt{\frac{k_s \rho_s c_{ps}}{k_m \rho_m c_{pm}}}\right\} = \frac{c_{ps}(T_f - T_0)}{L_f\sqrt{\pi}} = \frac{\text{Ste}}{\sqrt{\pi}} \tag{5.26}$$

where we have identified the Stefan number, Ste $= c_{ps}(T_f - T_0)/L_f$. Although this expression appears rather complicated, everything in Eq. (5.26) is known except ϕ. The left hand side of Eq. (5.26) is a monotonic function of ϕ, thus the solution is easily found, for example, by graphical methods. (See Example 5.1 below.)

Taken together, Eqs. (5.19) and (5.21-5.26) constitute the solution for the temperature and interface position we were looking for. Notice that the solution has developed naturally in terms of dimensionless groups: the Stefan number, and ratios of the effusivities and thermal diffusivities. An example will help to clarify how the solution is used.

Example 5.1 Solidification of aluminum in a graphite mold

Consider the solidification of pure aluminum in a graphite mold. The material properties for the mold, solid and liquid are tabulated below. The mold and liquid are assumed to be sufficiently thick for the assumption that the media are semi-infinite to be valid. The initial temperature for the liquid is 700°C, and the mold has an initial temperature of 25°C. The freezing point for Al is $T_f = 660$°C.

Material	Density kg m^{-3}	Specific heat J kg^{-1} K^{-1}	Thermal conductivity W m^{-1} K^{-1}	Heat of fusion J kg^{-1}
Graphite	2200	1700	100	–
Aluminum (solid)	2555	1190	211	3.98×10^5
Aluminum (liquid)	2368	1090	91	–

Begin by finding ϕ. Substituting the appropriate values for this example problem into Eq. (5.26) yields

$$\left\{\phi \exp\left(\phi^2\right) - 0.041\frac{\exp\left(-0.968\phi^2\right)}{\operatorname{erfc}\left(1.403\phi\right)}\right\} \cdot \left\{\operatorname{erf}\left(\phi\right) + 1.304\right\} = f(\phi) = 1.071$$

The solution $\phi = 0.522$ can be obtained graphically with the aid of the plot of $f(\phi)$ given below. Next, find the interface temperature $T_{ms} = 474°\mathrm{C}$ by substituting the problem parameters and newly found value for ϕ into Eq. (5.24). Finally, when using these two values, the interface position and temperatures in the various materials at any time are

$$x^*(t) = 8.7 \times 10^{-3}\sqrt{t}$$

$$T_m = 474 + 449 \operatorname{erf}\left(97.13\frac{x}{\sqrt{t}}\right)$$

$$T_s = 474 + 344.4 \operatorname{erf}\left(60.02\frac{x}{\sqrt{t}}\right)$$

$$T_\ell = 700 - 133.3 \operatorname{erfc}\left(84.17\frac{x}{\sqrt{t}}\right)$$

with all lengths given in m, the time in s and the temperature in °C. The result is shown for $t = 10$ s in the figure below.

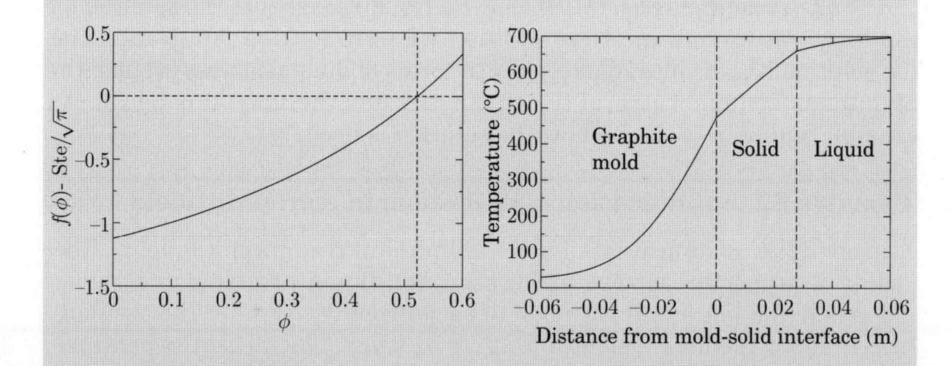

Notice that the slopes in the different materials do not match at the interfaces. This is due to the difference in conductivities between the mold and solid, and also to the liberation of latent heat at the solid-liquid interface. The temperature of the interface between the mold and the solid T_{ms} is also closer to the melting point of aluminum than to the initial temperature of the mold. This is mainly due to the fact that the effusivity of solid aluminum is large as compared to that of graphite.

Several dimensionless groups appear in the solution. There exist certain special cases of the general form, corresponding to asymptotic values of these dimensionless groups, which are useful to consider. Often, metals are cast with low superheat, in which case we can take $T_\infty \approx T_f$. One can partially compensate for the error in such an approximation by defining an effective latent heat L_{eff}, given by

$$L_{eff} = L_f + c_{p\ell}(T_\infty - T_f) \tag{5.27}$$

The general problem just considered can thus be somewhat simplified. Equation (5.24) remains unchanged. However, the solution for the temperature in the liquid is just $T_\ell = T_f$, and Eq. (5.26) also takes on a simpler form:

$$\phi \exp\left(\phi^2\right) \left\{ \operatorname{erf}(\phi) + \sqrt{\frac{k_s \rho_s c_{ps}}{k_m \rho_m c_{pm}}} \right\} = \frac{\mathrm{Ste}}{\sqrt{\pi}} \tag{5.28}$$

The assumption that $T_\ell = T_f$ is useful when estimating the solidification time of a slab of finite thickness. Once ϕ has been calculated, the freezing time t_f of a slab of thickness δ can be computed from Eq. (5.21) as

$$t_f = \frac{\delta^2}{4\phi^2 \alpha_s} \tag{5.29}$$

Note that this cannot be done when using the more general form, since the solution we found for the liquid assumes that the domain is semi-infinite. When there is no superheat, there is no conduction in the liquid, and Eq. (5.29) may be used.

Continuing under the assumption that $T_\infty = T_f$, there are two further special cases of interest. For some combinations of materials, the solid is a relatively poor conductor as compared to the mold, $k_s \ll k_m$. This might apply, for example, when a polymer solidifies in a metal mold, or when casting iron in a water-cooled metal mold. The density and specific heat of most materials are not all that different from each other, and thus $k_s \ll k_m$ implies that

$$\sqrt{\frac{k_s \rho_s c_{ps}}{k_m \rho_m c_{pm}}} \ll 1 \tag{5.30}$$

In this case, Eq. (5.24) can be simplified to $T_{ms} = T_0$, and Eq. (5.28) reduced even further to

$$\phi \exp\left(\phi^2\right) \operatorname{erf}(\phi) = \frac{\mathrm{Ste}}{\sqrt{\pi}} \tag{5.31}$$

Thus, for this case, the temperature in the mold and in the liquid are constant, and the temperature in the solid follows an error function form. Because the solidification rate is controlled by heat conduction in the solid, this case is sometimes referred to as "solid control."

The opposite case, where $k_m \ll k_s$, corresponds to a mold that is a relatively poor conductor as compared to the solid. An example is the casting of aluminum in a sand mold. For this case, Eq. (5.28) is first multiplied by the ratio $\sqrt{k_m \rho_m c_{pm}}/\sqrt{k_s \rho_s c_{ps}}$ ($\ll 1$), which gives

$$\phi \exp\left(\phi^2\right) \left\{ \operatorname{erf}(\phi) \sqrt{\frac{k_m \rho_m c_{pm}}{k_s \rho_s c_{ps}}} + 1 \right\} = \frac{c_{ps}(T_f - T_0)}{L_f \sqrt{\pi}} \sqrt{\frac{k_m \rho_m c_{pm}}{k_s \rho_s c_{ps}}} \tag{5.32}$$

Since $\operatorname{erf}(\phi)$ is of order one, the first term inside the brackets on the left hand side can be neglected for this case. The right hand side is also small, as it is multiplied by the small parameter, but it cannot be neglected without the entire solution being lost. However, for $\phi \exp(\phi^2) \ll 1$, one can approximate $\phi \exp(\phi^2) \approx \phi$ to obtain

$$\phi \approx \frac{c_{ps}(T_f - T_0)}{L_f \sqrt{\pi}} \sqrt{\frac{k_m \rho_m c_{pm}}{k_s \rho_s c_{ps}}} \tag{5.33}$$

Substituting this result into Eq. (5.21) gives

$$x^* = \frac{2(T_f - T_0)}{\rho_s L_f \sqrt{\pi}} \sqrt{k_m \rho_m c_{pm}} \sqrt{t} \tag{5.34}$$

Notice that the interface position still increases in proportion to $\sqrt{t}$. Dividing the numerator and denominator on the right hand side of Eq. (5.24) by $\sqrt{k_s \rho_s c_{ps}}$ and neglecting the diffusivity ratio leads to the result $T_{ms} = T_f$. Thus, in this case, the temperature in the solid and liquid are nearly constant, and the temperature distribution in the mold follows an error function. This type of solidification is sometimes referred to as "mold control."

5.2.2 Planar front solidification of a binary alloy

An analytical solution similar to the one developed in the previous section for the pure material can be found for the freezing of a binary alloy. Consider the solidification of a binary alloy, corresponding to composition C_0 in the phase diagram in Fig. 5.2. The alloy begins as a liquid at uniform temperature T_∞ above the liquidus temperature. The problem geometry is illustrated in Fig. 5.2. In order to highlight the new physics of the problem, take all of the thermal properties in the solid and liquid to be constant and equal, and replace conduction in the mold by a simple Dirichlet boundary condition at the mold-solid interface, $T(x = 0, t) = T_0$. This is not a requirement for the solution, but it does make it easier to follow. A closed form analytical solution exists for this problem, provided that $T_{sol} \leq T_0 < T_{liq}$.

The analysis begins by writing the equations to be solved for the temperature distributions in the solid and liquid phases. Once again, we assume that there is no fluid flow and no internal heat generation,

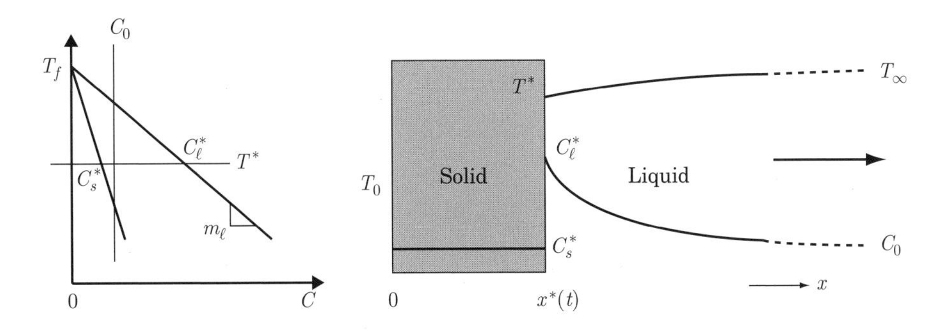

Fig. 5.2 A phase diagram, problem geometry and sketches of the expected temperature and composition profiles in the solid and liquid for one-dimensional solidification of a binary alloy.

leaving only conduction in both phases. We also assume that there is a flat interface between the two phases located at $x^*(t)$. Unlike the case of a pure material, the solid-liquid interface temperature T^* is unknown. However, we assume local equilibrium at the interface, so that the interfacial compositions of the solid and liquid correspond to the liquidus and solidus curves at temperature T^*, as shown in Fig. 5.2. Thus, the composition in the solid is given by $C_s^*(t) = k_0 C_\ell^*(t)$, where k_0 is the partition coefficient. Note that we have simply assumed that T^* is constant. If all of the boundary and initial conditions can be satisfied, then such a solution exists. Conversely, if no such solution exists, it will be impossible to satisfy all of the conditions. The heat conduction problem is then written as

$$\frac{\partial T_s}{\partial t} = \alpha \frac{\partial^2 T_s}{\partial x^2} \qquad 0 \leq x \leq x^*(t) \tag{5.35}$$

$$\frac{\partial T_\ell}{\partial t} = \alpha \frac{\partial^2 T_\ell}{\partial x^2} \qquad x^*(t) \leq x < \infty \tag{5.36}$$

$$T_s = T_0 \qquad x = 0 \tag{5.37}$$

$$T_s = T^* \qquad x = x^*(t) \tag{5.38}$$

$$\rho L_f \frac{dx^*}{dt} = k \left(\frac{\partial T_s}{\partial x} - \frac{\partial T_\ell}{\partial x} \right) \qquad x = x^*(t) \tag{5.39}$$

$$T_\ell = T_\infty \qquad x \to \infty \tag{5.40}$$

$$T_\ell = T_\infty \qquad t = 0 \tag{5.41}$$

$$x^* = 0 \qquad t = 0 \tag{5.42}$$

Note that since the thermal properties in the liquid and solid have been assumed to be equal, the subscripts on ρ, k, and α have been omitted.

For this problem, it is helpful to scale the governing equations in order to identify the important parameters. To that end, we define the following dimensionless variables. Note that there is no characteristic length in the problem, so we simply choose a symbol L and proceed with the scaling. It turns out that the length scale appears in each of the equations in such a way that it is eliminated.

$$\xi = \frac{x}{L}; \quad \xi^* = \frac{x^*}{L}; \quad \tau = \frac{\alpha t}{L^2}; \quad \theta = \frac{T - T_0}{T_\infty - T_0} \tag{5.43}$$

Two temperatures will become part of the solution: T_f, the equilibrium freezing temperature of the pure material, and T^*. (See Fig. 5.2.) Their scaled counterparts are defined as

$$\theta_f = \frac{T_f - T_0}{T_\infty - T_0}; \quad \theta^* = \frac{T^* - T_0}{T_\infty - T_0} \tag{5.44}$$

Substituting the scaled variables into the governing equations yields

$$\frac{\partial \theta_s}{\partial \tau} = \frac{\partial^2 \theta_s}{\partial \xi^2} \qquad 0 \le \xi \le \xi^*(\tau) \tag{5.45}$$

$$\frac{\partial \theta_\ell}{\partial \tau} = \frac{\partial^2 \theta_\ell}{\partial \xi^2} \qquad \xi^*(\tau) \le \xi < \infty \tag{5.46}$$

$$\theta_s = 0 \qquad \xi = 0 \tag{5.47}$$

$$\theta_s = \theta^* \qquad \xi = \xi^*(\tau) \tag{5.48}$$

$$\frac{1}{\text{Ste}} \frac{d\xi^*}{d\tau} = \frac{\partial \theta_s}{\partial \xi} - \frac{\partial \theta_\ell}{\partial \xi} \qquad \xi = \xi^*(\tau) \tag{5.49}$$

$$\theta_\ell = 1 \qquad \xi \to \infty \tag{5.50}$$

$$\theta_\ell = 1 \qquad \tau = 0 \tag{5.51}$$

$$\xi^* = 0 \qquad \tau = 0 \tag{5.52}$$

where the Stefan number is defined as $\text{Ste} = c_p(T_\infty - T_0)/L_f$.

Similarity solutions can be found for the temperatures in the solid and liquid. This time we simply write them down without derivation

$$\theta_s = \frac{\theta^*}{\text{erf}\,(\phi)} \text{erf}\left(\frac{\xi}{2\sqrt{\tau}}\right) \tag{5.53}$$

$$\theta_\ell = 1 - \frac{1 - \theta^*}{\text{erfc}\,(\phi)} \text{erfc}\left(\frac{\xi}{2\sqrt{\tau}}\right) \tag{5.54}$$

$$\xi^* = 2\phi\sqrt{\tau} \tag{5.55}$$

where ϕ is a constant to be determined later. Note also that the assumption that θ^* is constant requires that ξ^* be proportional to $\sqrt{\tau}$, as can be seen by setting $\xi = \xi^*$ in Eqs. (5.53) and (5.54). Note also that given $\xi^* = 2\phi\sqrt{\tau}$, one can write the temperature distributions in terms of ξ^*

$$\theta_s = \frac{\theta^*}{\text{erf}\,(\phi)} \text{erf}\left(\frac{\phi\xi}{\xi^*}\right) \tag{5.56}$$

$$\theta_\ell = 1 - \frac{1 - \theta^*}{\text{erfc}\,(\phi)} \text{erfc}\left(\frac{\phi\xi}{\xi^*}\right) \tag{5.57}$$

This is convenient for constructing graphical representations of the solutions. Substituting these expressions for θ_s and θ_ℓ into Eq. (5.49) gives one of the two necessary relations between the two unknowns θ^* and ϕ:

$$\theta^* = \text{erf}\,(\phi)\left(1 + \frac{\sqrt{\pi}}{\text{Ste}}\phi\exp(\phi^2)\text{erfc}\,(\phi)\right) \tag{5.58}$$

A second relation comes from the corresponding solute problem. The equations to be satisfied by the compositions in the solid and liquid were derived in Chap. 4:

$$\frac{\partial C_s}{\partial t} = D_s\frac{\partial^2 C_s}{\partial x^2} \qquad 0 \le x \le x^*(t) \tag{5.59}$$

$$\frac{\partial C_\ell}{\partial t} = D_\ell\frac{\partial^2 C_\ell}{\partial x^2} \qquad x^*(t) \le x < \infty \tag{5.60}$$

The boundary and initial conditions on the composition are

$$\frac{\partial C_s}{\partial x} = 0 \qquad x = 0 \tag{5.61}$$

$$C_s^* = k_0 C_\ell^* \qquad x = x^*(t) \tag{5.62}$$

$$D_s\frac{\partial C_s}{\partial x} - D_\ell\frac{\partial C_\ell}{\partial x} = C_\ell^*(1-k_0)\frac{dx^*}{dt} \qquad x = x^*(t) \tag{5.63}$$

$$C_\ell = C_0 \qquad x \to \infty \tag{5.64}$$

$$C_\ell = C_0 \qquad t = 0 \tag{5.65}$$

There is no initial condition on the composition in the solid, because the material is entirely liquid in its initial state.

The composition, since it is expressed as a mass fraction, is already dimensionless and order one, so we leave it as is. Applying the scaling given in Eq. (5.43), the dimensionless forms for the governing equations, boundary and initial conditions are

$$\frac{\partial C_s}{\partial \tau} = \frac{D_s}{\alpha}\frac{\partial^2 C_s}{\partial \xi^2} \qquad 0 \le \xi \le \xi^*(\tau) \tag{5.66}$$

$$\frac{\partial C_\ell}{\partial \tau} = \frac{D_\ell}{\alpha}\frac{\partial^2 C_\ell}{\partial \xi^2} \qquad \xi^*(\tau) \le \xi < \infty \tag{5.67}$$

$$\frac{\partial C_s}{\partial \xi} = 0 \qquad \xi = 0 \tag{5.68}$$

$$C_s^* = k_0 C_\ell^* \qquad \xi = \xi^*(\tau) \tag{5.69}$$

$$\frac{D_\ell}{\alpha}\left(\frac{D_s}{D_\ell}\frac{\partial C_s}{\partial \xi} - \frac{\partial C_\ell}{\partial \xi}\right) = C_\ell^*(1-k_0)\frac{d\xi^*}{d\tau} \qquad \xi = \xi^*(\tau) \tag{5.70}$$

$$C_\ell = C_0 \qquad \xi \to \infty \tag{5.71}$$

$$C_\ell = C_0 \qquad \tau = 0 \tag{5.72}$$

Typically, $D_s \ll D_\ell$ and $D_\ell \ll \alpha$. It is tempting, then, to neglect the right hand sides of both Eqs. (5.66) and (5.67). If we do so, however, the boundary conditions on C_ℓ cannot be satisfied. On the other hand, one *can* neglect the spatial terms in the solid, because $C_s = cst$ is a solution to the differential equation and boundary conditions. Further, the term involving D_s/D_ℓ in Eq. (5.70) can be neglected because the other term inside the parenthesis is order one. This decouples the problems for the composition in the solid and liquid, leaving the following system of equations to be solved for C_ℓ

$$\frac{\partial C_\ell}{\partial \tau} = \mathrm{Le}^{-1}\frac{\partial^2 C_\ell}{\partial \xi^2} \qquad \xi^*(\tau) \le x < \infty \tag{5.73}$$

$$-\mathrm{Le}^{-1}\frac{\partial C_\ell}{\partial \xi} = C_\ell^*(1-k_0)\frac{d\xi^*}{d\tau} \qquad \xi = \xi^*(\tau) \tag{5.74}$$

$$C_\ell = C_0 \qquad \xi \to \infty \tag{5.75}$$

$$C_\ell = C_0 \qquad \tau = 0 \tag{5.76}$$

where the *Lewis number* $\mathrm{Le} = \alpha/D_\ell$ represents the ratio of thermal diffusivity and diffusion coefficient. Finally, note that the phase diagram couples the temperature field to the composition field. Choosing (for the sake of convenience) a linear form for the liquidus curve, $T = T_f + m_\ell C_\ell$, and introducing the notation $C_\ell^* = C_\ell(\xi^*(\tau), \tau)$ gives

$$C_\ell^* = \frac{T_0 - T_\infty}{m_\ell}(\theta_f - \theta^*) \tag{5.77}$$

We now have a set of equations that can be solved for C_ℓ. Subsequently, C_s can be recovered from Eq. (5.69). The solution for the liquid composition satisfying Eqs. (5.73), (5.75) and (5.76) is

$$C_\ell = C_0 + \frac{C_\ell^* - C_0}{\mathrm{erfc}\left(\phi\sqrt{\mathrm{Le}}\right)}\mathrm{erfc}\left(\frac{\xi\sqrt{\mathrm{Le}}}{2\sqrt{\tau}}\right) \tag{5.78}$$

As noted earlier for the temperatures, Eq. (5.78) can be written in terms of ξ^* instead of τ,

$$C_\ell = C_0 + \frac{C_\ell^* - C_0}{\mathrm{erfc}\left(\phi\sqrt{\mathrm{Le}}\right)}\mathrm{erfc}\left(\frac{\phi\xi\sqrt{\mathrm{Le}}}{\xi^*}\right) \tag{5.79}$$

Combining Eqs. (5.58), (5.74) and (5.77) gives a transcendental equation to evaluate θ^*

$$\left\{\theta_f - \mathrm{erf}\,(\phi)\left(1 + \frac{\sqrt{\pi}}{\mathrm{Ste}}\phi\exp(\phi^2)\mathrm{erfc}\,(\phi)\right)\right\} \times$$
$$\left\{1 - \sqrt{\pi\mathrm{Le}}(1-k_0)\phi\exp\left(\phi^2\mathrm{Le}\right)\mathrm{erfc}\left(\phi\sqrt{\mathrm{Le}}\right)\right\} = \frac{m_\ell C_0}{T_0 - T_\infty} \tag{5.80}$$

Once again, although this equation appears complicated, the left hand side is a monotonic function of ϕ, which is the only unknown. The solution can thus proceed by first determining ϕ from Eq. (5.80), then θ^* from Eq. (5.58), and finally all of the temperature and composition solutions follow. This can be illustrated by an example.

Example 5.2 Solidification of an Fe-C alloy

Consider the solidification of an Fe-C alloy in a mold, corresponding to the geometry considered in this section. The material property data are tabulated below, along with the boundary and initial condition data, and computed values of the important dimensionless groups.

Substituting the values in the table into Eq. (5.80), and then solving graphically for ϕ, as in Example 5.1, yields $\phi = 0.1159$. Substituting this value for ϕ in Eq. (5.58) gives $\theta^* = 0.39$ and inserting in turn θ^* in Eq. (5.77) yields $C_\ell^* = 0.200$, and Eq. (5.69) gives $C_s = 0.034$. Once these values are known, one can plot the resulting temperature and composition profiles at various positions, using Eqs. (5.56), (5.57) and (5.79). The result appears below, where the temperatures and compositions in the solid and liquid are shown for several different interface positions. Notice that the composition gradient in the liquid ahead of the interface decreases with the distance from the wall, as a result of the decreased interface velocity there.

Material properties			
Nominal composition	C_0	0.05	mass %
Liquidus slope	m_ℓ	−81.1	K (mass %)$^{-1}$
Solidus slope	m_s	−478	K (mass %)$^{-1}$
Specific heat	c_p	820	J kg^{-1} K^{-1}
Density	ρ	7×10^3	kg m^{-3}
Latent heat	L_f	2.76×10^5	J kg^{-1}
Iron freezing point	T_f	1538	°C
Boundary and initial condition data			
Initial temperature	T_∞	1540	°C
Wall temperature	T_0	1510	°C
Derived quantities			
Segregation coefficient	k_0	0.17	–
Stefan number	Ste	0.089	–
Lewis number	Le	300	–
Dimensionless freezing point	θ_f	0.933	–

Computed dimensionless quantities			
Interface position parameter	ϕ	0.1159	–
Interface composition	C_ℓ^*	0.200	–
Solid composition	C_s	0.034	–
Interface temperature	θ^*	0.39	–

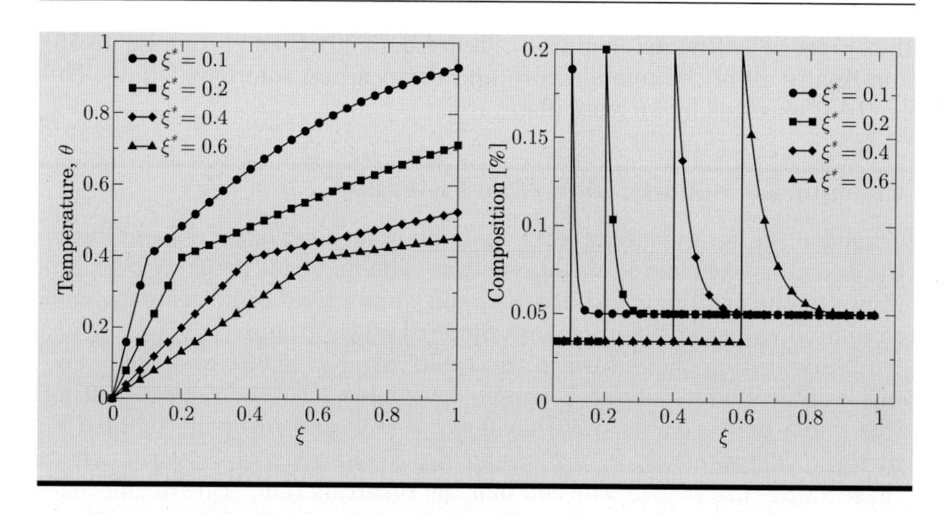

In later chapters, when we discuss microstructure development, two quantities of foremost importance will be the temperature gradient in the liquid at the interface and the interface velocity. These quantities can be evaluated from the solutions developed above. Differentiating Eq. (5.54) with respect to ξ and evaluating it at $\xi = \xi^*$ gives

$$\left.\frac{\partial \theta_\ell}{\partial \xi}\right|_{\xi^*} = \frac{1-\theta^*}{\text{erfc}(\phi)}\frac{2}{\sqrt{\pi}}\exp\left(-\phi^2\right)\frac{\phi}{\xi^*} \tag{5.81}$$

Similarly, differentiating Eq. (5.55) with respect to τ gives the dimensionless interface velocity

$$\frac{d\xi^*}{d\tau} = \frac{2\phi^2}{\xi^*} \tag{5.82}$$

The ratio of the temperature gradient in the liquid to the interface velocity turns out to be an important parameter in microstructure calculations. Using Eqs. (5.81) and (5.82) to compute this ratio gives

$$\frac{\partial\theta_\ell/\partial\xi|_{\xi^*}}{d\xi^*/d\tau} = \frac{1-\theta^*}{\sqrt{\pi}\text{erfc}(\phi)}\frac{\exp\left(-\phi^2\right)}{\phi} \tag{5.83}$$

It is remarkable that this ratio is constant for all locations of the interface. This is a consequence of the fact that the temperature gradient and velocity both decrease with distance in the same way, i.e., in proportion to $1/\xi^*$.

5.2.3 Transient solidification of a binary alloy at constant velocity

In the previous section, we considered a binary alloy freezing in contact with a wall maintained at constant temperature. The temperature gradient and velocity changed continuously as the solidification front advanced.

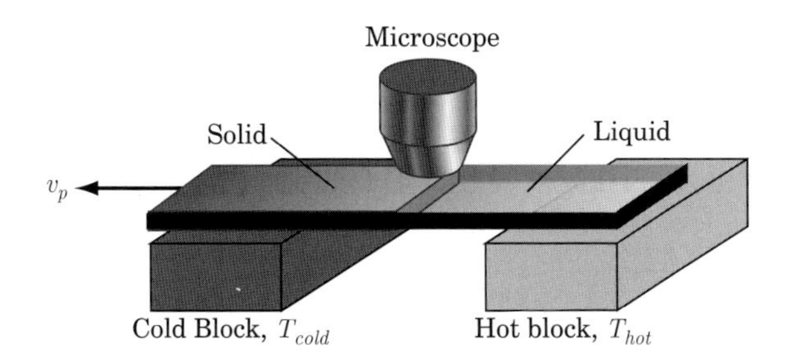

Fig. 5.3 A schematic drawing of a laboratory scale directional solidification apparatus.

Because these two quantities are important for microstructure development, a standard experiment has been devised in which they can be controlled independently. A temperature gradient is established across a sample, which is then drawn through the temperature field at constant velocity v_p. This experimental arrangement is usually referred to as *directional solidification*, and it is commonly used to grow crystals, particularly for semiconductors and aircraft turbine blades. The experiment can be adapted to transparent systems, using an apparatus such as that shown schematically in Fig. 5.3, to observe the solidification front during growth.

An analytical model of the directional solidification process was first presented by Smith, Tiller and Rutter [9]. These authors began their derivation by making the following assumptions:

1. The solidification process is one-dimensional, and the planar interface moves with a constant, prescribed velocity v_p.

2. All material properties are constant and equal in both the liquid and solid phases.

3. The sample is very long, in comparison to the size of the solute boundary layer ahead of the interface, and the domain is thus considered to be semi-infinite.

4. The temperature field is undisturbed by the movement of the interface. This is sometimes called the *frozen temperature approximation*, and is essentially a statement concerning the relative magnitudes of the terms in the Stefan condition, $\rho_s L_f v_n^* \ll k_{s,\ell} \nabla T_{s,\ell} \cdot \mathbf{n}$.

5. Diffusion in the solid is negligible. The justification for this assumption was presented in the preceding section via scaling analysis, and relies on the fact that $D_s \ll D_\ell$. It is also assumed that D_ℓ is constant.

6. There is no flow in the liquid. Note that this is consistent with the previous assumption that the densities of the solid and liquid are equal.

7. Local equilibrium applies at the planar interface, and the segregation coefficient k_0 is constant.

Although this may seem like many assumptions, note that they are quite reasonable in light of the previous analyses in this chapter.

The diffusion equation for the composition in the liquid must be solved:

$$\frac{\partial C_\ell}{\partial t} = D_\ell \frac{\partial^2 C_\ell}{\partial x^2} \tag{5.84}$$

Since the motion of the interface is prescribed as $x^* = v_p t$, it is convenient to recast this equation in the reference frame moving with the interface. We thus make the coordinate transformation $z = x - v_p t$, i.e., x is the absolute coordinate along the length of the specimen, whereas z is the relative position with respect to the interface position. In the new coordinate frame, Eq. (5.84) becomes

$$\frac{\partial C_\ell}{\partial t} - v_p \frac{\partial C_\ell}{\partial z} = D_\ell \frac{\partial^2 C_\ell}{\partial z^2} \tag{5.85}$$

The boundary conditions at the solid-liquid interface ($z = 0$) correspond to local thermodynamic equilibrium and solute balance. These are combined into one boundary condition on C_ℓ^*,

$$-D_\ell \frac{\partial C_\ell}{\partial z} = (C_\ell^* - C_s^*)v_p = C_\ell^*(1 - k_0)v_p \qquad z = 0 \tag{5.86}$$

where C_ℓ^* is the (unknown) liquid composition at the solid-liquid interface. The initial composition is $C_\ell = C_0$, and $C_\ell = C_0$ for $z \to \infty$.

Smith et al. derived an analytical solution to this system of equations using Laplace transforms. The development is easier to follow in scaled form, so we introduce the dimensionless variables:

$$\zeta = \frac{z}{L}; \quad \tau = \frac{D_\ell}{L^2}t; \quad L = \frac{D_\ell}{v_p} \tag{5.87}$$

The characteristic length scale $L = D_\ell / v_p$ is chosen to provide a simple form for the equations. Substituting the dimensionless variables into Eqs. (5.85) and (5.86) yields

$$\frac{\partial C_\ell}{\partial \tau} - \frac{\partial C_\ell}{\partial \zeta} = \frac{\partial^2 C_\ell}{\partial \zeta^2} \tag{5.88}$$

and

$$-\frac{\partial C_\ell}{\partial \zeta} = C_\ell^*(1 - k_0) \qquad \zeta = 0 \tag{5.89}$$

The Laplace transform $\bar{C}_\ell(s)$, and its inverse are defined as

$$\bar{C}_\ell(s) = \int_0^\infty C_\ell(\tau) e^{-s\tau} d\tau \; ; \qquad C_\ell(t) = \frac{1}{2\pi i} \lim_{M \to \infty} \int_{p-iM}^{p+iM} \bar{C}_\ell(s) e^{st} ds \tag{5.90}$$

Taking the Laplace transform of Eq. (5.88) (this is left as an exercise) yields an ordinary differential equation for $\bar{C}_\ell$

$$\frac{d^2\bar{C}_\ell}{d\zeta^2} + \frac{d\bar{C}_\ell}{d\zeta} - s\bar{C}_\ell = -C_0 \tag{5.91}$$

The initial condition $C_\ell = C_0$ has been used to obtain the right hand side. The Laplace transforms of the boundary condition at ∞ is

$$\bar{C}_\ell = \frac{C_0}{s} \qquad \zeta \to \infty \tag{5.92}$$

The solution for $\bar{C}_\ell$ can be obtained in a fairly straightforward manner by using the boundary conditions at the interface (Eq. (5.89)) and at infinity, with the result

$$\bar{C}_\ell = \frac{C_0}{s}\left[1 + \frac{2(1-k_0)}{2k_0 - 1 + \sqrt{1+4s}} \exp\left(-\frac{1+\sqrt{1+4s}}{2}\zeta\right)\right] \tag{5.93}$$

The inverse transform can be found in the tabulation of Campbell and Foster [2], given by

$$\begin{aligned} C_\ell = C_0 + 2C_0 \exp\left(-\frac{\zeta}{2}\right) \times \\ \left[\frac{1-k_0}{2k_0}\exp\left(-\frac{\zeta}{2}\right)\operatorname{erfc}\left(\frac{\zeta}{2\sqrt{\tau}} - \frac{\sqrt{\tau}}{2}\right) - \frac{1}{2}\exp\left(\frac{\zeta}{2}\right)\operatorname{erfc}\left(\frac{\zeta}{2\sqrt{\tau}} + \frac{\sqrt{\tau}}{2}\right)\right. \\ \left. + \frac{2k_0-1}{2k_0}\exp\left(\frac{2k_0-1}{2}\zeta + k_0(k_0-1)\tau\right)\operatorname{erfc}\left(\frac{\zeta}{2\sqrt{\tau}} + \frac{(2k_0-1)\sqrt{\tau}}{2}\right)\right] \end{aligned} \tag{5.94}$$

Redimensionalizing the solution in Eq. (5.94), using Eq. (5.87) yields, after some manipulation,

$$\begin{aligned} \frac{C_\ell}{C_0} = 1 + \frac{1-k_0}{2k_0}\exp\left(-\frac{v_p z}{D_\ell}\right)\operatorname{erfc}\left(\frac{z - v_p t}{2\sqrt{D_\ell t}}\right) - \frac{1}{2}\operatorname{erfc}\left(\frac{z + v_p t}{2\sqrt{D_\ell t}}\right) \\ + \left(1 - \frac{1}{2k_0}\right)\exp\left(-\frac{(1-k_0)v_p(z + k_0 v_p t)}{D_\ell}\right)\operatorname{erfc}\left(\frac{z + (2k_0-1)v_p t}{2\sqrt{D_\ell t}}\right) \end{aligned} \tag{5.95}$$

The composition in the solid at any position $x = z + v_p t$ can then be determined using the condition of local equilibrium at the interface,

$$\begin{aligned} C_s(x) &= k_0 \left. C_\ell \right|_{z=0} \\ &= \frac{C_0}{2}\left\{1 + \operatorname{erf}\left(\frac{1}{2}\sqrt{\frac{v_p x}{D_\ell}}\right)\right. \\ &\quad \left. + (2k_0 - 1)\exp\left(-\frac{k_0(1-k_0)v_p x}{D_\ell}\right)\operatorname{erfc}\left(\frac{2k_0-1}{2}\sqrt{\frac{v_p x}{D_\ell}}\right)\right\} \end{aligned} \tag{5.96}$$

An example of the development of the initial transient is shown in Fig. 5.4. The figure shows the composition profile in the liquid at several times near the beginning of solidification, as well as the composition profile in the solid (dashed line). The abscissa uses the scaled coordinate $v_p x/D_\ell$ since it appears naturally in Eq. (5.96). A typical value for directional solidification has $D_\ell/v_p \approx 1$ mm. Notice that the solid composition gradually increases from $C_s = k_0 C_0$ at $t = 0$ until it reaches the average alloy composition C_0 at steady state. This can also be seen from the analytical form, by examining the behavior of Eq. (5.96) for large x. As $x \to \infty$, the second (erf) term in the brackets goes to one, and the last term goes to zero, leaving $C_s = C_0$. One can examine a similar limit for the liquid composition in Eq. (5.95). It is more convenient to leave the variable z, and to examine the limit for $t \to \infty$. The last two terms go to zero, as both $\exp(-\infty)$ and $\mathrm{erfc}\,(\infty)$ are zero. In the second term, $\mathrm{erfc}\,(-\infty) \to 2$, thus leaving

$$\lim_{t\to\infty} C_\ell = C_0 \left\{ 1 + \frac{1-k_0}{k_0} \exp\left(-\frac{v_p z}{D_\ell} \right) \right\} \tag{5.97}$$

This result can also be determined by considering the steady state after the initial transient, i.e., Eq. (5.85) with the time derivative set to zero and the boundary condition in Eq. (5.86). This is left as an exercise. The important feature of the steady solution in Eq. (5.97), which will later be used for understanding microstructure formation, is the solute boundary layer ahead of the interface. Its characteristic length is D_ℓ/v_n^* (in this case, $v_n^* = v_p$). Note that in the steady solution at $z = 0$, $C_\ell = C_0/k_0$, which implies that $C_s = C_0$. This result is independent of v_p and D_ℓ, which means that a steady state planar front grows at the solidus temperature in the

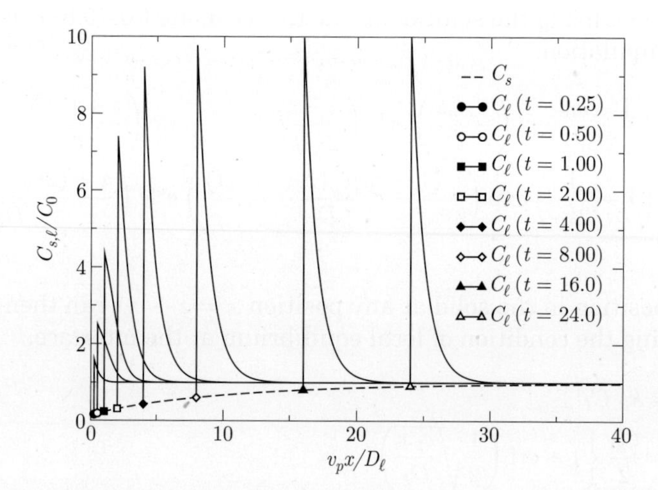

Fig. 5.4 The development of the initial transient composition profile for a binary alloy with $k_0 = 0.1$ solidified at constant velocity v_p.

equilibrium phase diagram. Furthermore, the solute gradient at the interface position, designated G_c, can be easily calculated (see Exercise 5.2).

$$G_c = \left(\frac{\partial C_\ell}{\partial z}\right)_{z=0} = \frac{C_0(1 - 1/k_0)}{D_\ell/v_p} \tag{5.98}$$

For a linear phase diagram, one therefore has $m_\ell G_c = \Delta T_0 (D_\ell/v_p)^{-1}$, where ΔT_0 is the equilibrium solidification interval of the alloy.

The extent of the initial transient can be estimated by examining the behavior of the third term within brackets in Eq. (5.96). For large values of x, and $k_0 \ll 1$, this term goes to zero approximately as $\exp(-k_0 v_p x/D_\ell)$. Thus, the contribution of this term becomes negligible at a distance of about $(3/k_0)(D_\ell/v_p)$. This observation is borne out by graphing C_s for various values of k_0, as shown in Fig. 5.5. The data are plotted on a logarithmic scale to emphasize the length of the initial transient. Note that for very small values of k_0, such as one finds in ice-salt systems, the length of the initial transient can be quite long.

In a finite length sample, say of length L, there is also a final transient region, corresponding to the end of solidification. When the diffusion tail ahead of the interface, with a length of roughly $3D_\ell/v_p$, reaches the end of the sample, the solute begins to build up ahead of the interface, since it has nowhere else to go. This makes the composition in the solid increase. Smith et al. presented the solution for this aspect of the problem, which they obtained by creating a set of "image" solute fields across the boundary $x = L$ to enforce the no-flux boundary condition. The result is given by

$$\frac{C_s}{C_0} = 1 + \sum_{n=1}^{\infty} (2n+1) \left(\frac{\prod_{m=1}^{n} (m - k_0)}{\prod_{m=1}^{n} (m + k_0)} \right) \exp\left(-\frac{n(n+1) v_p (L - x)}{D_\ell} \right) \tag{5.99}$$

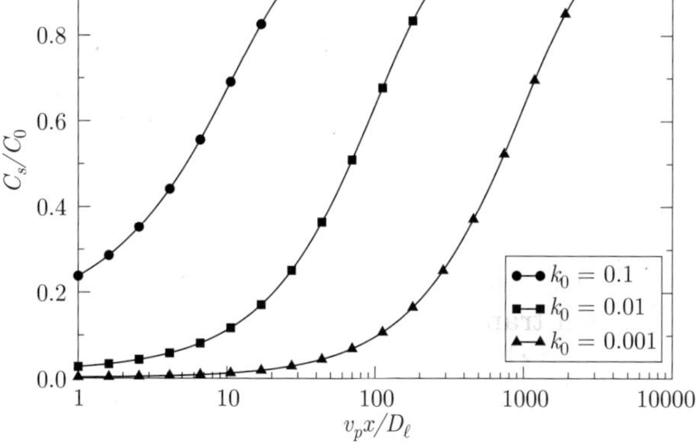

Fig. 5.5 The computed initial solid composition profile for binary alloys with different values of k_0, solidified at constant velocity.

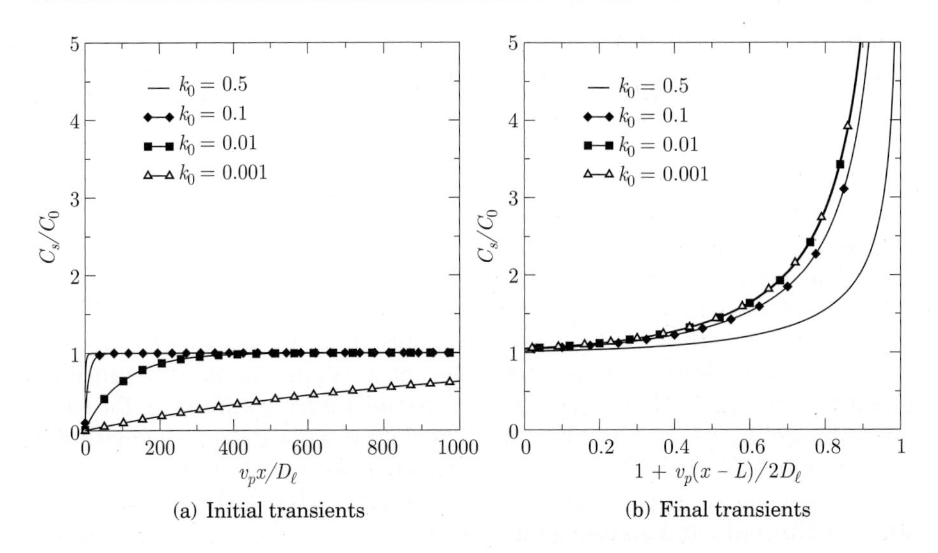

Fig. 5.6 The computed initial and final solid composition profiles for binary alloys with different values of k_0, solidified at constant velocity. The data in this figure correspond to $D_\ell/v_p = 1$ mm.

Notice that the length of the final transient should be roughly $3D_\ell/v_p$, and that only the magnitude of the composition, and not the length of the transient, should depend on k_0. A graph of the final transient can been seen in Fig. 5.6 for varying values of k_0. The corresponding initial transients are also shown for the sake of comparison. Note, however, that the length scales differ by a factor of 1000 between the two figures. According to the solution properties, the area under each curve over the entire length of the sample, i.e., the total solute content, must be equal to LC_0. The alloys with shorter initial transients (larger k_0) thus have smaller compositions in their final transients.

There is one more aspect of this problem that is very interesting to examine. Suppose that the solidification process is in the steady state regime, and that there is a sudden change of velocity from v_p to v_p'. Physically, what occurs is that the interface moves ahead faster than the solute is able to diffuse away, and as a result the composition in the liquid ahead of the interface increases. The solid composition increases in proportion. Eventually, a steady state is reached at the higher velocity. Since the steady state is characterized by $C_\ell = C_0/k_0$ and $C_s = C_0$ for all v_p, this implies that there is a transient "bump" in the composition, beginning at the interface position at the time of the speed change.

Smith et al. solved this problem in a similar way to the initial transient problem, by using Laplace transforms. This time, though, the initial condition for the composition in the liquid is given by Eq. (5.97) instead of a constant value of C_0. The solution for the solid composition is written in

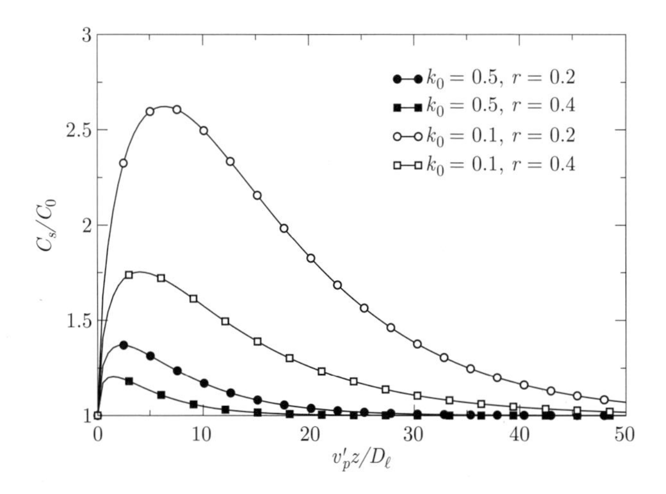

Fig. 5.7 The computed solid composition profile for binary alloys after a sudden increase in speed, for different values of k_0. The parameter $r = v_p/v'_p$.

terms of the parameters $r = v_p/v'_p$ and z, the distance measured from the interface position at the time of the speed change

$$\begin{aligned}\frac{C_s}{C_0} =& 1 - \frac{1}{2}\text{erfc}\left(\frac{1}{2}\sqrt{\frac{v'_p z}{D_\ell}}\right)\\ &+ \left(\frac{1-k_0}{2}\right)\left(\frac{1-2r}{k_0 - r}\right)\exp\left(-\frac{r(1-r)v'_p z}{D_\ell}\right)\text{erfc}\left(\frac{2r-1}{2}\sqrt{\frac{v'_p z}{D_\ell}}\right)\\ &+ \left(\frac{2k_0-1}{2}\right)\left(\frac{1-r}{k_0 - r}\right)\exp\left(-\frac{k_0(1-k_0)v'_p z}{D_\ell}\right)\text{erfc}\left(\frac{2k_0-1}{2}\sqrt{\frac{v'_p z}{D_\ell}}\right)\end{aligned} \tag{5.100}$$

The result is plotted in Fig. 5.7 for various values of k_0 and r. Notice that the perturbation eventually disappears, re-establishing the steady state. Such a phenomenon is responsible, for example, for "white bands" that appear in continuous casting of steels: a sharp increase in phosphorus content caused by a sudden increase in cooling rate due to sprays of water at the exit of the casting mold.

Variations on a theme ...

There are several other interesting problems in the same genre as those presented in detail earlier in this section. One such problem is called *zone melting*, in which a relatively short section (zone) of a solid rod is melted by application of local heating, after which the heater traverses the rod at constant velocity. This process is similar to the directional solidification process described in the previous section in that the solute is driven toward one end of the rod. The process is typically performed using multiple passes, which has the effect of moving the impurities to one end of the sample, while the rest of the sample can reach very high levels of purity.

The analysis is complicated, because the initial condition for each pass is the solute distribution at the end of the previous one. We leave as an exercise the calculation of the solute profile after the first pass, starting with a uniform composition alloy. The process was invented by Pfann [8], who presented computed curves showing the purification process for various values of the parameters. It is still commonly used to purify materials.

Both of the problems we considered for alloy solidification assumed that the diffusive transport in the solid could be neglected. This is not always the case, and there are many analyses where it is included. Such a phenomenon is normally referred to as *back diffusion*, and we will treat this subject in detail in Chap. 10. There may also be convection in the melt ahead of the interface. These phenomena interact to produce solid composition profiles that differ from those above, and there are several special cases of interest. The models are most useful for the study of segregation, and we therefore reserve their discussion for later chapters.

5.3 SOLIDIFICATION IN AN UNDERCOOLED MELT

In all of the preceding problems, the melt ahead of the interface was maintained at a higher temperature than that of the solid. Heat therefore flows from the liquid, through the interface, and through the solid. It is also possible for the melt to be *supercooled*, i.e., for the liquid to be at a temperature lower than the equilibrium freezing point. In this case, heat flows from the interface into the liquid. We consider several such problems in this section.

An example of this phenomenon occurs in *atomization* processes, where fine droplets of liquid are fractionated and cooled by a jet of air or an inert gas. Because of their small size, the temperature inside the droplets is nearly uniform. It is often observed that the droplets remain liquid down to temperatures substantially lower than the equilibrium freezing point. At some undercooled temperature, nucleation of the solid phase occurs, usually on the surface of the droplet. The temperature near the interface quickly recalesces to near the equilibrium freezing temperature, after which the solidification proceeds into the undercooled interior of the droplet.

5.3.1 Planar front growth

Consider the solidification of a pure material with a planar front growing into an undercooled melt. The physical system is illustrated in Fig. 5.8. The semi-infinite melt begins at temperature $T_\infty < T_f$, and at time $t = 0$

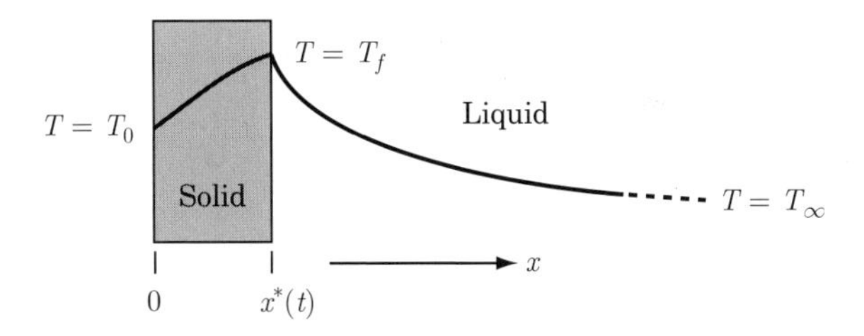

Fig. 5.8 A schematic drawing of the solidification of a pure material in an undercooled melt.

the left hand boundary ($x = 0$) is cooled to temperature T_s and solidification begins. In this case, heat flows away from the interface in both the solid and the liquid, as sketched in Fig. 5.8. The goal is to determine the evolution of the interface. As before, we assume that all of the material properties are constant and that there is no flow in the melt and that $\rho_s = \rho_\ell = \rho$.

The form of this problem should be familiar by now. We are to solve the heat conduction equation in both phases, subject to Dirichlet boundary conditions at the left and right hand boundaries, and the Stefan condition at the solid-liquid interface. Writing these out for the sake of completeness, we have

$$\frac{\partial T_s}{\partial t} = \alpha_s \frac{\partial^2 T_s}{\partial x^2} \qquad 0 \le x \le x^*(t) \tag{5.101}$$

$$\frac{\partial T_\ell}{\partial t} = \alpha_\ell \frac{\partial^2 T_\ell}{\partial x^2} \qquad x^*(t) \le x < \infty \tag{5.102}$$

$$T_s = T_0 \qquad x = 0 \tag{5.103}$$

$$T_\ell = T_\infty \qquad x \to \infty \tag{5.104}$$

$$T_s = T_\ell = T_f \qquad x = x^*(t) \tag{5.105}$$

$$k_s \frac{\partial T_s}{\partial x} - k_\ell \frac{\partial T_\ell}{\partial x} = \rho L_f \frac{dx^*}{dt} \qquad x = x^*(t) \tag{5.106}$$

Following the procedure from earlier sections, we can construct similarity solutions for T_s and T_ℓ,

$$T_s = T_0 + \frac{T_f - T_0}{\operatorname{erf}(\phi)} \operatorname{erf}\left(\frac{x}{2\sqrt{\alpha_s t}}\right) \tag{5.107}$$

$$T_\ell = T_\infty + \frac{T_f - T_\infty}{\operatorname{erfc}\left(\phi\sqrt{\alpha_s/\alpha_\ell}\right)} \operatorname{erfc}\left(\frac{x}{2\sqrt{\alpha_\ell t}}\right) \tag{5.108}$$

Once again, the interface position increases in proportion to $\sqrt{t}$,

$$x^*(t) = 2\phi\sqrt{\alpha_s t} \tag{5.109}$$

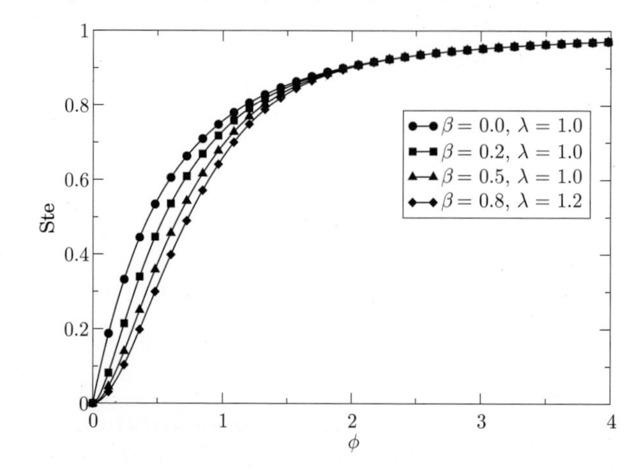

Fig. 5.9 Plots of the right hand side of Eq. (5.110) for various values of the parameters β and λ. The specific heats are taken to be equal for the two phases in this example.

Substituting the temperature solutions from Eqs. (5.107) and (5.108) into the Stefan condition (Eq. (5.106)) yields, after some rearrangement of terms,

$$\mathrm{Ste} = \frac{c_{p\ell}/c_{ps}}{\beta/\left[\sqrt{\pi}\phi\exp(\phi^2)\mathrm{erf}\,(\phi)\right] + 1/\left[\sqrt{\pi\alpha_T}\phi\exp(\alpha_T\phi^2)\mathrm{erfc}\left(\sqrt{\alpha_T}\phi\right)\right]} \tag{5.110}$$

where the parameters Ste, β and λ are defined as

$$\mathrm{Ste} = \frac{T_f - T_\infty}{L_f/c_{ps}}; \quad \beta = \frac{c_{ps}(T_f - T_0)}{c_{p\ell}(T_f - T_\infty)}; \quad \alpha_T = \frac{\alpha_s}{\alpha_\ell} \tag{5.111}$$

The right hand side of Eq. (5.110) is plotted in Fig. 5.9 for several values of the parameters. As usual, one computes the Stefan number, and determines from the graph the value of ϕ, from which the temperature solution and interface position follow. The temperature distribution in the solid and liquid at a particular instant in time for one set of parameters is shown in Fig. 5.10. Note, however, that there are no solutions for $\mathrm{Ste} \geq 1$. The physical explanation for this is that, under these conditions, called *hypercooling*, heat is conducted away from the interface too rapidly for the interface temperature to be maintained at T_f. At large undercooling, the interface velocity will be quite high (notice that $\phi \to \infty$ as $\mathrm{Ste} \to 1$), and so we might expect nonequilibrium kinetic effects to become important at the solid-liquid interface. This will be discussed in the next section.

Before proceeding to the consideration of kinetics, let us simplify the problem slightly by supposing that $T_0 = T_f$. This might occur, for example, if one were to insert a plate at temperature T_f into the undercooled melt. In fact, even if the plate were at a lower temperature than T_f, it would eventually heat up to T_f and remain at that temperature. The solution

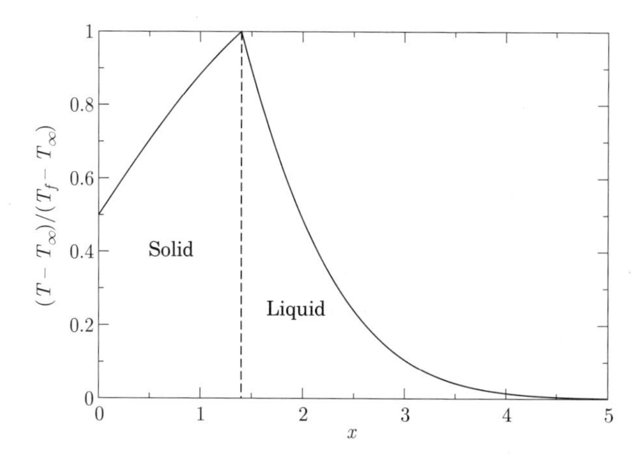

Fig. 5.10 The computed temperature distribution in the solid and liquid at $t = 1$ for $\phi = 0.7$.

for the solid temperature is then simply $T_s = T_f$. The liquid temperature distribution is the same, Eq. (5.108). However, now $\beta = 0$, so Eq. (5.110) is reduced to

$$\text{Ste} = \frac{c_{p\ell}}{c_{ps}}\sqrt{\pi\alpha_T}\phi\exp(\alpha_T\phi^2)\text{erfc}\left(\sqrt{\alpha_T}\phi\right) \tag{5.112}$$

If all of the thermal properties are equal in the liquid and solid, then $\alpha_T = 1$ and Eq. (5.112) is further reduced to

$$\text{Ste} = \sqrt{\pi}\phi\exp(\phi^2)\text{erfc}\left(\phi\right) \tag{5.113}$$

Eq. (5.113) also appears in Fig. 5.9, as the case where $\beta = 0$ and $\alpha_T = 1$.

Interface attachment kinetics

Next, we consider freezing for a case where the solid is isothermal, the melt is undercooled and interface attachment kinetics is considered. The problem then changes in two respects. Firstly, as shown in Chap. 2, the undercooling of the interface is proportional to the velocity. The system of equations for the temperature and interface motion with interface kinetics is given by

$$\frac{\partial T_\ell}{\partial t} = \alpha_\ell\frac{\partial^2 T_\ell}{\partial x^2} \qquad x^*(t) \le x < \infty \tag{5.114}$$

$$T_\ell(x^*(t), t) = T^* = T_f - \mu_k^{-1}\frac{dx^*}{dt} \qquad x = x^*(t) \tag{5.115}$$

$$\rho L_f\frac{dx^*}{dt} = -k_\ell\frac{\partial T_\ell}{\partial x} \qquad x = x^*(t) \tag{5.116}$$

$$T_\ell = T_\infty \qquad x \to \infty \tag{5.117}$$

$$T_\ell = T_\infty \qquad t = 0 \tag{5.118}$$

$$x^* = 0 \qquad t = 0 \tag{5.119}$$

where μ_k is the attachment kinetics coefficient relating the velocity of the interface v^* to its undercooling $\Delta T = (T_f - T^*)$. Notice that Eq. (5.115) can be directly inverted to express the velocity v^* in terms of T^*, although this quantity is as yet unknown.

$$v^* = \frac{dx^*}{dt} = \mu_k(T_f - T^*) \tag{5.120}$$

The second change is that the kinetic term introduces a length scale into the problem, which can be solved by scaling the equations. The following dimensionless variables are introduced:

$$\theta_\ell = \frac{T_\ell - T_\infty}{T_f - T_\infty} = \frac{T_\ell - T_\infty}{\Delta T}; \qquad \xi = \frac{x}{L}; \qquad \xi^*(\tau) = \frac{x^*(t)}{L}; \qquad \tau = \frac{\alpha_\ell}{L^2}t \tag{5.121}$$

By using these variables, Eqs. (5.114)-(5.119) become

$$\frac{\partial \theta_\ell}{\partial \tau} = \frac{\partial^2 \theta_\ell}{\partial \xi^2} \qquad \xi^*(\tau) \leq \xi < \infty \tag{5.122}$$

$$\theta_\ell = \theta^* \qquad \xi = \xi^*(\tau) \tag{5.123}$$

$$\frac{1}{\mathrm{Ste}}\frac{d\xi^*}{d\tau} = -\frac{\partial \theta_\ell}{\partial \xi} \qquad \xi = \xi^*(\tau) \tag{5.124}$$

$$\theta_\ell = 0 \qquad \xi \to \infty \tag{5.125}$$

$$\theta_\ell = 0 \qquad \tau = 0 \tag{5.126}$$

$$\xi^* = 0 \qquad \tau = 0 \tag{5.127}$$

where $\theta^* = (T^* - T_\infty)/(T_f - T_\infty)$. The length scale L cancels in each of these equations. However, it remains after scaling Eq. (5.120):

$$\frac{\alpha_\ell}{L\mu_k \Delta T}\frac{d\xi^*}{d\tau} = 1 - \theta^* \tag{5.128}$$

This means that there *is* now a length scale in the problem,

$$L = \frac{\alpha_\ell}{\mu_k \Delta T} \tag{5.129}$$

corresponding to the competition between attachment kinetics and diffusion.

We now seek a solution for which θ^* and $d\xi^*/dt = \mathrm{v}^*$ are constant. It is therefore convenient to transform to a frame translating with velocity v^*, i.e., define $\zeta = \xi - \mathrm{v}^*\tau$. The temperature field is steady in this frame, so Eq. (5.122) becomes

$$-\mathrm{v}^*\frac{\partial \theta_\ell}{\partial \zeta} = \frac{\partial^2 \theta_\ell}{\partial \zeta^2} \tag{5.130}$$

The boundary conditions at $\xi = \xi^*$ become boundary conditions at $\zeta = 0$, and it is therefore straightforward to apply the Dirichlet conditions at

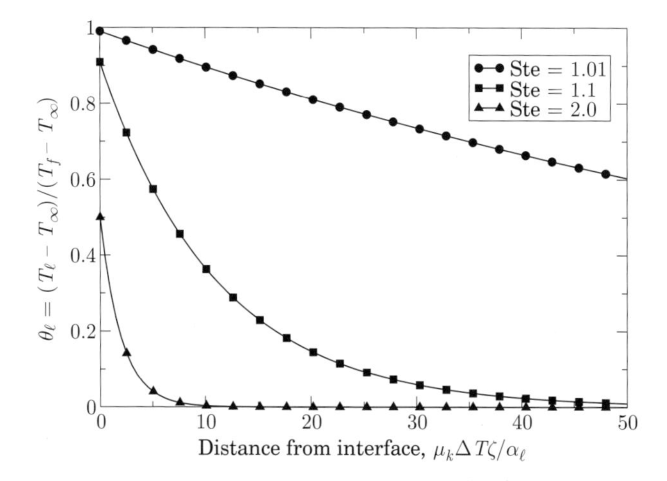

Fig. 5.11 The computed temperature distributions ahead of an advancing planar front with kinetic undercooling for several values of the Stefan number.

$\zeta = 0$ and $\zeta = \infty$ (Eqs. (5.123) and (5.125)) to obtain

$$\theta_\ell = \theta^* e^{-\mathrm{v}^* \zeta} \tag{5.131}$$

Substituting this result into the Stefan condition, Eq. (5.124) allows us to solve for θ^* and v^*

$$\theta^* = \frac{1}{\mathrm{Ste}}; \qquad \mathrm{v}^* = 1 - \frac{1}{\mathrm{Ste}} \tag{5.132}$$

This solution is valid for Stefan numbers strictly greater than one. Examples of the temperature solution are given in Fig. 5.11 for several values of Ste. Notice that as the undercooling increases, the interface temperature decreases proportionally. Note also that there is a boundary layer ahead of the interface with a thickness proportional to $(1 - 1/\mathrm{Ste})$.

5.3.2 Solidification of a paraboloid

One of the most common morphologies observed in solidification is a *dendrite*, i.e., a tree-like form that grows with a primary stalk and side branches. The primary and secondary branches reflect the underlying crystal symmetry, as well as the direction of heat flow. Analysis of dendritic growth is a very important problem in the development of microstructure; indeed, Chap. 8 is devoted to this topic. However, there is a basic problem relating to dendritic growth that we include here since it has an analytical solution.

Consider the growth of a parabolic shape in an undercooled melt. There are two related problems: a 2D parabolic platelet, and a 3D paraboloid of revolution, which we solve simultaneously. For convenience, we call both shapes paraboloids. This problem was first analyzed by

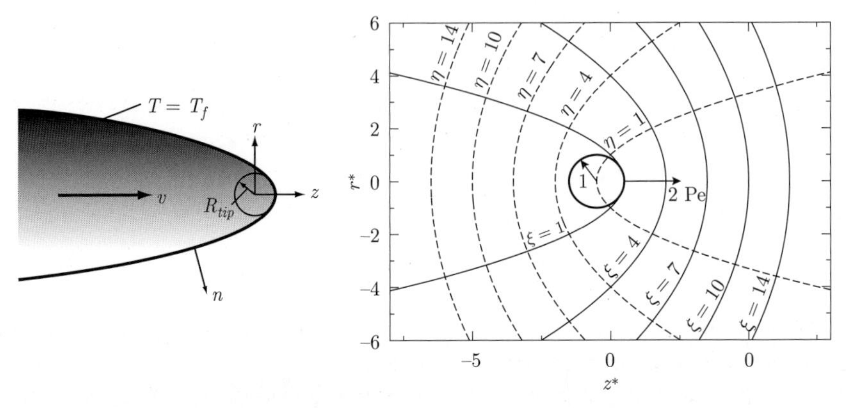

(a) A schematic view of paraboloid

(b) The paraboloidal coordinate system

Fig. 5.12 (a) Sketch of paraboloidal tip geometry; (b) parabolic coordinate system for the analysis. The curve $\xi = 1$ corresponds to the solid-liquid interface.

Ivantsov [7], and the analysis was later generalized by Horvay and Cahn. [6]. The presentation here follows the latter approach, omitting some of the mathematical details. The solid is assumed to grow at constant velocity v along its axis, in a shape-preserving way. The melt is taken to be pure and infinite with the far-field temperature designated by T_∞. The temperature at the solid-liquid interface is the equilibrium freezing temperature T_f. Note that this excludes the effect of curvature and kinetics at the interface. Although these effects are important, including them would preclude finding an analytical solution.

The problem geometry is illustrated in Fig. 5.12(a). The analysis is performed in a frame that translates with the dendrite velocity v in the $z-$direction. The surface of the dendrite is given by

$$z = \frac{R_{tip}}{2} - \frac{r^2}{2R_{tip}} = P(r; R_{tip}) \tag{5.133}$$

where R_{tip} is the radius of curvature at the tip ($r = 0$), and where $r = x$ in 2D, and $r = \sqrt{x^2 + y^2}$ in 3D. The notation $P(r; R_{tip})$ is a shorthand notation referring to the surface of the paraboloid. Note that the shape of the paraboloid can be specified by a single parameter, i.e., the tip radius R_{tip}. The equations governing the temperature fields in the liquid and solid are written as follows, and include the assumption that the fields are steady in the moving frame

$$-v\frac{\partial T_s}{\partial z} = \alpha_s \nabla^2 T_s \qquad z \le P(r; R_{tip}) \tag{5.134}$$

$$-v\frac{\partial T_\ell}{\partial z} = \alpha_\ell \nabla^2 T_\ell \qquad z \ge P(r; R_{tip}) \tag{5.135}$$

$$T_s = T_\ell = T_f \qquad z = P(r; R_{tip}) \tag{5.136}$$

$$\rho L_f v n_z = k_s \nabla T_s \cdot \mathbf{n} - k_\ell \nabla T_\ell \cdot \mathbf{n} \qquad z = P(r; R_{tip}) \tag{5.137}$$

$$T_\ell = T_\infty \qquad r^2 + z^2 \to \infty \tag{5.138}$$

where n_z is the component along the z-axis of the unit vector normal to the surface of the dendrite. Before going further, note that $T_s = T_f$ is a solution to Eqs. (5.134) and (5.136). It will also be possible to find a solution for the liquid temperature with $\nabla T_s = 0$. For this reason, the solution is sometimes called the "isothermal dendrite."

The analysis for the temperature in the liquid begins by scaling the lengths on R_{tip} and the temperatures on $\Delta T = T_f - T_\infty$:

$$\zeta = \frac{z}{R_{tip}}; \quad \varrho = \frac{r}{R_{tip}}; \quad \theta_\ell = \frac{T_\ell - T_\infty}{T_f - T_\infty} \tag{5.139}$$

Note that some authors, including Horvay and Cahn [6], whose treatment we follow here, choose instead to scale temperature as $(T - T_\infty)/(L_f/c_{p\ell})$. The form we chose in Eq. (5.139) is consistent with our approach of scaling the temperature to be order one. As a result, the scaled temperature we compute will correspond to that computed by Horvay and Cahn divided by the Stefan number. The solid-liquid interface is given by

$$\zeta = \frac{1}{2}(1 - \varrho^2) = \mathcal{P}(\varrho) \tag{5.140}$$

Now that the length has been scaled on R, the parametric dependence of the paraboloid has been eliminated in the scaled form. Substituting these scaled variables into Eqs. (5.135)-(5.138) gives

$$-2\text{Pe}\frac{\partial \theta_\ell}{\partial \zeta} = \nabla^2 \theta_\ell \qquad \zeta > \mathcal{P}(\varrho) \tag{5.141}$$

$$\theta_\ell = 1 \qquad \zeta = \mathcal{P}(\varrho) \tag{5.142}$$

$$\frac{2\text{Pe}}{\text{Ste}} n_z = -\nabla \theta_\ell \cdot \mathbf{n} \qquad \zeta = \mathcal{P}(\varrho) \tag{5.143}$$

$$\theta_\ell = 0 \qquad \zeta \to \infty \tag{5.144}$$

where $\text{Pe} = vR_{tip}/2\alpha_\ell$ is the Péclet number, and $\text{Ste} = C_{p\ell}\Delta T/L_f$ is the Stefan number. Notice that, since this problem has a natural length scale R_{tip}, two important dimensionless parameters appear.

It is convenient to transform the system to paraboloidal coordinates (ξ, η), conforming to the shape of the solid, defined as

$$\varrho = \xi\eta \quad ; \quad \zeta = \frac{1}{2}(\xi^2 - \eta^2) \tag{5.145}$$

$$\xi^2 = \zeta + \sqrt{\varrho^2 + \zeta^2} \quad ; \quad \eta^2 = -\zeta + \sqrt{\varrho^2 + \zeta^2} \tag{5.146}$$

The solid-liquid interface corresponds to $\xi = 1$. The coordinate system is shown in Fig. 5.12(b). Notice that curves of constant η and constant ξ are orthogonal. Thus, the vector normal to the solid-liquid interface is $\hat{\boldsymbol{\xi}}$, and in particular $\nabla\theta_\ell \cdot \mathbf{n} = \partial\theta_\ell/\partial\xi$. All of the boundary conditions are thus independent of η, whereas the solution depends only of ξ.

The temperature distribution in 2D is (see Horvay and Cahn [6] for details)

$$\theta_\ell(\xi) = \frac{\sqrt{\pi \text{Pe}}}{\text{Ste}} \exp(\text{Pe})\text{erfc}\left(\xi\sqrt{\text{Pe}}\right) \qquad 2\text{D}, \xi \geq 1 \tag{5.147}$$

Since we have neglected curvature and attachment kinetics at that point, the solid-liquid interface is at the melting point T_f, which means that $\theta_\ell^* = \theta_\ell(\xi = 1) = 1$. Therefore, the boundary condition at $\xi = 1$ relates the Stefan number to the Péclet number

$$\mathrm{Ste} = \sqrt{\pi \mathrm{Pe}}\exp(\mathrm{Pe})\mathrm{erfc}\left(\sqrt{\mathrm{Pe}}\right) \qquad \text{2D} \tag{5.148}$$

In 3D, the solution takes a slightly different form

$$\theta_\ell = \frac{\mathrm{E}_1(\mathrm{Pe}\xi^2)}{\mathrm{E}_1(\mathrm{Pe})} \qquad \text{3D}, \xi \geq 1 \tag{5.149}$$

where E_1 is the exponential integral, defined as

$$\mathrm{E}_1(u) = \int_u^\infty \frac{e^{-s}}{s} ds; \qquad \frac{\partial \mathrm{E}_1(u)}{\partial \xi} = -\frac{e^{-u}}{u}\frac{du}{d\xi} \tag{5.150}$$

Using the Stefan condition, Eq. (5.143) and the definition of the derivative of the exponential integral given above leads to the following relation between the Péclet and Stefan numbers in 3D

$$\mathrm{Ste} = \mathrm{Pe}\exp(\mathrm{Pe})\mathrm{E}_1(\mathrm{Pe}) \qquad \text{3D} \tag{5.151}$$

The right hand sides of Eqs. (5.148) and (5.151) are plotted in Fig. 5.13, along with the right hand side of Eq. (5.113), the corresponding equation relating the Stefan number to the growth parameter ϕ for planar front growth in an undercooled melt. Notice that, once again, there are no solutions for $\mathrm{Ste} \geq 1$. The solutions given in Eqs. (5.148) and (5.151) are sometimes given the shorthand form

$$\mathrm{Iv}_{2\mathrm{D}}(\mathrm{Pe}) \quad \text{or} \quad \mathrm{Ste} = \mathrm{Iv}_{3\mathrm{D}}(\mathrm{Pe}) \tag{5.152}$$

where $\mathrm{Iv}_{2\mathrm{D}}(\mathrm{Pe})$ and $\mathrm{Iv}_{3\mathrm{D}}(\mathrm{Pe})$ are called the *Ivantsov functions*, corresponding to 2D or 3D, respectively.

The temperature distributions for two different 3D paraboloids, corresponding to two values of the Stefan number, are plotted in Fig. 5.14. These solutions were computed by first determining the value of the Péclet number corresponding to the Stefan number from the graph of Eq. (5.151), and then substituting this result in Eq. (5.149). Notice that as the Stefan (and Péclet) numbers increase, the isotherms approach the interface, falling to zero at a normalized distance ahead of the tip given roughly by $1/\sqrt{\mathrm{Pe}}$.

There is another important aspect of the problem for the parabolic shape. In an experiment, one can set T_∞, but one cannot set the parabolic shape that we have characterized by the tip radius R. It is experimentally observed that the material selects a unique shape corresponding to each undercooling. However, our solution does not provide this result. The Stefan number is determined once T_∞ is chosen. As long as $\mathrm{Ste} < 1$,

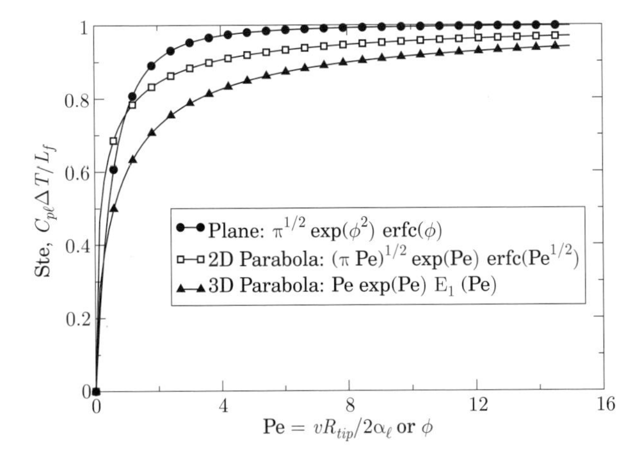

Fig. 5.13 Graphical representations of the various transcendental relations between the Stefan number and the velocity control parameter for solidification in various geometries in an undercooled melt.

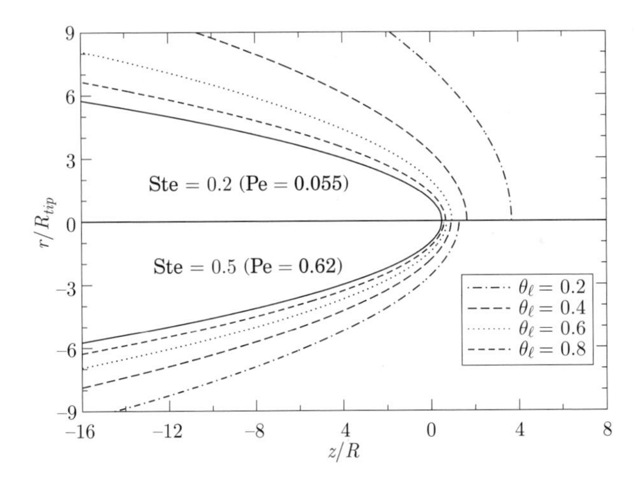

Fig. 5.14 Isotherms for 3D paraboloids of revolution growing with two different values of the Stefan (or Péclet) number. Upper half, $\mathrm{Ste} = 0.2$ and $\mathrm{Pe} = 0.055$; lower half, $\mathrm{Ste} = 0.5$ and $\mathrm{Pe} = 0.62$.

one can then find the Péclet number using the curves given in Fig. 5.13, and subsequently plot the temperature distribution using the appropriate solution. The curious result is that the heat transport solution determines $\mathrm{Pe} = vR_{tip}/2\alpha_\ell$, i.e., the product vR_{tip}, but not v and R_{tip} independently. In other words, by fixing the Stefan number (or the undercooling $\Delta T = T_f - T_\infty$), the system can produce in steady state a very sharp dendrite that grows rapidly, or a blunt one that grows slowly. The resolution of this matter will be discussed in Chap. 8.

Before leaving the paraboloidal dendrite, we should note that a similar analysis can be performed for alloy systems. For example, consider once again the solidification of alloy C_0, shown in Fig. 5.2. This time, let the melt be isothermal at temperature T^*, and assume that the solid grows as a paraboloid. The analysis presented above for the paraboloid of the pure material can be applied to this case, with some modification of the variables. The solid will now have a constant composition C_s, and the role of the Stefan number is taken by the *supersaturation* Ω,

$$\Omega = \begin{cases} \mathrm{Iv}_{2\mathrm{D}}(\mathrm{Pe}) \\ \mathrm{Iv}_{3\mathrm{D}}(\mathrm{Pe}) \end{cases}; \qquad \Omega = \frac{C_\ell^* - C_0}{C_\ell^* - C_s^*} \tag{5.153}$$

See Fig. 5.2 for the definitions of C_ℓ^*, C_0 and C_s^*. Note that, since we have neglected the curvature and kinetic undercoolings, the interface temperature T^* and the composition C_ℓ^* must lie on the liquidus curve of the phase diagram, i.e., $T^* = T_f + m_\ell C_\ell^*$. We will return to the discussion of this problem in Chap. 8.

5.4 THE EFFECT OF CURVATURE

One physical aspect of the problem that was not included in the analysis of the paraboloid, is the effect of curvature on the equilibrium temperature of the solid-liquid interface. This phenomenon was introduced in Chap. 2, where it was demonstrated that equilibrium at a curved interface of a pure substance led to the Gibbs-Thomson equation for the interface temperature T^*

$$T^* = T_f - 2\Gamma_{s\ell}\bar{\kappa} \tag{5.154}$$

where $\Gamma_{s\ell}$ is a material property, the Gibbs-Thomson coefficient, and $\bar{\kappa}$ is the mean curvature. The curvature correction makes no contribution to the problems with a planar front. Although it is important for the paraboloid, the inclusion of curvature precludes finding an analytical solution. We will thus continue the discussion of this geometry in Chap. 8. An analytical solution *can* be found for a sphere including curvature in the Gibbs-Thomson condition, a problem taken up in the next section.

5.4.1 Solidification of a sphere in an undercooled melt

Consider a solid sphere of radius $R(t)$ growing in an infinite undercooled melt. The initial radius of the sphere is R_i, and the temperature at infinity is T_∞. Assume that the system retains spherical symmetry and that it is made of a pure substance. The density and specific heats of the solid and liquid are taken to be constant and equal in both phases, but the thermal conductivity of the solid and liquid phases are allowed to differ from each other. The governing equations for the heat conduction problem, under the stated assumptions, are

$$\frac{\partial T_s}{\partial t} = \frac{\alpha_s}{r^2}\frac{\partial}{\partial r}\left(r^2\frac{\partial T_s}{\partial r}\right) \qquad 0 \le r \le R(t) \tag{5.155}$$

$$\frac{\partial T_\ell}{\partial t} = \frac{\alpha_\ell}{r^2}\frac{\partial}{\partial r}\left(r^2\frac{\partial T_\ell}{\partial r}\right) \qquad R(t) \le r < \infty \tag{5.156}$$

$$T_s < \infty \qquad r = 0 \tag{5.157}$$

$$T_s = T_\ell = T_f - 2\Gamma_{s\ell}\bar{\kappa} = T_f - \frac{2\Gamma_{s\ell}}{R(t)} \qquad r = R(t) \tag{5.158}$$

$$\rho L_f \frac{dR}{dt} = k_s\frac{\partial T_s}{\partial r} - k_\ell\frac{\partial T_\ell}{\partial r} \qquad r = R(t) \tag{5.159}$$

$$T_\ell = T_\infty \qquad r \to \infty \tag{5.160}$$

$$R = R_i \qquad t = 0 \tag{5.161}$$

$$T_\ell = T_\ell^0(r) \qquad t = 0 \tag{5.162}$$

$$T_s = T_s^0(r) \qquad t = 0 \tag{5.163}$$

where the functions $T_\ell^0(r)$ and $T_s^0(r)$ are initial temperature distributions, to be defined later.

As usual, the analysis begins by scaling the governing equations. Using the definition $\Delta T = T_f - T_\infty$,

$$\theta_s = \frac{T_s - T_f}{\Delta T}; \quad \theta_\ell = \frac{T_\ell - T_f}{\Delta T}; \quad \varrho = \frac{r}{R_i}; \quad \mathcal{R} = \frac{R}{R_i}; \quad \tau = \frac{t}{t_c} \tag{5.164}$$

where t_c is a characteristic time that will be chosen later. Notice that $-1 \le \theta_{s,\ell} \le 0$. This is a little unusual, but facilitates the application of the boundary conditions. Rewriting the governing equations in dimensionless form, and using the notation $k_T = k_s/k_\ell$ and $\alpha_T = \alpha_s/\alpha_\ell$ gives,

$$\left\{\frac{R_i^2}{\alpha_\ell t_c}\right\}\frac{\partial \theta_s}{\partial \tau} = \frac{\alpha_T}{\varrho^2}\frac{\partial}{\partial \varrho}\left(\varrho^2\frac{\partial \theta_s}{\partial \varrho}\right) \qquad 0 \le \varrho \le \mathcal{R}(\tau) \tag{5.165}$$

$$\left\{\frac{R_i^2}{\alpha_\ell t_c}\right\}\frac{\partial \theta_\ell}{\partial \tau} = \frac{1}{\varrho^2}\frac{\partial}{\partial \varrho}\left(\varrho^2\frac{\partial \theta_\ell}{\partial \varrho}\right) \qquad \mathcal{R}(\tau) \le \varrho < \infty \tag{5.166}$$

$$\theta_s < \infty \qquad \varrho = 0 \tag{5.167}$$

$$\theta_s = \theta_\ell = -\frac{2\Gamma_{s\ell}}{\Delta T R_i \mathcal{R}(\tau)} \qquad \varrho = \mathcal{R}(\tau) \tag{5.168}$$

$$\left\{\frac{\rho L_f R_i^2}{k_\ell \Delta T t_c}\right\}\frac{d\mathcal{R}}{d\tau} = k_T\frac{\partial \theta_s}{\partial \varrho} - \frac{\partial \theta_\ell}{\partial \varrho} \qquad \varrho = \mathcal{R}(\tau) \tag{5.169}$$

$$\theta_\ell = -1 \qquad \varrho \to \infty \tag{5.170}$$

$$\mathcal{R} = 1 \qquad \tau = 0 \tag{5.171}$$

$$\theta_\ell = \theta_\ell^0(\varrho) \qquad \tau = 0 \tag{5.172}$$

$$\theta_s = \theta_s^0(\varrho) \qquad \tau = 0 \tag{5.173}$$

Examination of Eqs. (5.165), (5.166) and (5.169) shows that there are two choices for the time scale t_c: either a diffusive time scale, $t_D = R_i^2/\alpha_\ell$, or a time scale associated with solidification $t_s = \rho L_f R_i^2/k_\ell \Delta T$. We can choose either t_s or t_D to scale the time, and the ratio of the two time scales then appears somewhere in the equations. Note that the ratio $t_D/t_s = \text{Ste}$, the Stefan number. Since we are interested in solidification, we choose $t_c = t_s$. The revised forms of Eqs. (5.165), (5.166) and (5.169) are

$$\text{Ste}\frac{\partial \theta_s}{\partial \tau} = \frac{\alpha_T}{\varrho^2}\frac{\partial}{\partial \varrho}\left(\varrho^2\frac{\partial \theta_s}{\partial \varrho}\right) \qquad 0 \leq \varrho \leq \mathcal{R}(\tau) \tag{5.174}$$

$$\text{Ste}\frac{\partial \theta_\ell}{\partial \tau} = \frac{1}{\varrho^2}\frac{\partial}{\partial \varrho}\left(\varrho^2\frac{\partial \theta_\ell}{\partial \varrho}\right) \qquad \mathcal{R}(\tau) \leq \varrho < \infty \tag{5.175}$$

$$\frac{d\mathcal{R}}{d\tau} = k_T\frac{\partial \theta_s}{\partial \varrho} - \frac{\partial \theta_\ell}{\partial \varrho} \qquad \varrho = \mathcal{R}(\tau) \tag{5.176}$$

In most of the previous problems, we have chosen to scale time on t_D, in which case the Stefan number appears in Eq. (5.176) instead. A typical undercooling ΔT for Al is less than 1K. The corresponding Stefan number is thus

$$\text{Ste} = \frac{c_p\Delta T}{L_f} = \frac{(1.19\times 10^3\ \text{Jkg}^{-1}\text{K}^{-1})(1\ \text{K})}{3.98\times 10^5\ \text{Jkg}^{-1}} = 0.003 \ll 1 \tag{5.177}$$

In view of this calculation, the analysis will proceed by allowing $\text{Ste} \to 0$ in Eqs. (5.174) and (5.175). The physical interpretation of this assumption is that the interface moves slowly in comparison to the diffusion of heat, and therefore the temperature distribution for any sphere diameter corresponds to the value found in steady state. This is sometimes called *quasi-steady state.*

The quasi-steady form of Eq. (5.175) is

$$\frac{1}{\varrho^2}\frac{\partial}{\partial \varrho}\left(\varrho^2\frac{\partial \theta_\ell}{\partial \varrho}\right) = 0 \qquad \mathcal{R}(\tau) \leq \varrho < \infty \tag{5.178}$$

which has the solution

$$\theta_\ell = c_1(\tau) + \frac{c_2(\tau)}{\varrho} \tag{5.179}$$

The integration "constants" c_1 and c_2 are allowed to vary with time in order to follow the series of quasi-steady states taken by the temperature in the liquid as the diameter of the sphere changes. The constants are evaluated by applying the boundary conditions given in Eqs. (5.168) and (5.170), and the solution for θ_ℓ is

$$\theta_\ell = -1 + \left(1 - \frac{2\Gamma_{s\ell}}{R_i\Delta T\mathcal{R}(\tau)}\right)\frac{\mathcal{R}(\tau)}{\varrho} \tag{5.180}$$

The solution for θ_s follows the same procedure, and has the same form as Eq. (5.179). However, the boundary conditions in the solid are different. In particular, since θ_s is bounded at $\varrho = 0$, the coefficient of the term involving $1/\varrho$ must be zero. Applying the boundary condition Eq. (5.168) gives

$$\theta_s = -\frac{2\Gamma_{s\ell}}{R_i \Delta T \mathcal{R}(\tau)} \tag{5.181}$$

Thus the temperature in the solid is uniform at any given time, i.e., independent of ϱ. Notice that one cannot satisfy arbitrary initial conditions on the temperature in the solid or liquid, Eqs. (5.172) and (5.173). This is a consequence of allowing Ste to go to zero, eliminating the time derivatives in Eqs. (5.174) and (5.175). The initial condition that *can* be satisfied is found by setting $\mathcal{R} = 1$ in Eqs. (5.180) and (5.181)

$$\theta_\ell^0 = -1 + \left(1 - \frac{2\Gamma_{s\ell}}{R_i \Delta T}\right)\frac{1}{\varrho}; \qquad \theta_s^0 = -\frac{2\Gamma_{s\ell}}{R_i \Delta T} \qquad \tau = 0 \tag{5.182}$$

We have not yet considered the Stefan condition, Eq. (5.176). Substituting the temperature solutions for the solid and the liquid into Eq. (5.176) gives

$$\frac{d\mathcal{R}}{d\tau} = \left(1 - \frac{2\Gamma_{s\ell}}{R_i \Delta T \mathcal{R}}\right)\frac{1}{\mathcal{R}} \tag{5.183}$$

To understand the behavior of the particle, first apply the initial condition $\mathcal{R}(0) = 1$

$$\left.\frac{d\mathcal{R}}{d\tau}\right|_{\tau=0} = 1 - \frac{2\Gamma_{s\ell}}{R_i \Delta T} \tag{5.184}$$

Notice that the initial growth rate changes sign for $R_i = 2\Gamma_{s\ell}/\Delta T$, which is called the *critical nucleus* radius. Particles with initial radii smaller than the critical radius will shrink and eventually disappear, and those with initial radii greater than the critical radius will grow. As an example, for $\Delta T = 1\,\mathrm{K}$ in Al, $R_i = 0.18\,\mu\mathrm{m}$. The concept of a critical nucleus is important for nucleation, and will be discussed further in Chap. 7. To make graphs of the temperature solution, let us choose the critical nucleus size $R_i = 2\Gamma_{s\ell}/\Delta T$. Then

$$\theta_\ell = -1 + \left(1 - \frac{1}{\mathcal{R}}\right)\frac{\mathcal{R}}{\varrho} \qquad \varrho \geq \mathcal{R} \tag{5.185}$$

$$\theta_s = -\frac{1}{\mathcal{R}} \tag{5.186}$$

Note that the only initial condition that can be satisfied by this solution is $\theta_s = \theta_\ell = -1$. The temperature distributions corresponding to several different values of $\mathcal{R}$ are shown in Fig. 5.15. Notice that the solid temperature approaches T_f as the sphere radius increases. It turns out that once the solid grows above a certain size, it becomes morphologically unstable.

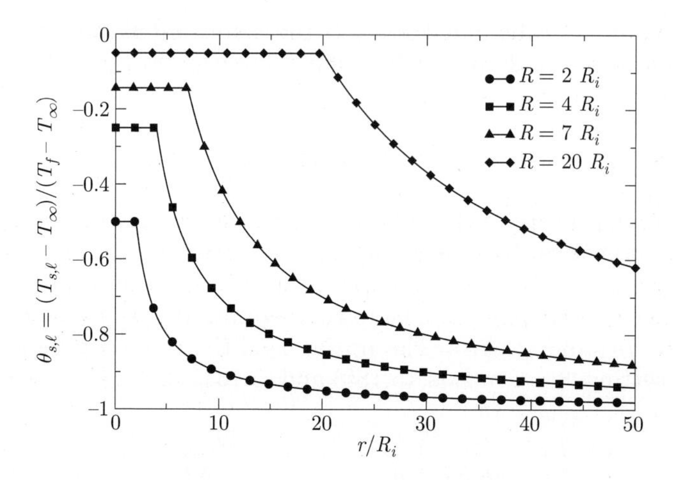

Fig. 5.15 Temperature distributions in the solid and liquid for a spherical particle growing in an undercooled melt. The initial particle diameter is the critical nucleus.

5.5 SUMMARY AND CONCLUSIONS

In this chapter, we have considered several basic problems in solidification, for which analytical solutions exist. Solutions were presented for pure materials and alloys solidifying with a planar front into a semi-infinite melt. For the case of an infinite superheated melt, solutions of the governing equations were provided in terms of a similarity variable, $x/(2\sqrt{\alpha t})$. The interface position grows in proportion to $\sqrt{t}$, and hence the velocity is proportional to $1/\sqrt{t}$. The constant of proportionality could be determined by solving a transcendental equation involving the Stefan number, $\text{Ste} = c_{ps}\Delta T/L_f$, where ΔT is a characteristic temperature difference. Solutions exist for all values of Ste.

Analysis of growth with a planar front into an undercooled melt followed a similar path. However, we found that similarity solutions exist only for $\text{Ste} < 1$. Adding interface kinetics to the problem allowed us to find solutions also for $\text{Ste} > 1$. We then examined growth of an isothermal paraboloid in an undercooled melt, and by analogy, an alloy in a supersaturated melt. Solutions once again could be found only for $\text{Ste} < 1$, and the solution had the curious property that the shape was indeterminate. In other words, the transport solution gives only the product $R_{tip}v^*$, where R_{tip} is the radius of curvature at the tip of the paraboloid and v^* its velocity, and not R_{tip} and v^* individually. We return to these problems in Chap. 8.

Finally, the solidification of a sphere in an undercooled melt was examined, including the effect of curvature on the equilibrium interface. A separation of time scales for solidification and diffusion was found, and a quasi-steady state analysis was performed. We determined that there is a critical nucleus size, above which particles can grow. This is the basis for understanding nucleation phenomena, which we discuss in Chap. 7.

The solutions presented in this chapter have several applications. The solutions for superheated melts can be used immediately as a guide to estimate, for example, the solidification characteristics of complex-shaped castings. More detailed analysis of such applications must be done numerically, but these analytical solutions provide good references and test cases for the numerical codes. The solutions for undercooled melts provide the basis for the understanding of microstructure development. For example, in Chap. 8, we examine the morphological stability of the basic solutions developed in this chapter, and find the development of length scales associated with this instability.

5.6 EXERCISES

Exercise 5.1. Freezing of a slab.
Consider the solidification of a slab of a pure material with a thickness $2L$, beginning with the liquid at the melting point, T_f. The slab surfaces are maintained at temperature T_0. Consider the problem to be one-dimensional. The thermal properties are constant and equal in both phases.

(a) Write the governing equations and boundary conditions for the solidification problem.

(b) Scale the equations, choosing to scale time on L^2/α. Identify any dimensionless groups that appear.

(c) Work out the solution for the solidification time of the slab for the case where $\mathrm{Ste} = 1$. (Answer: $\tau_f = 0.65$)

Exercise 5.2. Steady state directional solidification.
In Sect. 5.2.3 we presented Smith et al.'s transient solution for directional solidification at constant speed in a fixed temperature gradient. The steady state solution was found by considering the long time limit of the transient solution. If one is interested in only the steady state solution, it can be obtained in a more straightforward way. The solution we seek is steady in the frame fixed on the interface.

(a) Begin with the diffusion equation for the liquid composition given in Eq. (5.85), and modify it for the steady state.

(b) Write the general solution to this ordinary differential equation, and then apply the boundary conditions from Eq. (5.86) and the subsequent text to find the solution.

Exercise 5.3. Cryosurgery.
In crysosurgery, a probe is inserted into diseased tissue, and a cryogenic fluid, e.g., liquid nitrogen, is circulated through the probe in order to freeze the tissue. In this problem, we will model a simplified version of the process. We treat the cryoprobe as a line source of strength $\dot{R}_q^{lin}(\mathrm{Wm}^{-1}) < 0$ and assume that it is inserted into an infinite medium at a temperature T_∞ above the freezing point T_f. The object is to determine the time required to solidify a cylinder of tissue of radius R. This is the only set of conditions for which an analytical solution exists. Let all of the material properties in both the solid and liquid be constant, and assume that $\rho_s = \rho_\ell$, $c_{ps} = c_{p\ell}$, but that k_s and k_ℓ are unequal.

(a) Begin with the heat conduction equation in cylindrical coordinates and simplify it, making the usual set of appropriate assumptions. You should end up with a transient term, as well as terms involving derivatives with respect to r.

(b) Show by differentiation that the exponential integral, $\mathrm{E}_1(-r^2/4\alpha_\nu t)$ defined below, is a solution to this equation.

$$\mathrm{E}_1(u) = \int_u^\infty \frac{e^{-s}}{s} ds$$

Note that $\mathrm{E}_1(0) = -\infty$.

(c) Suppose that the solidified layer has thickness $R(t)$. Write the two appropriate boundary conditions at $r = R(t)$, and the boundary conditions at $r = 0$ and $r \to \infty$. Be sure to use a flux condition at the origin; there is no analytical solution for a fixed temperature there.

(d) Write the solution for the temperature in both the frozen and unfrozen tissue in the form $T \sim A + B\mathrm{E}_1(-r^2/4\alpha_\nu t)$. Show that the boundary condition $T = T_f$ at $r = R(t)$ implies that $R(t)$ is proportional to $\sqrt{t}$. To be consistent with the treatment in the chapter, choose $R(t) = 2\phi\sqrt{\alpha_s t}$.

(e) In order to enforce the boundary condition at $r = 0$, consider a small cylinder of radius r about the origin, set the total heat loss through the surface of that cylinder to $\dot{R}_q^{lin}$, and equate that to $-k_\ell \partial T/\partial r$. Then take the limit as the radius of the small cylinder goes to zero. This should allow you to solve for A and B in terms of ϕ. What is the temperature at $r = 0$, and what does this mean?

(f) Write a similar expression for the temperature outside the solid, $T_\ell = C + D\mathrm{E}_1(-r^2/4\alpha_\ell t)$. Use the boundary condition at $r \to \infty$ and the equilibrium condition to evaluate C and D, then use the Stefan condition to find a transcendental equation for ϕ.

(g) Taking the properties of the tissue to be the same as those for water, find the necessary flux to freeze a cylinder of radius 1 cm in 120 seconds. For simplicity, let $\alpha_s = \alpha_\ell$ for this part, and assume that $T_\infty = 37°$C. It may be helpful to know that Ei(-1.35) = 2.86.

Exercise 5.4. Zone melting.
The *zone melting* process is used to purify materials by taking advantage of the segregation of impurities to one end of the sample in directional solidification, as seen in Fig. 5.6. In zone melting, a segment of the sample of thickness δ is melted, for example using an induction heater. The coil then traverses the sample beginning at one end. The heater produces sufficient stirring in the melted section that its composition can be considered to be uniform. The composition profile at some time during the first pass is shown schematically in the figure below. The initial composition of the sample is C_0, and the composition in the liquid is designated C_ℓ^*.

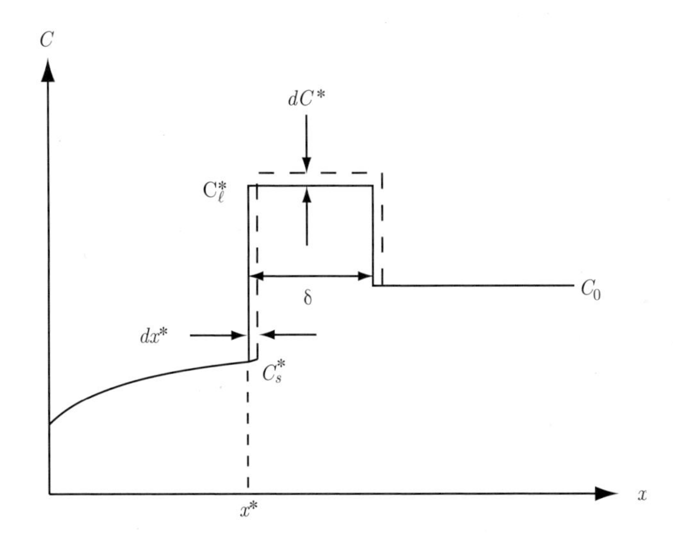

(a) Write a solute balance for the change in composition in the liquid when the interface moves by an increment dx^* as shown. Be sure to account for both the solute rejected at the solid-liquid interface, and the dilution associated with previously unmelted solid.

(b) Rearrange this equation so that it can be integrated from the initial condition, where the composition in the liquid is C_0 and the solid-liquid interface is located at $x^* = 0$.

(c) Perform the integration to obtain the result

$$\frac{C_\ell^*}{C_0} = \frac{1}{k_0}\left[1 - (1 - k_0)\exp\left(-\frac{k_0 x^*}{\delta}\right)\right]$$

(d) Real zone melting processes perform multiple passes. One cannot find an analytical solution for the later passes. Sketch the solute profile that you would expect to see after the second and tenth passes. What would happen if this were a multicomponent alloy, whose constituents had different segregation coefficients.

5.7 REFERENCES

[1] V. Alexiades and A. D. Solomon. *Mathematical modeling of melting and freezing processes*. Hemisphere, Washington, 1993.

[2] G. A. Campbell and R. M. Foster. *Fourier integrals for practical applications*. Van Nostrand, New York, 1948. Formula 827.1 with $\sigma^{1/2} = z^*, \rho = 1/4, \gamma = 0, \lambda = (2k_0 - 1)/2$.

[3] H. S. Carslaw and J. C. Jaeger. *Conduction of heat in solids*. Oxford University Press, London, 2nd edition, 1959.

[4] S. H. Davis. *Theory of solidification*. Cambridge University Press, Cambridge, 1999.

[5] M. C. Flemings. *Solidification processing*. McGraw-Hill, New York, 1974.

[6] G. Horvay and J. W. Cahn. Dendritic and spheroidal growth. *Acta. Metall.*, 9:695-705, 1961.

[7] G. P. Ivantsov. Temperature field around the spherical, cylindrical and needle-crystals which grow in supercooled melt. *Dokl. Akad. Nauk USSR*, 58:1113, 1947.

[8] W. G. Pfann. *Zone melting*. Wiley, New York, 2nd edition, 1966.

[9] V. G. Smith, W. A. Tiller, and J. W. Rutter. A mathematical analysis of solute redistribution during solidification. *Can. J. Phys.*, 33:723-745, 1955.

CHAPTER 6

NUMERICAL METHODS FOR SOLIDIFICATION

6.1 INTRODUCTION

The previous chapter presented several problems involving solidification in pure materials and alloys, all of which had analytical solutions. These problems are useful in several respects: they help to identify the important parameters governing solidification processes, and also provide basic states that are useful for examining the development of solidification microstructures. In addition, the analytical solutions can be used as benchmark solutions to test numerical schemes for modeling solidification in more complex situations, which is the subject of the present chapter.

Why pursue numerical approaches? The reason is that analytical solutions apply only under very restrictive conditions: the domain geometry must be particularly simple, often one-dimensional; only relatively restrictive boundary conditions may be applied; material properties must be constant; and in only very few problems can coupled phenomena, such as heat transfer and fluid flow, be considered. None of these conditions are met in real physical systems.

The analysis of more realistic solidification problems requires numerical methods. Numerous books are devoted to this subject, and only a small fraction of the topic can be covered here. The objective is to understand the use of numerical schemes, and in particular to focus on the *accuracy* of the methods, i.e., how close the numerical solution is to the exact solution. For this purpose, the analytical solutions of the preceding chapter provide very useful test problems. We are especially interested in the relationship between the accuracy of the solution and the parameters in the simulation, such as grid spacing and time step size.

Solidification problems present the same difficulties in numerical schemes that they do in analytical approaches: some means must be found to follow the phase boundaries through the domain as the material evolves from fully liquid to fully solid. Methods for this purpose are divided into two classes. *Fixed grid* methods, as their name implies, allow the interface to move through a fixed mesh, and special techniques are applied to ensure that the interfacial boundary conditions are satisfied. The interface location is recovered afterward. *Front tracking* methods follow the interface motion explicitly, ensuring that the grid conforms to the interface in order for the boundary conditions to be applied there. Examples of both types are included in the following sections.

6.2 HEAT CONDUCTION WITHOUT PHASE CHANGE

This section describes some of the basic methods for numerical analysis of heat conduction, covering in some detail the finite difference method. We focus on the relationship between grid size and accuracy of the solution. Several common algorithms for marching the solution forward in time from the initial condition are also given. Subsequently, the finite volume and finite element methods are presented, with the main objective of understanding how they work.

6.2.1 Finite difference method

In the finite difference method (FDM), time and space are divided into a structured grid, the continuous field variable is approximated by its value at discrete grid points, and the value of the field variable at intermediate points is obtained by interpolation. We begin in 1-D, and then extend the results to higher dimensions. The 1-D heat conduction equation is given by

$$\frac{\partial T}{\partial t} = \alpha \frac{\partial^2 T}{\partial x^2} \qquad 0 \leq x \leq L; \;\; t > 0 \tag{6.1}$$

The numerical solution of Eq. (6.1) starts with the discretization of the spatial and time domain on a grid such as the one illustrated in Fig. 6.1. The temperature field is represented at a series of grid points in space

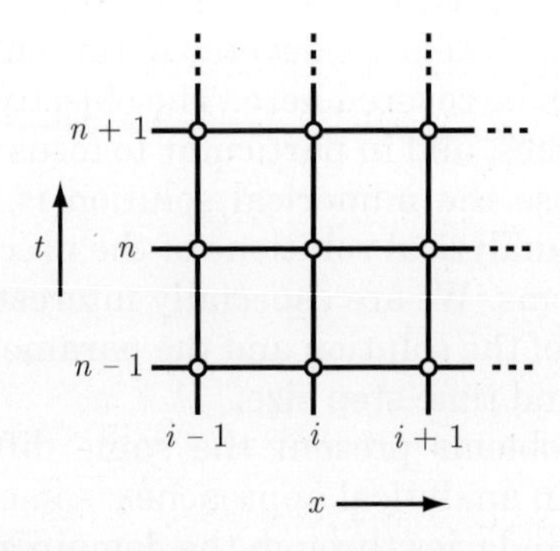

Fig. 6.1 A schematic representation of the discretization of time and space for one-dimensional problems.

called *nodes*, and the method follows the temperature history at these nodes through a sequence of discrete time steps. We define

$$T(x_i, t_n) = T_i^n = T((i-1)\Delta x, (n-1)\Delta t), \quad i = 1, \ldots I; \quad n = 1, 2, \ldots \tag{6.2}$$

where Δx and Δt are the spatial and time increments, respectively. For this discussion, both Δx and Δt are taken to be constant, although this is not a requirement. Approximations for spatial derivatives are constructed using Taylor series expansions about T_i^n at a certain fixed time t:

$$T_{i+1}^n = T_i^n + \Delta x \left.\frac{\partial T}{\partial x}\right|_{x_i,t} + \frac{\Delta x^2}{2}\left.\frac{\partial^2 T}{\partial x^2}\right|_{x_i,t} + \frac{\Delta x^3}{6}\left.\frac{\partial^3 T}{\partial x^3}\right|_{x_i,t} + \frac{\Delta x^4}{24}\left.\frac{\partial^4 T}{\partial x^4}\right|_{x_i,t} + \cdots \tag{6.3}$$

$$T_{i-1}^n = T_i^n - \Delta x \left.\frac{\partial T}{\partial x}\right|_{x_i,t} + \frac{\Delta x^2}{2}\left.\frac{\partial^2 T}{\partial x^2}\right|_{x_i,t} - \frac{\Delta x^3}{6}\left.\frac{\partial^3 T}{\partial x^3}\right|_{x_i,t} + \frac{\Delta x^4}{24}\left.\frac{\partial^4 T}{\partial x^4}\right|_{x_i,t} + \cdots \tag{6.4}$$

In order to make the presentation less cluttered, the x_i, t subscripts are dropped from this point on, but it is to be understood that the derivatives are evaluated at (x_i, t). Subtracting Eq. (6.4) from (6.3), and solving for $\partial T/\partial x$ gives

$$\frac{\partial T}{\partial x} = \frac{T_{i+1} - T_{i-1}}{2\Delta x} + \mathcal{O}(\Delta x^2) \tag{6.5}$$

The superscripts for T_{i+1} and T_{i-1} have been omitted since Eq. (6.5) is valid for any time t. Truncating the series after the first term on the right hand side of Eq. (6.5) gives the *central difference* form for the first derivative, accurate to order Δx^2. The order of the truncation error is significant in that it indicates how the scheme should converge to the exact solution as the grid spacing decreases. Following a similar procedure, but this time adding Eqs. (6.3) and (6.4), yields an expression for the second derivative:

$$\frac{\partial^2 T}{\partial x^2} = \frac{T_{i-1} - 2T_i + T_{i+1}}{\Delta x^2} + \mathcal{O}(\Delta x^2) \tag{6.6}$$

As before, truncating the series after the first term leads to an expression that is accurate to order Δx^2.

One can follow a similar procedure for the derivative with respect to time. Several schemes are obtained when the spatial derivatives are evaluated at different times $t_{n+\zeta}$, where $0 \le \zeta \le 1$. Expanding the Taylor series in time about the points (x_i, t_n) and (x_i, t_{n+1}) gives

$$T_i^{n+\zeta} = T_i^n + \frac{\partial T}{\partial t}\zeta\Delta t + \frac{1}{2}\frac{\partial^2 T}{\partial t^2}\zeta^2\Delta t^2 + \cdots \tag{6.7}$$

$$T_i^{n+\zeta} = T_i^{n+1} - \frac{\partial T}{\partial t}(1-\zeta)\Delta t + \frac{1}{2}\frac{\partial^2 T}{\partial t^2}(1-\zeta)^2\Delta t^2 + \cdots \tag{6.8}$$

Subtracting Eq. (6.7) from (6.8) and solving for $\partial T/\partial t$ gives

$$\frac{\partial T}{\partial t} = \frac{T_i^{n+1} - T_i^n}{\Delta t} + \frac{1-2\zeta}{2}\frac{\partial^2 T}{\partial t^2}\Delta t + \mathcal{O}(\Delta t^2) \tag{6.9}$$

To form a difference equation, Eq. (6.6) is first evaluated at time $t_{n+\zeta}$ using linear interpolation between the values at t_n and t_{n+1}, i.e.,

$$T_i^{n+\zeta} = (1-\zeta)T_i^n + \zeta T_i^{n+1} \tag{6.10}$$

Thus, Eq. (6.6) becomes

$$\frac{\partial^2 T}{\partial x^2} = (1-\zeta)\left(\frac{T_{i-1}^n - 2T_i^n + T_{i+1}^n}{\Delta x^2}\right) + \zeta\left(\frac{T_{i-1}^{n+1} - 2T_i^{n+1} + T_{i+1}^{n+1}}{\Delta x^2}\right) + \mathcal{O}(\Delta x^2) \tag{6.11}$$

Substituting Eqs. (6.9), and then (6.11) into Eq. (6.1), followed by separating the terms evaluated at t_n and t_{n+1} to opposite sides of the resulting equation, gives

$$\begin{aligned} T_i^{n+1} - \mathrm{Fo}_g\zeta\left(T_{i-1}^{n+1} - 2T_i^{n+1} + T_{i+1}^{n+1}\right) &= T_i^n + \mathrm{Fo}_g(1-\zeta)\left(T_{i-1}^n - 2T_i^n + T_{i+1}^n\right) \\ &\quad + \frac{1-2\zeta}{2}\frac{\partial^2 T}{\partial t^2}\Delta t + \mathcal{O}(\Delta t^2, \Delta x^2) \end{aligned} \tag{6.12}$$

where $\mathrm{Fo}_g = \alpha\Delta t/\Delta x^2$ is called the *grid Fourier number*.

Three common schemes can be obtained by truncating the series by eliminating the last term in Eq. (6.12) choosing different values of ζ: Explicit ($\zeta = 0$), Crank-Nicolson ($\zeta = 1/2$), and Fully Implicit or Backward Euler ($\zeta = 1$). Notice that for any value of $\zeta \neq 1/2$, the scheme is accurate to $\mathcal{O}(\Delta t, \Delta x^2)$. However, in the Crank-Nicolson scheme ($\zeta = 1/2$), the coefficient of the second time derivative error term vanishes, and the scheme becomes $\mathcal{O}(\Delta t^2, \Delta x^2)$. The *domain of dependence* for T_i^{n+1} is often illustrated by a *stencil*, as can be seen for these three common schemes in Fig. 6.2.

An in-depth discussion of the accuracy and stability of the various algorithms can be found in many texts devoted to numerical methods for partial differential equations. We present one example for the explicit scheme, and then simply state the corresponding results for the others.

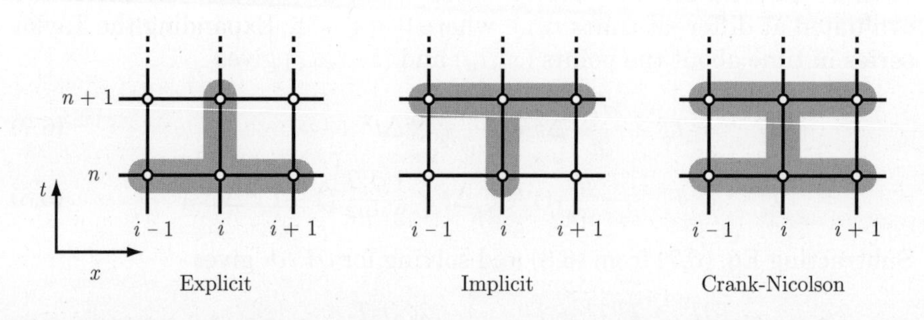

Fig. 6.2 Stencils for the three common finite difference schemes.

Consider the one-dimensional linear heat conduction problem

$$\frac{\partial T}{\partial t} = \alpha \frac{\partial^2 T}{\partial x^2} \qquad 0 \leq x \leq L \tag{6.13}$$

$$T = T_0 \qquad x = 0 \tag{6.14}$$

$$-k\frac{\partial T}{\partial x} = h_T(T - T_\infty) \qquad x = L \tag{6.15}$$

$$T = T(x) \qquad t = 0 \tag{6.16}$$

The explicit finite difference equations are obtained by setting $\zeta = 0$ in Eq. (6.12). At the interior nodes $i = 2, \ldots I - 1$, we have

$$T_i^{n+1} = \mathrm{Fo}_g T_{i-1}^n + (1 - 2\mathrm{Fo}_g)T_i^n + \mathrm{Fo}_g T_{i+1}^n + \mathcal{O}(\Delta t, \Delta x^2) \tag{6.17}$$

In order to complete the scheme, the boundary conditions need to be discretized. The boundary condition at $x = 0$ is simply

$$T_1^{n+1} = T_0 \tag{6.18}$$

To obtain the finite difference form for the boundary condition at $x = L$ ($i = I$), Eq. (6.5) is used to evaluate the first derivative at time t_n

$$\frac{T_{I+1}^n - T_{I-1}^n}{2k\Delta x} = h_T(T_I^n - T_\infty) + \mathcal{O}(\Delta x^2) \tag{6.19}$$

The grid point at $i = I + 1$ lies outside the domain. We resolve this problem by ensuring that, at node I, *both* the difference equation, Eq. (6.17), and the boundary condition, Eq. (6.19), are satisfied. Solving Eq. (6.19) for T_{I+1}^n and substituting the result in Eq. (6.17) gives

$$T_I^{n+1} = T_I^n + 2\mathrm{Fo}_g \left(T_{I-1}^n - \left(1 + \frac{h_T \Delta x}{k}\right) T_I^n + \frac{h_T \Delta x}{k} T_\infty \right) \tag{6.20}$$

The expression $h_T \Delta x / k$ is called the grid Biot number. However, in order to avoid introducing too many new symbols, this expression is left as is.

Equations (6.17), (6.18) and (6.20) constitute an explicit scheme for marching the solution forward in time from any specified initial condition, Eq. (6.16). Notice that we have been careful to ensure that the derivatives and boundary conditions are evaluated to the same level of accuracy, in order for the overall scheme to remain $\mathcal{O}(\Delta t, \Delta x^2)$. Coding of the explicit algorithm is trivial. For instance, it can be performed easily using a spreadsheet. This is shown in Exercise 6.2.

The accuracy of a numerical method can be determined by examining its behavior as the grid spacing and time step size decrease. This is done using the *modified equation* that can be thought of as the answer to the question: "To what differential equation does my difference equation correspond?" To that end, one should begin with the finite difference form of Eq. (6.17), and substitute the Taylor series expanded around T_i^n for each term. The result, after a certain amount of algebra, is

$$\frac{\partial T}{\partial t} - \alpha \frac{\partial^2 T}{\partial x^2} = \left(\frac{\alpha \Delta x^2}{12} \frac{\partial^4 T}{\partial x^4} - \frac{\Delta t}{2} \frac{\partial^2 T}{\partial t^2} \right) + \mathcal{O}(\Delta t^2, \Delta x^4) \tag{6.21}$$

The right hand side of Eq. (6.21) thus represents the truncation error in the difference equation. The behavior of the system with respect to changes in grid spacing becomes more clear when $\partial^2 T/\partial t^2$ is written in terms of $\partial^4 T/\partial x^4$. The result, obtained by multiple differentiation of Eq. (6.21) (see Exercise 6.3), is

$$\frac{\partial T}{\partial t} - \alpha \frac{\partial^2 T}{\partial x^2} = \frac{\alpha \Delta x^2}{2} \left(\frac{1}{6} - \mathrm{Fo}_g \right) \frac{\partial^4 T}{\partial x^4} + \mathcal{O}(\Delta t^2, \Delta x^4) \tag{6.22}$$

The modified equation shows that the explicit scheme is $\mathcal{O}(\Delta t, \Delta x^2)$, which, of course, we already knew. A closer examination of the first term on the right hand side provides some other interesting features. For $\mathrm{Fo}_g = 1/6$, the coefficient of $\partial^4 T/\partial x^4$ vanishes, and the scheme becomes $\mathcal{O}(\Delta t^2, \Delta x^4)$. One should also note that if Δt is fixed and a series of simulations is performed with $\Delta x \to 0$, there should be a minimum in the absolute error corresponding to $\mathrm{Fo}_g = 1/6$.

The explicit scheme is relatively easy to code, but suffers from the major disadvantage of being unstable for values of $\mathrm{Fo}_g > 1/2$, as shown in Exercise 6.1. In practice, one chooses the spatial resolution desired by selecting Δx, and the stability criterion then requires that $\Delta t \leq \Delta x^2/2\alpha$. The time step size dictated by the stability limit of the explicit scheme becomes unacceptable in most problems, especially in higher dimensions. Fortunately, better schemes can be constructed by choosing other values for ζ in Eq. (6.9). Selecting $\zeta = 1$ gives the Fully Implicit, or Backward Euler scheme, for which the difference equation at each interior node is given by

$$-\mathrm{Fo}_g T_{i-1}^{n+1} + (1 + 2\mathrm{Fo}_g) T_i^{n+1} - \mathrm{Fo}_g T_{i+1}^{n+1} = T_i^n + \mathcal{O}(\Delta t, \Delta x^2) \tag{6.23}$$

This scheme is referred to as *implicit* since the values at the next time step are interdependent. This leads to a system of simultaneous algebraic equations that must be solved for the values at the next time step. Boundary conditions are determined in the same way as for the explicit scheme. Let us continue with the same example described earlier in Eqs. (6.13)-

(6.15). Assembling all of the equations in the system from Eq. (6.23) into matrix form gives

$$\begin{bmatrix} 1 & 0 & 0 & \cdot & \cdot & \cdot & 0 \\ -\mathrm{Fo}_g & 1+2\mathrm{Fo}_g & -\mathrm{Fo}_g & 0 & \cdot & \cdot & \cdot \\ 0 & -\mathrm{Fo}_g & 1+2\mathrm{Fo}_g & -\mathrm{Fo}_g & 0 & \cdot & \cdot \\ \cdot & \ddots & \ddots & \ddots & \ddots & \ddots & \cdot \\ \cdot & \cdot & \ddots & \ddots & \ddots & \ddots & 0 \\ \cdot & \cdot & \cdot & 0 & -\mathrm{Fo}_g & 1+2\mathrm{Fo}_g & -\mathrm{Fo}_g \\ 0 & \cdot & \cdot & \cdot & 0 & -2\mathrm{Fo}_g & b_1 \end{bmatrix} \begin{Bmatrix} T_1^{n+1} \\ T_2^{n+1} \\ \cdot \\ \cdot \\ \cdot \\ \cdot \\ T_{I-1}^{n+1} \\ T_I^{n+1} \end{Bmatrix} = \begin{Bmatrix} T_0 \\ T_2^n \\ \cdot \\ \cdot \\ \cdot \\ \cdot \\ T_{I-1}^n \\ b_2 \end{Bmatrix} \tag{6.24}$$

where

$$b_1 = 1 + 2\mathrm{Fo}_g\left(1 + \frac{h_T \Delta x}{k}\right) \quad \text{and} \quad b_2 = T_I^n + 2\mathrm{Fo}_g T_\infty \frac{h_T \Delta x}{k} \tag{6.25}$$

came from the boundary condition at $x = L$. The matrix has a tridiagonal structure, originating from the form of the finite difference stencil. Linear equation solvers that take advantage of the tridiagonal form of the matrix are commonly available. As opposed to the explicit scheme, the Backward Euler scheme is unconditionally stable, thus permitting much larger time steps to be taken. One must nevertheless keep in mind that the truncation error remains $\mathcal{O}(\Delta t)$. The modified equation for this scheme is

$$\frac{\partial T}{\partial t} - \alpha \frac{\partial^2 T}{\partial x^2} = \left(\frac{\alpha \Delta x^2}{12} + \frac{\alpha^2 \Delta t}{2}\right) \frac{\partial^4 T}{\partial x^4} + \mathcal{O}(\Delta t^2, \Delta x^4) \tag{6.26}$$

The scheme is thus $\mathcal{O}(\Delta t, \Delta x^2)$, and there exists no choice of Fo_g that will improve the order of accuracy, as was the case for the explicit scheme. Note that, for this scheme, one would expect the absolute error to decrease monotonically as $\Delta x \to 0$, however the truncation error associated with Δt remains, even for very small Δx .

Finally, the Crank-Nicolson scheme is obtained by setting $\zeta = 1/2$, yielding the finite difference form

$$-\left(\frac{\mathrm{Fo}_g}{2}\right) T_{i-1}^{n+1} + (1 + \mathrm{Fo}_g)\, T_i^{n+1} - \left(\frac{\mathrm{Fo}_g}{2}\right) T_{i+1}^{n+1} = \left(\frac{\mathrm{Fo}_g}{2}\right) T_{i-1}^{n} + (1 - \mathrm{Fo}_g)\, T_i^{n} + \left(\frac{\mathrm{Fo}_g}{2}\right) T_{i+1}^{n} + \mathcal{O}(\Delta t^2, \Delta x^2) \tag{6.27}$$

Notice that this scheme is second order in both time and space. Moreover, it is an unconditionally stable implicit method, for which the matrix equation is similar in form to that of the Backward Euler scheme given in Eq. (6.24). The modified equation for this method is given by

$$\frac{\partial T}{\partial t} - \alpha \frac{\partial^2 T}{\partial x^2} = \frac{\alpha \Delta x^2}{12} \frac{\partial^4 T}{\partial x^4} + \mathcal{O}(\Delta t^2, \Delta x^4) \tag{6.28}$$

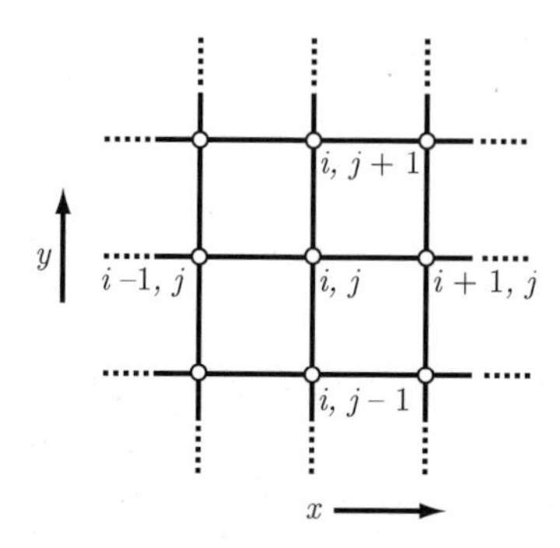

Fig. 6.3 A schematic representation of the spatial discretization for two-dimensional problems.

Extension to higher dimensions

The extension of the finite difference scheme to higher dimensions is straightforward. The only new matter consists in the evaluation of the finite difference representation of the Laplacian, as the time derivatives are evaluated in exactly the same way as in one dimension. Consider first a two-dimensional spatial grid, as illustrated in Fig. 6.3. Here, the x-direction discretization is labeled using i, as before, and the y-direction is labeled with j. For simplicity, we choose $\Delta x = \Delta y$, but that is not a requirement. Since the x- and y- directions are orthogonal, the extension of Eq. (6.6) is straightforward,

$$\begin{aligned} \nabla^2 T &= \frac{\partial^2 T}{\partial x^2} + \frac{\partial^2 T}{\partial y^2} \\ &= \frac{T_{i-1,j} - 2T_{i,j} + T_{i+1,j}}{\Delta x^2} + \frac{T_{i,j-1} - 2T_{i,j} + T_{i,j+1}}{\Delta x^2} + \mathcal{O}(\Delta x^2) \end{aligned} \quad (6.29)$$

This so-called five-point stencil is biased along the coordinate axes. For this reason, some analysts use a form where the representation given in Eq. (6.29) is averaged with one that is constructed by rotating the axes by 45°,

$$(\nabla^2 T)' = \frac{T_{i-1,j-1} + T_{i+1,j-1} + T_{i-1,j+1} + T_{i+1,j+1} - 4T_{i,j}}{2\Delta x^2} + \mathcal{O}(\Delta x^2) \quad (6.30)$$

Representing the Laplacian by the arithmetic mean of the expressions in Eqs. (6.29) and (6.30) results in a nine-point stencil, sometimes called the "isotropic Laplacian"

$$\begin{aligned} (\nabla^2 T)_{iso} \sim \frac{1}{4\Delta x^2} [& 2(T_{i-1,j} + T_{i+1,j} + T_{i,j-1} + T_{i,j+1}) \\ & + T_{i-1,j-1} + T_{i+1,j-1} + T_{i-1,j+1} + T_{i+1,j+1} - 12T_{i,j}] + \mathcal{O}(\Delta x^2) \end{aligned} \quad (6.31)$$

Similar forms can be derived for three dimensions. (See Exercise 6.4.) The time integration schemes remain unchanged in higher dimensions. Note, however, that the stability limit for the explicit algorithm becomes even more restrictive

$$\mathrm{Fo}_g^{2D} = \frac{\alpha \Delta t}{\Delta x^2 + \Delta y^2} \leq \frac{1}{4}; \qquad \mathrm{Fo}_g^{3D} = \frac{\alpha \Delta t}{\Delta x^2 + \Delta y^2 + \Delta z^2} \leq \frac{1}{6} \tag{6.32}$$

The matrix equation corresponding to Eq. (6.24) has a more complex form in higher dimensions; it becomes *banded*, which means that there are non-zero entries in a band of diagonals about the main diagonal. In one-dimension the bandwidth was two, i.e., there was only one filled diagonal above the main one. In higher dimensions, one eventually renumbers the nodes using a single index. The bandwidth is then larger, corresponding to the maximum difference in the number of nodes between neighboring rows in the grid. If the domain is rectangular in 2-D, or a simple box in 3-D, then the non-zero entries are confined to five or seven diagonals, respectively. In more complex domains, the non-zero entries in the coefficient matrix do not necessarily have a simple arrangement. The matrix is nevertheless *sparse*, indicating that there are many embedded zeros.

If a direct solver is used, the bandwidth affects the time required to solve the equations. Iterative solvers, particularly those designed for solving problems with sparse matrices, are less sensitive to bandwidth, and therefore find more frequent application for these problems. The reader is referred to texts on numerical methods for further details concerning equation solvers.

6.2.2 Finite volume method

The finite volume method (FVM) represents a different scheme for generating the discrete form of the governing equation at each node. In order to focus on the essentials of the method itself, rather than the details of the implementation, transient conduction is considered in two dimensions with all material properties taken as constant. A finite difference in time is still employed, thus leading to only the spatial derivatives being affected by the use of the FVM for discretization. Therefore, the same time integration schemes described previously still apply here. For the sake of simplicity, the equations are developed for the explicit scheme.

We begin with the energy balance equation written in integral form,

$$\int_V \rho c_p \frac{\partial T}{\partial t} dV + \int_A \boldsymbol{q} \cdot \boldsymbol{n} \, dA = 0 \tag{6.33}$$

where $\boldsymbol{q}$ is the heat flux vector and $\boldsymbol{n}$ is the normal to the surface. In the FVM, a grid is placed over the domain, and non-overlapping control volumes are defined around each node, as illustrated in Fig. 6.4. The integration given in Eq. (6.33) is subsequently carried out for each control volume, and the summation over all control volumes then represents Eq. (6.33).

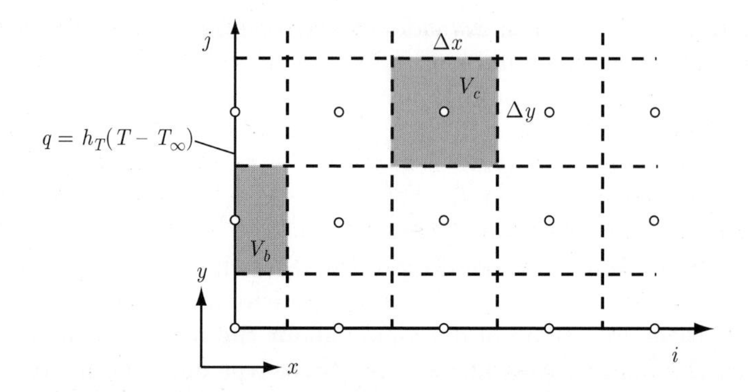

Fig. 6.4 A Cartesian grid with control volumes shown in gray in the interior and at the boundary.

The real power of the FVM derives from the fact that the volumes no longer have to conform to a structured grid. However, it is easier to illustrate the method and compare the results to the FDM when using a Cartesian grid. Consider first the interior control volume V_c, highlighted as a gray square in Fig. 6.4. The surface integral in Eq. (6.33) becomes a line integral in two dimensions, and, by convention, the path is traversed counter-clockwise. The choice of direction is not important, however it is essential that all volumes be traversed in the same sense to ensure that the fluxes are balanced in neighboring control volumes. In 2-D, the outward pointing normal vector is given by $\boldsymbol{n}\, dS = \hat{\mathbf{x}}\, dy - \hat{\mathbf{y}}\, dx$, and we have

$$\int_A \boldsymbol{q} \cdot \boldsymbol{n}\, dA = \oint (q_x dy - q_y dx) \tag{6.34}$$

The flux $\boldsymbol{q} = -k\nabla T$ (Eq. (4.123)) is evaluated at the midpoint of each face, with the result

$$\int_{A_c} \boldsymbol{q}\cdot\boldsymbol{n}\, dA_c = \left(k\frac{\partial T}{\partial x}\bigg|_{i-1/2,j} - k\frac{\partial T}{\partial x}\bigg|_{i+1/2,j} \right)\Delta y + \left(k\frac{\partial T}{\partial y}\bigg|_{i,j-1/2} - k\frac{\partial T}{\partial y}\bigg|_{i,j+1/2} \right)\Delta x \tag{6.35}$$

In the explicit scheme, each of the derivatives on the right hand side of Eq. (6.35) is evaluated at time t_n, using the central difference form given in Eq. (6.5) and the time derivative is evaluated using Eq. (6.9) with $\zeta = 0$, considering the nodal value $T_{i,j}$ to represent the entire volume. When these substitutions are performed, the following nodal equation is obtained for $T_{i,j}^{n+1}$:

$$\begin{aligned}(\rho c_p \Delta x \Delta y)&\frac{T_{i,j}^{n+1} - T_{i,j}^{n}}{\Delta t} \\ &= k\left(\frac{T_{i-1,j}^{n} - 2T_{i,j}^{n} + T_{i+1,j}^{n}}{\Delta x}\Delta y + \frac{T_{i,j-1}^{n} - 2T_{i,j}^{n} + T_{i,j+1}^{n}}{\Delta y}\Delta x \right)\end{aligned} \tag{6.36}$$

Dividing Eq. (6.36) by $\rho c_p \Delta x \Delta y$ yields an equation in which the right hand side is identical to the finite difference form for the Laplacian given in Eq. (6.29).

Boundary conditions are handled differently in the FVM than in the FDM. Consider the rectangular gray region V_b at the boundary of Fig. 6.4. For this control volume, the flux at the external boundary is evaluated from the boundary condition. The other faces are internal, and are thus evaluated as before. For V_b ($i = 1$), the result is

$$\rho c_p \left(\frac{T_{1,j}^{n+1} - T_{1,j}^{n}}{\Delta t} \right) \left(\frac{\Delta x \Delta y}{2} \right) = \left(k \frac{T_{2,j} - T_{1,j}}{\Delta x} - h_T \left(T_{1,j}^{n} - T_{\infty} \right) \right) \Delta y + \left(k \frac{T_{1,j+1} - T_{1,j}}{\Delta y} - k \frac{T_{1,j} - T_{1,j-1}}{\Delta y} \right) \frac{\Delta x}{2} \tag{6.37}$$

A little manipulation of Eq. (6.37), left as an exercise, will show that this form is identical to the FDM form of the boundary condition given in Eq. (6.20).

We have demonstrated that, when implemented in a rectangular Cartesian grid, the FVM approach produces the same nodal equations as its FDM counterpart. The real advantage of the FVM is that second order accuracy can be maintained even on irregular grids. An example of such a grid is illustrated in Fig. 6.5, where the interior and boundary control volumes are depicted in light gray for each node. The formulation of the nodal equations follows the same path as for the Cartesian grid, with the exception of the method for evaluating the area and boundary integrals. The FVM is closely related to the finite element method, described in the next section.

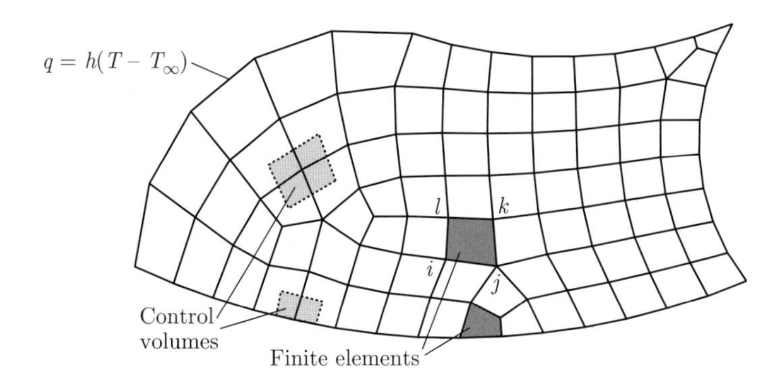

Fig. 6.5 An unstructured grid on an irregular shape, indicating control volumes in light gray and finite elements in dark gray.

6.2.3 Finite element method

The finite element method (FEM) is another common approach for solving heat conduction problems. This method is particularly well suited to irregular geometries. One begins by dividing the domain into discrete units, or *finite elements*, such as 2-dimensional triangles and quadrilaterals (see Fig. 6.5), or 3-dimensional tetrahedra, wedges and bricks. Each element is distinguished by the *connectivity* of its member nodes. For example, the interior element in Fig. 6.5 is defined as the quadrilateral consisting of nodes (i, j, k, l). The field variable, e.g., the temperature T, is interpolated within the element using a simple algebraic function, most often a polynomial.

Figure 6.6 gives an example of the representation of a 1-D temperature field using linear finite elements, which are developed further in order to illustrate the method. Let us focus our attention on element e, highlighted in gray, consisting of nodes i and j. As shown, there is a *shape function* N_i^e associated with node i in element e with the following properties: $N_i^e = 1$ at $x = x_i$; $N_i^e = 0$ at $x = x_j$, and $N_i^e = 0$ outside of element e. The shape function N_j^e associated with node j has similar properties. At any x inside of e, the shape functions sum to one. This is a general property that applies to any shape function in any number of dimensions, regardless of the interpolating polynomial. These properties ensure that any coordinate in the element can be written in terms of the element's nodal coordinates (x_i, x_j) as

$$x = N_i^e(x)\, x_i + N_j^e(x)\, x_j = [N_i^e \;\; N_j^e] \begin{Bmatrix} x_i \\ x_j \end{Bmatrix} = [\boldsymbol{N}^e]\{\boldsymbol{x}^e\} \tag{6.38}$$

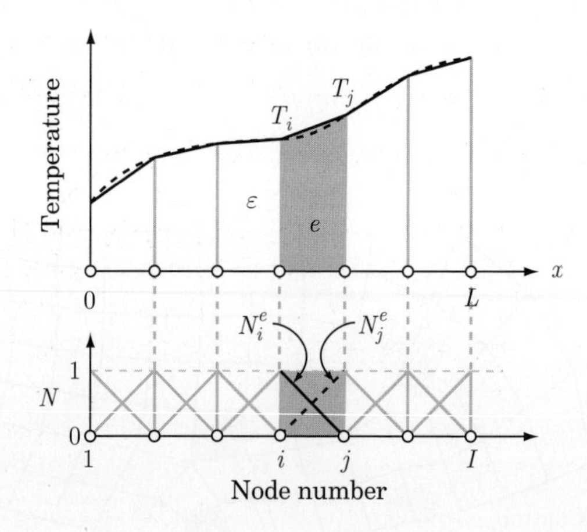

Fig. 6.6 An example of the approximation of a 1-D temperature distribution using linear finite elements. The shape functions associated with each node are depicted on the lower axes. A single element, consisting of nodes i and j is highlighted in gray.

The definition of the shape function matrix $[\boldsymbol{N}]$ and the vector of nodal coordinates $\{\boldsymbol{x}^e\}$ provides a convenient shorthand notation. It should be noted that more nodes are required to define the element and the shape functions if one uses a higher order polynomial for interpolation, e.g., quadratic interpolation in 1-D requires three nodes per element. The general properties given above, however, remain valid (see Exercise 6.5).

There are several versions of the FEM. The most common one, and the only one discussed here, is called the *isoparametric* form, in which the field variables are interpolated using the same shape functions as those used for the coordinates. In this case, we can write the temperature T in element e as

$$T(x) = N_i^e(x)\, T_i + N_j^e(x)\, T_j = [\boldsymbol{N}^e]\{\boldsymbol{T}^e\}; \qquad x_i \leq x \leq x_j \tag{6.39}$$

For the particular linear shape functions illustrated in Fig. 6.6, we have

$$N_i^e(x) = \frac{x_j - x}{x_j - x_i} = \frac{x_j - x}{\Delta x^e}; \qquad N_j^e(x) = \frac{x - x_i}{x_j - x_i} = \frac{x - x_i}{\Delta x^e} \tag{6.40}$$

and the linear form for the temperature

$$T = T_i + \frac{T_j - T_i}{x_j - x_i}(x - x_i); \qquad x_i \leq x \leq x_j \tag{6.41}$$

The spatial derivative of the temperature within the element is computed by differentiating Eq. (6.39) with respect to x,

$$\frac{\partial T}{\partial x} = \frac{d[\boldsymbol{N}^e]}{dx}\{\boldsymbol{T}^e\} = [\boldsymbol{B}^e]\{\boldsymbol{T}^e\} \tag{6.42}$$

where $[\boldsymbol{B}^e]$ contains the derivatives of the shape functions. For the 1-D linear element,

$$[\boldsymbol{B}^e] = \begin{bmatrix} \dfrac{dN_i^e}{dx} & \dfrac{dN_j^e}{dx} \end{bmatrix} = \begin{bmatrix} -\dfrac{1}{\Delta x^e} & \dfrac{1}{\Delta x^e} \end{bmatrix} \tag{6.43}$$

The time derivative is usually computed using a finite difference in time,

$$\frac{\partial T}{\partial t} = \frac{1}{\Delta t}[\boldsymbol{N}^e]\left(\{\boldsymbol{T}^e\}^{n+1} - \{\boldsymbol{T}^e\}^n\right) \tag{6.44}$$

With these preliminaries in hand, we are ready to derive the FEM form of the governing equations. We begin with the usual 1-D conduction problem

$$\rho c_p \frac{\partial T}{\partial t} - \frac{\partial}{\partial x}\left(k\frac{\partial T}{\partial x}\right) = 0 \tag{6.45}$$

In the FEM, Eq. (6.45) is satisfied in the weighted average sense in each element, i.e.,

$$\int_{x_i}^{x_j} W^e \rho c_p \frac{\partial T}{\partial t}\, dx - \int_{x_i}^{x_j} W^e \frac{\partial}{\partial x}\left(k\frac{\partial T}{\partial x}\right) dx = 0 \tag{6.46}$$

where W^e is a weighting function for the element, to be defined after a little manipulation of the equation. Clearly, summing Eq. (6.46) over all elements ensures that Eq. (6.45) is satisfied in the weighted average sense over the entire domain. Next, the second integral is expanded using integration by parts to obtain

$$\int_{x_i}^{x_j} W^e \rho c_p \frac{\partial T}{\partial t}\, dx + \int_{x_i}^{x_j} \frac{\partial W^e}{\partial x}\left(k\frac{\partial T}{\partial x}\right) dx = W^e k \frac{\partial T}{\partial x}\bigg|_{x_i}^{x_j} \tag{6.47}$$

The partial derivatives inside the integrals on the left hand side are now replaced by their FEM forms, using Eqs. (6.42) and (6.44).

$$\left(\int_{x_i}^{x_j} W^e \rho c_p [\boldsymbol{N}]\, dx\right) \frac{\{\boldsymbol{T}^e\}^{n+1} - \{\boldsymbol{T}^e\}^n}{\Delta t} + \left(\int_{x_i}^{x_j} \frac{\partial W^e}{\partial x} k[\boldsymbol{B}^e]\, dx\right)\{\boldsymbol{T}^e\}^{n+\zeta} = W^e k \frac{\partial T}{\partial x}\bigg|_{x_i}^{x_j} \tag{6.48}$$

Notice that, just as in the FDM, different schemes are produced by choosing to evaluate the spatial terms at different times $t_{n+\zeta}$. In order to compare the FEM results with the earlier results for FDM and FVM, we choose the explicit scheme, i.e., $\zeta = 0$. Also, note that the nodal temperature vectors have been taken outside of the integrals, as the nodal temperatures do not depend on location.

To go further, we must choose the weighting function W^e. Different versions of the FEM are obtained by selecting different forms for W^e. The most common form is known as *Galerkin's method*, in which one chooses $W^e = [\boldsymbol{N}^e]^T$, the transpose of the element shape function matrix. Carrying out this substitution in Eq. (6.48) gives

$$\left(\int_{x_i}^{x_j} [\boldsymbol{N}^e]^T \rho c_p [\boldsymbol{N}^e]\, dx\right) \frac{\{\boldsymbol{T}^e\}^{n+1} - \{\boldsymbol{T}^e\}^n}{\Delta t} + \left(\int_{x_i}^{x_j} [\boldsymbol{B}^e]^T k [\boldsymbol{B}^e]\, dx\right)\{\boldsymbol{T}^e\}^n = [\boldsymbol{N}^e]^T k \frac{\partial T}{\partial x}\bigg|_{x_i}^{x_j} \tag{6.49}$$

This equation can be written in more compact form as

$$[\boldsymbol{C}^e]\frac{\{\boldsymbol{T}^e\}^{n+1} - \{\boldsymbol{T}^e\}^n}{\Delta t} + [\boldsymbol{K}^e]\{\boldsymbol{T}^e\}^n = \{\boldsymbol{b}^e\} \tag{6.50}$$

where the element *capacitance matrix* $[\boldsymbol{C}^e]$, the element *conductance matrix* $[\boldsymbol{K}^e]$ (sometimes called the stiffness matrix due to a similar form appearing in elasticity problems) and the element *load vector* $\{\boldsymbol{b}^e\}$ are defined as

$$[\boldsymbol{C}^e] = \int_{x_i}^{x_j} [\boldsymbol{N}^e]^T \rho c_p [\boldsymbol{N}^e]\, dx; \quad [\boldsymbol{K}^e] = \int_{x_i}^{x_j} [\boldsymbol{B}^e]^T k [\boldsymbol{B}^e]\, dx; \quad \{\boldsymbol{b}^e\} = [\boldsymbol{N}^e]^T k \frac{\partial T}{\partial x}\bigg|_{x_i}^{x_j} \tag{6.51}$$

Recall that $N_i = 1$ and $N_j = 0$ at $x = x_i$ (and the converse at $x = x_j$). Also note that Fourier's law gives $k(\partial T/\partial x) = -q$, where q is the heat flux. Thus, the element load vector $\{\boldsymbol{b}^e\}$ takes the simpler form

$$\{\boldsymbol{b}^e\} = \left\{ \begin{array}{c} -q_i \\ q_j \end{array} \right\} \tag{6.52}$$

where the minus sign on q_i comes from the lower limit of integration. The FEM form for the entire domain is obtained by summing over all of the elements. Consider the element just to the left of Element e in Fig. 6.6 (which we call Element ε). The load vector in this element is

$$\{\boldsymbol{b}^\varepsilon\} = \left\{ \begin{array}{c} -q_{i-1} \\ q_i \end{array} \right\} \tag{6.53}$$

Thus, the fluxes at Node i exactly cancel when the load vectors for neighboring elements are summed. This is true at all interior nodes in the domain. Consequently, the summing of the load vectors produces a non-zero result at the boundaries, $x = 0, L$, and thus, summing Eq. (6.49) over all elements gives the FE form of Eq. (6.45):

$$\sum_e [\boldsymbol{C}^e] \frac{\{\boldsymbol{T}^e\}^{n+1} - \{\boldsymbol{T}^e\}^n}{\Delta t} + \sum_e [\boldsymbol{K}^e]\{\boldsymbol{T}^e\}^n = - \left. [\boldsymbol{N}]^T q_b \right|_0^L \tag{6.54}$$

It is left as an exercise to demonstrate that, for the 1-D linear element, the matrices $[\boldsymbol{K}^e]$ and $[\boldsymbol{C}^e]$ are computed from Eq. (6.49) as

$$[\boldsymbol{K}^e] = \frac{k}{\Delta x^e} \left[\begin{array}{cc} 1 & -1 \\ -1 & 1 \end{array} \right]; \qquad [\boldsymbol{C}^e] = \frac{\rho c_p \Delta x^e}{6} \left[\begin{array}{cc} 2 & 1 \\ 1 & 2 \end{array} \right] \tag{6.55}$$

The expression for $[\boldsymbol{C}^e]$ in Eq. (6.49) is called the consistent form. Often, and particularly for solidification problems, a "lumped" form of the capacitance matrix is instead used:

$$C_{i,j}^{lump} = \left\{ \begin{array}{ll} \sum\limits_j C_{i,j}^e & i = j \\ 0 & i \neq j \end{array} \right. \tag{6.56}$$

For the 1-D linear element, the lumped form is trivial to compute

$$[\boldsymbol{C}] = \frac{\rho c_p \Delta x^e}{2} \left[\begin{array}{cc} 1 & 0 \\ 0 & 1 \end{array} \right] \tag{6.57}$$

The lumped form has several attractive properties. As shown below, using the lumped form in 1-D instead of a consistent form produces identical equations to those of the FDM. However, this property does not carry over to higher dimensions. The lumped form of the mass matrix is diagonal, which makes the explicit form particularly simple. It is demonstrated in Exercise 6.6 that employing the lumped form is equivalent to a small decrease in the diffusivity. Recall that the stability criterion for the explicit scheme was $\Delta t < \Delta x^2/2\alpha$. Thus, employing the lumped form permits the use of larger time steps. A more detailed analysis of this topic can be found in the texts by Reddy, and by Gresho and Sani, listed at the end of this chapter.

The assembly of the global set of equations from the contributions of the individual element equations is essentially a bookkeeping exercise. One first defines *global* capacitance $[\boldsymbol{C}]$ and conductance $[\boldsymbol{K}]$ matrices as well as a global solution vector $\{\boldsymbol{T}\}$. The contributions of the individual elements are then mapped into the appropriate locations in the global matrices and summed. The assembly process for several linear elements of the same length Δx^e connected together with node numbers in numerical sequence is illustrated below:

$$[C] = \frac{\rho c_p \Delta x^e}{2} \begin{bmatrix} 1 & 0 & & & \\ 0 & 1+1 & 0 & & \\ & 0 & 1+1 & 0 & \\ & & 0 & 1+1 & \\ & & & & \ddots \end{bmatrix} \tag{6.58}$$

Notice that the contributions from neighboring elements are added where they overlap. A similar operation can be performed for the global conductance matrix, and the global equations after division by $(\rho c_p \Delta x^e/\Delta t)$ become

$$\begin{bmatrix} 1/2 & 0 & & \\ 0 & 1 & 0 & \\ & 0 & 1 & 0 \\ & & & \ddots \end{bmatrix} \begin{Bmatrix} T_1^{n+1} \\ T_2^{n+1} \\ T_3^{n+1} \\ \vdots \end{Bmatrix} = \begin{bmatrix} (1/2 - \mathrm{Fo}_g) & \mathrm{Fo}_g & & \\ \mathrm{Fo}_g & (1 - 2\mathrm{Fo}_g) & \mathrm{Fo}_g & \\ & \mathrm{Fo}_g & (1 - 2\mathrm{Fo}_g) & \mathrm{Fo}_g \\ & & \ddots & \end{bmatrix} \begin{Bmatrix} T_1^n \\ T_2^n \\ T_3^n \\ \vdots \end{Bmatrix} \tag{6.59}$$

Consider the row in Eq. (6.59) corresponding to the interior node i:

$$T_i^{n+1} = \mathrm{Fo}_g T_{i-1}^n + (1 - 2\mathrm{Fo}_g)T_i^n + \mathrm{Fo}_g T_{i+1}^n \tag{6.60}$$

Comparing Eq. (6.60) to Eq. (6.17) shows that the FEM produces nodal equations identical to those obtained in the FDM in 1-D. (The equations differ slightly when the consistent form is used instead of the lumped form of the capacitance matrix.) The true value of the FEM resides in its ability to handle complex geometries in higher dimensions. Nevertheless, it is useful to know that the methods produce the same nodal equations, and hence the discussion of accuracy of the FDM is valid also for the FEM.

Boundary conditions must be applied before Eq. (6.59) can be solved. If there is no flux at a node, there is nothing to be done as this is the natural boundary condition. To apply a Dirichlet boundary condition, e.g., $T_1 = T_b$, one replaces the first row of Eq. (6.59) by the boundary condition. Third kind boundary conditions are evaluated by determining the boundary flux in Eq. (6.54). This is demonstrated in any standard text on finite elements, to which the reader is referred for further details.

Extension to higher dimensions

The formulation of the FEM in higher dimensions, although fundamentally similar to that in 1-D, is sufficiently different as to warrant a more detailed explanation. The matrices and load vectors are defined as a generalization of the 1-D forms in Eq. (6.51).

$$[\boldsymbol{C}] = \int_V [\boldsymbol{N}]^T \rho c_p [\boldsymbol{N}]\, dV; \quad [\boldsymbol{K}] = \int_V [\boldsymbol{B}]^T k [\boldsymbol{B}]\, dV; \quad \{\boldsymbol{b}\} = \int_S [\boldsymbol{N}]^T k \nabla T \cdot \boldsymbol{n}\, dS \tag{6.61}$$

where the derivatives of the shape functions are given by $[\boldsymbol{B}] = \nabla[\boldsymbol{N}]$. The evaluation of the integrals is demonstrated in the context of an example. Let us first note that the element load vector represents the fluxes between elements. One must take care to ensure that the surface integrals are evaluated in the same sense in each element. This is guaranteed in most implementations by numbering all elements in the same sense, e.g., counterclockwise in 2-D.

Consider the 2-D four-noded quadrilateral element illustrated in Fig. 6.7. The power of the finite element method is its flexibility with regard to element shape. However, it can be difficult to perform the integrals needed for element matrices for an arbitrary shape. The typical process, as illustrated in Fig. 6.7, consists of carrying out the integration after mapping the element from physical (x, y) space to a reference space with coordinates (r, s), with $r(x, y) \in [-1, 1]$ and $s(x, y) \in [-1, 1]$. The method is called *isoparametric* when the field variables are interpolated using the

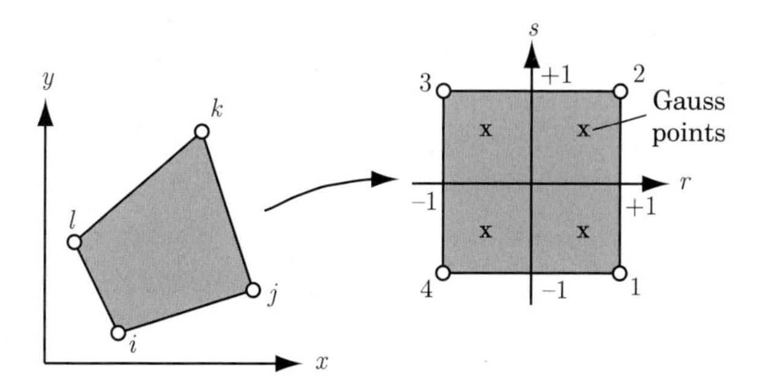

Fig. 6.7 The mapping of a quadrilateral element in physical (x, y) space to the reference (r, s) space.

same shape functions as the coordinates, i.e.,

$$\left\{ \begin{array}{c} x \\ y \end{array} \right\} = [\boldsymbol{N}]\{\boldsymbol{x}\} \; ; \quad T = [\boldsymbol{N}]\{\boldsymbol{T}\} \tag{6.62}$$

where $\{\boldsymbol{x}\}$ now represents the vectors of nodal coordinates in 2-D.

Because the shape is arbitrary, the shape functions are difficult to write in closed form in real space, whereas they are straightforward in the reference space. For the 2-D quadrilateral depicted in Fig. 6.7, the shape functions are

$$\begin{aligned} N_1 &= \frac{(1+r)(1-s)}{4} & N_2 &= \frac{(1+r)(1+s)}{4} \\ N_3 &= \frac{(1-r)(1+s)}{4} & N_4 &= \frac{(1-r)(1-s)}{4} \end{aligned} \tag{6.63}$$

Note that the shape functions are linear in the coordinate along the edges, but not in the interior of the element. The gradient requires the real space derivative of the shape functions, which can be written, using the chain rule, as

$$\left[\begin{array}{c} \partial N_m/\partial r \\ \partial N_m/\partial s \end{array} \right] = \left[\begin{array}{cc} \partial x/\partial r & \partial y/\partial r \\ \partial x/\partial s & \partial y/\partial s \end{array} \right] \left[\begin{array}{c} \partial N_m/\partial x \\ \partial N_m/\partial y \end{array} \right] = [\boldsymbol{J}] \left[\begin{array}{c} \partial N_m/\partial x \\ \partial N_m/\partial y \end{array} \right] \tag{6.64}$$

Here, $[\boldsymbol{J}]$ is the *Jacobian* of the transformation from real to reference coordinates. The various components of $[\boldsymbol{J}]$ are obtained by differentiating Eq. (6.62), e.g.,

$$\frac{\partial x}{\partial r} = \sum_{m=1}^{M} \frac{\partial N_m}{\partial r} x_m \tag{6.65}$$

where x_m is the x-coordinate of node m, and M is the total number of nodes in the element. The derivatives of the shape functions are then obtained by inverting Eq. (6.64),

$$\left[\begin{array}{c} \partial N_m/\partial x \\ \partial N_m/\partial y \end{array} \right] = [\boldsymbol{J}]^{-1} \left[\begin{array}{c} \partial N_m/\partial r \\ \partial N_m/\partial s \end{array} \right] \tag{6.66}$$

Consider now the conductance matrix, defined in Eq. (6.61). We first transform to the reference space

$$\iint_V [B]^T k [B] \; dxdy = \int_{-1}^{1} \int_{-1}^{1} [\tilde{B}]^T k [\tilde{B}] \det[\boldsymbol{J}] \; drds \tag{6.67}$$

Here, $\det[\boldsymbol{J}]$ is the determinant of the Jacobian, and the tilde in $[\tilde{B}]$ indicates that these are in (r, s) space. Recall that the inverse of the Jacobian also appears in the integral in the reference space, as shown in Eq. (6.66). It is not possible in general to find an analytical expression for $[\boldsymbol{J}]^{-1}$, and the integral in Eq. (6.67) is therefore evaluated numerically. We write this symbolically as

$$\int_{-1}^{1}\int_{-1}^{1} f(r,s)drds = \sum_{n=1}^{N} \mathcal{W}_n f(r_n, s_n) \tag{6.68}$$

where (r_n, s_n) are predefined coordinates, called integration points, for the evaluation of the integrand, and the $\mathcal{W}_n$ are weight functions. (Note that these weight functions are completely different from those in Eq. (6.46).) When the interpolating functions are polynomials, as is commonly the case in the FEM, the optimal placement of points is given by Gauss-Legendre quadrature. The locations are often called "Gauss points." For example, in the four-noded isoparametric quadrilateral element we have been considering, there are four Gauss points, located at the permutations of $(\pm\sqrt{3}/3, \pm\sqrt{3}/3)$, and the weights are all $\mathcal{W}_n = 1$. An advantage of this approach is that it is very natural to incorporate temperature-dependent material properties: one first evaluates the temperature, then the property and finally the integrand at each Gauss point.

Further details can be found in standard texts regarding the FEM. It is important for the reader to understand that the integration is typically carried out at just four points in the element. This has implications for solidification problems where the liquid-solid interface lies inside an element. The numerical schemes that are employed must be robust enough for correct solutions to be obtained in this situation. This point is discussed further in the next section.

Finally, we note that the extension to three dimensions is relatively straightforward. For example, the equivalent form to the four-noded isoparametric quadrilateral is the eight-noded "brick" element, for which eight Gauss points are defined. A mesh of the domain is built using such brick elements, and the various matrices are computed. As in the other methods that have been discussed, it can become expensive to solve the system of equations. Thus, more efficient schemes have been developed in order to solve for the nodal values. For details on this topic, the reader is referred to one of the numerous texts on the subject of equation solvers.

6.3 HEAT CONDUCTION WITH PHASE CHANGE

6.3.1 Fixed grid: Enthalpy methods

We now turn to the modeling of solidification problems using the numerical schemes presented in the preceding sections. For most metals, the ratio of latent heat to specific heat L_f/c_p is on the order of 300-400 K. For this reason, the primary difficulty is to accurately and efficiently handle the latent heat release during solidification. This section focuses on macroscopic modeling, e.g., the solidification of castings. The discretization will then take place at a length scale that is large relative to the microstructural features, but small with respect to the dimensions of the solidifying object.

Thus, the representative volume element associated with the length scale of the grid can contain both solid and liquid phases. (See Fig. 4.2.)

It is therefore appropriate to begin with the average form of the energy balance equation, Eq. (4.137)

$$\frac{\partial \langle \rho h \rangle}{\partial t} + \nabla \cdot \langle \rho h \boldsymbol{v} \rangle = \nabla \cdot \langle k \nabla T \rangle + \left\langle \rho \dot{R}_q \right\rangle \tag{6.69}$$

where $\langle k \nabla T \rangle = g_s k_s \nabla \langle T \rangle_s + g_\ell k_\ell \nabla \langle T \rangle_\ell$. In order to simplify the presentation, let us take $k_s = k_\ell = k$, and also assume that the volume is sufficiently small and the phases sufficiently intermixed that $\langle T \rangle_s = \langle T \rangle_\ell = T$. For the time being, focus is also placed on conduction, so that $\boldsymbol{v} = 0$, and internal heat generation is neglected. (It should be noted that this latter assumption indicates that $\dot{R}_q = 0$, and that it has nothing to do with the latent heat of fusion, which will appear in the term $\langle \rho h \rangle$.) Under these assumptions, Eq. (6.69) reduces to

$$\frac{\partial \langle \rho h \rangle}{\partial t} = \nabla \cdot (k \nabla T) \tag{6.70}$$

The right hand side of Eq. (6.70) is identical to that discussed in the preceding section. The new aspect for solidification is the handling of the enthalpy associated with the phase change, which constitutes the focus of this section.

The relationship between enthalpy and temperature was developed in Chap. 2

$$\langle \rho h \rangle = \int_{298}^{T} \langle \rho c_p \rangle \, dT' + \rho L_f g_\ell \tag{6.71}$$

where $\langle \rho c_p \rangle = \rho(c_{ps} g_s + c_{p\ell} g_\ell)$, and we have assumed that $\rho_s = \rho_\ell = \rho$. In many applications, the relationship between g_ℓ and temperature is considered to be known *a priori*. One such relation already encountered is the lever rule Eq. (3.11), which provides the solid fraction as a function of temperature when the solid and liquid phases are in complete equilibrium. Other forms are discussed in Chap. 10. Figure 6.8 presents an example of an enthalpy-temperature curve for an alloy.

In order to solve Eq. (6.70), it is possible to devise algorithms in which one keeps track of both the enthalpy and temperature at each node. The latent heat affects only the transient term, so all of the methods for spatial discretization discussed in the preceding sections lead to essentially the same algorithm, as they all use a finite difference in time:

$$\frac{\langle \rho h \rangle^{n+1} - \langle \rho h \rangle^{n}}{\Delta t} = \nabla \cdot (k \nabla T) \tag{6.72}$$

The right hand side can be evaluated using the FDM, FVM or FEM at any time $t_{n+\zeta}$ to obtain the various integration schemes discussed earlier. If the right hand side is evaluated at time t_n, i.e., employing an explicit scheme, a time-marching algorithm results in which the enthalpy at the next time step $\langle \rho h \rangle^{n+1}$ is computed using Eq. (6.72), after which the temperature

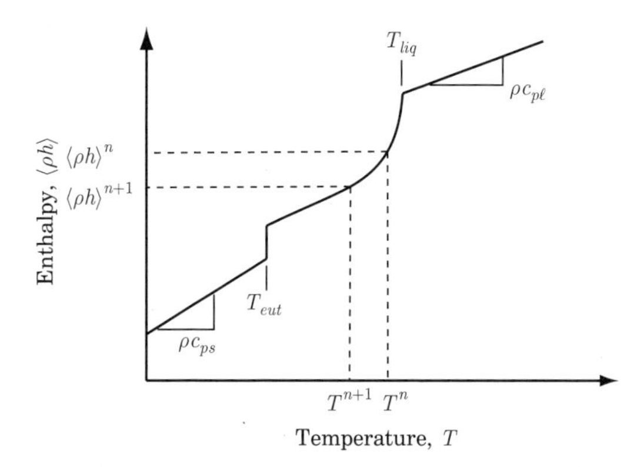

Fig. 6.8 Schematic enthalpy-temperature curve for an alloy exhibiting a eutectic reaction.

$\langle T \rangle^{n+1}$ is obtained from the enthalpy-temperature curve. Note that, in this scheme, the enthalpy can be a multi-valued function of temperature, as it would be for either a pure material or the invariant part of the curve for the alloy shown in Fig. 6.8. After the temperature has been updated, the next time step begins.

This version of the enthalpy method has certain inconvenient aspects. It can be difficult to estimate the relation between the size of the time step and the accuracy of the scheme, since the change in enthalpy varies with location as well as over the time of the simulation. Implementation of this algorithm in implicit form is difficult, as one needs to devise a means of ensuring that the proper relation between temperature and enthalpy is maintained in the solution. Failure to do so results in the loss of latent heat and/or artificial spreading of the phase boundary.

Another disadvantage is that this form of the enthalpy method is not implemented in very many commercial codes. Thus, users cannot take advantage of pre- and post-processors as well as other convenient features found in such codes. Most commercial packages implement instead a form where the left hand side of Eq. (6.70) is written in the form

$$\rho c_p^{eff}(T) \frac{\partial T}{\partial t} = \nabla \cdot (k \nabla T) \tag{6.73}$$

where $c_p^{eff}(T)$ is called the *effective specific heat*, and as noted, it is in general a function of temperature. Indeed, if $\langle \rho h \rangle$ depends only on temperature, then we can substitute Eq. (6.71) to find

$$\begin{aligned} \frac{\partial \langle \rho h \rangle}{\partial t} &= \frac{d \langle \rho h \rangle}{dT} \frac{\partial T}{\partial t} \\ &= \left(\langle \rho c_p \rangle + \rho L_f \frac{dg_\ell}{dT} \right) \frac{\partial T}{\partial t} \\ &= \rho c_p^{eff} \frac{\partial T}{\partial t} \end{aligned} \tag{6.74}$$

It turns out that this approach makes for a poor algorithm, because it is prone to large numerical errors if the grid and time steps are not sufficiently small. Equation (6.72) can be recast in a more robust form by manipulating its left hand side as follows. Since this scheme is readily implemented in implicit form, we derive it by evaluating the conduction terms at time t_{n+1}.

$$\frac{\langle \rho h \rangle^{n+1} - \langle \rho h \rangle^{n}}{\Delta t} = (\nabla \cdot (k \nabla T))^{n+1}$$
$$\left(\frac{\langle \rho h \rangle^{n+1} - \langle \rho h \rangle^{n}}{T^{n+1} - T^{n}} \right) \frac{T^{n+1} - T^{n}}{\Delta t} = (\nabla \cdot (k \nabla T))^{n+1} \tag{6.75}$$

We can then identify

$$\rho c_p^{eff} = \frac{\langle \rho h \rangle^{n+1} - \langle \rho h \rangle^{n}}{T^{n+1} - T^{n}} \tag{6.76}$$

This is sometimes called an *enthalpy-specific heat* method. It represents a very effective approach that is developed further below.

Equation (6.75) is non-linear since ρc_p^{eff} depends on the unknown temperature T^{n+1}. The solution of Eq. (6.75) begins by writing the equation in discrete form using any of the FD, FV or FE methods introduced in the preceding section. The evaluation of the right hand side of Eq. (6.75) at time $t_{n+\zeta}$ allows us to simultaneouly formulate both the explicit and implicit schemes, as in Eq. (6.12). Denoting the vector of average nodal temperatures by $\{\boldsymbol{T}\}^{n+\zeta}$, and generalizing Eq. (6.10), the components are evaluated by linear interpolation between the values at t_n and t_{n+1}

$$\{\boldsymbol{T}\}^{n+\zeta} = (1-\zeta)\{\boldsymbol{T}\}^{n} + \zeta\{\boldsymbol{T}\}^{n+1} \tag{6.77}$$

The system of equations to be solved for $\{\boldsymbol{T}\}^{n+1}$ has the same form, regardless of the method used to perform the spatial discretization:

$$\frac{1}{\Delta t}\left[\boldsymbol{C}\left(\{\boldsymbol{T}\}^{n+\zeta}\right)\right]\left(\{\boldsymbol{T}\}^{n+1} - \{\boldsymbol{T}\}^{n}\right) = \left[\boldsymbol{K}\left(\{\boldsymbol{T}\}^{n+\zeta}\right)\right]\{\boldsymbol{T}\}^{n+\zeta} \tag{6.78}$$

(For examples, see Eqs. (6.24) and (6.50).) Substituting Eq. (6.77) into Eq. (6.78) yields

$$\left(\frac{1}{\Delta t}\left[\boldsymbol{C}\left(\{\boldsymbol{T}\}^{n+\zeta}\right)\right] - \zeta\left[\boldsymbol{K}\left(\{\boldsymbol{T}\}^{n+\zeta}\right)\right]\right)\{\boldsymbol{T}\}^{n+1}$$
$$= \left(\frac{1}{\Delta t}\left[\boldsymbol{C}\left(\{\boldsymbol{T}\}^{n+\zeta}\right)\right] + (1-\zeta)\left[\boldsymbol{K}\left(\{\boldsymbol{T}\}^{n+\zeta}\right)\right]\right)\{\boldsymbol{T}\}^{n} \tag{6.79}$$

For the case $\zeta = 0$ (the explicit scheme), Eq. (6.79) becomes linear. This form is rarely used, however, both due to the stringent stability limits on Δt, and because the accuracy of the solution deteriorates rapidly as

Δt increases. The implicit forms are usually solved iteratively. One common algorithm, called *successive substitution*, implements the following procedure:

1. Estimate $\{\boldsymbol{T}\}^{n+1}$, perhaps from forward projection using $\{\boldsymbol{T}\}^{n}$ and $\{\boldsymbol{T}\}^{n-1}$. This initial estimate is denoted $\{\boldsymbol{T}\}_0^{n+1}$.

2. Evaluate $\{\boldsymbol{T}\}^{n+\zeta}$ using the current iterate (e.g., m) $\{\boldsymbol{T}\}_m^{n+1}$ in Eq. (6.77), then assemble the matrices $\left[\boldsymbol{C}\left(\{\boldsymbol{T}\}^{n+\zeta}\right)\right]$ and $\left[\boldsymbol{K}\left(\{\boldsymbol{T}\}^{n+\zeta}\right)\right]$.

3. Solve the resulting "linearized" system of equations for $\{\boldsymbol{T}\}^{n+1}$, designating this solution vector $\{\boldsymbol{T}\}_\star^{n+1}$.

4. Prepare a new estimate for $\{\boldsymbol{T}\}^{n+1}$. A common scheme is to use successive substitution with relaxation

$$\{\boldsymbol{T}\}_{m+1}^{n+1} = (1-\beta)\{\boldsymbol{T}\}_\star^{n+1} + \beta\{\boldsymbol{T}\}_m^{n+1} \tag{6.80}$$

   where β is referred to as a relaxation factor. When $\beta = 0$, this is called *Picard's method*.

5. The algorithm is then tested for convergence, as discussed further below. If the solution has converged, the time is incremented and the algorithm begins again at Step 1. If it is not converged, the algorithm returns to Step 2 and continues.

To test for convergence, one defines the *residual* vector for the $m+1$ iterate, $\{\boldsymbol{R}\}_{m+1}$, as

$$\begin{aligned}\{\boldsymbol{R}\}_{m+1} = &\left(\frac{1}{\Delta t}\left[\boldsymbol{C}\left(\{\boldsymbol{T}\}_{m+1}^{n+\zeta}\right)\right] - \zeta\left[\boldsymbol{K}\left(\{\boldsymbol{T}\}_{m+1}^{n+\zeta}\right)\right]\right)\{\boldsymbol{T}\}_{m+1}^{n+1} \\ &- \left(\frac{1}{\Delta t}\left[\boldsymbol{C}\left(\{\boldsymbol{T}\}_{m+1}^{n+\zeta}\right)\right] + (1-\zeta)\left[\boldsymbol{K}\left(\{\boldsymbol{T}\}_{m+1}^{n+\zeta}\right)\right]\right)\{\boldsymbol{T}\}^{n}\end{aligned} \tag{6.81}$$

Convergence of the iterative scheme is declared when both of the following two criteria are met:

$$\frac{\|\{\boldsymbol{T}\}_{m+1}^{n+1} - \{\boldsymbol{T}\}_m^{n+1}\|}{\|\{\boldsymbol{T}\}_m^{n+1}\|} < \epsilon_T \quad \text{and} \quad \frac{\|\{\boldsymbol{R}\}_{m+1}\|}{\|\{\boldsymbol{R}\}_0\|} < \epsilon_R \tag{6.82}$$

Here, the notation $\|\cdot\|$ indicates the L^2 norm of a vector, and ϵ_T and ϵ_R are preset error tolerances. Certain obvious programming issues related to the possibility of very small denominators exist but are not treated in detail here.

As an example of the application of the ideas presented thus far in this chapter, consider the 1-D freezing of a slab. The slab has unit thickness, and all properties are 1 (i.e., $\rho = c_p = L_f = k = T_f = 1$). Initially,

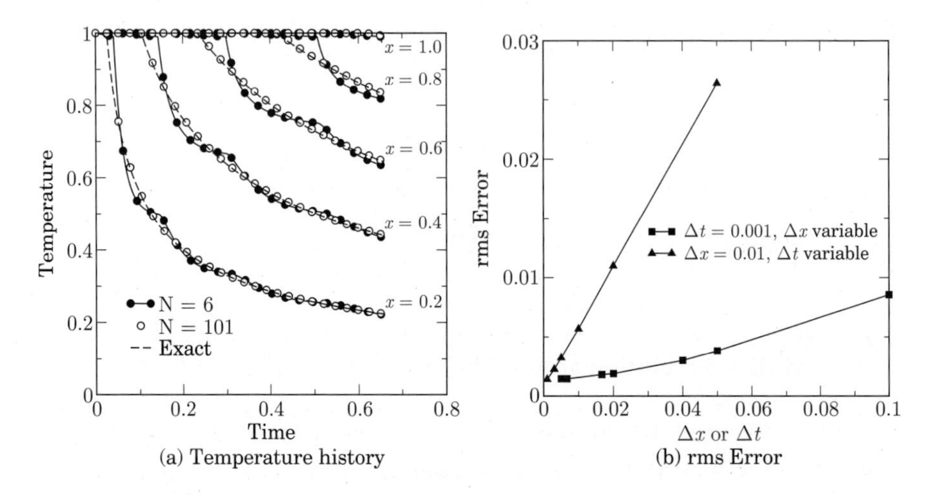

Fig. 6.9 The solution to a 1-D Stefan problem for various values of Δt and Δx.

the slab is a liquid at its melting point. At time zero, the face $x = 0$ is cooled to $T = 0$, while all other faces are insulated. This problem is readily solved analytically using the approach given in Chap. 5 to find a freezing time of 0.65. Figure 6.9(a) presents the results of the numerical solution to this problem using the enthalpy-specific heat method as the temperature history at nodes located at various distances from the surface. Results for two different grid spacings are shown, and it can be seen that the total solidification time is accurate even when there are as few as six nodes used across the domain. There are, however, internal fluctuations. The fluctuations disappear, and the numerical solution becomes indistinguishable from the exact solution, when 101 nodes are used. The rms error in the solution is displayed in Fig. 6.9(b) as a function of Δt and Δx. As expected, the rms error decreases linearly with Δt, and quadratically with Δx.

The successive substitution algorithm has a large radius of convergence, i.e., it is relatively tolerant of poor initial estimates. However, its terminal rate of convergence can be slow. A more advanced method is the *Newton-Raphson* algorithm. One begins by expanding the residual vector in a Taylor series about the current iterate

$$\{\boldsymbol{R}\}(\{\boldsymbol{T}\}^{n+1}) \approx \{\boldsymbol{R}\}(\{\boldsymbol{T}\}^{n+1}_m) + \left.\frac{\partial\{\boldsymbol{R}\}}{\partial\{\boldsymbol{T}\}^{n+1}}\right|_{\{\boldsymbol{T}\}^{n+1}_m} \{\Delta\boldsymbol{T}\}^{n+1} + \cdots \tag{6.83}$$

Truncating the series at the first term and setting the result to zero gives the linearized system

$$\left.\frac{\partial\{\boldsymbol{R}\}}{\partial\{\boldsymbol{T}\}^{n+1}}\right|_{\{\boldsymbol{T}\}^{n+1}_m} \{\Delta\boldsymbol{T}\}^{n+1} = -\{\boldsymbol{R}(\{\boldsymbol{T}\}^{n+1}_m)\} \tag{6.84}$$

The matrix $\partial\{\boldsymbol{R}\}/\partial\{\boldsymbol{T}^{n+1}\}$ is called the *tangent matrix*. After solving for $\{\Delta\boldsymbol{T}\}^{n+1}$, the solution is updated as

$$\{\boldsymbol{T}\}_{m+1}^{n+1} = \{\boldsymbol{T}\}_{m}^{n+1} + \{\Delta\boldsymbol{T}\}^{n+1}$$

The Newton-Raphson algorithm, although more complicated to code properly, offers quadratic terminal convergence. This algorithm is generally more sensitive to the initial estimate. However, in many transient solidification problems, after the first few time steps, the solution is very well behaved, and a good initial estimate can be made by extrapolation in time.

6.3.2 Fixed grid: Temperature recovery methods

A very useful variant of the enthalpy techniques is called the *temperature recovery method*. This algorithm uses separate vectors for the temperature, enthalpy and liquid fraction at each node. The calculations proceed in two substeps at each time step. In the first substep, the specific heat is assumed to be constant, i.e., no solidification occurs. Although the thermal conductivity may still be temperature dependent, this nonlinearity is usually much less severe than that associated with the latent heat of fusion, and the solution, designated $\{\bar{\boldsymbol{T}}\}^{n+1}$, can be obtained relatively easily. Separate vectors are maintained containing the average enthalpy $\langle \rho h \rangle$ associated with each node at the previous and current time steps, designated $\{\rho\boldsymbol{H}\}^{n}$ and $\{\rho\boldsymbol{H}\}^{n+1}$, respectively. After the first substep, the enthalpy vector is updated

$$\{\rho\boldsymbol{H}\}^{n+1} = \{\rho\boldsymbol{H}\}^{n} + \rho c_p \left(\{\bar{\boldsymbol{T}}\}^{n+1} - \{\boldsymbol{T}\}^{n}\right) \tag{6.85}$$

In the second substep, the change in enthalpy is resolved into the separate contributions from the latent heat associated with solidification and the specific heat associated with the change in temperature. The basis for the resolution is given in Eq. (6.71). When adapted to the enthalpy change in the substep at node i, Eq. (6.71) becomes

$$\langle \rho h \rangle_i^{n+1} - \langle \rho h \rangle_i^{n} = \int_{T_i^n}^{T_i^{n+1}} \rho c_p dT + \rho L_f \Delta \langle g_\ell \rangle_i \tag{6.86}$$

where $\Delta \langle g_\ell \rangle_i$ is the incremental change in liquid fraction at node i over the time step. Eq. (6.86) is then inverted so as to find the self-consistent set of temperatures and liquid fractions that gives the correct enthalpy change. The nodal temperature is subsequently "recovered" to the self-consistent value T_i^{n+1} and the simulation continues.

Alternatively, one can invert the enthalpy-temperature relation *a priori*, so as to have $T(\langle \rho h \rangle)$. The temperature T_i^{n+1} is then approximated as

$$T_i^{n+1} = T_i^n + \frac{dT}{d\langle \rho h \rangle}\left(\langle \rho h \rangle_i^{n+1} - \langle \rho h \rangle_i^{n}\right) \tag{6.87}$$

The derivative $(dT/d\langle\rho h\rangle)$ is approximated by $1/(\rho c_p)$. One can thus determine $\Delta\langle\rho h\rangle$ at each node, and then apply the relationship $T(\langle\rho h\rangle)$ to find a new temperature T_i^{n+1}. This procedure ensures that enthalpy is conserved in the algorithm.

The most important feature of this approach is the fact that it decouples the enthalpy, phase transformation and temperature calculations. As will be demonstrated in Chap. 11, one can introduce solidification models in which the enthalpy-temperature relationship is not known *a priori*. In such models, solid particles (nuclei) are introduced as the liquid undercools, and their number and growth rate are related to the local thermal conditions. This leads to a state where different locations in a macroscopic part follow different solidification paths, i.e., they traverse a variety of enthalpy-temperature curves. This approach is the basis of almost all of the so-called "macro-micro" methods for coupling the macroscale heat flow to microstructure formation. A further discussion of the approach can be found in Chap. 11.

6.3.3 Front tracking methods

As noted earlier, the most difficult and interesting aspect of solidification modeling is keeping track of the moving interface. The previous sections have been concerned with methods that obtain the interface position implicitly, inferring it from a temperature solution. There are also many techniques, known as *front tracking* methods, that follow the interface *explicitly*, with a local adaption of the grid in order to conform to the computed interface shape. As will be demonstrated in Chap. 8, the solid-liquid interface can develop very complex shapes. For this reason, front tracking methods are most appropriately applied using unstructured meshes, as in the FVM and FEM. These techniques are most applicable when the solid-liquid interface remains close to its initial position.

The methods differ most in the details of their implementation. The discussion here is confined to the basic framework applied to pure materials, leaving the extension to other cases in the references for the interested reader. The two boundary conditions that need to be satisfied at the interface are the interfacial energy balance, Eq. (4.133), and the Gibbs-Thomson condition neglecting attachment kinetics, Eq. (2.70). These equations are repeated here in a form that is more convenient for the current discussion:

$$v_n^* = \frac{(k_s \nabla T_s \cdot \boldsymbol{n})^* - (k_\ell \nabla T_\ell \cdot \boldsymbol{n})^*}{\rho_s L_f} \tag{6.88}$$

$$T^* = T_f - 2\Gamma_{s\ell}\bar{\kappa} \tag{6.89}$$

Suppose that the temperature and interface positions are known at a particular time t_n, and that the computational grid conforms to the interface at this time. Let us consider how to obtain the corresponding data at the next time step, t_{n+1}.

In almost all front tracking methods, this type of calculation proceeds in two substeps. In the first substep, the interface is treated as fixed, and a temperature solution is obtained using one of the two boundary conditions given in Eq. (6.88) and (6.89). Most often, the Gibbs-Thomson condition is employed as it provides a Dirichlet boundary condition. We confine our discussion to this particular case. Notice that Eq. (6.89) involves the curvature that has to be approximated from the representation of the interface. Such kinetic effects are disregarded for the remainder of this discussion. After computation of the temperature distribution in the first substep, the interface is relocated using the second boundary condition – in this case the interfacial energy balance given in Eq. (6.88). The right hand side of Eq. (6.88) is evaluated from the temperature distribution obtained in the first substep, providing the local interface velocity. The interface position is then updated in an appropriate way for the algorithm being used, and the computation continues.

Numerous algorithms that follow this general procedure have been developed. Quite a number of them work only in 1-D, and thus do not merit further discussion. We first describe an algorithm, illustrated in Fig. 6.10, that is easily visualized in one dimension, and that can be extended to higher dimensions. The method begins with an underlying fixed grid with spacing Δx. In general, the interface lies between nodes of the fixed grid, as illustrated. The temperature solution for the first substep is obtained in a straightforward manner using Eq. (6.89) as a boundary condition, coupled with any of the methods already described in the present chapter. Regardless of whether the FDM, FVM or FEM is used, the local nodal equations must account for the variations in grid spacing near the interface. The interface velocity is then obtained using Eq. (6.88), the position of the node that corresponds to the interface is updated, and the calculation proceeds to the next time step. There are certain implementation issues associated with the interface being near or crossing a node in the fixed mesh, as well as with the collision of two surfaces that we do not describe.

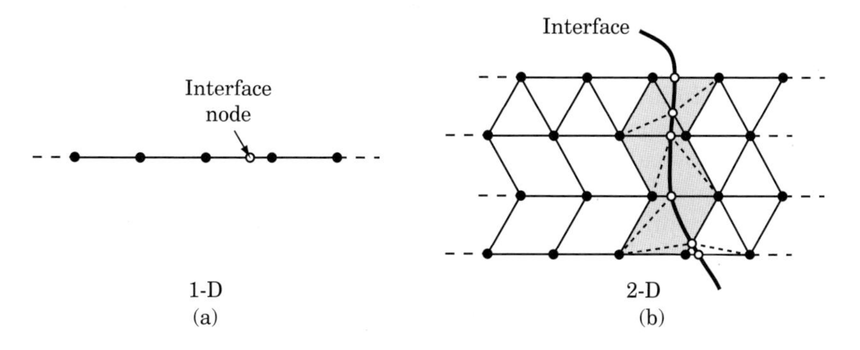

Fig. 6.10 An illustration of the addition of nodes at locations where the interface crosses the fixed grid in (a) 1-D and (b) 2-D.

In two dimensions, it is relatively easy to build unstructured meshes using triangles that conform locally to the interface (see Fig. 6.10(b).) Unfortunately, it is also relatively easy to produce highly distorted elements that generally give rise to poor solutions. Most of the computational effort in the method goes into dealing with this problem, e.g., by carrying out a local subdivision of the elements. An additional problem that appears in higher dimensions is the extraction of the local mean curvature $\bar{\kappa}$ to be used in the application of Eq. (6.89). This can be done, for instance, by making a sorted list of the interface nodes and first computing the local normal vector, followed by the curvature from the definition $\bar{\kappa} = \nabla \cdot \boldsymbol{n}$. Since this procedure involves derivatives of interpolated points, the curvature is generally much less accurate than the location of the points. The method can be extended to three dimensions, where tetrahedra are employed to complete the mesh as needed. This case contains nothing really new, but the geometric complexity is much greater in three dimensions than in two.

There are two other methods that should be mentioned in this context. The first is called a *Landau transformation*, in which the domain is divided into two regions, corresponding to the two phases. In the 1-D domain $0 \leq x \leq L$, one would define two variables, ξ and ζ, as

$$\xi(x,t) = \frac{x}{x^*(t)}; \quad \zeta(x,t) = \frac{x - x^*(t)}{L - x^*(t)} \tag{6.90}$$

The conduction equation is then rewritten in terms of these new variables as

$$\frac{\partial T_s}{\partial t} - \frac{v_\xi}{x^*}\frac{\partial T_s}{\partial \xi} = \frac{\alpha_s}{(x^*)^2}\frac{\partial^2 T_s}{\partial \xi^2} \qquad 0 \leq \xi \leq 1 \tag{6.91}$$

$$\frac{\partial T_\ell}{\partial t} - \frac{v_\zeta}{L - x^*}\frac{\partial T_\ell}{\partial \zeta} = \frac{\alpha_\ell}{(L - x^*)^2}\frac{\partial^2 T_\ell}{\partial \zeta^2} \qquad 0 \leq \zeta \leq 1 \tag{6.92}$$

where $v_\xi = \xi v^*$ and $v_\zeta = (1 - \zeta)v^*$. The interface velocity v^* is determined using Eq. (6.88). It is possible to extend this approach to higher dimensions under the condition that the interface position does not move very far from its initial location. It is thus particularly useful for steady problems, such as in the modeling of continuous casting.

Another very useful technique for precisely locating an interface that remains close to its initial location is illustrated in Fig. 6.11. An FE mesh is created with nodes placed on the interface, and as the interface moves, these nodes move along with it. A designated subset of nodes, illustrated as open circles in Fig. 6.11, are also allowed to move, whereas other nodes (shown as gray-filled circles) remain fixed. As long as the motion of the interface is not too large, the mesh retains the initial topology, i.e., nodes do not pass through element boundaries.

The fact that the nodes can move introduces another complexity into the formulation of the problem. The shape functions of the elements include information about the nodal coordinates, thus if the nodes move in time, the shape functions are no longer constant. Instead, they depend

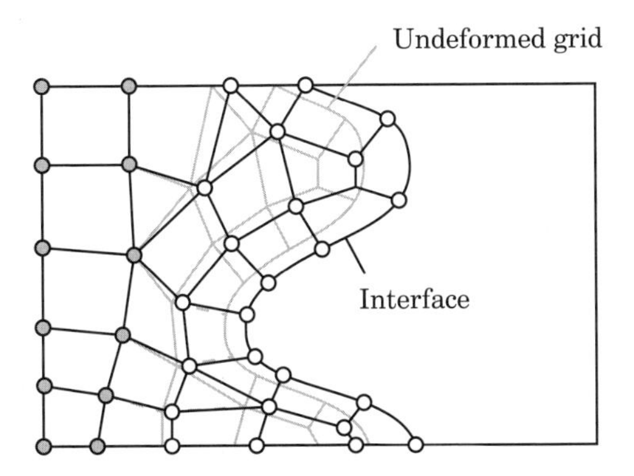

Fig. 6.11 An illustration of a deforming mesh following a moving interface. The original mesh is shown in gray. Nodes that are allowed to move have white centers, whereas the centers of the fixed nodes are gray. Notice that the movement of the interface affects the positions of nodes that are not connected to the interface.

on time through the movement of the nodes. This affects the transient term in the energy equation, as demonstrated below:

$$\begin{aligned}\int_V [\boldsymbol{N}]^T \rho c_p \frac{\partial T}{\partial t}\, dV &= \int_V [\boldsymbol{N}]^T \rho c_p \frac{\partial([\boldsymbol{N}]\{\boldsymbol{T}\})}{\partial t}\, dV \\ &= \left[\int_V [\boldsymbol{N}]^T \rho c_p [\boldsymbol{N}]\, dV\right] \frac{d\{\boldsymbol{T}\}}{dt} + \left[\int_V [\boldsymbol{N}]^T \rho c_p \frac{\partial [\boldsymbol{N}]}{\partial t}\, dV\right] \{\boldsymbol{T}\} \\ &= [\boldsymbol{C}] \frac{d\{\boldsymbol{T}\}}{dt} + \left[\int_V [\boldsymbol{N}]^T \rho c_p \frac{\partial [\boldsymbol{N}]}{\partial t}\, dV\right] \{\boldsymbol{T}\}\end{aligned} \tag{6.93}$$

The first term on the right hand side of Eq. (6.93) is the usual transient term, previously encountered in Eq. (6.49). The second term contains the contributions from the moving mesh, and can be evaluated by considering the shape function in the reference frame, $[\boldsymbol{N}(\boldsymbol{r})]$, where $\boldsymbol{r}$ is the vector of the reference coordinates (e.g., (r, s) in 2-D). Since the reference shape functions are independent of time, the total time derivative is zero in this frame. We thus have

$$\frac{d[\boldsymbol{N}(\boldsymbol{r})]}{dt} = 0 = \frac{\partial [\boldsymbol{N}]}{\partial t} + \frac{\partial [\boldsymbol{N}]}{\partial \boldsymbol{x}} \cdot \frac{d\boldsymbol{x}}{dt} \tag{6.94}$$

Now, introducing the transformation between the real space position $\boldsymbol{x}$ and the corresponding reference space position $\boldsymbol{r}$, and applying the chain rule gives

$$\frac{\partial [\boldsymbol{N}]}{\partial t} = -\frac{\partial [\boldsymbol{N}]}{\partial \boldsymbol{r}} \frac{\partial \boldsymbol{r}}{\partial \boldsymbol{x}} \frac{d\boldsymbol{x}}{dt} \tag{6.95}$$

Differentiating Eq. (6.62) and setting $d[\boldsymbol{N}]/dt = 0$ yields

$$\frac{d\boldsymbol{x}}{dt} = [\boldsymbol{N}]\frac{d\{\boldsymbol{x}\}}{dt} \tag{6.96}$$

where $\{x\}$ is the vector of nodal coordinates. It can also be recognized from Eq. (6.66) that $\partial \boldsymbol{r}/\partial \boldsymbol{x} = [\boldsymbol{J}]^{-1}$. Using these last two results in Eq. (6.95) gives

$$\begin{aligned}\frac{\partial[\boldsymbol{N}]}{\partial t} &= -\frac{\partial[\boldsymbol{N}]}{\partial \boldsymbol{r}}[\boldsymbol{J}]^{-1}[\boldsymbol{N}]\frac{d\{\boldsymbol{x}\}}{dt} \\ &= [\boldsymbol{B}][\boldsymbol{N}]\frac{d\{\boldsymbol{x}\}}{dt}\end{aligned} \tag{6.97}$$

where $[\boldsymbol{B}]$ is the matrix of spatial derivatives of $[\boldsymbol{N}]$. Finally, by substituting this expression into Eq. (6.93), one obtains

$$\int_V [\boldsymbol{N}]^T \rho c_p \frac{\partial T}{\partial t}\, dV = [\boldsymbol{C}]\,\frac{\partial\{\boldsymbol{T}\}}{\partial t} - [\boldsymbol{v}_{mesh}]\,\{\boldsymbol{T}\} \tag{6.98}$$

where the *mesh velocity matrix* $[\boldsymbol{v}_{mesh}]$, representing the advection of energy associated with the moving mesh, is defined as

$$[\boldsymbol{v}_{mesh}] = \int_V [\boldsymbol{N}]^T \rho c_p [\boldsymbol{B}][\boldsymbol{N}]\frac{d\{\boldsymbol{x}\}}{dt}\, dV \tag{6.99}$$

Note that it is only the *representation* of the energy that is advected with the mesh, not the energy itself. The conductance matrix and force vector are evaluated on the current mesh, and thus remain defined by Eq. (6.61). The modified form of the conduction equation for the deforming grid then becomes

$$[\boldsymbol{C}]\,\frac{\partial\{\boldsymbol{T}\}}{\partial t} - [\boldsymbol{v}_{mesh}]\,\{\boldsymbol{T}\} + [\boldsymbol{K}]\,\{\boldsymbol{T}\} = \{\boldsymbol{b}\} \tag{6.100}$$

In addition to the transport equation, an algorithm for moving the nodes must also be supplied. The movement of the interface nodes in a direction normal to the interface is of course specified from the solid-liquid interfacial energy balance. The concomitant motion of the other nodes in the mesh can be accomplished through several possible algorithms. One approach is to treat the domain as an elastic continuum, and to perform a linear elastic analysis to compute the displacement of the remaining nodes in the mesh.

6.3.4 Level set methods

A very useful extension of the above technique is the Level Set Method, first developed by Osher and Sethian [6]. The basic idea of this method is to define a function ϕ as the signed perpendicular distance from the solid-liquid interface. By convention, $\phi < 0$ corresponds to the solid, $\phi > 0$

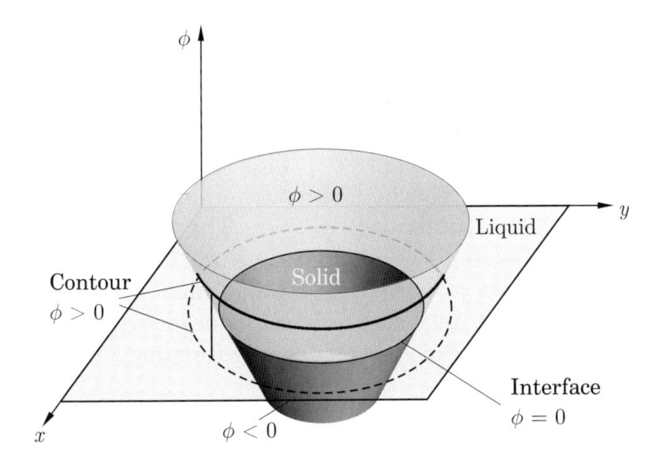

Fig. 6.12 An illustration of the definition of the distance function ϕ, and the level set $\phi = 0$ as the interface for a 2-D solidification problem.

represents to the liquid, and the "level set" $\phi = 0$ defines the interface. The distance function is illustrated in Fig. 6.12 for a 2-D solidification problem. In the level set algorithm, the interface is defined implicitly. Thus, when the portion of the domain used to form the nodal equations contains the interface, a special form of the discrete equations is employed, rather than explicitly placing new nodes on the interface. In the FDM, the interface is detected when ϕ changes sign between neighboring nodes, and a special stencil incorporating the implicit location of the interface is used. In the FEM, the interface is detected in an element if ϕ changes sign within it, in which case special shape functions are introduced. In both cases, the special handling ensures that the interfacial boundary condition is implicitly satisfied.

Suppose that the distance function ϕ and temperature T are known at a certain time. The level set algorithm for advancing them proceeds as follows:

1. The normal velocity of the interface is computed for all of the nodes and/or elements adjacent to the interface using Eq. (6.88) .

2. The interface is then implicitly "moved" by integrating in time the pure advection equation for ϕ given by

$$\frac{\partial \phi}{\partial t} + v_n |\nabla \phi| = 0 \tag{6.101}$$

   Particular care must be taken when solving Eq. (6.101) since numerical errors tend to spread the interface. Latent heat will be lost if this occurs.

3. After the computation of ϕ in the preceding step, the contours of ϕ are generally distorted, such that ϕ no longer represents the distance

to the interface. This property is restored by a process called "reinitialization" carried out through the integration of

$$\frac{\partial \phi}{\partial \tau} + S(\phi)\left(|\nabla \phi| - 1\right) = 0 \tag{6.102}$$

until steady state is reached. Inspection of Eq. (6.102) shows that steady state corresponds to $|\nabla \phi| = 1$. The reader is encouraged to make a simple sketch to demonstrate that $|\nabla \phi| = 1$ is equivalent to the statement that ϕ represents the distance from the interface. In Eq. (6.102), τ is an artificial time introduced simply to perform the reinitialization, and $S(\phi)$ is a "smearing function," typically given by

$$S(\phi) = \frac{\phi}{\sqrt{\phi^2 + \Delta x^2}} \tag{6.103}$$

Here, Δx is a smearing distance (usually the same as the typical grid spacing near the interface). This form ensures that the algorithm is well behaved near $\phi = 0$. The computational efficiency can be improved by confining the solution of Eq. (6.102) to a narrow region near the interface, which is the only place where ϕ is required to be accurate. This is called the "local level set method."

4. The temperature field is now computed, with the interface fixed and its location determined implicitly as the level set $\phi = 0$. Eq. (6.89) is satisfied implicitly as discussed earlier. The curvature is conveniently extracted from ϕ using

$$\boldsymbol{n} = \frac{\nabla \phi}{|\nabla \phi|}; \qquad \bar{\kappa} = \nabla \cdot \boldsymbol{n} \tag{6.104}$$

5. The computation then begins again at the next time step.

The level set method can be implemented readily using any of the discretization schemes presented earlier in this chapter. Since the interface is located implicitly, the method works equally well with both structured and unstructured grids. We will return to the level set method in Chap. 8 in the context of dendritic growth.

6.4 FLUID FLOW

Fluid flow can have a profound effect on solidification. Consider first the case where the solid is fixed in space during freezing i.e., $v_s = 0$. In this case, the evolution of the solid-liquid interface is affected by advection of heat in the liquid near the interface. This changes the local temperature gradients, thus altering the evolution of the solidification front. An example is shown schematically in Fig. 6.13. A second important mechanism

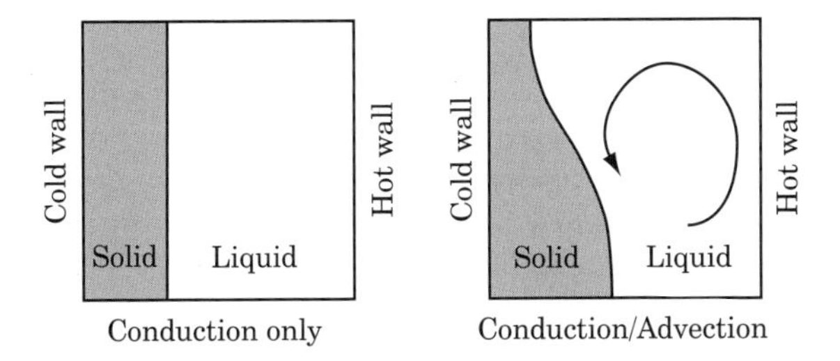

Fig. 6.13 An illustration of the effect of advection in the melt on the progress of the solid-liquid interface for a 2-D solidification problem.

by which fluid flow can alter the solidification pattern occurs when particles of the solid phase are transported along with the the fluid. This latter effect is very important in the evolution of microstructure and macrosegregation, topics discussed in Chaps 11 and 14. In this section, we focus on the first mechanism, and in particular, the numerical simulation of the fluid flow itself. There are many texts devoted to the numerical simulation of fluid flow, also called computational fluid dynamics (CFD). However it is beyond the scope of this text to treat the subject very deeply.

Most liquid metals are well represented as incompressible Newtonian fluids, and our discussion is confined to this case. For the purpose of the discussion, we consider pure materials, the properties of which are taken to be constant, in order for the algorithms to be developed more clearly. The governing equations include the mass balance, Eq. (4.48), and the Navier-Stokes equation with $\rho = cst$, Eq. (4.99), repeated here for the sake of convenience:

$$\nabla \cdot \boldsymbol{v} = 0 \tag{6.105}$$

$$\rho \left(\frac{\partial \boldsymbol{v}}{\partial t} + \nabla \cdot (\boldsymbol{v}\boldsymbol{v}) \right) = -\nabla p + \mu_\ell \nabla^2 \boldsymbol{v} + \rho \boldsymbol{b} \tag{6.106}$$

There are two important common cases where the flow is driven by the body force $\boldsymbol{b}$. In some processes, time-varying magnetic fields $\boldsymbol{B}$ are applied to drive and/or control the flow in a melt. These fields induce a current in the melt, represented by the current density $\boldsymbol{j}$. The interaction of the induced current with the magnetic field produces an electromagnetic body force $\boldsymbol{b}_{EM}$ (also called the *Lorentz force*) given by

$$\rho \boldsymbol{b}_{EM} = \boldsymbol{j} \times \boldsymbol{B} \tag{6.107}$$

The second important case is buoyancy-driven flows, where density differences owing to differences in temperature and/or composition produce flow. In Chap. 4, it was demonstrated that, when the density change associated with the characteristic temperature difference (i.e., $\beta_T \Delta T_c \ll 1$), we obtain the *Boussinesq form* of Eq. (6.106), given by

$$\rho_0 \left(\frac{\partial \boldsymbol{v}}{\partial t} + \boldsymbol{v} \cdot \nabla \boldsymbol{v} \right) = -\nabla \hat{p} + \mu_\ell \nabla^2 \boldsymbol{v} - \rho_0 \boldsymbol{g} \beta_T (T - T_0) \tag{6.108}$$

Here, T_0 is a reference temperature at which the density is ρ_0. This form creates a coupling between the energy and momentum balance equations, explored further in Sect. 6.4.3.

The two primary challenges in CFD include the non-linearity associated with the inertial term $\boldsymbol{v} \cdot \nabla \boldsymbol{v}$ in Eq. (6.106) and enforcing the constraint of the mass balance in Eq. (6.105). If the inertial terms are sufficiently large, it may also be necessary to add a model for turbulent flow. Here, however, we consider only laminar flows.

6.4.1 Finite difference method on staggered grids

The lessons learned from the study of heat conduction problems solved using finite differences, suggest that one should begin by discretizing Eq. (6.106) with central difference forms. It turns out that this results in unstable numerical schemes due to the gradient terms in Eqs. (6.105) and (6.106). In order to demonstrate the instability, let us write Eq. (6.105) in 2-D centered-space finite difference form at (x_i, y_j). (For the present section, the notations $v_x = u$ and $v_y = v$ are employed to make the expressions less cluttered.)

$$\begin{aligned} \nabla \cdot \boldsymbol{v} = 0 &= \frac{\partial u}{\partial x} + \frac{\partial v}{\partial y} \\ &\approx \frac{u_{i+1,j} - u_{i-1,j}}{2\Delta x} + \frac{v_{i,j+1} - v_{i,j-1}}{2\Delta y} \end{aligned} \tag{6.109}$$

Performing a similar operation for the pressure gradient that appears in Eq. (6.106), we obtain

$$\nabla p = \left\{ \begin{array}{c} \partial p/\partial x \\ \partial p/\partial y \end{array} \right\} \approx \left\{ \begin{array}{c} (p_{i+1,j} - p_{i-1,j})/2\Delta x \\ (p_{i,j+1} - p_{i,j-1})/2\Delta y \end{array} \right\} \tag{6.110}$$

Now consider the completely nonsensical flow field represented by the velocity field shown in Fig. 6.14(a). For this illustration, we take $u = v$ at each grid point, and assign to each the value shown in the figure. Nevertheless, Eq. (6.109) is exactly satisfied, i.e., the central difference form for this flow field gives $\nabla \cdot \boldsymbol{v} \equiv 0$. Similarly, there is no contribution to the pressure gradient from the oscillating pressure field shown in Fig. 6.14(b). Since such flow fields are solutions to the naive difference equations associated with Eq. (6.105) and (6.106), no stable and sensible solution can originate from this approach.

A very commonly employed method for stabilizing the solution is the use of a *staggered grid*, where the different degrees of freedom are solved

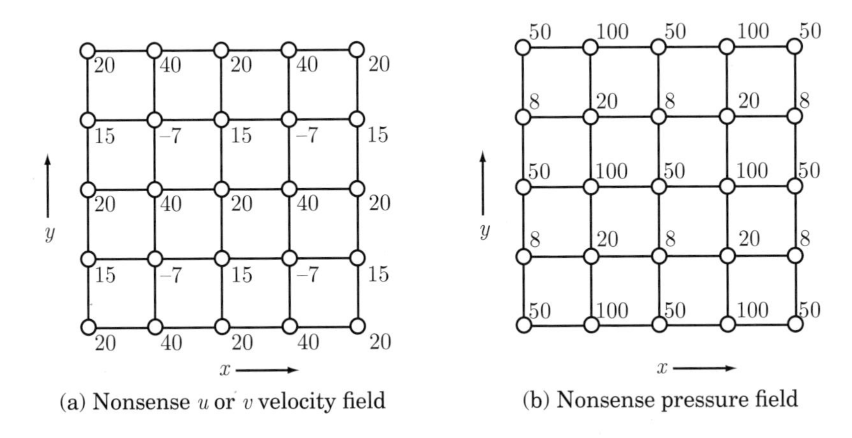

Fig. 6.14 (a) Nonsensical $u-$ or $v-$velocity field. (b) Nonsensical pressure field.

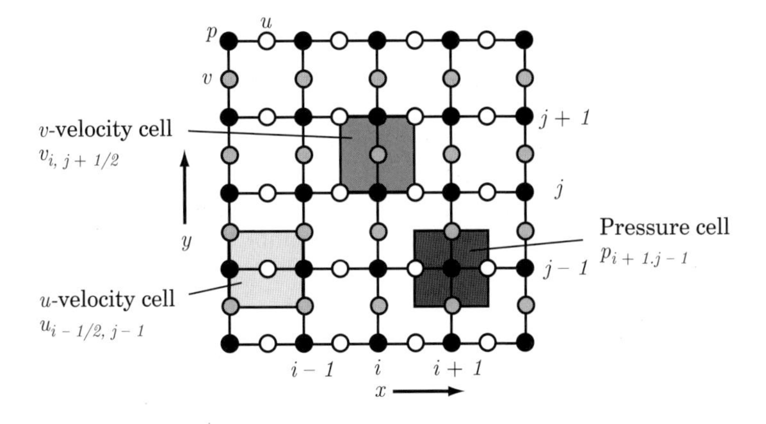

Fig. 6.15 A staggered grid. The pressure is solved at the black circles, the $u-$velocity is solved at the open circles, and the $v-$velocity is solved at the gray circles.

at various locations. An example is shown in Fig. 6.15. In this case, central difference forms use nearest neighbor grid points, and the instabilities described above for colocated grids, are suppressed.

Perhaps the most commonly used algorithm for solving this type of problem is the deceptively named SIMPLE algorithm, an acronym for Semi-Implicit Method for Pressure-Linked Equations. The velocities are calculated in a two-level scheme. At the first level, intermediate velocities $\tilde{u}$ and $\tilde{v}$ are computed using an estimated pressure field $\tilde{p}$. In general, the intermediate velocity field does not satisfy continuity, and a *pressure correction* equation is developed to provide an update to the velocity solution. The calculations are performed separately for the $u-$ and $v-$components of the velocity, by employing a "cell" around each grid point, as indicated in Fig. 6.15.

We present here only the equations and algorithms for the u–velocity, since the v–velocity solution is entirely analogous. The u–component of Eq. (6.106) is

$$\rho\left(\frac{\partial u}{\partial t}+\frac{\partial(u^2)}{\partial x}+\frac{\partial(uv)}{\partial y}\right)=-\frac{\partial p}{\partial x}+\mu_\ell\left(\frac{\partial^2 u}{\partial x^2}+\frac{\partial^2 u}{\partial y^2}\right) \tag{6.111}$$

Expanding the time derivative as a forward difference and all of the spatial derivatives as central differences gives

$$\begin{aligned}\rho u^{n+1}_{i+1/2,j} &= \rho u^{n}_{i+1/2,j}-\frac{\Delta t}{\Delta x}\left(p^n_{i+1,j}-p^n_{i,j}\right)\\ &+\mu_\ell\Delta t\left(\frac{u^n_{i+3/2,j}-2u^n_{i+1/2,j}+u^n_{i-1/2,j}}{\Delta x^2}+\frac{u^n_{i+1/2,j+1}-2u^n_{i+1/2,j}+u^n_{i+1/2,j-1}}{\Delta y^2}\right)\\ &-\rho\Delta t\left(\frac{(u^2)^n_{i+3/2,j}-(u^2)^n_{i-1/2,j}}{2\Delta x}+\frac{(uv)^n_{i+1/2,j+1/2}-(uv)^n_{i+1/2,j-1/2}}{\Delta y}\right)\end{aligned} \tag{6.112}$$

The intermediate values of u and v at the cell faces, corresponding to the indices $(i+1/2, j\pm 1/2)$ in the last term, are obtained by linear interpolation. For the sake of convenience, the two long terms in parentheses in Eq. (6.112) are lumped together into the symbol A, thus giving rise to the more compact form

$$\rho u^{n+1}_{i+1/2,j}=\rho u^{n}_{i+1/2,j}-\frac{\Delta t}{\Delta x}\left(p^n_{i+1,j}-p^n_{i,j}\right)+A\Delta t \tag{6.113}$$

The use of an analogous approach yields a similar equation for the v–velocity:

$$\rho v^{n+1}_{i,j+1/2}=\rho v^{n}_{i,j+1/2}-\frac{\Delta t}{\Delta x}\left(p^n_{i,j+1}-p^n_{i,j}\right)+B\Delta t \tag{6.114}$$

where B is defined similarly to A.

To generate an algorithm, Eqs. (6.113) and (6.114) are first expressed using the estimated pressure field $\tilde{p}$:

$$\rho\tilde{u}_{i+1/2,j}=\rho u^{n}_{i+1/2,j}-\frac{\Delta t}{\Delta x}\left(\tilde{p}_{i+1,j}-\tilde{p}_{i,j}\right)+A\Delta t \tag{6.115}$$

$$\rho\tilde{v}_{i,j+1/2}=\rho v^{n}_{i,j+1/2}-\frac{\Delta t}{\Delta x}\left(\tilde{p}_{i,j+1}-\tilde{p}_{i,j}\right)+B\Delta t \tag{6.116}$$

Since all terms on the right hand side are known, Eqs. (6.115) and (6.116) can be solved directly for $\tilde{u}$ and $\tilde{v}$. We now need to develop an equation for the pressure correction $p' = p^{n+1} - \tilde{p}$. Subtracting Eq. (6.115) from

Eq. (6.113), and Eq. (6.116) from Eq. (6.114), and then rearranging the terms gives

$$u^{n+1}_{i+1/2,j} = \tilde{u}_{i+1/2,j} - \frac{\Delta t}{\rho \Delta x}\left(p'_{i+1,j} - p'_{i,j}\right) \tag{6.117}$$

$$v^{n+1}_{i,j+1/2} = \tilde{v}_{i,j+1/2} - \frac{\Delta t}{\rho \Delta x}\left(p'_{i,j+1} - p'_{i,j}\right) \tag{6.118}$$

The continuity equation, Eq. (6.105), is now written in finite difference form for a pressure cell:

$$\frac{u^{n+1}_{i+1/2,j} - u^{n+1}_{i-1/2,j}}{\Delta x} + \frac{v^{n+1}_{i,j+1/2} - v^{n+1}_{i,j-1/2}}{\Delta y} = 0 \tag{6.119}$$

and Eqs. (6.117) and (6.118) are substituted into Eq. (6.119) to obtain:

$$\frac{p'_{i-1,j} - 2p'_{i,j} + p'_{i+1,j}}{\Delta x^2} + \frac{p'_{i,j-1} - 2p'_{i,j} + p'_{i,j+1}}{\Delta y^2} = \frac{\rho}{\Delta t}\left(\frac{\tilde{u}_{i+1/2,j} - \tilde{u}_{i-1/2,j}}{\Delta x} + \frac{\tilde{v}_{i,j+1/2} - \tilde{v}_{i,j-1/2}}{\Delta y}\right) \tag{6.120}$$

Eq. (6.120) is called a "pressure-Poisson" equation due to the left hand side being the standard discretization of $\nabla^2 p'$. The right hand side is simply the error in continuity of the intermediate solution $(\tilde{u}, \tilde{v})$, divided by $\rho/\Delta t$. The solution of Eq. (6.120) is normally obtained using an iterative solver. After determining p', the velocity can be updated using Eqs. (6.117) and (6.118).

The SIMPLE algorithm is $\mathcal{O}(\Delta t, \Delta x^2)$, as expected for the central difference formulation. Boundary conditions can sometimes be tricky to implement, and in order to illustrate the problem, let us assume that the left hand boundary of the domain shown in Fig. 6.15 is a no-slip boundary, i.e., $u = v = 0$. The implementation of the condition $v = 0$ is obvious, but what is to be done about u when there are no $u-$velocity nodes on the boundary? Consider as an example $u_{1,j}$, located inside the domain, at a distance of $\Delta x/2$ from the boundary. A relatively simple means of enforcing the boundary condition is to create a fictitious or "ghost" node outside the boundary, $u_{0,j} = -u_{1,j}$. This value can then be used to evaluate A in Eq. (6.115), as well as to determine the right hand side of Eq. (6.120). In order to ensure that the $u-$velocity maintains a value of zero at the boundary in the update procedure, Eq. (6.117) implies that the correct boundary condition on the pressure requires the normal derivative to vanish. This can be done by creating a ghost node also for the pressure, such that $p(0, j) = p(1, j)$.

This scheme, and its variants, are very commonly used for CFD. However, the main shortcoming of this method is that it relies on structured, staggered grids, which can cause problems in the implementation of boundary conditions. In addition to the problem described above, the method is difficult to apply correctly when the boundary is irregular, and one has to approximate the true shape with a "stair-step" that does not match the real boundary very well. For such problems, the FE and FV methods are much more convenient.

There is one other aspect that comes up fairly often in this scheme. When the velocities are sufficiently high, the solutions start to develop "wiggles" near boundaries wherever a boundary layer is insufficiently resolved. The common remedy is to use "upwind differencing" for the velocity derivatives, where the central difference is replaced by a difference biased in the flow direction. The reason that this works is that it actually adds additional "numerical diffusion" into the formulation. We return to this point in Sect. 6.4.3.

6.4.2 Finite element methods for CFD

The solution of fluid dynamics problems by finite element methods is less common than by finite difference approaches. However, the advantages of the FEM in terms of representing complex geometries make it the method of choice in many applications. The discrete equations are generated by applying Galerkin's method to the Navier-Stokes equations, using the same methods as those described in Sect. 6.2.3. The resulting equations are given without derivation. For further details, see, for example, the work by Gresho and Sani. [4] The equations, including buoyancy, are derived according to the Boussinesq form given in Eq. (6.108). For the sake of convenience, "pressure" will be used to mean the modified pressure, i.e., it will include the static pressure $\rho_0 \boldsymbol{g} h$.

We begin as in Sect. 6.2.3 with the interpolation functions for the velocity and the modified pressure in the element. It should be noted, however, that only the first derivative of the pressure appears in the momentum balance, whereas the second derivative of the velocity is found in the viscous terms. For this reason, the pressure is usually interpolated one order lower than the velocity, e.g., we have a linear interpolation for the velocity, and a constant discontinuous pressure; or a quadratic velocity and a linear continuous pressure. The interpolation scheme for these two fields is thus written as

$$\boldsymbol{v} = [\boldsymbol{N}]\{\boldsymbol{v}\}; \quad p = [\boldsymbol{N}_p]\{\boldsymbol{p}\} \tag{6.121}$$

where $[\boldsymbol{N}]$ and $[\boldsymbol{N}_p]$ are the shape functions for the velocity and pressure, respectively, and $\{\boldsymbol{v}\}$ and $\{\boldsymbol{p}\}$ represent the corresponding vectors of the nodal velocities and pressures.

The Galerkin finite element form of the continuity equation (Eq. (6.105)) is obtained with $[\boldsymbol{N}_p]^T$ as the weighting function, since this equation contains only the first spatial derivative of the velocity field

$$\int_V [\boldsymbol{N}_p]^T \nabla \cdot \boldsymbol{v} \, dV = 0 \tag{6.122}$$

After substituting the FEM form for v from Eq. (6.121), we have

$$\left[\int_V [\boldsymbol{N}_p]^T [\boldsymbol{B}]\, dV\right]\{\boldsymbol{v}\} = [\boldsymbol{K}_p]\{\boldsymbol{v}\} = 0 \tag{6.123}$$

where, $[\boldsymbol{B}]$ is the matrix of spatial derivatives of the shape functions $[\boldsymbol{N}]$. Note that the resulting matrix $[\boldsymbol{K}_p]$ is not symmetric. In some formulations, a small artificial compressibility is introduced on the right hand side of Eq. (6.123) in the form $-\varepsilon p$, where ε is called a *penalty parameter*, assessed to control the amount of permitted compressibility. The reason for doing so will become clear when we consider the momentum balance equation. The penalty formulation of Eq. (6.123) is then

$$[\boldsymbol{K}_p]\{\boldsymbol{v}\} = -\varepsilon\left[\int_V [\boldsymbol{N}_p]^T [\boldsymbol{N}_p]\, dV\right]\{\boldsymbol{p}\} = -\varepsilon\,[\boldsymbol{C}_p]\,\{\boldsymbol{p}\} \tag{6.124}$$

This equation can be solved to obtain the nodal pressures in terms of the nodal velocities

$$\{\boldsymbol{p}\} = -\frac{1}{\varepsilon}\,[\boldsymbol{C}_p]^{-1}\,[\boldsymbol{K}_p]\{\boldsymbol{v}\} \tag{6.125}$$

The Galerkin form of Eq. (6.108) is obtained using $[\boldsymbol{N}]$ as the weighting function,

$$\rho_0\int_V [\boldsymbol{N}]^T\frac{\partial \boldsymbol{v}}{\partial t}\, dV + \rho_0\int_V [\boldsymbol{N}]^T \boldsymbol{v}\cdot\nabla\boldsymbol{v}\, dV = -\nabla\int_V [\boldsymbol{N}]^T p\, dV + \int_V [\boldsymbol{N}]^T \mu_\ell \nabla^2 \boldsymbol{v}\, dV - \int_V [\boldsymbol{N}]^T \rho_0 \boldsymbol{g}\beta_T (T - T_0)\, dV \tag{6.126}$$

The temperature is interpolated using the same shape functions as for the velocity. The FEM form of the momentum balance equation can be derived by a straightforward application of the methods that were employed to derive the matrix form for the energy and continuity equations, and this derivation is therefore left as an exercise. The result is

$$[\boldsymbol{C}]\frac{\partial\{\boldsymbol{v}\}}{\partial t} + [\boldsymbol{A}(\boldsymbol{v})]\,\{\boldsymbol{v}\} + [\boldsymbol{K}_p]^T\{\boldsymbol{p}\} + ([\boldsymbol{K}_\mu] + (\mathrm{tr}[\boldsymbol{K}_\mu][\boldsymbol{I}]))\,\{\boldsymbol{v}\} + \beta_T \boldsymbol{g}[\boldsymbol{C}]\{\boldsymbol{T}\} = \{\boldsymbol{b}_T\} \tag{6.127}$$

where $[\boldsymbol{A}(\boldsymbol{v})]$ is the *advection matrix*, $[\boldsymbol{K}_\mu]$ is the *viscosity matrix*, and the vector $\{\boldsymbol{b}_T\}$ comes from buoyancy. The specific forms of these new terms are given below.

$$[\boldsymbol{C}] = \left[\rho_0\int_V [\boldsymbol{N}]^T[\boldsymbol{N}]\, dV\right];\quad [\boldsymbol{A}(\boldsymbol{v})] = \left[\rho_0\int_V [\boldsymbol{N}]^T \boldsymbol{v}\cdot[\boldsymbol{B}]\, dV\right]$$
$$[\boldsymbol{K}_\mu] = \left[\int_V [\boldsymbol{B}]^T \mu [\boldsymbol{B}]\, dV\right];\quad \{\boldsymbol{b}_T\} = \left[\int_V \rho_0\beta_T \boldsymbol{g}[\boldsymbol{N}]^T[\boldsymbol{N}] dV\right] \tag{6.128}$$

The matrix $[C]$ in the first term is the consistent form of the mass matrix, which is analogous to the capacitance matrix introduced in Eq. (6.49). The time derivative $\partial\{v\}/\partial t$ is handled in the usual way by a finite difference. Notice that the velocity appears within the matrix $[A]$, clearly showing the nonlinearity. The pressure can be eliminated from Eq. (6.127) using Eq. (6.125), thus reducing the number of degrees of freedom by one. The pressure can then be recovered from the velocity field with Eq. (6.125).

We must now include the advective term in the energy equation. The Galerkin form of the energy equation is obtained in the usual way, using $[N]^T$ as the weighting function, with the result

$$[C]\frac{\partial\{T\}}{\partial t} + [A_T(v)]\{T\} + [K]\{T\} = \{q_b\} + \{\rho\dot{R}_q\} \tag{6.129}$$

Here, $\{q_b\}$ and $\{\rho\dot{R}_q\}$ are the boundary fluxes and internal heat generation vectors, respectively. The *thermal advective matrix* $[A_T(v)]$ is given by

$$[A_T(v)] = \int_V \rho_0 c_p [N]^T v \cdot [B]\, dV \tag{6.130}$$

From this point, the evaluation of the various matrices and solution proceeds as described for the heat equation in Sect. 6.2.3. The essential nonlinearity embodied in the advection term in the momentum balance has a tendency to make the solution considerably more difficult than for the heat equation.

The Galerkin form of the FEM is also capable of producing "wiggles" in the solution when the advective terms are strong. As in the finite difference method, these terms are handled by introducing an artificial or "numerical" diffusion in the streamwise direction, a procedure called "upwinding" (see Gresho and Sani [4] for further details). In the next section, these methods are applied to examine a melting problem in the presence of buoyant convection, and to demonstrate the effect of the upwinding procedure on the solution.

6.4.3 Example: Melting of pure Ga

Let us investigate how all of these methods fit together. In an experiment performed by Gau and Viskanta [3], a block of solid Ga, initially at a uniform temperature of 28.3°C, was insulated on all sides with the exception of the right hand side, which was maintained at 28.3°C, and the left hand side, whose temperature was raised to 38°C. Since the melting point of Ga is 29.8°C, the block began to melt from the left hand side. The material properties are listed in Table 6.1, and the block used in the experiment had a height of 63.5 mm, a width of 88.9 mm, and a depth of 38.1 mm.

Although this problem concerns melting, rather than solidification, it provides a very instructive example of the use of the various modeling methods introduced in this chapter. If there were no buoyant convection, the temperature distribution in the solid and liquid, as well as the location of the interface, could be obtained readily at least for short times, by considering the block to be semi-infinite and using the analytical methods of

Table 6.1 Material properties for pure Ga.

Property	Value	Units
Density, ρ_0	6.093×10^3	kg m^{-3}
Viscosity, μ	1.81×10^{-3}	Pa s
Specific heat, c_p	381	J kg^{-1} K^{-1}
Conductivity, k_s	33.5	W m^{-1} K^{-1}
Conductivity, k_ℓ	32.0	W m^{-1} K^{-1}
Expansion coefficient, β_T	1.2×10^{-4}	K^{-1}
Latent heat of fusion, L_f	80.16×10^3	J kg^{-1}
Melting point, T_m	29.8	°C
Pr	2.16×10^{-2}	–

Chap. 5. Applying the boundary conditions and material property data for this problem in Eqs. (5.25) and (5.26), we find that the interface position $x^*(t) \approx 1.04 \text{ mm s}^{-1/2}\sqrt{t}$. This estimate will be useful for computing the relevant dimensionless groups.

Buoyant convection does exist, however, and it affects the shape and progress of the liquid-solid interface. We simulated the experiment in 2-D, using the enthalpy-specific heat method. The latent heat was spread uniformly over the temperature range [29.78°C, 29.98°C], and the viscosity was raised from the value given in Table 6.1 by a factor of 1×10^8 over the freezing range to effectively immobilize the solid. As discussed in Sect. 4.6, the Grashof and Prandtl numbers consitute two important dimensionless groups for this kind of flow. They are given by

$$\text{Gr} = \frac{g\beta_T \Delta T L^3}{\nu_0^2}; \qquad \text{Pr} = \frac{c_p \mu}{k} = \frac{\nu_0}{\alpha_0} \tag{6.131}$$

The characteristic temperature difference $\Delta T = 38 - 29.8$ K is defined by the boundaries of the melted region. The analytical solution for $x^*(t)$ provides a reasonable estimate for the length scale L. Substituting the expression for $x^*(t)$ for L into Gr in Eq. (6.131), gives

$$\text{Gr} = 1.1 \times 10^2 \text{mm}^{-3}\, x^*(t)^3 \tag{6.132}$$

The Grashof number is a coefficient that multiplies the nonlinear inertial terms in scaled form. The problem thus becomes increasingly difficult to solve numerically as Gr increases. For instance, when $x^* = 10\,\text{mm}$, $\text{Gr} = 10^5$. In other words, as melting proceeds, the buoyant flow becomes more intense, the interface is more strongly affected, and the solution is harder to obtain. Upwinding can be used to stabilize the solution on a grid that is too coarse for the equations to be solved in the standard form. Figures 6.16 and 6.17 display the melt flow streamlines and isotherms in the liquid for a series of times. The upper row shows results computed without upwinding, whereas the lower row presents results obtained with upwinding. The flow patterns are similar, but the flow velocities are higher,

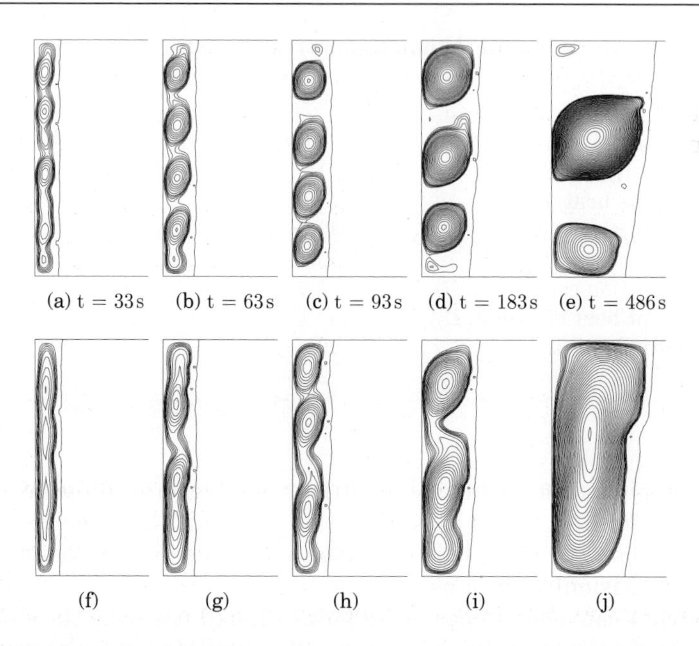

Fig. 6.16 Computed streamlines from the simulation of melting of pure Ga by means of natural convection. The simulation in the upper row (a)-(e) was carried out without upwinding, whereas the results in the lower row (f)-(j) were obtained when upwinding was included.

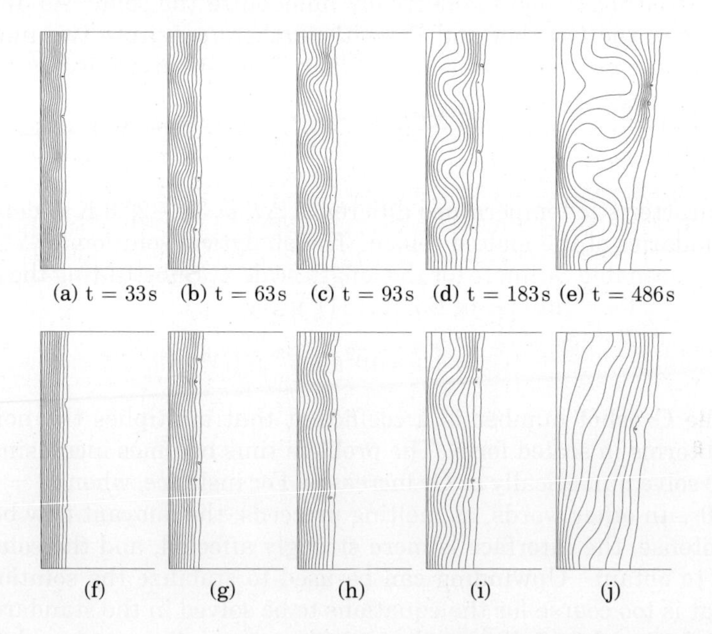

Fig. 6.17 Computed temperatures from the simulation of melting of pure Ga. The simulations in the upper row (a)-(e) were done without upwinding, whereas the results in the lower row (f)-(j) were obtained when upwinding was included. The isotherms are linearly interpolated between the boundary temperature 38°C to the left, and the freezing point, 29.8°C on the right in each panel.

and the internal cell structure better resolved, in the absence of upwinding. There are significant differences in the internal temperature distribution obtained using the two methods. These are shown in Fig. 6.17 for the same times as the streamlines in Fig. 6.16. Notice that the artificial diffusivity when upwinding is used wipes out much of the internal structure in the temperature field. On the other hand, the position of the melting isotherm is not very different between the two methods. Thus, as in many simulations, it is important to understand the effect of various assumptions and algorithms used, and also to decide what is most important.

6.5 OPTIMIZATION AND INVERSE METHODS

Up to this point, the focus of this chapter has been on numerical methods for solving the various equations that arise during modeling of solidification and associated phenomena. In particular, we specify the domain geometry, material properties and boundary conditions, after which a solution is constructed. However, it frequently occurs that the boundary conditions or material properties are not precisely known. In such cases, there is sometimes a limited amount of experimental data available, and one might like to use the analysis in combination with the experiments to determine the boundary conditions or material properties. This can be done by employing *inverse methods*.

Inverse methods are a subset of the more general field of numerical optimization, where one seeks to minimize (or maximize) a function, called an *objective*, which in certain ways represents the "ideal" solution to a particular problem. In the inverse problem, the objective might be a measure of the difference between a set of experimental data and the corresponding results obtained from the simulation. One could also ask the question: "Which process parameters should I choose in order to obtain a desired result?" This permits the use of simulations for design as well as analysis. Before getting into the mathematical details, let us consider an example.

Consider steady, one-dimensional heat conduction through a wall of thickness L with constant thermal conductivity k (see Fig. 6.18). At $x = 0$ the temperature is maintained at $T = 1$. At $x = L$, there is a convective boundary condition, i.e., $k\partial T/\partial x = -h_T(T(L) - T_\infty)$. We desire $T(L) = 0.5$. What values should be selected for h_T and T_∞ in order to achieve this result?

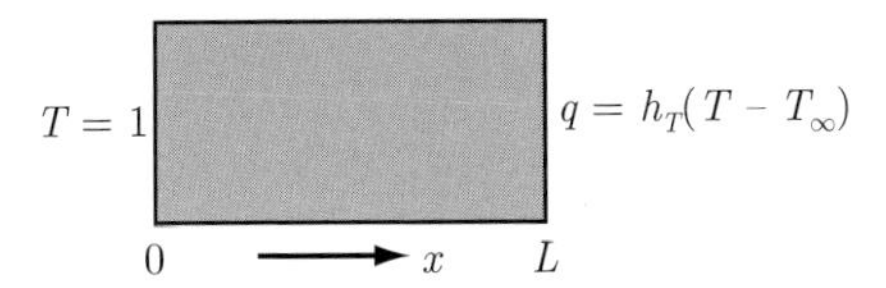

Fig. 6.18 An illustration of a 1-D heat conduction problem for optimization.

The problem is approached by first seeking the solution to the *forward problem*, i.e., solving the differential equation for a specified set of boundary conditions. The solution to the steady conduction problem is linear, and for the boundary conditions specified, the solution is

$$T(x) = 1 + \frac{\mathrm{Bi}}{1+\mathrm{Bi}}(T_\infty - 1)\frac{x}{L} \tag{6.133}$$

where Bi is the Biot number, $h_T L/k$. An objective function G that is minimized at the desired value of $T(L)$ can be defined as

$$G = [T(L) - 0.5]^2 \tag{6.134}$$

The optimal value corresponds to a minimum of G. The quadratic form is preferable to the absolute value since it is easily differentiated. Substituting $T(L)$, obtained from Eq. (6.133), into Eq. (6.134) yields

$$G = \left[0.5 + \frac{\mathrm{Bi}}{1+\mathrm{Bi}}(T_\infty - 1)\right]^2 \tag{6.135}$$

Figure 6.19(a) shows a contour plot of G as a function of the *design variables* h_T and T_∞.

The optimal solution is not unique! Any pair of h_T, T_∞ values on the contour $G = 0$ represents a solution. This occurs frequently in optimization problems. As a result, the specific optimal solution obtained may depend on the initial estimate and on the computational algorithm used to determine the optimum. This issue can be much more troublesome when the objective function has local minima, and one way around the problem

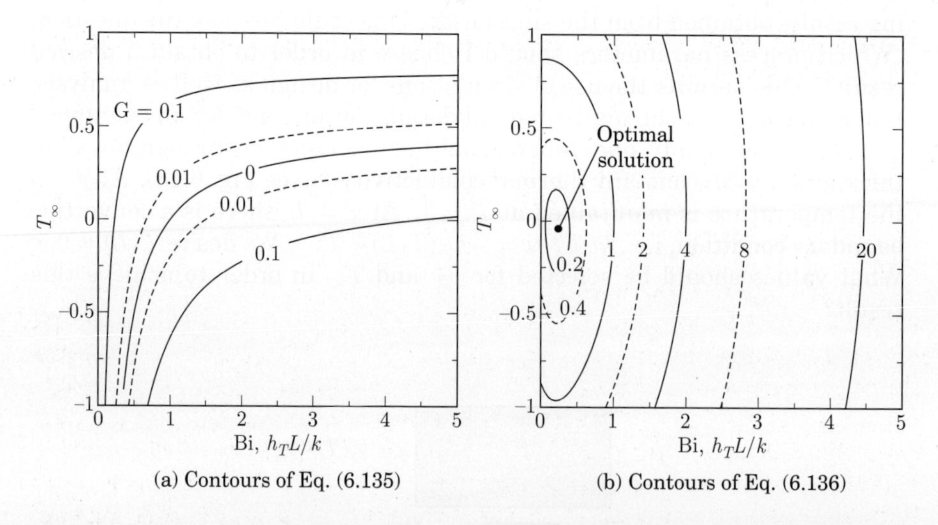

(a) Contours of Eq. (6.135)

(b) Contours of Eq. (6.136)

Fig. 6.19 Contours of the objective function for one-dimensional optimization problem (a) without regularization, and (b) with regularization $P = 1$.

is to modify the objective function employing a method called *regularization*. Suppose in our problem that there is a cost associated with providing large values of h_T and extreme values of T_∞. A regularized objective function could be constructed to reflect this by using the form

$$G_{reg} = \left[0.5 + \frac{\text{Bi}}{1+\text{Bi}}(T_\infty - 1)\right]^2 + P\left(\text{Bi}^2 + T_\infty^2\right) \tag{6.136}$$

Here, P is called a *penalty parameter*. Contours of G_{reg} for $P = 1$ are shown in Fig. 6.19(b), where it can be seen that there now exists a unique optimal value. The exact value is a weak function of P, with large values bringing the optimal solution closer to the origin, which is the minimum of the regularization term.

This is certainly one approach to optimization: representing the objective function graphically and visually selecting the minimum. This technique is not applicable for problems with more than two design variables, and more importantly, it is impractical when the cost of computing the objective function is high. The latter is often the case in solidification problems, where the evaluation of the objective function at each point requires a full numerical simulation. A systematic means of finding the optimum is therefore needed, and this requires a more fundamental approach to the problem.

Let us define the terms used above in a more formal manner. The process begins by the definition of several *design variables* $\boldsymbol{b} = \{b_1, b_2, \ldots\}$ that parameterize the system. These might be material properties, boundary conditions, part dimensions – anything that goes into the simulation. The design variables differ from the field variables discussed so far, quantities such as temperature, composition, etc, which will be generically referred to as $\Psi = \{\psi_1, \psi_2, \ldots\}$. Next, the *objective function* $G(\Psi; \boldsymbol{b})$ that is to be minimized (or maximized) is defined. There may also be *constraints* that come in three varieties:

- Inequality constraints of the form $f(\Psi; \boldsymbol{b}) \leq 0$
- Equality constraints of the form $h(\Psi; \boldsymbol{b}) = 0$
- Side constraints of the form $b_i^{min} \leq b_i \leq b_i^{max}$

Once again, optimization is a deep and complex subject, and only a brief introduction to it can be provided here. The basic idea of numerical optimization algorithms is to begin with an initial guess, and to systematically proceed to the optimum. Since, in our problems, the cost of evaluating a trial design is high, the focus is placed on algorithms that attempt to reach the optimum efficiently, i.e., they employ a relatively small number of trial designs. Some methods, such as genetic algorithms, require very large numbers of trial designs, and are thus inappropriate for the type of problem considered here. The optimal solution $\boldsymbol{b}^\star$ is obtained when the design vector is *feasible*, signifying that it satisfies all of the constraints, and when the following conditions (called the Kuhn-Tucker conditions) are satisfied [9]:

$$\lambda_j f_j(\Psi; \boldsymbol{b}^\star) = 0 \quad j = 1, \ldots J \quad \lambda_j \geq 0 \tag{6.137}$$

$$\nabla_b G(\Psi; \boldsymbol{b}^\star) + \sum_{j=1}^{J} \lambda_j \nabla_b f_j(\Psi; \boldsymbol{b}^\star) + \sum_{k=1}^{K} \lambda_{k+J} \nabla_b h_k(\Psi; \boldsymbol{b}^\star) = 0 \tag{6.138}$$

Here, the λ are Lagrange multipliers, and J and K respectively represent the total number of inequality and equality constraints, respectively. The notation ∇_b indicates derivatives with regard to the design variables, sometimes called the design gradient. When there are no constraints, the Kuhn-Tucker conditions reduce to the more familiar form from multivariable calculus, $\nabla_b G = 0$.

The most common algorithms start from an initial guess after which a sequence of searches is carried out in order to determine the optimal value. The algorithm first chooses a search direction in design space, and subsequently evaluates a series of trial designs (i.e., sets of b) along that line, searching for a local minimum in that direction. We will present further details on how the search directions are chosen shortly. There are several common techniques for finding the minimum along a one-dimensional curve. Typically, a few points are sampled to find an interval that contains a minimum. The minimum is then determined by one of several methods, such as fitting a simple function such as a parabola to the data, or using a systematic approach such as interval halving. Once the minimum is found in a given direction, a new search direction is chosen and the process continues. The search ends when the Kuhn-Tucker conditions are satisfied, or when successive minima differ by less than a preset tolerance.

Numerous algorithms employ this basic method of successive minimization in line searches. We have chosen to describe one of them, suitable for use with most commercial analysis codes, and the reader is referred to the literature for further details. A useful analogy to keep in mind during the description is the following: Imagine that you are placed somewhere in a mountainous region and given the task of finding the lowest point within a limited distance. It is completely dark, so you cannot see the region to decide which way to go. You have the following tools available: a compass that allows you to choose your direction of travel, a GPS unit that indicates your exact coordinates, and an altimeter to determine the altitude at any given location. You may consult the compass and GPS unit without charge, but you are required to pay a significant fee to determine the altitude. The coordinates are analogous to the design variables, i.e., your location in design space, the compass is analogous to a guide for following a search direction, and the altitude measurement is analogous to an evaluation of the objective function for a trial design.

Clearly, choosing a series of points and trying to build a contour map of the region could work, but is likely to be expensive. Such approaches are called "zero-order" methods. If the contours are relatively smooth, it is much more efficient to use "first-order" methods that employ information about the slope or gradient in design space. The gradient of the objective in design space is called the *design sensitivity*. There are techniques for extracting gradients analytically, described in some of the references at the end of this chapter. However, these are not commonly available for commercial packages, so for the present discussion we assume that only the objective is accessible, e.g., by reading and processing the results file from a simulation. The gradient in design space is then constructed by finite difference in design space, by first evaluating the function at the initial location $\boldsymbol{b}^0$, and then at nearby points, perturbing the design variables one at a time. The initial search direction $\boldsymbol{S}^0$ is subsequently chosen in the direction of the gradient, i.e., "downhill." The minimum is found along this direction as described earlier. Another gradient is computed at that point by finite difference, a new search direction is chosen, and the process repeats. In the *steepest descent* algorithm, the new search direction is simply the new gradient. This method tends to converge very slowly, but a slight modification called the *conjugate gradient* method converges very well. In the latter technique, the next search direction is made "conjugate" to the preceding ones by using the following expression.

$$\boldsymbol{S}^k = -\nabla G(\Psi; \boldsymbol{b}^k) + \beta_k \boldsymbol{S}^{k-1} \quad \text{where} \quad \beta_k = \frac{|\nabla G(\Psi; \boldsymbol{b}^k)|^2}{|\nabla G(\Psi; \boldsymbol{b}^{k-1}|^2} \tag{6.139}$$

This algorithm is sometimes called the Fletcher-Reeves method, and several other algorithms are available in typical optimization packages.

To illustrate the process, the one-dimensional heat conduction optimization problem presented earlier is solved with the Fletcher-Reeves algorithm. Figure 6.20(a) shows the initial estimate, $\boldsymbol{b}^0 = (\text{Bi}^0, T_\infty^0) = (2.5, -0.5)$, the line search directions and the minima along each search

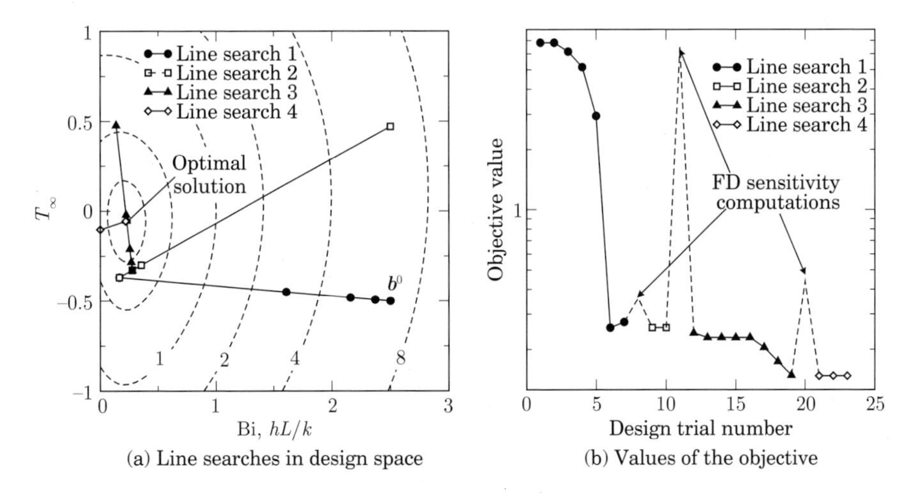

Fig. 6.20 The progress of the numerical optimization of Eq. (6.136) using the Fletcher-Reeves algorithm. Notice that the optimal solution is obtained after approximately 19 trials. The jumps in the objective on the right correspond to finite difference evaluations of the design derivative at the starts of new line searches.

direction. The optimal solution is $\boldsymbol{b}^{opt} = (0.22, -0.056)$. Figure 6.20(b) displays the history of the objective function, and the occasional peaks correspond to the finite difference evaluations of the gradient for new search directions.

The remainder of this section is devoted to discussing how to use optimization in applications of interest for solidification. It is surprising but true that one of the more difficult aspects turns out to be the construction of the objective function. The optimization code will lead to the "best" solution, but the user must sometimes be very specific about what is desired, and at certain times also must be somewhat creative in how to ask for it. This point is clarified by a few examples.

Suppose that, in a continuous casting operation, I thermocouples are placed at several locations in a casting mold, and those temperature data $\bar{T}_i$ are recorded during a typical operation. A plant engineer wishing to simulate this same process must specify the boundary conditions via heat transfer coefficients on both sides of the mold, material properties, etc., and the output of the simulation is the temperature at all locations in the mold. However, the boundary conditions are not known with great precision, and the engineer wants to use the experimental data to tune the simulation. Placed in the current context, the heat transfer coefficients would be the design variables, $\boldsymbol{b}$, and the temperatures at the thermocouple locations obtained in the simulation would represent the field variables, Ψ. In particular, ψ_i should be designated as the computed temperature at the location corresponding to thermocouple i. An objective function could then be defined as

$$G(\Psi; \boldsymbol{b}) = \sum_{i=1}^{I} \left(\psi_i - \bar{T}_i\right)^2 \tag{6.140}$$

We would subsequently seek to minimize G by a systematic variation of the design variables, i.e., the heat transfer coefficients. Naturally, these must be positive, and it may turn out that a certain amount of regularization is also needed. This is called a *parameter identification* problem.

There are certain pitfalls associated with this approach. In a problem such as the present one, the heat transfer coefficient probably varies with the position along the mold, and the user may want to divide the mold into several segments, each with a specific heat transfer coefficient. It is usually a bad idea to have discontinuous boundary conditions, and one should thus choose to represent the spatial variation within each segment with a functional representation that is at least continuous at the segment boundaries. A cubic spline representation is often employed, as it is easy to match the function and its first derivative at the segment boundaries. The coefficients of the splines in each segment then become the design variables. Note that the cost of carrying out the optimization increases with the number of design variables. If there are N design variables, then each gradient evaluation requires N additional simulations. This can get expensive if the user does not make carefully thought-out choices.

Another potential pitfall appears for large numbers of thermocouples. Convergence in the optimization routines is usually based on a certain percentage change in the objective. Thus, if the number of thermocouples is large, relatively significant discrepancies might remain at a few locations in the optimal solution. One means of dealing with this is to add a weighting function to the contribution to the terms in Eq. (6.140) at the problem locations.

How does one go about coding the optimization problem? Let us assume that there exists an available optimization library that can be summoned from within a program. The library routines typically take as arguments the current value of the design variables and objective function, and return new values for the design variables, requesting an evaluation of the objective. The user must then write a code to perform the following tasks:

- Write an input file to the simulation code using the new design variables;
- Run the simulation;
- Execute a program to read the output of the simulation to evaluate the objective.

This requires a certain amount of programming skill, but is usually not a formidable task.

When constructing the objective function, the user should also be careful with regard to units and scaling. In the example problem shown earlier, the design variables were Bi and T_∞. For this particular example, both variables were $\mathcal{O}(1)$. Although it was not specifically stated, the temperatures probably corresponded to °C. Assume that it was decided to work in Kelvin, i.e., that 273 should be added to each temperature value. The result should be the same. But if we look closely at Eq. (6.136), we find that if 273 is added to T_∞, the regularization attaches much more importance to T_∞ than it does to Bi. A similar sensitivity to units could arise if one were to choose h_T rather than Bi as the design variable. This can create problems in the optimization algorithm, especially when gradients are calculated. It is thus prefereble to scale the design variables such that they all have the same order of magnitude, and it would of course be even better to make them all order one.

This section concludes with another example. As part of a conference in 1988 [1], a large steel hammer was cast. The part is illustrated in Fig. 6.21. The casting and mold were instrumented with thermocouples, and the modeling community was challenged to simulate the casting and report the temperature histories at several locations. Numerous authors simulated the casting with various codes, based on the methods described earlier in this chapter, and essentially all of them were able to obtain accurate results. Figure 6.21(a) shows the outline of a half-section of the hammer, and the gray area encapsulates regions that are still partially liquid just at the time when the (poorly designed) riser freezes off

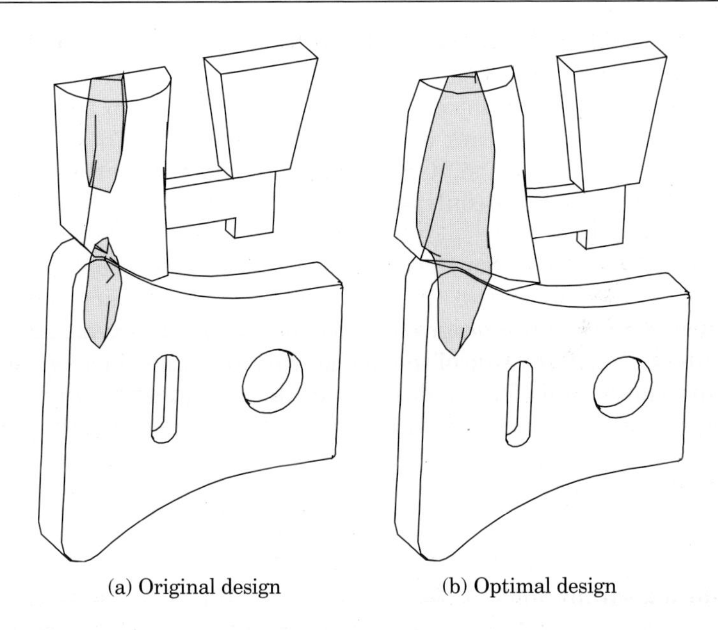

Fig. 6.21 (a) A half-section of the original design of a hammer casting. The gray area encapsulates material that is still partially liquid near the end of solidification, showing the failure of the riser to feed the casting. (b) An optimized design, showing uninterrupted feeding of the casting from the riser.

from the casting. The isolated region inside the casting is likely to have substantial porosity. In a later work, the riser design was modified using shape optimization so as to eliminate the isolated section [5]. The design variables consisted of a parameterization of the dimensions of the riser, and the objective function was defined as its total volume. There was also a constraint imposed to ensure directional solidification into the riser. We refer to the original paper for further details. The result is shown in Fig. 6.21(b). Notice that the optimization process removed material from the top of the riser, placing it at the bottom, where it helped keep open the feeding gate between the riser and casting. One of the interesting aspects to this problem is that it demonstrates that virtually anything going into the simulation, including the shape of the domain, can be made into a design variable.

6.6 SUMMARY

The present chapter has covered a wide variety of numerical methods that can be used for solidification problems. Finite difference, finite volume and finite element methods were examined for solving the governing equations. Analysis of the algorithms showed that with proper care, the methods converge to $\mathcal{O}(\Delta t, \Delta x^2)$, and on Cartesian grids, the three methods produce identical equation sets.

Several methods were introduced in order to handle the specific problem presented by solidification: a moving boundary. Methods suitable for fixed grids, using enthalpy-based approaches, were investigated, and will be further exploited in Chap. 11 for modeling the development of microstructure. Some front-tracking schemes, where the interface is fitted to the grid, were also studied.

Modeling methods for solving the fluid flow equations were presented, including both finite difference and finite element forms. Finally, optimization methods were considered, which provide a very powerful adjunct to process modeling.

6.7 EXERCISES

Exercise 6.1. Stability of the explicit scheme.
When an algorithm such as the one given in Eq. (6.17) is implemented on a computer, there will also be errors associated with finite precision arithmetic in the machine. Let us call the solution that comes from the computer $\bar{T}$. This solution can be represented as the sum of the exact solution to Eq. (6.17) (not Eq. (6.13)!) and an error that we will call ε, i.e.,

$$\bar{T} = T + \varepsilon$$

(a) Show by substitution of $\bar{T}$ into Eq. (6.17) that ε also satisfies Eq. (6.17).

(b) Assume that ε can be represented as a finite Fourier series over the domain, i.e.

$$\varepsilon(x,t) = \sum_{m=0}^{M} \varepsilon_m(x,t) = \sum_{m=0}^{M} e^{a_m t} e^{i k_m x}$$

where $2M$ is the number of grid points in the domain of length $2L$, and $k_m = m\pi/L$ is the wavenumber associated with mode m. If any component grows unbounded in time, the method is unstable, so it is sufficient to study the behavior of one mode. Show that for any mode m

$$e^{a_m (t+\Delta t)} e^{i k_m x} = e^{a_m t} e^{i k_m x} + r e^{a_m t} e^{i k_m x} \left(e^{i k_m \Delta x} - 2 + e^{-i k_m \Delta x} \right)$$

(c) Use trigonometric identities to reduce the right hand side to show that

$$\mathcal{A} = e^{a_m \Delta t} = 1 - 4r \sin^2 \frac{k_m \Delta x}{2}$$

Note that $\varepsilon_m(x, t+\Delta t) = \mathcal{A}\varepsilon_m(x,t)$, and for this reason, $\mathcal{A}$ is called the *amplification factor*. The condition for stability is $|\mathcal{A}| \leq 1$ for all m. Inspection shows that the algorithm is stable for $r \leq 1/2$.

Exercise 6.2. Spreadsheet solution of 1-D heat conduction.
The goal of this exercise is to learn to code the explicit algorithm for a one-dimensional heat conduction problem. We will also use the analytical solution to check the answer.

(a) The heat conduction equation in one dimension is given by

$$T_t = \alpha T_{xx}$$

Consider the domain $0 \leq x \leq L$ with boundary and initial conditions given by

$$\begin{aligned} T(x,0) &= T_i \\ T(0,t) &= T_0 \\ T_x(L,t) &= 0 \end{aligned}$$

Scale the differential equation, boundary and initial conditions using

$$x^* = \frac{x}{L},\ t^* = \frac{\alpha t}{L^2},\ \theta = \frac{T - T_0}{T_i - T_0}$$

(b) Use separation of variables and Fourier series to show that the solution is

$$\theta(x^*, t^*) = \frac{4}{\pi} \sum_{k=0}^{\infty} \frac{(-1)^k}{2k+1} \exp\left(-\frac{(2k+1)^2 \pi^2 t^*}{4}\right) \cos \frac{(2k+1)\pi x}{2}$$

(c) Start a spreadsheet with 12 columns, with Column A representing time, and Columns B-K corresponding to division of the domain $0 \leq x^* \leq 1$ into 10 equal segments, with $\Delta x^* = 0.1$. Note that we use the usual notation that rows are numbered and columns are marked with letters. Fill in Row 1 with the initial condition $\theta = 1$. Choose a cell, e.g. "1L" to hold the value of the parameter $\alpha \Delta t / \Delta x^2$, and insert the value 0.4 in that cell. Program Eq. (6.17) into cell "2C" and then propagate this formula across the row through cell "2J." Remember to refer to the value of Fo_g as "\$1\$L" in the formula. Then use Eqs (6.18) and (6.20) to define appropriate boundary conditions in cells "2B" and "2K". Propagate this row definition down the spreadsheet to integrate the equation in time.

(d) Make a plot of the temperature distribution at several times. In particular, plot $\theta(x^*, 0.8)$. Also plot the analytical solution at the same time. Note that only the first two or three terms should be necessary because of the exponential form.

(e) Choose a different value of $\alpha \Delta t / \Delta x^2$ and explore the results. Notice what happens if you choose a value larger than 0.5.

Exercise 6.3. Modified equation for heat conduction.
This exercise fills in the blanks of the derivation of Eq. (6.22). Begin with Eq. (6.21).

(a) Differentiate Eq. (6.21) with respect to t.

(b) Differentiate Eq. (6.21) twice with respect to x.

(c) Combine these two equations in such a way as to eliminate the term containing T_{txx}, and then rearrange the result to produce an expression for T_{tt}.

(d) Substitute this result into Eq. (6.21) to obtain Eq. (6.22)

Exercise 6.4. Isotropic Laplacian operator.
Follow the description of the 2-D isotropic Laplacian to develop a 3-D version of the operator, using a 27-noded stencil. Assume that the grid spacing is the same in all three directions, i.e., $\Delta x = \Delta y = \Delta z$.

Exercise 6.5. Quadratic 1-D finite elements.
Derive the shape functions for a quadratic 1-D element containing nodes i, j and k, where node k is at the midpoint between nodes i and j.

(a) Write the interpolation scheme in the form

$$x = a + bx + cx^2$$

and then write equations at each node to form a set of linear algebraic equations for a, b and c.

(b) Rearrange the equation in part (a) to have the form of Eq. (6.38), identifying the three shape functions N_i, N_j and N_k.

(c) Graph the three shape functions over the element.

Exercise 6.6. Lumped vs consistent capacitance matrices.
Begin with the lumped and consistent capacitance matrices from the 1-D linear element given in Eq. (6.55) and (6.57).

(a) Write the contribution to the nodal equation for an interior node i from neighboring elements for both forms of the capacitance matrix.

(b) Show by comparing the two formulations that the difference between the two formulations is proportional to the second derivative of the temperature.

(c) What is the effective diffusivity when using the lumped form of the capacitance matrix?

6.8 REFERENCES

[1] G. J. Abbaschian and A. F. Giamei, editors. *Modeling of casting and welding processes-IV*. Palm Coast, FL, 1988.

[2] M. S. Engelman, R. Sani, P. M. Gresho, and M. Bercovier. Consistent vs. reduced integration penalty methods for incompressible media using several old and new elements. *Intl. J. Num. Meth. Engr.*, 2:25-42, 1982.

[3] C. Gau and R. Viskanta. Melting and solidification of a pure metal on a vertical wall. *J. Heat Transfer*, 108:174-181, 1986.

[4] P. M. Gresho and R. L. Sani. *Incompressible flow and the finite element method*. Wiley, 1998.

[5] T. E. Morthland, P. E. Byrne, D. A. Tortorelli, and J. A. Dantzig. Optimal riser design for metal castings. *Metall. and Mat. Trans.*, 26B:871-885, 1995.

[6] S. Osher and J. A. Sethian. Fronts propagating with curvature-dependent speed: algorithms bases on Hamilton-Jacobi formulations. *J. Comp. Phys.*, 79:12-49, 1988.

[7] M. Rappaz, M. Bellet, and M. Deville. *Numerical modeling in materials science and engineering*. Springer-Verlag, Berlin, 2002.

[8] J. N. Reddy and D. K. Gartling. *The finite element method in heat transfer and fluid dynamics*. CRC Press, New York, 2nd edition, 2000.

[9] G. N. Vanderplaats. *Numerical optimization techniques for engineering design: with applications*. McGraw-Hill, New York, 1984.

Part II

Microstructure

Part II

Microstructure

The first part of this book dealt mainly with macroscopic behavior. In fact, Chapters 2 and 3, which cover thermodynamics and phase diagrams, make almost no reference to length scales at all. The following chapters were mostly concerned with modeling of macroscale behavior. In this part of the book, we extend our view to the microscopic scale, focusing on the formation of *microstructure* from the liquid state, which we define as the patterns observed in the range of 10 nm to 100 μm. These structures are most commonly seen using conventional optical microscopy that reveals compositional inhomogeneities after chemical etching. They can also exhibit distinct crystallographic orientations that can be mapped using Electron Back-Scattered Diffraction (EBSD) in a scanning electron microscope.

The tools that we have developed in the first part of this book still apply at these length scales. In particular, we use the concepts of thermodynamic equilibrium (developed in Chaps. 2 and 3), as well as the balance equations from Chap. 4, to model the formation and evolution of microstructures during solidification. We also need to account for deviations from equilibrium at the solid-liquid interface, in particular the contribution associated with its curvature (see Sect. 2.4.2). Another important phenomenon that has yet to be discussed is the *interface stability*. The concept of stability is essential to the understanding of how patterns are selected during microstructure evolution.

We begin our study in Chap. 7, where we consider the first solid to form and the role that thermodynamics and impurities play in the initial stages of solidification. We see there that curvature has a major role in the selection of initial length scales. From there, we move on to study cells and dendrites in Chap. 8. These are the microscale patterns that evolve in most materials after nucleation of the solid phase. It is in this chapter that we introduce the concept of *morphological stability*, i.e., whether a particular shape, such as a sphere or planar front, is stable with respect to perturbations in morphology. The result of the instability is the microstructural pattern. These concepts are introduced in the context of single phase alloys, or at least alloys whose microstructures are dominated by a single primary phase. Another very important class of alloys inludes the eutectics and peritectics, discussed in Chap. 9. In these alloys, two (or more) phases solidify simultaneously. Nevertheless, the analytical methods used to understand these alloys are in many ways similar to those used to study cells and dendrites. In the same chapter, the competition of phases and/or microstructures is also presented. Finally, we discuss microsegregation in Chap. 10. This chapter deals with the chemical segregation patterns associated with the formation of microstructures.

CHAPTER 7

NUCLEATION

7.1 INTRODUCTION

In this text, we focus our attention on crystalline solids that form from the melt. The process begins with the creation of a cluster of atoms of crystalline structure, which may occur due to random fluctuations. This chapter describes the processes by which clusters form, and the conditions that affect the rate of their production and survival. Clusters that are too small to survive will be called *embryos*, whereas those that are sufficiently large to be stable are termed *nuclei*. Thermodynamics plays an important role, with the clusters becoming more plentiful and viable as the temperature decreases below the equilibrium freezing point. The concept of nucleation also applies in other contexts, such as the first appearance of voids associated with gas porosity and other defects. This is discussed in Part III.

We begin with the classical theory of *homogeneous nucleation*, where the clusters are assumed to appear spontaneously in a melt free of any impurities. This theory is the easiest to understand, but it leads to predictions for nucleation rates that are far different from those observed in practice. This leads to the concept of *heterogeneous nucleation*, introduced in Sect. 7.3, in which the clusters form preferentially on foreign particles in the melt, or at interfaces such as those between the melt and its container. Particles may be present simply as a result of inevitable impurities, or they may be intentionally added in a process called *inoculation*, which promotes the formation of nuclei just below the equilibrium melting point, i.e., at small undercooling. Relatively recent experimental results show that the heterogeneous nucleation rate depends, in this case, on the size distribution of foreign particles in the melt, as discussed in Sect. 7.4.

7.2 HOMOGENEOUS NUCLEATION

7.2.1 Embryos and nuclei

Consider a homogeneous, pure liquid at uniform temperature above the equilibrium melting point T_f. The structure of the liquid, which can be measured for example by neutron diffraction, has short range order over distances of a few atomic radii, but is disordered over longer distances. The atomic mobility is also high. This permits the formation of clusters of a few atoms with a crystalline structure, but it is not energetically favorable for the crystal structure to persist in such small clusters. This was discussed in Chap. 2, and we refer the reader to Fig. 2.2 for a graphical description. Indeed, for $T > T_f$, the Gibbs free energy of the solid G_s^m is higher than that of the liquid G_ℓ^m. When $T < T_f$, the free energy diagram shows that $G_s^m < G_\ell^m$, however this simple balance does not account for the surface energy contribution associated with the formation of the solid-liquid interface (Eq. (2.60)). Nor does it reveal the rate at which solid nuclei appear. This is the subject of the present chapter.

Despite the fact that we discuss clusters consisting of a relatively small number of atoms, the liquid and solid phases are considered to be continuous media, and the interface between the two phases is considered to be a sharp surface of separation. For the sake of simplicity in the presentation, the solid cluster is assumed to be a sphere of radius R containing N atoms. We treat the solid and liquid as continua, which implies that R is much larger than the actual dimensions of the interface, which is typically a few atomic diameters thick. We assume that both phases are homogeneous, with molar free energies G_s^m and G_ℓ^m in the solid and liquid phases, respectively, and that the interfacial energy is given by $\gamma_{s\ell}$. The free energy of a liquid containing a solid particle was described in detail in Sect. 2.4.2, and the results from those derivations are reused here. The difference in free energy between a system containing a solid particle of volume V_s and surface area $A_{s\ell}$ in contact with the liquid, and an entirely liquid system is given by

$$\Delta G = V_s \frac{G_s^m - G_\ell^m}{V^m} + A_{s\ell}\gamma_{s\ell} \tag{7.1}$$

It was shown in Eq. (2.21) that for a small undercooling ΔT, $G_s^m - G_\ell^m \approx -\Delta S_f^m \Delta T$, where ΔS_f^m is the molar entropy of fusion. Furthermore, from Eq. (2.62) we have $\rho\Delta s_f = \Delta S_f^m / V^m$. Substituting these results into Eq. (7.1), and replacing V_s and $A_{s\ell}$ by appropriate values for a sphere of radius R, yields

$$\Delta G = -\frac{4}{3}\pi R^3 \rho \Delta s_f \Delta T + 4\pi R^2 \gamma_{s\ell} \tag{7.2}$$

Figure 7.1 shows the three terms in Eq. (7.2), for the particular case of pure Al, when $\Delta T > 0$, i.e., $T < T_f$. The first term on the right hand side is proportional to R^3 and negative, indicating that energy is released by solidification when $\Delta T > 0$. The second term is quadratic in R and positive, corresponding to the energy "penalty" associated with the creation of a surface. For small R, the surface energy penalty exceeds the volumetric liberation of energy, whereas the volumetric contribution dominates at larger radii. This creates a maximum in ΔG called the *homogeneous nucleation barrier* ΔG_n^{homo}, as indicated in Fig. 7.1. The radius R_c at which this

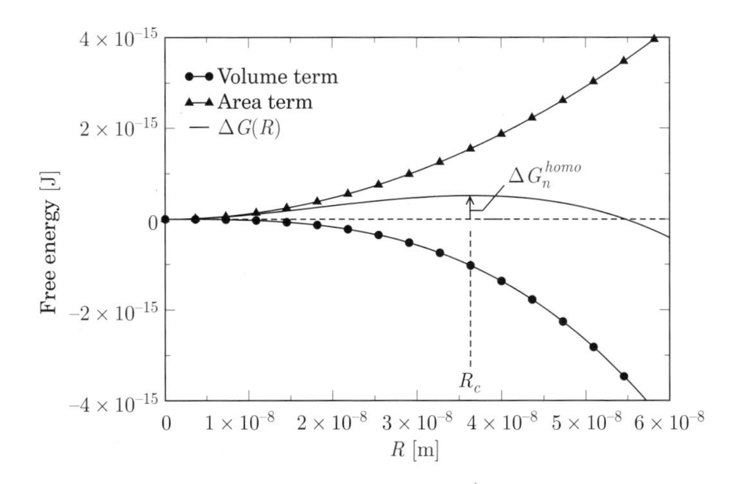

Fig. 7.1 Surface, bulk and total free energies of a spherical solid as functions of its radius for a fixed undercooling $\Delta T = 5\,\text{K}$. Property data for Al are tabulated in Table 7.1.

occurs is determined by differentiating Eq. (7.2) with respect to R and setting the result equal to zero:

$$R_c = \frac{2\gamma_{s\ell}}{\rho \Delta s_f \Delta T} = \frac{2\Gamma_{s\ell}}{\Delta T} \tag{7.3}$$

where $\Gamma_{s\ell} = \gamma_{s\ell}/(\rho_s \Delta s_f)$ is the Gibbs-Thomson coefficient. Substituting this result into Eq. (7.2) gives

$$\Delta G_n^{homo} = \frac{4\pi\gamma_{s\ell}R_c^2}{3} = \frac{16\pi}{3}\frac{\gamma_{s\ell}^3}{(\rho\Delta s_f)^2 \Delta T^2} \tag{7.4}$$

An embryo of radius R_c is called a *critical nucleus*, since it is energetically favorable for nuclei with $R < R_c$ to melt, and for $R > R_c$ to grow. Note that for $\Delta T < 0$, corresponding to temperatures above T_f, both terms on the right hand side are positive for all values of R, and any embryo of the solid that forms should always remelt. Let us note that the exact geometry of the embryo is not important – it changes the prefactor in Eq. (7.4), but not in such a way as to affect the conclusions.

We have actually already encountered this result twice in earlier chapters. First, in Chap. 2, we found the relationship between the undercooling and curvature at equilibrium, and Eq. (7.3) simply corresponds to Eq. (2.61) with the mean curvature $\bar{\kappa} = R_c^{-1}$. We also found this result in Sect. 5.4.1, where an analysis of the growth rate of a spherical solid in an undercooled melt revealed that the sphere had a critical radius of $R_c = 2\Gamma_{s\ell}/\Delta T$. (See Eq. (5.184) and the subsequent discussion.) The important point is that all of these analyses are tied together: the contribution of curvature to thermodynamic equilibrium leads to a condition

Table 7.1 Material properties for Al.

Property	Value	Units
T_f	933	K
$\gamma_{s\ell}$	0.093	$\mathrm{J\ m^{-2}}$
$\rho\Delta s_f$	1.02×10^6	$\mathrm{J\ m^{-3}\ K^{-1}}$
$\Gamma_{s\ell}$	9.12×10^{-8}	K m
V^m	1.138×10^{-5}	$\mathrm{m^3\ mol^{-1}}$

of "equilibrium," where a particle of a given radius R grows or shrinks, depending on its size relative to R_c. This equilibrium point is unstable because $d^2(\Delta G)/dR^2 < 0$, indicating that the Gibbs free energy decreases for both $R < R_c$ and $R > R_c$. Clusters of atoms smaller than the critical nucleus are called *embryos*, while those larger than R_c are called *nuclei*.

Example 7.1

Consider a melt of pure Al, with the properties tabulated in Table 7.1. Determine the critical radius R_c, the number of atoms N_c, contained in a critical nucleus, and the free energy change upon solidification as a function of the undercooling.

The solution is a straightforward application of the equations just derived. We have:

$$R_c = \frac{2\Gamma_{s\ell}}{\Delta T} = \frac{2(9.12 \times 10^{-8}\ \mathrm{K\ m})}{\Delta T} = \frac{1.82 \times 10^{-7}\ \mathrm{K\ m}}{\Delta T}$$

$$N_c = \frac{4\pi N_0}{3V^m} R_c^3$$
$$= \frac{(4\pi)(6.02 \times 10^{23}\ \mathrm{atoms\ mol^{-1}})(1.82 \times 10^{-7}\ \mathrm{K\ m})^3}{3(11.38 \times 10^{-6}\ \mathrm{m^3\ mol^{-1}})\Delta T^3}$$
$$N_c = \frac{1.33 \times 10^9\ \mathrm{atoms\ K^3}}{\Delta T^3}$$

For convenience in plotting, ΔG_n^{homo} is normalized by the characteristic energy $k_B T$ associated with the thermal fluctuations of one atom, k_B being Boltzmann's constant:

$$\frac{\Delta G_n^{homo}}{k_B T} = \frac{16\pi}{3k_B T} \frac{\gamma_{s\ell}^3}{(\rho\Delta s_f)^2 \Delta T^2}$$
$$= \frac{(16\pi)(0.093\ \mathrm{J\ m^{-2}})^3}{(3)(1.38 \times 10^{-23}\ \mathrm{J\ K^{-1}})(1.02 \times 10^6\ \mathrm{J\ m^{-3}\ K^{-1}})^2} \frac{1}{T\Delta T^2}$$
$$\frac{\Delta G_n^{homo}}{k_B T} = \frac{9.39 \times 10^8\ \mathrm{K^3}}{T\Delta T^2}$$

The results are plotted below.

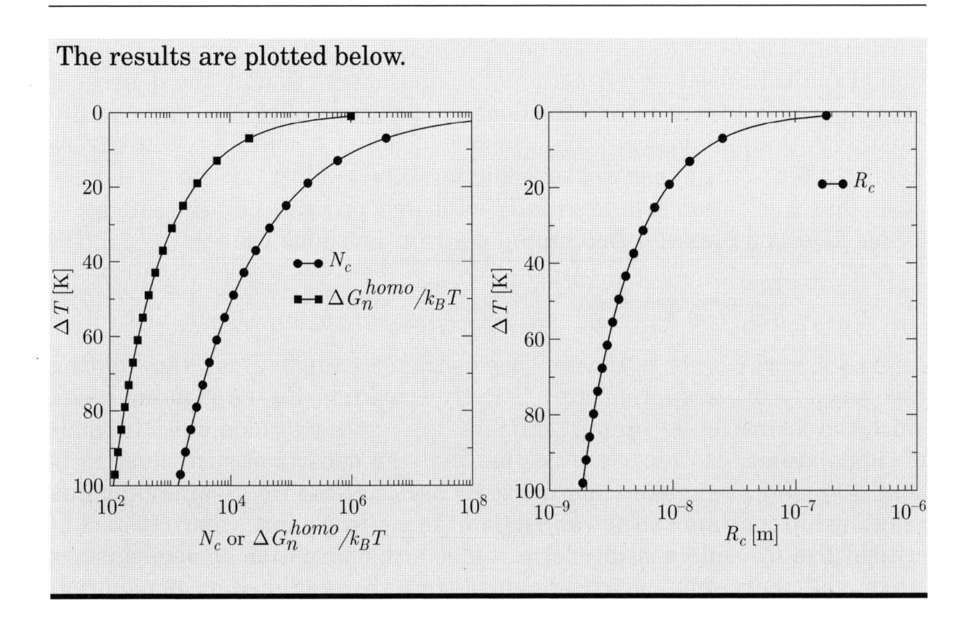

Example 7.1 calculates the critical nucleus size and the magnitude of the homogeneous nucleation barrier as a function of the undercooling for pure Al. A closer examination of the results leads us to doubt the validity of the homogeneous nucleation theory. These doubts become even more apparent in the next section. For undercooling values up to about 20 K, the critical radius is larger than 10 nm, which is sufficiently large for the assumptions of continuum behavior and a sharp interface to be reasonable. However, a cluster of this size contains more than $100,000$ atoms, and thus the formation of a critical nucleus of even this small size requires an unlikely amount of organization in the liquid. On the other hand, if we consider a cluster containing a more probable number of atoms i.e. approximately 1000, the corresponding undercooling is almost 100 K. Such large undercoolings are rarely achieved in practice, unless very rapid solidification techniques are used. Further, since the critical nucleus size then approaches atomic dimensions, the assumptions made in this theory are unlikely to apply. These inconsistencies in the theory of homogeneous nucleation lead us to propose alternative mechanisms for nucleation in practical applications.

7.2.2 Nucleation rate

The relationship between cluster size and free energy is important as it allows an estimation of the rate at which clusters of a given size will appear. We have seen that in an undercooled melt ($\Delta T > 0$), the formation of a cluster or embryo of size R requires an additional free energy $\Delta G(R)$, which exhibits a maximum of height ΔG_n^{homo} at $R = R_c$, as shown in Fig. 7.1. On the other hand, $\Delta G(R)$ is an increasing function of R for a superheated melt ($\Delta T < 0$). We now proceed to compute the rate of formation of critical nuclei as a function of temperature. Denote the density of atoms in the liquid as n_ℓ, and the density of clusters of radius R in equilibrium with the liquid as n_R. The cluster of radius R contains N atoms, so we also refer to the density of such clusters as n_N. Most nucleation models posit that nucleation is a *thermally activated* process, i.e., the energy of atoms follows a Maxwell-Boltzmann distribution, such that

$$\frac{n_R}{n_\ell} = \exp\left(-\frac{\Delta G(R)}{k_B T}\right) \quad \text{or} \quad \frac{n_N}{n_\ell} = \exp\left(-\frac{\Delta G(N)}{k_B T}\right) \tag{7.5}$$

As $\Delta G(N=1) = 0$ for a "cluster" consisting of a single atom, the limiting case $n_{N=1} = n_\ell$ is appropriate. Strictly speaking, Eq. (7.5) is not exact since the formation of dimers, trimers, etc., naturally depletes the melt of single atoms. However we assume that the density of such clusters is small. Further, we make a more drastic assumption: we assume that this relationship continues to be valid below the melting point, when clusters of critical size R_c could possibly form. Indeed, the formation of such clusters which will then grow clearly should distort the equilibrium distribution given by Eq. (7.5). A more detailed discussion of this point can be found in the text by Christian [1]. For the current discussion, we nonetheless assume that the density of clusters of critical size $R = R_c$ is small with respect to n_ℓ, and given by Eq. (7.5).

The distribution of embryo sizes is shown schematically in Fig. 7.2. For values of $R > R_c$, the growth of nuclei invalidates Eq. (7.5), indicated in the figure by a dashed line that continues the distribution in this region. Thus, we are interested in finding n_c, the density of embryos that reach the critical radius R_c, which can be found by setting $\Delta G(R) = \Delta G_n^{homo}$ in Eq. (7.5), and using Eq. (7.4) to obtain

$$\frac{n_c}{n_\ell} = \exp\left(-\frac{\Delta G_n^{homo}}{k_B T}\right) = \exp\left(-\frac{16\pi}{3}\frac{\gamma_{s\ell}^3}{(\rho\Delta s_f \Delta T)^2 k_B T}\right) \tag{7.6}$$

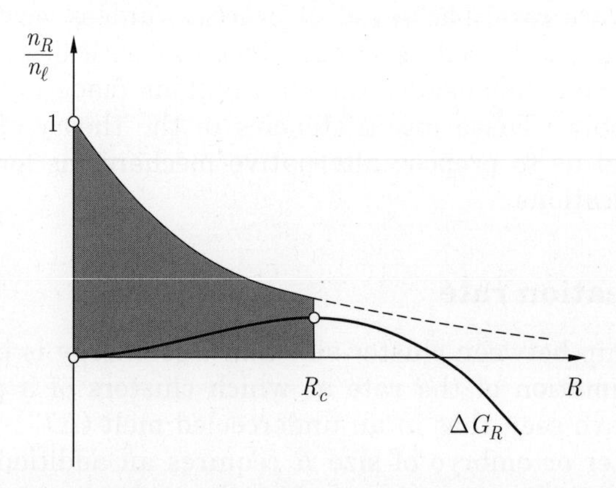

Fig. 7.2 The density of embryos as a function of their radius.

A nucleus of critical size R_c will grow if it manages to add one more atom. The rate at which this occurs is proportional to the atomic vibration frequency ν_0 and the probability of capturing an atom at the surface, p_c. Thus, the rate of formation of homogeneous nuclei, I^{homo}, is

$$I^{homo} = \nu_0 p_c n_c = \underbrace{\nu_0 p_c n_\ell}_{I_0^{homo}} \exp\left(-\frac{16\pi}{3}\frac{\gamma_{s\ell}^3}{(\rho\Delta s_f \Delta T)^2 k_B T}\right) \tag{7.7}$$

The prefactor I_0^{homo} is not strongly temperature-dependent, and its value can be estimated from its three component terms. As an example, we use the data for Al given in Table 7.1 to compute $n_\ell = N_0/V^m \approx 6 \times 10^{28}$ atoms m^{-3}. The atomic vibration frequency being about 10^{13} s^{-1} at T_f, let us assume that there is no difficulty in attaching an atom to the surface of the nucleus, i.e., $p_c \approx 1$. Substituting these values into Eq. (7.7) yields

$$I^{homo} = \left(6 \times 10^{41}\ \mathrm{m^{-3}s^{-1}}\right) \exp\left(-\frac{16\pi}{3}\frac{\gamma_{s\ell}^3}{(\rho\Delta s_f \Delta T)^2 k_B T}\right) \tag{7.8}$$

The nucleation rate depends strongly on temperature through the two competing terms in the denominator of the exponential in Eq. (7.8). The decrease in the nucleation energy barrier contributes the term ΔT^2, which tends to make the nucleation rate increase very strongly with increasing undercooling. The Maxwell-Boltzmann distribution contributes the term T, corresponding to the decreasing mobility of atoms at low temperatures.

Example 7.2

Use the data from Example 7.1 to compute the nucleation rate and the time to reach one nucleus per cm^3 as a function of temperature.

The solution is again a straightforward substitution. From Eq. (7.8) we have

$$\begin{aligned} I^{homo} &= \left(6 \times 10^{41}\ \mathrm{m^{-3}s^{-1}}\right) \exp\left(-\frac{16\pi}{3}\frac{\gamma_{s\ell}^3}{(\rho\Delta s_f \Delta T)^2 k_B T}\right) \\ &= (6 \times 10^{41}\ \mathrm{m^{-3}s^{-1}}) \times \\ &\exp\left(-\frac{16\pi(93 \times 10^{-3}\ \mathrm{J\,m^{-2}})^3}{3[(1.02 \times 10^6\ \mathrm{J\,m^{-3}\,K^{-1}})(933 - T)]^2(1.38 \times 10^{-23}\ \mathrm{J\,K^{-1}})T}\right) \end{aligned}$$

Subsequently, the nucleation time can be computed as simply

$$t_n = \frac{10^6\ \text{nuclei m}^{-3}}{I^{homo}}$$

The results in the figure below show the formation of critical nuclei for homogeneous nucleation as a function of the absolute temperature. To

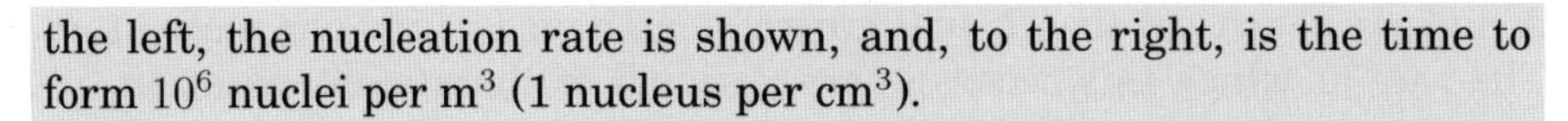
the left, the nucleation rate is shown, and, to the right, is the time to form 10^6 nuclei per m^3 (1 nucleus per cm^3).

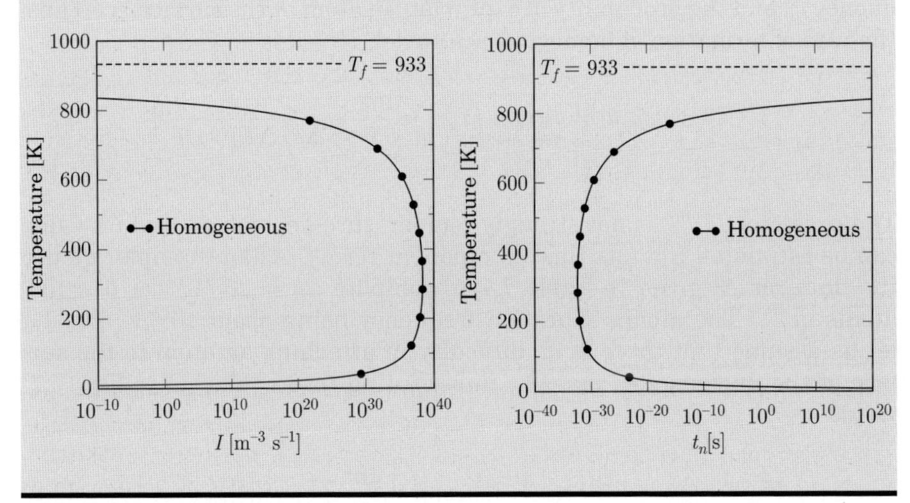

Example 7.2 illustrates the competition between the two phenomena through the calculation of the homogeneous nucleation rate as a function of temperature for Al. The nucleation rate increases by over 20 orders of magnitude as the temperature decreased from about 833 K to 800 K. Example 7.2 also shows the time required to form one nucleus per cm^3 as a function of temperature. The curve has the characteristic shape of a Time-Temperature-Transformation (TTT) curve. Note also the huge time scale difference in this diagram: at 800 K, one crystal per cm^3 forms in about 0.1 μs, whereas at 833 K, this takes more than 450, 000 years!

Before going any further, we should note that the predictions of the homogeneous nucleation model do not agree very well with observations under typical experimental conditions. The example suggests that it would take almost forever to observe nucleation in Al at temperatures less than 100 K below the equilibrium melting point. However, a simple experiment, performed by placing a thermocouple into a melt and allowing it to freeze, shows that solidification begins with undercooling of at most a few K. The solution to this dilemma is that, in practice, solidification is initiated by *heterogeneous nucleation*, where solidification begins on foreign particles suspended in the melt or at external surfaces. We deal with this phenomenon in the next section. The form of Eq. (7.8), where $\gamma_{s\ell}$ is raised to the third power inside the exponential suggests that one might be able to observe homogeneous nucleation in systems where $\gamma_{s\ell}$ is weak, i.e., when the structures of the solid and liquid phases are very similar. Such seems to be the case in $Ga_{24}Mg_{36}Zn_{40}$, for which very small grains containing $MgZn_2$ dendrites have been observed in droplets of about 120 μm solidified in a gas-filled drop tube (see Spaepen and Fransaer [12]).

On the other hand, if one wants to obtain a glassy material, nucleation must be avoided. Example 7.2 shows that very rapid cooling is required to avoid the "nose" of the TTT-diagram. In pure metals, this is probably impossible due to the nucleation time being so short. In materials with more complex structures, such as SiO_2, glasses can be obtained relatively easily at low cooling rates since the attachment kinetics of the molecules at the interface, as well as their mobility, is low. Glasses can be obtained in some metallic alloys by very rapid cooling of the melt. The earliest observations of this phenomenon implemented rapid heat extraction by using specimens where at least one dimension was small (e.g., ribbons or droplets) and massive heat extraction. More recently, bulk metallic glasses (BMG) have been obtained under almost conventional solidification conditions in alloys exhibiting a eutectic reaction at very low temperature (so-called "deep eutectic forming" alloys). In such systems, the liquidus of both phases is decreased so much that diffusion becomes severely limited at temperatures where the driving force (undercooling) is appreciable. Such BMG's exhibit fairly unique mechanical and corrosion properties. [3]

Finally, if we consider homogeneous nucleation in alloys, the composition of the nucleus has to be taken into account. Figure 7.3 shows schematically the equilibrium Gibbs free energy curves for the liquid and solid phases in a binary alloy of nominal composition $X_{B\ell}$ at a certain temperature T smaller than $T_{liq}(X_{B\ell})$. As for Fig. 2.20, the two solid curves correspond to the liquid and solid Gibbs free energies for a solid-liquid interface of zero curvature (curves labeled G_ℓ^m and $G_s^{m\infty} = G_s^m(R = \infty)$, respectively). However, we have seen in Chap. 2 (Eq. (2.60)) that the free energy $G_s^m(R)$ of a spherical solid of finite radius R is higher than $G_s^m(\infty)$ by an amount $2V^m\gamma_{s\ell}R^{-1}$. Thus, there is a family of free energy curves, obtained by translating the curve $G_s^{m\infty}$ along the vertical axis, each corresponding

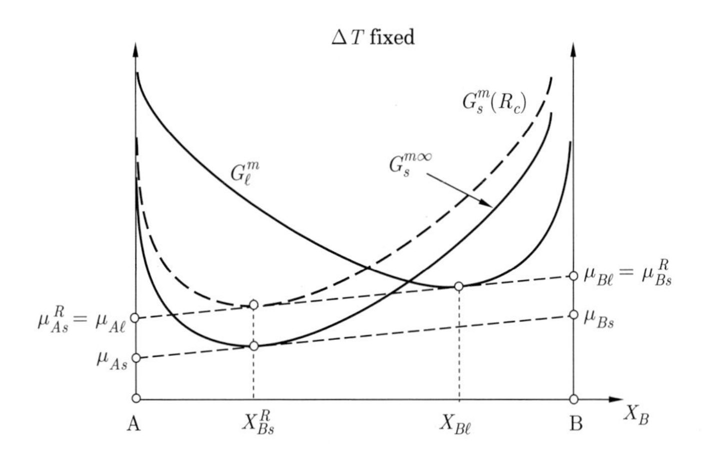

Fig. 7.3 Gibbs free energies of the liquid and solid phase at a certain undercooling, ΔT, and the construction that provides the maximum variation of energy for the formation of a first nucleus.

to a different value of the radius R. The problem is to determine which of these free energy curves gives us the solid composition X_{Bs}^{R} and the size of the critical nucleus R_c in equilibrium with the melt.

In order to solve this problem, we note that the composition of the melt is $X_{B\ell}$ when the nucleus first appears. If we assume that the solid nucleus is in equilibrium with this liquid, we thus obtain, taking into account its curvature,

$$\mu_{As}(R_c) = \mu_{A\ell}(X_{B\ell}); \quad \text{and} \quad \mu_{Bs}(R_c) = \mu_{B\ell}(X_{B\ell}) \tag{7.9}$$

The chemical potentials can be determined from the free energy curves using the tangent construction illustrated in Fig. 7.3. (See Sect. 2.3 and Eq. (2.36) for a detailed discussion). Equation (7.9) is satisfied when we determine the unique free energy curve for the solid, designated $G_s^m(R_c)$ in Fig. 7.3, that forms a common tangent with the composition $X_{B\ell}$ on the liquid free energy curve. The critical radius R_c can then be determined using Eq. (2.60)

$$G_s^m(R_c) - G_s^m(\infty) = \frac{2V^m\gamma_{s\ell}}{R_c} \tag{7.10}$$

This construction also gives the composition of the nucleus. Note that this approach requires knowledge of the composition-dependence of the surface energy $\gamma_{s\ell}$. Thompson and Spaepen [13] obtained a similar result by hypothesizing that the composition of the nucleus was such as to give the largest energy difference between the liquid and solid states. It is then possible to show (see Exercise 7.2) that this occurs when $\Delta\mu_A = \mu_{A\ell} - \mu_{As} = \Delta\mu_B = \mu_{B\ell} - \mu_{Bs}$, i.e., when one draws a tangent to the solid curve $G_s^m(\infty)$ that is parallel to the tangent to the liquid free energy curve at $X_{B\ell}$. This construction is also illustrated in Fig. 7.3.

7.3 HETEROGENEOUS NUCLEATION

7.3.1 Motivation

In the previous section, we introduced the thermodynamics of homogeneous nucleation. Embryos appeared in the melt by random fluctuations, and the rate at which the embryos became viable was computed based on the energy difference between the embryo and the melt. The nucleation rate thus depended strongly on the temperature. In Example 7.2, we showed that the homogeneous nucleation theory predicts that there is essentially no nucleation for less than 100 K of undercooling. We also noted that this is contrary to common observations in the laboratory.

Some other mechanism is clearly needed to explain the behavior of real systems. In this section, we introduce the theory of *heterogeneous nucleation*, where solidification is initiated on a *foreign surface*. By foreign surface, we mean the surface of the mold containing the melt, the oxide skin forming at the free surface of the melt in contact with ambient air, or foreign particles suspended in the melt. These foreign particles may be simply unintentional impurities, or deliberate additions for the purpose of controlling the microstructure. The theory we have developed to describe homogeneous nucleation will be extended to describe this new phenomenon. Prior to this, however, let us anticipate one of the results in order to discuss a very important experiment that demonstrates the phenomenon of heterogeneous nucleation.

Consider again the TTT curve in Example 7.2. The time for one nucleus to appear in one cm^3 of material drops from about $450,000$ years at $T = 833$ K to 0.1 μs at about 800 K. This shows that homogeneous nucleation occurs abruptly over a fairly narrow temperature interval. Let us assume for the moment (and demonstrate later) that the same phenomenon occurs for nucleation of the solid on a distribution of foreign impurities suspended in the melt, each impurity being characterized by a single nucleation temperature. Of course, these temperatures are higher than those characterizing homogeneous nucleation, otherwise these particles would have no effect on nucleation. For each particle, no nucleation occurs above its characteristic temperature, whereas it is almost instantaneous just below. The nucleation temperature characterizing each foreign particle is representative of what is called its *nucleation potency*. We demonstrate later that this model is indeed reasonable.

Consider now the implications of such a distribution of foreign particles on the solidification of a small droplet undergoing heat extraction at a constant rate. As time proceeds, the temperature of the droplet falls linearly in proportion to the heat extraction rate divided by the specific heat. In the absence of nucleation, the droplet becomes undercooled once the temperature falls below T_f. Eventually the temperature reaches the characteristic nucleation temperature of one of the particles in suspension. After solidification begins, latent heat is liberated, and the temperature of the droplet may actually increase, even though the overall heat content decreases. This phenomenon is called *recalescence*, and the path through enthalpy-temperature space under these conditions is shown schematically in Fig. 7.4. The implication of recalescence for our model with multiple nuclei of varying potency is that once the temperature begins to increase, the particles with characteristic nucleation temperatures below the minimum are never activated.

An experiment that provides deeper insight into this phenomenon was performed by Perepezko and coworkers. [7] Droplets of high purity Sn were dispersed in an inert solid matrix. For that purpose, tin and the matrix, immiscible in the liquid state, were mixed with a high speed stirrer before the mixture was allowed to solidify. Depending on the stirring conditions, tin droplets of various sizes were obtained in the inert matrix. The mixture was then placed in a Differential Thermal Analysis (DTA) apparatus. DTA measures the temperature difference between a small specimen undergoing phase transformations and a reference (e.g., an empty crucible), both placed inside a furnace of uniform temperature which is slowly cooled. Peaks in the DTA output therefore correspond to phase transformations; in our case, small temperature differences induced

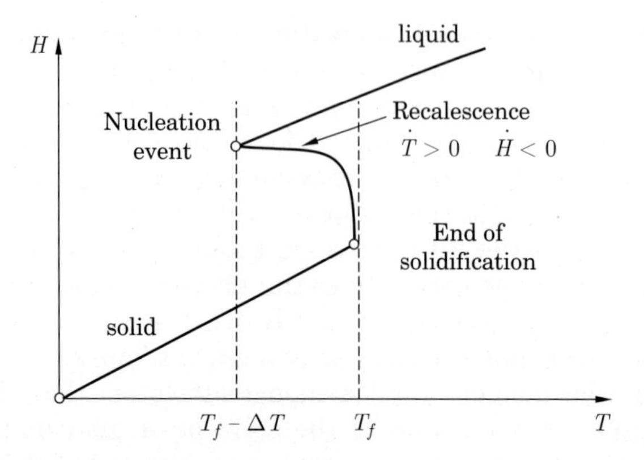

Fig. 7.4 A schematic representation of the enthalpy-temperature relationship after nucleation at a certain undercooling, ΔT.

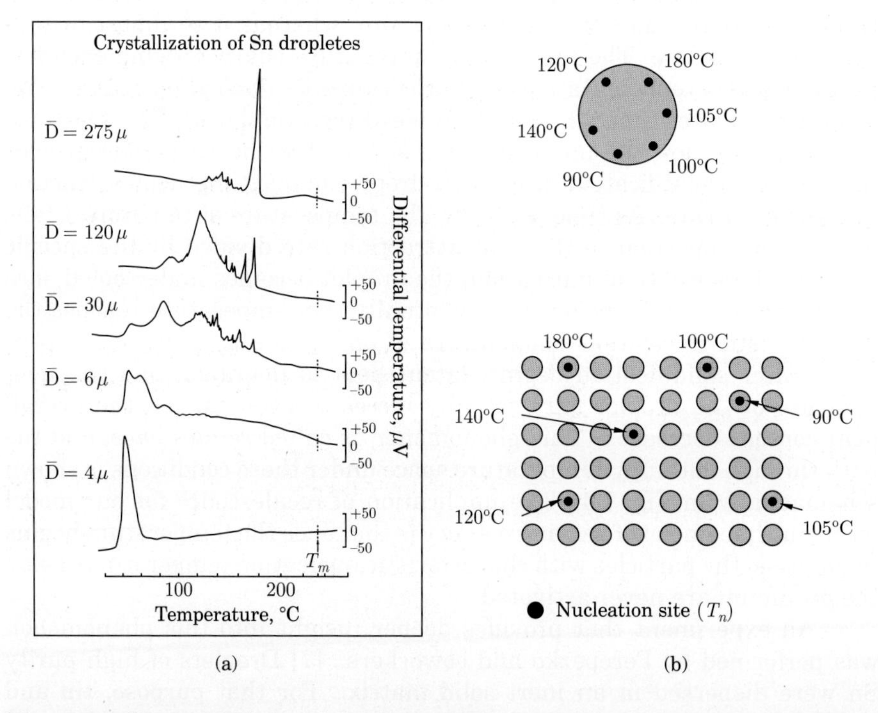

Fig. 7.5 Differential Thermal Analysis (DTA) spectra measured on (a) small tin droplets (courtesy of J. Perepezko, Univ. of Wisconsin-Madison). (b) A schematic illustration of the isolation of potent heterogeneous nucleation sites as the average droplet size is decreased.

by latent heat releases of various solidification events. Figure 7.5 shows some of the results from these experiments, as a series of DTA traces for various droplet size distributions.

For relatively large droplets, there is a sharp peak at about 180°C, along with a few smaller peaks at lower temperatures. The equilibrium melting point (232°C) is indicated with small vertical lines, and the mean droplet diameter is indicated to the left. As the average droplet size decreases, the peak shifts to lower temperatures, until settling at about 50°C. The explanation behind the results is illustrated schematically on Fig. 7.5(b). A distribution of heterogeneous nucleation sites with varying potencies exists in the droplets. When the critical temperature for a particular particle is reached, solidification begins as described earlier. When the droplets are large, nearly all of them contain one of the most potent nuclei, with a critical temperature of about 180°C. Thus, almost all of the droplets freeze at this temperature. As the melt becomes more finely divided, the most potent nuclei are segregated to fewer and fewer droplets, corresponding to a decreasing volume of the total melt. Solidification of these droplets still occurs at around 180°C, but now is stopped by the matrix, and the latent heat release is too small to be detected. Other, less potent, nuclei then become active in a larger number of droplets, which explains the shift in the DTA peak to lower temperatures. The final value that the peak reaches for average droplet sizes of a few μm probably corresponds to nucleation on the interface between the droplets and the matrix. (It is left as an exercise to show that the homogeneous nucleation rate for Sn at this temperature is extremely small.) With this motivation, we now proceed to examine the thermodynamics of heterogeneous nucleation.

7.3.2 Basic theory

The essential physics for the analysis of heterogeneous nucleation is illustrated schematically in Fig. 7.6. When a foreign substrate has a structure and chemistry that match sufficiently well those of the solid, it is energetically favorable to form a solid nucleus on the surface, as shown. In the analysis, we assume that the embryo takes the form of a spherical cap. This implies that the surface energies are all isotropic, and that gravitational effects can be neglected. These phenomena can be included in the analysis, but are not essential for understanding the concepts.

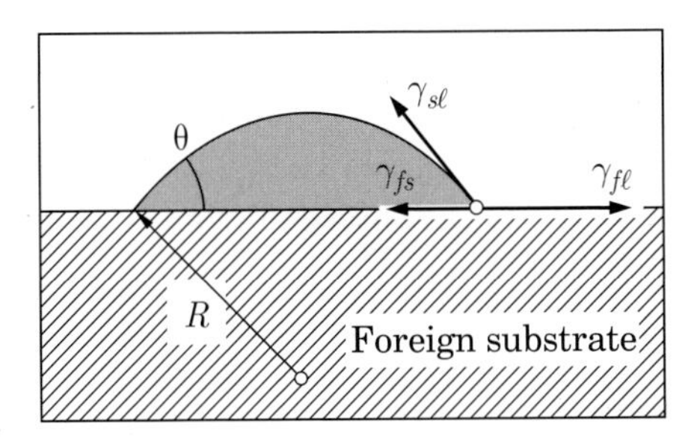

Fig. 7.6 The nucleation of a spherical solid cap at a liquid-substrate interface.

Summation of the forces parallel to the surface in Fig. 7.6 relates the various surface energies to θ, the equilibrium *contact angle* (also called the *wetting angle*) in the Laplace-Young equation

$$\gamma_{f\ell} = \gamma_{fs} + \gamma_{s\ell} \cos\theta \tag{7.11}$$

where the two subscripts on the surface energies indicate the two substances in contact, and the index "f" represents the foreign substrate. There are two cases where there is no value for θ that satisfies Eq. (7.11):

- When $\gamma_{f\ell} > \gamma_{fs} + \gamma_{s\ell}$, it is energetically favorable for the solid layer to completely cover the substrate, effectively separating the liquid from the substrate. This case is discussed further in Sect. 7.3.3.
- Similarly, when $\gamma_{fs} > \gamma_{f\ell} + \gamma_{s\ell}$, there is no possible gain of energy by forming a solid layer on the foreign particle surface. Such particles do not participate in the heterogeneous nucleation process and might even be pushed away from a solid-liquid front coming from other nucleation centers due to this very unfavorable wetting condition.

The primary case of interest is when there is a solution for the contact angle θ requiring that the surface energies satisfy

$$\left| \frac{\gamma_{f\ell} - \gamma_{fs}}{\gamma_{s\ell}} \right| \leq 1 \tag{7.12}$$

When $0 \leq \theta \leq \pi/2$, the solid is said to "wet" the surface, whereas contact angles greater than $\pi/2$ are called "non-wetting." To analyze the case where Eq. (7.12) is satisfied, we follow the derivation for homogeneous nucleation given in Sect. 7.2. The difference in volumetric Gibbs free energy of the liquid/substrate system due to the appearance of the solid spherical cap is given by

$$\Delta G = \rho V_s (g_s - g_\ell) + A_{s\ell}\gamma_{s\ell} + A_{fs}(\gamma_{fs} - \gamma_{f\ell}) \tag{7.13}$$

where V_s is the volume of the solid nucleus, in this case a spherical cap, $A_{s\ell}$ is the area of the new surface created between the solid and liquid, and A_{fs} is the area on the surface of the foreign substrate where the liquid is replaced by the solid. We have taken the density to be the same in the solid and liquid for simplicity. The necessary volume and surface areas for a spherical cap with radius R are obtained from trigonometry as

$$\begin{aligned} V_s &= \frac{\pi R^3}{3}\left(2 - 3\cos\theta + \cos^3\theta\right) = \frac{\pi R^3}{3}(2+\cos\theta)(1-\cos\theta)^2 \\ A_{s\ell} &= 2\pi R^2(1-\cos\theta) \\ A_{fs} &= \pi\left(R\sin\theta\right)^2 = \pi R^2(1-\cos^2\theta) \end{aligned} \tag{7.14}$$

Equation (7.13) can be simplified by substituting the expressions in Eq. (7.14), using Eq. (2.21) to replace $\rho(g_s - g_\ell)$ by $-\rho\Delta s_f \Delta T$, and then finally applying Eq. (7.11) to yield

$$\Delta G = \left(-\frac{4\pi R^3}{3}\rho\Delta s_f \Delta T + 4\pi R^2 \gamma_{s\ell} \right) f(\theta) \tag{7.15}$$

where $f(\theta) \in [0, 1]$ is a geometric factor given by the ratio of the volumes of the spherical cap and a full sphere of identical radius

$$f(\theta) = \frac{V_s}{4\pi R^3/3} = \frac{(2+\cos\theta)(1-\cos\theta)^2}{4} \tag{7.16}$$

A comparison of Eq. (7.15) with Eq. (7.2) reveals a remarkable result: the same form that we had for homogeneous nucleation appears for the free energy in this case, but is now multiplied by $f(\theta)$. Minimization of ΔG with respect to R therefore gives exactly the same result here as for the case of homogeneous nucleation

$$R_c = \frac{2\gamma_{s\ell}}{\rho\Delta s_f \Delta T} = \frac{2\Gamma_{s\ell}}{\Delta T} \tag{7.17}$$

However, the free energy corresponding to R_c, i.e., the heterogeneous nucleation energy barrier is lower by a factor $f(\theta)$

$$\Delta G_n^{heter} = \Delta G_n^{homo} f(\theta) = \frac{16\pi}{3} \frac{\gamma_{s\ell}^3}{(\rho\Delta s_f)^2 \Delta T^2} f(\theta) \tag{7.18}$$

In fact, this result is really not that surprising. Fixing the undercooling determines R_c, the radius of curvature of the solid-liquid interface, via the Gibbs-Thompson Eq. (2.63), regardless of whether this is a full sphere or just a spherical cap. The height and contact angle θ must then be adjusted so as to satisfy the Laplace-Young Eq. (7.11) at the three-phase junction. This, in turn, determines the number of atoms required to form the critical nucleus, $N^{heter}(R_c)$, which is simply the product $N^{homo}(R_c)f(\theta)$.

Thus, we see that the nature of the heterogeneous nucleation is determined by the function $f(\theta)$. Figure 7.7 shows a graph of $f(\theta)$, and also illustrates the equilibrium shape corresponding to three interesting cases:

- For $\theta = \pi$, we have $f(\theta) = 1$, which corresponds to complete non-wetting of the solid on the substrate. This case is equivalent to homogeneous nucleation, since the cap is now the entire sphere.

- When $\gamma_{fs} = \gamma_{f\ell}$, then $\theta = \pi/2$ and $f(\theta) = 0.5$. In this case, the cap forms a hemisphere. Although there is no reduction in the surface energy when the solid embryo forms on the substrate, the presence of the substrate does reduce by half the number of atoms (and thus the nucleation energy barrier) required to form a critical nucleus.

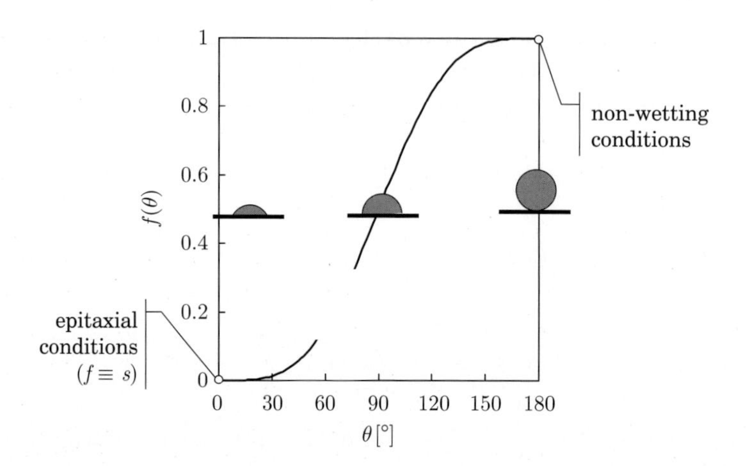

Fig. 7.7 The function $f(\theta)$ for heterogeneous nucleation.

- For $\theta = 0$, the function $f(\theta) = 0$. This case is that of perfect wetting of the solid phase on the substrate, i.e., of perfect compatibility between these two phases. An example is when the solid itself is the substrate, a situation for which there is clearly no nucleation barrier and solidification is limited only by growth. An example of this case is when solidification occurs on fragments of the solid, such as on dendrite arms detached by partial remelting, which then act as new "nucleation"sites.

It should be noted that the assumptions of this theory are not valid for very small values of θ, i.e., when the substrate is very well wetted by the solid. To illustrate this point, consider the nucleation of Al at an undercooling of 1 K on a substrate where $\theta = \pi/30$. Referring to Example 7.1, we see that this undercooling corresponds to $R_c \approx 0.2\ \mu$m. A simple geometric construction shows that the height of the spherical cap is given by $h_{cap} = R_c(1 - \cos\theta)$, which then evaluates to about 1 nm. The extent of the cap, given by $R_c \sin\theta$, is approximately 20 nm. Such a solid nucleus is better represented by a few atomic layers at a surface than a spherical cap. We see later that a realistic treatment requires that other phenomena be included in the analysis.

Now that the free energy barrier for nucleation has been determined, we consider the rate of nucleation. In contrast to homogeneous nucleation, where every atom in the liquid is a potential nucleation site, the number of sites for heterogeneous nucleation depends on the amount and type of foreign substrates. Equation (7.7) can thus be updated for heterogeneous nucleation

$$I^{heter} = \underbrace{\nu_0 p_c n_p}_{I_0^{heter}} \exp\left(-\frac{16\pi}{3}\frac{\gamma_{s\ell}^3}{(\rho\Delta s_f \Delta T)^2 k_B T} f(\theta)\right) \tag{7.19}$$

where n_p is now the density of particles in the melt (m^{-3}) or the density of nucleation sites at the surface of the melt (m^{-2}) that provide heterogeneous nucleation conditions with a wetting angle θ. Note that in contrast to the case for homogeneous nucleation, where the number of potential sites was equal to n_ℓ, we expect that $n_p \ll n_\ell$. Let us assume the same frequency for the atomic attachment as in the case of homogeneous nucleation, $\nu_0 p_c = 10^{13}\ s^{-1}$. Moreover, for the sake of estimation, we also assume that there is one foreign particle per mm^3, i.e., $n_p = 10^9\ m^{-3}$. We then compute $I_0^{heter} = 10^{22}\ m^{-3}s^{-1}$, which is much smaller than the value we used for homogeneous nucleation, $I_0^{homo} = 6 \times 10^{41}\ m^{-3}s^{-1}$. However, the argument of the exponential is also modified by the factor $f(\theta)$, which can be much more important, depending on the value of θ.

Figure 7.8 shows the rate of formation of critical nuclei in an Al melt as a function of the undercooling for several values of θ. The figure also displays the TTT curve associated with the formation of one critical nucleus per cm^3 and per s. These are the same quantities that we computed for the case of homogeneous nucleation in Example 7.2, and the figure also presents those data for comparison (curve with filled circles). Note first that the computed curves for $\theta = \pi$ are parallel to those for homogeneous nucleation, merely displaced due to the difference in the pre-exponential factors. If a liquid melt is quenched very rapidly and then maintained at low temperature (e.g., 500 K), it can be seen that homogeneous nucleation is likely to occur over a very short time, due to the large pre-exponential factor. However, if the same procedure is carried out at a quench temperature greater than 800 K, heterogeneous nucleation prevails provided that the wetting angle θ is smaller than $\pi/4$ (curves with filled triangles, filled squares and open circles). This is due to the factor $f(\theta)$ appearing

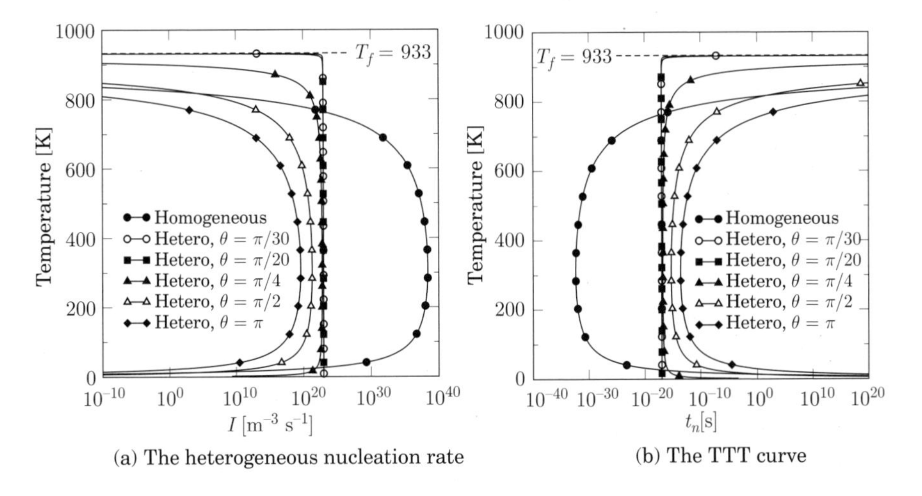

(a) The heterogeneous nucleation rate

(b) The TTT curve

Fig. 7.8 The formation of critical nuclei for heterogeneous nucleation as a function of the absolute temperature for various values of the contact angle θ. The time t_n is defined as the time to form one nucleus per cm^3.

in the nucleation barrier which, despite the reduced pre-exponential term, largely favors heterogeneous nucleation.

In other words, whenever easily wetted foreign particles are present in the melt, they induce heterogeneous nucleation under normal cooling conditions. We can clarify this point by plotting the undercooling required to produce a particular nucleation rate for various values of θ. This corresponds to the construction of a vertical line at a particular value of I in Fig. 7.8 and plotting the intersections of the various curves with the line. (Note that the intersection with the curves near 0 K is ignored in this discussion). The points for this graph can be approximated by solving Eq. (7.19) for $\Delta T^{heter}(I, \theta)$ assuming that $T \approx T_f$

$$\Delta T^{heter}(I, \theta) = \left(-\frac{16\pi}{3} \frac{\gamma_{s\ell}^3 f(\theta)}{(\rho \Delta s_f)^2 k_B T \ln(I/I_0^{heter})} \right)^{1/2} \tag{7.20}$$

The result is shown in Fig. 7.9, where it is clear that the characteristic undercooling required for nucleation decreases to just a few K for small values of θ, consistent with experimental observations. Foreign particles for which the characteristic nucleation undercooling is increasingly small are referred to as increasingly "potent."

Since the number of available heterogeneous nucleation sites is limited, the extinction of sites should be accounted for. Indeed, considering Fig. 7.8(a), it can be seen that the maximum nucleation rate (10^{22} m^{-3}s^{-1}) is reached as soon as $T \lesssim 800$ K when $\theta < \pi/4$. This would mean that all the sites are "activated" in less than 10^{-13} s! To include the extinction of

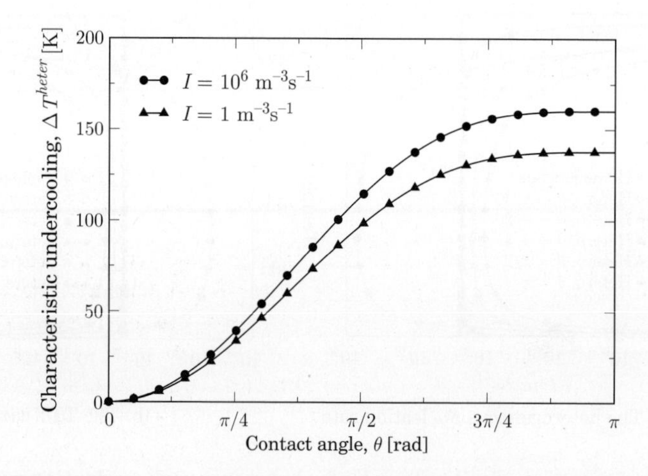

Fig. 7.9 The characteristic heterogeneous nucleation undercooling for Al using two values of the nucleation rate, I, plotted as functions of the contact angle.

sites, we write the instantaneous nucleation rate in terms of the number of remaining nuclei $n(t)$ as

$$I^{heter}(t) = \frac{dn(t)}{dt} = (n_p - n(t))\nu_0 p_c \exp\left(-\frac{16\pi}{3}\frac{\gamma_{s\ell}^3 f(\theta)}{(\rho\Delta s_f \Delta T)^2 k_B T}\right) \tag{7.21}$$

Since each nucleus is expected to grow into a macroscopic grain, $n(t)$ is referred to as the *grain density*. To provide a concrete example, suppose that the temperature decreases linearly with time, i.e., $\Delta T(t) = -\dot{T}t$. If we maintain the approximation $T \approx T_f$, valid for small undercooling, the grain density becomes

$$n(t) = n_p \left[1 - \exp\left\{-\nu_0 p_c \int_0^t \exp\left(-\frac{16\pi}{3}\frac{\gamma_{s\ell}^3 f(\theta)}{(\rho\Delta s_f \dot{T})^2 k_B T_f \tau^2}\right) d\tau\right\}\right] \tag{7.22}$$

This expression, evaluated for $\theta = \pi/20$, cooling rate $\dot{T} = -1\ \mathrm{K\,s^{-1}}$, and the parameters previously given for Al, is shown graphically in Fig. 7.10. The grain density demonstrates a nearly instantaneous increase at about $t = 2\,\mathrm{s}$ ($\Delta T = 2\,\mathrm{K}$), from its initial value of zero to saturation of all available sites $n_p = 10^9\ \mathrm{m^{-3}}$. The undercooling for which this burst occurs turns out to be almost independent of the cooling rate, and is in fact given by Eq. (7.20).

Up to this point, we have considered nucleation to be a thermally-activated process, i.e., the probability distribution of critical nuclei follows the Maxwell-Boltzmann law. We found that the assumption of homogeneous nucleation led to the prediction that nucleation occurs at a significant rate only when the undercooling reaches a value that is much larger than that observed in experiments. Introducing the concept of

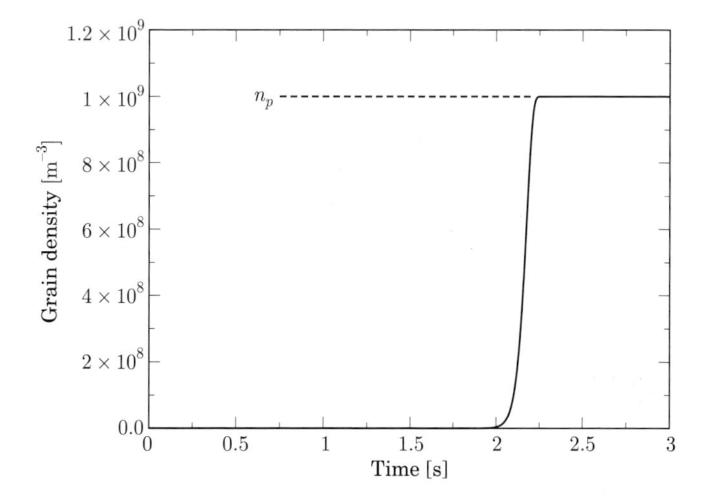

Fig. 7.10 The evolution of the density of grains corresponding to a wetting angle of $\pi/20$ for the conditions of Fig. 7.9 and a cooling rate of -1 K/s.

heterogeneous nucleation leads to predictions of nucleation at more realistic values of undercooling. However, the theory still has shortcomings that preclude the description of other behaviors observed in real castings. The most important of these shortcomings is illustrated by the result just obtained, shown in Fig. 7.10. This result predicts that the number of viable nuclei is equal to the number of potential sites, which implies that the grain density in the final casting is identical for all cooling conditions. However, in real castings, the grain density increases markedly at higher cooling rates. Further, the observation that the result shown in Fig. 7.10 is nearly independent of cooling rate implies that all castings should solidify with identical undercooling. This is also contrary to observation, where the undercooling increases with increasing cooling rate. In the next section, we extend the theory to include the ability to model these two missing behaviors.

7.3.3 Instantaneous or athermal nucleation

For thermally-activated nucleation, the rate of formation of critical nuclei depends on temperature and time. However, the last result obtained in the preceding section shows that, in this model, the number of nuclei is almost immediately increased from zero to the saturation of all sites n_p as soon as the necessary undercooling is reached. In this section, we treat the nucleation as instantaneous, and posit that there exists a distribution of nucleants of varying potency. Each variant has its own density of possible sites and its own characteristic temperature at which the sites become active. Thus, there is a distribution $n_p(\Delta T)$ (or $n_p(\theta)$) of sites. We then explore the implications of this approach and demonstrate how it can be used to model the behavior of real castings. Since $n_p(\Delta T)$ is not affected by the Maxwell-Boltzmann distribution, i.e., it is not thermally activated, this approach is sometimes called "athermal" nucleation, which has its origins in the theory of martensite formation. This term is somewhat misleading in the current context, because the main result is that the nucleation rate depends on temperature, and not on time. Nevertheless, since it is the commonly used term, we use it to describe this model.

In order to show that there is a physical basis for this model, we first describe solidification in a cavity, where we find that the nucleation behavior depends on the *size of the cavity*, whereas it is independent of time. The geometry is illustrated in Fig. 7.11. The cavity is modeled as a conical indentation with cone angle 2α, as shown. The cavity becomes very effective in promoting a new solid nucleus if the contact angle θ is less than $\pi/2 - \alpha$. Indeed, the solid-liquid interface in this case is concave toward the melt, i.e., the curvature is negative, and thus the Gibbs-Thomson equation tells us that the equilibrium temperature is *higher* than T_f. The thermodynamics of this situation is now described, following the work of Turnbull [14].

As in the previous section, consider first the difference in free energy when the solid forms in the cavity, as compared to when the cavity is filled

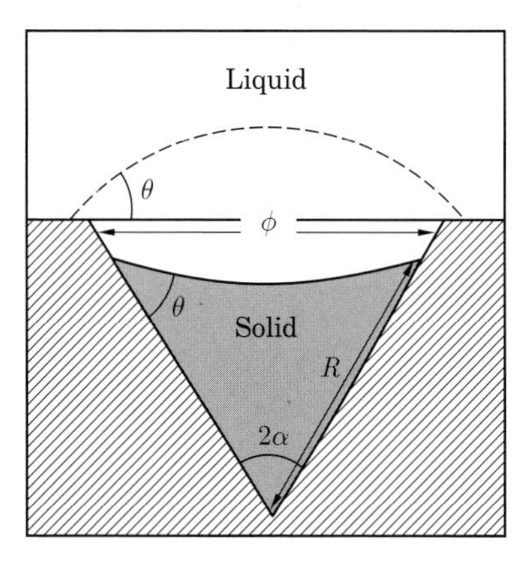

Fig. 7.11 A schematic of heterogeneous nucleation in a conical groove of exit diameter ϕ and aperture angle 2α.

with liquid. For that purpose, it is necessary to calculate the volume of solid nucleus and the surfaces of the various interfaces (see Exercise 7.6). When using the Laplace-Young equation at the triple junction, one obtains

$$\Delta G^{cone} = \frac{4\pi R^3}{3}\rho(g_s - g_\ell)f_V^g + \pi R^2 \gamma_{s\ell} f_A^g \tag{7.23}$$

The factor f_V^g represents the volume of the solid nucleus, i.e., a conical pyramid of edge R from which a spherical cap is removed, normalized by the volume of the sphere of identical radius (the index "g" has been added to indicate that a groove is being considered). The surface factor f_A^g is obtained similarly to the derivation made for a flat surface (Eq. (7.13)): we consider the two newly created interfaces ("fs" for the cone and "$s\ell$" for the spherical cap) at the expense of the previous "$f\ell$" interface. Using the Laplace-Young Eq. (7.11) at the triple junction permits the elimination of γ_{fs} and $\gamma_{f\ell}$, leading to the final result (see Exercise 7.6)

$$\begin{aligned} f_V^g(\alpha,\theta) &= \sin^3\alpha\left(\frac{1}{4\tan\alpha} - \frac{f(\alpha+\theta-\pi/2)}{\cos^3(\alpha+\theta)}\right) \\ f_A^g(\alpha,\theta) &= \frac{2\sin^2\alpha}{1+\sin(\alpha+\theta)} - \sin\alpha\cos\theta \end{aligned} \tag{7.24}$$

Notice that the same function we encountered earlier for heterogeneous nucleation (Eq. (7.16)) appears in f_V^g, this time as $f(\alpha+\theta-\pi/2)$, because of the way the angles are defined in Fig. 7.11.

The interesting aspect of this case is that when $T > T_f$, the volumetric term in Eq. (7.23) is positive (i.e., the solid has a higher bulk free energy than the liquid), but for $\theta < \pi/2 - \alpha$, the contribution of the surface term is negative (see Fig. 7.12). The total Gibbs free energy thus goes through a *minimum* at some radius R, corresponding to a stable embryo

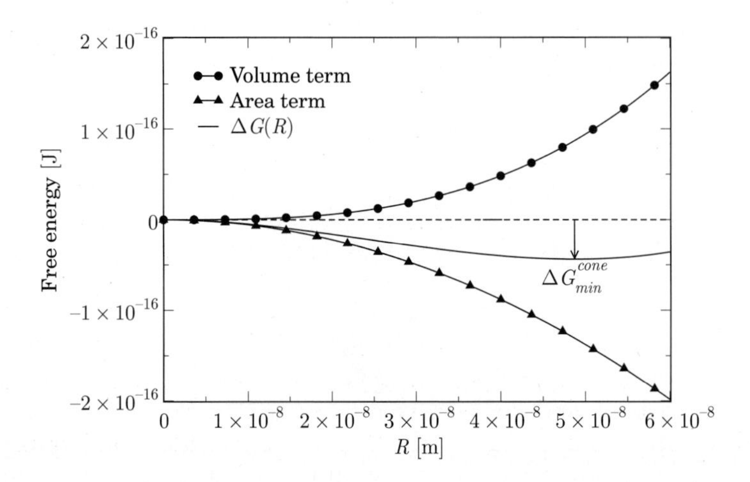

Fig. 7.12 Energy terms associated with the formation of an embryo in a conical cavity, using data for Al and $\alpha = \pi/3$, $\theta = \pi/10$ and $T = 938\,\text{K}$.

that can exist above the melting point simply due to the geometry. We call this solid an "embryo," rather than a nucleus, since it cannot grow unless the temperature decreases.

Let us now consider what happens to such an embryo as the temperature gradually decreases from its initial value above T_f. The difference between the solid and liquid free energies $\rho(g_s - g_\ell)$ decreases to zero at T_f, and then changes sign as undercooling begins. This alters the balance between the volume and area terms in Eq. (7.23), shifting the equilibrium point to larger values of R. At a certain temperature, still larger than T_f, the base of the embryo reaches the mouth of the cavity. The embryo must stop growing at this point as it cannot extend onto the flat part of the substrate while retaining the negative curvature required to be in equilibrium with the melt. Instead, the ends remain pinned at the mouth of the cavity, and the curvature gradually decreases as the temperature falls, until finally, at $T = T_f$, the embryo is flat. When the temperature falls below T_f, the undercooling becomes positive and the solid grows as a convex cap of fixed circular base until it reaches a radius of curvature, R_c, such that the angle θ with the flat substrate satisfies the Laplace-Young equation. This occurs when

$$R_c = \frac{\phi}{2\sin\theta} \quad \text{and} \quad \Delta T_g = \frac{4\Gamma_{s\ell}\sin\theta}{\phi} \tag{7.25}$$

where ϕ is the diameter of the mouth of the cavity and $\Gamma_{s\ell}$ the Gibbs-Thomson coefficient. At that point the solid can grow onto the flat part of the substrate (see the dashed line in Fig. 7.11), thus increasing its radius and decreasing the curvature undercooling.

This might seem like the same nucleation theory as in the previous sections, but there is a major difference. In the standard theory, a critical nucleation radius exists at *any* temperature: if fluctuations of atoms at a given temperature are such that they can overcome the corresponding energy barrier, nuclei will form at a given rate (and then grow) at this temperature. In the case of the cavity, provided that $\theta < \pi/2 - \alpha$, the morphology of the cavity (ϕ and α) completely *dictates the temperature and undercooling* at which the stable embryo will grow out of it. Thus, a new grain is guaranteed to grow from the cavity once the critical undercooling defined by Eq. (7.25) is reached. The key difference in this case is that we do not consider the thermal activation of atoms, i.e., the vibration and the probability of capture, thus producing athermal nucleation.

Another important situation where growth is determined by the geometry of the substrate occurs when $\gamma_{fs} + \gamma_{s\ell} < \gamma_{f\ell}$. Recall that there is no value of θ that satisfies the Laplace-Young equation for this case. In fact, the substrate and melt are so dissimilar that the system can reduce the total interfacial energy if the original substrate-liquid interface is replaced by two interfaces, one between the substrate and solid, and the other between the solid and liquid (see top part of Fig. 7.13). Once again, this may occur above the melting temperature T_f for certain combinations of the free and surface energies, as we now show. The thickness, δ, of the solid layer can be computed by starting with an overall energy balance

$$\Delta G = -A_{sub}\delta(\rho\Delta s_f)\Delta T + A_{sub}(\gamma_{fs} + \gamma_{s\ell} - \gamma_{f\ell}) \tag{7.26}$$

where A_{sub} is the area of the substrate exposed to the melt, and where we have substituted the now familiar expression $\rho\Delta s_f\Delta T$ for the volumetric part of the free energy. Notice the implicit dependence on the size of the foreign substrate – this becomes important later. By assumption, the interfacial energy contribution for this case is negative, thus implying that a solid layer δ can form above the melting point, i.e., for $\Delta T < 0$. The

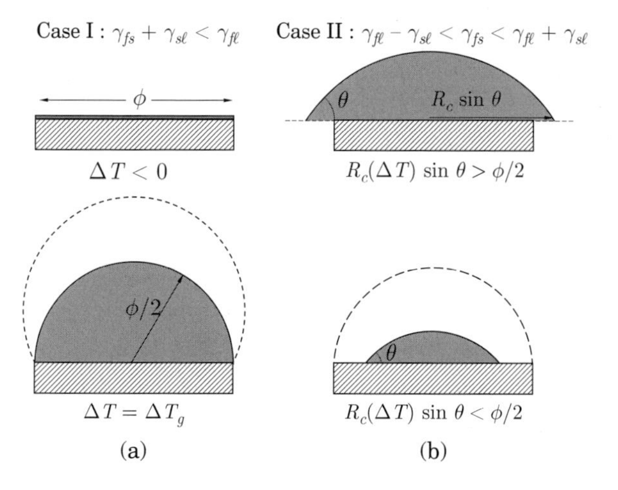

Fig. 7.13 A schematic of heterogeneous nucleation on a foreign platelet particle (a) when $\gamma_{f\ell} > \gamma_{fs} + \gamma_{s\ell}$ and (b) when $\gamma_{f\ell} - \gamma_{s\ell} < \gamma_{fs} < \gamma_{f\ell} + \gamma_{s\ell}$.

equilibrium thickness δ is found by differentiating Eq. (7.26) with respect to A_{sub} and setting the result to zero

$$\delta = \frac{\gamma_{fs} + \gamma_{s\ell} - \gamma_{f\ell}}{\rho \Delta s_f \Delta T} > 0 \quad \text{when} \quad \Delta T < 0 \tag{7.27}$$

Analyzing the behavior of the system as the temperature decreases toward T_f, i.e., $\Delta T \to 0$, Eq. (7.27) shows that the thickness of the solid layer increases. (The limit of $\Delta T = 0$ is singular, and, as one can see from Eq. (7.27), the thickness is indeterminate at this point.) Once the melting point is reached, the process becomes controlled by growth, provided that the area of the foreign substrate is large enough. But what happens if the area of the substrate, A_{sub}, is limited? This situation is shown schematically in Fig. 7.13(a). Let us assume that the solid can form only on the surface of a circular disk-shaped substrate, as illustrated. Once again, the geometry is chosen for ease of presentation, and the detailed shape is not important for the understanding of the phenomena. As the temperature decreases below T_f, the solid begins to adopt a curved surface to satisfy the Gibbs-Thomson equation. The curvature increases as T decreases, until it reaches a critical value given by the hemisphere, $R = \phi/2$, where ϕ is the diameter of the disk. This occurs at the geometrically limited undercooling ΔT_g, given by

$$\Delta T_g = \frac{4\Gamma_{s\ell}}{\phi} \tag{7.28}$$

This is the maximum undercooling that is seen by this particle, since further growth of the solid leads to a decrease in the curvature, as illustrated by the dashed line in the lower part of Fig. 7.13(a). Thus, once again the growth of the solid is determined entirely by the temperature and the geometry of the nucleant particle.

When $\gamma_{f\ell} - \gamma_{s\ell} < \gamma_{fs} < \gamma_{f\ell} + \gamma_{s\ell}$, i.e., the case we discussed extensively in the previous section on heterogeneous nucleation, nucleation can still be affected by the size of the foreign substrate particles. We continue to use the circular disk with diameter ϕ as an example. A critical nucleus with radius R_c given by Eq. (7.17) can form on the particle if $R_c \sin\theta < \phi/2$ (see the bottom part of Fig. 7.13(b)). However, if $\phi < 2R_c \sin\theta$, nucleation on such a particle is not possible since the substrate is not large enough to fit the spherical cap of the required size (see top part of Fig. 7.13(b)). The disk cannot support a critical nucleus until the undercooling increases sufficiently such that $\phi \geq 2R_c \sin\theta$.

Heterogeneous nucleation can also be limited by the particle size in another way. Suppose that the undercooling is such that $R_c(\Delta T)\sin\theta < \phi/2$ and that heterogeneous nucleation has occurred on such a particle (see bottom part of Fig. 7.13). This is the same heterogeneous nucleation event that we discussed previously. The solid can continue to grow along the substrate until it reaches the edge of the particle. At that point, the curvature must increase in order for further growth to occur. This provides a geometric limit to the growth of the solid, such that if $\Delta T > \Delta T_g$ given by Eq. (7.28), the solid can grow past the maximum curvature given by

$R = \phi/2$, whereas if $\Delta T < \Delta T_g$, it will stop at the edge of the particles. In this case, this stable embryo will be able to grow only if the undercooling is larger than ΔT_g.

The various situations we have described for geometrically limited heterogeneous nucleation on a flat circular particle are summarized in Fig. 7.14. The diagram is divided into three regions: The curve $\Delta T/\Delta T_g = \sin\theta$ separates the region where a critical nucleus can form from the region where it cannot. The vertical line $\Delta T/\Delta T_g = 1$ separates the region where the nucleus is able to form but is stopped at the edge of the particle from the region where the nucleus can continue to grow ($\Delta T > \Delta T_g$). Further details on athermal nucleation can be found in Quested and Greer [8].

Let us now see how we can use the concept of athermal nucleation to understand the development of grain structures in castings. Suppose that the melt contains foreign particles constituted of a material where the solid can form easily, i.e., the two cases we have just described for athermal nucleation with a small value of the contact angle θ. If the particles are sufficiently large for heterogeneous nuclei to form, but at the same time small enough to geometrically limit the growth of the solid, then the heterogeneous nucleation is athermal. This corresponds to the gray zone in Fig. 7.14. At such small wetting angles (and undercoolings), the formation of the first nuclei on these particles occurs very rapidly, as shown in Fig. 7.10, at a typical nucleation undercooling, $\Delta T^{heter}(\theta)$, which is at first independent of the size of the particle. Grains will then grow up to the limit of the particle edge: at that point, grains can continue to grow only if ΔT increases beyond $\Delta T_g(\phi)$.

Next, suppose that there exists a distribution of sizes of such particles. We can classify the distribution by the density Δn_i of particles

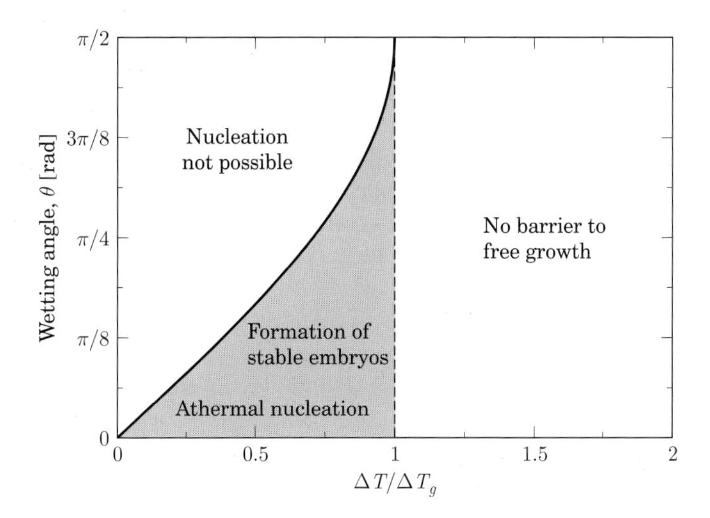

Fig. 7.14 Regimes of nucleation on a foreign particle as functions of the wetting angle and normalized undercooling (after Quested and Greer [8]).

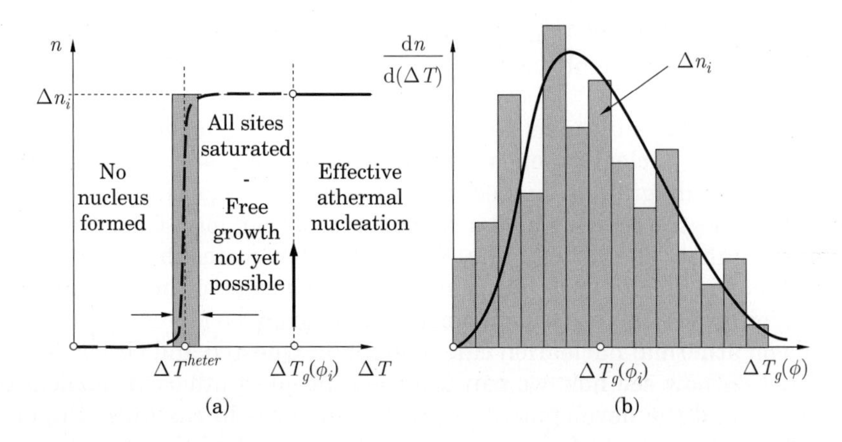

Fig. 7.15 (a) The evolution of the density of grains as a function of the undercooling for a constant cooling rate when only one class of particles is considered. (b) The construction of a continuous distribution of nucleation sites becoming active as the undercooling is increased.

of average particle size $\phi_i \leq \phi < \phi_i + \Delta\phi_i$. For each ϕ_i, heterogeneous nucleation occurs essentially instantaneously at a characteristic undercooling, $\Delta T^{heter}(\theta)$, given by Eq. (7.20), and independently of the particle size (provided that $R_c(\Delta T^{heter}) \sin\theta < \phi_i/2$). This is shown schematically in Fig. 7.15(a) for a particular class, $[\phi_i, \phi_i + \Delta\phi_i]$. The nuclei formed on these particles cannot grow until $\Delta T \geq \Delta T_g(\phi_i)$. Therefore, the discrete distribution of particle sizes results in a discrete distribution of undercoolings, $\Delta T_g(\phi_i)$, for which each class of particles becomes active and produces new grains. For modeling purposes, the discrete distribution of particles can be replaced by a continuous distribution, $dn/d(\Delta T)$, represented by the continuous curve in Fig. 7.15(b). This distribution represents dn, the number of particles per unit volume that become active growth sites between ΔT and $\Delta T + d(\Delta T)$.

The idea of using such a continuous distribution of nucleation sites was first introduced in 1966 by Oldfield to explain grain structures in gray cast iron. [6] Oldfield observed experimentally in slowly cooled specimens of varying size, that the density of grains, n, was related to the maximum undercooling reached at recalescence ΔT by

$$n \approx A\Delta T^2 \tag{7.29}$$

Differentiating this expression with respect to ΔT gives the athermal distribution law

$$\frac{dn}{d(\Delta T)} = 2A\Delta T \tag{7.30}$$

Other forms of the distribution are sometimes used to model the formation of dendritic grains. The parameters of the distribution are then fitted to observations of the distribution of grain sizes in experimental castings. One example is the Gaussian distribution illustrated in Fig. 7.16

$$\frac{dn}{d(\Delta T)} = \frac{n_{max}}{\Delta T_\sigma \sqrt{2\pi}} \exp\left[-\frac{1}{2}\left(\frac{\Delta T - \Delta T_0}{\Delta T_\sigma}\right)^2\right] \tag{7.31}$$

where n_{max} is the total density of available particles, ΔT_0 is the mean value of the distribution, and ΔT_σ is the standard deviation. The Gaussian distribution extends from $-\infty$ to $+\infty$, which is clearly impossible. A more physical distribution comes from the assertion that the particle sizes ϕ follow a log-normal distribution, given by

$$\frac{dn}{d\phi} = \frac{n_{max}}{\phi_\sigma \sqrt{2\pi}} \frac{1}{\phi} \exp\left[-\frac{1}{2}\left(\frac{\ln\phi - \ln\phi_0}{\phi_\sigma}\right)^2\right] \tag{7.32}$$

where the mean value of the distribution is $\ln\phi_0$ and the standard deviation is ϕ_σ. Equation (7.28) can be used to convert the particle size distribution into a distribution of the associated nucleation undercooling

$$\frac{dn}{d\Delta T} = \frac{n_{max}}{\Delta T_\sigma \sqrt{2\pi}} \frac{1}{\Delta T} \exp\left[-\frac{1}{2}\left(\frac{\ln\Delta T - \ln\Delta T_0}{\Delta T_\sigma}\right)^2\right] \tag{7.33}$$

This distribution is also shown in Fig. 7.16. We note that regardless of the functional form chosen for the distribution, the parameters must be measured experimentally. In the next section, we discuss such experiments, and in particular show that the distribution of grain density and undercooling can be correlated with the particle size of inoculants intentionally added to the melt to promote grain refinement.

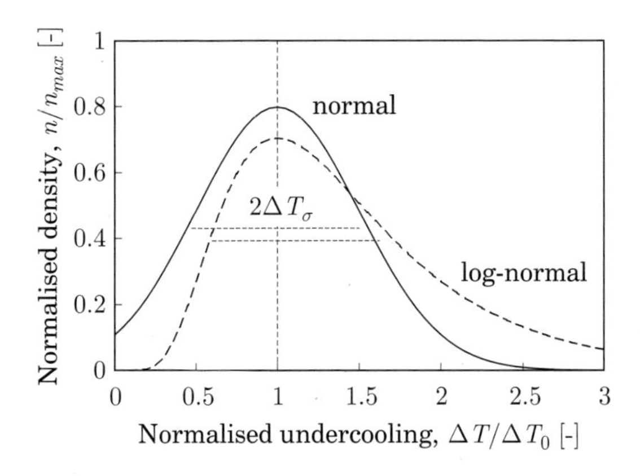

Fig. 7.16 Gaussian (continuous line) and log-normal (dashed line) distributions of heterogeneous nucleation particles (sites). $\Delta T_\sigma = 0.5$

7.4 MECHANISMS FOR GRAIN REFINEMENT

In most casting applications, the desired microstructure is most often represented by fine equiaxed grains, as they exhibit more isotropic and uniform mechanical properties. Such structures also show less tendency for hot tear formation, or decreased centerline macrosegregation in continuous casting processes as well as other desirable properties for further processing of materials (e.g., extrusion, stamping, etc.). As we show in Chap. 11, such structures can be obtained by using a relatively high cooling rate while keeping a low thermal gradient. One can also promote the formation of equiaxed microstructures by two additional mechanisms:

- *Inoculation* of the melt by intentional addition of *inoculant* particles, on which heterogeneous nucleation is easy.
- *Agitation* of the melt, e.g., by mechanical means, by application of controlled electromagnetic fields or by ultrasound. Movement of the liquid induces partial remelting of existing dendritic grains, thus creating fragments of solid from which growth can start.

The first mechanism, i.e., inoculation, is described here, and discussion of the second is reserved for Chap. 11.

Inoculation of the melt is a common practice for most Al alloys and is also possible for other alloy systems. Despite many scientific studies, the addition of particles to a melt in order to promote heterogeneous nucleation is still an art, and the mechanisms have yet to be well understood. Some inoculants may be effective when added to a pure substance, but ineffective when certain solute elements are present. A good example of this phenomenon is found in Zn-Al alloys, which are used extensively as coatings on sheet steel in order to protect against corrosion. In the coating process, the steel sheets pass through a melt at a controlled rate so as to form a coating 20 to 40 μm thick. A typical bath composition of Zn-0.2wt%Al (also containing iron picked up from the steel) produces coatings with a grain size of approximately 1 mm. However, in the presence of even very small amounts of Pb, Sb or Bi, the grains are much larger (typically 1 cm).

This phenomenon is illustrated by the microstructures in Fig. 7.17. Two specimens with nominal composition Zn-3wt%Al were produced by directional solidification under controlled thermal conditions (thermal gradient, solidification speed) in a Bridgman furnace; one with an intentional addition of 0.11wt%Pb, and the other Pb-free. A longitudinal section through the specimen without lead shows relatively fine equiaxed grains (Fig. 7.17(a)). On the other hand, the sample with the Pb addition (Fig. 7.17(b)) is made up of columnar grains. Careful analysis of experiments performed for various Al compositions indicated that the addition of Pb increased the temperature at which heterogeneous nucleation was observed from about $0.5 - 0.9$ K in the sample without Pb to about 2.3 K

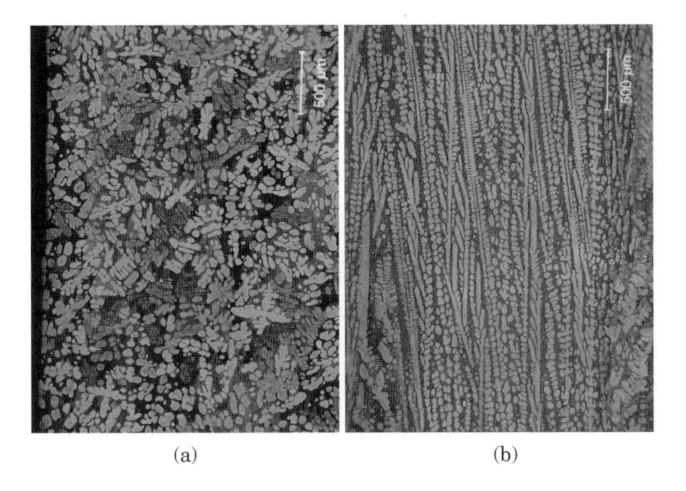

Fig. 7.17 The grain structure in a longitudinal section of a Zn-3wt%Al solidified under Bridgman conditions without (a) and with (b) 0.11wt%Pb. The vertical thermal gradient (23 K/cm) and the solidification speed (66.7 μm/s) were identical in both specimens (courtesy of A. Sémoroz [11]).

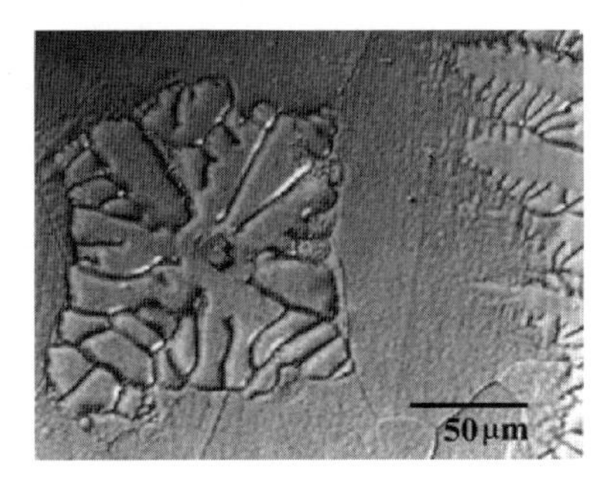

Fig. 7.18 The formation of an equiaxed grain around a TiN particle in a ferritic stainless steel during welding (courtesy of H. Kerr [4]).

undercooling with Pb. Unfortunately the detailed mechanism by which this occurs is still unknown.

Figure 7.18 shows a micrograph from a ferritic stainless steel specimen welded with a gas tungsten arc (GTA), where a TiN particle is clearly located at the center of a dendritic grain. However, TiN is not commonly used as an inoculant due to the improvement in microstructure seemingly being offset by the reduction of mechanical properties induced by the presence of TiN particles. In superalloy components produced by investment casting, a fine layer of cobalt aluminate ($CoAl_2O_4$), also called "cobalt blue" for its use in the pigment industry, is first deposited on the wax pattern to promote nucleation. Such castings are called "equiaxed" because of their surface appearance, but a cross-section often reveals a columnar structure. In all of these cases, the detailed mechanism of heterogeneous nucleation is still largely unknown. A typical rule of thumb, based on coherent or semi-coherent interfaces in solids, is to find particles with a crystallographic structure that matches that of the solid at least on one plane, in order to minimize the interfacial energy, γ_{fs}, and thus the wetting angle. However, this is not sufficient, as demonstrated by the analysis of inoculation in aluminum alloys.

In the aluminum industry, particularly in the direct chill (DC) semi-continuous casting process, it is very common to add 0.1 to 1wt% of Al-3wt%Ti-1wt%B or Al-5wt%Ti-1wt%B to the liquid melt as an inoculant. The starting materials typically contain Al_3Ti and TiB_2 particles. The Al_3Ti particles are thought to be unstable at concentration levels lower than 0.15wt% Ti. On the other hand, TiB_2 particles are stable at these concentrations, but can act as inoculants for the aluminum primary phase only if excess titanium is present in the melt. TiB_2 can also promote the formation of other phases, such as Al_2Cu in Cu-bearing alloys on the prismatic faces.

Although the detailed mechanisms are not yet fully understood, recent work by Greer and co-workers [2; 8] has shed some much-needed light on the subject. The nucleation phenomena we have described occur on very small length scales in a liquid, and are therefore difficult to observe. This research group had the idea to first form a metallic glass by melt spinning an alloy with composition $Al_{85}Y_8Ni_5Co_2$ containing inoculant particles, and then observe crystallization of the glass around the inoculant particles *in situ* using High Resolution Transmission Electron Microscopy.

Figure 7.19 contains an image from one such experiment. The glassy phase appears white, the heterogeneous nucleant particle, whose composition is nominally TiB_2, is gray, and there are two phases that appear in black: Al_3Ti and Al. One can see that the Al does not form directly on the TiB_2 surface. Instead, a thin layer of Al_3Ti first grows epitaxially (e.g., the thin black layer visible on the left boundary), and then many islands of Al

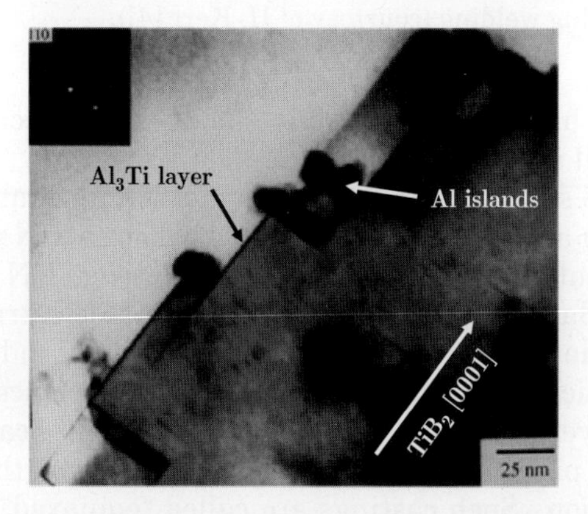

Fig. 7.19 An electron micrograph of small aluminum islands nucleating at the surface of a TiB_2 particle in $Al_{85}Y_8Ni_5Co_2$ (after Greer et al. [2]).

form on this Al_3Ti layer. This occurs only on the basal plane of the TiB_2 particles. Some islands can be seen even on the small platform on the left side of the particle, but not on the re-entrant corner with the vertical faces. This indicates that such faces are characterized by a fairly high interfacial energy with crystalline aluminum, or conversely with a low interfacial energy when Al is glassy.

The crystallographic structures of the three phases involved in the nucleation of aluminum are shown in Fig. 7.20. The orientation relationships between them were identified by electron diffraction as

$$\{0001\}\langle 11\bar{2}0\rangle_{\mathrm{TiB_2}} \quad \| \quad \{112\}\langle 201\rangle_{\mathrm{Al_3Ti}} \quad \| \quad \left\{ \begin{array}{c} \{112\}\langle 110\rangle_{\mathrm{Al}} \\ \text{or} \\ \{111\}\langle 110\rangle_{\mathrm{Al}} \end{array} \right\} \tag{7.34}$$

Note that the ordered Al_3Ti phase is tetragonal, however, it has a c-axis twice that of the a-axis, i.e., it is equivalent to two stacked FCC units. The $\{112\}$ plane of coincidence outlined in light gray for Al_3Ti therefore corresponds to the $\{111\}$ plane of the cubic unit (and the $\langle 201\rangle$ direction to $\langle 101\rangle$).

It is interesting to note that the angle θ of the Al nuclei formed on the Al_3Ti layer is not small, despite the fact that they preferentially nucleate on this plane at very low undercooling. The nuclei appear to grow epitaxially from this surface with the orientation relationship given in Eq. (7.34). Since the nuclei have the same crystallographic relationship with the particle, which is itself a single crystal, once all of the solid islands meet to form a single layer, they produce only one grain. This layer then grows according to the athermal nucleation mechanism described previously. The process is shown schematically in Fig. 7.21.

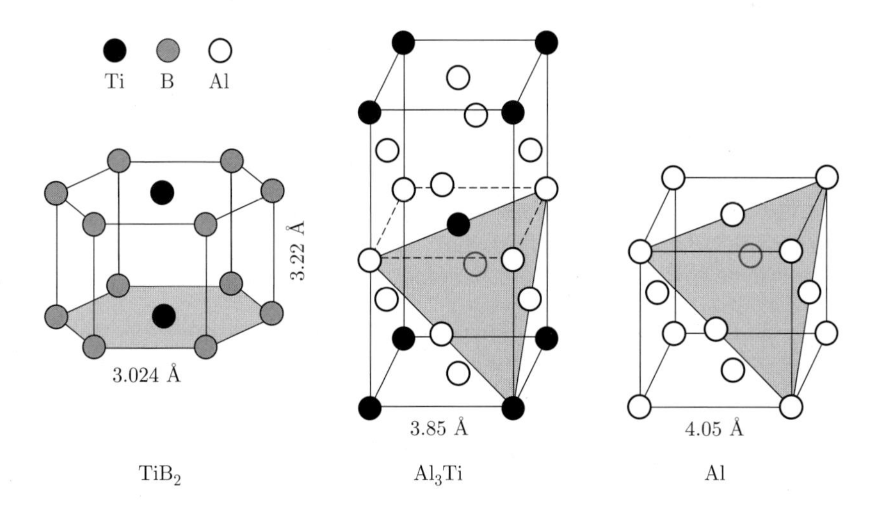

Fig. 7.20 Crystallographic structures of TiB_2, Al_3Ti and Al with indication of the main planes of coincidence involved in heterogeneous nucleation.

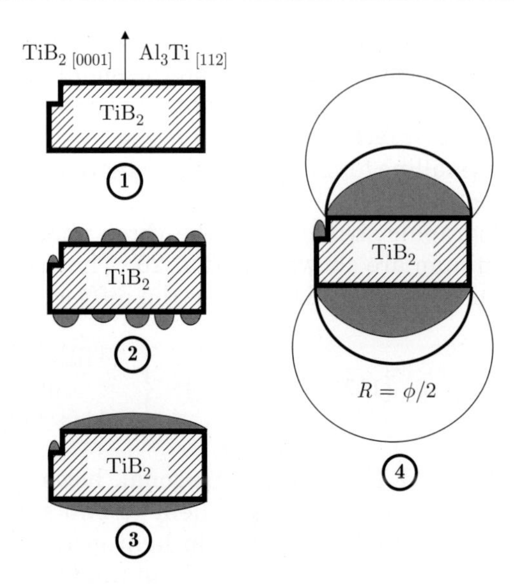

Fig. 7.21 Steps of nucleation-growth of a new aluminum grain on a TiB_2 particle (after Greer et al. [2; 8]).

Although this type of analysis has not yet been able to explain all the phenomena encountered under industrial conditions, it has brought much deeper insight into some of the important mechanisms. These include the following observations:

- The double layer mechanism, i.e., the formation of a layer of Al_3Ti around the TiB_2 particles, explains why an excess of titanium is required for the nucleation to be effective.

- The role of Si in foundry alloys is more complex and not fully understood. Below about 3wt%Si, an increase in the Si-content decreases the grain size, as the growth of dendrites is made more difficult (see Chap. 11), but beyond that, the grain size increases again. This has been attributed to silicon "poisoning" the nucleation sites. By using $Al_{80}Ni_{10}Cu_8Si_2$ glasses, it was found that over extended periods of contact with the melt, $TiSi_2$ particles formed on the surface of TiB_2 particles instead of Al (and to some extent Al_2Cu).

- It has been observed that the presence of Zr eliminates the inoculating properties of TiB_2. This is explained by noting that Zr is known to destroy the Al_3Ti layer through the formation of Al_3Zr.

- The fact that only a small portion of the inoculant particles are effective in the nucleation of Al alloys is explained very well by the athermal model. Nucleation is not governed by the formation of caps (or islands) on the TiB_2 particles, which occurs for a very small undercooling, but rather by the ability to grow afterwards. The few largest

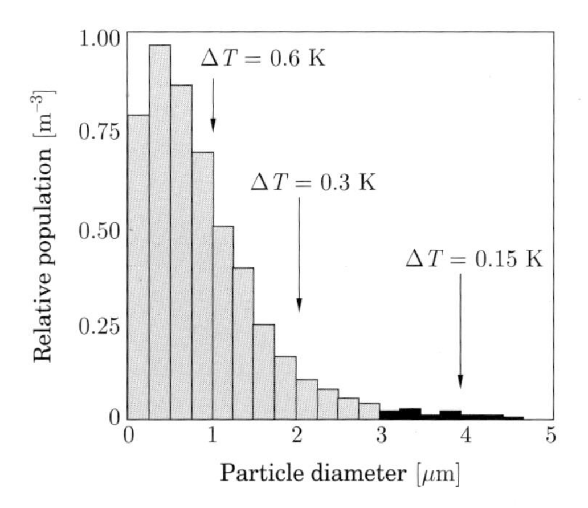

Fig. 7.22 The size distribution of TiB_2 particles with the particles effectively active during nucleation of aluminum grains indicated in black (after Greer et al. [2; 8]).

particles are characterized by the smallest ΔT_g undercooling, and are thus the first to become active. Once they have promoted the growth of new grains, they grow and entrap other finer particles. Figure 7.22 shows the portion of the distribution of particles that is truly active in aluminum. If the magnitude of the cooling rate is increased, more particles can be activated before being entrapped by growth. Ideally, it would be preferable to have the narrowest possible particle size distribution in order for all the particles to become active at about the same undercooling. This model also explains why the agglomeration of particles is detrimental to effective inoculation.

Numerous investigations have been carried out concerning the nucleation characteristics of a variety of compounds, and in the presence of numerous other elements. The details are beyond the scope of this text, where we strive to understand the general mechanisms for structure formation, leaving the details of the particular alloy systems to the practitioner.

7.5 SUMMARY

In this chapter, we have explored the processes by which the first solid appears in a melt. Thermodynamics shows that the balance between volume and surface energies dictates that there exist a minimum particle size in order for an embryo of the solid phase to be stable. This minimum size decreases as the undercooling increases. Analysis of the expected rate of nucleation for a homogeneous melt demonstrated that this mechanism was not sufficient to explain the observation that melts tend to nucleate within a few K of their equilibrium melting temperature.

The analysis was extended to consider nucleation on the surface of foreign particles in the melt, a process called heterogeneous nucleation. In this case, nucleation was found to occur almost instantaneously at an undercooling in roughly the correct temperature range, i.e., close to T_f. The exact value of the nucleation temperature was shown to depend on the potency of the foreign particles, as characterized by the contact angle between the solid, substrate and liquid. However, this model predicts that one will see the same number of critical nuclei, regardless of the other experimental conditions, such as cooling rate.

We then considered athermal nucleation, in which the nucleation characteristics depended not only on the thermodynamics of the substrate and melt, but also on the geometry of the individual particles. By positing the existence of a distribution of sizes and potencies of the nucleating particles, we were able to provide a model that can be used to predict the grain structure in real castings. We return to this model in Chap. 11 for such analysis.

Finally, we examined some of the experimental evidence that supports the presented nucleation models. Experimental evidence for the nucleation of Al on TiB_2 particles showed that a thin epitaxial layer of Ti_3Al first forms on the surface of the particle, and Al solid is then created on this layer. This together with other experimental evidences provide a complete picture of nucleation. In the next chapter, we deal with the question of growth of solid from this initial stage.

7.6 EXERCISES

Exercise 7.1. Critical radius and energy barrier.
Using Eq. (7.2), obtain the expressions for the radius of the critical nucleus and for the associated energy barrier (Eqs (7.3) and (7.4)).

Exercise 7.2. Nucleation in alloys.
Assuming that the critical nucleus formed in a liquid of nominal composition X_ℓ is the one that induces the largest variation in Gibbs free energy, derive the construction shown in Fig. 7.3.

Exercise 7.3. Nucleation in alloys.
Considering the construction shown in Fig. 7.3 and assuming constant thermo-physical properties, how should one expect the critical nucleation temperature for a given nucleation rate to vary with increasing solute content in a binary alloy?

Exercise 7.4. Competition of phases induced by nucleation.
Using the results from the previous exercise, describe the competition between the δ and γ phases in Fe-C (see diagram below) as a function of

the carbon content. For that purpose, calculate the critical nucleation temperature for a given nucleation rate of the two phases. Use the following data for δ and γ iron.

Quantity		δ	γ	Units
Melting point	T_f	1809	1763	K
Slope of liquidus	m_ℓ	−849	−800	K
Partition coefficient	k_0	0.17	0.3	–
Interfacial energy	$\gamma_{s\ell}$	0.204	0.235	$\mathrm{J\,m^{-2}}$
Entropy of fusion	$\rho\Delta s_f$	1.07×10^6	1.23×10^6	$\mathrm{J\,m^{-3}\,K^{-1}}$

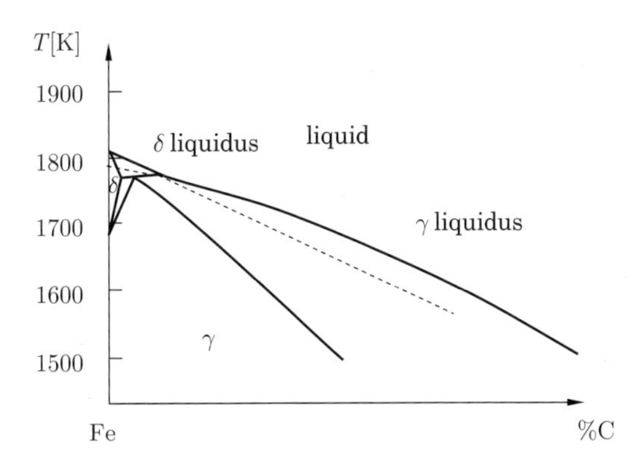

Exercise 7.5. Nucleation of a spherical cap.
Calculate the volume and surface area of a hemispherical cap (Eq. (7.14)) and show that the Gibbs free energy barrier of such a cap is equal to that of a sphere (i.e., homogeneous nucleation) multiplied by the function $f(\theta)$.

Exercise 7.6. Nucleation in a conical groove.
Calculate the volume and surface areas of interest of an embryo of radius R formed with a wetting angle θ in a conical groove of aperture angle 2α. Using these relationships and the Laplace-Young equation (7.11), derive relationships (7.23) and (7.24).

Exercise 7.7. Nucleation in a scratch and in a cylindrical cavity.
Calculate the volume and surface areas for an embryo of radius R formed with a wetting angle θ in a scratch (i.e, linear groove) of aperture 2α. Using these relationships and the Laplace-Young equation (7.11), derive the conditions for heterogeneous nucleation in this geometry. Perform a similar analysis for a cylindrical cavity of radius R_0.

Exercise 7.8. Determination of nucleation parameters.
In order to measure the nucleation distribution parameters in a given alloy under particular inoculation conditions, one usually cast several small

specimens under conditions where the sample temperature is nearly uniform. Varying the cooling rate $\dot{T}$, explain why one would obtain cooling curves such as those shown in the accompanying figure, where the temperature history is non-monotonic just below the liquidus. How do you think the grain density varies in these specimens? Describe how you would use these data to determine the distribution of the heterogeneous nucleation sites.

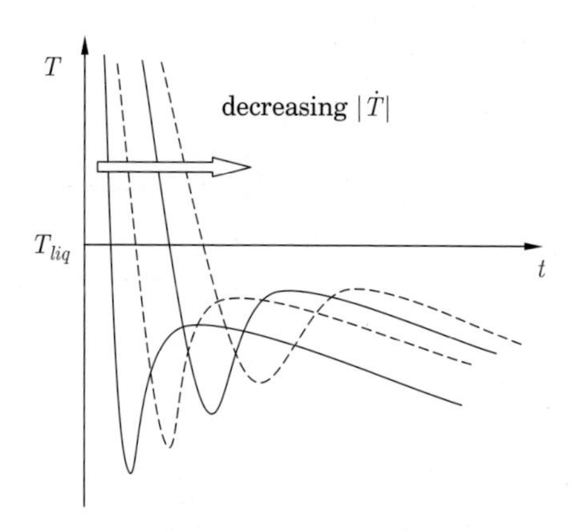

Cooling curves recorded for a small specimen of uniform temperature solidified at various cooling rates.

Exercise 7.9. Heterogeneous nucleation of Al on TiB_2 particles.
Considering the crystallographic structures of TiB_2, Al_3Ti and Al (Fig. 7.20) and the relations of coincidence (7.34), sketch the atomic positions on the planes of interest.

Exercise 7.10. Snowflakes with 12 branches.
Snowflakes usually exhibit 6 branches, following the symmetry of the hexagonal lattice of ice (see Chap. 8). However, in certain cases, they can have 12 symmetric branches distributed every 30°. Can you find an explanation for this observation related to nucleation?

7.7 REFERENCES

[1] J. W. Christian. *The Theory of transformations in metals and alloys*. Pergamon Press, 2nd edition, 1975.

[2] A. L. Greer, P. S. Cooper, M. W. Meredith, W. Schneider, P. Schumacher, J. A. Spittle, and A. Tronche. Grain refinement of aluminium alloys by inoculation. *Adv. Engng Mater.*, 5:81, 2003.

[3] W. L. Johnson, A. Inoue, and C. T. Liu, editors. *Bulk metallic glasses*, volume 554. Materials Research Society, 1999. ISBN 1-55899-460-2.

[4] H. Kerr. Columnar to equiaxed transitions in welds. In M. Rappaz and R. Trivedi, editors, *Solidification microstructures: 2nd Zermatt Workshop*. EPFL, 1998.

[5] B. S. Murty, S. A. Kori, and M. Chakraborty. Grain refinement of aluminium and its alloys by heterogeneous nucleation and alloying. *Int. Mater. Rev.*, 47:3, 2002.

[6] W. Oldfield. Quantitative approach to casting solidification – freezing of cast iron. *Trans. ASM*, 59:945-961, 1966.

[7] J. H. Perepezko and M. J. Uttormark. Nucleation controlled solidification kinetics. *Met. and Mat Trans.*, 27A:553-547, 1996.

[8] T. E. Quested and A. L. Greer. Athermal heterogeneous nucleation of solidification. *Acta Mater.*, 53:2683, 2005.

[9] M. Rappaz. Modelling of microstructure formation in solidification processes. *Int. Mater. Rev.*, 34:93, 1989.

[10] T. Sato, W. Kurz, and K. Ikawa. Experiments on dendrite branch detachment in the succinonitrile-camphor alloy. *Trans. Jap. Inst. Metals*, 28:1012, 1987.

[11] A. Sémoroz, L. Strezov and M. Rappaz. Formation of large grains in hot-dipped galvanized coatings: Assessment of the nucleation and growth contribution by Bridgman solidification experiments. In M. Lamberights, editor, *Galvatech 2001*. Verlag Stahleisen, 2001, p. 612.

[12] F. Spaepen and J. Fransaer. Advances in modeling of crystal nucleation from the melt. *Adv. Engng Mater.*, 2:593, 2000.

[13] C. V. Thompson and F. Spaepen. Homogeneous crystal nucleation in binary metallic melts. *Acta Metall.*, 31, 1983.

[14] D. Turnbull. Kinetics of heterogeneous nucleation. *J. Chem. Phys.*, 18:198, 1950.

CHAPTER 8

DENDRITIC GROWTH

8.1 INTRODUCTION

The preceding chapter focused on nucleation, i.e., the first appearance of solid in the melt, and in this context the nucleus was treated as a sphere or a spherical cap. As the solid grows, the spherical morphology eventually becomes unstable, which leads to the formation of *dendrites*. This is the topic of the present chapter. Figure 8.1 shows several examples that are discussed further in Sect. 8.2.

We divide the subject into a discussion of *free growth*, the evolution of one isolated solid particle in an infinite undercooled melt, and a discussion of *constrained growth*, where many solid structures grow in concert under an imposed thermal gradient. The important physical phenomena are described, with a particular focus on the development of microstructural length scales by the interaction between capillarity and transport of heat and solute. We begin each section with general observations regarding microstructural features, and how their formation is related to processing conditions. For certain simple shapes, these qualitative descriptions are augmented by a formal stability analysis of the base growth morphology, a sphere for free growth and a planar front for constrained growth. These analyses help us derive expressions relating the processing conditions to material properties and microstructure. Such models can be used to predict microstructures in more complex geometries and processing conditions, as is demonstrated in subsequent chapters. We also discuss the effect of melt convection on the final microstructure.

8.2 FREE GROWTH

8.2.1 General observations

Let us begin the discussion where we left off in the previous chapter, i.e., with a spherical solid nucleus of radius greater than the critical radius,

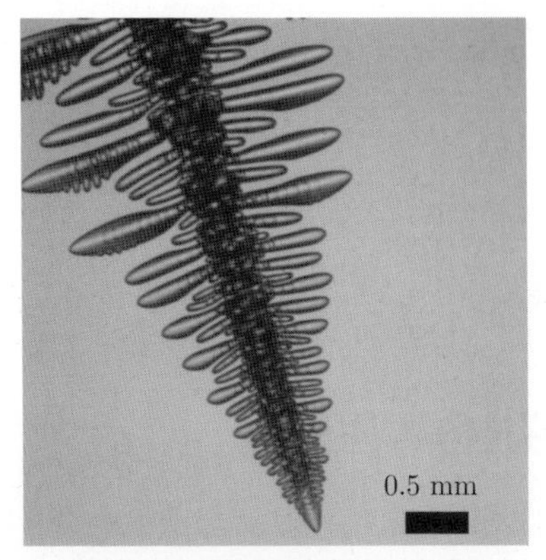

(a) Succinonitrile in free growth

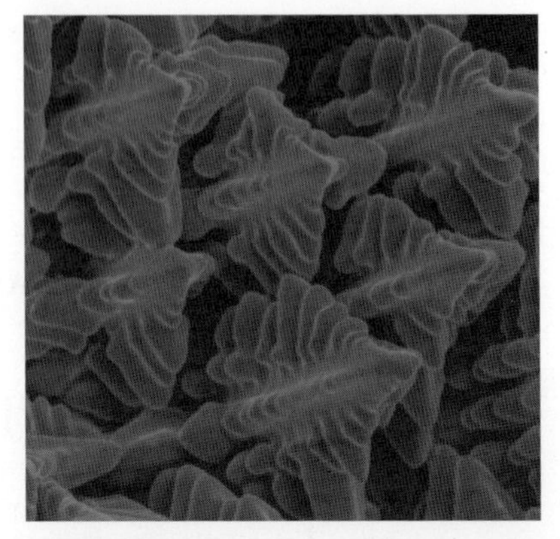

(b) Ni dendrites in superalloy spotweld

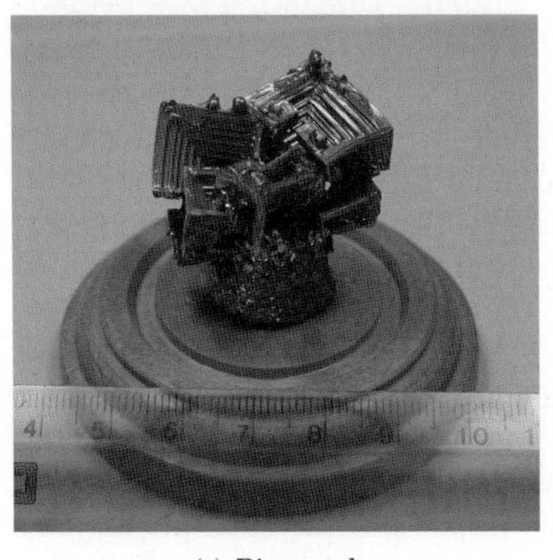

(c) Bi crystal

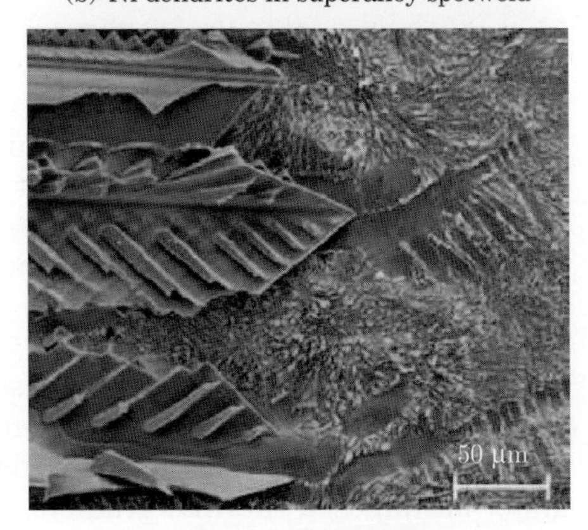

(d) Faceted Si dendrite

Fig. 8.1 (a) A micrograph of a pure succinonitrile dendrite solidifying in an undercooled melt with $\Delta T = 0.8$ K. (Reproduced with permission from C. Beckermann) (b) A spotweld in an Ni-based superalloy. (Reproduced with permission from ref. [10]) (c) A photograph of a Bi crystal grown from a chill inserted into an undercooled melt. (d) A micrograph of a quenched interface in a directionally solidified Si-Al alloy, displaying faceted Si primary dendrites (Reproduced with permission from ref. [34]).

$R_c = 2\Gamma_{s\ell}/\Delta T$, growing in a pure, infinite and undercooled melt. The temperature at infinite distance from the nucleus is $T_\infty = T_f - \Delta T$, where ΔT is the undercooling. We refer to this configuration as *free growth*. It was demonstrated in Chap. 7 that it is energetically favorable for the sphere to continue to grow as soon as $R > R_c$. One might then expect that under these conditions, the sphere would continue to grow until reaching macroscopic dimensions. Indeed, in Sect. 5.4.1, we computed the temperature profiles in the solid and liquid assuming this to be the case. However, as the sphere grows, the spherical morphology becomes unstable with respect to perturbations in shape. A formal stability analysis is given later to help understand the details of the emergence of the microstructure pattern during growth.

Once disturbances to the spherical shape can grow, i.e., after the spherical morphology is unstable, the solid shape begins to express the preferred growth directions of the underlying crystal. This preference is derived from anisotropy; whether it be in the surface energy of the solid-liquid interface, in the ease of attachment of atoms to the interface on different crystallographic planes, or both. These phenomena were discussed in Chap. 2. The fully developed structure looks quite different from the initial sphere. Figure 8.1(a) shows a crystal of succinonitrile (SCN) growing in an undercooled melt. This material, which crystallizes with a BCC structure, has been extensively used in solidification studies due to the fact that it behaves similarly to metals, as well as it being transparent. The solid takes on a tree-like form, called a *dendrite*, consisting of a primary trunk growing along one of the six equivalent $\langle 100 \rangle$ directions of the crystal, with secondary arms in the four conjugate $\langle 100 \rangle$ directions appearing just behind the advancing dendrite tip. The $\langle 100 \rangle$ preferred growth direction is typical of BCC crystals, and reflects the underlying surface energy anisotropy, even though the anisotropy represents only a few percent of the actual value of the surface energy. Preferred growth directions for other crystal structures are tabulated in Table 8.1. Figure 8.1(b) shows a regular array of Ni-rich dendrites formed during solidification of a superalloy weldment. This configuration is referred to as *constrained growth*, and is discussed in detail in Sect. 8.3.

Table 8.1 Preferred crystallographic directions for dendritic growth.

Crystal structure	Preferred growth direction	Examples
FCC	$\langle 100 \rangle$	Al, Cu, Ni, γ-Fe
BCC	$\langle 100 \rangle$	δ-Fe, Succinonitrile (SCN), NH_4Cl (CsCl-type)
Tetragonal	$\langle 110 \rangle$	Sn
HCP	$\langle 10\bar{1}0 \rangle$	Zn, H_2O

In most metallic systems, where interface attachment kinetics is negligible, the fact that the solid grows along preferred crystallographic directions can be understood as an attempt by the system to minimize the area of those surfaces with the highest surface energy. Consequently, as the anisotropy of the solid-liquid interfacial energy $\gamma_{s\ell}$ increases, assuming that all other quantities remain the same, the dendrite will exhibit a sharper and sharper tip. If the anisotropy in $\gamma_{s\ell}$ is large enough, the dendrites will present a *faceted* morphology. As two examples of this morphology, Fig. 8.1(c) shows a crystal of pure Bi grown from an undercooled melt, and Fig. 8.1(d) displays an array of faceted Si dendrites growing during directional solidification in an Si-Al alloy. In the Bi crystal, one can clearly see *facets* of the rhombohedral crystal growing in a spiral pattern, probably originating from a screw dislocation on the initial surface. In such materials, the kinetics of atom attachment at the interface also plays a significant role.

Let us further consider dendrites in metals, for which the anisotropy in $\gamma_{s\ell}$ controls the growth direction, at least at low undercooling. It is difficult to measure the surface energy itself accurately, and even more so for its anisotropy. It has been observed that materials with high values of surface energy anisotropy also tend to have relatively large entropies of fusion, leading to the rule of thumb that faceted crystals form in materials for which $\Delta S_f^m/\mathcal{R} > 2$, where $\Delta S_f^m = \mathcal{M}L_f/T_f$ is the molar entropy of fusion and $\mathcal{R}$ is the ideal gas constant (see Chap. 2). Table 8.2 gives examples of entropies of fusion for several common materials.

Since the surface energy and its anisotropy play such important roles in dendritic growth, it is worth discussing these properties in greater detail. The topic was introduced in Sect. 2.4.2, and we first recall certain points from that discussion for the current context. In 2-D, one often represents the surface energy in the form

$$\gamma_{s\ell} = \gamma_{s\ell}^0 \left[1 + \varepsilon_n \cos(n\phi)\right] \tag{8.1}$$

where ϕ is the azimuthal angle measured from a reference direction, ε_n represents the strength of the anisotropy and n is the degree of symmetry. For example, $n = 4$ for cubic symmetry in a (100) plane, and $n = 6$ for hexagonal symmetry in a (0001) plane. The surface stiffness $\Psi_{s\ell}$, defined in Eq. (2.64), is given in 2-D by

$$\Psi_{s\ell} = \gamma_{s\ell} + \frac{d^2\gamma_{s\ell}}{d\phi^2} \tag{8.2}$$

Thus, for a fourfold symmetry in 2-D, we have

$$\Psi_{s\ell} = \gamma_{s\ell}^0 \left[1 - 15\varepsilon_4 \cos(4\phi)\right] \tag{8.3}$$

Table 8.2 Entropies of fusion and structures for some common materials.

Material	L_f (J g^{-1})	$\mathcal{M}$ (g mol^{-1})	T_f (K)	$\Delta S_f^m/\mathcal{R}$
Non-faceting materials				
Al	373	26.982	933	1.30
Cu	211	63.546	1356	1.19
Ni	298	58.693	1728	1.22
Sn	59.2	118.71	504	1.68
SCN	43.8	80.1	331	1.27
Zn	112	65.409	693	1.27
Faceting materials				
Bi	54	208.98	544	2.50
Si	1790	28.086	1687	3.59
H_2O	334	18	273	2.65

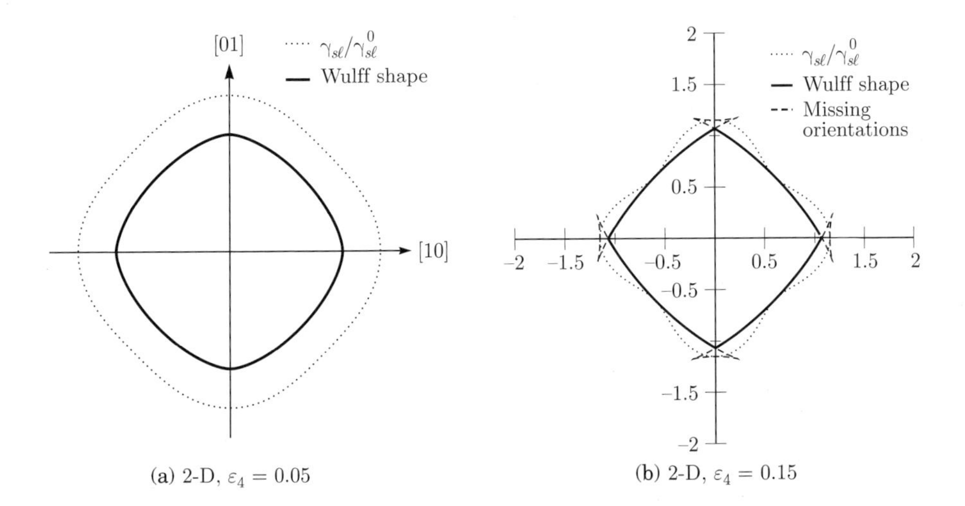

Fig. 8.2 The equilibrium shape of fourfold symmetric 2-D crystals. (a) $\varepsilon_4 = 0.05$. The Wulff shape has been drawn inside of $\gamma_{s\ell}$ for clarity. (b) $\varepsilon_4 = 0.15$. Notice the "ears" on the coordinate axes, corresponding to the missing orientations.

Notice that, if $\varepsilon_4 > 1/15$, there are some angles, ϕ, for which the surface stiffness becomes negative. This is forbidden by thermodynamics, and results in a range of orientations of the surface normal with respect to the crystal axis that cannot appear. The equilibrium shape of the crystal, computed using the Wulff construction (see Sect. 2.4.2), is illustrated in Fig. 8.2 for two values of ε_4 above and below this transition. The "ears" along the axes at the cusp of the Wulff shape for $\varepsilon_4 = 0.15$ correspond to the orientations that are missing in the equilibrium crystal shape.

These concepts can be extended to 3-D, but one must be somewhat more formal. For crystals with a cubic symmetry, the surface energy can be expanded in terms of cubic harmonics, i.e., orthogonal combinations of spherical harmonics exhibiting a cubic symmetry:

$$\gamma_{s\ell}(\boldsymbol{n}) = \gamma^0_{s\ell}(1 + a_1[Q_4 - 3/5] + a_2[3Q_4 + 66S_4 - 17/7] + \ldots) \tag{8.4}$$

where a_1 and a_2 parameterize the strength of the anisotropy. The functions Q_4 and S_4 are written in Cartesian and spherical form as

$$Q_4 = \begin{cases} n_x^4 + n_y^4 + n_z^4 \\ \sin^4\theta\left(\cos^4\phi + \sin^4\phi\right) + \cos^4\theta \end{cases} \tag{8.5}$$

$$S_4 = \begin{cases} n_x^2 n_y^2 n_z^2 \\ \sin^4\theta\cos^2\theta\sin^2\phi\cos^2\phi \end{cases} \tag{8.6}$$

The polar angle θ is measured from the [001] direction, and the azimuthal angle ϕ is measured as in 2-D, from the [100] direction in the (001) plane. The Cartesian components of the unit normal $\boldsymbol{n}$ in this reference frame correspond to n_x, n_y and n_z. We further note that it is possible to choose a_1 and a_2 such that the expression in Eq. (8.4) becomes equivalent to Eq. (8.1) in the (001) plane (see Exercise 8.2). The Wulff shape is computed using the formalism of Cahn and Hoffman [9], as the surface swept by the vector $\boldsymbol{\xi}$, defined as

$$\boldsymbol{\xi} = \nabla[r\gamma_{s\ell}(\hat{\boldsymbol{n}})] = \gamma_{s\ell}\hat{\boldsymbol{n}} + \frac{\partial \gamma_{s\ell}}{\partial \theta}\hat{\boldsymbol{\theta}} + \frac{1}{\sin\theta}\frac{\partial \gamma_{s\ell}}{\partial \phi}\hat{\boldsymbol{\phi}} \tag{8.7}$$

where $r = (x^2 + y^2 + z^2)^{1/2}$. Examples of the shape of $\gamma_{s\ell}$ for various values of the parameters, equivalent in the (001) plane to those in Fig. 8.2, are given in Fig. 8.3, along with their corresponding Wulff shapes. The "ears" that were present in 2-D, and which become "flaps" along some edges in 3-D, again indicate a band of forbidden orientations. They have been removed in Fig. 8.3(d) for the sake of clarity. One can see that the behavior in 3-D is considerably more complicated in comparison to 2-D. The Wulff

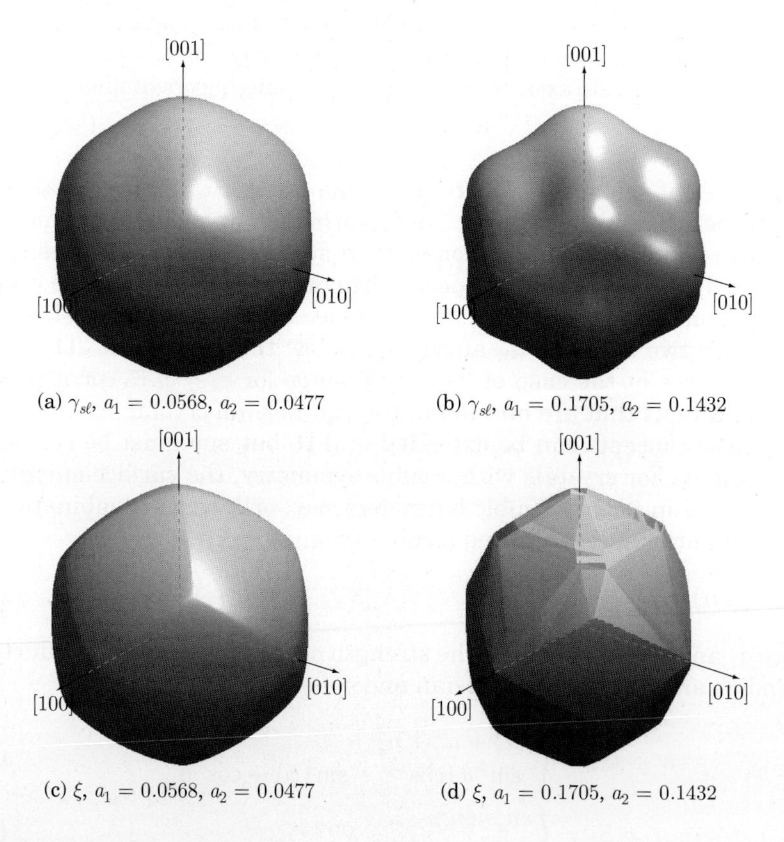

Fig. 8.3 (a)-(b) The surface energy $\gamma_{s\ell}$ of 3-D crystals with levels of cubic anisotropy comparable with those in Fig. 8.2. (c)–(d) The corresponding $\boldsymbol{\xi}$-vector surfaces, equivalent to equilibrium shape crystals. The "ears" and "flaps" corresponding to missing orientations have been removed in (d).

construction can be used to deduce the anisotropy of the surface energy from an experiment in which one produces a sample of the material of interest which contains small droplets of liquid.

8.2.2 Stability and scale selection for a freely growing sphere of a pure material

This section examines in more detail the evolution of the shape of an initially spherical solid. We analyze the *morphological stability* of such a sphere solidifying in an infinite undercooled melt, with the objective of understanding how the microstructural length scale evolves during growth. We will find that it is the interaction of diffusional transport with surface energy that determines the shape. Stability analysis is concerned with understanding how a system reacts to disturbances to a base state. An example is the simple pendulum with its two equilibrium positions: the one usually observed where the weight is parallel to gravity, and the inverted position. Experience tells us that the first position is a state of *stable equilibrium*, signifying that a small disturbance from the initial state produces bounded motions about the initial position which eventually decay under friction to the base state. Similarly, practical experience shows us that the inverted position is *unstable*, where any disturbance of the equilibrium state produces large motions that do not return to the initial configuration.

One can formalize this experience-based observation, turning it into a mathematical procedure known as *linear stability analysis*, in which small perturbations to the base state are introduced, and the governing equations are used to determine whether the disturbance will decay or grow in time. In this context, "linear" refers to neglecting any terms arising in the analysis that are products of perturbations. This is sometimes called linearization. If the disturbance decays in time, the base state is said to be stable. Conversely, if the perturbation grows in time, the base state is called unstable. If the disturbances neither grow nor decay, the system is referred to as *neutrally stable*. We now apply this procedure to analyze the stability of a sphere growing in an undercooled melt. We omit some of the mathematical details of the calculation in order to focus on the important physics of the problem.

Section 5.4.1 considered in detail the solidification of a sphere in an infinite undercooled melt. The results of the analysis that are required in the current context are restated here, and the reader is encouraged to review that section before proceeding, if these results seem unfamiliar. We derived a spherically symmetric temperature solution to this problem in terms of the following dimensionless variables:

$$\varrho = \frac{r}{R_i}\ ;\ \tau = \frac{t}{R_i^2/\alpha_s}\frac{1}{\text{Ste}}\ ;\ \mathcal{R}^*(\tau) = \frac{R^*(t)}{R_i}\ ;\ \theta_{s,\ell} = \frac{T_{s,\ell} - T_f}{T_f - T_\infty} = \frac{T_{s,\ell} - T_f}{\Delta T} \tag{8.8}$$

where r is the radial coordinate measured from the center of the sphere, R_i is the initial radius of the sphere, t is the time, $\mathrm{Ste} = c_p\Delta T/L_f$, $\Delta T = T_f - T_\infty$ is the undercooling, and $R^*(t)$ is the radius of the sphere at time t. We also introduced the parameters $\alpha_T = \alpha_s/\alpha_\ell$ and $k_T = k_s/k_\ell$, which are close to one for most materials. It was demonstrated in Sect. 5.4.1 that, for small values of Ste, the movement of the solid-liquid interface is very slow as compared to heat conduction, and that the temperature field is then quasi-steady, signifying that the temperature in the solid and liquid phases depends on time only through the radius of the sphere R^*:

$$\theta_\ell^0 = -1 + \left(1 - \frac{2\Gamma_{s\ell}}{\Delta T R_i \mathcal{R}^{0*}(\tau)}\right)\frac{\mathcal{R}^{0*}(\tau)}{\varrho} \tag{8.9}$$

$$\theta_s^0 = -\frac{2\Gamma_{s\ell}}{\Delta T R_i \mathcal{R}^{0*}(\tau)} \tag{8.10}$$

The index "0" on θ_ℓ^0, θ_s^0 and $\mathcal{R}^{0*}$ has been added to indicate that these quantities constitute the base state solution whose stability we wish to examine. The temperature solutions are shown schematically in Fig. 8.4, where one can see that the temperature in the solid is independent of position, whereas the temperature in the liquid falls off with the distance as $1/r$. We also obtained an expression for the growth rate of the sphere by substituting Eqs. (8.9) and (8.10) into the Stefan condition at the surface of the sphere, with the result

$$\frac{d\mathcal{R}^{0*}}{d\tau} = -\left(1 - \frac{2\Gamma_{s\ell}}{\Delta T R_i \mathcal{R}^{0*}(\tau)}\right)\frac{1}{\mathcal{R}^{0*}(\tau)} \tag{8.11}$$

From this equation, it was inferred that the critical nucleus size for the sphere is $R^* = 2\Gamma_{s\ell}/\Delta T$, above which the sphere will grow (the sign of the right-hand side is positive), and below which it will shrink.

The linear stability of this solution was first examined by Mullins and Sekerka [33], who introduced small, non-symmetric perturbations θ_s^1,

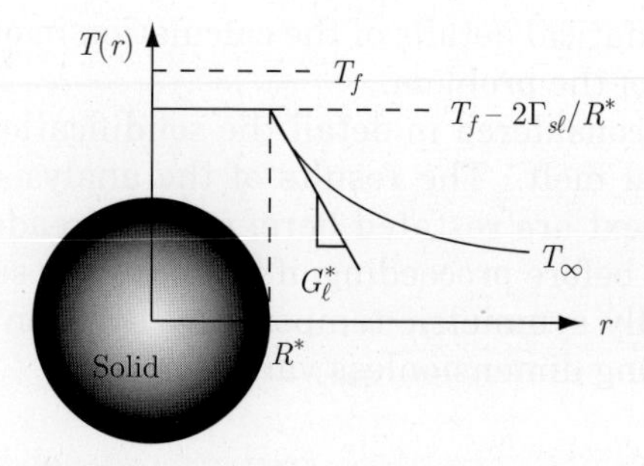

Fig. 8.4 A schematic representation of the temperature distribution near a solidifying sphere of a pure material in an undercooled melt.

θ_ℓ^1 and $\mathcal{R}^{1*}$ to the base state in the form

$$\theta_s = \theta_s^0 + \theta_s^1(\varrho, \Theta, \Phi, \tau) \tag{8.12}$$

$$\theta_\ell = \theta_\ell^0 + \theta_\ell^1(\varrho, \Theta, \Phi, \tau) \tag{8.13}$$

$$\mathcal{R}^* = \mathcal{R}^{0*} + \mathcal{R}^{1*} \tag{8.14}$$

Here, we have used Θ and Φ to denote the orthogonal polar angular directions to avoid confusion with the symbol θ that refers to the dimensionless temperature. After substituting these forms into the governing equations, the perturbed temperatures in the solid and liquid can be determined by separation of variables, giving solutions of the form

$$\theta_\ell^1 = \frac{A_\ell(\tau)}{\varrho^{n+1}} Y_{nm}(\Theta, \Phi) \ ; \quad \theta_s^1 = A_s(\tau)\varrho^n Y_{nm}(\Theta, \Phi) \tag{8.15}$$

where Y_{nm} is a spherical harmonic, n is an integer called the *mode* of the harmonic, and m is also a positive integer. The coefficients $A_s(\tau)$ and $A_\ell(\tau)$ are evaluated from the boundary conditions at the solid-liquid interface. The details of this calculation are omitted. The Stefan condition yields an equation for the growth rate of the perturbation $\mathcal{R}^{1*}$, given by

$$\frac{d\mathcal{R}^{1*}}{d\tau} = \left[1 - \frac{\Gamma_{s\ell}}{\Delta T R_i \mathcal{R}^{0*}}\{(n+1)(n+2) + 2 + k_T n(n+2)\}\right] \frac{n-1}{(\mathcal{R}^{0*})^2}\mathcal{R}^{1*} \tag{8.16}$$

Equation (8.16) is less complicated than it at first appears to be. This is because for any given value of $\mathcal{R}^{0*}$, everything on the right hand side is constant, with the exception of $\mathcal{R}^{1*}$. Let us then write Eq. (8.16) and its solution in the shorthand form

$$\frac{d\mathcal{R}^{1*}}{d\tau} = s_n \mathcal{R}^{1*} \quad \Rightarrow \quad \mathcal{R}^{1*} \sim \exp(s_n \tau) \tag{8.17}$$

One can thereby infer the stability of the spherical morphology to disturbance of mode n simply by considering the sign of s_n: if $s_n > 0$, the perturbation grows exponentially in time, and if $s_n < 0$, the perturbation shrinks exponentially in time. The sphere is morphologically stable if $s_n < 0$ for *all* modes of n, and, conversely, the sphere is morphologically unstable if $s_n > 0$ for *any* value of n. Finally, the case $s_n = 0$ corresponds to neutral stability. Returning now to the original form, i.e., Eq. (8.16), one can see that the mode $n = 1$ is neutrally stable, since the factor $(n-1)$ causes the right-hand side to be zero. Stability is thus determined by the mode $n = 2$, for which we have

$$\frac{d\mathcal{R}^{1*}}{d\tau} = \left[1 - \frac{2\Gamma_{s\ell}(4k_T + 7)}{\Delta T R_i \mathcal{R}^{0*}}\right] \frac{\mathcal{R}^{1*}}{(\mathcal{R}^{0*})^2} \tag{8.18}$$

The perturbations grow when the term inside the square brackets is positive, i.e., when

$$\mathcal{R}^{0*} > (4k_T + 7)\frac{2\Gamma_{s\ell}}{R_i \Delta T} \tag{8.19}$$

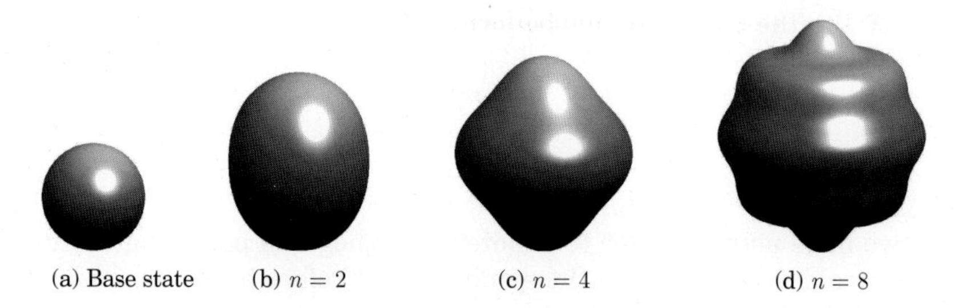

(a) Base state (b) $n = 2$ (c) $n = 4$ (d) $n = 8$

Fig. 8.5 (a) Base state sphere. (b)-(d) Perturbed spheres, $n = 2, 4, 8$.

Recognizing that $2\Gamma_{s\ell}/\Delta T = R_c$, i.e., the radius of a critical nucleus, and that $R_i\mathcal{R}^{0*} = R^*$, the criterion for instability can be written in a more compact form as

$$R^* > (4k_T + 7)R_c \tag{8.20}$$

Note that, as the sphere grows, higher modes also become unstable, and these modes will in fact have larger growth rates than the $n = 2$ mode. (see Exercise 8.4). The mode shapes for $n = 2, 4, 8$ are shown in Fig. 8.5.

It is particularly revealing to estimate the wavelength of the instability in terms of the temperature gradient G_ℓ^* in the liquid at the interface. Differentiating Eq. (8.9) with respect to ϱ and evaluating at $\varrho = \mathcal{R}^{0*}$ gives

$$\left.\frac{\partial\theta_\ell^0}{\partial\varrho}\right|_{\mathcal{R}^{0*}} = -\left(1 - \frac{2\Gamma_{s\ell}}{R_i\Delta T\mathcal{R}^{0*}}\right)\frac{1}{\mathcal{R}^{0*}} \tag{8.21}$$

Rewriting Eq. (8.21) in terms of the dimensional temperature gradient G_ℓ^* and radius R^*, leads to

$$\frac{G_\ell^* R_i}{\Delta T} = -\left(1 - \frac{R_c}{R^*}\right)\frac{R_i}{R^*} \tag{8.22}$$

It is left as an exercise to demonstrate that combining Eqs. (8.20) and (8.22) yields the criterion.

$$(R^*)^2 > -\frac{2(4k_T + 6)\Gamma_{s\ell}}{G_\ell^*} \tag{8.23}$$

Note that for growth in an undercooled melt, $G_\ell^* < 0$. The unstable mode $n = 2$ corresponds to a disturbance on the surface of the sphere for which the wavelength $\lambda = \pi R^*$ (see Fig. 8.5). We thus have

$$\lambda = \pi\sqrt{2(4k_T + 6)\left(-\frac{\Gamma_{s\ell}}{G_\ell^*}\right)} \tag{8.24}$$

For most pure materials, $k_T \approx 1$, whereupon Eq. (8.24) gives

$$\lambda \approx 4.5\pi\sqrt{-\Gamma_{s\ell}/G_\ell^*} \tag{8.25}$$

This result shows that the length scale of the microstructure is selected based on the interplay of surface energy and diffusion. A similar characteristic wavelength, with a prefactor between 2π and 5π, shows up repeatedly in the stability analysis of solidification problems, and plays a major role in calculations of microstructural length scales. It should be noted that the factor 4.5 in Eq. (8.24) is reduced by a factor $2/n$ as higher order modes become unstable (see Exercise 8.4).

To make the physical interpretation of Eq. (8.25) even more clear, consider the Stefan condition

$$\rho L_f \frac{dR^*}{dt} = k_s \frac{\partial T_s}{\partial r} - k_\ell \frac{\partial T_\ell}{\partial r} \qquad r = R^*(t) \tag{8.26}$$

In the base state, the solid is isothermal (Eq. (8.10)), causing the first term on the right-hand side to be zero. By replacing the interface velocity dR^*/dt by the symbol v^* and the temperature gradient in the liquid by G_ℓ^*, we obtain

$$\rho L_f v^* = -k_\ell G_\ell^* \tag{8.27}$$

Solving for G_ℓ^* and substituting the result into Eq. (8.25) gives

$$\lambda \approx 4.5\pi\sqrt{d_0(\alpha_\ell/v^*)} \tag{8.28}$$

where $d_0 = c_{p\ell}\Gamma_{s\ell}/L_f$ is called the *thermal capillary length*, a material property that characterizes the surface energy. Thus, Eq. (8.28) relates the wavelength of the disturbance to the geometric mean of two length scales: the capillary length d_0 and the characteristic length scale for the diffusion field ahead of the moving interface, α_ℓ/v^*. This helps us to understand how the microstructural length scale is set by the balance between surface energy and diffusion. This is the most important result of the analysis, and we will see that it has far-reaching consequences for the modeling of microstructure in a large variety of processes.

Figure 8.6 shows the results of a phase-field simulation (described in Sect. 8.6) of the growth of a small spherical seed of a pure material with constant $\gamma_{s\ell}$ solidifying in an infinite undercooled pure melt. The

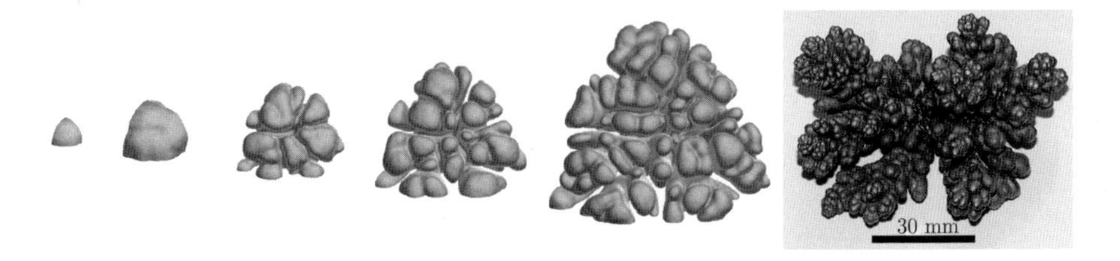

Fig. 8.6 The evolution of an initially spherical solid particle in an infinite undercooled melt. The simulations in the first five images were performed in one octant of the domain, with $\mathrm{Ste} = 0.5$. The final panel shows a photograph of an Ni sample grown from solution in an electrochemical cell.

morphological instability becomes apparent in the second image. As the solid evolves, one sees the continued breakdown of the smooth surface, eventually producing what is known as a "dense-branched" morphology. Notice that the length scale of each branch is roughly the same as that of the initial instability, indicating that the same mechanism continues to determine the microstructure well after the initial simple shape has disappeared. The final image is a photograph of pure Ni grown from solution in a plating bath, showing a remarkably similar morphology, although the growth mechanism is somewhat more complicated in the case of the Ni sample. We note that these shapes do not bear a close resemblance to the dendrites shown in Fig. 8.1. It turns out that the key missing feature in our analysis is the anisotropy of the surface energy, which so far has been neglected. The discussion of this topic, however, is deferred to Sect. 8.4.

8.2.3 Extension to binary alloys

The preceding analysis applies to a pure material solidifying in an undercooled melt. It can be adapted to the solidification of a binary metallic alloy in a relatively straightforward way by noting that the disparity in diffusivities ($D_s \ll D_\ell \ll \alpha_\ell \lesssim \alpha_s$) leads to a large separation in length and time scales for heat and solute transport, which permits us to treat the melt as isothermal. (The problem in which both heat and solute diffusion are considered can be solved, but this increases the complexity of the presentation without adding any essential phenomena.) The problem definition is given schematically in Fig. 8.7, which also displays some of the important notation. Note that we have chosen an alloy for which the partition coefficient $k_0 < 1$. The results are applicable for $k_0 > 1$ as well.

The analysis proceeds in a manner similar to that for the pure material. For this reason, most of the details are omitted here, and are instead left to the reader in Exercise 8.5. The quantity that is analogous to the

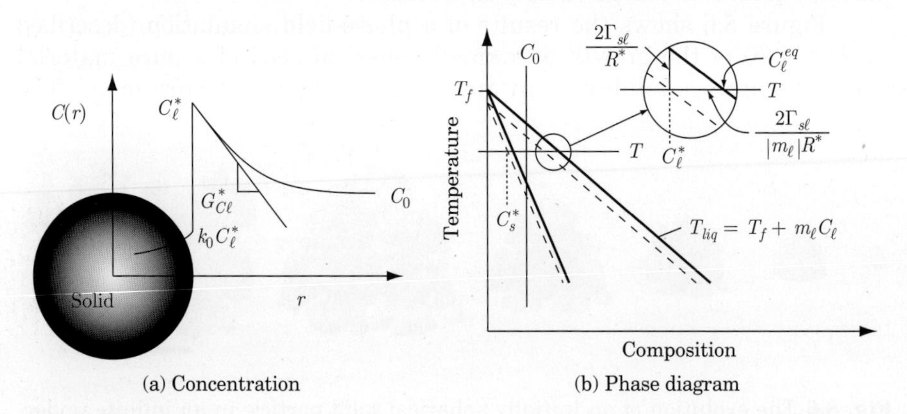

Fig. 8.7 (a) A schematic representation of the composition distribution near a solidifying sphere; (b) A segment of a binary alloy phase diagram, showing the solidus and liquidus curves corrected for the curvature of a spherical particle of radius R^*.

Stefan number in the thermal problem is called the *supersaturation* Ω, and is defined as

$$\Omega = \frac{C_\ell^{eq} - C_0}{C_\ell^{eq}(1 - k_0)} \tag{8.29}$$

It should be noted that, in Chap. 2, the symbol C_ℓ^∞ was used to indicate the composition for $R^* \to \infty$. Here, it is changed to C_ℓ^{eq} to avoid confusion with the boundary condition at $r \to \infty$, where $C_\ell \to C_0$. To simplify the analysis, we assume that the supersaturation $\Omega \ll 1$ and that $C_\ell^* - C_s^* \approx C_\ell^{eq} - C_s^{eq}$. Under these assumptions, the governing equations and boundary conditions for the liquid composition take an identical form to those in the thermal problem (see Exercise 8.5), and we thus simply write down the key results. The radius of the critical nucleus $R_c = 2\Gamma_{s\ell}/\Delta T$ for the alloy sphere is obtained by setting $\Delta T = m_\ell(C_0 - C_\ell^*) \approx m_\ell(C_0 - C_\ell^{eq})$ (see Fig. 8.7). In analogy to Eq. (8.19), the alloy sphere becomes unstable for

$$R^* > \left(4\frac{D_s}{D_\ell} + 7\right) R_c = \left(4\frac{D_s}{D_\ell} + 7\right) \frac{2\Gamma_{s\ell}}{m_\ell(C_0 - C_\ell^{eq})} \tag{8.30}$$

Solving for the composition gradient in the liquid, and following the same steps as in the case of the pure material yields the criterion for instability for alloys at small supersaturation

$$R^{*2} > \left(4\frac{D_s}{D_\ell} + 6\right) \frac{2\Gamma_{s\ell}}{m_\ell G_{C\ell}^*} \tag{8.31}$$

where $G_{C\ell}^*$ is the composition gradient in the liquid at the interface (see Fig. 8.7(a)). Notice that the ratio of the thermal conductivities k_s/k_ℓ is replaced in this problem by the ratio of the chemical diffusivities D_s/D_ℓ, which is typically very close to zero. By making this approximation and, once again, equating the wavelength λ to πR^*, we have

$$\lambda \approx 3.5\pi\sqrt{\frac{\Gamma_{s\ell}}{m_\ell G_{C\ell}^*}} \tag{8.32}$$

Note the similarity of this form to Eq. (8.25). We can also write the analogous form to Eq. (8.28), setting $D_s/D_\ell \approx 0$, to obtain

$$\lambda \approx 3.5\pi\sqrt{d_0^C (D_\ell/v^*)} \tag{8.33}$$

where the *chemical capillary length* d_0^C is defined as

$$d_0^C = -\frac{\Gamma_{s\ell}}{m_\ell C_\ell^{eq}(1 - k_0)} \tag{8.34}$$

Thus, we see once again that the wavelength is selected by a competition between surface energy and diffusion. The composition gradient in the liquid is sometimes represented by the so-called "Zener approximation"

$$G_{C\ell}^* = \left.\frac{\partial C_\ell}{\partial r}\right|_{R^*} \approx \frac{C_0 - C_\ell^{eq}}{R^*} \tag{8.35}$$

Substituting Eq. (8.35) into Eq. (8.30) and rearranging terms yields

$$R^{*2} > \left(4\frac{D_s}{D_\ell} + 7\right) \frac{2\Gamma_{s\ell}}{m_\ell G_{C\ell}^*} \tag{8.36}$$

in which case the prefactor in Eqs. (8.32) and (8.33) becomes 3.7 instead of 3.5.

In Sect. 8.4.2, we consider the corresponding form for the mode $n = 4$, for which we have

$$(R_4^*)^2 > \left(32 + 24\frac{D_s}{D_\ell}\right) \frac{\Gamma_{s\ell}}{m_\ell G_{C\ell}^*} \tag{8.37}$$

Here, the subscript "4" indicates the mode $n = 4$. Taking the limit $D_s/D_\ell \to 0$ and noting that the wavelength corresponding to the mode $n = 4$, is $\lambda_4 = \pi R_4^*/2$, we have

$$\lambda_4 \approx \frac{\pi R_4^*}{2} = 2\pi\sqrt{2}\sqrt{\frac{\Gamma_{s\ell}}{m_\ell G_{C\ell}^*}} \tag{8.38}$$

The morphological stability analysis of the sphere yields the important result that the length scale for the instability is proportional to the geometric mean of the length scales associated with diffusion (α/v^* or D_ℓ/v^*) and capillarity (d_0 or d_0^C). The constant of proportionality depends to a certain extent on whether one has an alloy or pure material, and the particular mode that one chooses to examine, nevertheless, the essential physics are the same.

8.3 CONSTRAINED GROWTH

8.3.1 General observations

The concept of free growth of a single dendrite in an infinite undercooled melt is very useful for understanding the important phenomena in dendritic growth, and can be applied to equiaxed growth in realistic situations. A more common configuration is solidification from a chill surface, such as when a liquid is poured into a cold mold. Chapter 5 presented several analytical solutions for the progress of a planar solidification front for such a case, considering both pure materials and alloys. For the pure material, the temperature gradient G_ℓ^* at the solid-liquid interface is positive for this configuration, and the interface is stable. In alloys, we still have $G_\ell^* > 0$, however, the composition gradient in the liquid at a planar front

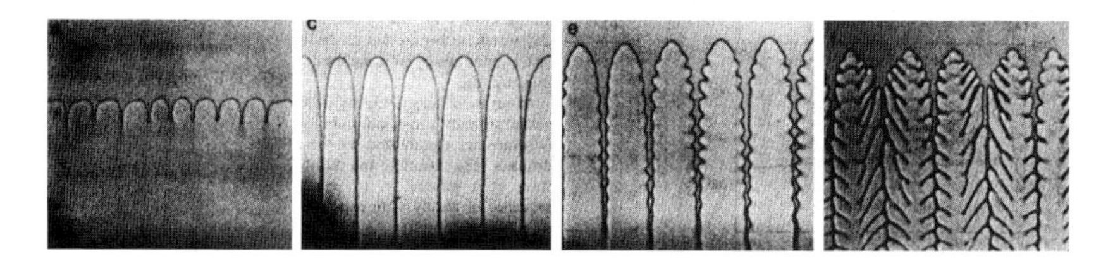

Fig. 8.8 Microstructures observed in an SCN-0.66 wt% Salol alloy, directionally solidified with $G = 4.5$ K mm^{-1}. The pulling speed from left to right, is 0.57, 2.0, 5.7, and 7.6 μm s^{-1}. (Reproduced with permission from ref. [31].)

$G^*_{C\ell}$ is negative (see Fig. 5.2 and the surrounding discussion). We demonstrate below that this configuration is *conditionally stable*, depending on the temperature gradient G_ℓ and interface speed v^*.

Figure 8.8 shows some examples of microstructures observed during the directional solidification of succinonitrile (SCN)-Salol alloys. In this experiment, a thin sample of a dilute alloy is confined between two microscope slides that are placed on a microscope stage between a hot and cold block (see Fig. 5.3). The sample is then translated at constant pulling speed v_p, and the microstructure is observed. The sequence of micrographs was obtained using a fixed temperature gradient, G, and various pulling speeds, v_p. At very low velocities (not shown), the interface is planar. Above a critical velocity, which we compute below, shallow *cells* develop, separated by grooves parallel to G. The cells do not follow any particular crystallographic orientation, and are characterized by a large radius of curvature. As the solidification velocity increases, the grooves become deeper and the cell spacing decreases. At higher velocity, there is a gradual transition to dendrites characterized by much sharper tips and by trunks whose orientation follows the preferred growth direction, and by sidebranches.

An example showing the temporal evolution of an array of dendrites in directional solidification is given in Fig. 8.9. The experiment begins with an alloy of SCN and acetone confined between two glass plates, held stationary in a temperature gradient to establish the initial planar interface. The remaining figures show the development of the dendritic microstructure when the sample is pulled at constant speed v_p. The time sequence shows that, during the early stages of growth, the initially planar interface (to the left) begins to develop perturbations, which then grow into fully developed dendrites as time proceeds (to the right in the figure). Upon closer examination, one sees that the perturbations appear first at irregularities, such as at the edges of the sample and a pre-existing grain boundary approximately 20% above the lower edge of the sample. In the third panel of the sequence, nearly sinusoidal perturbations appear everywhere along the previously planar front between the initial disturbances. The succeeding panels show the development of aligned primary dendrites competing with each other, growing like trees, with primary trunks and secondary branches on the side. The spacing between the primary trunks

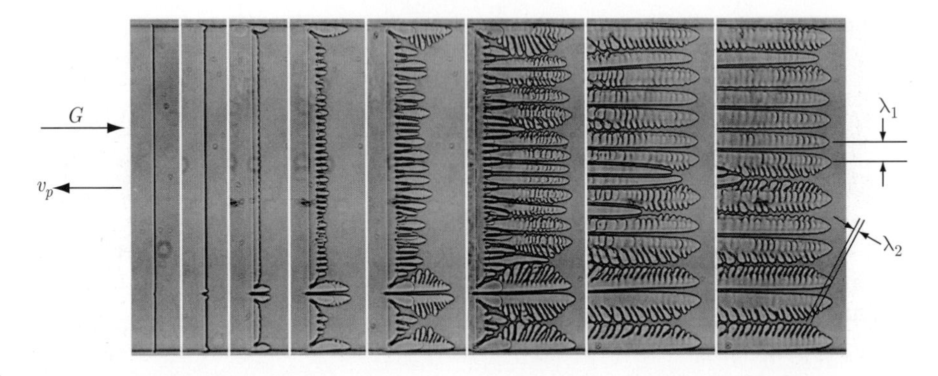

Fig. 8.9 A time series from the directional solidification of a succinonitrile-acetone alloy. The initial condition is a solid at rest in a temperature gradient, giving the planar interface in the left-hand panel. Solidification occurs by pulling the alloy at constant speed v_p to the left, as indicated. (Reproduced with permission from W. Losert.)

is called the *primary dendrite arm spacing*, and is labeled λ_1. The spacing between the secondary branches, called the *secondary dendrite arm spacing*, is labeled λ_2. The nearly parabolic tips of the primary dendrites are very similar to those observed in free dendrites, also growing in the $\langle 100 \rangle$ direction. The secondary arms tend to grow at a small angle to the preferred growth direction when the anisotropy is not very strong and the thermal gradient is high. Due to the dendrites being constrained to grow along the direction of the thermal gradient, this situation is called *constrained growth*.

The dendrites in the array shown in Fig. 8.9 are oriented such that their $\langle 100 \rangle$ preferred growth direction is aligned almost perfectly with the temperature gradient and growth direction. Of course, this is not always the case. When several grains of differing orientation grow in a columnar form, a competition takes place between the grains, eventually leading to the elimination of those least favorably oriented. This phenomenon is explored further in Chap. 11.

Surface energy anisotropy plays as important a role in constrained growth as it does in free growth. We provide an example to illustrate this phenomenon: Gonzales, et al. [15] performed directional solidification experiments on a range of Al-Zn alloys, measuring the misorientation of the primary dendrite trunks with respect to the $\langle 100 \rangle$. The results are summarized in Fig. 8.10, where it can be observed that the primary growth direction shifts gradually from $\langle 100 \rangle$ to $\langle 110 \rangle$ as the Zn content increases. This is due to the effect that Zn, with its hcp crystal structure, has on the weak anisotropy of Al ($\varepsilon_4 \approx 0.01$). At the limits of the transition region ($C_0^{Zn} \approx 25\%$ and $C_0^{Zn} \approx 55\%$), the anisotropy in $\gamma_{s\ell}$ becomes very small, leading to the appearance of *seaweed* structures, which represent the analogous form in constrained growth to the dense-branched morphology found

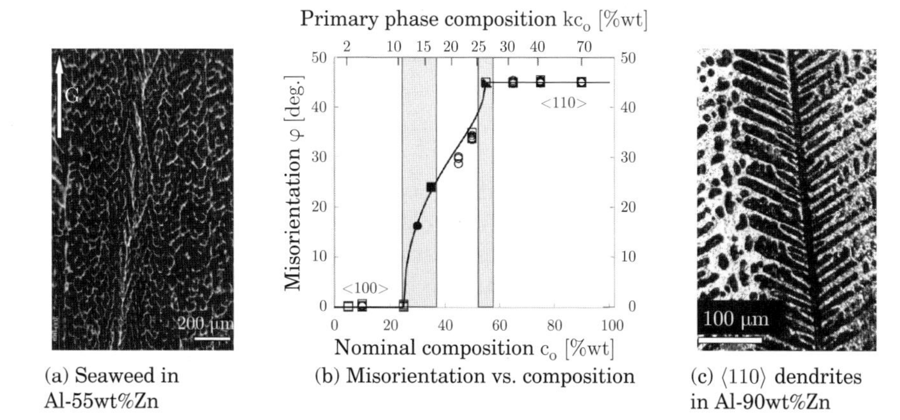

(a) Seaweed in Al-55wt%Zn (b) Misorientation vs. composition (c) ⟨110⟩ dendrites in Al-90wt%Zn

Fig. 8.10 Solidification microstructures of Al-Zn alloys observed in longitudinal sections after constrained growth. (a) "Seaweed" structure in Al-55 wt%Zn specimen. (b) The angle between the ⟨100⟩ and either the orientation of the primary dendrite trunk, or the strongest texture component for seaweed structures, plotted as a function of the zinc content. (c) ⟨110⟩ dendrites in an Al-90wt%Zn specimen. (Reproduced with permission, see ref. [15].)

in free growth. The seaweed structure nevertheless exhibits a texture whose average orientation follows the continuous curve shown in Fig. 8.10.

8.3.2 Length scales and pattern selection in constrained growth

The phenomenological description of dendritic growth leads us to ask two key questions: Under what conditions is the solid-liquid interface morphologically unstable, and how are the length scales selected by the solidifying interface? We begin with a simple model that addresses only the first of the two questions.

Consider the planar front directional solidification of a binary alloy at constant pulling speed, v_p, in a constant temperature gradient, $G > 0$. For such a planar front growth, the pulling speed and interface speed are the same, i.e., $v^* = v_p$. This model problem was considered in Sect. 5.2.3, where it was shown that after an initial transient, the system reached a steady state in which the composition in the liquid is given by Eq. (5.97),

$$C_\ell = C_0 \left\{ 1 + \frac{1-k_0}{k_0} \exp\left(-\frac{v_p z}{D_\ell} \right) \right\} \tag{8.39}$$

Here, z is the distance measured from the interface and C_0 is the nominal composition of the alloy. The interface temperature is the solidus temperature of the alloy, T_{sol}; the solid and liquid compositions at the interface are $C_s^* = C_0$ and $C_\ell^* = C_0/k_0$, respectively. The composition gradient in the liquid, $G^*_{C\ell}$, ahead of the interface is obtained by differentiating Eq. (8.39) and evaluating the result at the interface ($z = 0$)

$$G^*_{C\ell} = \left.\frac{dC_\ell}{dz}\right|_{z=0} = -\frac{C_0(1-k_0)v_p}{k_0 D_\ell} == -\frac{\Delta C_0 v_p}{D_\ell} \tag{8.40}$$

where $\Delta C_0 = C_0(1-k_0)/k_0$ is the difference in composition between the liquid and solid phases at the interface. For convenience of presentation, we have assumed a linearized phase diagram, as shown in Fig. 8.11. The liquidus temperature corresponding to the composition at any point z ahead of the interface is then given by

$$T_{liq}(z) = T_f + m_\ell C_\ell(z) = T_{sol} + \Delta T_0 \left[1 - \exp\left(-\frac{v_p z}{D_\ell}\right)\right] \tag{8.41}$$

for which $\Delta T_0 = -m_\ell \Delta C_0 = T_{liq} - T_{sol}$ is the nominal freezing range of the alloy.

Based on this solution and the sketch in Fig. 8.11, Tiller et al. [45] presented a heuristic explanation of the mechanism by which the planar interface becomes unstable for a binary alloy. They argued that if the *actual temperature* $T_\ell(z)$, ahead of the interface is greater than the local liquidus temperature, $T_{liq}(C_\ell(z))$, then any perturbation to the interface will be remelted. In other words, the planar interface is stable for this condition. Conversely, if the actual temperature anywhere ahead of the interface is below the local equilibrium liquidus temperature, a perturbation will continue to grow, and the interface will be unstable. For constant G, the actual temperature in the sample is given by

$$T_\ell(z) = T_{sol} + Gz \tag{8.42}$$

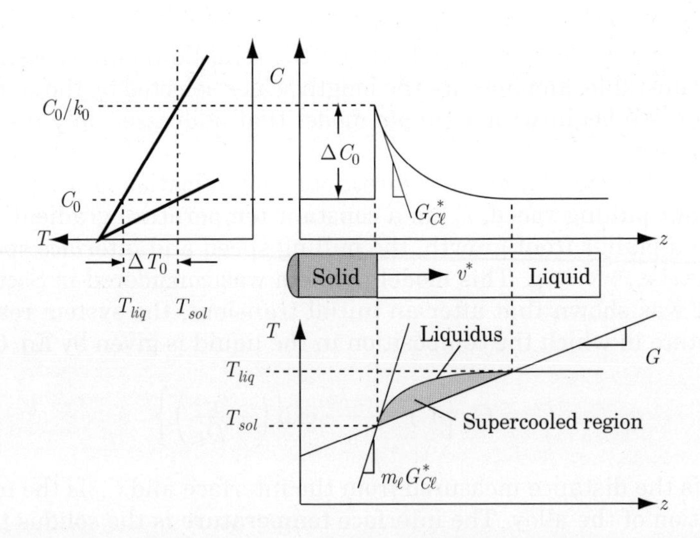

Fig. 8.11 An illustration of the constitutional supercooling criterion for morphological stability of a planar interface for a binary alloy under steady-state unidirectional solidification at constant velocity.

Since $T^* = T_\ell(z = 0) = T_{liq}(C_0/k_0) = T_{sol}(C_0)$, the stability condition reduces to

$$G \geq m_\ell G^*_{C\ell} = \frac{\Delta T_0 v_p}{D_\ell} \tag{8.43}$$

If the inequality is reversed, i.e., $G < m_\ell G^*_{C\ell}$, the liquid ahead of the interface is said to be *supercooled* due to its composition (or constitution), and for this reason, the criterion goes by the name *constitutional supercooling*. The situation is illustrated in Fig. 8.11, in which the relationship between the phase diagram, the composition profile in the liquid, and the local temperature field is demonstrated. It is convenient to separate the processing variables, G and v_p, from the material properties, D_ℓ and ΔT_0, in Eq. (8.43) to write the stability criterion as

$$\frac{G}{v_p} \geq \frac{\Delta T_0}{D_\ell} \tag{8.44}$$

8.3.3 Stability of planar front growth in binary alloys

The constitutional supercooling criterion helps us understand why the planar interface can become unstable during the directional solidification of alloys, but reveals nothing about the length scale of the microstructure that develops from the instability. As we saw in the case of free growth, the key missing ingredient is the surface energy, and its incorporation requires that we once again perform a stability analysis. In order to demonstrate the procedure, the analysis is presented in some detail for this case. Readers who are less interested in mathematical details may safely skip ahead to the key results, beginning at Eq. (8.67).

The base state solution is the steady solution given by Eq. (8.39). Using this solution as a guide, we choose to scale length on the characteristic diffusion length $L = D_\ell/v_p$. We scale time on the characteristic diffusion time $t_c = L^2/D_\ell = D_\ell/(v_p)^2$. Note that in the base state, the interface speed $v^* = v_p$, however, this is not the case when the interface is perturbed. Let us define the following dimensionless variables:

$$\xi = \frac{x}{L};\ \zeta = \frac{z}{L};\ \tau = \frac{t}{t_c};\ \theta_{s,\ell} = \frac{T_{s,\ell} - T_{sol}(C_0)}{GL};\ \mathcal{C}_{s,\ell} = \frac{C_0/k_0 - C_{s,\ell}}{\Delta C_0} \tag{8.45}$$

Substituting these definitions into the governing equations for temperature and solute (given in Sect. 5.2.3) yields

$$\frac{D_\ell}{\alpha_s}\left(\frac{\partial \theta_s}{\partial \tau} - \frac{\partial \theta_s}{\partial \zeta}\right) \approx 0 = \nabla^2 \theta_s \tag{8.46}$$

$$\frac{D_\ell}{\alpha_\ell}\left(\frac{\partial \theta_\ell}{\partial \tau} - \frac{\partial \theta_\ell}{\partial \zeta}\right) \approx 0 = \nabla^2 \theta_\ell \tag{8.47}$$

$$\frac{\partial \mathcal{C}_\ell}{\partial \tau} - \frac{\partial \mathcal{C}_\ell}{\partial \zeta} = \nabla^2 \mathcal{C}_\ell \tag{8.48}$$

$$\left(\frac{\partial \mathcal{C}_s}{\partial \tau} - \frac{\partial \mathcal{C}_s}{\partial \zeta}\right) = \frac{D_s}{D_\ell}\nabla^2 \mathcal{C}_s \approx 0 \tag{8.49}$$

where we have applied the usual observation that $D_s \ll D_\ell \ll \alpha_{s,\ell}$. Eqs. (8.46) and (8.47) imply that the temperature field does not vary with time, sometimes referred to as the "frozen temperature approximation", which can be seen to be a result of scaling the equations. The dimensionless form of the base state steady solution for the composition in the liquid becomes

$$\mathcal{C}_\ell^0 = 1 - e^{-\zeta} \tag{8.50}$$

When the interface is flat, the temperature field is one-dimensional, and Eqs. (8.46) and (8.47) can be easily integrated to show that the temperature field is linear. The slope must be the imposed temperature gradient G. The Gibbs-Thomson condition for the flat interface reveals that $T_s(z=0) = T_\ell(z=0) = T_f + m_\ell C_0/k_0$. We thereby have

$$T = T_f + \frac{m_\ell C_0}{k_0} + Gz \tag{8.51}$$

Now, substituting $\zeta = z/L$ and rearranging terms gives

$$\theta_s = \theta_\ell = \frac{T - T_f - m_\ell C_0/k_0}{GL} = \zeta \tag{8.52}$$

Suppose now that the interface is perturbed from its base state $z = 0$ to $z = H^*(x,t)$, as shown in Fig. 8.12(b). The perturbations are assumed to be infinitesimal, and simply magnified for clarity. The interface velocity of the perturbed interface is given by

$$v^* = v_p + \frac{\partial H^*}{\partial t} \tag{8.53}$$

In scaled form, the definition of $\mathcal{H}^* = H^* v_p/D_\ell$, and $\mathbf{v}^* = v^*/v_p$ gives

$$\mathbf{v}^* = 1 + \frac{\partial \mathcal{H}^*}{\partial \tau} \tag{8.54}$$

Scaling the equations revealed the temperature field as being independent of time. Moreover, the solid composition simply follows that of the

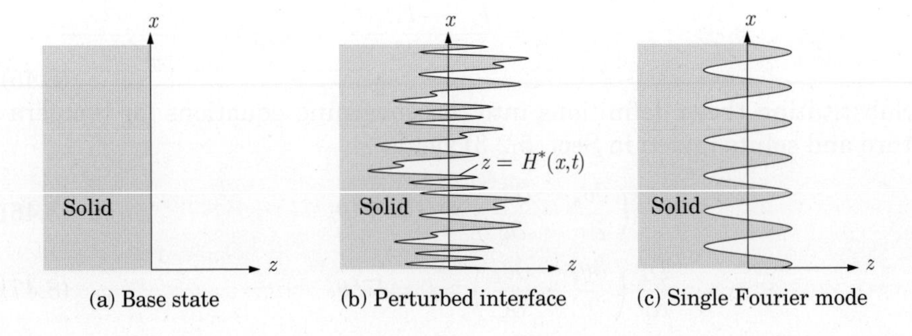

Fig. 8.12 A schematic view of the interface. (a) A flat interface in the base state. (b) A perturbed interface consisting of multiple Fourier modes. (c) A single Fourier mode.

liquid. It is therefore necessary to examine only the composition field in the liquid and the interface position, which we now write as we did for the sphere, in the form of a sum of the base state and a small perturbation

$$\mathcal{C}_\ell = \mathcal{C}_\ell^0 + \mathcal{C}_\ell^1 \tag{8.55}$$

$$\mathcal{H}^* = 0 + \mathcal{H}^{1*} \tag{8.56}$$

Note that the interface is flat in the base state, which implies that $\mathcal{H}^{0*} = 0$. Substituting Eq. (8.55) into Eq. (8.48) and eliminating $\mathcal{C}_\ell^0$ gives

$$\frac{\partial \mathcal{C}_\ell^1}{\partial \tau} - \frac{\partial \mathcal{C}_\ell^1}{\partial \zeta} = \nabla^2 \mathcal{C}_\ell^1 = \frac{\partial^2 \mathcal{C}_\ell^1}{\partial \xi^2} + \frac{\partial^2 \mathcal{C}_\ell^1}{\partial \zeta^2} \tag{8.57}$$

Since the base state solution satisfies the boundary condition $\mathcal{C}_\ell^0 \to 1$ as $\zeta \to \infty$, we require that

$$\mathcal{C}_\ell^1(\infty) = 0 \tag{8.58}$$

Boundary conditions on the interface are obtained by expanding the various entities and boundary conditions in a Taylor series, and then linearizing by neglecting products of perturbations. The details are left as an exercise. The result of this process is

$$\mathcal{C}_\ell^*(\mathcal{H}^*) = \mathcal{C}_\ell^1(0) + \mathcal{H}^* \tag{8.59}$$

$$\left.\frac{\partial \mathcal{C}_\ell^1}{\partial \zeta}\right|_0 = (k_0 - 1)\mathcal{C}_\ell^1(0) + k_0\mathcal{H}^* + \frac{\partial \mathcal{H}^*}{\partial \tau} \tag{8.60}$$

where Eq. (8.50) has been used to evaluate the base state and its derivative. Notice that the terms involving $\mathcal{C}_\ell^1$ are evaluated on $\zeta = 0$, which is sometimes referred to as "transferring" the boundary condition to the original interface. It is actually a result of neglecting higher order terms that appear as products of perturbations, a process known as linearization. The linearized form of the Gibbs-Thomson condition becomes

$$\mathcal{C}_\ell^1(0) = \left(\frac{1}{M} - 1 - S\frac{\partial^2}{\partial \xi^2}\right)\mathcal{H}^* \tag{8.61}$$

where the *morphological number*, M, and the *surface number*, S, are given by

$$M = \frac{\Delta T_0 v_p}{G D_\ell}; \quad S = \frac{\Gamma_{s\ell} v_p^2}{M G D_\ell^2} = -\frac{\Gamma_{s\ell} v_p k_0}{D_\ell m_\ell C_0 (1 - k_0)} = \frac{d_0^C}{D_\ell / v_p} \tag{8.62}$$

Note that M is always positive, regardless of the sign of m_ℓ. One can see that S represents the ratio of two length scales: the capillary length $d_0^C = \Gamma_{s\ell}/\Delta T_0$, and the characteristic length for diffusion D_ℓ/v_p. As an example, consider an alloy of Al-2 wt%Cu, solidifying at a velocity of $10\,\mu\text{m s}^{-1}$ in a temperature gradient of 100 K m^{-1}. In this case, we would obtain $d_0^c = 7.5 \times 10^{-9}$ m and $D_\ell/v_p = 3 \times 10^{-4}$ m. The dimensionless groups are $S = 2.5 \times 10^{-5}$ and $M = 5.3$. We continue to use this alloy in the examples as we take the analysis further.

Since any perturbation on the interface can be written as a Fourier series, and the governing equations are linear, it is sufficient to consider the behavior of a single Fourier mode, such as the one illustrated in Fig. 8.12(c). The solution in terms of such "normal modes" can thus be expressed as

$$\mathcal{C}_\ell^1 = A(\zeta)\exp\{\sigma_n\tau + in\xi\} \tag{8.63}$$
$$\mathcal{H}^* = \mathcal{H}_1\exp\{\sigma_n\tau + in\xi\} \tag{8.64}$$

where the wavenumber n determines the length scale of the perturbations for mode n, i.e., $\lambda_n = 2\pi/n$. As we saw in the stability analysis of the sphere, stability is indicated by the sign of σ_n. When any $\sigma_n > 0$, the perturbation grows exponentially in time, and the base state solution is unstable. Similarly, $\sigma_n < 0$ for all n indicates stability of the base state, and for $\sigma_n = 0$, the system is neutrally stable. Substituting the normal mode forms into Eq. (8.57) yields an ordinary differential equation for $A(\zeta)$, which can be solved along with the boundary conditions to give

$$\mathcal{C}_\ell^1 = B\exp\left[\left(-\frac{1}{2} - \frac{1}{2}\sqrt{1 + 4(\sigma_n + n^2)}\right)\zeta\right]\exp\{\sigma_n\tau + in\xi\} \tag{8.65}$$

Next, substituting $\mathcal{C}_\ell^1$ from Eq. (8.65) into Eqs. (8.60) and (8.61) leads to the following system of linear algebraic equations for B and $\mathcal{H}_0$:

$$\begin{bmatrix} 1 & -\left(\frac{1}{M} - 1 + Sn^2\right) \\ \frac{1}{2}\left(1 - \sqrt{1 + 4(\sigma_n + n^2)}\right) - k_0 & -(k_0 + \sigma_n) \end{bmatrix}\begin{Bmatrix} B \\ \mathcal{H}_1 \end{Bmatrix} = \begin{Bmatrix} 0 \\ 0 \end{Bmatrix} \tag{8.66}$$

The existence of a nontrivial solution to this homogeneous system requires the determinant of the coefficient matrix to be zero. It is most revealing to consider the *neutral stability curve*, where $\sigma_n = 0$, which leads to

$$\frac{1}{M} = 1 - Sn^2 + \frac{2k_0}{1 - 2k_0 - \sqrt{1 + 4n^2}} \tag{8.67}$$

Figure 8.13(a) shows a graph of $1/M$ versus n for several values of S, for an Al-2wt%Cu alloy. For each value of S, the stable region ($\sigma_n < 0$) corresponds to the zone above the associated neutral stability curve, and the instability corresponds to combinations of $1/M$ and n that fall below it, distinguished by the gray shading in the figure. Since M is always positive, the region $1/M < 0$ is physically unattainable. We keep this region in the graph for reasons that will be clear momentarily. At the limit of zero surface energy, $S = 0$, there are possible unstable states for all wavenumbers. Consequently, in order to ensure stability of the interface, we require $1/M > 1$. This condition is identical to the constitutional supercooling

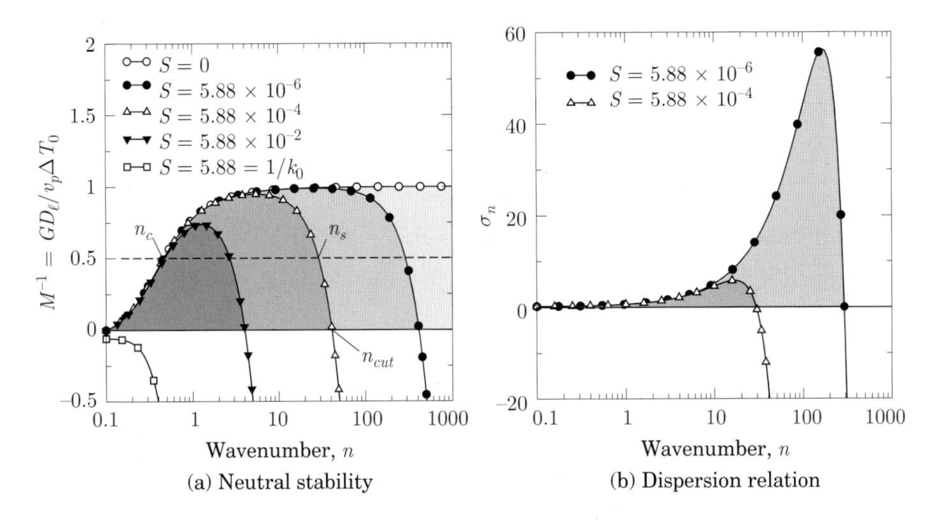

Fig. 8.13 Stability curves for planar front growth and various values of S for Al-2 wt%Cu. (a) Neutral stability curves. The shaded regions correspond to values of $1/M$ that are unstable. Negative values of $1/M$ are physically unattainable. Absolute stability corresponds to $S \geq 1/k_0$. (b) The dispersion relation, portraying the value of the growth exponent σ_n as a function of the wavenumber for two values of S. Again, the shaded regions correspond to the unstable modes.

criterion given in Eq. (8.44). As S increases, the range of unstable wavenumbers becomes smaller, and eventually for a large enough S, there are no positive values of $1/M$. This is called the *absolute stability limit*. By differentiating Eq. (8.67) twice, one can demonstrate that the absolute stability limit occurs for $S = 1/k_0$, or $v_p^a = D_\ell \Delta T_0 / k_0 \Gamma_{s\ell}$.

It is also instructive to compute the *dispersion relation*, defined as the growth rate of the disturbance σ_n as a function of the wavenumber n, for fixed values of M and S. This can be obtained fairly easily by setting the determinant of the coefficient matrix to zero in Eq. (8.66). Assuming that $n^2 \gg \sigma_n$, we have

$$\sigma_n \approx k_0 - n \left(\frac{1}{M} - 1 + S n^2 \right) \tag{8.68}$$

The result is shown in graphical form in Fig. 8.13(b) for the Al-2wt%Cu alloy, for $M = 2$ and two different values of S. Note that the numerical results validate the assumption that $n^2 \gg \sigma_n$.

Let us now try to make the results less abstract. If one selects particular values for G and v_p in an experiment, then for any given alloy, the values of M and S can be computed with Eq. (8.62). As an example, suppose that G, v_p and the alloy are chosen such that $1/M = 0.5$ and $S = 5.88 \times 10^{-4}$. Figure 8.13(a) shows that there is a range of unstable wavenumbers between n_c and n_s. Let us now estimate the corresponding range of unstable wavelengths. The value of n_c is very close to the constitutional supercooling limit ($S = 0$), so we set $S = 0$ in Eq. (8.67) and solve for n_c, which gives

$$n_c = \frac{1}{2}\left[\left(1 - \frac{2k_0}{1-M}\right)^2 - 1\right]^{1/2} \tag{8.69}$$

The significance of this result becomes more apparent if it is written in terms of the dimensional wavelength

$$\lambda_{max} = \frac{2\pi}{n_c}\frac{D_\ell}{v_p} = \frac{4\pi D_\ell}{v_p}\left[\left(1 - \frac{2k_0}{1-\Delta T_0 v_p/(GD_\ell)}\right)^2 - 1\right]^{-1/2} \tag{8.70}$$

We use here the subscript "max" to indicate that the minimum wavenumber $n = n_c$ corresponds to the maximum unstable wavelength. The short wavelength limit, λ_{min}, corresponds to n_s, which for most values of S occurs at $n_s \gg 1$. In that case, we can neglect the last term in Eq. (8.68) and solve for n_s, which gives the following result:

$$n_s = \sqrt{\frac{M-1}{MS}}; \quad \lambda_{min} = \frac{2\pi}{n_s}\frac{D_\ell}{v_p} = 2\pi\sqrt{\frac{\Gamma_{s\ell}}{m_\ell G^*_{C\ell} - G}} \tag{8.71}$$

The set of unstable wavelengths is plotted in Fig. 8.14, using values for an Al-2wt%Cu alloy and $G = 10\ \mathrm{K\,mm^{-1}}$. There is a minimum velocity $v_p^c = GD_\ell/\Delta T_0$ corresponding to the constitutional supercooling limit, and a maximum velocity $v_p^a = D_\ell \Delta T_0/k_0\Gamma_{s\ell}$ that represents absolute stability.

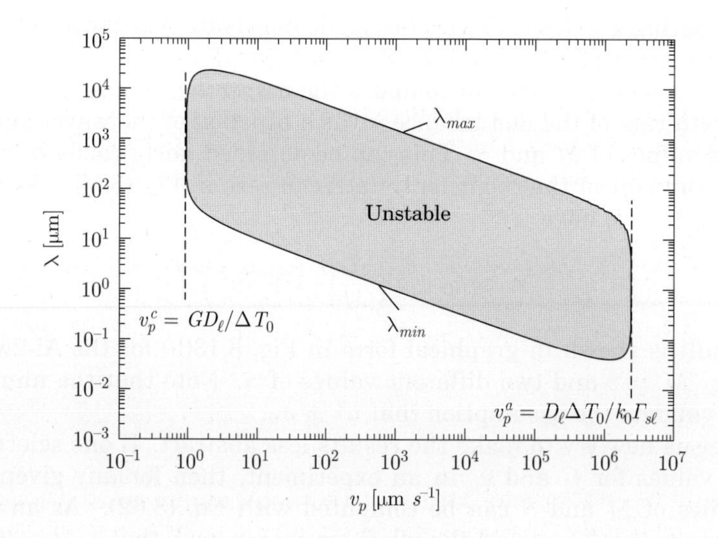

Fig. 8.14 The computed region of unstable wavelengths as a function of the pulling speed for an Al-2wt%Cu alloy. $\Delta T_0 = 32.91$ K, $D_\ell = 3 \times 10^{-9}\mathrm{m^2/s}$, $k_0 = 0.17$, $\Gamma_{s\ell} = 2.4 \times 10^{-7}$ mK and $G = 10^4\ \mathrm{Km^{-1}}$.

The curve has two branches, corresponding to λ_{min} and λ_{max}. Note that the approximations leading to the computation of λ_{min} and λ_{max} are no longer valid near absolute stability, where $n \to 0$. To obtain that portion of the curve, one can solve Eq. (8.68) numerically. It is worth noting that for our Al-2wt% Cu alloy, $v_p^a = 2.42\,\mathrm{m\,s^{-1}}$, which is indeed a very high velocity, usually attainable only in very thin samples quenched by a massive chill.

Finally, we note that there is a cutoff wavenumber n_{cut} corresponding to the zero-crossing of the neutral stability curve, which, for values of $S \ll 1/k_0$, i.e., far from the absolute stability limit, occurs at $n_{cut} \gg 1$ (see Fig. 8.13a). We may then compute n_{cut} by neglecting the first term (k_0) term on the right-hand side of Eq. (8.68) and setting both $1/M$ and the left-hand side to zero, with the result

$$n_{cut} \approx \frac{1}{\sqrt{S}} = \sqrt{\frac{D_\ell/v_p}{d_0^c}} \tag{8.72}$$

The corresponding wavelength λ_{cut} is given by

$$\lambda_{cut} = \frac{2\pi}{n_{cut}}\left(\frac{D_\ell}{v_p}\right) = 2\pi\sqrt{d_0^C\,(D_\ell/v_p)} \tag{8.73}$$

Thus, the analysis once again gives a characteristic length scale that is the geometric mean of the two important characteristic length scales, d_0^c (surface energy) and D_ℓ/v_p (diffusion).

Figure 8.14 provides a succinct summary of all of the results regarding the stability of directional solidification of a binary alloy with a planar front. For a fixed temperature gradient G:

- The planar interface is stable below a minimum velocity given by the constitutional supercooling limit $v_p^c = GD_\ell/\Delta T_0$.

- As the pulling speed, v_p, is increased, the interface may become unstable over a range of wavelengths that depends on v_p. The minimum wavelength λ_{min}, given by Eq. (8.71), corresponds to the balance between the destabilization of the interface due to composition and its stabilization by surface energy. The long wavelength limit λ_{max}, given by Eq. (8.70), corresponds to constitutional supercooling ahead of the interface in the limit of zero surface energy. The magnitude of the unstable wavelengths becomes smaller as the velocity increases.

- At a sufficiently high velocity, $v_p^a = D_\ell\Delta T_0/k_0\Gamma_{s\ell}$, referred to as absolute stability, the interface is once again stable. In this case, stability is due to the wavelength decreasing more slowly than D_ℓ/v_p.

The analysis presented here was based on numerous assumptions, such as the thermal conductivity being constant and equal in both phases, and neglecting latent heat. These phenomena can be included in the analysis, but the corresponding result would present essentially the same physics as the simplified problem.

8.4 GROWTH OF A NEEDLE CRYSTAL

8.4.1 General observations

The analysis in the previous sections was confined to the initial stages of the instability of either a spherical particle or a planar interface. As discussed in Sect. 8.1, as growth proceeds beyond the initial instability, the preferred crystallographic growth directions become dominant, eventually forming fully-developed dendrites. In this section, we focus on the dynamics of the dendrite tip, with the objective of understanding the relationship between the growth conditions, as measured by the undercooling, and the resultant microstructural features, such as the tip radius and the growth rate. The discussion is here in the context of thermal dendrites, but the extension to isothermal growth of alloy dendrites is straightforward, and follows the same path as the one taken when analyzing the sphere. Section 8.4.2 provides an approximate analysis of the problem that includes both thermal and solutal transport. The purpose of the present section is to provide a qualitative description of the theory of dendrite growth, to reveal the relationship between processing conditions, physical parameters and microstructural length scales. It is useful to follow the chronological development when studying this problem, as it helps us understand the role of the various physical phenomena.

The basic model is the constant velocity solidification of a paraboloidal "needle crystal," first introduced in Sect. 5.3.2. The reader may find it useful to review this section, in particular Fig. 5.12, before continuing. The principal assumptions and results include the following: (i) The solid is a parabola (in 2-D) or a paraboloid of revolution (in 3-D) of tip radius R_{tip}, growing in a shape-preserving manner at constant velocity v^* into an infinite undercooled melt, whose far-field temperature, T_∞, is constant. In the simplest model, the effect of curvature on the melting temperature is neglected, leading to the solid being isothermal with temperature T_f; and (ii) The transport solution for this problem, derived in Sect. 5.3.2, is given by

$$\theta_\ell = \frac{\sqrt{\pi \mathrm{Pe}}}{\Delta} \exp(\mathrm{Pe}) \mathrm{erfc}\left(\xi\sqrt{\mathrm{Pe}}\right) \qquad \text{2-D}, \xi \geq 1 \qquad (8.74)$$

$$\theta_\ell = \frac{\mathrm{E}_1(\mathrm{Pe}\xi^2)}{\mathrm{E}_1(\mathrm{Pe})} \qquad \text{3-D}, \xi \geq 1 \qquad (8.75)$$

where $\Delta = c_p(T_f - T_\infty)/L_f$ is the dimensionless undercooling, $\mathrm{Pe} = R_{tip}v^*/2\alpha_\ell$ is the Péclet number, ξ is the paraboloidal coordinate that follows the shape of the interface, and $E_1(\cdot)$ is the exponential integral. The interface corresponds to $\xi = 1$, and evaluating the boundary condition there gives rise to the following results in 2-D and 3-D:

$$\Delta = \mathrm{Iv}_{2D}(\mathrm{Pe}) = \sqrt{\pi \mathrm{Pe}}\ \exp(\mathrm{Pe}) \mathrm{erfc}\left(\sqrt{\mathrm{Pe}}\right) \qquad \text{2-D} \qquad (8.76)$$

$$\Delta = \mathrm{Iv}_{3D}(\mathrm{Pe}) = \mathrm{Pe} \exp(\mathrm{Pe}) \mathrm{E}_1(\mathrm{Pe}) \qquad \text{3-D} \qquad (8.77)$$

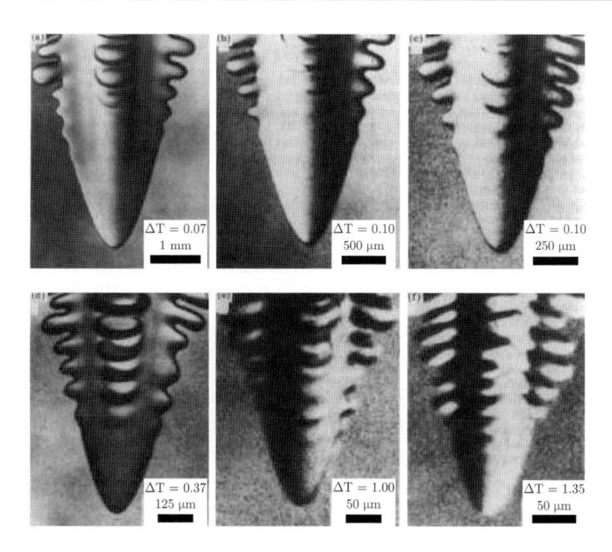

Fig. 8.15 Experimental observations of pure succinonitrile solidifying in an undercooled melt. The undercooling ranges form 0.07 K in the upper left to 1.35 K in the lower right, and each image has been scaled by a factor proportional to its undercooling. (Reproduced with permission from ref. [14].)

The puzzling aspect of the Ivantsov solutions is that they do not yield unique values for v^* and R_{tip}. Rather, a solution is obtained only in terms of their product, expressed in the Péclet number, Pe. Thus, for any given undercooling ΔT, there is an entire family of solutions, $v^* R_{tip} = \text{constant}$. The indeterminate nature of the solution contradicts experimental observations. Figure 8.15 shows a series of micrographs taken from carefully controlled experiments by Glicksman and coworkers [14; 18], who observed pure succinonitrile growing into an undercooled melt. For a given undercooling, the dendrites always select a particular growth velocity and tip radius. As the undercooling increases, the velocity increases and the tip radius decreases.

One phenomenon that is missing from the solutions in Eqs. (8.74)-(8.77) is the surface energy. The Ivantsov solution was obtained by assuming that the solid is isothermal at T_f, which clearly does not satisfy the Gibbs-Thomson condition. The correction is largest near the tip, where the curvature is maximum. The incorporation of curvature in the analysis is difficult because, when it is included, the shape is no longer a paraboloid, and the transport solution, which relied on the transformation to paraboloidal coordinates no longer applies. This fact delayed progress on this problem for quite some time.

A number of researchers tried to “patch” the Ivantsov solution, gradually adding more physics to the problem in an attempt to explain the experimental observations. Let us follow a timeline, in order to observe how the various phenomena affect the solution, and also to determine how the combination of analysis and computer simulation can increase our understanding of a complex problem.

Temkin [44] developed an approximate solution in which he retained the basic paraboloid shape, characterized by the tip radius R_{tip}. He then applied the Gibbs-Thomson condition to compute the reduced tip temperature, $T_{tip} = T_f - 2\Gamma_{s\ell}/R_{tip}$. This allowed him to evaluate the flux condition at the tip, and to compute the velocity based on the reduced undercooling $T_{tip} - T_\infty$. The analysis leads to a minimum tip radius $R_{tip}^{min} = 2\Gamma_{s\ell}/\Delta T$, for which $T_{tip} = T_\infty$, where the velocity goes to zero. The locus of possible pairs (v^*, R_{tip}) is often illustrated in a log-log plot, as shown in Fig. 8.16. As can be seen, the Ivantsov solution, for which Pe is constant, is a straight line with slope (−1) in such a plot. All (v^*, R_{tip}) pairs on the heavy line represent solutions to the transport equations. Temkin's correction, presented as a long-dashed line on the same graph, produces the extremum indicated in Fig. 8.16. This extremum was thought to be the selected "operating state" for the dendrite until careful experiments revealed that the actual operating state was quite far (several orders of magnitude in both v^* and R_{tip}) from this value, as indicated by the symbol marked "Experiment."

Nash and Glicksman [36] noted that Temkin's solution did not satisfy the Gibbs-Thomson condition everywhere, and also that when this condition was applied, the dendrite could no longer be isothermal. Consequently, conduction in the solid must also be considered. They developed an integral equation solution using the Green's function for the heat equation, satisfying the Gibbs-Thomson equation with isotropic surface energy on the interface, and assuming that, far from the tip, where curvature is

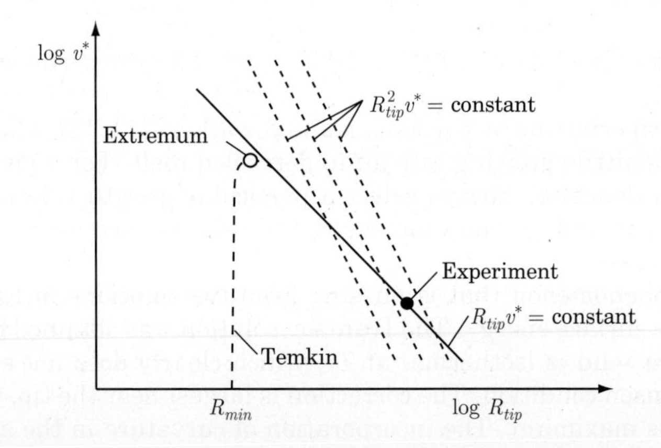

Fig. 8.16 A schematic representation of the various models for determining the "operating state" of a needle crystal. The heavy solid line shows the Ivantsov transport solution $R_{tip}v^* = \text{constant}$. The dashed line corresponds to Temkin's correction to this solution, characterized by an extremum marked with an open circle. The filled circle indicates experimental observations in succinonitrile. The set of short-dashed lines corresponds to the predictions of various stability theories, which indicate that $R_{tip}^2 v^* = \text{constant}$.

negligible, the solution converges to the Ivantsov solution. The method iteratively generates a self-consistent interface shape and thermal field. Unfortunately, their results were only marginally different from those of Temkin, still resulting in a continuum of $v^* - R_{tip}$ pairs, with a minimum at a certain value of R_{tip}. The integral equation approach will turn out to be very important a little later.

Oldfield [37] was the first to attempt to investigate the dynamics of the growth of the parabolic needle crystal with numerical simulation. He began with an Ivantsov parabola as the initial shape, then integrated forward in time using the Gibbs-Thomson condition on the interface and the Ivantsov solution far away as boundary conditions to find the temperature field. He then advanced the interface using the Stefan condition. Note that this is essentially a numerical implementation in transient form of the Nash-Glicksman approach, but it also permits the interface shape to evolve. He found the surprising result that the needle crystal shape with surface energy was unstable. Rather than continuously propagating with self-similar shape, the tip continually split in the simulations. Oldfield then presented a heuristic stability argument based on the balance between the perturbation due to surface energy in the Gibbs-Thomson equation and the disturbance to the external temperature solution, which had the form $R_{tip}^2 v^* = cst$. This provides a second set of lines in the plot of R_{tip} vs. v^*, shown in Fig. 8.16 as short-dashed lines. Each such curve has a unique intersection point with the transport solution line. There was as yet no theory to indicate which of the dashed lines to choose.

Langer and Müller-Krumbhaar [29] formalized and extended Oldfield's work, demonstrating that the needle crystal with surface energy is unstable for all finite values of (isotropic) surface energy. Based on the observation in other solidification problems (see for example the preceding section and the exercises at the end of this chapter) that the competition of surface energy and diffusion produces a characteristic length scale, Langer and Müller-Krumbhaar conjectured that

$$R_{tip} = \frac{1}{\sigma^*}\sqrt{d_0 \frac{\alpha_\ell}{v^*}} \tag{8.78}$$

where $d_0 = c_{p\ell}\Gamma_{s\ell}/L_f$ is the capillary length, introduced earlier in the stability analysis for the sphere in an undercooled melt. The constant σ^* is usually called the *selection constant*. By squaring both sides and rearranging terms, we obtain a relation of the form given by Oldfield:

$$\frac{R_{tip}^2 v^*}{d_0 \alpha_\ell} = \left(\frac{1}{\sigma^*}\right)^2 \tag{8.79}$$

Langer and Müller-Krumbhaar set $\sigma^* = 1/(2\pi)$, motivated by the value obtained for the minimum wavelength found in the morphological stability analysis of planar front growth, Eq. (8.73). This value turns out to be quite close to that obtained in Glicksman's experiments. Langer and Müller-Krumbhaar's analysis goes by the name of *marginal stability*, and Eq. (8.78), with $\sigma^* = 1/(2\pi)$, is referred to as the *marginal stability criterion*.

There are various attractive features for this theory: it includes a result from a stability analysis (although that analysis is for a planar front rather than a paraboloid), it provides an analytical expression that is easy to evaluate, and it is in reasonable agreement with experimental data. However, it is based on a solution that was demonstrated to be unconditionally unstable, which is of course a source of concern. The theory also fails to explain the fact that, in practice, needle crystals always grow in particular crystallographic directions. The principal difficulty for making further progress resided in the mathematics of the full diffusion problem in an infinite domain with anisotropic surface energy being quite complicated, and it was therefore difficult to find a basic state to test for stability.

Two groups independently developed approximate models that were more tractable, yet had similar dynamics to the real problem. Ben-Jacob, et al. [4; 5] used a boundary layer model, restricting the diffusion to a thin layer near the interface, whereas Kessler et al. [8; 25] used a more geometric model. There is no stable solution in either model for isotropic surface energy, as Oldfield and Langer and Müller-Krumbhaar found. However, when the surface energy is *anisotropic*, both models find that, instead of a continuous spectrum of solutions depending only on the Péclet number, there is a family of solutions corresponding to discrete (v^*, R_{tip}) pairs. Further, of these solutions, only one is stable, represented by the fastest growing mode. Both models also lead to an expression similar to the one given by Oldfield and Langer and Müller-Krumbhaar, namely that $R_{tip}^2 v^* = cst$, albeit from significantly different approaches.

The extension of the stability analysis to a model including the full diffusion equation was done numerically by Kessler and Levine [26]. They employed an analytical technique first developed by Vanden-Broeck [47] for the related problem of "fingers" developing as one fluid invades another having a different viscosity, the so-called Saffman-Taylor problem. The mathematics of this analysis is beyond the needs of the present text, but the essentials of the argument can be understood as a simple extension of the Nash-Glicksman method to the case of anisotropic surface energy. One imposes the Ivantsov solution at a large (infinite) distance from the tip, and then determines the tip shape using the Nash-Glicksman integral form with anisotropic surface energy. The shape computed in this way generally displays a cusp (non-zero slope) at the tip. At a unique (v^*, R_{tip}) pair, the cusp reduces to a smooth shape with zero slope at the tip. This is called the *microscopic solvability* condition, and gives the theory its name. The calculations find only one stable dendrite, with a unique operating state v^* and R_{tip}. By performing a series of numerical experiments, Kessler and Levine also showed that the microscopic solvability theory leads to a relationship of the form $R_{tip}^2 v^* = C(\varepsilon)$, where ε is a measure of the anisotropy of the surface energy. Similar calculations were performed by Saito, et al. [41] and by Barbieri and Langer [2], showing that σ^* is roughly proportional to ε. This result has not yet been validated experimentally, and thus

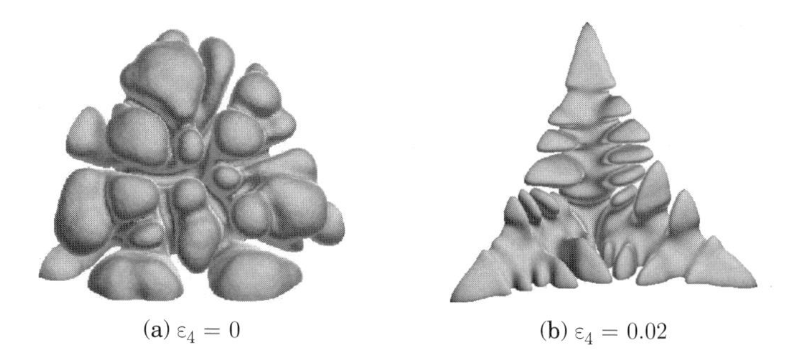

(a) $\varepsilon_4 = 0$ (b) $\varepsilon_4 = 0.02$

Fig. 8.17 Computed crystals starting from a spherical one in an undercooled melt with $\Delta = 0.5$ for (a) ε_4=0 and (b) $\varepsilon_4 = 0.02$. The calculations are performed in the octant $(x, y, z) > 0$, with zero-flux boundary conditions on all faces.

remains somewhat of an open question. It is thought that interface attachment kinetics or melt convection, not included in the microscopic solvability model, or errors in the experimental measurement of the surface energy anisotropy, may be responsible for the discrepancy between theory and experiment. To illustrate the effect of the surface energy anisotropy, Fig. 8.17 shows dendritic crystals computed using the phase-field method (described in Sect. 8.6), beginning with a small sphere in an undercooled melt with $\Delta = 0.5$, varying the strength of the cubic anisotropy ε_4. One can clearly see that even a very small amount of crystalline anisotropy changes the structure from the dense-branched morphology observed at $\varepsilon_4 = 0$ to a branched dendrite with cubic symmetry. Sidebranches form behind the tip due to the amplification of noise.

Finally, we note that this body of theoretical work can be directly applied to solutal dendrites in a similar way to that used in the analysis of the growth of spherical particles in Sect. 8.2.2. If the length scales for heat and solute transport are sufficiently disparate, the melt can be treated as isothermal, and the analysis proceeds in a similar way, with the thermal capillary number d_0 replaced by the chemical capillary number d_0^c, and the dimensionless undercooling Δ replaced by the supersaturation $\Omega = (C_\ell^{eq} - C_0)/(C_\ell^{eq}(1 - k_0))$. The problem becomes somewhat more complicated when both solute and heat transport are included, and an approximate treatment for this case is presented in Sect. 8.4.2.

Let us summarize the most important results to this point: A dendrite growing in an undercooled melt can be approximated as a "needle crystal" that is nearly parabolic in shape, except for a small but important correction at the tip, and that grows in a shape-preserving manner at constant velocity. The thermal (or solutal) field obtained by solving the transport equations admits a continuum of solutions corresponding to pairs of the growth velocity v^* and tip radius R_{tip} such that $R_{tip}v^*$ is a constant. When surface energy is included, no stable solution exists unless the surface energy has some non-zero anisotropy. This leads to a second "solvability" condition stating that $R_{tip}^2 v^*$ is constant. Together, these two criteria provide a unique value for the tip shape and growth velocity. This is particularly important because these same tip dynamics apply even when the melt is finite. This permits the solution to be used to analyze real microstructures.

8.4.2 Approximate models for growth at the dendrite tip

The body of work described in the preceding section provides a good understanding of the physics of dendritic growth. It is not in a particularly convenient form, however, for it to be applied to the prediction of microstructure formation under less ideal conditions, e.g., where the melt is not infinite, where both solute and thermal diffusion are important, etc. Lipton et al. [30] developed a model that incorporates most of the essential phenomena, and for which there is a relatively simple analytical procedure for computing a unique pair of (v^*, R_{tip}) for a given fixed undercooling, or a unique pair $(R_{tip}, \Delta T)$ with an imposed velocity, such as in constrained growth.

The model is shown schematically in Fig. 8.18. The solid dendrite is considered to be a paraboloid of revolution, with temperature T^* and composition C_ℓ^* in the liquid at the tip. The temperature and composition are unknown *a priori*, but are assumed to correspond to a point on the liquidus curve for the alloy, reduced by the undercooling required to maintain the tip curvature, $\Delta T_R = 2\Gamma_{s\ell}/R_{tip}$. The temperature and solute fields are given by their respective Ivantsov solutions, Eq. (8.75), and are

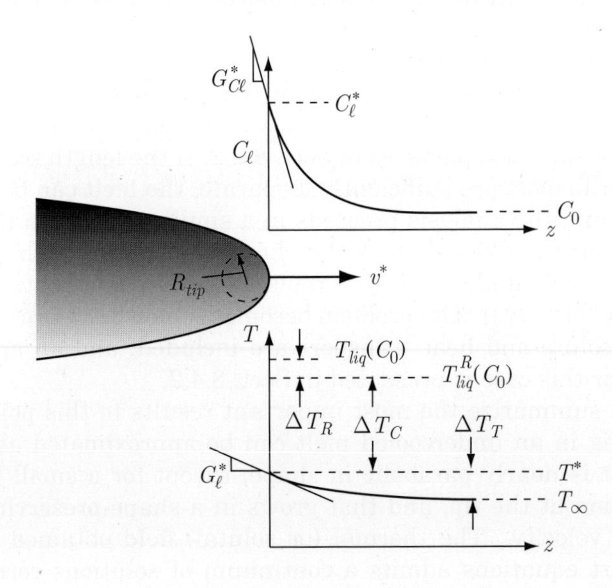

Fig. 8.18 A schematic illustration of the LGK model, portraying the curvature undercooling, ΔT_R, solutal undercooling, ΔT_C, and thermal undercooling, ΔT_T.

repeated here using the notation of Fig. 8.18 to place them in the current context

$$\theta_\ell = \frac{T_\ell - T_\infty}{T^* - T_\infty} = \frac{\mathrm{E}_1(\mathrm{Pe}_T\xi^2)}{\mathrm{E}_1(\mathrm{Pe}_T)}; \qquad \mathcal{C}_\ell = \frac{C_\ell - C_0}{C_\ell^* - C_0} = \frac{\mathrm{E}_1(\mathrm{Pe}_C\xi^2)}{\mathrm{E}_1(\mathrm{Pe}_C)} \tag{8.80}$$

Here, ξ is the paraboloidal coordinate parallel to the interface, scaled by the tip radius, R_{tip}, the thermal Péclet number, $\mathrm{Pe}_T = v^* R_{tip}/2\alpha_\ell$, and the solutal Péclet number, $\mathrm{Pe}_C = v^* R_{tip}/2D_\ell$. It should be noted that although T^* and C_ℓ^* are as yet unknown, they have been assumed to be constant, in order for them to be used for scaling the variables. The flux conditions relate the undercooling and supersaturation to their respective Péclet numbers

$$\Delta = \frac{T^* - T_\infty}{L_f/c_p} = \mathrm{Pe}_T \exp(\mathrm{Pe}_T)\mathrm{E}_1(\mathrm{Pe}_T) = \mathrm{Iv}_{3D}(\mathrm{Pe}_T) \tag{8.81}$$

$$\Omega = \frac{C_\ell^* - C_0}{C_\ell^*(1-k_0)} = \mathrm{Pe}_C \exp(\mathrm{Pe}_C)\mathrm{E}_1(\mathrm{Pe}_C) = \mathrm{Iv}_{3D}(\mathrm{Pe}_C) \tag{8.82}$$

Next, the total undercooling $\Delta T = T_{liq}(C_0) - T_\infty$ is written as the sum of three contributions: the thermal undercooling, ΔT_T, the solutal undercooling, ΔT_C, and the curvature undercooling, ΔT_R. In this model, valid for small undercooling, the undercooling associated with attachment kinetics is neglected. An expression for ΔT is obtained using Fig. 8.18 and Eqs. (8.81) and (8.82):

$$\begin{aligned}\Delta T &= \Delta T_T + \Delta T_C + \Delta T_R \\ &= \frac{L_f}{c_p}\mathrm{Iv}_{3D}(\mathrm{Pe}_T) - m_\ell C_0 \left[\frac{(1-k_0)\mathrm{Iv}_{3D}(\mathrm{Pe}_C)}{1-(1-k_0)\mathrm{Iv}_{3D}(\mathrm{Pe}_C)}\right] + \frac{2\Gamma_{s\ell}}{R_{tip}}\end{aligned} \tag{8.83}$$

Notice that we have at this point the usual problem of one equation with two unknowns, v^* and R_{tip}.

Lipton et al. [30] resolved this issue in the same way as Langer and Müller-Krumbhaar, i.e., by assuming that R_{tip} can be estimated from the expression for λ_{min} derived in the linear stability analysis of the planar front given in Eq. (8.71). In other words, they took $R_{tip} \approx \lambda_{min} = 2\pi\sqrt{\Gamma_{s\ell}/(m_\ell G_{C\ell}^* - G_\ell^*)}$. They also found better agreement with experimental observations in succinonitrile-acetone alloys by changing the prefactor from 2π to $2\pi\sqrt{2}$. This gives

$$R_{tip} \approx 2\pi\sqrt{2}\sqrt{\frac{\Gamma_{s\ell}}{(m_\ell G_{C\ell}^* - G_\ell^*)}} \tag{8.84}$$

In order to use Eq. (8.84), we compute the temperature gradient G_ℓ^* and the solute gradient $G_{C\ell}^*$ at the dendrite tip by first differentiating the solutions for θ_ℓ and $\mathcal{C}_\ell$ in Eq. (8.80) with respect to ξ, and then evaluating at

$\xi = 1$. We then apply Eqs. (8.81) and (8.82), and finally express the result in dimensional form as

$$G_\ell^* = -\frac{2\text{Pe}_T L_f}{c_p R_{tip}}; \qquad G_{C\ell}^* = -\frac{2\text{Pe}_C C_\ell^*(1-k_0)}{R_{tip}} \tag{8.85}$$

Substituting these expressions into Eq. (8.84) and performing some algebraic manipulation gives

$$R_{tip} = 8\pi^2\Gamma_{s\ell}\left(\frac{2\text{Pe}_T L_f}{c_p} - \frac{2\text{Pe}_C m_\ell C_0(1-k_0)}{(1-(1-k_0)\text{Iv}_{3D}(\text{Pe}_C))}\right)^{-1} \tag{8.86}$$

One should note that Lipton et al. kept the prefactor 2π in Eq. (8.84), and instead obtained the necessary factor of 2 by taking half of $G_{C\ell}^*$ in Eq. (8.85). Nonetheless, the results are essentially the same.

Equations (8.83) and (8.86) can be solved iteratively for any given value of ΔT to obtain a unique pair (v^*, R_{tip}) corresponding to each composition C_0. Lipton et al. found good agreement with experimental measurements in dilute succinonitrile-acetone alloys when using the model. Since our interest in later chapters is focused more on metallic systems, we present in Fig. 8.19 similar calculations for Al-Cu alloys, displaying computed values of v^* and R_{tip} as functions of the alloy composition C_0 for two undercoolings, $\Delta T = 0.5\,\text{K}$ and 1.5K. The results show that, as C_0 increases, v^* decreases monotonically, whereas R_{tip} passes through a minimum at a very small value of C_0 before increasing monotonically

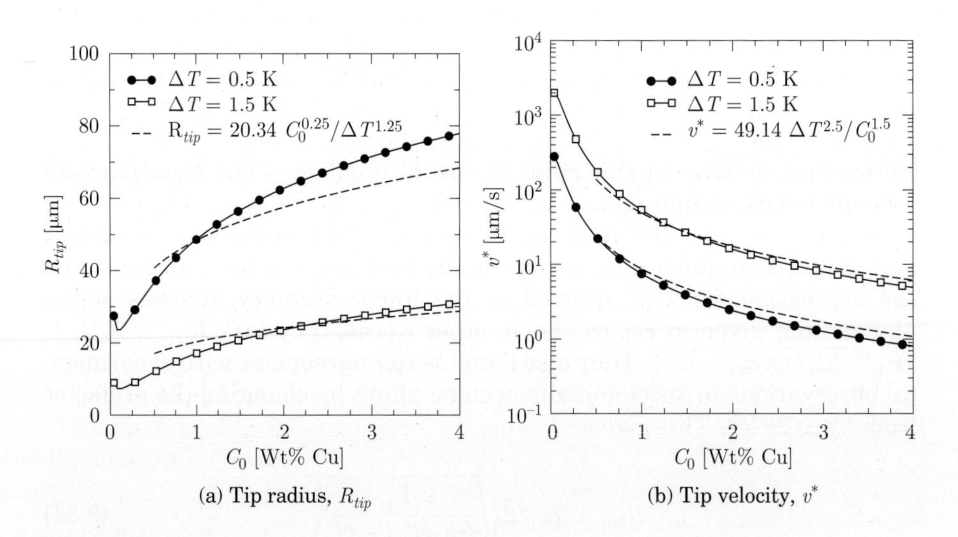

Fig. 8.19 Values of R_{tip} and v^*, computed with Eqs. (8.83) and (8.86), for Al-Cu alloys at two undercoolings. The dashed lines represent the approximate forms considering only the solute, given in Eqs. (8.91) and (8.92), valid at low undercooling for compositions that are not too small.

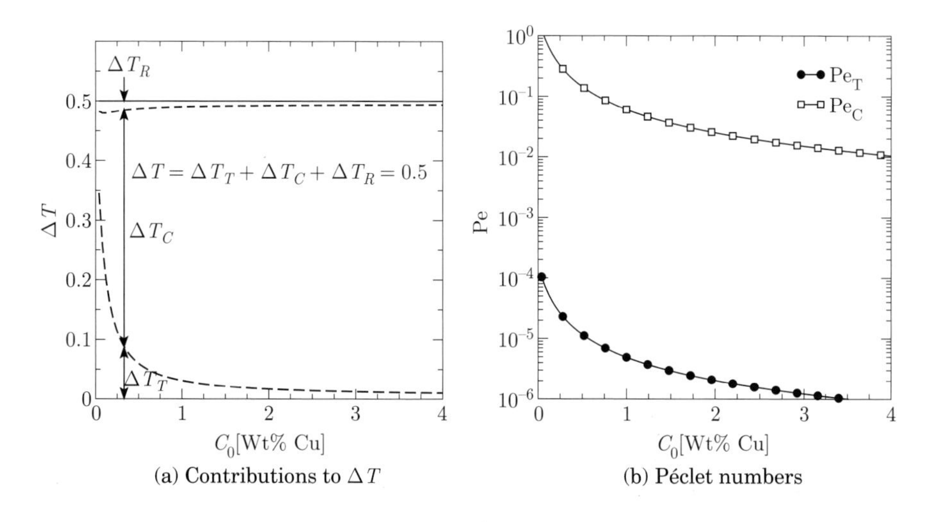

(a) Contributions to ΔT

(b) Péclet numbers

Fig. 8.20 Calculation details for Al-Cu alloys for $\Delta T = 0.5$ K. (a) Contributions to the total undercooling. (b) Péclet numbers Pe_T and Pe_C.

thereafter. This minimum corresponds to the transition from thermal to solutal dendrites for $C_0 \gtrsim 0.1$ wt%Cu. The figures also show a further approximation, represented by dashed curves, which is developed next.

The LGK model is very commonly used in microstructure modeling since it provides a compact analytical form for obtaining the operating state. However, the repeated iterative solutions of Eqs. (8.83) and (8.86) is too costly to be used for computation in simulations of macroscale castings, where there may be millions of grains. To motivate a further approximation, the graph in Fig. 8.20(a) shows the individual contributions of ΔT_T, ΔT_C and ΔT_R to the total undercooling, evaluated for $\Delta T = 0.5$ K. One can clearly see that for $C_0 \gtrsim 0.5$ wt%Cu, ΔT_C accounts for more than 90% of the total undercooling. Moreover, Fig. 8.20(b) shows that $\mathrm{Pe}_T \to 0$ and that $\mathrm{Pe}_C \ll 1$ in this composition range. These approximations hold true for most cases of practical interest, where ΔT is limited to just a few Kelvin. (In rapid solidification processes, ΔT can be much larger, and this approximations cannot be made.)

These results suggest that we can develop a simpler form of the LGK model, valid for small undercoolings, by neglecting the contributions of thermal and curvature undercooling, i.e., the solidification is dominated by solute transport. To formalize this approximate form, we take the limit $\mathrm{Pe}_T \to 0$ and $\mathrm{Pe}_C \ll 1$, which also implies that $\mathrm{Iv}_{3D}(\mathrm{Pe}_C) \ll 1$. Applying these approximations to Eq. (8.86) and rearranging terms yields the much simpler form

$$R_{tip}^2 v^* = -\frac{8\pi^2 \Gamma_{s\ell} D_\ell}{m_\ell C_0 (1 - k_0)} \tag{8.87}$$

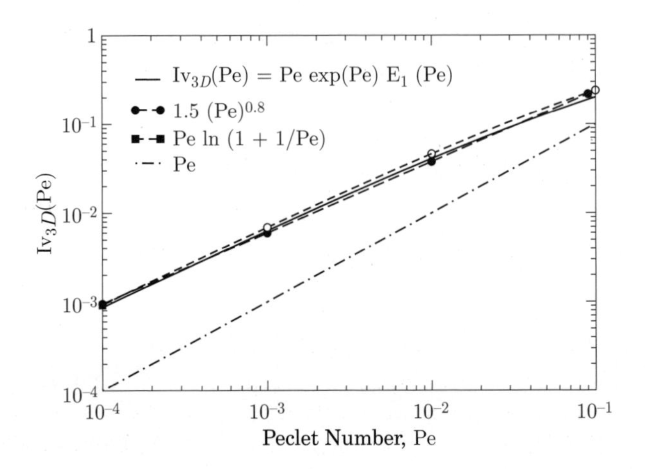

Fig. 8.21 A plot of $Iv_{3D}(Pe)$ for small Pe, along with several approximate expressions.

Equation (8.83) also takes on a less complex form after neglecting the terms associated with thermal and curvature undercooling

$$\Delta T = -m_\ell C_0(1 - k_0)Iv_{3D}(Pe_C) \tag{8.88}$$

In order to produce an even more convenient form, we would like to have an analytical expression for $Iv_{3D}(Pe_C)$ allowing v^* and R_{tip} to be more easily separated. Figure 8.21 presents several candidates, from which we choose one that is both simple and accurate for $Pe_C \lesssim 0.1$

$$Iv_{3D}(Pe_C) \approx 1.5 Pe_C^{0.8} \tag{8.89}$$

Substituting this expression into Eq. (8.88) and rearranging the terms gives

$$R_{tip} v^* = 2D_\ell \left(-\frac{2\Delta T}{3m_\ell C_0(1 - k_0)} \right)^{1.25} \tag{8.90}$$

Notice that Eqs. (8.87) and (8.90) are in exactly the same form as the sketch in Fig. 8.16. One can easily solve Eqs. (8.87) and (8.90) to obtain the following approximate expressions for R_{tip} and v^*, which are plotted as dashed lines in Fig. 8.19:

$$R_{tip} = 6.64\pi^2 \Gamma_{s\ell}(-m_\ell(1 - k_0))^{0.25} \left(\frac{C_0^{0.25}}{\Delta T^{1.25}} \right) \tag{8.91}$$

$$v^* = \frac{D_\ell}{5.51\pi^2(-m_\ell(1 - k_0))^{1.5}\Gamma_{s\ell}} \left(\frac{\Delta T^{2.5}}{C_0^{1.5}} \right) \tag{8.92}$$

One should note that slightly different forms for Eqs. (8.91) and (8.92) are often presented in the literature. These are obtained by using the approximate form $\mathrm{Iv}_{3D}(\mathrm{Pe}) \approx \mathrm{Pe}$ instead of Eq. (8.89). This derivation is left to the reader in Exercise 8.8.

8.4.3 Primary dendrite arm spacing in constrained growth

One of the most important applications for the models developed in this chapter is the prediction of microstructure in castings. For equiaxed grain morphologies resulting from free growth, metallurgists measure the mean grain size and the secondary dendrite arm spacing (DAS), λ_2. For the columnar grains that develop during constrained growth, they measure the spacing between the primary trunks, λ_1, as well as, λ_2, (see Fig. 8.9). The DAS provides a sort of "frozen record" of the solidification process, because the microstructure usually does not change much in the solid state during cooling after solidification.

The combination of nucleation and growth determines the grain size in equiaxed solidification; an issue that is addressed in Chap. 11. In columnar growth, the competition between adjacent dendrites decides the primary spacing, λ_1. The model developed in the previous section can be used to predict λ_1 with the aid of a simple geometrical construct, illustrated in Fig. 8.22. The model introduces an "envelope" around the solid dendrite that encloses all of the solid. We use this concept in several places in subsequent chapters. For constrained growth, the envelope surrounding the dendrite is modeled as an ellipsoid, with major and minor axes a and b, respectively. The behavior of the tip of the ellipsoid is assumed to follow

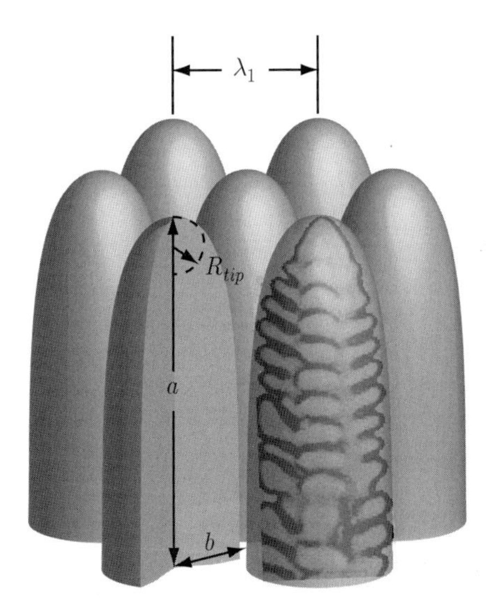

Fig. 8.22 The geometric model of a hexagonal dendrite array, displaying the model of the dendrite envelope as an ellipsoid.

the kinetics described by the simplified LGK model, obtained by assuming that the thermal and curvature undercooling are negligible. Thus, Eq. (8.87) applies. In constrained growth, the tip velocity v^* is known, so one can solve Eq. (8.87) directly for R_{tip}:

$$R_{tip} = \left(-\frac{8\pi^2 \Gamma_{s\ell} D_\ell}{m_\ell C_0 (1-k_0) v^*} \right)^{1/2} = \left(\frac{8\pi^2 \Gamma_{s\ell} D_\ell}{k_0 \Delta T_0 v^*} \right)^{1/2} \tag{8.93}$$

The length of the major axis, a, is determined by the temperature gradient and the freezing range of the alloy

$$a = \frac{T_{tip} - T_{base}}{G} \tag{8.94}$$

where T_{base} is the temperature at the base of the dendrites. Under equilibrium conditions, and provided that the composition of the alloy is below the maximum solubility of the primary phase, the freezing range is simply $T_{tip} - T_{base} = \Delta T_0$. Under non-equilibrium conditions (see. Chap. 10) and/or when C_0 exceeds the solubility limit, solidification ends by some other means, e.g., a eutectic reaction. In that case, $T_{base} = T_{eut}$. Bearing this in mind, we write $T_{tip} - T_{base} = \Delta T_0'$, and Eq. (8.94) becomes

$$a = \frac{\Delta T_0'}{G} \tag{8.95}$$

Figure 8.22 shows the dendrites represented as a hexagonal array of identical ellipsoids whose minor axes b are defined, as shown, as the distance from the axis of the dendrite to the center of the interstitial space formed between three neighboring dendrites. The primary spacing, λ_1, is defined as the distance between the axes of nearest neighbors. It is straightforward to show that $b = \lambda_1/\sqrt{3}$.

The model is completed by computing the radius of curvature at the tip of the ellipsoid

$$R_{tip} = \frac{b^2}{a} = \frac{\lambda_1^2 G}{3\Delta T_0'} \tag{8.96}$$

Substituting the expression for R_{tip} from Eq. (8.93) and solving for λ_1 gives

$$\lambda_1 = \left(-\frac{72\pi^2 \Gamma_{s\ell} D_\ell \Delta (T_0')^2}{k_0 \Delta T_0} \right)^{1/4} (v^*)^{-1/4} G^{-1/2} \tag{8.97}$$

For $\Delta T_0' \approx \Delta T$, this expression reduces to

$$\lambda_1 = \left(-\frac{72\pi^2 \Gamma_{s\ell} D_\ell \Delta T_0}{k_0} \right)^{1/4} (v^*)^{-1/4} G^{-1/2} \tag{8.98}$$

Since all variables within the parentheses are known, this term is simply a constant for any given alloy. The important aspect of Eq. (8.98) is that it predicts a measurable microstructure feature, λ_1, in terms of alloy properties, as well as a specific set of solidification conditions, v^* and G. It is common practice to correlate microstructural features with the *local solidification time* $t_f = \Delta T_0'/|\dot{T}|$, where $|\dot{T}| = Gv^*$ is the cooling rate. Thus, t_f is easily measured, e.g., by placing thermocouples in test castings. According to Eq. (8.98), such a correlation is impossible for the primary dendrite arm spacing due to G and v^* occurring with different exponents. On the other hand, the secondary dendrite arm spacing correlates well with t_f, as demonstrated in the next section.

8.4.4 Secondary dendrite arm spacing: Coarsening

Figure 8.23 portrays the region near the tips of SCN dendrites during directional solidification of a SCN-acetone alloy. Let us now focus on the evolution of λ_2, i.e., the spacing between secondary dendrite arms. It is clear from the figure that secondary arms begin to develop a few tip radii behind the tip itself. They then increase in amplitude with the distance from the tip. In this region, the solute layers associated with the primary dendrite tips do not overlap. Secondary arms therefore grow to fill the space by a mechanism that, along with natural thermal and solutal fluctuations, leads to arms of various sizes at the time the solute layers overlap. From that point, the secondary spacing changes by elimination of certain sidebranches and growth of others. This process continues all the way to the root of the primary dendrites, where one reaches the final dendrite arm spacing, λ_2. The same phenomenon occurs in equiaxed growth over the entire solidification time.

This phenomenon, where the length scale of the microstructure increases over time, is called *coarsening*. We begin by presenting a simple model developed by Kattamis and Flemings [24], illustrated in the right-hand panel of Fig. 8.23, to help explain the important phenomena in the process and identify the significant parameters. The model focuses on three neighboring sidebranches, presented as a white box on the micrograph and idealized as shown to the right. The sidebranches are treated

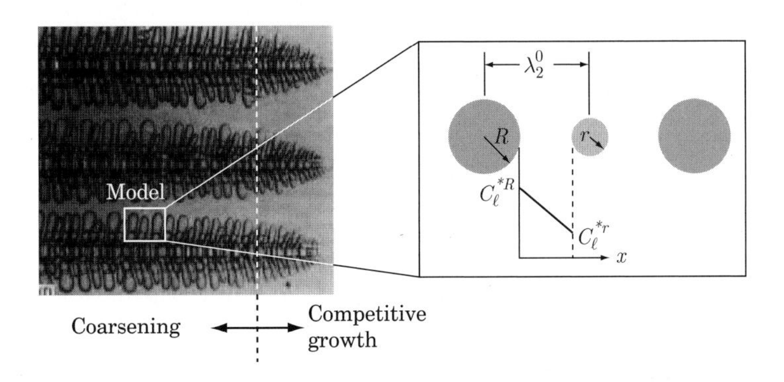

Fig. 8.23 Fully developed dendrites during directional solidification of a SCN-acetone alloy (left). (Reproduced with permission from ref. [46].) A simple model for the coarsening of neighboring dendrite arms (right).

as cylinders that have evolved to two different radii, R and r. We assume that the temperature is uniform at the length scale of this elementary volume, and that the solid and melt are in thermodynamic equilibrium. The Gibbs-Thomson equation then implies that the composition in the liquid must be different at the surface of the neighboring dendrite arms, due to them being at the same temperature but having different curvatures (see Fig. 8.7). The composition at the larger arm is expressed as C_ℓ^{*R}, and that at the smaller arm as C_ℓ^{*r}. This arrangement produces a gradient in composition, as illustrated, which in turn gives rise to a solute flux from the larger arm to the smaller one. In order to maintain local equilibrium at each interface, the solute flux arriving at the smaller arm dissolves (remelts) some of the solid whereas the larger dendrite arm produces this flux by growing. As a result, the difference $(C_\ell^{*R} - C_\ell^{*r})$ increases, and the process continues until the smaller arm is completely remelted.

For the sake of simplicity, yet retaining the essential physics, we assume the composition profile to be linear between the neighboring arms, and treat the diffusion as one-dimensional. The flux in the liquid is then given by

$$j = -D_\ell \frac{\partial C_\ell}{\partial x} \approx D_\ell \frac{C_\ell^{*R} - C_\ell^{*r}}{\lambda_2^0} \tag{8.99}$$

Taking the slope of the liquidus curve to be constant, and equating the temperature at both cylinders gives

$$T_f + m_\ell C_\ell^{*R} - \frac{\Gamma_{s\ell}}{R} = T_f + m_\ell C_\ell^{*r} - \frac{\Gamma_{s\ell}}{r} \tag{8.100}$$

Solving Eq. (8.100) for $(C_\ell^{*R} - C_\ell^{*r})$ and substituting the result in Eq. (8.99) leads to

$$j = -\frac{D_\ell \Gamma_{s\ell}}{m_\ell \lambda_2^0}\left(\frac{1}{r} - \frac{1}{R}\right) \tag{8.101}$$

This solute flux dissolves the smaller dendrite arm at a rate that can be computed from the interfacial flux balance, Eq. (4.152), which takes the following form in the context of the present problem:

$$j = -C_\ell^{*r}(1 - k_0)\frac{dr}{dt} \tag{8.102}$$

Equating the fluxes from Eqs. (8.101) and (8.102) and solving for dr/dt yields

$$\frac{dr}{dt} = \frac{D_\ell \Gamma_{s\ell}}{\lambda_2^0 m_\ell C_\ell^{*r}(1 - k_0)}\left(\frac{1}{r} - \frac{1}{R}\right) \tag{8.103}$$

Note that for $m_\ell < 0$, the expression on the right-hand side is negative, indicating dissolution of the smaller arm. The larger arm should also increase in size, but in this model we assume that R is approximately constant, which is not necessarily inconsistent if $R \gg r$. Considering the very small curvature undercooling, it is also reasonable to assume that $C_\ell^{*R} \approx C_\ell^{*r} \approx C_\ell$. This assumption is not inconsistent with the basic model, since even a small difference between C_ℓ^{*R} and C_ℓ^{*r} can produce a large flux, as the secondary arms are so close together.

Next, we assume that the composition of the liquid, C_ℓ, varies linearly from C_0 to a final composition, e.g., C_{eut}, over the solidification time, t_f, i.e.,

$$C_\ell^{*r} = C_0 + (C_{eut} - C_0)\frac{t}{t_f} \tag{8.104}$$

We now try to determine the time, $t_{r=0}$, required for the smaller arm to dissolve. By substituting Eq. (8.104) into Eq. (8.103) and integrating, we obtain

$$\int_{r_0}^{0} \frac{r dr}{1 - r/R} = \frac{\Gamma_{s\ell} D_\ell}{\lambda_2^0 m_\ell (1 - k_0)} \int_0^{t_{r=0}} \frac{dt}{C_0 + (C_{eut} - C_0)(t/t_f)} \tag{8.105}$$

The result of the integration, after some rearrangement of terms, is

$$R^2 \lambda_2^0 = \left(\frac{1}{r_0/R + \ln(1 - r_0/R)} \right) \left(-\frac{\Gamma_{s\ell} D_\ell \ln(C_\ell(t_{r=0})/C_0)}{m_\ell (1 - k_0)(C_{eut} - C_0)} \right) t_f \tag{8.106}$$

where $C_\ell(t_{r=0})$ is the liquid composition at time $t_{r=0}$. After this time, as a result of the middle arm disappearing, the initial DAS has doubled. If we choose as an example the initial values $R = \lambda_2^0/2$ and $r_0/R = 0.5$, the time at which $\lambda_2 = 2\lambda_2^0$ is given by

$$\lambda_2(t_{r=0}) = 5.5 \left(-\frac{\Gamma_{s\ell} D_\ell \ln(C_\ell(t_{r=0})/C_0)}{m_\ell (1 - k_0)(C_{eut} - C_0)} \right)^{1/3} t_{r=0}^{1/3} \tag{8.107}$$

Finally, when extending this procedure to the end of solidification, $t = t_f = \Delta T_0'/Gv^*$, we obtain

$$\lambda_2(t_f) = 5.5 \left(-\frac{\Gamma_{s\ell} D_\ell \ln(C_{eut}/C_0)}{m_\ell (1 - k_0)(C_{eut} - C_0)} \right)^{1/3} \left(\frac{\Delta T_0'}{Gv^*} \right)^{1/3} \tag{8.108}$$

This expression should be compared with Eq. (8.98), where the exponents of G and v^* are $-1/2$ and $-1/4$, respectively.

Example 8.1

We develop a map of primary and secondary dendrite arm spacings in an alloy of Al-4 wt%Cu. Contours for particular values of λ_1 are computed from Eq. (8.98), and for λ_2 from Eq. (8.108). The plots also reveal the region where planar front growth is expected to be stable, as computed from Eq. (8.43). The material properties include:

$\Gamma_{s\ell} = 2.4 \times 10^{-7}$ mK, $D_\ell = 3.0 \times 10^{-9}$ m^2 s^{-1}, $m_\ell = -3.37$ K wt%$^{-1}$, $k_0 = 0.17$ and $\Delta T_0' = 98.5$ K.

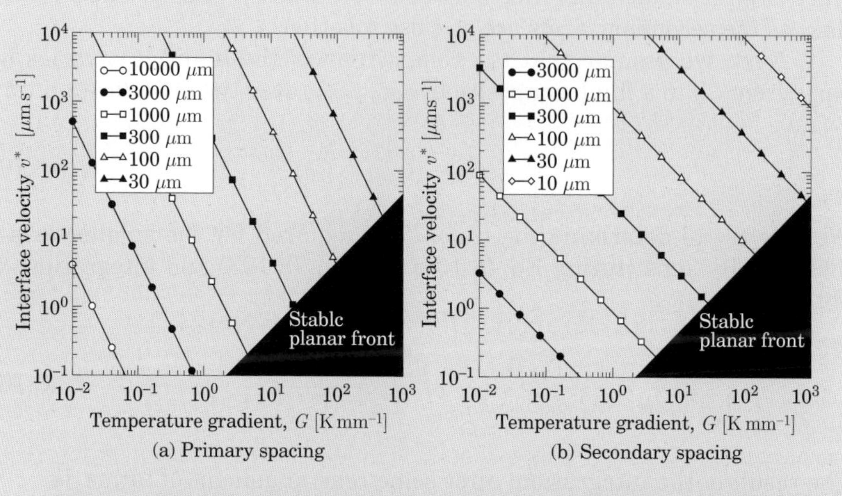

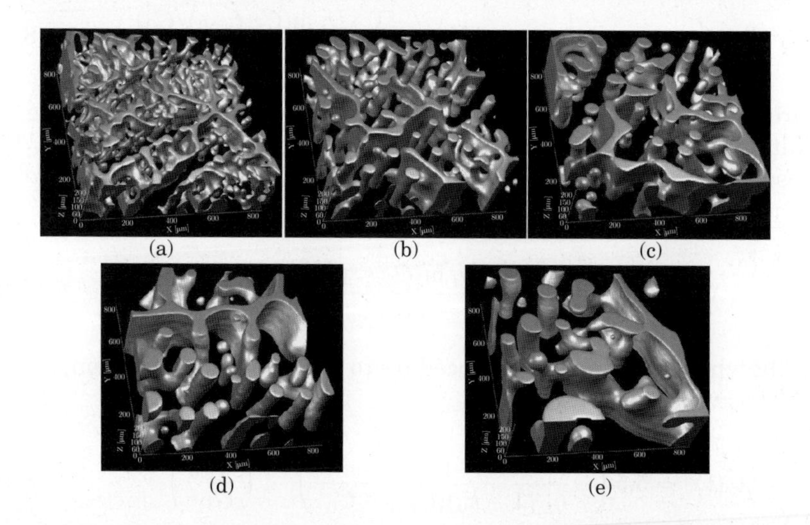

Fig. 8.24 A series of micrographs portraying the coarsening of the primary phase in an Al-15wt%Cu alloy maintained slightly above the eutectic temperature for various amounts of time. (Reproduced with permission from P. Voorhees, see ref. [32].)

Coarsening in real microstructures is far more complicated, due to interactions between many solid segments, not just adjacent arms. To provide a specific example, Fig. 8.24 shows a sequence of microstructures obtained after isothermal coarsening of an Al-15wt%Cu alloy for various times. [32] The sample was first directionally solidified, then reheated to a temperature just above T_{eut}, where it was maintained for the indicated amounts of time. The micrographs were reconstructed from a stack of images obtained by serial sectioning of quenched samples. The portion of the sample that was liquid at the time of the quench was made transparent.

Kammer and Voorhees [21] analyzed the microstructure and determined that the specific surface S_V was proportional to $t^{-1/3}$. This can be thought of as a generalization of the scaling found in our simplified model (see Eq. (8.107)). They also studied the evolution of the principal curvatures as the sample coarsened, and came to the conclusion that the microstructure evolved toward a collection of cylinders, i.e., where one of the principal curvatures is zero. This is probably due to the directional solidification process that produced the original microstructure.

8.5 CONVECTION AND DENDRITIC GROWTH

8.5.1 Convection and free growth

We have seen that the scale of the microstructure is set by competition between heat/solute diffusion and surface energy. Thus far, we have considered only conduction and diffusion as mechanisms for transport. Advection, i.e., the transport of heat and solute by fluid flow, can play a very significant, or even dominant role in transport near solidifying interfaces. Since we discuss the origins of fluid flow in other chapters, we simply recall here that flows can be induced by pouring of the melt into a mold, buoyancy-driven flows due to thermal and solutal gradients (see Chap. 14), or perhaps intentionally driven by electromagnetic forces. The present section concentrates on the effect of such flows on dendritic microstructures.

Figure 8.25(a) shows a crystal of NH_4Cl growing while settling in the melt due to its higher density. We see that the growth of the

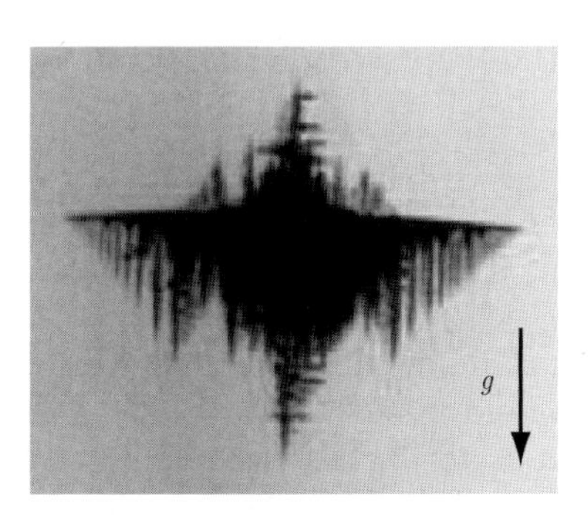

(a) Settling NH_4Cl crystal

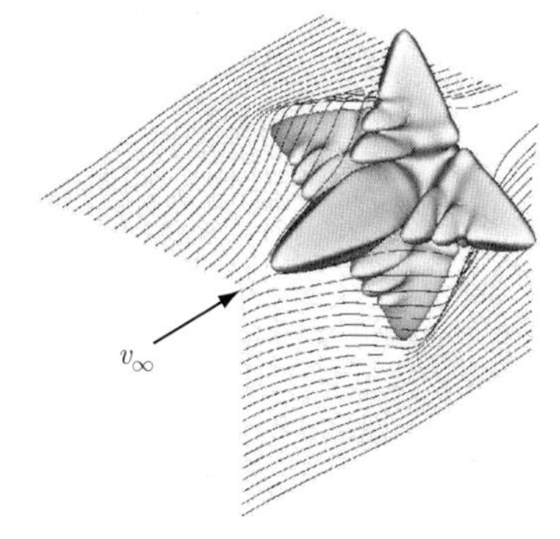

(b) Computed dendrite

Fig. 8.25 The effect of fluid flow on dendrite growth. (a) The settling of a NH_4Cl crystal in aqueous solution. (Reproduced with permission from C. Beckermann) (b) A dendritic crystal, computed using a phase-field method [19]. The particle paths illustrate the nature of the fluid flow. Note the enhanced growth of the primary and secondary branches in the flow direction in both panels.

downward-facing primary arm is much greater than that of its upward-facing counterpart. Moreover, the secondary arm growth is much more significant on the leading edges of the horizontal primary arms, as compared to their trailing edges. The mechanism for this is the increased transport on the leading edge, and the corresponding decrease of transport on the trailing edges, as heat and solute are advected by the flow over the dendrite arms. It is important to realize that this mechanism makes the phenomenon inherently three-dimensional. In 3-D, fluid may easily pass around a dendrite arm, whereas in 2-D, this is much more difficult.

It is useful to perform some scaling in order to understand which phenomena are of most importance. Suppose that the induced flow has a characteristic speed $v_\infty \approx 10$ mm s^{-1}. Let us also take some typical values for dendritic growth of pure SCN, (a thermal dendrite), with $R_{tip} \approx 10\ \mu$m, and $v^* \approx 10\ \mu$m s^{-1}. We then have

$$\begin{aligned}
\mathrm{Re} &= \frac{v_\infty R_{tip}}{\nu} = \frac{(10\ \mathrm{mm\ s^{-1}})(10^{-2}\ \mathrm{mm})}{2.67\ \mathrm{mm^2\ s^{-1}}} = 0.038 \\
\mathrm{Pe}_T &= \frac{v^* R_{tip}}{2\alpha_\ell} = \frac{(10^{-2}\ \mathrm{mm\ s^{-1}})(10^{-2}\ \mathrm{mm})}{2(0.116\ \mathrm{mm^2\ s^{-1}})} = 4.3 \times 10^{-4} \\
\mathrm{Pe}_V &= \frac{v_\infty R_{tip}}{2\alpha_\ell} = \frac{(10\ \mathrm{mm\ s^{-1}})(10^{-2}\ \mathrm{mm})}{2(0.116\ \mathrm{mm^2\ s^{-1}})} = 0.43
\end{aligned} \tag{8.109}$$

From this simple calculation, one can anticipate that inertial effects are negligible, that the approximation of a "low undercooling" for the dendrite transport is appropriate, and that thermal transport in the external fluid depends on both the conduction and advection of heat. Saville and Beaghton [42] analyzed this configuration by modeling the transport problem for a pure material solidifying as a paraboloid of revolution in an infinite undercooled melt, anti-parallel to a uniform flow with velocity v_∞. The details of the solution are not of great interest to us here; we simply note that they found a modified form of the Ivantsov solution given by

$$\Delta = \mathcal{B}(\mathrm{Pe}_T, \mathrm{Pe}_V, \mathrm{Pr}) \tag{8.110}$$

where $\mathrm{Pr} = \nu/\alpha_\ell$ is the Prandtl number for the material. Boissou and Pelcé then considered the stability of this solution, and concluded the tip dynamics to be a weak function of v_∞. [7]

Jeong et al. [19; 20] performed phase-field calculations to study this same problem. They employed a phase-field model including flow which was originally developed by Beckermann et al. [3] It should be noted that this is a challenging computational problem. The phenomena require modeling in 3-D, where there are six degrees-of-freedom at each computational node (three velocities, pressure, temperature and the phase-field parameter, described in the next section). Furthermore, one must have high grid resolution, at least near the interface, in order to resolve the phase boundary. Figure 8.25(b) shows the evolved dendrite, along with particle paths that clearly demonstrate the low Reynolds number nature of the flow, as well as the preferential growth of the primary and secondary arms in the upstream direction. Comparing this result to that of the NH_4Cl dendrite in Fig. 8.25(a), shows that the essential phenomena are captured in the simulation.

These results help us understand the role of transport in the evolution of the microstructure. Consider the four dendrite arms perpendicular to the primary arm. One can clearly see that the secondary arms on these branches grow preferentially into the flow. It is important to recognize, however, that the strong $\langle 100 \rangle$ crystalline anisotropy is still expressed by the sidebranches. The average solidification direction tilts upstream, but the underlying crystal symmetry remains unchanged.

8.5.2 Convection and columnar growth

What happens if the flow is present during the grain selection phase in constrained growth? To answer this question, consider the structures that form during twin-roll casting of an Fe-3wt%Si alloy illustrated in Fig. 8.26. In this process, used for fabricating thin strips of steel alloys, a pair of water-cooled rolls constitutes the mold. Solidification begins on the surface of the rolls, which move at relatively high speed (typically 0.5-1 m/s). The fluid is dragged by the solid shell, thus creating a boundary layer. Near the dendrite tips, the solid sees the fluid moving in an anti-parallel manner to the casting direction. It is demonstrated in Chap. 11 that, in the absence of convection, grains with their $\langle 100 \rangle$ direction most favorably aligned with the thermal gradient are selected from a random population nucleated at the surface. In twin roll casting, melt convection biases the dendrite selection such that the $\langle 100 \rangle$ tilts into the upstream direction, resulting in the structure shown in Fig. 8.26(b). Figure 8.26(c) shows this

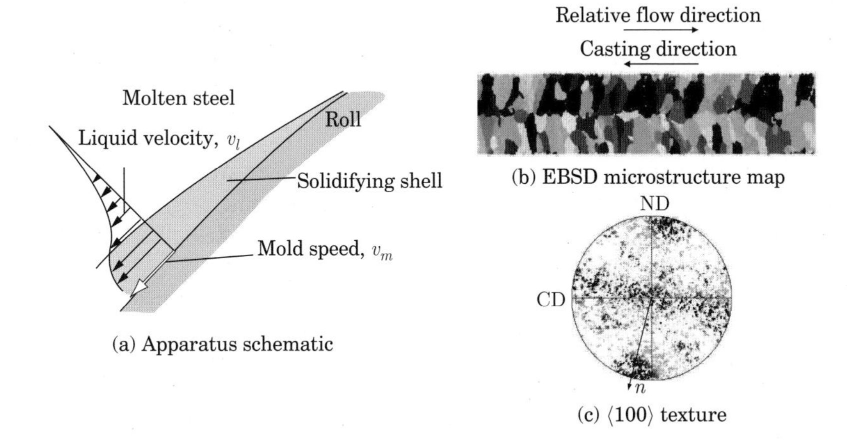

Fig. 8.26 Microstructure development in twin roll casting. (a) A schematic of solidification on the moving roll. (b) An EBSD image showing the tilting of selected grains in the upstream direction. (c) A pole figure for the upper half of the strip, showing preferred orientation of grains with their $\langle 100 \rangle$ tilted about 15° in the upstream direction. (Reproduced with permission from M. Rappaz, see ref. [43].)

result more quantitatively in the form of a $\langle 100 \rangle$ texture for the upper half of the strip. In the absence of fluid flow, the $\langle 100 \rangle$ direction of the selected grains would be mostly aligned with the normal to the strip, indicated as ND in Fig. 8.26(c). Due to the flow, we see instead a rotation of the expected texture by about 15°.

8.6 PHASE-FIELD METHODS

The models described up until this point have all assumed that the shape of the growing solid phase is either known *a priori* or computed as a relatively small perturbation from such a shape. This limitation can be overcome only by computational methods that are sufficiently robust so as to be able to handle arbitrarily complex interface shapes. The *phase-field method* has become the approach of choice for this purpose because it provides a systematic means for solving, on a fixed grid, the complete system of equations governing both diffusion and the evolution of the interface, irrespective of the complexity of the shape. In the development given below, we provide a general description of the method for a thermal dendrite, and then briefly present the extension to binary alloys. The reader is referred to the articles listed at the end of the chapter for a more in-depth description.

The phase-field method introduces an *order parameter* $\psi \in [-1, 1]$ to identify the phase, where $\psi = -1$ corresponds to the liquid, and $\psi = +1$ corresponds to the solid. Intermediate values of ψ represent the liquid-solid interface, and one normally identifies a specific value, e.g., $\psi = 0$ as "the" location of the interface.[1] Thus, the usual "sharp" interface between the solid and liquid phases now becomes diffuse, and the "phases" turn into a continuous field, hence the name phase-field. The model is designed to confine the interface to a narrow region of width W, as described further below. A free energy $\mathcal{F}(\psi, T)$ is introduced, and the system dynamics ensure that the total energy decreases with time, by using the Ginzburg-Landau form,

$$\tau \frac{\partial \psi}{\partial t} = -\frac{\delta \mathcal{F}}{\delta \psi} + \Upsilon(\mathbf{x}, t) \tag{8.111}$$

Here, τ is a characteristic time scale, and $\Upsilon(\mathbf{x}, t)$ is a stochastic variable representing thermal noise. The expression $\delta\mathcal{F}/\delta\psi$ on the right-hand side indicates the variational derivative. The formulation is completed by coupling the evolution of the phase-field to the thermal field by writing

$$\frac{\partial \theta}{\partial t} = \alpha \nabla^2 \theta + \frac{1}{2} \frac{\partial h(\psi)}{\partial t} \tag{8.112}$$

where $\theta = c_p(T - T_f)/L_f$ is the scaled temperature. We define $h(\psi)$ momentarily.

[1]Some authors use $\psi \in [0, 1]$ and locate the interface at $\psi = 1/2$. There is no fundamental difference.

The key to the model is the free energy, chosen in such a way that three criteria are satisfied: (i) the integration of the free energy over a volume that includes an interface gives the correct value for the surface energy; (ii) there are two stable states, corresponding to $\psi = \pm 1$; and (iii) the preferred state is coupled to the temperature field in such a way that the transition occurs at the equilibrium melting point for a flat interface. We write

$$\mathcal{F}(\psi, \theta) = \int_V F(\psi, \theta) dV = \int_V \left[\frac{1}{2} (W |\nabla \psi|)^2 + f(\psi) + \lambda \theta g(\psi) \right] dV \tag{8.113}$$

where the free energy density $F(\psi, \theta)$ has been introduced. We ascribe physical meaning to the terms in the free energy shortly. Since the temperature is a continuous function, and phases are distinguished only implicitly through the value of ψ, there is no subscript "s" or "ℓ" on θ. The above three criteria are guaranteed to be satisfied if one chooses the function $f(\psi)$ to be a "double well" potential with minima at $\psi = \pm 1$, and $g(\psi)$ to be an odd function with $g'(\pm 1) = 0$. Commonly used functional forms for $f(\psi)$ and $g(\psi)$ are

$$f(\psi) = -\frac{\psi^2}{2} + \frac{\psi^4}{4}; \quad g(\psi) = \psi - 2\frac{\psi^3}{3} + \frac{\psi^5}{5} \tag{8.114}$$

Other forms often used in the literature are given in Table 8.3 (the constants β_1 and β_2 are defined in Eq. (8.120)). Anisotropy can be included in the model by allowing τ and W to depend on the local normal vector to the interface $\boldsymbol{n} = \nabla \psi / |\nabla \psi|$ as [22]

$$\tau(\boldsymbol{n}) = \tau_0 A^2(\boldsymbol{n}); \quad W = W_0 A(\boldsymbol{n}) \tag{8.115}$$

where $A(\boldsymbol{n})$ expresses the anisotropy. For example, for cubic symmetry, one can use Eq. (8.4). Substituting the various expressions for F, τ, W, $f(\psi)$ and $g(\psi)$ into Eq. (8.111) yields the evolution equation for ψ

$$\begin{aligned} \tau_0 A^2(\boldsymbol{n}) \frac{\partial \psi}{\partial t} =& \nabla \cdot (W_0^2 A^2(\boldsymbol{n}) \nabla \psi) + \psi - \psi^3 - \lambda \theta (1 - \psi^2)^2 \\ &+ W_0 \frac{1}{2} \nabla \cdot \left[|\nabla \psi|^2 \frac{\partial A^2(\boldsymbol{n})}{\partial (\nabla \psi)} \right] + \Upsilon(\mathbf{x}, t) \end{aligned} \tag{8.116}$$

To illustrate the physical meaning of the free energy, consider the case of a stationary, planar interface. This makes the problem one-dimensional

Table 8.3 Functional forms for several phase-field models. [27]

Model	$\partial g(\psi)/\partial \psi$	$h(\psi)$	β_1	β_2
1	$1 - \psi^2$	ψ	$1/\sqrt{2}$	$5/6$
2	$(1 - \psi^2)^2$	ψ	$5/4\sqrt{2}$	$48/75$
3	$(1 - \psi^2)^3$	ψ	1.0312	0.52082
4	$(1 - \psi^2)^4$	ψ	1.1601	0.45448
5	$(1 - \psi^2)^2$	$15(\psi - 2\psi^3/3 + \psi^5/5)/8$	$5/4\sqrt{2}$	0.39089

and steady, and also implies that $A(\boldsymbol{n}) \to 1$ and $T = T_f$ ($\theta = 0$). Under these restrictions, Eq. (8.116) reduces to

$$0 = W_0^2 \frac{d^2\psi_1}{dx^2} + \psi_1 - \psi_1^3 \tag{8.117}$$

where the subscript on ψ_1 indicates that it is in 1-D. The solution (see Exercise 8.9) thus becomes

$$\psi_1 = \tanh\left(\frac{x}{W_0\sqrt{2}}\right) \tag{8.118}$$

The associated free energy density is given by

$$F(\psi_1, \theta = 0) = \frac{1}{4}\cosh^{-4}\left(\frac{x}{W_0\sqrt{2}}\right) - \frac{1}{2}\tanh^2\left(\frac{x}{W_0\sqrt{2}}\right) + \frac{1}{4}\tanh^4\left(\frac{x}{W_0\sqrt{2}}\right) \tag{8.119}$$

These results for $\psi_1(x)$ and $F(\psi_1, \theta = 0)$ are plotted in Fig. 8.27(a). One can observe that the interface extends over a width of approximately $8W_0$, and that ψ_1 is essentially constant outside this region. Positive values of x/W_0 correspond to the solid phase, $\psi = 1$, and negative values of x/W_0 represent the liquid phase, $\psi = -1$. The free energies of the solid and liquid phases are equal, as one would expect for $\theta = 0$. The additional contribution to the free energy within the interfacial region, arising from the first two terms on the right-hand side of Eq. (8.113), is thus identified as proportional to the surface energy. The area under the free energy curve, shaded in gray, is consequently $\gamma_{s\ell}$, and one can see that it depends on the value of W_0.

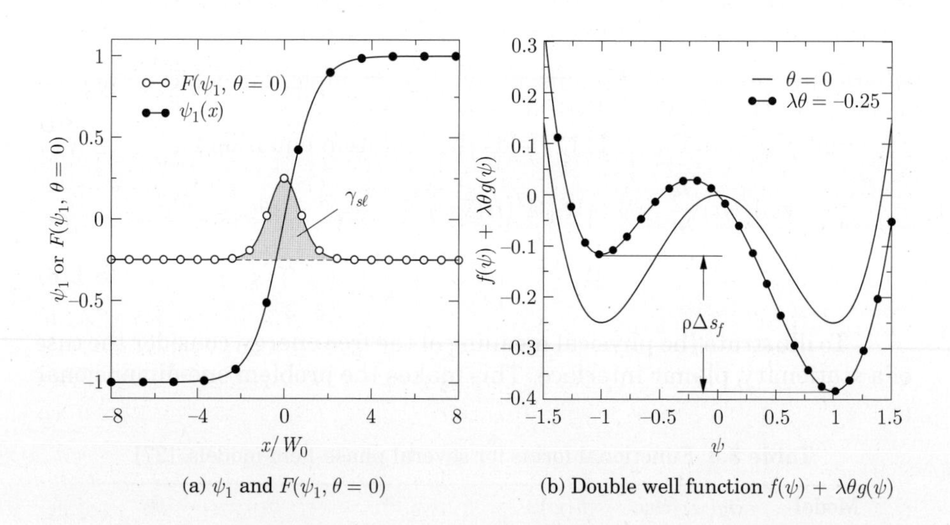

Fig. 8.27 The phase-field, and various contributions to the free energy, near the interface. (a) The stationary phase-field and the associated free energy density. (b) The double well potential, displaying the tilting of the potential toward the solid phase $\psi = 1$ for $\theta < 0$. The difference in energy between the two local minima is proportional to $\Delta s_f \Delta T$.

To understand the physical meaning of the third term in the free energy density, $\lambda\theta g(\psi)$, one should recall from Eq. (2.21) that $\Delta g \approx \rho \Delta s_f \Delta T$. The parameter λ is thus proportional to Δs_f, as demonstrated graphically in Fig. 8.27(b). When $\theta = 0$, the two phases have equal free energy. This is manifested by the two energy wells at $\psi = \pm 1$ having the same energy. As θ decreases, i.e., as the melt becomes undercooled, the wells tilt toward the solid. The energy difference between the wells is proportional to Δs_f. Thus, the physical quantity Δs_f is represented by the model parameter λ. One can relate the phase-field model parameters to the kinetic coefficient, μ_k, by considering the case of a one-dimensional interface moving at constant speed. (Cf. Boettinger et al. [6]).

The exact form of the relationships between the model parameters W_0, λ, and τ_0 and the physical quantities $\gamma_{s\ell}$, Δs_f, and μ_k depends on the specific functional forms chosen for $f(\psi)$ and $g(\psi)$. Karma and Rappel [22] performed an asymptotic analysis of the phase-field method which provided the following formulae for selecting the model constants for a given set of physical parameters

$$d_0 = \beta_1 \frac{W_0}{\lambda}; \quad \frac{1}{\mu_k} = \frac{\beta_1 \tau_0}{W_0 \lambda} \left(1 - \beta_2 \lambda \frac{W_0^2}{\alpha \tau_0}\right) \tag{8.120}$$

Here, β_1 and β_2 are constants determined from the functional forms used in formulating the phase-field model. Their values are tabulated in Table 8.3 for several combinations of $g(\psi)$ and $h(\psi)$, with $f(\psi)$ given by Eq. (8.114) for all cases. Karma and Rappel's analysis also revealed that the use of these formulae ensures that the phase-field model solution converges to the corresponding sharp interface problem in the limit where the characteristic interface width, W_0, is small in comparison to the thermal diffusion field, α/v^*. This permits quantitative modeling of dendritic growth in 3-D using only modest computational resources. Kim et al. [27] showed that all of the choices for $g(\psi)$ and $h(\psi)$ given in Table 8.3 result in identical solutions for the computed shape of the dendrite as well as its velocity, and further that this solution gives the same operating state (R_{tip}, v^*) as predicted by microscopic solvability theory, described in Sect. 8.4.

Before considering any more simulations, we should discuss some of the computational issues. It is a relatively straightforward exercise to code Eqs. (8.116) and (8.112) using the methods given in Chap. 6. The presence of the highly nonlinear terms in Eq. (8.116) usually leads to the choice of an explicit finite difference in time, at least for the ψ-equation. In certain cases, a different time stepping scheme is employed for the conduction problem (Eq. (8.112)), using either a larger time step, an implicit scheme, or both. The spatial terms may be evaluated using either finite difference or finite element methods. A practical difficulty arises during the resolution of the multiple length scales that appear in the problem. There are three physical length scales of interest: d_0, R_{tip} and $\delta_T = \alpha_\ell/v^*$, in addition to three model/computational length scales: W_0, the characteristic grid spacing, Δx and the size of the computational domain, L. Clearly, any resolved calculation must depend on the first set of physical length scales, and not on the second (computational) length scales.

The asymptotic analysis makes it possible to work in the regime where W_0 is as large as $R_{tip}/10$, and one can normally expect to resolve the interface properly with on the order of 10 grid points within the interface. The problem in implementation arises when simultaneously trying to resolve the diffusion field, whose characteristic length v^*/α_ℓ is typically much larger than R_{tip}, especially at low undercooling, where v^* is small. The solution to this problem is to employ adaptive techniques that resolve the smallest length scales near the interface, coarsening to much larger length scales far away, in order to resolve the diffusion field. Provatas et al [40; 39] developed such methods based on adaptive mesh refinement approaches in 2-D, which were later extended to 3-D by Jeong et al. [19; 20] and by Narski and Picasso [35]. Plapp and Karma [38] were able to obtain a similar computational efficiency by coupling a finite difference calculation in an inner region, near the growing dendrite, to a Monte Carlo calculation of the far-field diffusion.

The phase-field method described above for pure materials can also be extended to alloys. The first modification consists in adding terms to the free energy associated with the composition, e.g., the regular solution model (see Sect. 2.3.2). Diffusion equations for each solute element are then coupled to a phase-field evolution equation such that the solid and liquid phases are in equilibrium, as specified by the equilibrium phase diagram. There are several important implementation issues, not described here. Instead, we refer the reader to the original papers, as well as to a review by Boettinger et al. [6].

Figure 8.28 shows an example of a simulation performed with such a model of the directional solidification of a dilute SCN-acetone alloy. The important phenomena associated with constrained growth are apparent: the instability of an initially planar interface, the competitive selection of primary spacing by elimination of certain dendrites due to solute rejection from neighbors, and secondary arms appearing once the primary dendrites are sufficiently far apart. Figure 8.28 also displays a micrograph from an experiment by Trivedi and Somboonsuk [46], revealing a remarkably similar structure. We should note that it takes a very long time to establish steady state conditions in both the experiments and simulations, which makes it very difficult to go beyond such qualitative comparisons.

To illustrate the power of the phase-field method, we present some results from the work of Haximali et al. [17]. In their study, the model was employed to investigate the effect of surface energy anisotropy in Al-Zn alloys. A collection of experimental results for this system were presented in Fig. 8.10. Haximali et al. performed a series of phase-field simulations of freely growing alloy dendrites for varying combinations of the parameters a_1 and a_2 in Eq. (8.4), and the results are shown in Fig. 8.29. The surface stiffness, shown in Fig. 8.29(a), contains distinct regions, separated by a dashed line, where the maxima of the stiffness inverse Ψ^{-1} correspond to $\langle 100 \rangle$ and $\langle 110 \rangle$. Along the dashed line, the anisotropy is degenerate in the (001). Figure 8.29(b) shows the results of a series of

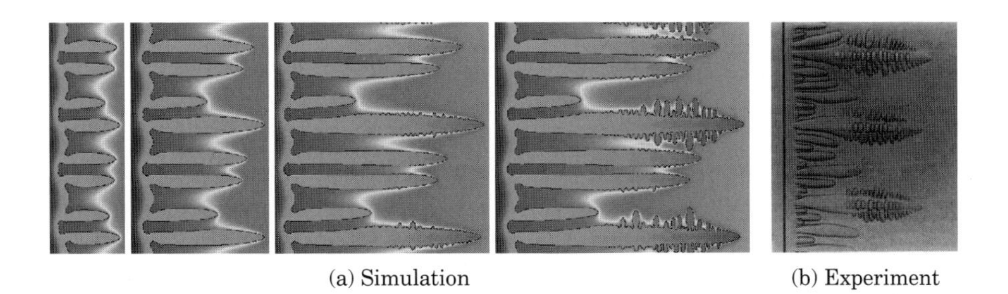

Fig. 8.28 (a) A 2-D simulation of directional solidification. (b) An experimental observation of SCN-acetone dendrites. (Reproduced with permission from ref. [46].)

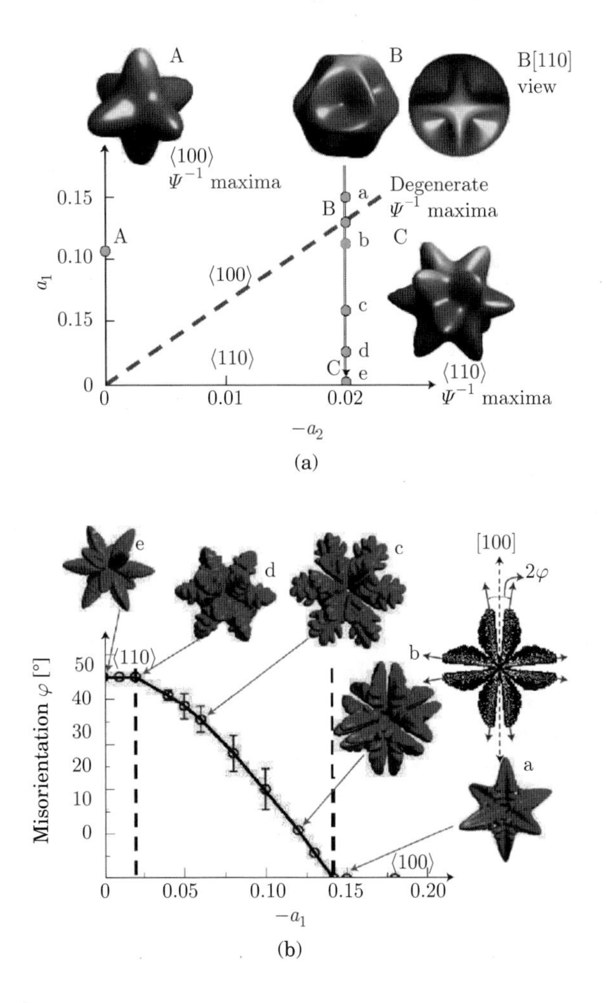

Fig. 8.29 (a) A map of the spatial variation of the minimum stiffness as a function of the anisotropy parameters a_1 and a_2. (b) Phase-field computations of free dendrite growth for various values of the anisotropy parameters. (Reproduced with permission from ref. [17].)

phase-field simulations for these sets of parameter, clearly indicating that a rich variety of dendrite shapes are possible for various combinations of the anisotropy parameters. Thus, the phase-field method provides a very powerful tool to be used in conjunction with experimental observations, with the aim of increasing the understanding of the fundamental aspects of microstructure development.

8.7 SUMMARY

This chapter has been dedicated to dendritic growth, one of the most prevalent solidification microstructures. We showed that simple geometries, such as spherical grains or planar fronts, are morphologically unstable with respect to shape perturbations. As such perturbations grow, dendrites evolve as the expression of the underlying crystalline anisotropy of the solid phase. If the anisotropy is large enough, some orientations are thermodynamically forbidden and facets appear. We presented stability analyses for planar fronts and spherical growth, and found that the length scale selected by the instability originates from a balance between the destabilizing forces due to transport, and the stabilizing effect of capillary forces.

One of the most important topics we considered is the relationship between dendrite growth velocity, length scale (as measured by the tip radius) and undercooling or supersaturation. We found a curious result, namely that the solution to the transport equations was insufficient to determine the length scale, instead yielding only a relation between the undercooling and the Péclet number, $\text{Pe} = R_{tip}v^*/2\alpha_\ell$. A second relation was developed by considering the morphological stability of the steady dendrite shape, which gave a second condition of the form $R_{tip}^2 v^* = \text{cst}$, thus producing a unique operating state (v^*, R_{tip}) for any given undercooling, ΔT, in free growth. In a similar way, one can find a unique pair $(R_{tip}, \Delta T)$ for constrained growth, where v^* is known. The growth kinetics models providing $v^*(\Delta T)$ are put to use in Chap. 11, when we discuss the coupling of microstructure formation to macroscopic heat flow.

We also examined the evolution of secondary arm spacing due to a coarsening of the initial microstructure; a section that was concluded with a description of the effect of convection on the evolution of microstructure. Finally, the phase-field method was introduced, and examples demonstrating this powerful tool were given.

8.8 EXERCISES

Exercise 8.1. Negative anisotropy coefficient.
Construct plots similar to those in Fig. 8.2 for negative values of the anisotropy coefficient, $\varepsilon_4 = -0.05, \ -0.15$. Describe the results.

Exercise 8.2. Forms for anisotropy.

(a) A number of articles in the literature use the following expression for the surface energy anisotropy in 2-D:

$$1 - 3\epsilon_4 + 4\epsilon_4(\cos^4\phi + \sin^4\phi)$$

Demonstrate that this expression is equivalent to Eq. (8.1).

(b) Follow a similar procedure to find values for a_1 and a_2 such that the expansion in terms of the first two cubic harmonics in Eq. (8.4) reduces in 2-D to Eq. (8.1) in the (001) plane.

Exercise 8.3. Wavelength for planar front growth.
Begin with Eq. (8.67) and utilize the definitions of M and S in Eq. (8.62) to show that, for large values of the wavenumber n, the wavelength $\lambda_{\min}$ can be approximate by Eq. (8.71).

Exercise 8.4. Higher modes for the perturbed sphere.
In the analysis of the morphological stability of the solidifying sphere, we considered the initial instability, where the mode $n = 2$ is the only unstable mode. However, for larger radii, other modes become unstable as well. Consider Eq. (8.16), and compare the rate of growth of various modes as the sphere radius increases.

(a) Recast the equation in the following form:

$$\frac{\dot{\mathcal{R}}^{1*}}{\mathcal{R}^{1*}} = f(n, k_T, R_c/R^*)$$

(b) Plot the growth rate obtained for various n at the following values of R_c/R^*: 11, 25, 50 and 100. Make two sets of graphs: one for $k_T = 1$ and the other for $k_T = 0$.

(c) What can you conclude with regard to the probable evolution of the morphology as the particle continues to grow?

(d) How would this affect Eq. (8.25)?

Exercise 8.5. Free growth of a binary alloy.
This exercise demonstrates the analogy between the thermal and solutal models for the growth of a sphere. Write the governing equations for isothermal growth of a binary alloy particle.

(a) Substitute the scaled variables defined as follows

$$\mathcal{C}_\ell = \frac{C_\ell - C_\ell^\infty}{C_\ell^\infty - C_0}; \quad \mathcal{C}_s = \frac{C_s - C_\ell^\infty}{C_\ell^\infty - C_0}; \quad \varrho = \frac{r}{R_i}; \quad \mathcal{R}^* = \frac{R^*(t)}{R_i}; \quad \tau = \frac{t}{t_c}$$

into the dimensional form of the governing equations. For the moment, leave t_c unspecified.

(b) Make the approximation $|2\Gamma_{s\ell}/m_\ell R^*| \ll C_\ell^{eq}$ in the governing equation, and choose a value for t_c such that the coefficient in front of dR^*/dt in the flux condition goes to one. Then, massage the equations into the appropriate form.

Exercise 8.6. Planar front growth of a pure material.
This problem examines the stability of directional solidification of a pure material. Consider the case where the material solidifies at a constant velocity v_n into an infinite undercooled melt, where the temperature at infinity is T_∞. Note that this problem differs from the one considered in Sect. 5.3.1, because the interface velocity in that problem was not constant.

(a) Begin by finding the base state solution for the growth of a planar front. Write the governing equation for conduction in the solid and liquid phases, along with the boundary conditions. Use a coordinate system where z is measured from the moving interface. Simplify the governing equations as far as possible, assuming that the solution is steady in the frame moving with the interface, and that all material properties are constant. Also assume that the density of the solid and liquid are equal.

(b) The solution for the temperature in the solid is trivial: $T_s = T_f$. It is convenient to scale T_ℓ as follows:

$$\theta_\ell = \frac{T_\ell - T_\infty}{L_f/c_p}; \qquad \Delta = \frac{T_f - T_\infty}{L_f/c_p}$$

This actually becomes clearer if we leave z dimensional. Make this substitution, rewrite the governing equations and boundary conditions for the liquid temperature in terms of the dimensionless variable.

(c) Use the two boundary conditions on the temperature at $z = 0$ and $z \to \infty$ to find θ_ℓ. (Hint: assume a solution of the form $\theta_\ell \sim A + B\exp(sz)$.)

(d) Now, substitute this solution into the Stefan (flux) condition and show that solutions exist only for $\Delta = 1$. What is the physical interpretation of this result?

(e) We now consider the stability of this solution, closely following the treatment in Sect. 8.3.3 for the binary alloy. Write the temperature solution as

$$\theta_\ell = \theta_\ell^0 + \theta_\ell^1$$

where θ_ℓ^0 was obtained in the previous part of this problem. Develop the governing equation and boundary condition at $z \to \infty$ for θ_ℓ^1. Be careful to note that the problem is now 2-D.

(f) Assume a solution of the form $\theta_\ell^1 \sim A(z)\exp(\sigma t + inx)$. Substitute this form into the differential equation, to find an ordinary differential equation for $A(z)$. Using a trial solution of the form $A(z) \sim \exp(\beta z)$, find the two roots for β, i.e.,

$$A(z) = A_1 \exp(\beta_1 z) + A_2 \exp(\beta_2 z)$$

where A_1 and A_2 are constants. Finally, use the boundary condition at $z \to \infty$ to eliminate one of these terms.

(g) Following the procedure outlined in Sect. 8.3.3, linearize the Gibbs-Thomson condition, assuming a normal mode form for the interface shape, i.e., $h(x,t) = h_0 \exp(\sigma t + inx)$. This should leave you with the following equation:

$$d_0 h_0 n^2 - \frac{v_n \Delta}{\alpha} h_0 + A_1 = 0$$

where $d_0 = c_p \Gamma_{s\ell}/L_f$ is the capillary number for a pure material.

(h) Linearize the Stefan condition to obtain the result

$$\left(\frac{\sigma}{\alpha_\ell} + \frac{v_n^2}{\alpha_\ell^2}\right) h_0 - \left[\frac{v_n}{2\alpha_\ell} + \frac{1}{2}\sqrt{\left(\frac{v_n}{\alpha_\ell}\right)^2 + 4\left(\frac{\sigma}{\alpha_\ell} + n^2\right)}\right] A_1 = 0$$

(i) Solve the homogeneous system presented by the results of the previous two parts of this problem. Before doing so, simplify the expressions by assuming that $n \gg v_n/\alpha_\ell$ and $\sigma \ll n^2 \alpha_\ell$.

(j) Sketch the dispersion relation (σ vs. n), and find the cutoff value n_{max}, and then use that value to find $\lambda_{\min}$.

Exercise 8.7. Unstable planar growth in Al-4wt%Cu.
Consider an alloy of Al-4wt%Cu. Develop a plot similar to Fig. 8.14 for the directional solidification of the alloy in a temperature gradient $G = 5000$ K/m.

(a) Compute ΔT_0 for this alloy.

(b) Compute the minimum pulling speed, v_p^c, that will result in an unstable interface.

(c) Compute the interface speed corresponding to the absolute stability, v_p^a.

(d) Compute and plot Eqs. (8.70) and (8.71) using these values.

(e) You will not find the proper result near v_p^a in the previous part of this exercise. Write a program to solve Eq. (8.67) numerically in this region. A suggestion for the algorithm is as follows:

(i) Separate the solution into upper and lower branches, similar to λ_{min} and λ_{max}, for numerical solution.

(ii) Loop through values of v_p, and by using either Eq. (8.70) or (8.71) as appropriate, find a range of n-values containing the root to Eq. (8.67).

(iii) Find the root by interval halving.

Exercise 8.8. Alternate form of the simplified LGK model.
One often finds in the literature a form of the LGK model where the approximation $\mathrm{Iv}_{3D}(\mathrm{Pe}) \approx \mathrm{Pe}$ is used instead of Eq. (8.89). Develop the corresponding forms to Eqs. (8.91) and (8.92) when this approximation is made.

Exercise 8.9. One-dimensional phase field solution.
Show by substitution that Eq. (8.118) is a solution to Eq. (8.117).

8.9 REFERENCES

[1] S. Akamatsu, , G. Faivre, and M. Ihle. Symmetry-broken double fingers and seaweed patterns in thin-film directional solidification. *Phys. Rev. E*, 51(5):4751-4753, 1995.

[2] A. Barbieri and J. S. Langer. Prediction of dendritic growth rates in the linearized solvability theory. *Phys. Rev. A*, 39:5314-5325, 1989.

[3] C. Beckermann, H.-J.Diepers, I. Steinbach, A. Karma, and X. Tong. Modeling melt convection in phase-field simulation of solidification. *J. Comp. Phys.*, 154:468, 1999.

[4] E. Ben-Jacob, N. Goldenfeld, J. S. Langer, and G. Schön. Dynamics of interfacial pattern formation. *Phys. Rev. Lett.*, 51:1930-1932, 1983.

[5] E. Ben-Jacob, N. Goldenfeld, J. S. Langer, and G. Schön. Boundary-layer model of pattern formation in solidification. *Phys. Rev. A*, 29(1):330-340, 1984.

[6] W. J. Boettinger, J. A. Warren, C. Beckermann, and A. Karma. Phase-field simulation of solidification. *Annu. Rev. Mater. Res.*, 32:163-194, 2002.

[7] P. Bouissou and P. Pelcé. Effect of a forced flow on dendritic growth. *Phys. Rev. A*, 40: 6673-6680, 1989.

[8] R. C. Brower, D. A. Kessler, J. Koplik, and H. Levine. Geometrical models of interface evolution. *Phys. Rev. A*, 29:1335-1342, 1984.

[9] J. A. Cahn and D. W. Hoffman. A vector thermodynamics for anisotropic surfaces. II curved and faceted surfaces. *Acta Met.*, 22:1205-1214, 1974.

[10] S. A. David, S. S. Babu, and J. M. Vitek. Welding: Solidification and microstructure. *J. Metals*, pages 14-20, 2003.

[11] S. H. Davis. *Theory of solidification*. Cambridge University Press, Cambridge, 1999.

[12] S. Eck, J. Mogeritsch, and A. Ludwig. Experimental observation of convection during equiaxed solidification of transparent alloys. *Materials Science Forum*, 508:157-162, 2006.

[13] H. Esaka. *Dendrite growth in succinonitrile-acetone alloys*. PhD thesis, EPFL, Lausanne, 1986. Thesis no. 615.

[14] M. E. Glicksman. Free dendritic growth. *Mat. Sci. Engrg.*, pages 45-55, 1984.

[15] F. Gonzales and M. Rappaz. Dendrite growth directions in aluminum-zinc alloys. *Met. Mater. Trans*, 37A:2797-2806, 2006.

[16] Quantum Chemistry Group, 2007. `http://web.uniovi.es/qcg/harmonics/harmonics.html`.

[17] T. Haxhimali, A. Karma, F. Gonzales, and M. Rappaz. Orientation selection in dendritic evolution. *Nature Materials*, 5:660-664, 2006.

[18] S.-C. Huang and M. E. Glicksman. Overview 12: Fundamentals of dendritic solidification-II development of sidebranch structure. *Acta Metallurgica*, 29(5):717-734, 1981.

[19] J.-H. Jeong, J. A. Dantzig, and N. Goldenfeld. Dendritic growth with fluid flow in pure materials. *Met. Mater. Trans.*, 34A(3):459-466, 2003.

[20] J.-H. Jeong, N. Goldenfeld, and J. A. Dantzig. Phase field model for three-dimensional dendritic growth with fluid flow. *Phys. Rev. E*, 64:041602-1 to 041602-14, 2001.

[21] D. Kammer and P. Voorhees. The morphological evolution of dendritic microstructures during coarsening. *Acta Mat.*, 54:1549-1558, 2006.

[22] A. Karma and W.-J. Rappel. Quantitative phase-field modeling of dendritic growth in two and three dimensions. *Phys. Rev. E.*, 57:4323-4349, 1998.

[23] A. Karma. Phase-field formulation for quantitative modeling of alloy solidification. *Phys. Rev. Lett.*, 87:115701-1:4, 2001.

[24] T. Z. Kattamis and M. C. Flemings. *Trans. AIME*, 233:992-999, 1965.

[25] D. A. Kessler, J. Koplik, and H. Levine. Pattern selection in fingered growth phenomena. *Adv. Phys.*, 37:255-339, 1988.

[26] D. A. Kessler and H. Levine. Velocity selection in dendritic growth. *Phys Rev B*, 33: 7867-7870, 1986.

[27] Y.-T. Kim, N. Provatas, N. Goldenfeld, and J. A. Dantzig. Universal dynamics of phase field models for dendritic growth. *Phys. Rev. E*, 59(3):2546-2549, 1999.

[28] W. Kurz and D. Fisher. *Fundamentals of solidification*. Trans. Tech. Publ., Aedermansdorf, Switzerland, 4th edition, 2005.

[29] J. S. Langer and H. Müller-Krumbhaar. Theory of dendritic growth - I. elements and stability analysis. *Acta Met. et Mater.*, 26:1681-1688, 1978.

[30] J. Lipton, M. E. Glicksman, and W. Kurz. Equiaxed dendrite growth at small supercooling. *Metall. Trans.*, 18A:341-345, 1987.

[31] L. X. Liu and J. S. Kirkaldy. Thin-film forced velocity cells and cellular dendrites. 1. experiments. *Acta Metall. et Mat.*, 43(8):2891-2904, 1995.

[32] R. Mendoza, J. Alkemper, and P. W. Voorhees. Three-dimensional morphological characterization of coarsened microstructures. *Zeitschrift für metallkunde*, 96:155-160, 2005.

[33] W. W. Mullins and R. F. Sekerka. Morphological stability of a particle growing by diffusion or heat flow. *J. Appl. Phys.*, 34:323-329, 1963.

[34] R. Napolitano, H. Meco, and C. Jung. Faceted solidification morphologies in low-growth-rate Al-Si eutectics. *J. Metals*, 56:16-21, 2004.

[35] J. Narski and M. Picasso. Adaptive finite elements with high aspect ratio for dendritic growth of a binary alloy including fluid flow with shrinkage. *Comp. Meth. Appl. Mech. and Engrg.*, 196:3562-3576, 2007.

[36] G. E. Nash and M. E. Glicksman. Capillarity-limited steady-state dendritic growth - 1. theoretical development. *Acta. Metall*, 22:1283-1290, 1974.

[37] W. Oldfield. Computer model studies of dendritic growth. *Materials Science and Engineering*, 11:211-218, 1973.

[38] M. Plapp and A. Karma. Multiscale finite-difference-diffusion Monte-Carlo method for simulating dendritic solidification. *J. Comp. Phys.*, 165 :592-619, 2000.

[39] N. Provatas, J. A. Dantzig, and N. Goldenfeld. Adaptive mesh refinement computation of solidification microstructures using dynamic data structures. *J. Comp. Phys.*, 148: 265-290, 1999.

[40] N. Provatas, J. Dantzig, and N. Goldenfeld. Efficient computation of dendritic microstructures using adaptive mesh refinement. *Phys. Rev. Lett.*, 80:3308-3311, 1998.

[41] Y. Saito, G. Goldbeck-Wood, and H. Müller-Krumbhaar. Numerical simulation of dendritic growth. *Phys. Rev. A*, 38:2148-2157, 1988.

[42] D. A. Saville and P. J. Beaghton. Growth of needle-shaped crystals in the presence of convection. *Phys. Rev. A*, 37:3423-3430, 1988.

[43] H. Takatani, C.-A. Gandin, and M. Rappaz. EBSD characterisation and modelling of columnar dendritic grain growing in the presence of fluid flow. *Acta Mater.*, 48:675-88, 2000.

[44] D. E. Temkin. Growth velocity of a needle crystal in a supercooled melt. *Dokl. Akad. Nauk SSSR*, 132:1307-1310, 1960.

[45] W. A. Tiller, K. A. Jackson, J. W. Rutter, and B. Chalmers. The redistribution of solute atoms during the solidification of metals. *Acta Metallurgica*, 1:428-437, 1953.

[46] R. Trivedi and K. Somboonsuk. Constrained dendritic growth and spacing. *Mat. Sci. and Engg.*, 65:65-74, 1984.

[47] J.-M. Vanden-Broeck. Fingers in a Hele-Shaw cell with surface tension. *Physics of Fluids*, 26(8):2033-2034, 1983.

[48] J. A. Warren and W. J. Boettinger. Prediction of dendritic growth and microsegregation patterns in a binary alloy using the phase-field method. *Acta Met.*, 43:689-703, 1995.

CHAPTER 9

EUTECTICS, PERITECTICS AND MICROSTRUCTURE SELECTION

9.1 INTRODUCTION

In the previous chapter, we saw that the growth of a single phase α from a binary melt usually occurs in the form of equiaxed or columnar dendrites at some undercooling ΔT below the liquidus temperature T_{liq}. As the solidification proceeds, the liquid becomes richer in solute (if $k_0 < 1$), and, at some point, its composition reaches an invariant point, typically a eutectic or a peritectic. In order for the invariant reaction to proceed, a second phase β must nucleate and grow concurrently with, or at the expense of, the α-phase.

In the case of eutectic growth, the exchange of solute between the two solid phases occurs via transport in the liquid phase. The α-phase rejects solute B, whereas the β-phase rejects A. We demonstrate in Sect. 9.2 that the spatial distribution of the solute, and the curvature of the $\alpha - \ell$ and $\beta - \ell$ interfaces, combine to make the eutectic front grow at an undercooling ΔT below the equilibrium eutectic temperature T_{eut}. Section 9.2 also presents the various types of eutectic morphologies: regular, irregular, nodular, divorced and multi-component/multi-phase eutectics.

We describe the morphologies associated with the peritectic transformation $\alpha + \ell \rightarrow \beta$ in Sect. 9.3. This transformation also occurs below the equilibrium peritectic temperature T_{per} due to solute transport being required. We distinguish two mechanisms besides the solidification of β from the melt. The *peritectic reaction* occurs in the region where all three phases α, β and ℓ are in contact. The peritectic β-phase can also expand at the expanse of α via the *peritectic transformation*, a solid state transformation at the $\alpha - \beta$ interface involving diffusion through the peritectic β-phase. At very low speed, when the α and β planar fronts are normally stable, new growth mechanisms resulting in novel peritectic microstructures will be seen to operate.

Finally, Sect. 9.4 describes the competition between various phases and/or morphologies. Although nucleation kinetics determines which phase forms first, it will become clear that growth kinetics can confer an advantage to a particular phase or growth morphology. In a sample transforming at a nearly uniform temperature, the phase or microstructure that grows the fastest wins, whereas in directional solidification, the phase that grows at the highest temperature is the winner. Illustrative examples include the competition between austenite and ferrite dendrites in Fe-Ni-Cr steels, and between graphite and cementite in eutectic cast iron. We will also discuss the competition between dendritic and eutectic morphologies in the *coupled zone* of a eutectic phase diagram.

The notation in this chapter becomes somewhat complicated, since we need to distinguish between various interfaces, e.g., $\alpha - \ell$ or $\beta - \ell$. Therefore, in addition to the "$*$" superscript used throughout this book for the interfacial compositions in a given phase, e.g., C_ℓ^* for the interfacial liquid composition, another superscript is added for the phase with which it is in contact. For example, $C_\ell^{*\alpha}$ and $C_\beta^{*\ell}$ represent the interfacial composition in the liquid ahead of the $\alpha - \ell$ interface and the interfacial composition of β at the $\beta - \ell$ interface, respectively.

9.2 EUTECTICS

9.2.1 General considerations

Recall that the eutectic reaction is the direct transformation of the liquid into two solid phases, $\ell \rightarrow \alpha + \beta$. We first treat the invariant eutectic reaction in binary alloys using the hypothetical phase diagram shown in Fig. 9.1(a). Consider the hypoeutectic alloy C_0 solidifying in a temperature gradient, as illustrated in Fig. 9.1(b). As discussed in Chap. 8, the dendrite

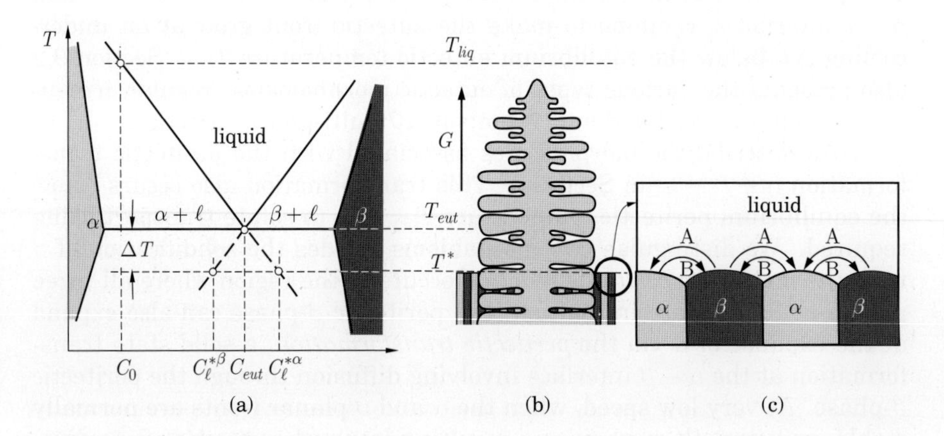

Fig. 9.1 (a) A typical eutectic phase diagram and (b) the associated microstructure that is formed during solidification in a vertical thermal gradient. (c) A schematic diagram of the coupled growth of α- and β-lamellae during the eutectic reaction.

tip will be found at a temperature $T^*_{tip} < T_{liq}(C_0)$, and the interdendritic liquid becomes increasingly enriched with increasing distance behind the tip. Eutectic growth occurs at a certain undercooling ΔT below T_{eut} when the liquid composition is close to C_{eut}. Let us now explore the mechanisms by which the two solid phases α and β grow simultaneously, as sketched in Fig. 9.1(c).

In a hypoeutectic alloy of composition $C_0 \lesssim C_{eut}$, the α-phase is likely to nucleate first. As the primary phase grows, the liquid is enriched in B. At some point, the liquid composition at the $\alpha - \ell$ interface, indicated as $C^{*\alpha}_\ell$ in Fig. 9.2, becomes larger than C_{eut}. Thus, the enriched liquid is undercooled by an amount ΔT^β_n with respect to the β-phase liquidus temperature for that composition. If we assume that the $\alpha - \ell$ interface is at equilibrium, the undercooling ΔT^β_n is given by $(T^\beta_{liq}(C^{*\alpha}_\ell) - T^\alpha_{liq}(C^{*\alpha}_\ell))$ (see Fig. 9.2). At a certain critical value of ΔT^β_n, the β-phase nucleates somewhere on the $\alpha - \ell$ interface, as shown in Fig. 9.3. The nucleus usually has a particular crystallographic relationship with the parent phase. The β-phase then spreads across the $\alpha - \ell$ interface. Akamatsu et al. [2] observed this phenomenon in transparent organic systems, in which they found that the lateral spreading can become unstable, forming fingers or cells of β-phase as the interface advances. The α-phase is still able to continue its growth in between these β-cells, naturally establishing an

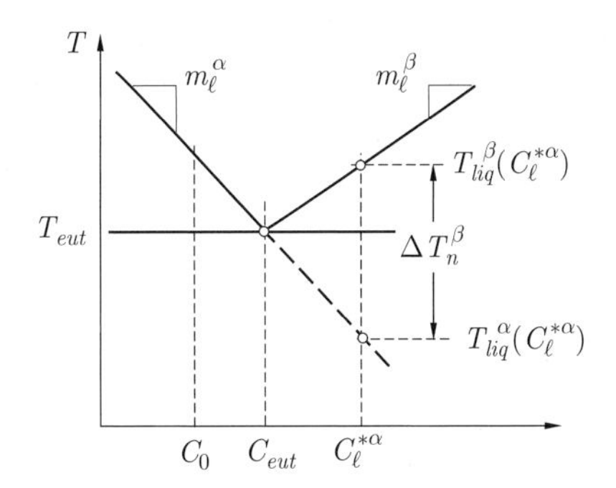

Fig. 9.2 Nucleation of the second phase in a hypoeutectic alloy.

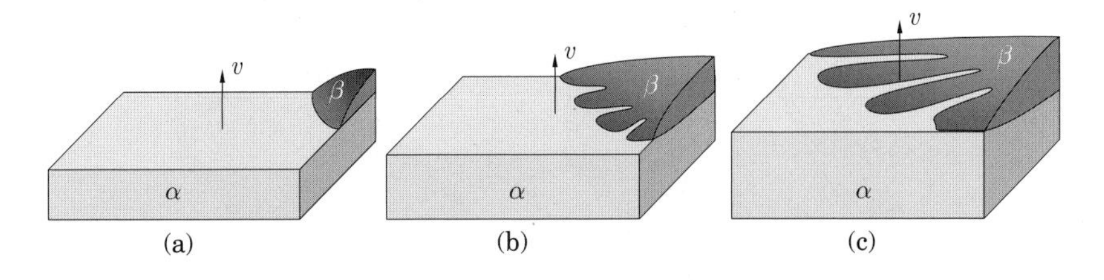

Fig. 9.3 (a) Nucleation of a β-phase at a certain point of a growing $\alpha - \ell$ interface. (b) Lateral growth of the β-phase is unstable, leading to the formation of cells. (c) The α-phase continues to grow in between the β-phase cells, leading to a periodic arrangement of lamellae. (After Akamatsu et al. [2]).

alternating sequence of α- and β-phases. This form persists during *coupled growth*, characterized by the presence of triple junctions between the α, β and ℓ phases. If both solid phases are at approximately the same temperature, corresponding to the same position in a thermal gradient, we call this mode *isothermal coupled growth*.

Consider next the alloy C_0 close to the eutectic composition shown in the idealized phase diagram in Fig. 9.4. Suppose that there are two solid phases, α and β, each growing with a planar front at constant velocity v^*, but that they are unaware of each other's presence. The solute profiles in the liquid ahead of α and β in this "thought experiment" are illustrated in Fig. 9.4(b). The liquid interfacial compositions ahead of the phases are obtained by extending the liquidus and solidus curves, as shown in Fig. 9.4(a). The composition in the liquid at the α-ℓ interface is enriched, $C_\ell^{*\alpha} = C_0/k_{0\alpha}$ ($k_{0\alpha} < 1$), whereas ahead of the β-phase the solute is depleted $C_\ell^{*\beta} = 1 - (1 - C_0)/k_{0\beta}$. Note that the partition coefficient $k_{0\beta} < 1$, and thus, if one were to redraw the phase diagram swapping elements A and B, the results would be similar. This requires that we define $k_{0\beta} = (1 - C_s^\beta)/(1 - C_\ell^\beta)$ when the composition is measured as the mass

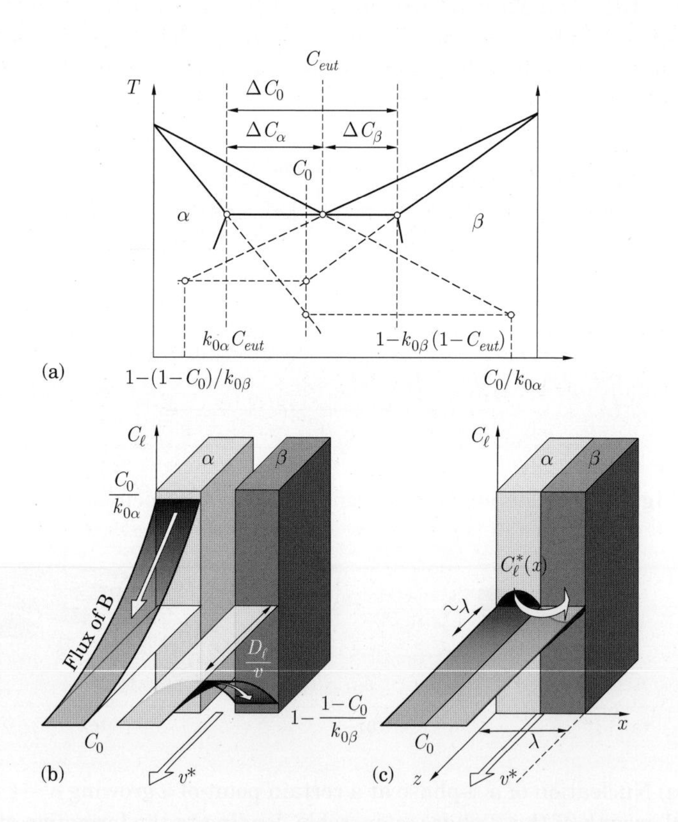

Fig. 9.4 (a) A linearized eutectic phase diagram. Schematic diagrams showing the steady-state solute field in the liquid ahead of (b) two independent α and β planar fronts, and (c) two lamellae α and β growing in a coupled manner.

fraction of element B. Note also that the two planar fronts in this thought experiment would not be at the same temperature.

Coupled growth provides a much more efficient mechanism for solute transport than the independent planar growth of each phase. Since the α-phase rejects B while the β-phase rejects A, the coupling of α and β, illustrated in Fig. 9.4(c), allows a *lateral* diffusion in the liquid, resulting in much smaller maximum and minimum compositions ahead of the two solid phases. The speed of the lateral diffusion increases as the lamellar spacing λ decreases. We will see in Sect. 9.2.3 that surface tension requires the interfaces to be curved, and the lamellar spacing is determined by a balance between transport and surface tension, similarly to what was found for dendritic growth in Chap. 8.

9.2.2 Coupled eutectic growth morphologies

Recall from Chap. 8 that materials for which $\Delta S_f^m/\mathcal{R} > 2$ usually also have a large anisotropy in solid-liquid interfacial energy and are therefore faceted, whereas those for which $\Delta S_f^m/\mathcal{R} < 2$ are non-faceted. The various combinations of such materials lead to different morphologies for the $\alpha - \beta$ eutectic. The eutectic microstructure is also affected by the volume fractions of the two phases. The present section describes the observed morphologies for coupled growth, as shown schematically in Fig. 9.5.

Faceted-faceted eutectics

If the two phases α and β are both faceted, they grow along well-defined directions. Unless a very specific orientation relationship occurs right from

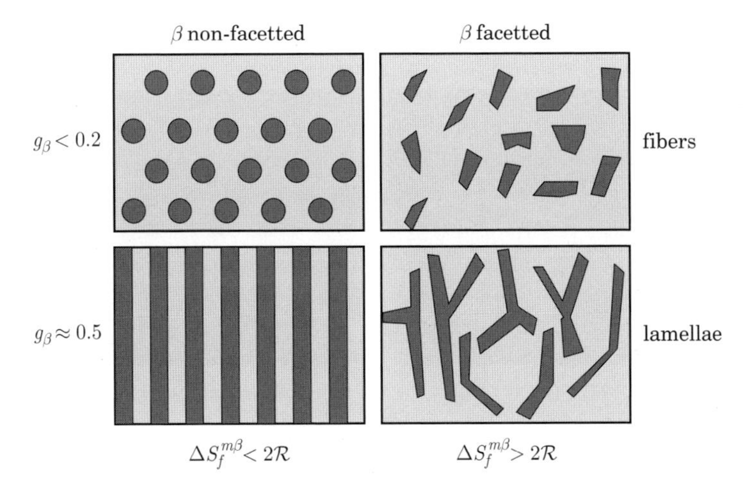

Fig. 9.5 Eutectic interface morphologies that can be obtained when the α-phase is non-faceted and the β-phase is either non-faceted (left) or faceted (right). This is shown for two volume fractions of the β-phase. The eutectic is growing in a thermal gradient perpendicular to the page.

the start of growth, the faceted needles or plates cannot maintain steady triple junctions. They will appear to grow "independently" from each other, although they still exchange solute via the liquid phase. Therefore, no coupled growth is possible in this case.

Irregular eutectics

When one of the two phases, e.g., α, grows with a non-faceted morphology while the β-phase is faceted, triple junctions can be maintained: the non-faceted phase follows the needles or plates of the faceted phase. As discussed in Sect. 2.4.2, the faceted phase grows along well-defined directions with the help of defects such as twins or screw dislocations. Therefore, the resulting eutectic structure is very complex and irregular, giving rise to an *irregular eutectic* morphology, as sketched schematically on the right-hand side of Fig. 9.5. This type of eutectic is found in metal-non metal systems such as Fe-C (gray cast iron) or Al-Si (Fig. 9.6b), and in various organics such as the borneol-succinonitrile alloy (Fig. 9.6d).

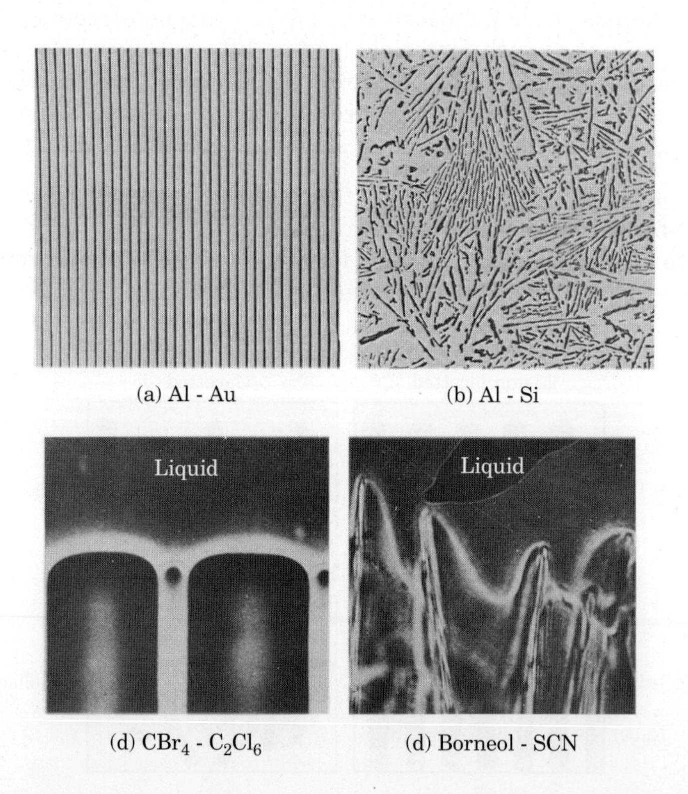

Fig. 9.6 Regular and irregular eutectics: (a) regular Al-Au eutectic observed in a longitudinal section; (b) irregular Al-Si eutectic observed in a transverse section; (c) regular CBr_4-C_2Cl_6 eutectic observed during growth; (d) an irregular borneol-succinonitrile eutectic observed during growth. The faceted borneol phase leads the growth of succinonitrile in between. (After Kurz and Fisher [9].)

Regular eutectics

When both solid phases are non-faceted, which is typical of two metallic phases, triple junctions can be maintained in almost any growth direction (Fig. 9.5, left). Some sort of crystallographic relationship usually exists between the α- and β-lamellae or fibers as a result of the nucleation sequence described above. However, during growth, the $\alpha - \beta$ interface becomes incoherent and can accommodate any growth direction. This leads to a *regular eutectic*, examplified by for instance Pb-Sn, Al-Zn, Fe-Fe_3C and Al-Au (Fig. 9.6a). Figure 9.6(c) shows the regular eutectic structure that forms during the growth of a C_2Cl_6-CBr_4 eutectic.

Let us consider further the crystallographic relationship between the α- and β-phases established at the nucleation stage. Figure 9.7 shows a transverse section of two grains from a DS sample of Al-Zn eutectic. The orientation of the lamellae in this regular eutectic clearly reveals the grain boundary. The figure also shows Al $\langle 100 \rangle$ and Zn $\langle 11\bar{2}0 \rangle$ pole figures for the lamellae of each grain. One can clearly see that, in each grain, the (111) plane of fcc Al (line drawn through the $\langle 110 \rangle$ directions in the pole figure) coincides with the (0001) plane of hcp Zn (line drawn through the $\langle 11\bar{2}0 \rangle$ directions in the pole figure). Furthermore, a $\langle 110 \rangle$ direction of Al coincides with a $\langle 11\bar{2}0 \rangle$ direction of Zn at the center of the pole figure, which corresponds to the direction of the thermal gradient. However, one can see by comparing the pole figures and the micrographs that the $\alpha - \beta$ interfaces of the lamellae are not exactly parallel to the planes of coherency.

In addition to the faceted or non-faceted character of the phases, the eutectic morphology is also affected by their relative volume fractions. In order to minimize the interfacial energy created during eutectic growth, the system develops either lamellae or fibers as illustrated in Fig. 9.5. For

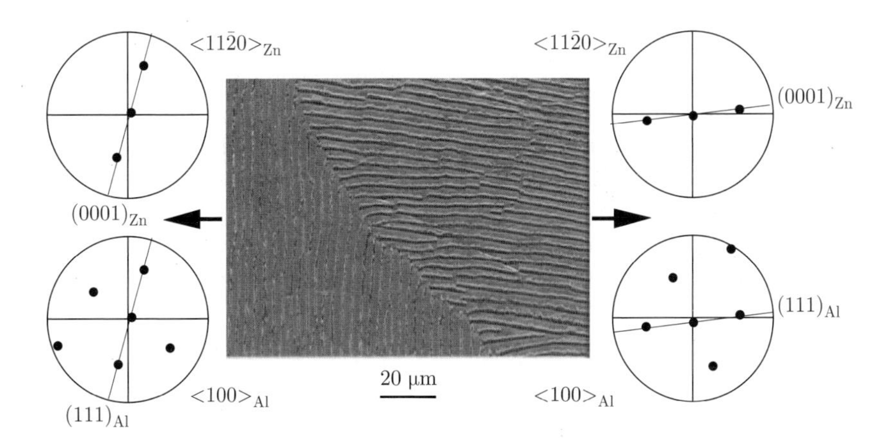

Fig. 9.7 Two grains of the lamellar Al-Zn regular eutectic with the corresponding pole figures for the fcc Al and hcp Zn phases, transverse cross section. (After Rhême et al. [33].)

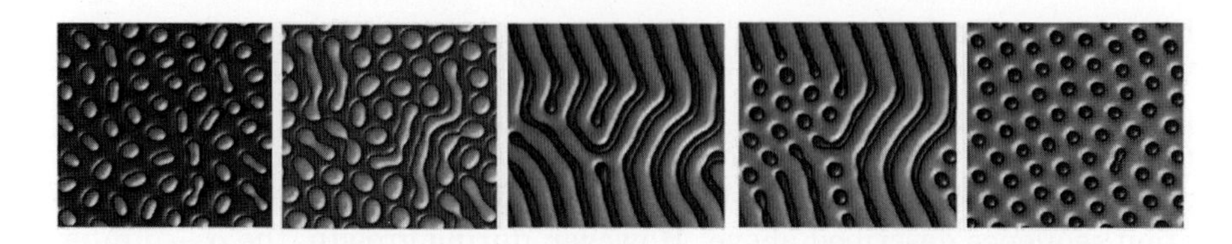

Fig. 9.8 Phase-field simulations of the interface in a regular eutectic for several values of g_β, increasing from left to right. The structure consists of β fibers (light gray) in a α-matrix at low g_β (left). At high g_β, α-fibers are embedded in a β-matrix. At intermediate values of g_β, a labyrinth-like lamella-structure is produced. (Courtesy of A. Parisi and M. Plapp [28].)

a volume fraction g_β close to 0.5, lamellae are preferred, whereas small values of g_β favor the formation of fibers. Depending on the arrangement of the fibers (e.g., square or hexagonal lattices), it can be shown that the transition between fibers and lamellae occurs for $g_\beta \approx 28 - 37\%$ (see Exercise 9.1).

Figure 9.8 shows computed interface morphologies from 3D phase-field simulations of regular eutectic growth in a model system. The α (dark gray) and β (light gray) phases grow towards the viewer. Moving from left to right in the figure, the eutectic point C_{eut} is gradually varied from low to high compositions of B, thus producing an evolution of the eutectic structure from low to high g_β. The system evolves from a fairly random arrangement of β fibers to a lamellar structure when $g_\beta \approx 0.5$. As g_β continues to increase, the lamellae of α "fragment" into a fairly regular hexagonal arrangement of nearly cylindrical fibers. In the simulations, the transition from random fibers to lamellae occurs at $g_\beta \approx 30\%$, whereas the transition from lamellae to a regular arrangement of fibers takes place at $g_\alpha \approx 20\%$.

The fully lamellar structure shown in the middle frame of Fig. 9.8 has a sinuous morphology, whereas the lamellae schematically represented in Fig. 9.5 or observed in Figs. 9.6(a) or 9.7 in Al-Au and Al-Zn, respectively, are fairly straight. Using phase field simulations, Plapp has shown that the "straightening" of the lamellae is the result of slightly curved isotherms. A small horizontal component of the thermal gradient tends to propagate the lamellae across the solid-liquid interface as growth proceeds, and to orient them along this component.

Following this qualitative description of eutectic structures, the next section develops the theory of Jackson and Hunt for the calculation of the lamellae spacing λ as a function of the growth rate in a regular lamellar eutectic [16]. This theory provides a basic explanation for the phenomena involved in coupled isothermal regular eutectic growth, which will then help in understanding more complex eutectic structures. This model extends a model previously derived by Hillert for eutectoids [13] to eutectic and off-eutectic alloy solidification with lamellae and fibers.

9.2.3 Jackson-Hunt analysis for regular eutectics

Consider a periodic arrangement of α- and β-lamellae with a characteristic spacing λ, solidifying at steady state with a constant velocity v^*. Symmetry permits us to restrict the analysis to the domain $x \in [0, \lambda/2]$ and $z \in [0, \infty[$, where $x = 0$ is positioned in the center of the α-lamella, and $z = 0$ corresponds to the eutectic front (see Fig. 9.9a). For the moment, it is assumed that the solid-liquid interface is flat, i.e., we neglect variations in the z-position of the $\alpha - \ell$ and $\beta - \ell$ interfaces. The triple junction $\alpha - \beta - \ell$ is therefore located at $x = g_\alpha \lambda/2$. The composition field in the liquid $C_\ell(x, z)$ satisfies the steady advection-diffusion equation:

$$\frac{\partial^2 C_\ell}{\partial x^2} + \frac{\partial^2 C_\ell}{\partial z^2} + \frac{v^*}{D_\ell}\frac{\partial C_\ell}{\partial z} = 0 \tag{9.1}$$

where D_ℓ is the diffusion coefficient in the liquid. The boundary conditions for the problem are given by:

$$\begin{aligned} D_\ell \frac{\partial C_\ell}{\partial x} &= 0 \qquad\qquad x = 0, \lambda/2 \\ C_\ell &= C_0 \qquad\qquad z \to \infty \end{aligned} \tag{9.2}$$

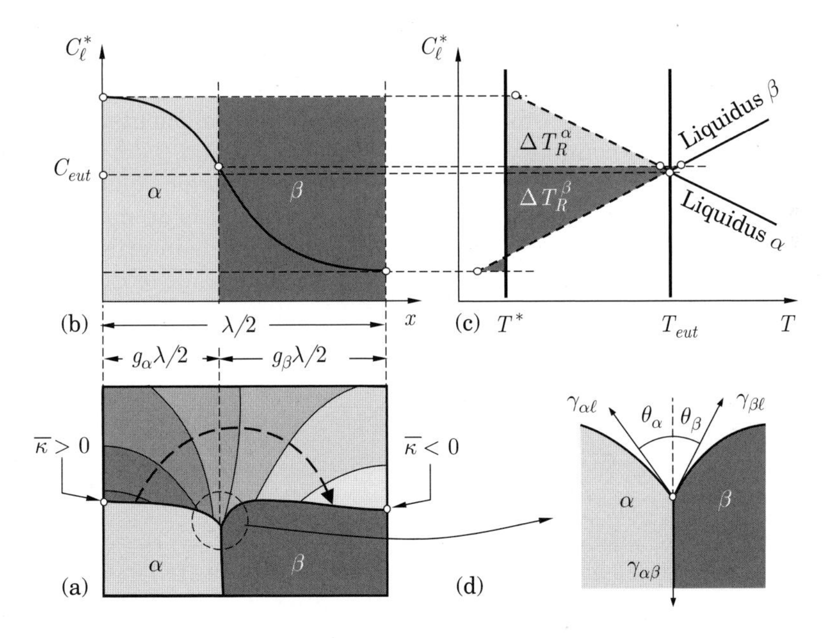

Fig. 9.9 (a) The solute profile $C_\ell(x, z)$ ahead of two lamellae α and β and (b) the corresponding solute profile in the liquid at the interface, $C_\ell^*(x)$. (c) Assuming an isothermal eutectic front at a temperature T^*, the difference between the corresponding liquidus $T_{liq}^\nu(C_\ell^{*\nu}(x)), \nu = \alpha, \beta$, and T^* is a direct measure of the curvature undercooling $\Delta T_R(x)$. (d) The Laplace-Young condition at the triple junction $\alpha - \beta - \ell$.

where C_0 is the nominal composition of the alloy, not necessarily equal to C_{eut}. Finally, at the eutectic interface $z = 0$, the flux condition is obtained according to:

$$D_\ell \left(\frac{\partial C_\ell}{\partial z}\right)^* = -v^* C_\ell^*(1 - k_{0\alpha}) \qquad z = 0, \quad 0 \leq x \leq g_\alpha \frac{\lambda}{2}$$

$$D_\ell \left(\frac{\partial C_\ell}{\partial z}\right)^* = v^*(1 - C_\ell^*)(1 - k_{0\beta}) \quad z = 0, \quad g_\alpha \frac{\lambda}{2} \leq x \leq \frac{\lambda}{2} \tag{9.3}$$

The x-periodicity of the eutectic structure implies that the solution to Eq. (9.1) can be obtained by a Fourier series (see Exercise 9.2)

$$C_\ell(x, z) = C_0 + \sum_{n=0}^{\infty} B_n \cos\left(\frac{2n\pi x}{\lambda}\right) \exp(-b_n z) \tag{9.4}$$

where the decay coefficients b_n that appear in the exponential are given by

$$b_n = \frac{v^*}{2D_\ell} + \left[\left(\frac{v^*}{2D_\ell}\right)^2 + \left(\frac{2n\pi}{\lambda}\right)^2\right]^{1/2} \qquad n \geq 0 \tag{9.5}$$

Note that the mode $n = 0$ corresponds to the familiar diffusion profile of a planar front growing at a velocity v^*, i.e., $b_0 = v^*/D_\ell$. As the eutectic spacing λ is much smaller than the planar front solute diffusion layer D_ℓ/v^*, the coefficients $b_n \simeq 2n\pi/\lambda$ for $n \geq 1$. The Fourier coefficients B_n of the series satisfying the boundary conditions at the $\alpha - \ell$ and $\beta - \ell$ interfaces (Eq. (9.3)) are determined from a system of linear algebraic equations (see Donaghey and Tiller [7]). However, if the eutectic front is at small undercooling, valid at low velocity, the liquid composition at the interface is close to C_{eut}. Replacing C_ℓ^*, which appears on the right-hand side of Eq. (9.3), by C_{eut}, permits the direct determination of the coefficients B_n of the Fourier series (see Exercise 9.2):

$$B_0 = \Delta C_\alpha g_\alpha - \Delta C_\beta g_\beta = C_{eut}(1 - k_{0\alpha})g_\alpha - (1 - k_{0\beta})(1 - C_{eut})g_\beta$$

$$B_n = \frac{1}{(n\pi)^2} \frac{v^*\lambda}{D_\ell} \Delta C_0 \sin(n\pi g_\alpha) \qquad n \geq 1 \tag{9.6}$$

where $\Delta C_0 = (\Delta C_\alpha + \Delta C_\beta) = [(1 - k_{0\beta}) + (k_{0\beta} - k_{0\alpha})C_{eut}]$ corresponds to the composition extension of the eutectic plateau as indicated in Fig. 9.4(a). The approximate solution to the diffusion equation with an arbitrary nominal composition $C_0 \neq C_{eut}$, valid for small undercooling, is thus

$$C_\ell(x, z) = C_0 + (\Delta C_\alpha g_\alpha - \Delta C_\beta g_\beta) \exp\left(-\frac{v^* z}{D_\ell}\right)$$
$$+ \Delta C_0 \frac{v^*\lambda}{D_\ell} \sum_{n=1}^{\infty} \frac{\sin(n\pi g_\alpha)}{(n\pi)^2} \cos\left(\frac{2n\pi x}{\lambda}\right) \exp\left(-\frac{2n\pi z}{\lambda}\right) \tag{9.7}$$

The first term on the left hand side is associated with long-range diffusion ahead of the eutectic front, whereas the summation corresponds to short-range diffusion at the scale of the lamellar spacing. As one would expect, the Péclet number based on the eutectic spacing $\mathrm{Pe} = v^*\lambda/D_\ell$ appears in this latter contribution.

Let us consider the behavior of the solution under the assumption of a small undercooling, where g_α and g_β are given by the lever rule regardless of the nominal composition C_0. We then have $g_\alpha = (1-(1-C_{eut})k_{0\beta} - C_0)/\Delta C_0$ and $g_\beta = (C_0 - k_{0\alpha}C_{eut})/\Delta C_0$. In that case, the term $B_0 = (\Delta C_\alpha g_\alpha - \Delta C_\beta g_\beta)$ is simply equal to $(C_{eut} - C_0)$. Therefore, for a nominal composition of the alloy $C_0 = C_{eut}$, $B_0 = 0$ and the long-range diffusion term, with a length scale D_ℓ/v^*, vanishes, leaving only cross diffusion between the lamellae. If the nominal composition of the alloy $C_0 = k_{0\alpha}C_{eut}$, i.e., the maximum solubility of the α-phase, we obtain $g_\alpha = 1$ and $g_\beta = 0$. This gives rise to $B_0 = C_{eut}(1-k_{0\alpha}) = C_0/k_{0\alpha}(1-k_{0\alpha})$ and $B_n = 0$ for $n \geq 1$. Substituting this result into Eq. (9.7) recovers the solution for a steady planar α-front growing at the solidus (Eq. (5.97)).

Figure 9.9(a) shows isopleths calculated using the Jackson-Hunt solution for $C_0 = C_{eut}$, along with one flux line perpendicular to the isopleths (thick dash line). Figure 9.10 presents the composition profiles in the liquid calculated for the same phase diagram and two different nominal compositions.

The composition at the interface in the liquid is obtained by setting $z = 0$ in Eq. (9.7):

$$C_\ell(x, z=0) = C_\ell^*(x) = C_0 + (\Delta C_\alpha g_\alpha - \Delta C_\beta g_\beta) + \Delta C_0 \frac{v^*\lambda}{D_\ell} F(x)$$

$$\text{where} \qquad F(x) = \sum_{n=1}^{\infty} \frac{\sin(n\pi g_\alpha)}{(n\pi)^2} \cos\frac{2n\pi x}{\lambda} \tag{9.8}$$

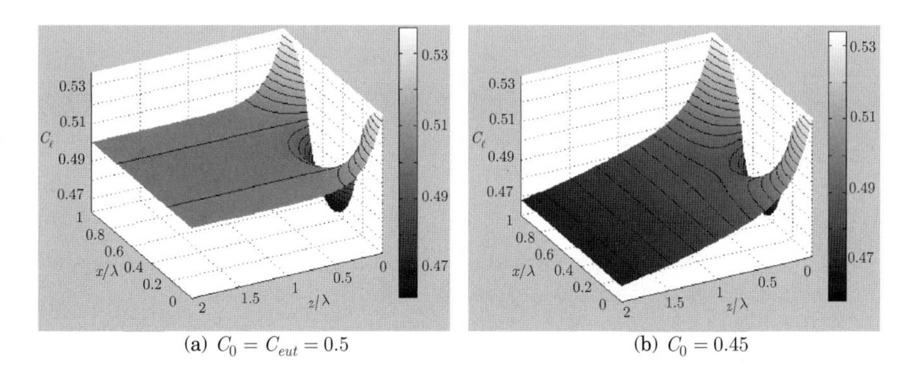

Fig. 9.10 A 3D view of the Jackson-Hunt solute profiles for two cases, as a function of the normalized coordinates x/λ and z/λ for $C_{eut} = 0.5$, $\mathrm{Pe} = 0.5$, $k_{0\alpha} = 0.2$, $k_{0\beta} = 0.2$. (a) $C_0 = C_{eut} = 0.5$; (b) $C_0 - 0.45$. Note in this case the long range diffusion with a normalized boundary layer $\delta/\lambda = 1/\mathrm{Pe} = 10$.

As shown schematically in Fig. 9.9(b), the interfacial liquid composition varies continuously from a maximum at the center of the α-lamella to a minimum at the center of the β-lamella. The corresponding liquidus temperature $T^{\nu}_{liq}(C^*_\ell(x))$ for each phase $\nu = \alpha, \beta$, is shown in Fig. 9.9(c), in which it can be seen to vary as a function of x: it is close to the eutectic temperature near the triple junction and is at its minimum at the center of the lamellae. However, for regular eutectics, both the α- and β-lamellae are at approximately the same position in the thermal gradient (see Fig. 9.6c). For example, if the triple junction is located 0.1μm behind the center of the lamellae in a thermal gradient of 10^4 Km^{-1}, the maximum temperature difference is 1 mK. Therefore, if we assume an isothermal front at a temperature T^*, the difference $(T^{\nu}_{liq}(C^*_\ell(x)) - T^*)$ represents directly the *curvature undercooling* $\Delta T^{\nu}_R(x)$ of the $\nu - \ell$ interface. This undercooling is maximum at the triple junction, where the α, β and ℓ phases have to satisfy the Laplace-Young equation (Fig. 9.9d). This is shown in gray in Fig. 9.9(c). Note that the curvature of the β-liquid interface was set arbitrarily to be negative at the center of the lamellae which also makes the curvature undercooling negative in this region. It also should be noted that the triple junction does not necessarily correspond to the eutectic composition.

As we have taken the eutectic front to be isothermal and neglected the attachment kinetics, the sum of the solute and curvature undercoolings, $\Delta T = \Delta T_C(x) + \Delta T_R(x)$, must be constant. The curvature contribution is shape-dependent and can only be predicted with a numerical technique such as front-tracking or phase-field. Jackson and Hunt simplified the analysis by computing the average solute and curvature undercoolings along the interface. Averaging the composition at the $\alpha - \ell$ and $\beta - \ell$ interfaces gives

$$\overline{C_\ell}^{*\alpha} = \frac{2}{g_\alpha \lambda} \int_0^{g_\alpha \lambda/2} C^*_\ell(x)dx = C_0 + (\Delta C_\alpha g_\alpha - \Delta C_\beta g_\beta) + \text{Pe}\frac{\Delta C_0}{g_\alpha}\bar{F}$$

$$\overline{C_\ell}^{*\beta} = \frac{2}{g_\beta \lambda} \int_{g_\alpha \lambda/2}^{\lambda/2} C^*_\ell(x)dx = C_0 + (\Delta C_\alpha g_\alpha - \Delta C_\beta g_\beta) - \text{Pe}\frac{\Delta C_0}{g_\beta}\bar{F}$$

where $$\bar{F} = \sum_{n=1}^{\infty} \frac{\sin^2(n\pi g_\alpha)}{(n\pi)^3} \tag{9.9}$$

Therefore the average solutal undercooling of each lamella, measured with respect to the eutectic temperature, can be expressed as:

$$\overline{\Delta T}^{\alpha}_C = |m_{\ell\alpha}|(\overline{C_\ell}^{*\alpha} - C_{eut}) = |m_{\ell\alpha}|\,(C_0 - C_{eut} + \Delta C_\alpha g_\alpha - \Delta C_\beta g_\beta) + \text{Pe}\frac{|m_{\ell\alpha}|\Delta C_0}{g_\alpha}\bar{F}$$

$$\overline{\Delta T}^{\beta}_C = |m_{\ell\beta}|(C_{eut} - \overline{C_\ell}^{*\beta}) = |m_{\ell\beta}|\,(C_{eut} - C_0 - \Delta C_\alpha g_\alpha + \Delta C_\beta g_\beta) + \text{Pe}\frac{|m_{\ell\beta}|\Delta C_0}{g_\beta}\bar{F} \tag{9.10}$$

On the other hand, the curvature of the $\alpha - \ell$ or $\beta - \ell$ interfaces is proportional to the second derivative of the interface shape $z(x)$ shown in Fig. 9.9(d). Averaging the curvature undercooling term for each phase (see Exercise 9.2) gives

$$\overline{\Delta T}_R^\alpha = \Gamma_{\alpha\ell} \frac{2}{g_\alpha \lambda} \int_0^{g_\alpha \lambda/2} \frac{z''(x)dx}{(1+z'^2(x))^{3/2}} = \frac{2\Gamma_{\alpha\ell}\cos\theta_\alpha}{g_\alpha \lambda}$$

$$\overline{\Delta T}_R^\beta = \Gamma_{\beta\ell} \frac{2}{(g_\beta)\lambda} \int_{g_\alpha \lambda/2}^{\lambda/2} \frac{z''(x)dx}{(1+z'^2(x))^{3/2}} = \frac{2\Gamma_{\beta\ell}\cos\theta_\beta}{g_\beta \lambda} \tag{9.11}$$

where θ_α and θ_β are the angles at the triple junction between the $\alpha - \beta$ interface and the tangents to the $\alpha - \ell$ and $\beta - \ell$ interfaces, respectively (see Fig. 9.9d). The total average undercooling $\overline{\Delta T}^\nu$ of phase ν is given by the sum of $\overline{\Delta T}_C^\nu$ and $\overline{\Delta T}_R^\nu$ obtained in Eqs. (9.10) and (9.11), respectively. Since we assumed that the eutectic front is isothermal, one has $\overline{\Delta T}^\alpha = \overline{\Delta T}^\beta$. Following Jackson and Hunt, an average of $\overline{\Delta T}^\alpha$ and $\overline{\Delta T}^\beta$ is obtained with the weight factors $|m_{\ell\alpha}|^{-1}$ and $|m_{\ell\alpha}|^{-1}$, respectively.

$$\overline{\Delta T} = \frac{\overline{\Delta T}^\alpha |m_{\ell\alpha}|^{-1} + \overline{\Delta T}^\beta |m_{\ell\beta}|^{-1}}{|m_{\ell\alpha}|^{-1} + |m_{\ell\beta}|^{-1}} \tag{9.12}$$

This simplifies the expression by canceling the first term in Eqs. (9.10). It turns out that this contribution is very small; it is in fact zero under the assumption that the fractions of α and β phases follow the lever rule regardless of C_0 (since $B_0 = \Delta C_\alpha g_\alpha - \Delta C_\beta g_\beta = (C_{eut} - C_0)$ in this case). The total undercooling of the eutectic front is then given by the key relationship:

$$\Delta T = A_C \text{Pe} + \frac{A_R}{\lambda} = A_C \frac{v^* \lambda}{D_\ell} + \frac{A_R}{\lambda} \tag{9.13}$$

with

$$A_C = \frac{\Delta C_0}{g_\alpha g_\beta} \frac{|m_{\ell\alpha}||m_{\ell\beta}|}{|m_{\ell\alpha}| + |m_{\ell\beta}|} \sum_{n=1}^{\infty} \frac{\sin^2(n\pi g_\alpha)}{(n\pi)^3}$$

$$A_R = \frac{|m_{\ell\alpha}||m_{\ell\beta}|}{|m_{\ell\alpha}| + |m_{\ell\beta}|} \left(\frac{2\Gamma_{\alpha\ell}\cos\theta_\alpha}{|m_{\ell\alpha}|g_\alpha} + \frac{2\Gamma_{\beta\ell}\cos\theta_\beta}{|m_{\ell\beta}|g_\beta} \right) \tag{9.14}$$

For clarity of presentation, we omit the overbar from $\overline{\Delta T}$. The expression for the undercooling of the eutectic front ΔT in Eq. (9.13) separates the contributions of diffusion and capillarity. The former is proportional to the Péclet number Pe = $v^* \lambda / D_\ell$ and thus increases with spacing λ and velocity v^*. The latter is inversely proportional to λ with a coefficient proportional to a weighted sum of the Gibbs-Thomson coefficients of the two phases. The relationship $\Delta T(\lambda)$ is plotted in Fig. 9.11 for a hypothetical alloy, at three interface velocities. For any given velocity, the undercooling

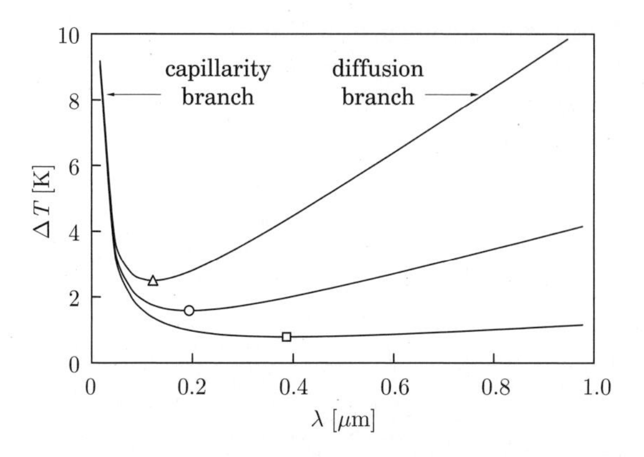

Fig. 9.11 Undercooling of the eutectic front ΔT as a function of the lamellar spacing λ for 3 growth velocities: $v^* = 100$ (square), 400 (circle) and 1000 μm s^{-1} (triangle). The symbol indicates the position of the minimum undercooling. $|m_{\ell\alpha}| = |m_{\ell\beta}| = 250$ K, $\Gamma_{\alpha\ell} = \Gamma_{\beta\ell} = 10^{-7}$ Km, $\theta_\alpha = \theta_\beta = 3\pi/8$, $g_\alpha = g_\beta = 0.5$, $\Delta C_0 = 0.5$, $D_\ell = 3 \times 10^{-9}$ m^2 s^{-1}.

has a minimum at a particular value λ_{ext}, represented by a symbol on the curve.

It is straightforward to show (see Exercise 9.2) that this minimum is given by

$$\lambda_{ext} = \sqrt{\frac{A_R}{A_C}\frac{D_\ell}{v^*}} \tag{9.15}$$

Note that A_R/A_C is essentially a weighted average capillarity length, whereas D_ℓ/v^* is the thickness of the boundary layer ahead of a steady state planar front. It is interesting to note that at the extremum, $\Delta T_R = \Delta T_C$, i.e., the average curvature and solutal undercoolings provide equal contributions to the total undercooling. Thus, λ_{ext} is the geometric mean of these two length scales. The undercooling, growth rate and lamellae spacing of eutectic structures at the extremum are interrelated via the following equations:

$$\lambda_{ext}^2 v^* = \frac{A_R}{A_C} D_\ell \quad ; \quad \Delta T = 2\sqrt{\frac{A_R A_C}{D_\ell}}\sqrt{v^*} \quad ; \quad \lambda_{ext}\Delta T = 2A_R \tag{9.16}$$

Note that these expressions are independent of the thermal gradient G. Equations (9.16) determine the spacing and the undercooling of the eutectic lamellae when the interface velocity is set. These equations can also be used for equiaxed solidification, where one assumes that the local undercooling, rather than the velocity, is fixed. In this case, the extremum criterion corresponds to a maximum velocity of the eutectic structure for a given undercooling (see Fig. 9.12).

The theory of coupled eutectic growth has been extended to rod-like structures by Jackson and Hunt [16] as well as to rapid solidification

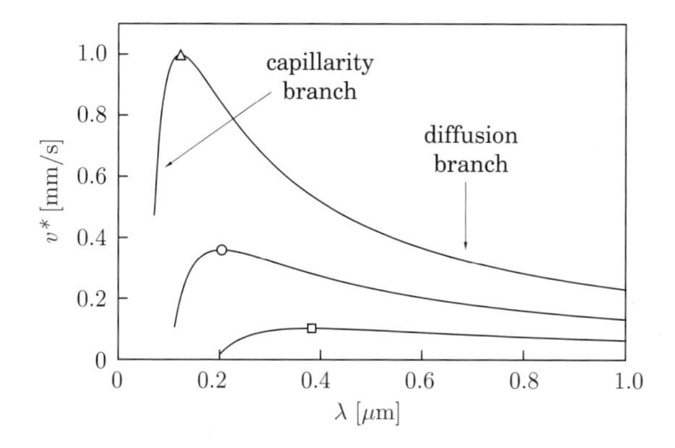

Fig. 9.12 The velocity of the eutectic front v^* as a function of the lamellar spacing λ for 3 undercoolings: $\Delta T = 0.8$ K (square), 1.5 K (circle) and 2.5 K (triangle). The symbol indicates the position of the maximum velocity. The other parameters are the same as those given in Fig. 9.11.

conditions by Trivedi et al. [35] (see also the book of Kurz and Fisher [21].). We now consider the stability of the structures that have just been computed.

9.2.4 Operating point and stability of regular eutectic

Experimental observations show that regular eutectics grow close to the extremum condition defined by Eq. (9.16). It should be noted that this contrasts with dendrites, where the operating state is found to be far from the extremum (see Sect. 8.4). One does find, however, that the lamellae or fibers exhibit fluctuations in their spacing and arrangement, especially when the velocity changes. Figure 9.13 illustrates the mechanisms for adaptation of lamellar spacing under such changes in velocity in the transparent carbon tetrabromide (CBr_4)-hexachlorethane (C_2Cl_6) eutectic alloy. Let us assume that the steady-state lamellar spacing $\lambda_{old} = \lambda(v^*_{old})$ before the velocity change corresponds to a point close to the extremum given by the open circle in Fig. 9.11. When the velocity decreases from v^*_{old} to v^*_{new}, the spacing λ_{old} is smaller than the new extremum $\lambda_{new} = \lambda(v^*_{new})$, given for instance by the open square in Fig. 9.11. In order to adapt to the new conditions, the system eliminates lamellae by an overgrowth mechanism, as evidenced by the dark gray lamella at mid-height of Fig. 9.13 being overgrown by its two light gray neighbors. The system then readapts the other lamellar spacings by lateral adjustment at the remaining triple junctions. This elimination of lamellae occurs as a result of the system being forced to move below λ_{ext}, i.e., onto the capillarity branch of the $\Delta T(\lambda, v^*_{new})$ curve shown in Fig. 9.11.

When the velocity increases, the steady-state lamellar spacing λ_{old} is larger than the new extremum spacing λ_{new} (e.g., the open triangle in

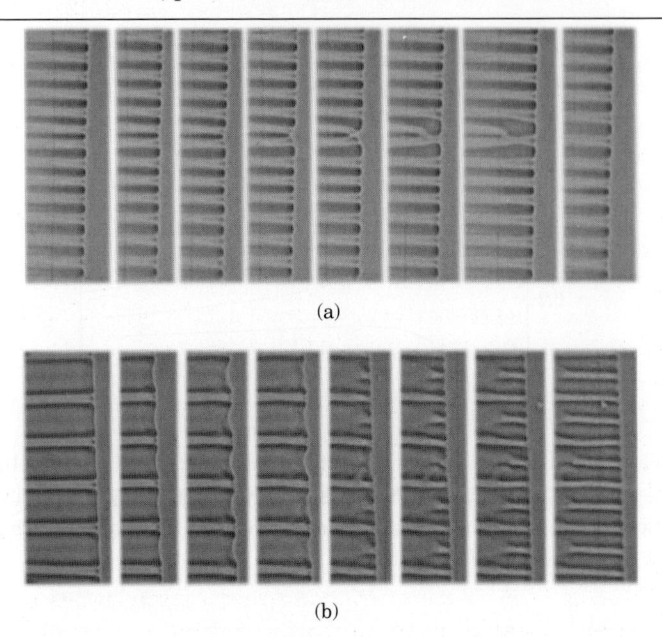

Fig. 9.13 Adaptation of lamellar spacing due to speed change in CBr_4-C_2Cl_6 eutectic alloy. (a) Speed reduction leading to elimination of lamellae. (b) Speed increase leading to creation of lamellae. (Sequences of pictures assembled from the movie of Jackson and Hunt).

Fig. 9.11). This means that the system is located to a larger degree on the diffusion branch of the $\Delta T(\lambda, v^*_{new})$ relationship. As a result, solute builds up ahead of the α-lamellae and the solvent does the same in front of the β-lamellae. As can be seen in Fig. 9.13(b), the wider lamellae (i.e., α) begin to develop a negative curvature at their centers in order to accommodate the extra solute. When the curvature becomes negative, even more solute builds up, until finally new β-lamellae form by nucleation in the middle. In thin specimens, such as those shown in Fig. 9.13, this occurs at about $v^*_{new} = 4v^*_{old}$ as a result of the spacing being approximately halved during lamellae creation and $\lambda^2 v^*$ being constant at the extremum (Eq. (9.16)).

These observations led Jackson and Hunt to conclude that regular eutectics cannot grow below λ_{ext} on the capillarity branch (otherwise a lamella is eliminated). However, they can grow up to about $2\lambda_{ext}$ on the diffusion branch (at which point the spacing is halved by formation of new lamellae). This would give an average value of about 1.5λ in 3D specimens, without taking into account stereological effects (lamellae that do not grow precisely along the thermal gradient appear wider and with larger spacing in a transverse metallographic section). However, recent theoretical analyses, phase field simulations and experimental observations have demonstrated that the situation is not that simple, as we now explore.

Since regular eutectics do not grow exactly at the extremum given by Eq. (9.16), we introduce the dimensionless eutectic spacing $\Lambda = \lambda/\lambda_{ext}(v^*)$, where λ is the actual spacing and $\lambda_{ext}(v^*)$ is defined by Eqs. (9.15). For $\Lambda < 1$, the curvature undercooling is greater than the solutal undercooling, and the situation is reversed for $\Lambda > 1$. Jackson and Hunt conjectured that regular eutectics grow with Λ in the range $[\Lambda_R, \Lambda_C]$, where $\Lambda_R = 1$ is imposed by capillarity (termination of one lamella) and $\Lambda_C \approx 2$ is due to the creation of new lamellae, as previously described. Through a series of well-controlled experiments on very thin CBr_4-C_2Cl_6 specimens (typically 15 μm thick, and thus approximately 2D), Faivre and co-workers determined that Λ_R depends on the thermal gradient, and can be as small as 0.7 for typical solidification conditions. Under more regular 3D solidification conditions, it seems that Λ_R is closer to 1.

Along the diffusion branch, the situation is more complicated and several new mechanisms have been identified in addition to the one seen in Fig. 9.13(b). The mechanism selected depends on the alloy composition, thermal gradient and geometry (2D or 3D). In 2D, experimental observations [2; 3] and boundary integral calculations performed by Karma and Sarkissian [18] revealed multiple transitions in regular eutectic systems, illustrated in Fig. 9.14. The symmetric eutectic pattern described in the previous section is stable up to a critical spacing $\Lambda_{2\lambda}$, beyond which an oscillatory mode appears, called 2λ since the periodicity of the phase α is still λ but that of the β-phase is 2λ. A periodic tilted lamellar structure appears when $\Lambda > \Lambda_{tilt}$, as a result of the associated undercooling ΔT being smaller than that of periodic or oscillatory modes. The tilt angle between the lamellae and the thermal gradient increases with $(\Lambda - \Lambda_{tilt})$. Outside the stable growth regime of tilted states (shown as a thick line in Fig. 9.14), other oscillatory modes such as 1λ-oscillations have been observed. Further details can be found in Akamatsu et al. [3].

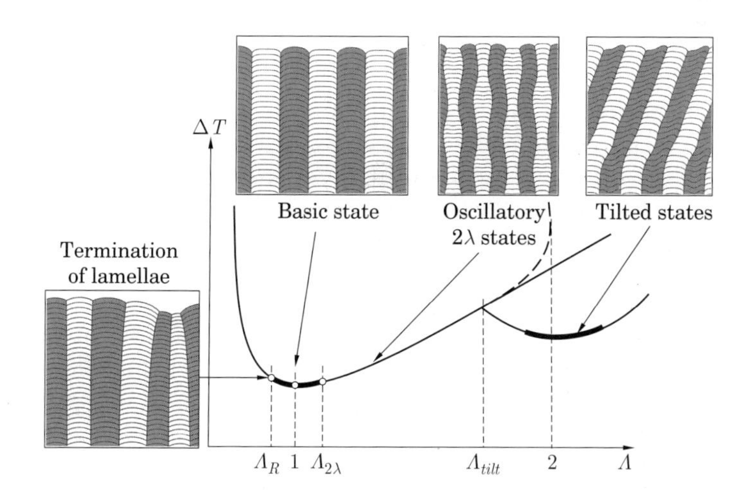

Fig. 9.14 Stability diagram for regular eutectics, redrawn from Faivre, [2; 3] along with the calculations of Karma and Sarkissian [18]. Note that the curve $\lambda - \Delta T$ of the basic symmetric state has a cut-off at a value of Λ slightly larger than 2.

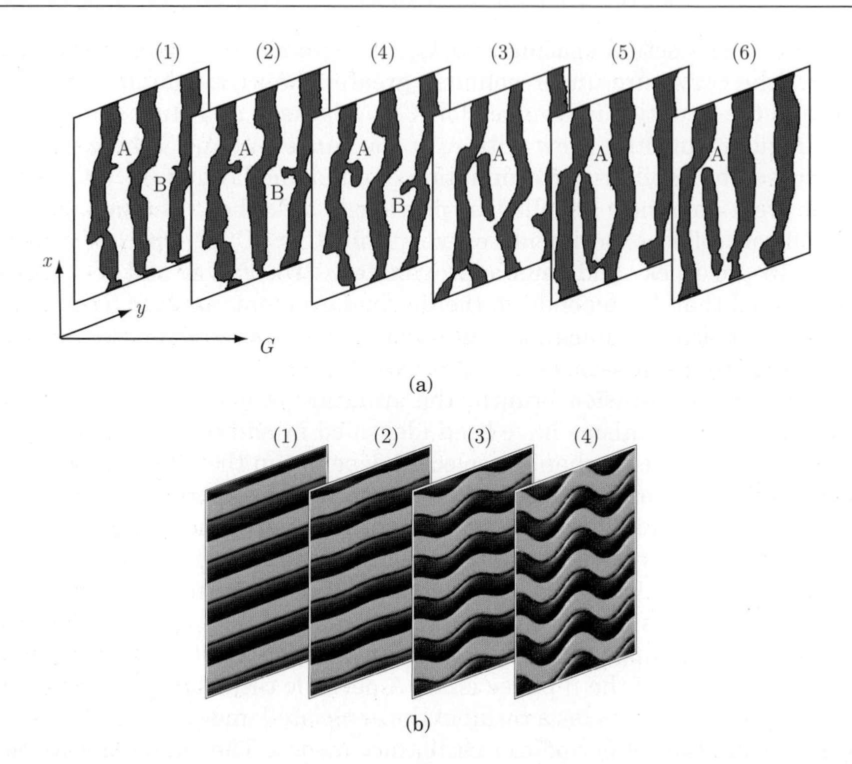

Fig. 9.15 3D mechanisms for decreasing the lamellar eutectic spacing: (a) Lateral instability leading to the formation of a new lamella in the Al-Al_2Cu system, as observed after solidification in serial sections labeled (1) to (6) (Adapted from Walker et al. [36]) (b) Development of a zigzag structure calculated using a multiphase field model for $\Lambda = 1.3$ (steady state is achieved in (4)). (Adapted from Parisi and Plapp [28].)

In 3D specimens, the situation is markedly different, and eutectic morphologies can accommodate a velocity increase by a wide variety of mechanisms, two of which are illustrated in Fig. 9.15. Lamellae can remain more or less straight, but they exhibit lateral instabilities when Λ exceeds a critical value Λ_{crit} (Fig. 9.15a). Walker et al. [36] observed the development of such instabilities in small diameter Al-Cu specimens, illustrated by the serial sections shown in Fig. 9.15(a). A new lamella starts as a small initial protrusion, labeled A in section (1), which develops and then grows parallel to the other lamellae. It eventually detaches from its parent lamella to form a new one (Sect. 6). Note that during this process, the initial protrusion B could not develop because of its proximity to protrusion A. Walker et al. also found the growth, coalescence and subsequent detachment of an array of protrusions to be another mechanism of lamella creation. The critical value Λ_{crit} for the formation of a new lamella in 3D was determined to be around 1.2, instead of the value 2 found in the 2D experiments of Jackson and Hunt (Fig. 9.13b).

Another mechanism for reducing the lamellar spacing is the formation of a zigzag structure, as illustrated in Fig. 9.15(b). These four sections show the calculated evolution of the structure upon a velocity increase corresponding to a change from $\Lambda = 1$ to $\Lambda = 1.3$. Straight lamellae first develop a lateral oscillation that then becomes stabilized into a zigzag array. The diffusion distance between lamella centers is decreased by a factor $\sin\theta$ in the straight part of the zigzag, where θ is the angle between the zigzag and the original direction of the lamellae. This allows the system to operate closer to the extremum value. The images shown in sections (1)-(4) of Fig. 9.15(b) came from simulations by Parisi and Plapp [28] using a multi-phase field model. Such structures have been also observed *in-situ* in thick transparent systems by means of long-distance oblique microscopy [1]. At smaller values of Λ, extensions to 3D of 2λ- and 1λ-oscillation modes have also been observed in phase-field simulations. At large values of Λ, labyrinth-like structures similar to that shown in the middle panel in Fig. 9.8 can occur.

9.2.5 Irregular eutectics

When one of the two phases is faceted, the eutectic becomes *irregular* since the faceted phase is capable of growing only along well-defined planes and/or directions. For example, the lamellae or flakes of Si in Al-Si grow as (111) planes, and the graphite in Fe-C grows as (0001) planes. The faceted phase lamellae "lead" the eutectic morphology with the non-faceted phase surrounding them, as shown in Fig. 9.6(d) for the borneol-succinonitrile system. As discussed in Chap. 2, the faceted phase grows with the help of defects such as twins or screw dislocations. It can also develop instabilities along the edges of lamellae, and attachment kinetics also may play a significant role. Figure 9.16 illustrates various defect-assisted mechanisms operating during irregular eutectic growth. In gray cast iron (Fig. 9.16a), the (0001) graphite flakes can develop a rotational defect along the $[10\bar{1}0]$ during solidification, resulting in very intricate and complex 3D graphite morphologies that resemble leaves of a cabbage. In the Al-Si eutectic (Fig. 9.16b), (111) twins form flakes with a typical spacing of about $0.4 - 1\ \mu$m under normal solidification conditions. The flakes, seen here in cross section, look fairly random (Fig. 9.6b), although they are interconnected in 3D and usually can be traced back to a single nucleation center.

As the growth mechanisms of the faceted phase are fairly complex, the resulting microstructure of irregular eutectics can be affected in various ways, for example by the growth rate and/or by the addition of "reactive" elements. For Al-Si alloys, the addition of Na, Sr, Ba and certain rare earth elements refines the Si phase to a point where it becomes non-faceted and fibrous. The mechanism is thought to be an enhancement of twinning in the Si phase by the adsorption of elements with large atomic radii. Similar modification can be obtained if the growth rate is increased to about 1 mm s^{-1}. The Na-modification of Al-Si foundry alloys, discovered by accident in the 1920's, provided an incentive to search for similar

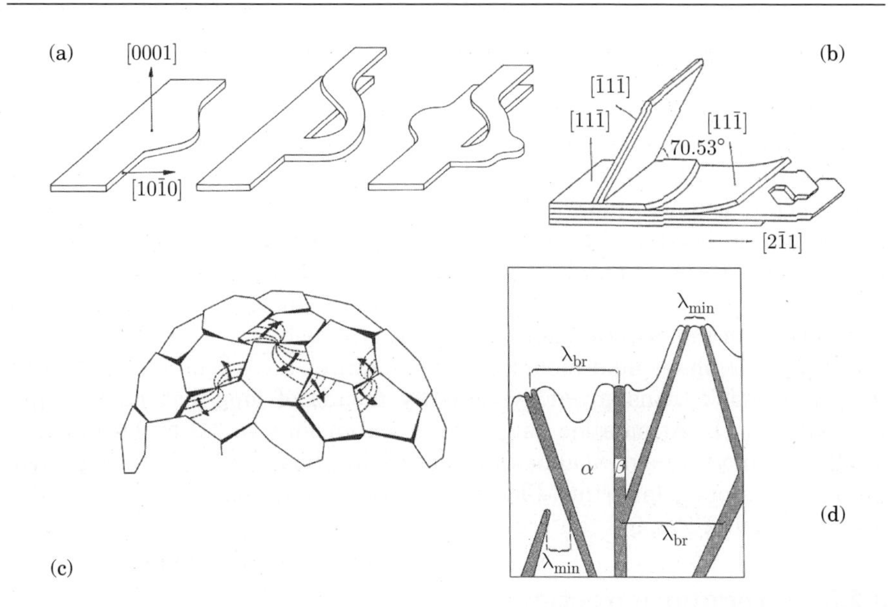

Fig. 9.16 Possible branching mechanisms in irregular eutectic growth: (a) A rotation boundary defect along $[10\bar{1}0]$ for Fe-C (after Minkoff and Lux [26]); (b) Twin formation in Al-Si (after Lu and Hellawell [24]); (c) Screw dislocation growth mechanism for Fe-C nodular cast iron (after Double and Hellawell [8]); (d) Schematic idealized irregular eutectic growth (after Fisher and Kurz [9]).

modifiers for gray cast iron. Just after the Second World War, it was found that when sulphur was eliminated from the melt, the addition of Ca or Mg transformed the graphite flakes into spherulites, also called nodules. This peculiar morphology, shown in Fig. 9.19(a), is very important since it renders the graphite-iron composite much less brittle, for which reason nodular cast iron is also called "ductile iron".

Similar nodular morphologies can be obtained in Ni-C alloys without any melt treatment, provided that cooling is fast enough. This structure can also be obtained by keeping molten gray iron in vacuum for very long periods of time (several days, although this is clearly not a commercially viable process!) in order to remove volatile elements such as O, P or S. The Mg treatment of cast iron seems to have an indirect effect on the growth mechanism of graphite. Double and Hellawell suggested that Mg traps trace solute elements that would otherwise be adsorbed at the edges of the basal plane and promote flake graphite through a complex growth mechanism. They suggested that, in very clean Fe-C, graphite does not grow along the basal plane but rather perpendicular to it along the c-axis. The natural graphite morphology would therefore not be flakes, but rather spherulites with an interface bounded by the basal plane. Figure 9.16(c) illustrates this mechanism, showing a graphite nodule consisting of a family of conical crystals, each with an axis parallel to $[0001]$ and bounded at their base by a (0001) plane. Note that the presence of screw dislocation created steps on the (0001) surface, enabling growth perpendicular to this plane.

The growth mechanisms of the faceted phase naturally control the growth patterns in irregular eutectics. The schematic diagram of Fisher and Kurz shown in Fig. 9.16(d) summarizes this process well: the faceted phase leads the eutectic reaction, and is constrained to grow along specific crystallographic planes/directions, often with the help of defects. When two faceted lamellae converge, the spacing decreases and the behavior is similar to the case of regular eutectics when $\Lambda < 1$ (Fig. 9.13a): one of the lamellae becomes overgrown. On the other hand, when two lamellae of the faceted phase diverge, $\Lambda > 1$ and the non-faceted phase starts to build up solute at the center (C in gray cast iron, Si in Al-Si). The local liquidus temperature therefore decreases (with $m_\ell < 0$) and the interface recedes in the temperature gradient. For a particular spacing, called λ_{br}, the driving force for the growth of the faceted phase in this region is large enough for branching to occur. In this sense, the range $[\lambda_{min}, \lambda_{br}]$ is conceptually similar to the range $[\lambda_R, \lambda_C]$ seen in regular eutectics, but appears to be much wider due to the growth restrictions on the faceted phase.

Note that the schematic diagram of Fig. 9.16(d) is 2D, and therefore it does not consider faceted lamellae originating from other sections that are able to grow in the most undercooled regions ahead of the non-faceted phase. Nevertheless, it can be concluded for irregular eutectics that: (i) the eutectic front is not truly isothermal, as the non-faceted phase can exhibit large depressions; (ii) the eutectic spacing varies. When $\lambda \approx \lambda_{min}$, a lamella is stopped, whereas when $\lambda \approx \lambda_{br}$, branching occurs. Results from experiments performed by Jones and Kurz on Fe-C eutectics are presented in Fig. 9.17. These authors measured the minimum, maximum and average eutectic spacings in samples directionally solidified using various combinations of G and v. The diagram also indicates the extremum spacing for a regular Fe-C eutectic by an open circle on the corresponding Jackson and Hunt expression, as well as the measured undercooling of the front, indicated by a horizontal arrow. It should be noted that the measured eutectic front temperature seemed to be very close to the maximum spacing λ_{br}.

As can be seen in Fig. 9.17, the value λ_{min} was close to λ_{ext} whereas λ_{br} was substantially larger (in a few cases, by nearly one order of magnitude). Assuming that both spacings lie on the $\Delta T(\lambda)$ curve calculated using the theory of Jackson and Hunt, and defining an average spacing $\bar{\lambda}$ as:

$$\bar{\lambda} = \frac{\lambda_{min} + \lambda_{br}}{2} = \phi\lambda_{ext} \tag{9.17}$$

it can be demonstrated that the growth kinetics of the irregular eutectic is given by (cf. Exercise 9.3):

$$\begin{aligned}
\Delta T\bar{\lambda} &= A_R\left(1+\phi^2\right) \\
\Delta T &= \left(\phi^{-1}+\phi\right)\sqrt{\frac{A_R A_C}{D_\ell}}\sqrt{v} \\
\bar{\lambda}^2 v &= \phi^2\frac{A_R}{A_C}D_\ell
\end{aligned} \tag{9.18}$$

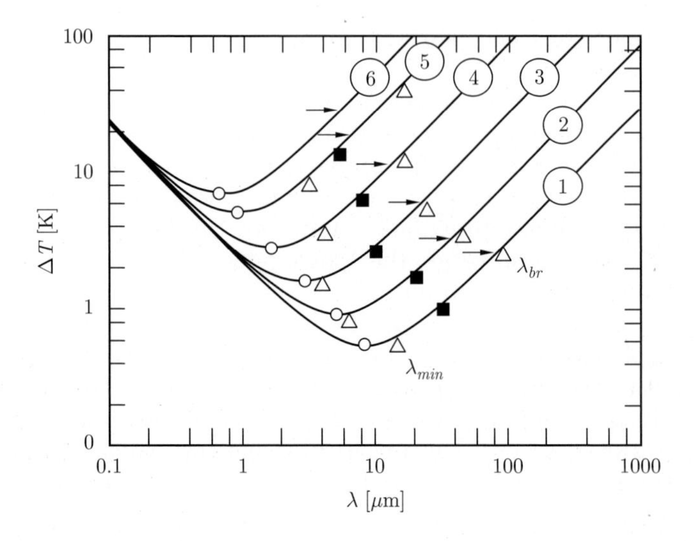

Fig. 9.17 Minimum and maximum eutectic spacings (open triangles) as well as average measured spacings (filled squares) in Fe-C irregular eutectics observed under various growth conditions: (1) $G = 65$ K cm^{-1}, $v = 0.41$ μm s^{-1}, (2) $G = 69$ K cm^{-1}, $v = 1.11$ μm s^{-1}, (3) $G = 72$ K cm^{-1}, $v = 3.47$ μm s^{-1}, (4) $G = 70$ K cm^{-1}, $v = 10.69$ μm s^{-1}, (5) $G = 71$ K cm^{-1}, $v = 435.2$ μm s^{-1}, (6) $G = 100$ K cm^{-1}, $v = 64.7$ μm s^{-1}. In each case, the measured undercooling is indicated by a horizontal arrow, the calculated Jackson-Hunt relation for regular eutectics by a continuous curve, and the extremum by an open circle. (After Jones and Kurz [17].)

As the eutectic front is no longer isothermal, these relationships become dependent on the thermal gradient (cf. Exercise 9.3).

9.2.6 Other eutectic morphologies

Divorced eutectics

Under certain conditions, hypoeutectic alloys may solidify with what is called a *divorced eutectic* structure. This typically occurs for hypoeutectic alloys with a fairly low solute composition, i.e., a relatively high volume fraction of the primary phase at the eutectic point. In this case, the remaining liquid is in the form of very thin films or even isolated droplets. Such a limited volume of liquid makes nucleation of the second phase more difficult, because nucleation has to occur within each individual droplet if the coalescence of the primary phase is well advanced. Since this requires a relatively large nucleation undercooling, the primary phase can continue to grow before the secondary phase appears, thus, further restricting the available space for the secondary phase to grow. When the diffusion distance between the primary and secondary phases is comparable to, or smaller than that required for coupled eutectic growth, there is no energy to be gained from forming triple junctions. The secondary phase

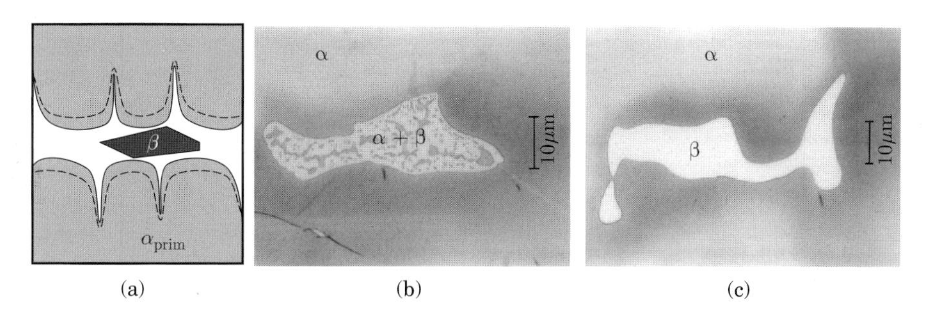

Fig. 9.18 (a) A schematic diagram of divorced eutectic solidification. (b) Partially-divorced and (c) fully divorced eutectic in AZ91E (Mg-9%Al-1%Zn) alloy, directionally solidified under conditions that are similar to those representative of the center and surface of a permanent mold casting. (Courtesy of A. Dahle, Australia [27].)

then grows independently while the primary phase continues its growth. This is shown schematically in Fig. 9.18(a).

Divorced or partially-divorced eutectics are common in Mg alloys. Depending on the cooling conditions, a binary Mg-Al alloy of eutectic composition (33 wt% Al) typically solidifies with a regular lamellar or fibrous morphology comprised of the hcp Mg primary phase and $Mg_{17}Al_{12}$ intermetallic phase. As the composition of Al decreases, there is an increased tendency to form a divorced eutectic, because the volume fraction of the primary phase increases. This tendency is reinforced by the addition of small amounts of Zn, a substance that, compared to Al, is more strongly segregated into the liquid and also causes the temperature along the eutectic monovariant line to decrease rapidly. The tendency to form a divorced eutectic is further reinforced by a high cooling rate. Figure 9.18(b) and (c) display micrographs of a commercial AZ91E alloy (Mg-9 wt%Al-0.8 wt%Zn) that has been solidified under cooling conditions that reproduce those encountered near the center of a permanent mold casting (cooling rate typically $-20\,\mathrm{K\,s^{-1}}$) and near the surface (cooling rate $-80\,\mathrm{K\,s^{-1}}$), respectively. For the slower cooling rate, the Mg primary phase (zone α) appears as light gray, and gradually becomes darker, corresponding to higher Al and Zn content, as we approach the zone where a final coupled eutectic zone appears (labeled $(\alpha+\beta)$ in Fig. 9.18(b)). Note the very small size of this last liquid region where the partially divorced (or partially coupled) eutectic has formed. In Fig. 9.18(b), the $Mg_{17}Al_{12}$ phase appears as a much lighter hue all along the perimeter of the liquid pocket, with small Mg islands inside. Figure 9.18(c) shows a fully divorced eutectic region. In this case , the magnesium phase reaches a higher Al content near the $Mg_{17}Al_{12}$ single phase intermetallic particle (bright zone β). Further details on divorced eutectics in Mg alloys can be found in Nave et al. [27].

Nodular cast iron

As already pointed out in Sect. 9.2.5, Mg and Ca can inhibit the growth of graphite on its basal plane, promoting the formation of spherulites.

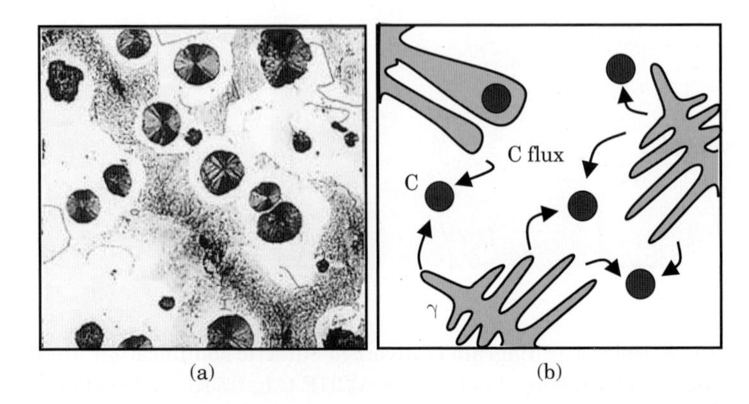

Fig. 9.19 (a) A micrograph of a nodular cast iron specimen (Courtesy of G. Lesoult). (b) Schematic of the initial stage of nodular cast iron solidification, displaying graphite nodules and austenite dendrites.

Each spherulite is comprised of an ensemble of cones whose axis coincides with the c-axis of graphite. Each cone base is therefore bounded by a (0001) plane that grows with the help of screw dislocations (see Fig. 9.16c). The microstructure of nodular or spheroidal gray iron (SGI) is shown in Fig. 9.19(a), and Fig. 9.19(b) describes schematically the solidification of a hypoeutectic Fe-C alloy. After nucleation and growth of austenite phase dendrites, the first graphite nodules nucleate in the remaining liquid. As described for divorced eutectics, C diffuses from the primary austenite toward the graphite, and Fe rejected by the graphite diffuses toward the austenite dendrites. At a fairly early stage during solidification, each nodule is surrounded by an austenite shell, and the growth then proceeds by diffusion of C atoms through this shell. (This growth mode is similar to the peritectic transformation described in Sect. 9.3). The Fe-C phase diagram and the solute composition profiles in the graphite nodule, austenite shell and surrounding liquid for a representative spherical domain are presented in Fig. 9.20. The driving force for the growth of the eutectic is given by $(C_\gamma^{*\ell} - C_\gamma^{*G})$, which corresponds to the composition difference across the austenite shell. Once the last liquid solidifies, austenite transforms into ferrite (surrounding the graphite nodules), and eventually to pearlite upon further cooling.

Several models for the special growth kinetics of SGI have been developed over the years, the most recent and complete one being that of Lesoult et al. [23]. Some models also include solid state transformations [22]. In order to simplify the presentation, we have chosen not to consider the formation of primary austenite dendrites, but focus our attention instead on the stage of solidification controlled by solid state diffusion. There are three phases: graphite G of radius $R_G(t)$, austenite γ of radius $R_\gamma(t)$ and liquid ℓ of radius $R_{g0}(t)$. This last quantity is given on average by $(3/(4\pi n_g))^{1/3}$, where n_g is the density of graphite nodules (or grains). Our domain is a representative volume element of a micro-model of

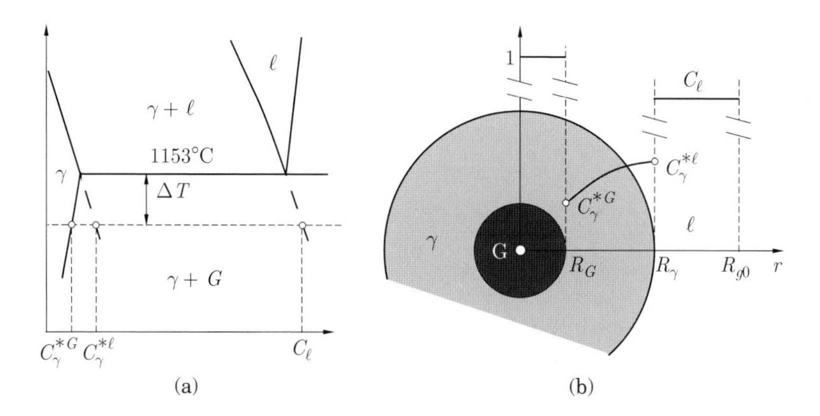

Fig. 9.20 (a) Fe-C phase diagram and (b) schematic of the carbon solute profile in the graphite nodule, in the surrounding austenite shell and in the intergranular liquid after the nodule has been captured by the austenite phase.

solidification (see Chap. 11). Note that all of the radii, including R_{g0}, vary over time as the lower density graphite has a tendency to expand while austenite contracts. All of the densities are taken to be constant (but not equal), and the solute is assumed to be completely mixed in the liquid. By further assuming that the solute profile is steady in the austenite shell, it can be shown that (see Exercise 9.4)

$$C_\gamma(r) = \frac{C_\gamma^{*\ell} R_\gamma (R_G/r - 1) + C_\gamma^{*G} R_G (1 - R_\gamma/r)}{R_G - R_\gamma} \tag{9.19}$$

Applying the mass and solute conservation equations at the graphite-austenite interface, one can deduce the growth rate of the graphite nodule, v_G (see Exercise 9.4)

$$v_G = \dot{R}_G = \frac{dR_G}{dt} = D_\gamma \frac{\rho_\gamma}{\rho_G} \frac{R_\gamma}{R_G(R_\gamma - R_G)} \frac{C_\gamma^{*\ell} - C_\gamma^{*G}}{1 - C_\gamma^{*G}} \tag{9.20}$$

The equivalent procedure at the austenite-liquid interface gives the growth rate v_γ of the austenite shell (see Exercise 9.4)

$$\begin{aligned} v_\gamma = \dot{R}_\gamma = \frac{dR_\gamma}{dt} = D_\gamma \frac{R_G}{R_\gamma(R_\gamma - R_G)} \frac{C_\gamma^{*\ell} - C_\gamma^{*G}}{C_\ell - C_\gamma^{*\ell}} \left(1 + \frac{\rho_\gamma/\rho_G - 1}{1 - C_\gamma^{*G}}\right) \\ - \frac{\rho_\ell}{\rho_\gamma} \frac{dC_\ell}{dt} \frac{1}{C_\ell - C_\gamma^{*\ell}} \frac{R_{g0}^3 - R_\gamma^3}{3R_G^2} \end{aligned} \tag{9.21}$$

The last term inside the parentheses of Eq. (9.21) derives from the graphite expansion that pushes the austenite, assumed incompressible in the present derivation. The second line corresponds to the solute variation in the liquid, which we assume to be completely mixed. The time variation of C_ℓ can be directly related to the temperature evolution by using the liquidus curve for austenite, shown in Fig. 9.20.

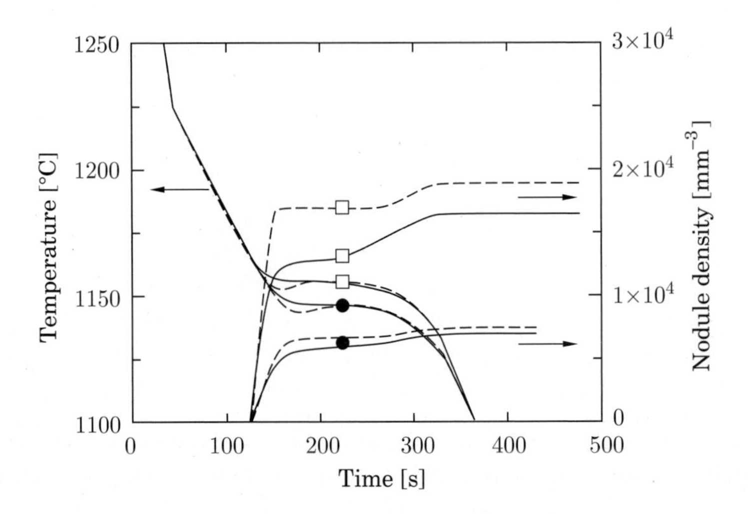

Fig. 9.21 Simulated cooling curves and the evolution of nodule density for a hypoeutectic nodular iron (3wt% C-2.6wt%Si) cast at 1450°C in a mold with a volume/surface ratio equal to 0.0086 m. The solid and dashed lines correspond to simulations performed with initial radii of the graphite nodule equal to 0.5 μ and 5 μm, respectively. The curves with filled circles correspond to $A = 10^{11}$ m^{-3}K^{-1} and those with open squares to $A = 4 \times 10^{11}$ m^{-3}K^{-1}. (Redrawn from Lesoult et al. [23].)

Equations (9.20) and (9.21) are solved to obtain R_G and R_γ, from which the latent heat evolution can be computed. This can be coupled with a heat balance on the volume element to predict the overall cooling curve of a specimen subjected to a given heat extraction. The procedures for this coupling are described in detail in Chap. 11. The cooling curves shown in Fig. 9.21 were obtained in simulations of SGI specimens with varying graphite nodule densities. Nucleation was assumed to be athermal (see Chap. 7, Eq. (7.30)) with a law given by

$$\frac{dn_g}{d(\Delta T)} = 2A\Delta T \tag{9.22}$$

where dn_g is the increment of the graphite nodule density in the remaining liquid over an increment of undercooling $d(\Delta T)$ with respect to the graphite liquidus. The evolution of the nodule density is presented on the same diagram, using the scale shown to the right. As can be seen, the specimen with the higher nodule density produces less undercooling. However, as the kinetics of SGI are controlled by solid state diffusion and are therefore slow, the undercooling is larger than that calculated for the lamellar gray cast iron under similar conditions of cooling rate and final grain size. The temperature difference between the beginning and end of SGI solidification is also much larger. Three other interesting features appear in this diagram: (i) By starting the simulation with smaller initial graphite nodules (dashed curves), the amplitude of the recalescence is reduced; (ii) Graphite nodules nucleate primarily during the early stage of solidification, but a few can still form when the undercooling increases again near the end of solidification; (iii) Solidification of primary dendrites (not discussed in this section) is evident from the change of slope of the cooling curves near $T = 1225°C$.

Eutectic colonies

Most industrial alloys have more than two components. The question therefore arises as to the influence of a third element on eutectic growth. Let us consider the simplest possible ternary eutectic phase diagram, first introduced in Chap. 3, for which the liquidus surface projection is shown in Fig. 9.22(a). The A-B, B-C and C-A binary phase diagrams are all simple eutectics. The three invariant eutectic points of the binary alloys become monovariant lines inside the ternary phase diagram, and these lines converge to the ternary eutectic point indicated by a black dot.

Consider now an initial composition in this alloy system residing on the $\alpha - \beta$ monovariant line, denoted by the open square in Fig. 9.22a. The liquid composition is in equilibrium with the compositions of the α- and β-phases identified by the open circles at the corners of the tie-line triangle. Considering the $\alpha - \beta$ eutectic valley and tie-lines, it can be seen that both lamellae reject element C. This has the consequence that a planar eutectic front consisting of α- and β-lamellae builds up a solute layer of C ahead of the interface. As described in Chap. 8 for the planar front of a single phase, this may lead to a morphological instability of the eutectic front

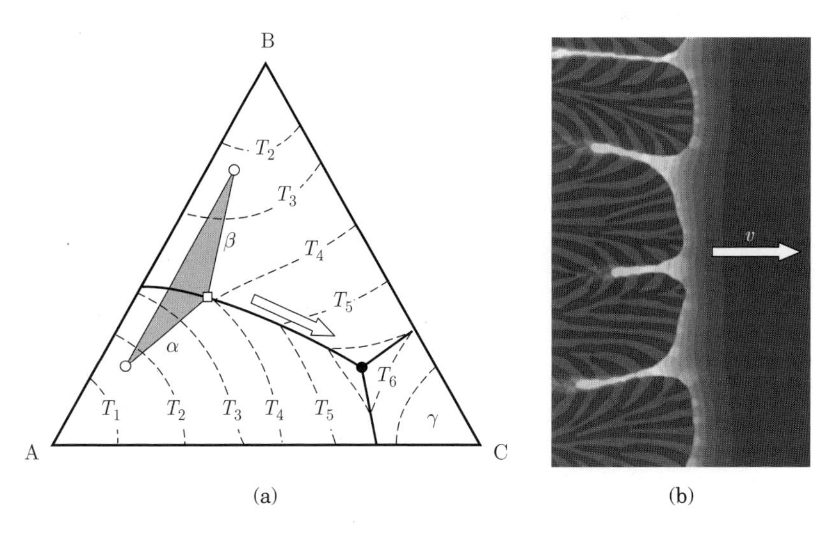

Fig. 9.22 (a) A generic eutectic ternary phase diagram in which isotherms are indicated by dashed lines ($T_1 > T_2 > T_3 > ...$). A tie-line triangle is presented in gray. (b) Phase-field computation of the coupled growth of α- and β-lamellae both rejecting solute element C. (From Plapp and Karma [31].)

and the formation of eutectic cells, or even eutectic dendrites, called *eutectic colonies*. Figure 9.22(b) shows the formation of such eutectic colonies in a ternary alloy, as calculated by Plapp and Karma [31]. The development of a solute layer of C is clearly visible ahead of the $\alpha - \beta$ lamellar front, along with a coupled diffusion of A and B elements near the lamellae appearing as small "bubbles" between them. Eutectic colonies observed experimentally in an Al-Cu-Ag alloy are shown in Fig. 9.23.

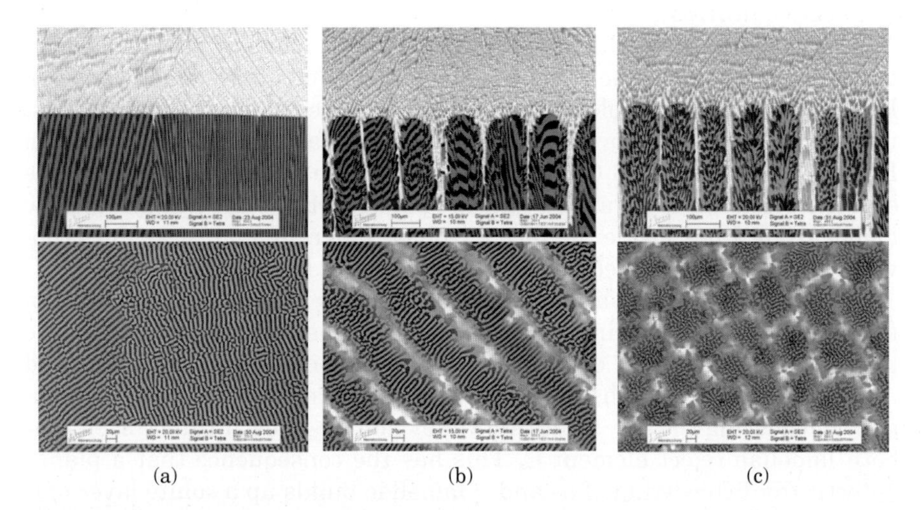

Fig. 9.23 Typical morphologies at the quenched solid/liquid interfaces of an Al-16.1wt%Cu-5wt%Ag alloy in longitudinal (top) and transverse (bottom) sections for (a) planar coupled growth, (b) cellular coupled growth with elongated cells, and (c) cellular coupled growth with regular cells. (From Hecht et al. [11].)

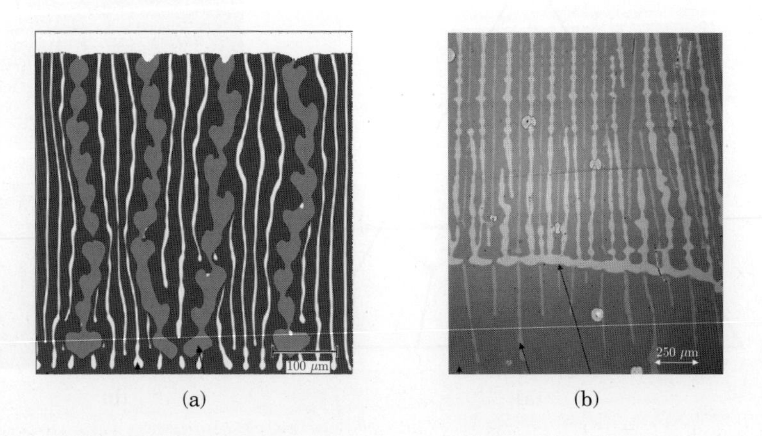

Fig. 9.24 (a) Phase-field simulation and (b) experimentally observed microstructure in a longitudinal section of a directionally solidified In-Bi-Sn ternary eutectic. In both cases, the α-$BiIn_2$, β-In and γ-Sn phases appear as dark gray, wavy intermediate gray and thin light gray regions, respectively. (From Boettger et al. [5].)

For an alloy having the ternary eutectic composition, indicated by the filled circle in Fig. 9.22(a), the α-, β- and γ-phases are expected to grow simultaneously from the liquid. This type of 3-phase eutectic microstructure is shown in Fig. 9.24 for the In-Bi-Sn system. The similarities between the phase field simulation shown in Fig. 9.24(a) and the actual microstructure observed in Fig. 9.24(b) are striking.

9.3 PERITECTICS

9.3.1 General considerations

Many important industrial alloys exhibit a peritectic transformation during solidification. The most important systems are steel (Fe-C), bronze (Cu-Sn) and brass (Cu-Zn). Peritectic invariants are also found in Al-Ti or Al-Ni and high-T_c YBaCuO superconductors. Recall from Chaps. 2 and 3 that the invariant binary peritectic reaction is $\alpha + \ell \rightarrow \beta$, which occurs at the peritectic temperature T_{per} and for the peritectic composition C_{per} (see Fig. 9.25a). Notice that C_{per} lies at the intersection of the β-phase solvus on the left and β-phase solidus on the right . Thus, we may write $C_{per} = C^{\beta}_{sol}(T_{per})$, where $C^{\beta}_{sol}(T)$ indicates the β-phase solidus. The equilibrium compositions of α and liquid at T_{per} are given by $C^{\alpha}_{sol}(T_{per}) = C^{per}_{\alpha}$ and $C^{\alpha}_{liq}(T_{per}) = C^{\beta}_{liq}(T_{per}) = C^{per}_{\ell}$, respectively.

Consider the equilibrium solidification path of a hypoperitectic alloy C_0 whose composition lies in the range $C^{per}_{\alpha} < C_0 < C_{per}$ (see Fig. 9.25a). The primary α-phase solidifies first and the β-phase appears at the $\alpha - \ell$ interface at $T = T_{per}$, at the expense of the liquid which disappears.

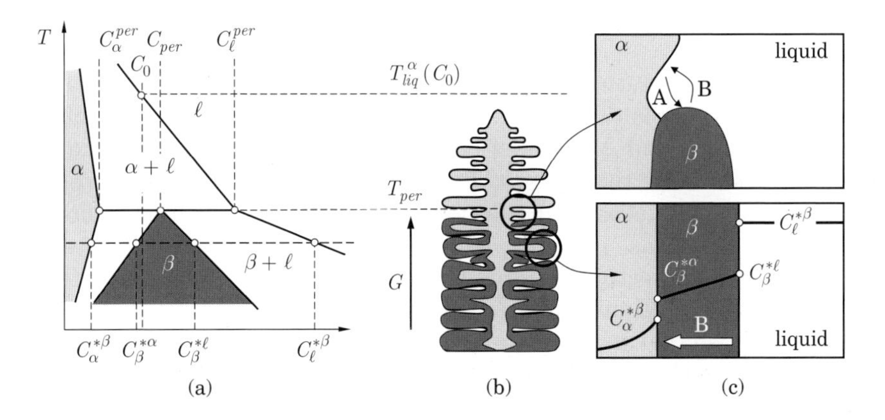

Fig. 9.25 (a) A typical peritectic phase diagram, showing the equilibrium compositions at the peritectic temperature and at some undercooling during the peritectic transformation. (b) Formation of the peritectic microstructure in a vertical thermal gradient under normal solidification conditions. (c) Schematic diagrams of the peritectic reaction (top) and of the peritectic transformation (bottom).

However, the transformation is not complete when the temperature falls below T_{per}, since the alloy is now in the $(\alpha + \beta)$ two-phase domain. Upon further cooling, the β-phase continues to consume the α-phase until the phase boundary separating the $(\alpha + \beta)$- and β-domains is reached. We refer to the part of the transformation that occurs in the solid state as the *peritectic transformation*. Hyperperitectic alloys, whose compositions lie between C_{per} and C_ℓ^{per} follow, a path similar to the hypoperitectic alloy until T_{per} is reached. Below T_{per}, the α-phase disappears and β now solidifies from the melt until the β-solidus is reached. Please note that these solidification or transformation paths depend strongly on the shape of the β domain, in particular the slopes of the β-solvus and β-solidus lines.

Under non-equilibrium solidification conditions, nucleation and growth of the peritectic β-phase have an important effect on the microstructure that forms, as we discuss in Sect. 9.3.2 and 9.3.3. In Sect. 9.3.4, we describe the competition that occurs between the primary α- and the peritectic β-phases at a very low solidification rate. More details on peritectic solidification can be found in the review article of Kerr and Kurz [19].

9.3.2 Nucleation

Recall that, in eutectic alloys, the α- and β-phase liquidus curves meet at the eutectic composition C_{eut} (see Fig. 9.1). Therefore, the α-phase is expected to nucleate first in hypoeutectic alloys, whereas the β-phase should be the first to form in hypereutectic alloys. In peritectic alloys, the α- and β-phase liquidus curves meet at the composition C_ℓ^{per} (see Fig. 9.25a). Thus, this composition, and not C_{per}, marks the transition between the compositions where the α- and β-phases are expected to nucleate first. In fact, for this reason, some authors identify the composition C_ℓ^{per} as the peritectic composition, but we prefer to define C_{per} as $C_{sol}^{\beta}(T_{per})$.

In order to determine which phase forms first in practice for a peritectic alloy of composition C_0, the characteristic nucleation undercooling of each phase, ΔT_n^α and ΔT_n^β, must be considered in addition to the liquidus temperatures of the two phases, $T_{liq}^\alpha(C_0)$ and $T_{liq}^\beta(C_0)$. For the purpose of this discussion, let us assume that the characteristic nucleation undercooling ΔT_n^ν of each phase ν is independent of the composition C_0. This allows us to construct "effective liquidus curves" for each phase, parallel to, but ΔT_n^ν below the original liquidus, as shown by the thick dashed lines in Fig. 9.26(a). We then conclude that the α-phase will nucleate before β only if the critical nucleation temperature $T_n^\alpha = (T_{liq}^\alpha(C_0) - \Delta T_n^\alpha)$ is higher than $T_n^\beta = (T_{liq}^\beta(C_0) - \Delta T_n^\beta)$. In general $\Delta T_n^\alpha \neq \Delta T_n^\beta$, and thus the composition $C_n^{\alpha\rightarrow\beta}$ at which the transition from α- to β-nucleation occurs is shifted with respect to C_ℓ^{per}. It is given by:

$$C_n^{\alpha\rightarrow\beta} = C_\ell^{per} + \frac{\Delta T_n^\alpha - \Delta T_n^\beta}{m_\ell^\alpha - m_\ell^\beta} \tag{9.23}$$

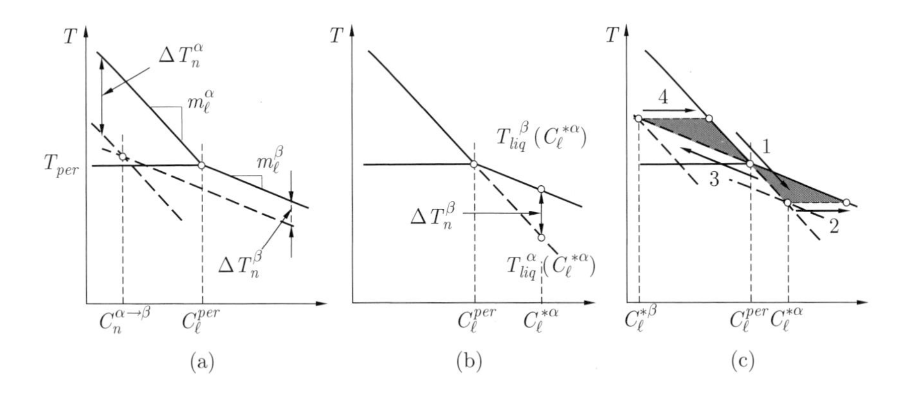

Fig. 9.26 Various situations of nucleation in peritectic systems: (a) The nucleation competition between α and β for an alloy of nominal composition near C_ℓ^{per}; (b) The nucleation of β on an existing α-phase; (c) The nucleation-growth cycle at low growth rate.

where m_ℓ^{α} and m_ℓ^{β} are the slopes of the two liquidus lines (both negative in the phase diagram shown in Fig. 9.25(a) and with $|m_\ell^{\alpha}| > |m_\ell^{\beta}|$).

Consider now a hyperperitectic alloy of composition close to the peritectic composition C_{per} shown in Fig. 9.25(a). Nucleation and growth of the α-phase are assumed to occur before the system reaches the peritectic temperature. Below T_{per}, when β is not yet present, the liquid composition at the $\alpha - \ell$ interface, $C_\ell^{*\alpha}$, follows the metastable α-liquidus curve represented by the thick dashed line in Fig. 9.26(b). Note that we assume the $\alpha - \ell$ interface to be at equilibrium, i.e., that the curvature and kinetic contributions are negligible. Therefore, the undercooling for the nucleation of β for $T < T_{per}$ is given by $\Delta T_n^{\beta} = (T_{liq}^{\beta}(C_\ell^{*\alpha}) - T_{liq}^{\alpha}(C_\ell^{*\alpha})) = (m_\ell^{\beta} - m_\ell^{\alpha})(C_\ell^{*\alpha} - C_\ell^{per})$, as indicated by the vertical line in Fig. 9.26(b). This situation is similar, with just a change of sign of m_ℓ^{β}, to the situation already described for nucleation in a hypoeutectic alloy (see Fig. 9.2). Note that the undercooling is measured with respect to the reference liquidus temperature of β, $T_{liq}^{\beta}(C_\ell^{*\alpha})$, and not T_{per}.

Since the α-phase is already present, the α-liquid interface is the most probable site for nucleation of β. This mechanism is similar to that described for eutectics (Figs. 9.3 and 9.7). There is usually a coherency relationship between the phases. For example, austenite (fcc) nucleating on existing ferrite (bcc) in steel follows the Kurdjumov-Sachs relationship:

$$\{111\}_{\text{fcc}} \parallel \{110\}_{\text{bcc}} \quad \langle 110\rangle_{\text{fcc}} \parallel \langle 111\rangle_{\text{bcc}} \tag{9.24}$$

The same coherency relationship has been observed for the nucleation of β (bcc) at the surface of α (fcc) in Cu-Sn alloys. The need to follow such a coherency relationship explains the observation that the undercooling required to nucleate the peritectic phase increases with the lattice mismatch between the primary and peritectic phases.

9.3.3 Solidification of peritectics at normal speed

In this section, we consider the microstructures that develop in peritectic alloys solidified under typical casting conditions. If the solidification rate is extremely slow, as in some laboratory directional solidification experiments, different microstructures can appear. We address the latter case in the next section. Examples of peritectic microstructures formed in two alloys under normal solidification conditions are shown in Fig. 9.27, along with their associated phase diagrams.

Figure 9.27(b) presents a quenched peritectic microstructure observed in Sn-bronze, with the β-phase completely surrounding the primary α-phase (the dark zone corresponds to the quenched liquid (QL)).

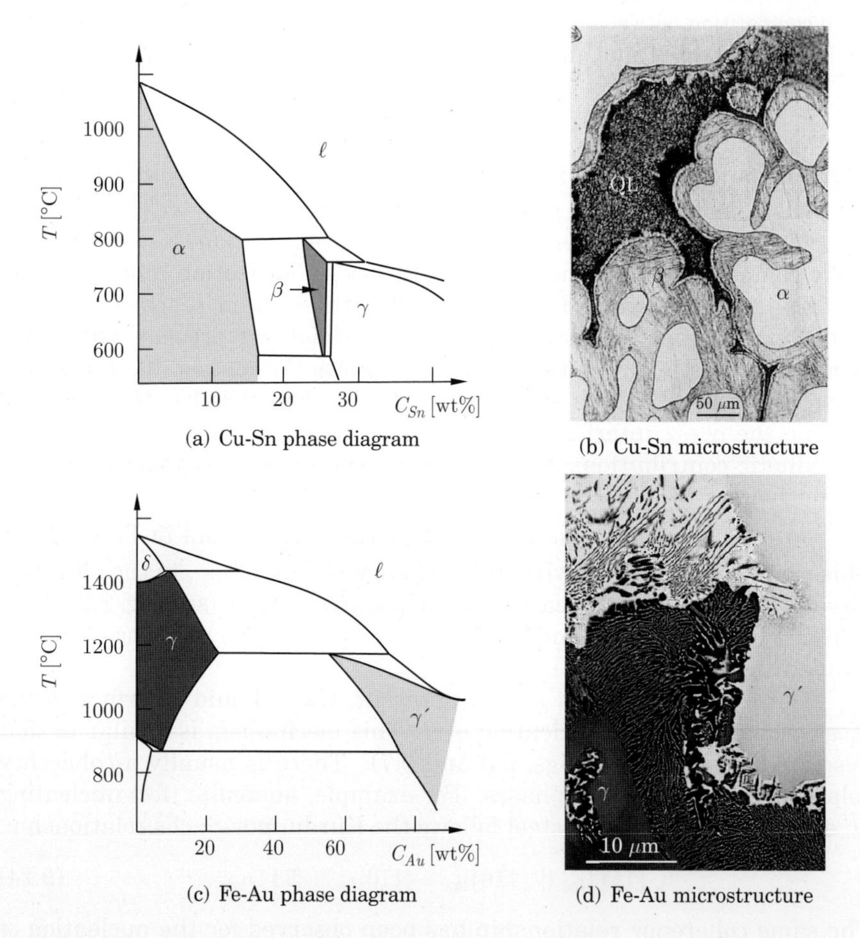

(a) Cu-Sn phase diagram

(b) Cu-Sn microstructure

(c) Fe-Au phase diagram

(d) Fe-Au microstructure

Fig. 9.27 Examples of peritectic microstructures and their associated phase diagrams: (a) The Cu-Sn phase diagram and (b) microstructure observed in a quenched Cu-20wt%Sn specimen [19]. (c) The Fe-Au phase diagram and (d) microstructure in a quenched Fe-50wt%Au specimen. (Courtesy, D. Favez.)

The corresponding phase diagram is shown in Fig. 9.27(a) with the α- and β-single phase domains involved in the peritectic transformation highlighted in gray. Unlike the schematic peritectic phase diagram shown in Fig. 9.25, the solvus curves delimiting the $(\alpha + \beta)$ domain in Cu-Sn both have a negative slope. It is left as an exercise to determine the consequence of this fact on the fraction of phases $g_\alpha(T)$ and $g_\beta(T)$ (see Exercise 9.5). In other systems, such as Zn-Ni, the transformation might be incomplete in the sense that the peritectic phase only partially covers the primary phase. Incomplete peritectic reactions have been attributed to the difficulty of nucleation and growth of the peritectic phase [19].

In some alloy systems, the peritectic phase has the same crystal structure as the primary phase. Figure 9.27(d) shows a micrograph of an Fe-50wt%Au alloy with the corresponding phase diagram in (c). Austenite (in dark gray) and the peritectic Au-rich phase (in light gray) have the same fcc structure. During cooling, the primary γ-phase (dark gray) solidifies first, with a gradual enrichment in Au as g_γ increases. Once the liquid composition exceeds C_ℓ^{per}, a thin layer of γ'-Au phase (light gray) forms around γ due to the peritectic reaction. As the temperature continues to decrease below T_{per}, the solvus lines of γ and γ' delimiting the $(\gamma + \gamma')$ domain show that the solubility of Au in γ-Fe and of Fe in γ'-Au decreases. This causes γ' to precipitate in γ, and γ to precipitate in γ'. Since γ and γ' are actually a single fcc structure, these reactions can occur by a *discontinuous* or *cellular transformation*: fine γ'-lamellae grow in γ and fine γ-lamellae grow in γ'. Two fronts, similar to those found in a eutectoid, are clearly visible in this figure: The first one separates the untransformed, supersaturated γ' Au-rich phase (in light gray) and the region of γ(black)-γ'(white) lamellae, whereas the second marks the transition between the supersaturated γ Fe-rich phase (in dark gray) and a γ'(white)-γ(black) lamellae region.

In order to understand the microstructure evolution in peritectic alloys, let us consider again the idealized peritectic phase diagram of Fig. 9.25(a). For the whole peritectic range $C_\alpha^{per} < C < C_\ell^{per}$, the α-dendrites are assumed to grow at some undercooling below the liquidus. The β-phase, which has nucleated at the periphery of the α-dendrites, grows below T_{per} by three mechanisms: (i) Direct solidification; (ii) *Peritectic reaction*; (iii) *Peritectic transformation*. The first mechanism is not detailed here as it is similar to the solidification of a primary phase, i.e., the $\beta - \ell$ interface solidifies with rejection of B into the liquid. The other two mechanisms, which are illustrated schematically in Fig. 9.25(c), are detailed further below.

The peritectic reaction occurs near the triple junction between the α-, β- and ℓ-phases (Fig. 9.28). For the sake of simplicity, we will assume equal surface energies for the three interfaces, i.e., $\gamma_{\alpha\beta} = \gamma_{\alpha\ell} = \gamma_{\beta\ell}$. The three interfaces thus each form an angle of $2\pi/3$. Moreover, the slow growth of β inside α by the peritectic transformation is neglected. This implies that the $\alpha - \beta$ interface is vertical. These geometric constraints

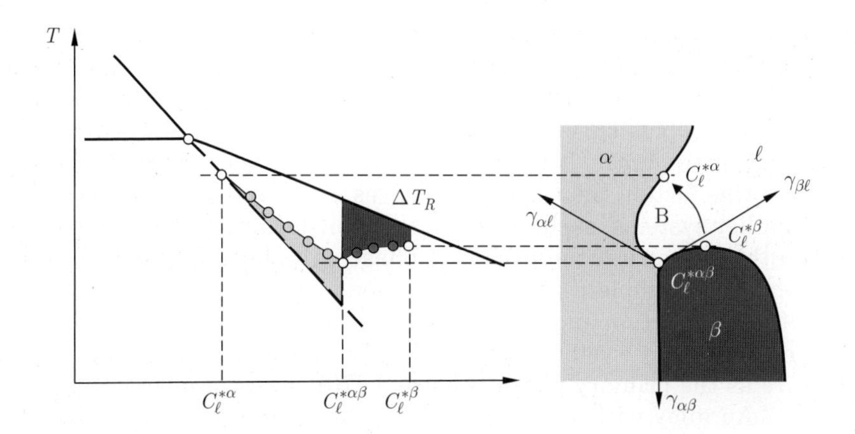

Fig. 9.28 A schematic diagram of diffusion during a peritectic reaction and of the corresponding curvature undercooling contribution ΔT_R.

require the $\alpha - \ell$ interface to be *concave* near the triple junction. Accordingly, α locally remelts and then re-solidifies with an incoming solute flux originating from β. Indeed, below T_{per}, the liquidus line of β is located above that of the α-phase and therefore $C_\ell^{*\beta} > C_\ell^{*\alpha}$ for a small region of nearly uniform temperature surrounding the triple junction. Taking into account the positive curvature of β and zero curvature of α at the inflection point (corresponding to a simplified 2D geometry), the liquid composition-temperature couples of the interfacial points can be drawn schematically on the peritectic phase diagram as we did for the coupled growth of eutectic lamellae in Fig. 9.9. A few such points are represented by gray-shaded circles with a hue identical to that of the corresponding solid phase (light/dark gray for $\alpha/\beta - \ell$ interfaces). The difference between these points and the corresponding liquidus defines the curvature undercooling contribution ΔT_R (gray region in the phase diagram of Fig. 9.28). Further details on the kinetics of the peritectic reaction can be found in the review paper by Kerr and Kurz [19]. Figure 9.29 shows this phenomenon as observed *in-situ* by confocal microscopy, revealing δ-ferrite remelting ahead of a growing γ-austenite lamella [29].

At temperatures below the peritectic reaction, the peritectic transformation thickens the β-layer by partially dissolving the α-phase. This solid state reaction is naturally very dependent on the shape of the $(\alpha + \beta)$ and β domains in the phase diagram. It is also much slower than the direct solidification of β, since it is governed by the diffusion of solute through the peritectic layer. We consider this case again in Sect. 10.2.2, where we perform a 1D solute balance over the three phases, α, β and liquid, in order to deduce the evolution of the $\alpha - \beta$ and $\beta - \ell$ interfaces. It is left as an exercise to show that the growth kinetics of the phases depends on the difference $(C_\beta^{*\ell} - C_\beta^{*\alpha})$, i.e., the difference between the β-solidus and the β-solvus compositions (see Exercise 9.6).

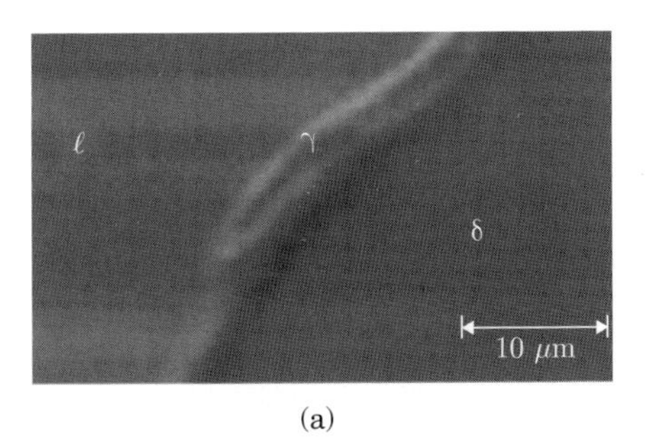

(a)

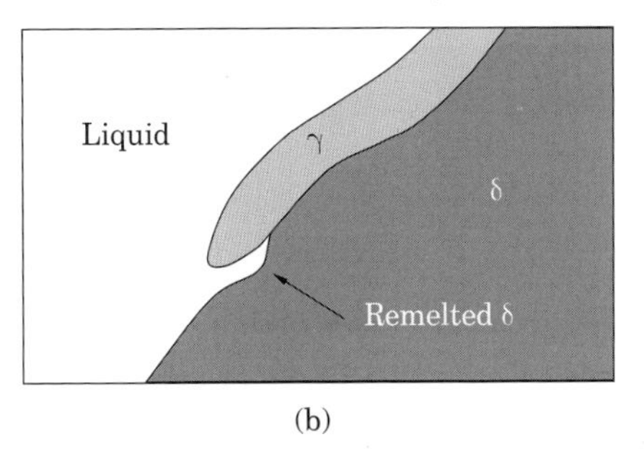

(b)

Fig. 9.29 (a) An image by confocal microscopy and (b) a sketch of the local remelting that occurs in δ-ferrite ahead of a lamella of γ-austenite. (Courtesy of R. Dipenaar.)

9.3.4 Solidification of peritectics at low speed

We now consider peritectic alloys directionally solidified in a thermal gradient G at such a low speed that either the α- or the β-planar front would be stable if the other phase were not present. We use the linear peritectic phase diagram presented in Fig. 9.25(a) for the discussion. Since $\Delta T_{0\alpha}(C_0) > \Delta T_{0\beta}(C_0)$, it is sufficient to consider the stability condition for the α-phase: $v^* < D_\ell G/\Delta T_{0\alpha}$ (see Eq. (8.44)). Consider now the solidification of a hypoperitectic alloy of composition C_0 with $C_\alpha^{per} < C_0 < C_{per}$. It can be demonstrated that neither the α-, nor the β-front is stable under such conditions.

During the initial transient of planar front growth of α (see Sect. 5.2.3), the composition in the liquid ahead of the interface, $C_\ell^{*\alpha}$, increases from C_0 to $C_0/k_{0\alpha}$, following the liquidus line (1) in Fig. 9.26(c). Since $C_0 > C_\alpha^{per}$, i.e., $C_0/k_{0\alpha} > C_\ell^{per}$, the composition $C_\ell^{*\alpha}$ necessarily exceeds C_ℓ^{per} during this transient, at which point it follows the metastable α-liquidus line (dashed line in Fig. 9.26c). At some undercooling ΔT_n^β, the β-phase might form at the $\alpha - \ell$ interface, as indicated by the horizontal arrow (2) in Fig. 9.26(c). It will then grow both laterally and parallel to the thermal gradient, since the liquid close to the $\alpha - \ell$ interface is undercooled with respect to β. The lateral growth of β is rapid, so that this phase overgrows the α-phase. Now, as the new $\beta - \ell$ interface tries to reach steady-state, the composition $C_\ell^{*\beta}$ tends toward $C_0/k_{0\beta}$, which is less than C_ℓ^{per}. Thus, $C_\ell^{*\beta}$ decreases and the temperature of the $\beta - \ell$ interface increases, following line (3) in Fig. 9.26(c). Above T_{per}, the liquid is undercooled with respect to α, so that nucleation of the α-phase occurs at the $\beta - \ell$ interface when some undercooling ΔT_n^α is reached. The α-phase then spreads rapidly in the lateral and longitudinal directions (horizontal arrow (4) in Fig. 9.26c), leading to the α-phase overgrowing the β-phase and the cycle beginning again. This mechanism explains the observation of alternating α- and β-*bands* essentially parallel to the isotherms, such as those displayed in Fig. 9.30(a).

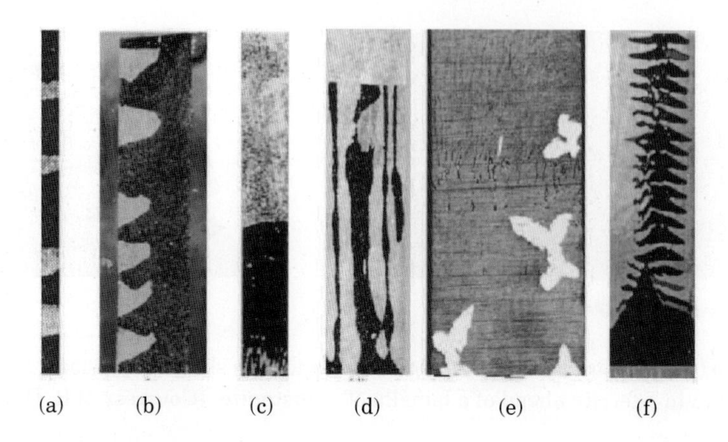

Fig. 9.30 A summary of peritectic microstructures observed at very low speed during solidification in a vertical thermal gradient: (a) Bands; (b) Islands; (c) Single transition from α to β; (d) Coupled growth of α- and β-lamellae; (e) Nucleation of α in the liquid ahead of a $\beta - \ell$ planar front; (f) Helicoidal structure. (a-d) Hypoperitectic compositions; (e-f) Hyperitectic compositions. (After Boettinger et al. [6].)

Depending on the nucleation conditions (nucleant density and undercooling), growth conditions and alloy composition, the overlaying of one phase on the planar front of the other phase might not be complete, leading to the formation of *islands* of one phase inside a matrix of the other phase as depicted in Fig. 9.30(b). On the other hand, the overlaying may occur only once (Fig. 9.30c), where the α-phase grows until the interfacial composition $C_\ell^{*\alpha}$ exceeds C_ℓ^{per} and the β-phase forms at some undercooling ΔT_n^β. If the composition ahead of the steady-state $\beta - \ell$ front, $C_0/k_{0\beta}$, is such that $(T_{liq}^\alpha(C_0/k_{0\beta}) - T_{liq}^\beta(C_0/k_{0\beta})) < \Delta T_n^\alpha$, the α-phase will be unable to re-nucleate. It is left as an exercise to demonstrate that nucleation of α can occur in the liquid of hyperperitectic alloys such as the one shown in Fig. 9.30(e), but not directly at the interface of a steady-state β planar front (Exercise 9.8).

Since all of these microstructures form at a very low growth rate (typically a few μm s^{-1}), convection can play an important role. The microstructure of the hyperperitectic alloy shown in Fig. 9.30(f) looks like symmetric islands of dark phase similar to those seen in Fig. 9.30(b). However, careful sectioning shows that the dark phase is, in fact, a continuous helicoidal entity surrounded by light phase. This morphology is associated with the convection pattern induced in this specimen.

Figure 9.30(d) shows a structure similar to those found in coupled eutectic growth, comprising alternating δ- and γ-lamellae growing cooperatively from the liquid (the quenched liquid is at the top of the figure). A higher magnification view of such a "coupled" lamellar structure in peritectic alloys is shown in Fig. 9.31(a). Although both phases reject solute, they can establish a cooperative mode in which the most rejecting phase

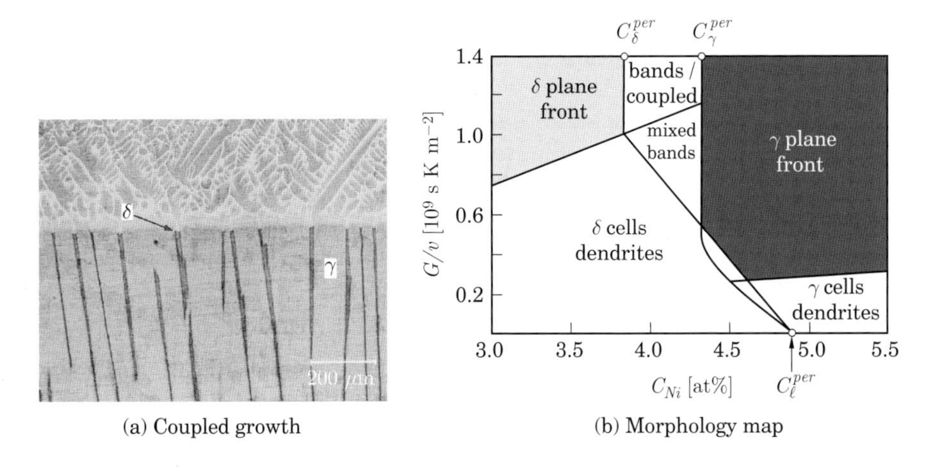

(a) Coupled growth (b) Morphology map

Fig. 9.31 (a) Coupled (or cooperative) growth of γ and δ phases in Fe-Ni. (b) The $G/v - C_0$ diagram showing the various morphologies that can be observed in Fe-Ni. (After Hunziker et al. [15].)

(δ in this case) is helped by the less-rejecting γ-phase. The solute profile in this case consists of a long-range diffusion layer D_ℓ/v^*, as for a eutectic alloy of off-eutectic composition (see Sect. 9.2.3, Fig. 9.10b), but the diffusion field corresponding to the exchange of solute between the two phases (i.e., other terms of the Fourier series) is not as pronounced in the case of coupled peritectic growth. The conditions under which planar front, cells-dendrites, bands or islands and coupled peritectic growth have been observed in the Fe-Ni system are summarized in the microstructure map of Fig. 9.31(b).

9.4 PHASE SELECTION AND COUPLED ZONE

In many systems under real solidification conditions, the phase predicted by the equilibrium phase diagram for a given composition does not necessarily appear, due to the nucleation and/or the growth kinetics of an otherwise metastable phase being more favorable than the stable one. This competition is detailed in Sect. 9.4.1. In Sect. 9.4.2, we describe the competition between solidification of a single phase vs. coupled growth of the two phases in eutectic alloys. Other forms of competition include morphology competition, such as the columnar-to-equiaxed tansition discussed in Chap. 11.

9.4.1 Phase competition

We saw in the previous section that, for peritectic systems, the transition between α- and β-nucleation does not occur for a nominal composition

C_ℓ^{per} if $\Delta T_n^\alpha \neq \Delta T_n^\beta$. Taking the example of Fig. 9.26(a), nucleation of β is expected for a composition $C_n^{\alpha \to \beta} < C < C_\ell^{per}$, whereas the equilibrium phase diagram predicts that α will be the primary phase. Assume now that both phases can nucleate "easily" in a given process. What happens during growth? The phase diagram suggests that C_ℓ^{per} marks the transition between primary α-dendrites, which then partially transform into β at their periphery, and β-dendrites solidifying directly from the melt. Whether or not this is what actually occurs depends on the respective growth kinetics of the dendrites, i.e., $v_\alpha^*(T)$ and $v_\beta^*(T)$.

Consider the ternary Fe-Ni-Cr system that forms the basis for stainless steels. Cr (bcc) favors the growth of δ-ferrite and the corresponding binary Fe-Cr phase diagram represents a complete solid solution of δ above 1400°C. Mo, Si and Nb have a similar effect. On the other hand, Fe-Ni alloys exhibit a very narrow peritectic invariant around 5 wt% Ni, above which composition they solidify as primary γ austenite. C and Mn also favor the formation of austenite. In the ternary Fe-Ni-Cr phase diagram, the peritectic composition C_ℓ^{per} of the Fe-Ni binary becomes a monovariant line separating the two liquidus surfaces associated with γ and δ. However, this line gradually transforms into a eutectic valley and finally joins the eutectic point of the Cr-Ni binary alloy ($C_{eut} \cong 53$ wt%). In order to handle the complex solidification of multi-component iron alloys into austenite and ferrite, it is common to define chromium- and nickel-equivalent compositions, C_{Ni}^{eq} and C_{Cr}^{eq}, as:

$$\begin{aligned} C_{Cr}^{eq} &= C_{Cr} + 1.5C_{Si} + C_{Mo} + 0.5C_{Nb} \\ C_{Ni}^{eq} &= C_{Ni} + 0.5C_{Mn} + 30C_C \end{aligned} \tag{9.25}$$

In a 2D map where these equivalent compositions represent the axes, Schaeffler's diagram allows a prediction of the amount of ferrite, austenite and martensite required for "standard" welding conditions.

However, such a diagram is unable to predict whether ferrite dendrites grow first with the formation of a surrounding shell of austenite occurring later, or whether austenite dendrites grow directly from the melt. In the latter case, some ferrite might form between the austenite dendrites when reaching the peritectic-eutectic monovariant line. In order to predict the competition between γ- and δ-phases in an Fe-Ni-Cr alloy of composition (C_0^{Ni}, C_0^{Cr}), it is necessary to take into account their corresponding liquidus temperatures, $T_{liq}^\gamma(C_0^{Ni}, C_0^{Cr})$ and $T_{liq}^\delta(C_0^{Ni}, C_0^{Cr})$, as well as their growth kinetics, $v^{*\gamma}(\Delta T)$ and $v^{*\delta}(\Delta T)$. Figure 9.32(a) shows the dendrite tip temperatures calculated for the two phases, $T_\delta^*(v) = (T_{liq}^\delta - \Delta T^\delta(v^*))$ and $T_\gamma^*(v) = (T_{liq}^\gamma - \Delta T^\gamma(v^*))$. The growth kinetics of both δ- and γ-dendrite tips were computed using the marginal stability dendrite growth model, presented in Chap. 8, extended to multi-component alloys.

Assuming that the phase that forms first is the one with the highest temperature, it can be seen that δ-dendrites, subsequently surrounded by a γ-shell, form at low growth velocity. This is shown schematically as a

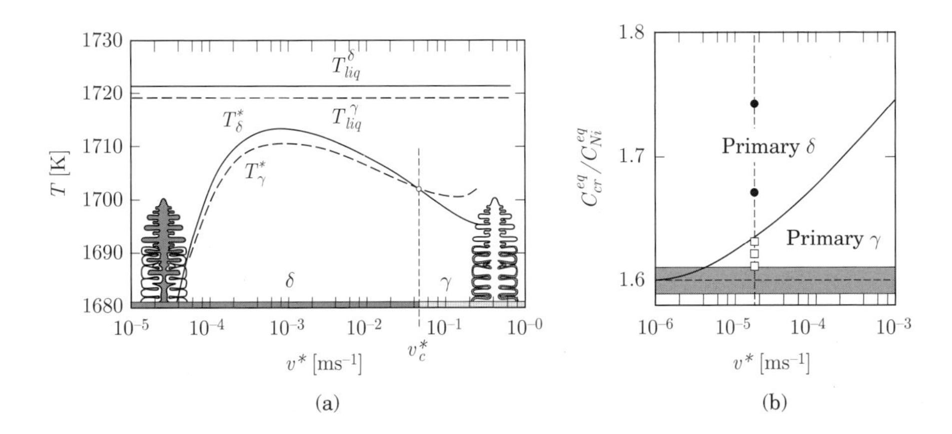

Fig. 9.32 (a) Dendrite tip temperatures T^* calculated as a function of the velocity for the γ and δ phases in Fe-18.2wt%Cr-10.7wt%Ni. The thermal gradient $G = 4 \times 10^5 \mathrm{Km}^{-1}$. (Redrawn from Fukumoto and Kurz. [10]). (b) Cr $\div$ Ni equivalent composition ratio marking the transition between primary γ- and primary δ-phases in Fe-Ni-Cr alloys as a function of the velocity (C_{Cr} = 18.5wt%; γ-inducing elements: 0.024 wt% $< C_C <$ 0.039 wt%, 0.013 wt% $< C_N <$ 0.015 wt%). The continuous curve was calculated using a dendrite growth model, whereas the symbols correspond to the data from directional solidification experiments. The gray rectangle indicates the equilibrium value given by the phase diagram. (Redrawn from Bobadilla et al. [4].)

gray dendrite surrounded by a white austenite layer. This morphology is consistent with the equilibrium phase diagram that gives $T^{\delta}_{liq} > T^{\gamma}_{liq}$. However, beyond a critical velocity v^*_c of a few cm s^{-1}, the computed tip temperature of γ-dendrites T^*_γ is higher than that of δ-dendrites, T^*_δ. Austenite dendrites are therefore expected, with minor amounts of ferrite formed in between, when the monovariant line is reached (as represented by a white dendrite surrounded by a thin gray ferrite layer). Figure 9.32(b) illustrates the critical velocity v^*_c separating γ and δ primary solidification as a function of the Cr-Ni equivalent composition ratio (continuous line). The experimental points, measured by Bobadilla et al. [4], are presented as open squares for γ-dendrites, and filled circles for δ-solidification.

Steel is just one of many examples of polymorphic materials encountered in nature. Other systems such as lipids or calcium carbonate can solidify in more than one crystalline phase (e.g., calcite or aragonite for $CaC0_3$). If its nucleation and growth kinetics are more favorable than the stable phase, a metastable phase may form. An industrially important example of this phenomenon is the competition between gray and white cast irons. The Fe-C phase diagram exhibits two eutectic invariants: a stable one $\ell \rightarrow (\gamma + \mathrm{graphite})$ at 1153°C and a metastable one $\ell \rightarrow (\gamma + \mathrm{Fe_3C})$ at 1147°C. The stable gray cast iron is therefore expected under equilibrium conditions, which is the case when the cooling rate is low. However, the stable eutectic is irregular due to the faceted morphology of the graphite lamellae. On the other hand, metastable white cast iron grows with a regular eutectic morphology. This fact gives white cast iron an advantage with respect to the growth kinetics, because regular eutectics grow closer to the extremum than irregular eutectics, and its undercooling is thus lower. Furthermore, the eutectic range ΔC_0 of the eutectic plateau for austenite-graphite is much larger than that of austenite-Fe_3C (graphite is nearly pure whereas cementite only has 25at%C). Consequently, there is much less solute (and solvent) exchange between austenite and cementite, thus giving a second growth advantage to white cast iron. The growth competition between these two morphologies can be assessed quantitatively using the theories developed in Sect. 9.2. A shown in Fig. 9.33(b), the $\Delta T - \sqrt{v^*}$ relationship for gray cast iron has a much steeper slope than that of white cast iron. Therefore, the two curves $T^*_{\gamma+\nu} = T^{\gamma+\nu}_{eut} - \Delta T_{\gamma+\nu}(v^*)$ associated with the two possible eutectic morphologies (ν = graphite (C) or Fe_3C) cross each other at a velocity v^*_c.

The transition between white and gray cast iron does not occur at exactly v^*_c. There is a significant hysteresis, related to nucleation of the new graphite or cementite phase. Let us assume that (irregular) gray cast iron is growing at low speed in a Bridgman furnace. As the withdrawal rate is increased, the liquid becomes undercooled with respect to the white cast iron once the temperature somewhere along the irregular eutectic front is below 1147°C. If the nucleation of cementite requires an undercooling $\Delta T_n^{Fe_3C}$, the new phase will form once $T^*_{\gamma+C} < T^{\gamma+Fe_3C}_{eut} - \Delta T_n^{Fe_3C}$. Once cementite forms, white cast iron regular eutectic rapidly overgrows the gray cast iron. If the withdrawal speed of the specimen is subsequently decreased, the process does not revert to the original state along

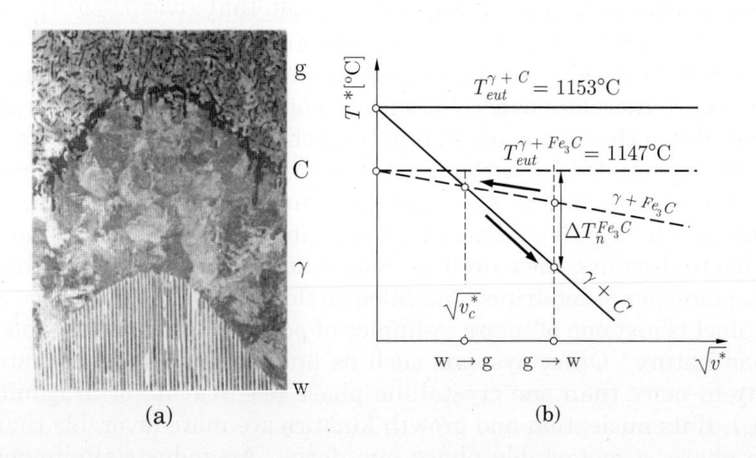

Fig. 9.33 (a) The white to gray cast iron transition in Fe-C (+100 ppm P) during directional solidification at 8 μm s^{-1}. (b) The eutectic front temperature T^* versus the square root of the front velocity $\sqrt{v^*}$ for (γ + C) gray and (γ + Fe_3C) white cast iron, with indications of the observed white-to-gray (w→g) and gray-to-white (g→w) transitions. (After Magnin and Kurz [25].)

the same path. Graphite can nucleate ahead of the white cast iron front if $(T_{eut}^{\gamma+C} - \Delta T_n^C) > T_{\gamma+Fe_3C}^*(v^*)$. Since the eutectic temperature of $(\gamma + C)$ is 6°C higher than that of $(\gamma + Fe_3C)$, nucleation of graphite can occur even for $v^* > v_c^*$. However, if it does, the gray cast iron has a larger undercooling and will be overgrown by the white cast iron. When this occurs just above v_c^*, graphite nucleates just ahead of the white cast iron growth front. As it grows, it depletes the melt of carbon, and at some point, this stops the growth of white cast iron. An example of this phenomenon is shown in Fig. 9.33(a). In the C-depleted zone between the lamellar white cast iron and the graphite "moustache," we see a region consisting entirely of austenite. This region is similar in some ways to the halos observed around graphite nodules in Fig. 9.19. Further growth of gray cast iron can occur only if the associated front temperature is slightly higher than that of white cast iron, or equivalently if its growth rate is comparable to (or faster than) that of white cast iron. The competition of white and gray cast iron is very commonly observed in practice: White cast iron forms at the surface of a casting where the cooling rate is high, whereas gray cast iron is found at the center of the component (see for instance the review by Stefanescu. [34]).

9.4.2 Coupled zone

The final section of this chapter focuses on the *coupled zone* in eutectic systems resulting from the competition between a primary phase and a eutectic morphology. Such a competition can also be found in solid state transformations, such as pearlite (the α-Fe_3C eutectoid) in homogenized Fe-C alloys which forms over a range of compositions. This range becomes wider as the cooling rate is increased. The formation of a primary phase in off-eutectic alloys can be similarly suppressed depending on the cooling conditions, resulting in a fully coupled growth structure.

Consider first the eutectic phase diagram shown in Fig. 9.34(a). We assume that both phases α and β formed from the melt are non-faceted. Since the diagram is fairly symmetrical, the expected fractions of α and β in an alloy of eutectic composition are close to 0.5. According to what we learned in Sect. 9.2.2, a regular lamellar eutectic structure should form in this system. For off-eutectic compositions, the fractions of primary phase dendrites and interdendritic eutectic will depend on composition and growth rate. Consider for example a hypereutectic alloy of composition slightly higher than C_{eut}, solidified at a speed below the stability limit for a β-planar front, i.e., for $v^* < D_\ell G/\Delta T_{0\beta}$, where $\Delta T_{0\beta}$ is the solidification interval of β. If the eutectic does not appear, the $\beta - \ell$ interface temperature corresponds to that of the β-solidus for C_0, $T_{sol}^\beta(C_0) = T_{liq}^\beta(C_0/k_{0\beta})$, which is much lower than the eutectic temperature T_{eut}. Therefore, rather than growing a β-planar front at this velocity, the alloy is more likely to develop a coupled growth eutectic morphology, with fractions of α and β determined roughly by the lever-rule, $g_\beta = (C_0 - k_{0\alpha}C_{eut})/\Delta C_0$, as described in Sect. 9.2.3. If we apply similar logic to an alloy far from the eutectic composition, e.g., equal to or greater than the solubility limit of β, $C_0 \geqslant (1-(1-C_{eut})k_{0\beta})$, its solidus temperature will be greater than or equal to T_{eut}. Therefore, for such compositions, there is no longer an advantage to growth as a coupled eutectic front as opposed to planar front of β at low velocity.

If we now consider a velocity larger than the stability limit, β-dendrites would be expected to grow, with a tip temperature given by $T^*_\beta(v^*) = (T^\beta_{liq}(C_0) - \Delta T^\beta(v^*, C_0))$ as discussed in detail in Chap. 8. On the other hand, we have seen in this chapter that eutectic coupled growth will occur at a temperature $T^*_{eut}(v^*) = (T_{eut} - \Delta T_{eut}(v^*, C_0))$. These two competing relationships are plotted schematically as functions of v^* on the right-hand side of Fig. 9.34(a), with continuous and dashed-dotted lines, respectively. A similar curve for the growth of α-dendrites is also shown with a dashed line. However, since $T^*_\alpha(v^*)$ is below both of the other two curves, α-dendrites are not expected to appear. The $T^*_{eut}(v^*)$ curve decreases monotonically with v^*, whereas the curve associated with the β-dendrites exhibits a maximum. At low speed, the temperature increase from the β-solidus to just below the β-liquidus, corresponds to the transition from a planar front to cells and dendrites. Beyond the maximum, in the dendritic regime, the temperature of the β-dendrite tips decreases as discussed in Chap. 8. In directional solidification, where the interface velocity is fixed,

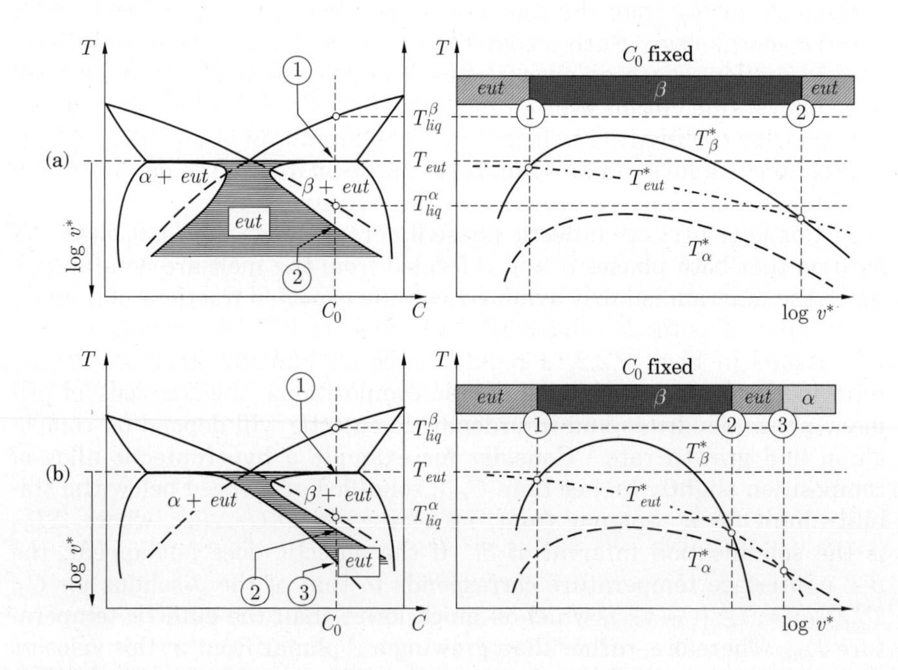

Fig. 9.34 (a) A symmetric and (b) asymmetric coupled zone. (After Kurz and Fisher [20].)

the microstructure that forms is the one that has the highest temperature. On this basis, we can distinguish three regions:

- At low speed, the temperature of the eutectic front is higher than that of a β-planar front or of the tips of β-cells. Therefore, a fully coupled eutectic structure is expected, as already discussed.
- At intermediate speed, the temperature of the β-dendrite tips is higher than the eutectic front temperature and the structure will thus consist of β-dendrites and interdendritic eutectic, as shown schematically in Fig. 9.1.
- As the growth velocity increases further, the undercooling of the β-dendrite tips increases more rapidly than that of the eutectic front, thus resulting in the two curves crossing each other again. Beyond this critical velocity, where $T^*_\beta(v^*_c) = T^*_{eut}(v^*_c)$, dendrites no longer grow ahead of the coupled eutectic front and only the coupled morphology remains in the final structure.

The velocities at which the eutectic front temperature T^*_{eut} is equal to that of the β-dendrites (or cells or planar front), T^*_β, are labeled (1) and (2) in Fig. 9.34(a). Since the two temperatures at which this occurs are smaller than T_{eut}, they can be reported directly on the phase diagram at the corresponding composition C_0, as indicated on the left of Fig. 9.34(a). The temperature scale below T_{eut} can be changed to a velocity scale, with an axis pointing downward since ΔT increases with v^*. In between these two points, β-dendrites form with some amount of eutectic in between, whereas outside this range, a fully coupled eutectic structure is expected. By making similar constructions for various alloy compositions, the coupled zone of this system can be computed and is indicated by the dashed region in the figure. In this region, no primary phase is expected to form ahead of the eutectic front.

We now apply the same reasoning to a eutectic system in which the β-phase is faceted (Fig. 9.34b). This modifies the problem in two ways: Firstly, the eutectic will be irregular and the growth kinetics is therefore slower (see Sect. 9.2.5). Secondly, the growth kinetics of the faceted phase alone also differs from that of a non-faceted morphology. Indeed, the attachment kinetics can play a significant role in this case and the tip morphology is more like that of a plate rather than that of a needle. Both contributions tend to make the growth kinetics of the faceted phase more sluggish, i.e., they require an undercooling $\Delta T^\beta(v^*)$ that increases faster with velocity. This is shown schematically in Fig. 9.34(b) where the curvature (in absolute value) of the curves $T^*(v^*)$ for the α-phase, irregular eutectic ($\alpha + \beta$) and β-phase increases in that order at high velocity. This has the consequence that the α-phase can become the leading phase at high velocity in a hypereutectic alloy, i.e., it becomes the phase with the highest temperature ($T^*_\alpha > \max(T^*_{eut}, T^*_\beta)$, even though the corresponding equilibrium temperature is the lowest ($T^\alpha_{liq} < \min(T_{eut}, T^\beta_{liq})$). Three transitions, labeled (1), (2) and (3) in Fig. 9.34(b), define four regions. Below velocity (1), a fully coupled eutectic structure is observed. Between (1) and (2), β-dendrites and interdendritic eutectics are expected. Between (2) and (3), a fully coupled eutectic region is found once again, and above velocity (3), α-dendrites with interdendritic eutectics are predicted for this hypereutectic alloy. All together, the resultant coupled zone, where only the eutectic phase is observed, is *skewed* toward the faceted phase, as shown in Fig. 9.34(b).

Unlike regular eutectic systems, it is difficult to predict the skewed coupled zone of an irregular eutectic, since the growth kinetics of both the faceted β-phase and the irregular eutectic cannot be readily quantified. Kurz and Fisher [20] fit the growth kinetics of the faceted α- and non-faceted β-phases with equations of the form:

$$\Delta T^{\alpha} = \frac{GD_{\ell}}{v^*} + A^{\alpha}\sqrt{v^*} \quad ; \quad \Delta T^{\beta} = \frac{GD_{\ell}}{v^*} + A^{\beta}\sqrt{v^*} \tag{9.26}$$

with two adjustable parameters A^{α} and A^{β}. The growth kinetics of the eutectic is similar to that given in Eq. (9.18) but also takes into account the influence of the thermal gradient on the growth rate of the irregular eutectic, written as

$$\Delta T^{eut} = A^{eut}\sqrt{\frac{v^*}{G}} \tag{9.27}$$

Exercise 9.9 describes the calculation of the skewed coupled zone for Al-Si binary alloys.

The concept of a skewed coupled zone in a faceted/non-faceted system helps to explain the Al-Si eutectic microstructures shown in Fig. 9.35. In such an alloy solidified at fairly low speed (Fig. 9.35a), a fully eutectic structure is observed. When solidified at higher cooling rate (Fig. 9.35b), a

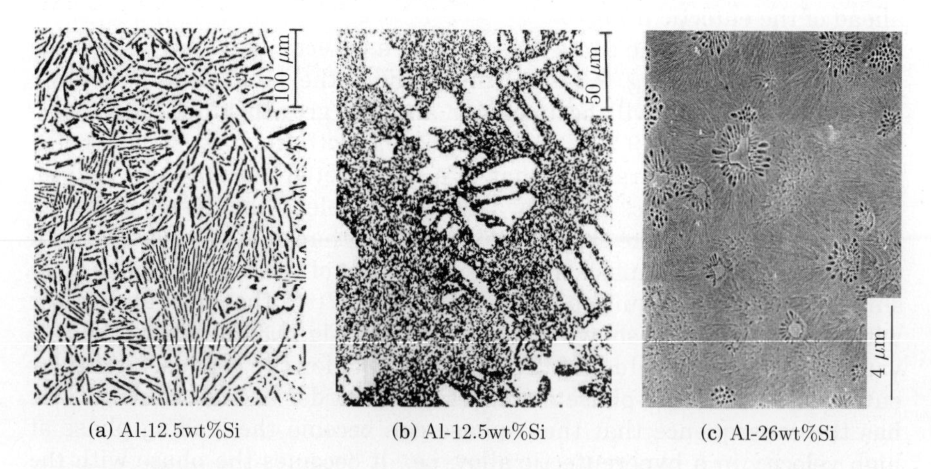

Fig. 9.35 Microstructures of various Al-Si alloys: (a) A eutectic composition, slow cooling; (b) A eutectic composition, fast cooling; (c) A hypereutectic composition, laser remelted at $0.1\ \mathrm{m\,s^{-1}}$. ((a,b) after Hellawell [12], (c) Pierantoni et al. [30].)

few aluminum dendrites form first, before the eutectic, as if the alloy composition was hypoeutectic. Put in the context of the model described above, the corresponding point falls in the ($\alpha + eut$)-region of Fig. 9.34(b). Similarly, for an Al-Si of hypereutectic composition (Fig. 9.35c), a fully eutectic microstructure can be expected if the growth rate is large enough, i.e., if the corresponding point falls in the skewed coupled zone. The micrograph of the laser-solidified alloy remelted at high speed, shows a structure that is slightly more complex. An "equiaxed" structure is formed by nucleation of the silicon phase, as opposed to columnar growth. As the silicon phase grows, the surrounding melt becomes depleted in Si and Al nucleates. It grows as a dendritic microstructure for a short distance, before coupled growth with a very fine lamellar spacing completes the solidification of each grain.

9.5 SUMMARY

The present chapter has demonstrated how two solid phases can grow together in eutectic alloys, such as the foundry alloys Al-Si and cast iron, or the soldering alloys Pb-Sn and Sn-Cu. Near-eutectic composition alloys have a rather narrow solidification interval and are thus easier to cast. The classical theory of Jackson and Hunt for regular lamellar eutectics has been derived in detail, demonstrating how solute exchanges between various phases can contribute to decreasing the required undercooling. This basic theory also facilitates the understanding of the mechanisms involved in the solidification of irregular eutectics, divorced eutectics, eutectic cells and nodular cast iron. Subsequently, peritectic solidification was addressed. Important industrial alloys in this class include bronze and steel. Whereas the peritectic reaction and solidification of the peritectic phase in hypoperitectic alloys occur close to the peritectic temperature, the transformation of the primary α-phase into β is usually incomplete, as it takes place through solid state transformation. Finally, we have seen that the competition of various phases and/or morphologies depends upon the equilibrium phase diagram, but is also strongly influenced by nucleation and growth kinetics. Considering these phenomena, a general criterion for phase selection was identified. The phase or morphology that is observed is the one that has the highest temperature when the velocity is imposed, or the fastest kinetics when the temperature is fixed.

9.6 EXERCISES

Exercise 9.1. Lamellar-fiber transition.
Consider two non-faceted phases α and β having an isotropic interfacial energy $\gamma_{\alpha\beta}$ and forming a regular eutectic. Based on an interfacial energy minimum criterion, calculate the transition volume fraction between fibers

and lamellae. The answer depends on the arrangement of the cylindrical fibers, so make three assumptions. (1) For cylindrical fibers arranged in a square lattice of spacing λ, where λ corresponds to the spacing of lamellae, show that the transition occurs at $g_{\beta,crit} = 1/\pi \approx 32\%$. (2) For fibers arranged in a hexagonal network with the same spacing λ of the lamellae, $g_{\beta,crit} = \sqrt{3}/(2\pi) \approx 28\%$. (3) For fibers arranged in a hexagonal network but having originated from the "fragmentation" of lamellae, i.e., spacing $a = 2\lambda/\sqrt{3}$, $g_{\beta,crit} = 2/(\sqrt{3}\pi) \approx 37\%$. Compare these values with those that could be deduced from Figs. 9.8(b) and (d).

Exercise 9.2. Jackson-Hunt model.

(a) By using the method of separate variables, i.e., by writing $C_\ell(x,z) = X(x)Z(z)$, show that the general solution to Eq. (9.1) is given by Eq. (9.4).

(b) Under the assumptions that $D_\ell/v \gg \lambda$ and $C_\ell(x, z = 0) \simeq C_0$, show that the coefficients B_n of the Fourier series in Eq. (9.4) are given by Eq. (9.6).

(c) Develop an expression for the average curvature contribution of a lamella starting from the integral of Eq. (9.11). For this purpose, apply two successive changes of variable: $y'(x) = u(x)$ and then $u(x) = \tan t(x)$.

(d) Finally, assuming that regular eutectic lamellae structures grow at the minimum undercooling for a fixed velocity v, show that the undercooling of the front $\Delta T(v^*)$ and the lamellae spacing $\lambda(v^*)$ are given by Eqs. 9.16.

Exercise 9.3. Irregular eutectics.
Assuming that $\overline{\lambda} = \phi\lambda_{ext}$ and that λ_{ext} is given by the theory of Jackson and Hunt, derive Eqs. (9.18). How can you explain that, for a fixed growth rate, the average spacing of irregular eutectic is a decreasing function of the thermal gradient?

Exercise 9.4. Nodular cast iron solute profile and boundary conditions.

(a) Beginning with the steady-state diffusion equation in spherical coordinates

$$\frac{1}{r^2}\frac{d}{dr}\left(r^2\frac{dC_\gamma}{dr}\right) = 0 \tag{9.28}$$

show that the solute profile in the austenite shell surrounding a graphite nodule is given by Eq. (9.19). Derive the solute flux at the two boundaries R_G and R_γ.

(b) Use the mass and solute balances at the graphite-austenite interface $r = R_G$, to derive the following relationship:

$$\rho_\gamma D_\gamma \left(\frac{\partial C_\gamma}{\partial r}\right)_{R_G} = \rho_G \dot{R}_G (1 - C_\gamma^{*g}) \tag{9.29}$$

By using the steady-state solution (Eq. (9.19)), show that Eq. (9.20) is obtained.

(c) Follow the same procedure for the austenite-liquid interface in order to obtain the following solute balance:

$$\rho_\gamma (v_\gamma(R_\gamma) - \dot{R}_\gamma)(C_\ell^{*\gamma} - C_\gamma^{*\ell}) = \rho_\ell D_\ell \left(\frac{\partial C_\ell}{\partial r}\right)_{R_\gamma} - \rho_\gamma D_\gamma \left(\frac{\partial C_\gamma}{\partial r}\right)_{R_\gamma} \tag{9.30}$$

Develop the last term in this equation with the help of the steady state solute diffusion solution (Eq. (9.19)). On the other hand, by assuming complete mixing in the liquid, show that the first term on the right-hand side is equal to:

$$\rho_\ell D_\ell \left(\frac{\partial C_\ell}{\partial r}\right)_{R_\gamma} = \rho_\ell \frac{dC_\ell}{dt} \frac{R_\ell^3 - R_\gamma^3}{3R_\gamma^2} \tag{9.31}$$

where $R_\ell(t)$ is the radius of the sphere available for the growth of a eutectic nodule. Its evolution is given by the overall mass balance:

$$\rho_G R_G^3 + \rho_\gamma \left(R_\gamma^3 - R_G^3\right) + \rho_\ell \left(R_\ell^3 - R_\gamma^3\right) = \rho_\ell R_0^3 \tag{9.32}$$

The radius R_0 is simply equal to $(3/(4\pi n_0))^{1/3}$ where n_0 is the density of graphite nodules.

(d) Finally, assuming that the γ phase is incompressible, its velocity $v_\gamma(R_\gamma)$ at the interface with the liquid is equal to $v_\gamma(R_G)(R_G/R_\gamma)^2$, where $v_\gamma(R_G)$ is its speed at the interface with the graphite nodule. Considering the graphite expansion, the previous mass balance at this interface is given by: $v_\gamma(R_G) = (1 - \rho_G/\rho_\gamma)\dot{R}_G$. With all these developments, show that the growth rate of the austenite nodule is given by Eq. (9.21).

Exercise 9.5. Influence of β domain on peritectic solidification. Consider the two peritectic alloys shown in the accompanying figure. In (a), the slopes of the β-solvus and β-solidus have the same sign and, in (b), the primary γ- and peritectic γ'-phases have the same crystallographic structures. Describe in both cases the progress of solidification at regular speed for two alloys with $C_0 < C_{per}$ and $C_0 > C_{per}$.

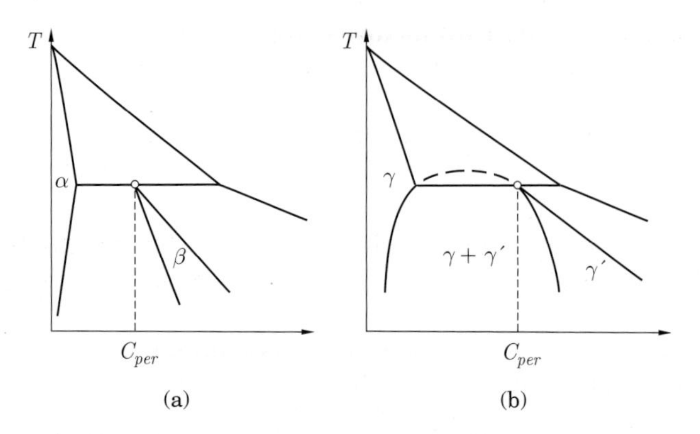

Two peritectic phase diagrams.

Exercise 9.6. Peritectic transformation kinetics.
In Sect. 10.2.2, a solute balance was applied to a peritectic phase diagram in order to deduce the evolution of the $\alpha-\beta$ and $\beta-\ell$ interfaces (Eqs. (10.39) and (10.40)). It was shown that the velocity of the two interfaces, v_α^* and v_β^*, depends in particular on the solute gradient in the peritectic phase, $D_\beta(\partial C_\beta/\partial x)$, taken at each interface. Assume that the solute profile is linear in the β-phase, and show that the expressions for v_α^* and v_β^* become:

$$v_\alpha^* = \frac{dx_\alpha^*}{dt} = \frac{1}{\left(C_\alpha^{*\beta} - C_\beta^{*\alpha}\right)} \left[D_\beta \frac{C_\beta^{*\ell} - C_\beta^{*\alpha}}{x_\beta^* - x_\alpha^*} - D_\alpha \left.\frac{\partial C_\alpha}{\partial x}\right|_{x_\alpha^*} \right]$$

$$v_\beta^* = \frac{dx_\beta^*}{dt} = \frac{1}{\left(C_\beta^{*\ell} - C_\ell^{*\beta}\right)} \left[\frac{dC_\ell^{*\beta}}{dt} \left(\frac{\lambda_2}{2} - x_\beta^*\right) - D_\beta \frac{C_\beta^{*\ell} - C_\beta^{*\alpha}}{x_\beta^* - x_\alpha^*} \right]$$

Neglecting diffusion in the α-phase abd the composition variation in the liquid, integrate these two equations to obtain $x_\alpha^*(t)$ and $x_\beta^*(t)$, and thus the evolution of the thickness of the peritectic phase.

Exercise 9.7. Solidification at low growth rate of a hypoperitectic alloy.
Consider a hypoperitectic alloy of nominal composition $C_\alpha^{per} < C_0 < C_{per}$ with a phase diagram such as the one shown in Fig. 9.25(a). By assuming steady-state and stable α-planar front growth, draw the composition profile in the liquid $C_\ell(z)$, the associated liquidus temperature profiles, $T_{liq}^\alpha(C_\ell(z))$ and $T_{liq}^\beta(C_\ell(z))$, and the actual temperature profile $T(z)$. Do the same for a β-planar front. Show graphically that an undercooled region exists at the solid-liquid interface in both cases.

The cycle illustrated in Fig. 9.26(c) is dependent on the undercooling ΔT_n^α and ΔT_n^β, besides the composition C_0. Fixing ΔT_n^α and ΔT_n^β, what is the range of nominal compositions allowing this cycle to operate?

Exercise 9.8. Solidification at low growth rate of a hyperperitectic alloy.
Consider a hyperperitectic alloy of nominal composition $C_{per} < C_0 < C_\ell^{per}$ with a phase diagram such as the one shown in Fig. 9.25(a) and a stable steady-state β planar front growing at velocity v^* in a thermal gradient G. Draw the composition profile in the liquid $C_\ell(z)$, the associated liquidus temperature profiles $T_{liq}^{\beta}(C_\ell(z))$ and $T_{liq}^{\alpha}(C_\ell(z))$, and the actual temperature profile $T(z)$. Show that nucleation of the α phase can take place only in the bulk liquid and not at the $\beta - \ell$ interface. Determine the conditions under which this can occur.

Exercise 9.9. Skewed coupled zone of Al-Si.
Consider the binary Al-Si alloy phase diagram, making the following assumptions for the eutectic invariant and liquidus lines (compositions in wt%, temperatures in °C):

$$T_{liq}^{Al} = 577 - 6.8 \times (C_\ell - 12.2) \quad ; \quad T_{liq}^{Si} = 577 + 9.53 \times (C_\ell - 12.2)$$
$$T_{eut} = 577 \quad ; \quad C_{eut} = 12.2 \tag{9.33}$$

The growth kinetics parameters that appear in Eqs. 9.26 and 9.27 are [20]:

$$A^{Al} = 20\ ^\circ\mathrm{Cs}^{1/2}\mathrm{mm}^{-1/2} \quad ; \quad A^{Si} = 60\ ^\circ\mathrm{Cs}^{1/2}\mathrm{mm}^{-1/2}$$
$$A^{eut} = 100\ ^\circ\mathrm{C}^{3/2}\mathrm{s}^{1/2}\mathrm{mm}^{-1} \tag{9.34}$$

Compute the skewed coupled zone of Al-Si for a thermal gradient $G = 10$ K/mm and $D_\ell = 3 \times 10^{-3}$ mm²/s. Compare the results with the microstructures shown in Fig. 9.35

9.7 REFERENCES

[1] S. Akamatsu, S. Bottin-Rousseau, and G. Faivre. Experimental evidence for a zigzag bifurcation in bulk lamellar eutectic growth. *Phys. Rev. Lett.*, 93:175701, 2004.

[2] S. Akamatsu, S. Moulinet, and G. Faivre. The formation of lamellar-eutectic grains in thin samples. *Met. Mater. Trans.*, 32A:2039, 2001.

[3] S. Akamatsu, M. Plapp, G. Faivre, and A. Karma. Overstability of lamellar eutectic growth below the minimum-undercooling spacing. *Met. Mater. Trans.*, 35A:1815, 2004.

[4] M. Bobadilla, J. Lacaze, and G. Lesoult. Influence des conditions de solidification sur le déroulement de la solidification des aciers inoxydables austénitiques. *J. Cryst. Growth*, 89:531, 1988.

[5] B. Boettger, V. Witusiewicz, and S. Rex. Phase-field method coupled to Calphad: Quantitative comparison between simulation and experiments in ternary eutectic In-Bi-Sn. In C.-A. Gandin and M. Bellet, editors, *Modeling of Casting, Welding and Advanced Solidification Processes - XI*, page 425, Warrendale, PA, USA, 2006. TMS Publ.

[6] W.J. Boettinger, S.R. Coriell, A.L. Greer, A. Karma, W. Kurz, M. Rappaz, and R. Trivedi. Solidification microstructures: Recent development, future directions. *Acta Mater.*, 48: 43, 2000.

[7] L.F. Donaghey and W.A. Tiller. On the diffusion of solute during the eutectoid and eutectic transformations, Part I. *Mater. Sci. Engrg*, 3:231, 1968/69.

[8] D. D. Double and A. Hellawell. The nucleation and growth of graphite-the modification of cast iron. *Acta Metall. Mater.*, 43:2435, 1995.

[9] D. J. Fisher and W. Kurz. A theory of branching limited growth of irregular eutectics. *Acta Met.*, 28:777, 1980.

[10] S. Fukumoto and W. Kurz. Predictions of the δ to γ transitions in austenitic stainless steels during laser treatment. *ISIJ Intl.*, 38:71, 1998.

[11] U. Hecht, V. T. Witusiewicz, A. Drevermann, B. Boettger, and S. Rex. Eutectic solidification of ternary Al-Cu-Ag alloys: Coupled growth of α(Al) and Al_2Cu in univariant reaction. *Mater. Sci. Forum*, 508:57, 2006.

[12] A. Hellawell. The growth and structure of eutectics with silicon and germanium. *Prog. Mater. Sci.*, 15:1, 1970.

[13] M. Hillert. The role of interfacial energy during solid-state phase transformations. *Jernkont. Annal.*, 141:757, 1957.

[14] L.M. Hogan, R.W. Kraft, and F.D. Lemkey. Eutectic grains. *Adv. Mater. Res.*, 5:83, 1971.

[15] O. Hunziker, M. Vandyoussefi, and W. Kurz. Phase and microstructure selection in peritectic alloys close to the limit of constitutional undercooling. *Acta Mater.*, 46:6325, 1998.

[16] K.A. Jackson and J.D. Hunt. Lamellar and rod eutectic growth. *Trans. Metal. Soc. AIME*, 236:1129, 1966.

[17] H. Jones and W. Kurz. Relation of interphase spacing and growth temperature to growth velocity in Fe-C and Fe-Fe_3C eutectic alloys. *Z. Metall.*, 72:792, 1981.

[18] A. Karma and A. Sarkissian. Morphological instabilities of lamellar eutectics. *Met. Mat. Trans. A*, 27:635, 1996.

[19] H.W. Kerr and W. Kurz. Solidification of peritectic alloys. *Intl. Mater. Rev.*, 41:129, 1996.

[20] W. Kurz and D. J. Fisher. Dendritic growth in eutectic alloys: the coupled zone. *Intl. Metals Rev.*, 5-6:177, 1979.

[21] W. Kurz and D. J. Fisher. *Fundamentals of Solidification*. Trans. Tech. Publ., Aedermansdorf, Switzerland, 4th edition, 1998.

[22] J. Lacaze and V. Gerval. Modeling of the eutectoid reaction in spheroidal graphite Fe-C-Si alloys. *ISIJ Intl.*, 38:714, 1998.

[23] G. Lesoult, M. Castro, and J. Lacaze. Solidification of spheroidal graphite cast irons - Parts I/II. *Acta Mater.*, 46:983/997, 1998.

[24] S.-Z. Lu and A. Hellawell. Modification of Al-Si alloys: microstructure, thermal analysis, and mechanics. *J. Metals*, 47:38–40, 1995.

[25] P. Magnin and W. Kurz. Competitive growth of stable and metastable Fe-C-X eutectics: Parts I / II. *Metal. Trans. A*, 19:1955/1969, 1988.

[26] I. Minkoff. *The physical metallurgy of cast iron*. John Wiley and Sons Ltd, Chichester, UK, 1983.

[27] M. D. Nave, A. K. Dahle, and D. H. StJohn. Eutectic growth morphologies in magnesium-aluminium alloys. In H. I. Kaplan, J. N. Hryn, and B. B. Clow, editors, *Magnesium Technology 2000*, page 233; see also 243, Warrendale, PA, USA, 2000. TMS Publ.

[28] A. Parisi and M. Plapp. Stability of lamellar eutectic growth. *Acta Mater.*, 56:1348–1357, 2008.

[29] D. Phelan, M. Reid, and R. Dippenaar. Kinetics of the peritectic reaction in an Fe-C alloy. *Mater. Sci. Engrg A*, 477:226, 2008.

[30] M. Pierantoni, M. Gremaud, P. Magnin, D. Stoll, and W. Kurz. The coupled zone of rapidly solidified Al-Si alloys in laser treatment. *Acta Metall. Mater.*, 40:1637, 1992.

[31] M. Plapp and A. Karma. Eutectic colony formation: a stability analysis. *Phys. Rev. E*, 60:6865, 1999.

[32] M. Plapp. Three-dimensional phase-field simulations of directional solidification. *J. Cryst. Growth*, 303:49, 2007.

[33] M. Rhême, F. Gonzales, and M. Rappaz. Growth directions in directionally solidified Al-Zn and Zn-Al alloys near eutectic compositions. *Scripta Mater.*, 59:440, 2008.

[34] D. Stefanescu. Modeling of cast iron solidification: the defining moments. *Met. Mater. Trans. A*, 38:1433, 2007.

[35] R. Trivedi, P. Magnin, and W. Kurz. Theory of eutectic growth under rapid solidification conditions. *Acta Metal.*, 35:971, 1987.

[36] H. Walker, S. Liu, J. H. Lee, and R. Trivedi. Eutectic growth in three dimensions. *Met. Mater. Trans.*, 38A:1417, 2007.

CHAPTER 10

MICROSEGREGATION AND HOMOGENIZATION

10.1 INTRODUCTION

The preceding chapters have described the formation of various types of microstructures, including cells, dendrites, eutectics and peritectics. The microstructure can be thought of as a combination of morphology and length scales. These factors are important for determining material properties, both in service and during processing. The present chapter explores the latter, focusing on how the solidification path relates to the evolution of solid fraction and composition in the microstructure. These results are, in turn, coupled to the evolution of latent heat, which is the essential link between microstructure and macroscale heat transfer, which is the subject of Chap. 11.

We focus on the segregation profile that develops within the microstructure due to the interaction of solute rejection by solidification with solute diffusion in the liquid and solid phases. One such example has already been encountered in Chap. 5, where the steady state, and the initial and final transients in planar growth of a binary alloy were considered. In that case, however, the main concern was segregation at the length scale of an entire crystal, whereas the present chapter is focused on segregation at the scale of the microstructure.

We will demonstrate that, for almost all cases of interest, both heat and solute transport at the scale of the microstructure are sufficiently rapid for the temperature and liquid composition to be considered uniform. Solute transport in the solid state, however, can vary significantly depending on the local solidification conditions. The first part of this chapter describes relatively simple one-dimensional models for solute transport for binary alloys in the solid state. We then move on to cases of higher complexity, including multi-component alloys and more realistic morphologies.

10.2 1-D MICROSEGREGATION MODELS FOR BINARY ALLOYS

Consider the segment of a dendritic array shown in Fig. 10.1. The micrograph on the left-hand side comes from a directional solidification experiment in a succinonitrile – acetone alloy [18]. An idealization of the structure suitable for modeling is presented to the right. The region of the secondary dendrite arms is modeled as a 2-D periodic array of unit cells, with lateral dimension of the secondary dendrite arm spacing λ_2 and length L. We neglect any variation in the third dimension for this part of the development. It is also assumed (for now) that there are only two components and that the densities of the solid and liquid are constant and equal.

For the sake of simplicity, the thermal properties of the solid and liquid are taken as equal. (Scaling will show that the temperature can be safely considered as uniform.) The governing equations for heat and solute transport in the unit cell are then

$$\frac{\partial T}{\partial t} = \alpha \left(\frac{\partial^2 T}{\partial x^2} + \frac{\partial^2 T}{\partial z^2} \right) \tag{10.1}$$

$$\frac{\partial C_\ell}{\partial t} = D_\ell \left(\frac{\partial^2 C_\ell}{\partial x^2} + \frac{\partial^2 C_\ell}{\partial z^2} \right) \tag{10.2}$$

$$\frac{\partial C_s}{\partial t} = D_s \left(\frac{\partial^2 C_s}{\partial x^2} + \frac{\partial^2 C_s}{\partial z^2} \right) \tag{10.3}$$

We now introduce the following scaled variables

$$\xi = \frac{x}{\lambda_2/2}; \; \zeta = \frac{z}{L}; \; \tau = \frac{t}{t_f} \tag{10.4}$$

where t_f is the local solidification time. Substituting the scaled variables into Eqs. (10.1)-(10.3) yields the semi-scaled equations

$$\frac{\partial T}{\partial \tau} = \frac{4\alpha t_f}{\lambda_2^2} \left(\frac{\partial^2 T}{\partial \xi^2} + \frac{\lambda_2^2}{L^2} \frac{\partial^2 T}{\partial \zeta^2} \right) \approx \mathrm{Fo_T} \frac{\partial^2 T}{\partial \xi^2} \tag{10.5}$$

$$\frac{\partial C_\ell}{\partial \tau} = \frac{4D_\ell t_f}{\lambda_2^2} \left(\frac{\partial^2 C_\ell}{\partial \xi^2} + \frac{\lambda_2^2}{L^2} \frac{\partial^2 C_\ell}{\partial \zeta^2} \right) \approx \mathrm{Fo}_\ell \frac{\partial^2 C_\ell}{\partial \xi^2} \tag{10.6}$$

$$\frac{\partial C_s}{\partial \tau} = \frac{4D_s t_f}{\lambda_2^2} \left(\frac{\partial^2 C_s}{\partial \xi^2} + \frac{\lambda_2^2}{L^2} \frac{\partial^2 C_s}{\partial \zeta^2} \right) \approx \mathrm{Fo_s} \frac{\partial^2 C_s}{\partial \xi^2} \tag{10.7}$$

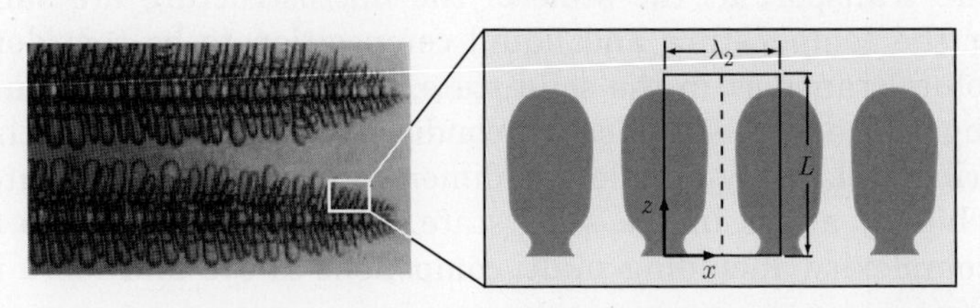

Fig. 10.1 (Left) A micrograph of a directionally solidified succinonitrile-acetone alloy. (Right) An idealized segment, displaying a periodic arrangement of secondary dendrite arms.

We have invoked the assumption that $\lambda_2 \ll L$, as suggested by the micrograph in Fig. 10.1. This renders the problem one-dimensional. The dimensionless groups that appear in each equation are Fourier numbers, each representing the ratio of the local solidification time t_f to the characteristic time for the diffusion of heat or solute. The meaning of the subscripts should be clear.

In order to proceed, experimental correlations between λ_2 and t_f need to be introduced. For Al alloys, it is found that, over a very wide range of solidification times, $0.1 \text{ s} < t_f < 10^7 \text{ s}$, the following correlation provides a fairly accurate estimate for λ_2:

$$\lambda_2 \approx \left(10^{-5} \text{ m s}^{-1/3}\right) t_f^{1/3} \tag{10.8}$$

The following property values for Al-Cu alloys are used as an example:

$$\alpha \approx 4 \times 10^{-5} \text{ m}^2 \text{ s}^{-1}; \quad D_\ell \approx 3 \times 10^{-9} \text{ m}^2 \text{ s}^{-1}; \quad D_s \approx 3 \times 10^{-13} \text{ m}^2 \text{ s}^{-1} \tag{10.9}$$

One should note the large disparity in the magnitude of the various diffusivities. Substituting these data into the expressions for the Fourier numbers gives

$$\text{Fo}_\text{T} \approx \left(1.6 \times 10^6 \text{ s}^{-1/3}\right) t_f^{1/3}; \quad \text{Fo}_\ell \approx \left(120 \text{ s}^{-1/3}\right) t_f^{1/3}; \quad \text{Fo}_\text{s} \approx \left(0.012 \text{ s}^{-1/3}\right) t_f^{1/3}; \tag{10.10}$$

Now, for all of the solidification times in the given range $0.1 - 10^7$ s, we have both $\text{Fo}_\text{T} \gg 1$ and $\text{Fo}_\ell \gg 1$. In this case, one can rewrite Eqs. (10.5) and (10.6), neglecting terms that are small compared to unity, to obtain

$$\frac{\partial^2 T}{\partial \xi^2} \approx 0 \tag{10.11}$$

$$\frac{\partial^2 C_\ell}{\partial \xi^2} \approx 0 \tag{10.12}$$

The solution of these two equations with symmetry boundary conditions at $\xi = 0, 1$ $(x = 0, \lambda_2/2)$ is trivial, giving simply $T = T(t)$ and $C_\ell = C_\ell(t)$. In other words, both T and C_ℓ are independent of spatial position, but may vary with time.

There are two regimes for the composition in the solid. For $t_f > 1000$ s, both sides of Eq. (10.7) are of comparable order, and the transient diffusion in the solid must be simulated. This case is called *back diffusion*, and several models for this phenomenon are described beginning in Sect. 10.2.1. For more typical casting conditions, where $t_f < 1000$ s, we have $\text{Fo}_\text{s} \ll 1$ and Eq. (10.7) becomes

$$\frac{\partial C_s}{\partial \tau} \approx 0 \tag{10.13}$$

This implies that the composition does not change with time in the solid, or that the diffusion is negligible there. The result that was just obtained via scaling analysis of the transport equations was used as a starting point for a model first presented by Gulliver [5] and later by Scheil [11]. The completion of the model requires an appropriate expression for the solute balance at the moving interface.

Consider a solute balance over the unit cell. Since the densities of the solid and liquid phases have been assumed to be equal, and the unit cell is a closed system, we have

$$\int_0^{x^*(t)} C_s(x,t)\,dx + \int_{x^*}^{\lambda_2/2} C_\ell(x,t)\,dx = C_0 \tag{10.14}$$

where x^* is the location of the solid-liquid interface. For all of the microsegregation models that we will consider, we assume the liquid composition to be uniform in space and equal to the interface composition C_ℓ^*, so that the second integral can be directly evaluated, giving

$$\int_0^{x^*(t)} C_s(x,t)\,dx + \left(\frac{\lambda_2}{2} - x^*\right) C_\ell^* = C_0 \tag{10.15}$$

Differentiating Eq. (10.15) with respect to time gives

$$C_s^* \frac{dx^*}{dt} + \int_0^{x^*(t)} \frac{\partial C_s(x,t)}{\partial t} dx + \left(\frac{\lambda_2}{2} - x^*\right) \frac{dC_\ell^*}{dt} - C_\ell^* \frac{dx^*}{dt} = 0 \tag{10.16}$$

where we have used $C_s(x^*,t) = C_s^*$. The remaining integral is resolved by first applying the one-dimensional form of Eq. (10.3) to Eq. (10.16). We also impose thermodynamic equilibrium at the interface, i.e, $C_s^* = k_0 C_\ell^*$, and combine the terms involving the interface velocity to obtain

$$-(1-k_0) C_\ell^* \frac{dx^*}{dt} + \int_0^{x^*(t)} D_s \frac{\partial^2 C_s(x,t)}{\partial x^2} dx + (\lambda_2/2 - x^*) \frac{dC_\ell^*}{dt} = 0 \tag{10.17}$$

The integration can be performed easily, with the result

$$-(1-k_0) C_\ell^* \frac{dx^*}{dt} + D_s \frac{\partial C_s(x^*,t)}{\partial x} + (\lambda_2/2 - x^*) \frac{dC_\ell^*}{dt} = 0 \tag{10.18}$$

The symmetry boundary condition at $x = 0$ was also imposed. Finally, we divide by $\lambda_2/2$, and recognize that $g_s = x^*/(\lambda_2/2)$ to obtain

$$-(1-k_0) C_\ell^* \frac{dg_s}{dt} + \frac{2D_s}{\lambda_2} \frac{\partial C_s(x^*,t)}{\partial x} + (1-g_s) \frac{dC_\ell^*}{dt} = 0 \tag{10.19}$$

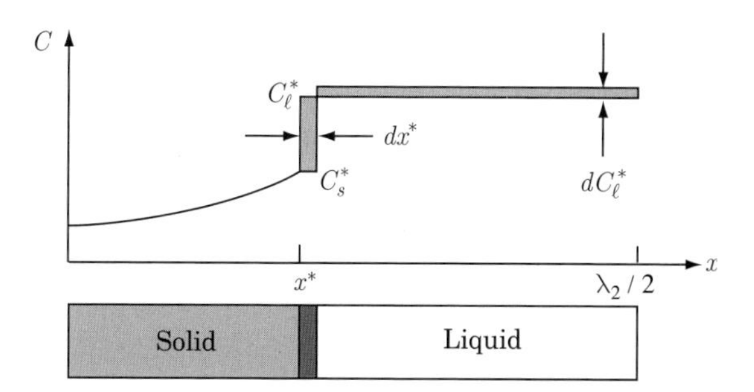

Fig. 10.2 A construct demonstrating the partitioning of liquid from the interface to the liquid in the Gulliver-Scheil model of microsegregation.

For values of $\mathrm{Fo_s}$ where diffusion in the solid is negligible, as in the Gulliver-Scheil model, Eq. (10.19) attains a simpler form by setting D_s to zero, leaving

$$(1 - k_0)C_\ell^* \frac{dg_s}{dt} = (1 - g_s)\frac{dC_\ell^*}{dt} \tag{10.20}$$

This form of the boundary condition is illustrated in Fig. 10.2, and one can see that the solute rejected by the moving interface in time dt is exactly balanced by the increase in solute content of the liquid. Equation (10.20) integration can be rearranged for integration

$$\int_1^{g_\ell} \frac{dg_\ell}{g_\ell} = -\frac{1}{1 - k_0} \int_{C_0}^{C_\ell^*} \frac{dC_\ell^*}{C_\ell^*} \tag{10.21}$$

where the initial conditions $C_\ell^* = C_0$ when $f_\ell = 1$ has been invoked. Integration and rearrangement of terms yields the *Gulliver-Scheil* equation,

$$C_\ell^* = C_0 g_\ell^{k_0 - 1} \tag{10.22}$$

The microsegregation profile in the solid is obtained by setting $C_s^* = k_0 C_\ell^*$. It is also convenient to have the liquid fraction in terms of temperature. Making the usual assumption that the liquidus curve is given by the straight line $T = T_f + m_\ell C_\ell$, it is straightforward to show that

$$g_s = 1 - g_\ell = 1 - \left(\frac{T - T_f}{T_{liq} - T_f}\right)^{1/(k_0 - 1)} \tag{10.23}$$

This equation can be compared to its counterpart derived for the lever rule, Eq. (3.11), which is repeated here for convenience

$$g_s = \frac{1}{1 - k_0} \frac{T - T_{liq}}{T - T_f} \tag{10.24}$$

Recall that the lever rule was derived as a solute balance assuming a uniform spatial composition in each phase, and that, for a constant density, $g_s = f_s$.

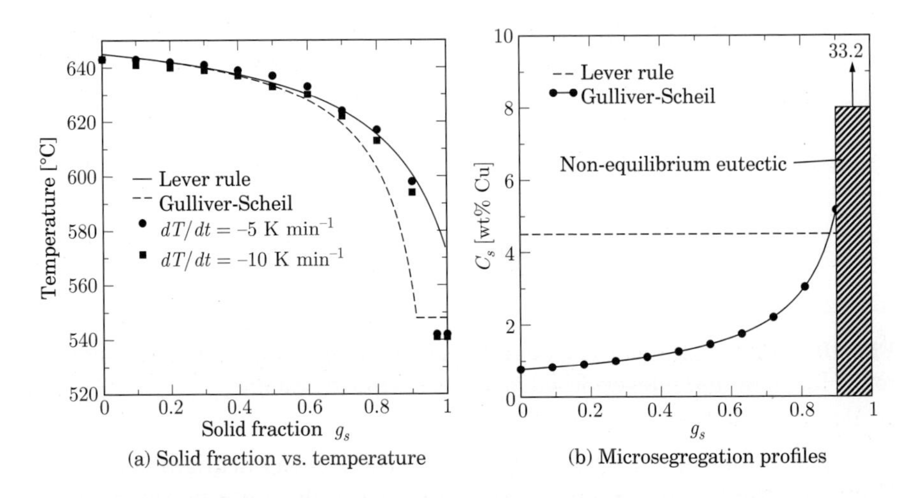

Fig. 10.3 (a) Computed fraction solid vs. temperature curves for an Al-4.5 wt% Cu alloy, using the lever rule and the Gulliver-Scheil equation. The experimental data were obtained by DTA experiments. [7] (b) The corresponding microsegregation patterns for the two solidification models.

Figure 10.3 compares the solid fraction calculated using the Gulliver-Scheil and lever rule equations for an Al-4.5 wt% Cu alloy. This system forms a eutectic at a composition of 33 wt% Cu at 548°C. The Gulliver-Scheil model predicts that the solidification will be incomplete at the eutectic temperature. It is usually assumed that any liquid that remains at the eutectic point will solidify as a eutectic. This would yield approximately 10% non-equilibrium eutectic for this alloy in the Gulliver-Scheil model. The lever rule, on the other hand, predicts that no eutectic will form. Figure 10.3 also includes data obtained by DTA experiments at various cooling rates [7]. At a low cooling rate of 5 K min^{-1}, which corresponds to a solidification time t_f of approximately 20 min, the measured solid fraction vs. temperature is close to that obtained using the lever rule. As the cooling rate increases to $-10\,\text{K}\,\text{min}^{-1}$, the solid fraction curve moves in the direction of the Gulliver-Scheil model. Note that the experimental data in both cases shows about 4% nonequilibrium eutectic.

The experimental observations are consistent with the scaling analysis presented earlier. The cooling rate of 5 K min^{-1} represents a solidification time $t_f \approx 1200$ s, and a corresponding value of $\text{Fo}_\text{s} \approx 14 \gg 1$. Thus, one would expect to find a solid composition that is almost independent of spatial position, i.e., the lever rule would apply. On the other hand, if the cooling rate were sufficiently high that $\text{Fo}_\text{s} \approx\ll 1$, then the Gulliver-Scheil model would be valid. This represents the opposite extreme from the lever rule. For the intermediate cooling rate of $-10\,\text{K}\,\text{min}^{-1}$, $\text{Fo}_\text{s} \approx 7 \gtrsim 1$. For this case, we need to consider diffusion in the solid state, which will be addressed in the next section.

Before taking up this latter case, we note that solid fraction calculations such as these can be used as input into macroscopic models. If the solidification time t_f is known to fall in the range where the Gulliver-Scheil model is valid ($\mathrm{Fo_s} \ll 1$, $\mathrm{Fo}_\ell \gg 1$, $\mathrm{Fo_T} \gg 1$), then one can write the enthalpy-temperature relationship used in Eq. (6.71) directly, using

$$\langle \rho h(T) \rangle = \int_{298}^{T} \rho c_p dT + \rho L_f (1 - g_s(T)) \tag{10.25}$$

as discussed in Chap. 6. If the values of the Fourier numbers are not appropriate for the Gulliver-Scheil model, e.g. $\mathrm{Fo}_\ell \sim 1$, then a back-diffusion model is needed, as described in the next section.

10.2.1 Microsegregation with diffusion in the solid state

The two models presented thus far represent two extremes: the lever rule is applicable when there is complete mixing in both the solid and the liquid, whereas for the Gulliver-Scheil model there is complete mixing only in the liquid and no diffusion at all in the solid. These models are the simplest, since in both cases, all of the transport equations reduce to a single term (cf., Eqs. (10.11)-(10.13)). This effectively eliminates the length and time scale from the problem, leading to expressions such as Eqs. (10.23) and (10.24), where length and time are absent. The present section examines several models representing intermediate behavior, where there is a limited amount of diffusion in the solid phase. Length and time then resurface, and the models must account for the microstructure.

The example of the Al-4.5 wt%Cu alloy, given in the preceding section, demonstrated that at a cooling rate of about $-10\,\mathrm{K\,min^{-1}}$, neither the lever nor Gulliver-Scheil model matched the experimental data very well. A common application where diffusion in the solid must almost always be accounted for is the solidification of Fe-C alloys, such as steel or cast iron. The diffusivity of C in solid iron is smaller than that of the liquid by a factor of only about 100. Therefore, performing a calculation similar to that in Eq. (10.10) for the various Fourier numbers would reveal that $\mathrm{Fo_T} \gg 1$ and $\mathrm{Fo}_\ell \gg 1$, but that for such alloys $\mathrm{Fo_s} \sim \mathcal{O}(1)$. Therefore, one must account for finite diffusion in the solid. The models presented in the remainder of this section handle solute diffusion in the solid in a variety of approximate ways, all with the goal of providing a relatively simple analytical expression relating the composition, solid fraction and temperature. Note that $\mathrm{Fo_s}$ is sometimes called the "back diffusion parameter."

Brody-Flemings model

The first treatment of this problem was presented by Brody and Flemings [2], who retained the assumption that the temperature and liquid composition are uniform in space. They then considered a solute balance at the interface, using Eq. (10.18). The term representing diffusion in the solid is approximated in this model by applying the chain rule and thermodynamic equilibrium, such that

$$\frac{\partial C_s^*}{\partial x^*} \sim k_0 \frac{dC_\ell^*/dt}{dx^*/dt} \sim k_0 \frac{dC_\ell^*}{dx^*} \tag{10.26}$$

Equation (10.19) then becomes

$$C_\ell^*(1-k_0)\frac{dx^*}{dt} = D_s k_0 \frac{dC_\ell^*}{dx^*} + \left(\frac{\lambda_2}{2} - x^*\right)\frac{dC_\ell^*}{dt} \tag{10.27}$$

The approximation in Eq. (10.26) is illustrated schematically in Fig. 10.4, where the positions of the interface and the compositions in the solid and liquid phases are presented at times t and $t+\Delta t$. One can see from the construction that Eq. (10.26) does not account for the change in solute in the solid phase over the time interval, with the exception of at the interface. Therefore, the solute illustrated in the black triangle is "lost" in the calculation. Thus, this model does not conserve solute, and this shortcoming led other researchers to extend the model as described below. Before continuing with these other approaches, let us first complete the Brody-Flemings analysis.

In order to integrate Eq. (10.27), some assumption needs to be made regarding the relation between x^* and t. In their model, Brody and Flemings assumed "parabolic growth," i.e.,

$$x^* = \frac{\lambda_2}{2}\sqrt{\frac{t}{t_f}}; \quad \frac{dx^*}{dt} = \frac{\lambda_2^2}{8t_f x^*} \tag{10.28}$$

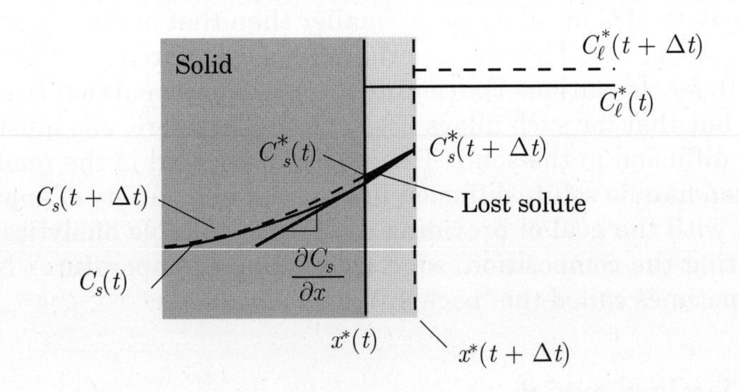

Fig. 10.4 A schematic illustration of the approximation for the solute balance at the interface in the Brody-Flemings model.

Using this expression to replace the denominator in the first term on the right hand side of Eq. (10.27), and rearranging the terms yields

$$\left[\left(2k_0\left\{\frac{4D_s t_f}{\lambda_2^2}\right\}-1\right)x^*+\frac{\lambda_2}{2}\right]\frac{dC_\ell^*}{dt}=C_\ell^*(1-k_0)\frac{dx^*}{dt} \tag{10.29}$$

Notice that the Fourier number now appears in the first term inside the square brackets, bringing with it the connotation of length and time scales.

Equation (10.29) can be integrated directly, noting that $C_\ell^* = C_0$ when $x^* = 0$. Using $x^*/(\lambda_2/2) = g_s$ as before, the result becomes

$$C_\ell^* = C_0\left[1-(1-2k_0\mathrm{Fo_s})\,g_s\right]^{-(1-k_0)/(1-2k_0\mathrm{Fo_s})} \tag{10.30}$$

Once again assuming that the liquidus is a straight line, Eq. (10.30) can be manipulated into a relationship between g_s and T, given by

$$g_s = \frac{1}{1-2k_0\mathrm{Fo_s}}\left[1-\left(\frac{T-T_f}{T_{liq}-T_f}\right)^{-(1-2k_0\mathrm{Fo_s})/(1-k_0)}\right] \tag{10.31}$$

One would expect to recover the Gulliver-Scheil and lever rule forms from this model for extreme values of $\mathrm{Fo_s}$. Fast solidification, meaning that $t_f \ll 4\lambda_2^2/D_s$, corresponds to $\mathrm{Fo_s} \to 0$. Substituting $\mathrm{Fo_s} = 0$ into Eq. (10.30) or (10.31) reduces them to the Gulliver-Scheil forms, Eqs. (10.22) and (10.23). One would also anticipate that $\mathrm{Fo_s} \to \infty$ would produce the lever rule, however this is not the case. This limit, e.g., in Eq. (10.30), is undefined. It is left as an exercise to show that, in fact, setting $\mathrm{Fo_s} = 0.5$ recovers the lever rule. This is where the liability that the model does not conserve solute in the solid phase appears. Thus, the solution is most accurate for $\mathrm{Fo_s} < 0.1$, where back-diffusion is less important.

Despite its shortcomings, we have chosen to present this model in some detail for several reasons. It is still widely used by many engineers, although more accurate models (discussed below) are available. Also, it forms the basis for several other models that apply various means to rectify the problem of solute conservation. Before turning to these other models, let us first consider an example of microsegregation in 0.8%C steel.

Example 10.1 Segregation in 0.8%C steel

Consider the solidification of 0.8%C steel, for which a portion of the phase diagram is shown below. Notice that the primary phase is γ-Fe. In order to employ the above-presented formulas that use T_f, the liquidus and solidus curves must be projected back to find the point (C', T_f') where they meet, and the value for T_f' must be used instead of T_f. This calculation is straightforward, and starts by computations of the slopes of the solidus and liquidus curves in order to find

$$m_\ell = -90.5 \text{ K(wt\%)}^{-1}; \quad m_s = -175.8 \text{ K(wt\%)}^{-1}$$

Using these values and taking the intersections at the peritectic temperature gives two simultaneous equations that can be solved to find $T_f' = 1562°\text{C}$ and $C' = -0.211$ wt%. The fact that C' is negative is not a problem, since it is simply a geometric construct and not a real composition. Caution must be taken with regard to the calculation of k_0. In order to be consistent with the definitions in the model, one should use $k_0 = m_\ell/m_s = 0.51$. Another option would be to use $k_0 = (C_s - C')/(C_\ell - C')$.

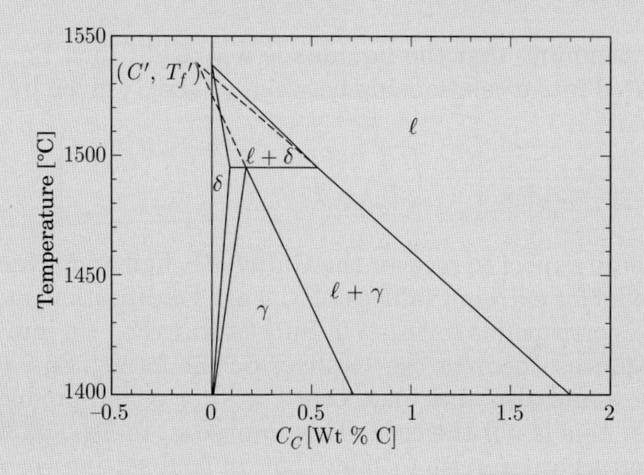

With these results in hand, we can proceed to calculate the relation between the solid fraction, composition and temperature for several values of Fo_s, as shown below.

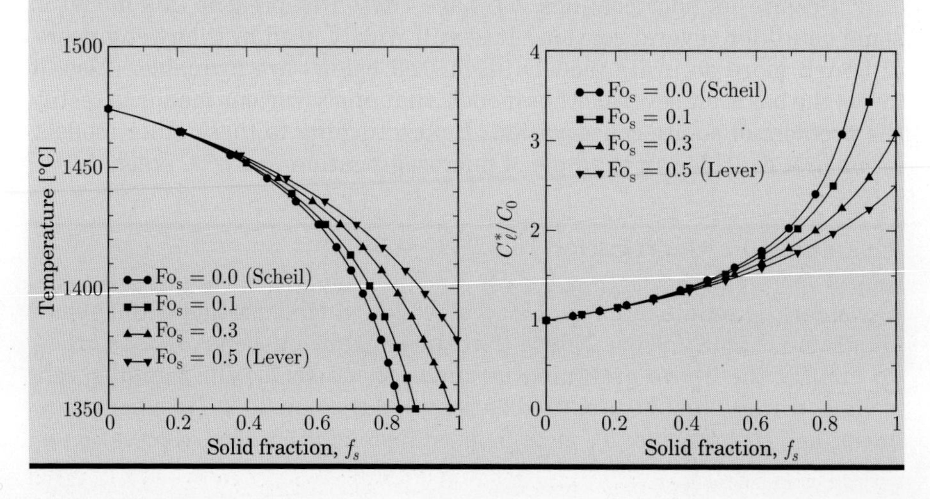

Derivative models

Numerous models have been developed to extend and improve the Brody-Flemings model. The first of these was presented by Clyne and Kurz [4], who introduced an empirical expression based on the Fourier number given by

$$f(\mathrm{Fo_s}) = \mathrm{Fo_s}\left[1 - \exp\left(-\frac{1}{\mathrm{Fo_s}}\right)\right] - 0.5\exp\left(-\frac{1}{2\mathrm{Fo_s}}\right) \tag{10.32}$$

Replacing $\mathrm{Fo_s}$ by $f(\mathrm{Fo_s})$ in Eqs. (10.30) and (10.31) causes them to approach the lever rule in the limit as $\mathrm{Fo_s} \to \infty$. The corresponding forms for the composition and solid fraction are given by

$$C_s = k_0 C_0 \left[1 - (1 - 2k_0 f(\mathrm{Fo_s}))\, g_s\right]^{-(1-k_0)/(1-2k_0 f(\mathrm{Fo_s}))} \tag{10.33}$$

and

$$g_s = \frac{1}{1 - 2k_0 f(\mathrm{Fo_s})}\left[1 - \left(\frac{T - T_f}{T_{liq} - T_f}\right)^{-(1-2k_0 f(\mathrm{Fo_s}))/(1-k_0)}\right] \tag{10.34}$$

Because of its simplicity, this model is widely used. However, it treats only the symptom of the problem in the Brody-Flemings model (incorrect behavior for large values of $\mathrm{Fo_s}$), rather than the underlying cause, which is that the solute is not conserved because diffusion in the solid is not handled consistently.

Ohnaka [8] extended the Brody-Flemings model by assuming a parabolic form for the solute distribution in the solid, expressed as $C_s = C_s^* + c(x^2 - x^{*2})$, where C_s^* is the value at the interface and c is a constant that is determined by enforcing a solute balance in the unit cell. To complete the model, Ohnaka invoked the parabolic growth assumption as in the Brody-Flemings model, Eq. (10.28). He also generalized the model to take into account both the planar geometry that has been considered so far, and a circular geometry inscribed in a hexagon, representing columnar growth. After substituting these assumptions and integrating the resulting differential equation, the following expression is obtained:

$$C_s = k_0 C_0 \left(1 - \gamma g_s\right)^{(k_0-1)/\gamma} \tag{10.35}$$

where

$$\gamma = 1 - \frac{n k_0 \mathrm{Fo_s}}{1 + n\mathrm{Fo_s}} \tag{10.36}$$

where $n = 2, 4$ for the planar and circular geometries, respectively. It is easy to demonstrate that Eq. (10.35) reduces to the Gulliver-Scheil equation for $\mathrm{Fo_s} \to 0$ and to the lever rule for $\mathrm{Fo_s} \to \infty$. Ohnaka also solved the same model system assuming a linear time evolution of the interface, $x^* = t/t_f$, producing a more complex expression that is omitted here since an exact solution will be provided for it by the following model.

Kobayashi [6] took Ohnaka's model for microsegregation in columnar dendrites, i.e., the cylindrical geometry inscribed in a hexagon, one step further by actually solving the diffusion equation in the solid for C_s, rather than assuming a parabolic distribution. He also assumed that the position of the interface in the circular geometry increased linearly with time, $r^* = t/t_f$. The result, obtained by separation of variables, is

$$C_s = k_0 C_0 \sum_{n=0}^{\infty} A_n \mathfrak{L}_n \left(-\frac{r^2}{4r^{*2}\mathrm{Fo_s}} \right) g_s^n \tag{10.37}$$

where $\mathfrak{L}_n$ is the Laguerre polynomial of order n, and the series coefficients A_n are given by

$$A_n = \frac{1}{\mathfrak{L}_n(-1/(4\mathrm{Fo_s}))} \prod_{m=0}^{n-1} \left[1 - \frac{k}{(m+1)\mathfrak{L}_m(-1/(4\mathrm{Fo_s}))} \sum_{j=0}^{\infty} \mathfrak{L}_j(-1/(4\mathrm{Fo_s})) \right] \tag{10.38}$$

Kobayashi also demonstrated that Ohnaka's solution could be obtained as an approximate form of this exact solution. Although this solution is exact, albeit to an approximate problem where $g_s = t/t_f$, the complexity of the functional form seems to have prevented its widespread use in the literature. In the same article, Kobayashi extended the model to consider a case where the solid fraction is an arbitrary function of time. For this case, the solution was applied in incremental form, and an iterative scheme was used to numerically solve for a self-consistent set of temperature, composition and solid fraction parameters.

The Brody-Flemings model and its derivative analytical expressions are summarized in Table 10.1. The major attraction of these models is the fact that each provides some sort of analytical expression for the microsegregation profile when driven by an externally applied, time-dependent change in the solid fraction. This can be implemented in a numerical code with particular ease. Once one decides to solve the problem numerically, however, it becomes possible to include many phenomena that otherwise have to be excluded since they preclude analytical solutions. Some of

Table 10.1 Analytical microsegregation models with prescribed movement of the interface and assumed solid composition profiles.

Model	Solid Diffusion Approximation	Eq. No.	Comment	Ref.
Lever Rule	Complete	(10.24)	$\mathrm{Fo_s} \to \infty$	Chap. 3
Scheil-Gulliver	None	(10.22)	$\mathrm{Fo_s} = 0$	[5; 11]
Brody-Flemings	Limited	(10.30)	Solute not conserved	[2]
Clyne-Kurz	Limited	(10.33)	Empirical expression	[4]
Ohnaka	Parabolic	(10.35)	Approximate C_s	[8]
Kobayashi	Laguerre	(10.37)	Exact solution	[6]

these phenomena include temperature- and composition-dependent properties, arbitrary temperature histories, possible deviations from thermodynamic equilibrium (especially undercooling of the eutectic), and coarsening of secondary dendrite arms. Several models involving some or all of these phenomena have been developed, but we have chosen to mention just two.

Roosz and Exner [9] and Sundarraj and Voller [13] simultaneously developed similar models including the use of actual liquidus and solidus curves from phase diagrams (i.e., non-constant k_0), arbitrary thermal histories and solute conservation via integral balances. In both models, the size of the one-dimensional domain increases with time, following an input coarsening law, such that $\lambda_2 \sim t^{1/3}$. Note that this is consistent with the coarsening laws derived in Chap. 8. The analytical models considered earlier all assumed that $\rho_s = \rho_\ell = cst$. The problem becomes more complex when density is allowed to vary with the composition and temperature in each phase, as mass conservation requires additional fluid to enter the domain in order to compensate for the change in density. To be able to treat this problem properly, one must include the motion of the bulk fluid. Such models are presented in Chap. 14.

The two models addressed at present assume that the fluid entering the domain has the mean composition of the liquid, which is appropriate for solidification of secondary dendrite arms. The two studies differ mainly in the manner that the equations are solved. Roosz and Exner used a "node jumping" scheme, where the time step was adjusted such that the interface always jumps from one nodal point to the next, thereby ensuring that boundary conditions at the solid-liquid interface can be readily enforced. Sundarraj and Voller, on the other hand, used a Landau transformation to divide the domain into two parts at the moving solid-liquid interface. (See Eq. (6.90) and the surrounding discussion.) This facilitated the handling of the variable domain size associated with coarsening and density changes. Both sets of authors compared their results to a dataset given by Sarreal and Abbaschian [10], and found reasonable agreement at a low cooling rate, but significant deviations at higher cooling rates. In a later paper, Sundarraj and Voller [19] presented an extension of their model including eutectic undercooling, which enabled them to match the reported experimental data very closely. The results are summarized in Fig. 10.5. Note that the experimental data point at the largest cooling rate was obtained under very different experimental conditions, and is less reliable than the others. Thus, one should not overemphasize the agreement between the models and experimental results!

10.2.2 Peritectic alloys

Peritectics represent a very important class of alloys where microsegregation plays a dominant role in the microstructure formation. Their behavior is sufficiently different from that of the eutectic alloys as to warrant a separate section. The solidification of these alloys was discussed in detail in

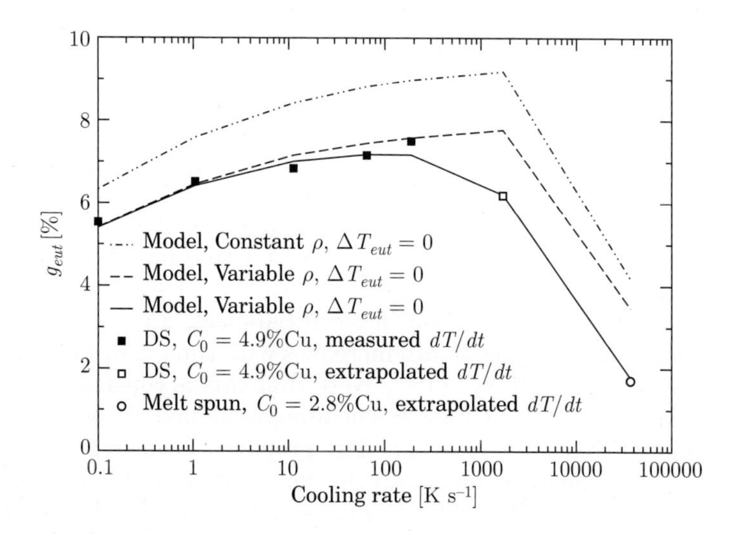

Fig. 10.5 A comparison of results from the microsegregation calculations of Sundarraj [19] to the experimental data of Sarreal and Abbaschian for Al-Cu alloys.

Chap. 9, so the discussion here is limited to the essentials relating to the subject of this chapter. The equilibrium solidification path for an alloy of composition C_0 in the range $C_\alpha^{per} \leq C_0 \leq C_\ell^{per}$ begins with the formation of a primary α-phase. At the peritectic temperature T_{per}, the peritectic reaction $\ell + \alpha \rightarrow \beta$ begins. In practice, such reactions rarely go to completion, since the initial reaction product forms a solid layer encapsulating the α-phase. Moreover, further reaction requires the diffusion in the solid state through this layer. Fig. 10.6 gives an example of such a microstructure in the Fe-Ni system, along with a measured composition profile across a dendrite arm. The simplest model capable of explaining this microsegregation

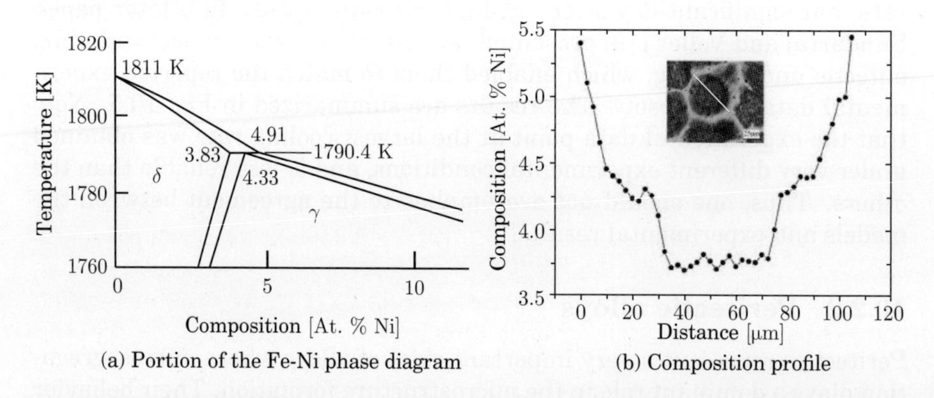

Fig. 10.6 (a) The Fe-rich portion of the Fe-Ni phase diagram. (b) The measured composition profile across a dendrite arm in an alloy of peritectic composition [3].

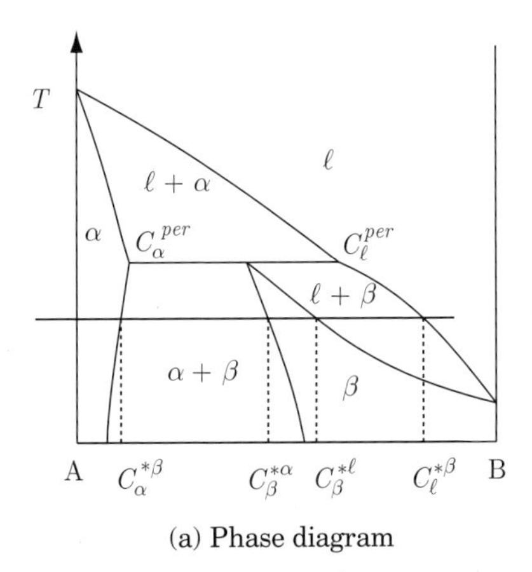

(a) Phase diagram

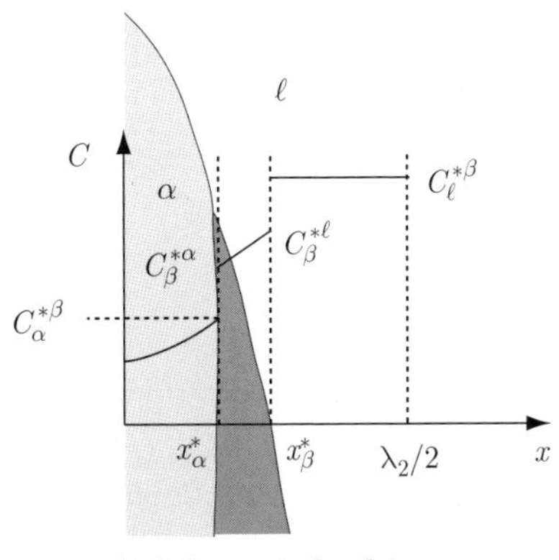

(b) Schematic dendrite arm

Fig. 10.7 (a) An idealized binary peritectic phase diagram. (b) A schematic drawing of the composition profile behind the dendrite tip where peritectic solidification has begun.

profile is one that begins with solidification of the α-phase according to the Gulliver-Scheil model until T_{per} is reached, after which solidification the β-phase solidifies on the α core, again following the Gulliver-Scheil model, but now for the β-phase.

A more realistic model must include diffusion through the solid $\beta-$phase. Indeed, back diffusion not only modifies the solidification of the $\beta-$phase, but also consumes the pre-existing $\alpha-$phase. This phenomena is illustrated in Fig. 10.7 by a dendrite segment growing in a temperature gradient. As shown, there are two moving interfaces to consider: one between the $\alpha-$ and $\beta-$phases, and another between the $\beta-$phase and the liquid. Writing the usual solute balances at the interfaces gives the boundary conditions for their motion

$$\left(C_\alpha^{*\beta} - C_\beta^{*\alpha}\right) \frac{dx_\alpha^*}{dt} = D_\beta \left.\frac{\partial C_\beta}{\partial x}\right|_{x_\alpha^*} - D_\alpha \left.\frac{\partial C_\alpha}{\partial x}\right|_{x_\alpha^*} \tag{10.39}$$

$$\left(C_\beta^{*\ell} - C_\ell^*\right) \frac{dx_\beta^*}{dt} = \frac{dC_\ell^*}{dt}\left(\frac{\lambda_2}{2} - x_\beta^*(t)\right) - D_\beta \left.\frac{\partial C_\beta}{\partial x}\right|_{x_\beta^*} \tag{10.40}$$

Thus, a complete model of microsegregation in peritectic alloys begins with one of the microsegregation models described earlier for solidification of the $\alpha-$phase until the temperature T_{per} is reached. Subsequently, simultaneous solidification of the $\beta-$phase from the melt takes place along with the solid state peritectic transformation at the $\alpha - \beta$ interface.

Fig. 10.8 provides an example of such a calculation for an alloy in the Fe-Au system. The figure shows the computed composition profile at several temperatures. At 1232°C, only the α and liquid phases are present. When the temperature drops below the peritectic temperature of 1173°C,

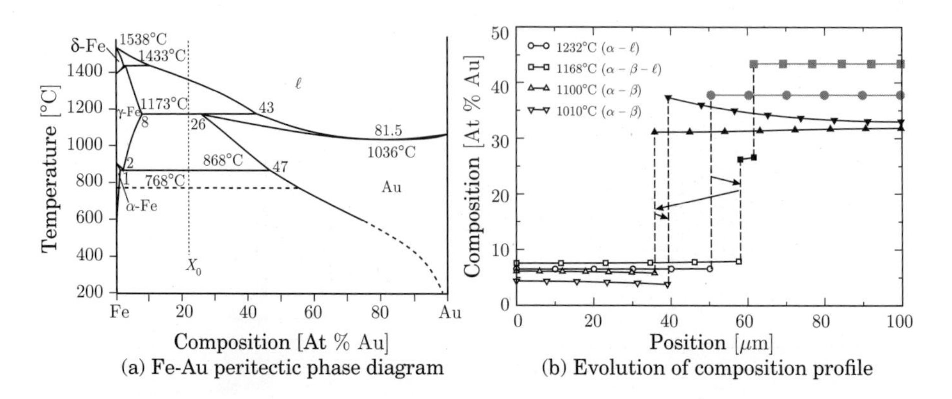

Fig. 10.8 Solidification in the peritectic Fe-Au system. (a) Equilibrium phase diagram. (b) Evolution of the composition profile across a dendrite arm in an alloy of peritectic composition. Filled gray symbols indicate the liquid phase, filled black symbols the $\beta-$phase, and open symbols the $\alpha-$phase. The path from high to low temperatures is indicated by arrows.

the $\beta-$phase appears at a position of $x \approx 60\ \mu$m. As the temperature decreases further, the β-phase region expands in both directions due to solidification from the melt to the right, and the peritectic transformation to the left. Eventually, just below 1135°C, the last liquid is consumed. As cooling continues, the compositions continue to change due to variations in the equilibrium compositions of the α and $\beta-$phases.

10.2.3 Volume averaged model

The main shortcoming of the models presented in the previous sections is the fact that they are all one-dimensional. Even the numerical models, which are capable of including a wide variety of physical phenomena, are limited to 1-D. This is a particular problem when coarsening and non-constant density are included, because the models require the composition of the solute surrounding the 1-D domain when liquid is added to compensate for solidification shrinkage. Further, the way that coarsening is included in these 1-D models is not physically correct. It was demonstrated in Chap. 8 that coarsening takes place through the disappearance of individual dendrite branches, rather than through a gradual separation of neighboring dendrite arms. Treating the coarsening with a law that evolves the average spacing $\langle \lambda_2 \rangle$ over time makes more sense in a model that averages over the microstructure.

Wang and Beckermann [20] developed a more general approach to the microsegregation problem; an approach based on averaging over the volume of a dendritic grain and its surrounding liquid. This enabled them to consider more complex geometries, and to include the effect of undercooling on dendritic growth. They began with the concept of a "grain envelope," first introduced by Thévoz and Rappaz [15; 16] and illustrated in

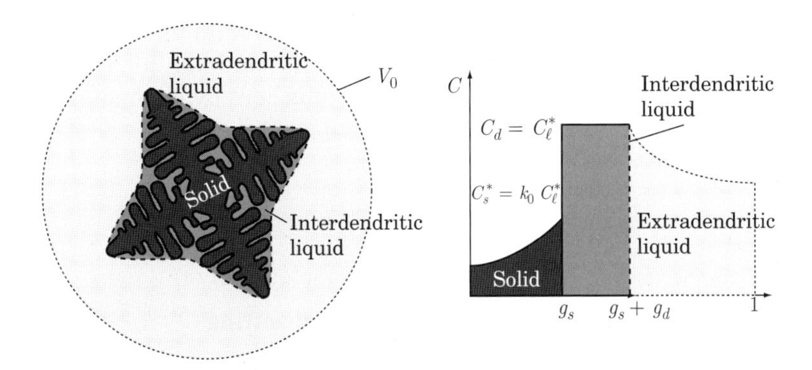

Fig. 10.9 The physical model of an equiaxed dendrite enclosed by an envelope. The fluid inside the envelope is assumed to be well mixed, i.e., of constant composition, and the extradendritic liquid composition decreases with distance, as indicated.

Fig. 10.9. As can be seen, the solid phase has a dendritic microstructure with varying composition. An imaginary envelope is drawn surrounding the solid, inside of which the "interdendritic liquid phase" is assumed to be well mixed, with constant composition C_ℓ^*. Outside the envelope, in the "extradendritic liquid phase," the composition falls rapidly to a far-field value. The quotes above indicate that although the solid, interdendritic and extradendritic liquids are treated as separate phases, there are actually only two phases, solid and liquid. For the remainder of the discussion, however, the quotes will be dropped. The model assumes that the domain is isothermal, that there is no advection, that the densities of the solid and liquid are equal and constant, and that there are only two components. The solid, interdendritic and extradendritic phases are labeled s, d and e, respectively, in the development.

The model consists of a set of four simultaneous coupled ordinary differential equations, three of which express the solute balance in each phase. We consider a binary alloy and assume that there is no velocity. The solute balance for phase j is then written in integral form as

$$\int_{V_j} \frac{\partial C_j}{\partial t}\, dV = \int_{A_j} D_j \nabla C_j \cdot \boldsymbol{n}\, dA \tag{10.41}$$

The volumes of the solid and of the grain envelope are normalized by the total volume V_0 in which the grain can grow (see Fig. 10.9). This transforms the integrals over the phase volumes into integrals over the volume fractions of the solid fraction g_s, interdendritic liquid fraction g_d, and extradendritic liquid fraction $(1 - g_s - g_d)$. For the solid, we have

$$\int_0^{g_s(t)} \frac{\partial C_s}{\partial t}\, dg = \mathcal{S}_V^{sd} D_s \left.\frac{\partial C_s}{\partial n}\right|_{g_s} \tag{10.42}$$

where $\mathcal{S}_V^{sd}$ is the specific surface area between the solid and interdendritic liquid, i.e., normalized by the total volume V_0 available for the grain to grow. Note also that the solute flux at $g_s = 0$ is zero by symmetry. Taking the time derivative outside the first integral gives us

$$\frac{d}{dt}\int_0^{g_s(t)} C_s\, dg_s = C_s^* \frac{dg_s}{dt} + \mathcal{S}_V^{sd} D_s \left.\frac{\partial C_s}{\partial n}\right|_{g_s} \tag{10.43}$$

Writing the left hand side in terms of $\langle C\rangle_s$, and setting $C_s^* = k_0 C_\ell^*$ provides the solute balance for the solid phase

$$\frac{d\left(g_s \langle C\rangle_s\right)}{dt} = k_0 C_\ell^* \frac{dg_s}{dt} + \mathcal{S}_V^{sd} D_s \left.\frac{\partial C_s}{\partial n}\right|_{g_s} \tag{10.44}$$

Following the same procedure for the interdendritic liquid yields

$$\frac{d}{dt}\int_{g_s(t)}^{g_s+g_d} C_d\, dg + C_\ell^* \frac{dg_s}{dt} - C_\ell^* \frac{d(g_s+g_d)}{dt} = \mathcal{S}_V^{de} D_d \left.\frac{\partial C_d}{\partial n}\right|_{g_s+g_d} - \mathcal{S}_V^{sd} D_d \left.\frac{\partial C_d}{\partial n}\right|_{g_s} \tag{10.45}$$

The specific surface area of the envelope between the inter- and extradendritic phases has been designated $\mathcal{S}_V^{de}$. The right hand side of Eq. (10.45) cannot be evaluated directly because $C_d = C_\ell^*$ is assumed to be constant. It is demonstrated in Exercise 10.2 that these two terms, representing the solute flux at the respective boundaries, can be written in terms of gradients of C_s and C_e, and the average velocity of the solid-liquid phase boundary. With these results and after performing some algebra, Eq. (10.45) becomes

$$\frac{d\left(g_d C_\ell^*\right)}{dt} = C_\ell^* \frac{dg_d}{dt} + (1-k_0) C_\ell^* \frac{dg_s}{dt} + \mathcal{S}_V^{de} D_\ell \left.\frac{\partial C_e}{\partial n}\right|_{g_s+g_d} - \mathcal{S}_V^{sd} D_s \left.\frac{\partial C_s}{\partial n}\right|_{g_s} \tag{10.46}$$

In order to simplify the computations, the gradients are replaced by approximate expressions written in terms of differences between the interface and average compositions, i.e.,

$$\left.\frac{\partial C_s}{\partial n}\right|_{g_s} \approx \frac{k_0 C_\ell^* - \langle C\rangle_s}{\delta_s}\;; \quad \left.\frac{\partial C_e}{\partial n}\right|_{g_s+g_d} \approx -\frac{C_\ell^* - \langle C\rangle_e}{\delta_e} \tag{10.47}$$

where δ_s and δ_e are characteristic diffusion lengths in the the solid and extradendritic phases, respectively. Equation (10.46) then takes on the simpler form

$$\frac{d\left(g_d C_\ell^*\right)}{dt} = C_\ell^* \frac{dg_d}{dt} + (1-k_0) C_\ell^* \frac{dg_s}{dt} - \mathcal{S}_V^{de} D_\ell \frac{C_\ell^* - \langle C\rangle_e}{\delta_e} - \mathcal{S}_V^{sd} D_s \frac{k_0 C_\ell^* - \langle C\rangle_s}{\delta_s} \tag{10.48}$$

Finally, applying the same procedure one more time for the extradendritic liquid gives the result

$$\frac{d\left(g_e \langle C\rangle_e\right)}{dt} = -C_\ell^* \frac{d(g_s + g_d)}{dt} - S_V^{de} D_\ell \left.\frac{\partial C_e}{\partial n}\right|_{g_s+g_d} = C_\ell^* \frac{dg_e}{dt} - S_V^{de} D_\ell \left.\frac{\partial C_\ell}{\partial n}\right|_{g_s+g_d} \tag{10.49}$$

Once again, by using the approximations in Eq. (10.47), Eq. (10.49) becomes

$$\frac{d\left(g_e \langle C\rangle_e\right)}{dt} = C_\ell^* \frac{dg_e}{dt} + S_V^{de} D_\ell \frac{C_\ell^* - \langle C\rangle_e}{\delta_e} \tag{10.50}$$

It is easy to see that the sum of Eqs. (10.44), (10.46) and (10.49) is

$$\frac{d\left(g_s \langle C\rangle_s\right)}{dt} + \frac{d\left(g_d C_\ell^*\right)}{dt} + \frac{d\left(g_e \langle C\rangle_e\right)}{dt} = 0 \tag{10.51}$$

The last equation needed for the model is obtained by setting the growth rate of the envelope equal to the dendrite tip velocity, determined from the LGK model given in Sect. 8.4.2

$$\frac{d(g_s + g_d)}{dt} = \frac{S_V^{de} D_\ell m_\ell (k_0 - 1) C_\ell^*}{\pi^2 \Gamma_{s\ell}} \left[\mathrm{Iv}^{-1}(\Omega)\right]^2 \tag{10.52}$$

Here, $\mathrm{Iv}^{-1}(\Omega)$ is the inverse of the Ivantsov function for the supersaturation $\Omega = (C_\ell^* - \langle C\rangle_e)/[(1 - k_0)C_\ell^*]$.

Equations (10.44), (10.46), (10.49) and (10.52) provide a set of four coupled, non-linear ordinary differential equations in five unknowns: g_s, g_d, C_ℓ^*, $\langle C\rangle_s$, $\langle C\rangle_e$. To complete the specification of the problem, an additional condition is applied relating to the rate of solidification ($\dot{g}_s$), the cooling rate ($\dot{T}$), or the rate of heat loss, which can be written as a combination of the first two. The solution of the coupled set of equations is discussed in further detail in Chap. 11. Once the initial condition is specified, standard numerical techniques can be employed to integrate forward in time and evolve the microsegregation profile. Wang and Beckermann presented several examples, including one where they compared their computational results to the experiments of Sarreal and Abbaschian shown in Fig. 10.5. The fit of their model was comparable to that obtained by Sundarraj and Voller, except for the data point at the highest undercooling.

10.3 HOMOGENIZATION AND SOLUTION TREATMENT

Microsegregation can have either beneficial or deleterious effects on the properties of cast products. In shape castings of Al-Si or Al-Mg-Si alloys, the additional second-phase particles can improve the wear characteristics of the part in question, which is useful in certain automotive applications. In many other cases, however, the desired microstructure has a more uniform distribution of precipitates (or none at all) in order to

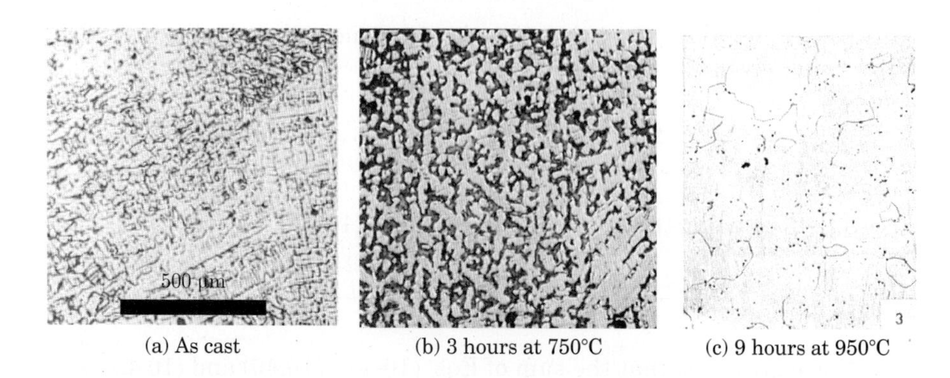

(a) As cast (b) 3 hours at 750°C (c) 9 hours at 950°C

Fig. 10.10 Modification of the microstructure by homogenization treatment in an Cu-15 wt%Ni alloy. (Reproduced with permission from ref. [1])

obtain improved mechanical properties. Further, the chemical inhomogeneity associated with microsegregation usually leads to poor corrosion resistance. For these reasons, cast alloys are often subject to a *homogenization treatment*, so as to reduce or eliminate residual segregation patterns, and/or to a *solution treatment* in order to re-dissolve non-equilibrium secondary phases produced by microsegregation. We do not consider the possibility of precipitation.

10.3.1 Homogenization

Homogenization is first considered in an isomorphous system, such as Cu-Ni, illustrated in Fig. 10.10. The as-cast microstructure to the left clearly shows the dendritic structure due to the microsegregation profile. Only the smallest length scales were coarsened after a heat treatment of 3 hours at 750°C, Fig. 10.10(b), and as can be seen in Fig. 10.10(c). Complete homogenization of the as-cast structure was obtained after heating for 9 hours at 950°C. In this case, the final invariant reaction would be the solidification of pure Cu, which takes place at 1083°C. Note that $k_0 > 1$ for this case, and we assume the Gulliver-Scheil model to obtain the initial microsegregation profile.

The analysis of homogenization consists of modeling the diffusion process in the same idealized periodic array in which the segregation was modeled, illustrated in Fig. 10.1(b). The analysis proceeds in a similar fashion to that of the segregation in the preceding sections, but there is now only diffusion in the solid phase to consider. We thus begin with a slightly modified form of Eq. (10.7), retaining the assumption that $\lambda_2 \ll L$. Clearly, the solidification time t_f is not an appropriate time scale for homogenization, so let us simply designate the characteristic time t_c. Equation (10.7) then becomes

$$\frac{\partial C_s}{\partial \tau} = \frac{4D_s t_c}{\lambda_2^2}\frac{\partial^2 C_s}{\partial \xi^2} \quad 0 \leq \xi \leq 1 \tag{10.53}$$

where $\tau = t/t_c$ and $\xi = 2x/\lambda_2$. The obvious choice for t_c is $t_c = \lambda_2^2/4D_s$. If the only matter of concern is to estimate the time for complete homogenization, computing the value of t_c might be sufficient. However, it may be of interest to track the evolution of the homogenization, e.g., to examine a partial homogenization treatment. The boundary conditions are the same as in the segregation analysis, zero flux at $\xi = 0, 1$. The initial condition for this analysis is the final composition profile from the microsegregation analysis. The Gulliver-Scheil solution is taken as an example, since it represents an extreme case. The scaled governing equation, boundary and initial conditions are

$$\frac{\partial C_s}{\partial \tau} = \frac{\partial^2 C_s}{\partial \xi^2} \qquad 0 \leq \xi \leq 1 \tag{10.54}$$

$$\frac{\partial C_s}{\partial \xi} = 0 \qquad \xi = 0, 1 \tag{10.55}$$

$$C_s = k_0 C_0 (1-\xi)^{k_0 - 1} \qquad \tau = 0 \tag{10.56}$$

The solution to this system of equations can be found in the form of a Fourier cosine series, given by

$$C_s(\xi, \tau) = \sum_{m=0}^{\infty} A_m \cos(m\pi\xi) e^{-m^2\pi^2\tau} \tag{10.57}$$

The coefficients A_m are obtained from the initial condition given in Eq. (10.56). Although it is possible to obtain a solution using the Gulliver-Scheil equation as the initial condition, the result in terms of incomplete elliptic integrals is not particularly intuitive. Instead, the Gulliver-Scheil solution is approximated as a parabola, preserving the values of the Gulliver-Scheil solution at $\xi = 0, 1$ as well as the total solute, determined by integrating $C_s(\tau = 0)$ from $\xi = 0$ to $\xi = 1$. The result is

$$C_s(\xi, \tau = 0) \approx k_0 C_0 + (6 - 4k_0) C_0 \xi + (3k_0 - 6) C_0 \xi^2 \tag{10.58}$$

Using Eq. (10.58) as the initial condition, instead of Eq. (10.56), the A_m can be readily computed, which gives the solution

$$C_s(\xi, \tau) = C_0 + \sum_{m=1}^{\infty} \frac{4C_0}{m^2\pi^2} \left[(k_0 - 3)\cos(m\pi\xi) + 2k_0 - 3 \right] e^{-m^2\pi^2\tau} \tag{10.59}$$

To illustrate the properties of this solution, we choose values close to those of the Cu-Ni system, with $C_0 = 0.2$, and assume $k_0 = 1.8$ to be constant. The solution given in Eq. (10.59) is plotted in Fig. 10.11(a) for various values of τ. The parabolic initial condition is seen to be a very reasonable approximation to the Gulliver-Scheil solution, also displayed in the figure. Notice that homogenization is essentially complete at $\tau = 0.5$, i.e., when

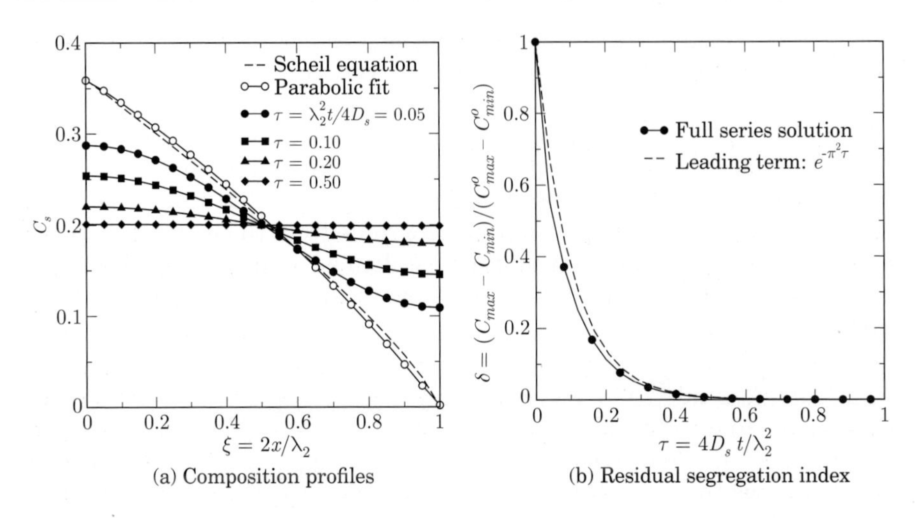

Fig. 10.11 The computed progress of the homogenization of a Cu-Ni alloy. (a) The evolution of the composition profile. (b) The residual segregation index, defined in Eq. (10.60).

$t \approx t_c/2$. Some authors define the following *residual segregation index* δ_r as a measure of the progress of homogenization

$$\delta_r = \frac{C_{max} - C_{min}}{C^0_{max} - C^0_{min}} \tag{10.60}$$

where the superscript "0" refers to the initial values. Figure 10.11(b) plots the time evolution of δ_r for the solution given in Eq. (10.59). We also display the residual segregation index that one would compute by maintaining only the first term in the Fourier series, which is clearly a good approximation for $\tau \gtrsim 0.3$. We use this observation to our advantage when analyzing the following solution heat treatment.

Finally, to illustrate the practical application of the solution, we apply it to the Cu-Ni alloy whose micrographs are shown in Fig. 10.10. From the length scale, we can estimate the dendrite arm spacing $\lambda_2 \approx 10\ \mu$m. The diffusion coefficient of Ni in Cu is given approximately by

$$D_{Ni} = \left(2.7 \times 10^{-5}\ \text{m}^2/\text{s}\right) \exp \frac{-256\ \text{kJ/mol}}{RT} \tag{10.61}$$

Using these data to compute the dimensionless time corresponding to the two heat-treated microstructures shown in Fig. 10.10, we find that $\tau = 9.76 \times 10^{-4}$ for the sample heat treated for 3 hours at 750°C, and that $\tau = 0.40$ for the sample heat treated for 9 hours at 950°C. Thus, the minimal homogenization evident from Fig. 10.10(b), and the complete homogenization seen in Fig. 10.10(c) are consistent with our analysis. It should be noted that the characteristic length scale should actually increase with time as the microstructure coarsens, but the present analysis both illustrates the phenomenon, and also provides a reasonable estimate of the necessary time of homogenization.

10.3.2 Solution heat treatment

This section addresses the solution heat treatment of alloys to re-dissolve non-equilibrium second phases. The basic model, illustrated in Fig. 10.12, is similar to the previous 1-D models at the scale of the dendrite arms. The initial condition reflects the segregation profile produced by solidification, and in particular the volume fraction of the primary α-phase is given by g_α^p. The fraction of eutectic formed is thus $g_{eut} = 1 - g_\alpha^p$. The heat treatment takes place at a temperature between the solvus temperature for the alloy and the eutectic temperature, shown as a dashed line in Fig. 10.12(a). It is assumed that the eutectic region is at local equilibrium, which implies that the composition of the α-phase in the eutectic region is given by the solvus value C_α, and the composition of the β-phase is given by C_β. The volume fraction of β-phase is designated as g_β, and the volume fraction of α-phase in the $\alpha + \beta$ region thus becomes $1 - g_\alpha^p - g_\beta$. The composition of the α-phase in the eutectic is assumed to remain constant, even as the β-phase dissolves. This is a reasonable assumption when g_β is small.

The scaling procedure is identical to that for homogenization. The governing equation and boundary conditions for this problem under the stated assumptions are

$$\frac{\partial C_s}{\partial \tau} = \frac{\partial^2 C_s}{\partial \xi^2} \qquad 0 \le \xi \le g_\alpha \tag{10.62}$$

$$\frac{\partial C_s}{\partial \xi} = 0 \qquad \xi = 0 \tag{10.63}$$

$$C_s = C_\alpha \qquad \xi = g_\alpha \tag{10.64}$$

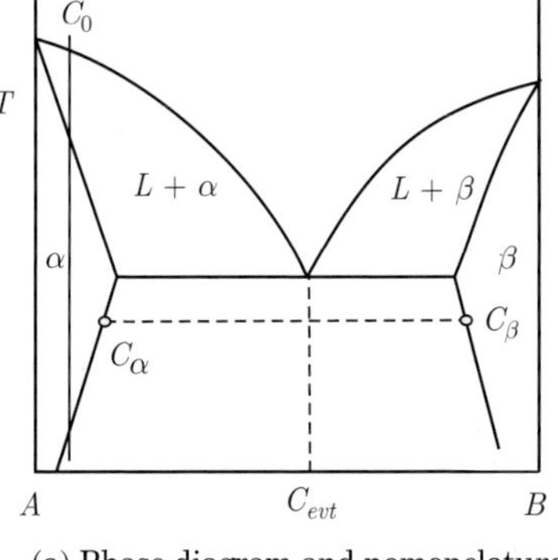

(a) Phase diagram and nomenclature

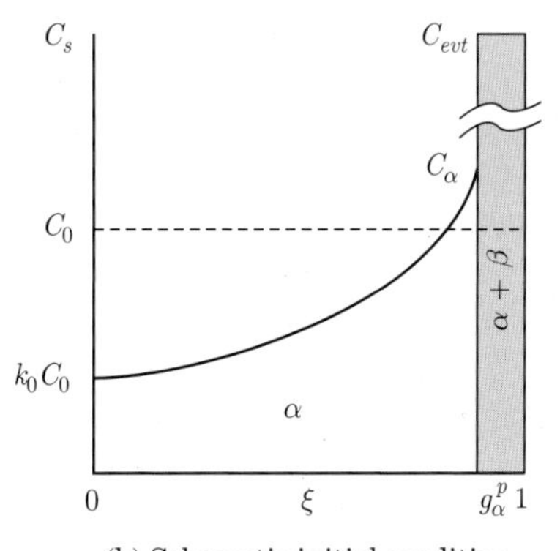

(b) Schematic initial condition

Fig. 10.12 A schematic view of (a) the phase diagram and (b) initial condition for solution treatment analysis.

The solution can be written once again as a cosine series. The boundary condition at $\xi = g_\alpha$ is not the same as in homogenization, and thus the form differs slightly from Eq. (10.57). The solution is given by

$$C_s(\xi, \tau) = C_\alpha + \sum_{m=1}^{\infty} A_m \cos\left(\frac{(2m-1)\pi\xi}{2g_\alpha}\right) e^{-(2m-1)^2\pi^2\tau/4g_\alpha^2} \tag{10.65}$$

An initial condition is required to evaluate the coefficients A_m. Our experience in the previous problem, however, demonstrated that the initial condition quickly disappears, and only the first term in the series survives beyond very small times. In fact, since the series in Eq. (10.65) contains only odd values of m, the terms beyond the first go to zero even faster than they did in the homogenization problem. Thus, it is quite reasonable to choose an approximate initial condition of the form

$$C_s(\xi, 0) \approx C_\alpha + A_0 \cos\left(\frac{\pi\xi}{2g_\alpha}\right) \tag{10.66}$$

Note that, with this simple form, it is not possible to choose a single constant A_0 that satisfies the boundary condition on C_s at $\xi = 0$ and $\xi = g_\alpha^p$, and to enforce a total solute balance all at the same time. It will be demonstrated, however, that the exact choice of A_0 is unimportant for the analysis, and we thus choose A_0 to ensure that the total solute content is equal to C_0. With this choice of the initial condition, only the leading term in the series has a non-zero coefficient, and the solution is given by

$$C_s(\xi, \tau) = C_\alpha + A_0 \cos\left(\frac{\pi\xi}{2g_\alpha}\right) e^{-\pi^2\tau/4g_\alpha^2} \tag{10.67}$$

Next, we perform a solute balance over the domain at $\tau = 0$.

$$\int_0^{g_\alpha} C_s(\xi, 0)\, d\xi + C_\alpha(1 - g_\alpha - g_\beta^0) + g_\beta^0 C_\beta = C_0 \tag{10.68}$$

where g_β^0 is the fraction of β-phase at $\tau = 0$. Substituting Eq. (10.66) and carrying out a little manipulation gives

$$\int_0^{g_\alpha} A_0 \cos\left(\frac{\pi\xi}{2g_\alpha}\right) d\xi = -(C_\alpha - C_0) - (C_\beta - C_\alpha)g_\beta^0 \tag{10.69}$$

Computing a similar solute balance at time τ yields

$$\left(\int_0^{g_\alpha} A_0 \cos\left(\frac{\pi\xi}{2g_\alpha}\right) d\xi\right) e^{-\pi^2\tau/4g_\alpha^2} = -(C_\alpha - C_0) - (C_\beta - C_\alpha)g_\beta \tag{10.70}$$

Dividing Eq. (10.70) by Eq. (10.69) provides a relation between the volume fraction of β-phase and time

$$\frac{g_\beta + (C_\alpha - C_0)/(C_\beta - C_\alpha)}{g_\beta^0 + (C_\alpha - C_0)/(C_\beta - C_\alpha)} = e^{-\pi^2\tau/4g_\alpha^2} \tag{10.71}$$

Note that Eq. (10.71) takes a particularly simple form when the heat treatment is done at the solvus temperature of the alloy, where $C_\alpha = C_0$.

To illustrate the result, consider an alloy of Al-4 wt% Cu. The density is assumed to be constant, so the mass and volume fractions are identical. Let us assume that the Gulliver-Scheil equation can be applied to compute the initial microsegregation of this alloy system. For this alloy system, we have $k_0 = 0.17$, and the compositions of the α- and β-phases at the eutectic temperature are 5.65 and 52 wt% Cu, respectively. The composition of the eutectic is 33 wt% Cu. A simple calculation will show that the fraction of non-equilibrium eutectic is 0.08. We then compute

$$g_\beta^0 = 0.08 \times \frac{33 - 5.65}{52 - 5.65} = 4.72 \times 10^{-2} \tag{10.72}$$

Suppose that the heat treatment takes place at a temperature for which $C_\alpha = 5$ and $C_\beta = 52$. The result of substituting these values into Eq. (10.71) is shown in Fig. 10.13. We have plotted the ratio g_β/g_β^0, since this is usually the result of most interest. The increase in solute content in the α-phase comes from the dissolution of the β-phase. The solution is valid only up to the time where $g_\beta = 0$, which turns out to be $\tau = 0.403$ for the given values; a very similar result to that obtained for the homogenization problem. This similarity is not very common, however, as it depends on the

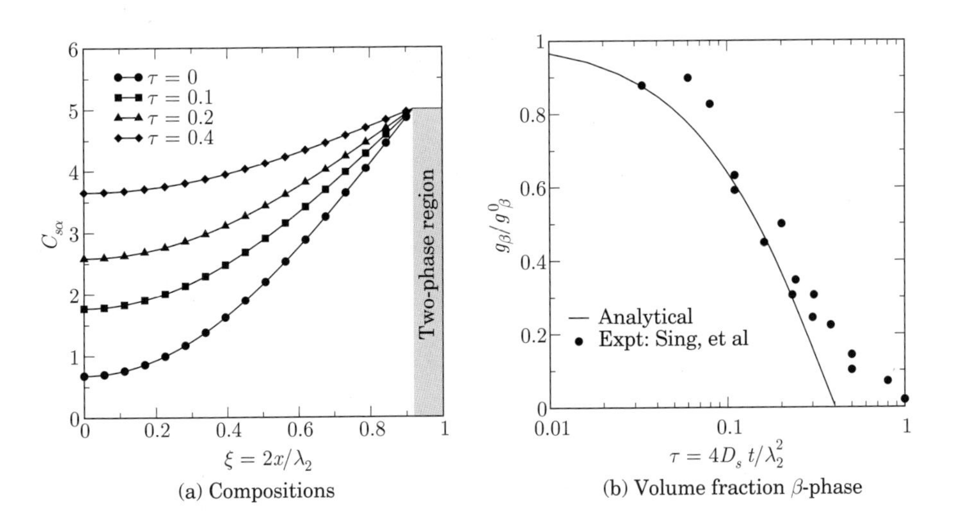

Fig. 10.13 Computed values of (a) compositions in the primary phase and (b) the residual volume fraction β-phase for an Al-4Cu alloy.

specific values of the compositions. Finally, Fig. 10.13 also shows experimental data obtained by Singh et al. [12] for an Al-4.5 Cu alloy. These data agree very well with the analysis, with the exception of at very long times. The discrepancy there is due to longer range diffusion as secondary arms coarsen, and the appropriate length scale thus increases to become the primary spacing.

10.4 MULTICOMPONENT ALLOYS

Most important industrial alloys contain more than two components. This section describes how to extend the models developed for binary alloys to multicomponent systems. The first observation that can be made concerns the increased complexity of multicomponent systems. To illustrate this, we return to the ternary eutectic system first discussed in Chap. 3. Figure 10.14 displays two views of the phase diagram, simplified so as to emphasize the solidification path of alloy x. As shown, solidification begins with primary α, the composition of which slides down the solidus surface until it reaches the dashed line corresponding to the three-phase equilibrium between α, β and ℓ. At this point, the liquid composition first reaches the monovariant curve between the $\ell + \alpha$ and $\ell + \beta$ phase regions. As the temperature decreases, the liquid composition follows the monovariant curve, finally terminating at the ternary eutectic point. Notice that the solidification path does not follow a straight line – a consequence of

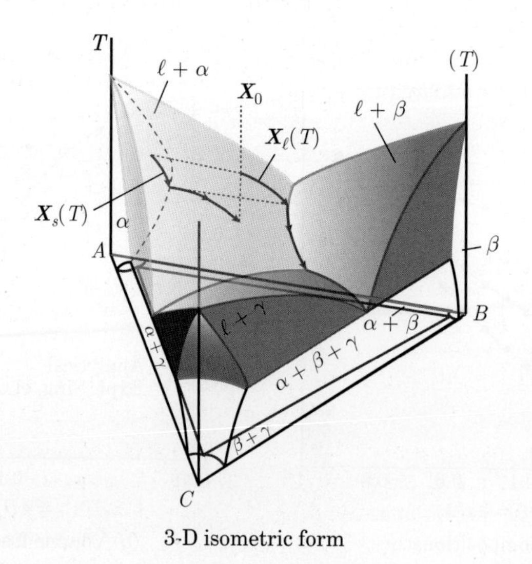

Fig. 10.14 A ternary eutectic equilibrium phase diagram, displaying the solidification path for alloy X_0 in for equilibrium solidification.

the fact that the tie-lines do not necessarily lie in the the same vertical plane over the series of temperatures found during solidification. These tie-line data are essential to be able to determine the solidification path.

The condition where the solid and liquid phases are in complete equilibrium at all temperatures corresponds to what was referred to as lever rule solidification in the context of microsegregation. In this case, solidification ends when the composition of the solid is equal to $\boldsymbol{X}_0$. Thus, if the tie-lines are known, one can compute *a priori* the phases that exist, as well as their compositions and the fraction of each phase present at each temperature. These data could then be stored in tabulated form, for example, and be used in other computations, such as in the modeling of the solidification of a casting. Nevertheless, one generally cannot write a simple analytical expression for multicomponent alloys, as can be done in the case of binary alloys.

It is also possible to provide a similar tabulation for the Gulliver-Scheil model. In this model, the composition of each increment of solid remains unchanged after solidification. The solidification path extends all the way to the ternary eutectic. In order to create a tabulation of phases, fractions, compositions and temperatures for the Gulliver-Scheil model as described for the lever rule, one must solve the solute balance in incremental form by generalizing Eq. (10.20)

$$(C^*_{J\ell} - C^*_{Js})dg_s = (1 - g_s)dC^*_{J\ell} \quad J = 1, \ldots N_c \tag{10.73}$$

The data relating the solid and liquid compositions C^*_{Js} and $C^*_{J\ell}$ at the interface in Eq. (10.73) originate from the tie-lines. One thus begins with the initial condition, solves Eq. (10.73) at a series of increments corresponding to specified temperature changes, and then tabulates the results in the same form as described for the case of the lever rule. Once again, a simple analytical expression is generally not possible.

Furthermore, it is possible that the diffusivities of the two species will be quite different. A very common, and extremely important example is the solidification of Fe-C-X alloys, where X is a species that diffuses by substitution, rather than interstitially, as is the case for C. In this case, the diffusion of C may follow a lever rule, whereas species X obeys a Gulliver-Scheil model. In such a case, the solidification path for the alloy does not follow any simple rule.

Finally, let us consider the case where there is a limited amount of back diffusion in the solid. The evolution of the phases is then no longer independent of the history, and one cannot tabulate the phase data *a priori*. For this case, Eq. (10.73) can still be applied at the interface, but the model must also simultaneously solve the diffusion problem in the solid phase before proceeding to the next step. Thus, it is essential to couple the solution of the macroscopic heat and solute field equations to a thermodynamic database that supplies the interface compositions at each temperature. This has been carried out in the literature for a few cases, but it is indeed a very slow process. The essential problem is that one cannot predict beforehand what solidification path the alloy will traverse. The thermodynamic equilibrium calculations thus must be performed at each step for each of the current interface compositions. A number of commercial packages carry out such calculations.

There exists a further complication that we now explore in the simpler context of solid state diffusion. Let us suppose that an alloy solidifies in such a way that a non-equilibrium solute profile is formed, similar to the one illustrated in Fig. 10.14. As described for binary alloys, this profile can be altered by homogenization and/or a solution treatment at elevated temperature. The microsegregation profile formed during the solidification then becomes the initial condition for the heat treatment process. It is thus necessary to solve the multicomponent diffusion equation.

We begin with Fick's law, given in Eq. (4.148) as

$$\dot{j}_K = -\rho \sum_{J=1}^{N_c} D_{KJ} \nabla C_J + \frac{M_K C_K \mu_K}{T^2} \nabla T \tag{10.74}$$

If the heat treatment process takes place at constant temperature, the second term can be neglected. Substituting the reduced form into the solute balance, Eq. (4.146), and taking the velocity in the solid to be zero, gives Fick's second law

$$\frac{\partial(\rho C_K)}{\partial t} = \nabla \cdot \left(\rho \sum_{J=1}^{N_c} D_{KJ} \nabla C_J \right) \tag{10.75}$$

If the diffusivity tensor is diagonal, i.e., $D_{KJ} = 0$ for $K \neq J$, then the solution for the diffusion of the two species is uncoupled, and all of the methods discussed earlier for binary alloys can be separately applied to the individual species. The problem becomes somewhat more interesting when there are interactions between the various species, i.e., when the diffusivity tensor is not diagonal.

The most interesting new aspect of the problem is the interaction of the species, and an example that focuses on this aspect is thus developed. We consider dilute alloys of Cr and Al in Ni, and use the experimental data of Thompson et al [17], who measured the diffusivity tensor as

$$[D_{Al \cdot Cr}] = \begin{bmatrix} 22.0 & 7.6 \\ 7.8 & 12.6 \end{bmatrix} \times 10^{-15} \text{ m}^2 \text{ s}^{-1} \tag{10.76}$$

for the composition range (in atomic percent) $5.0 \leq X_{Al} \leq 9.5$ and $8 \leq X_{Cr} \leq 19$ at 1100°C. Note that this formulation is rather inconsistent with our definition of Fick's law, which was written in terms of mass fraction, and that the conversion between the two is not simply a matter of replacing C_J by X_J when the molecular weights of the species are significantly different. However, since the measurements were all made in terms of atomic fraction, a modified form of Fick's law is adopted for this analysis, where Eq. (10.75) is replaced by

$$\frac{\partial X_K}{\partial t} = \nabla \cdot \left(\sum_{J=1}^{N_c} D_{KJ} \nabla X_J \right) \tag{10.77}$$

To isolate the interaction of the two species, diffusion couples are formed consisting of a pair of very long rods, each with a uniform initial composition. The two rods are brought into intimate contact (e.g., by welding), after which the couple is heated to a high temperature where diffusion is allowed to take place for a fixed period of time. The couple is then quenched, sectioned and the composition is measured as a function of the position in the couple.

A series of such experiments was performed for a variety of pairs of initial compositions. In the first experiment, one of the rods had an initial composition of Ni-5 at%Al-8 at%Cr, and the other had an initial composition of Al-5 at%Al-17 at%Cr. In the second experiment, the corresponding values were Ni-5 at%Al-8 at%Cr and Al-10 at%Al-8 at%Cr. Before solving Eq. (10.77) for these two cases, the diffusion tensor is first rewritten in the form

$$[D_{Al \cdot Cr}] = 22 \times 10^{-15}\ \mathrm{m^2 s^{-1}} \begin{bmatrix} 1.0000 & 0.3455 \\ 0.3545 & 0.5727 \end{bmatrix} = D_{max}[D'_{Al \cdot Cr}] \tag{10.78}$$

Since the domain is of infinite extent, there is no natural length scale. The symbol L is chosen and dimensionless variables are defined as

$$\xi = \frac{x}{L} \qquad \tau = \frac{D_{max} t}{L^2} \tag{10.79}$$

The resulting scaled equation is

$$\frac{\partial X_K}{\partial \tau} = \sum_{J=1}^{2} D'_{KJ} \nabla^2 X_J \tag{10.80}$$

where we have taken $N_c = 2$ and assumed that D'_{KJ} are constants. This pair of equations is then solved for the composition using zero-flux boundary conditions at $\xi \to \pm\infty$. The numerical solution of Eq. (10.80) is straightforward, employing the methods developed in Chap. 6.

Figure 10.15 shows the composition profiles computed for these two diffusion couples at several times. Notice that, after the initial condition, the profile is essentially the same at each time, simply stretched along the ξ direction. This is typical of a similarity solution. It is, however, very untypical that the species that initially had no composition gradient actually develops one in both cases. The explanation for this behavior is the interaction between the species with no gradient and the other species. This phenomenon is sometimes called *cross-diffusion*. Even more surprising is that if one looks carefully at each of these two cases, there is a continual transport of the species that was originally constant to one side of

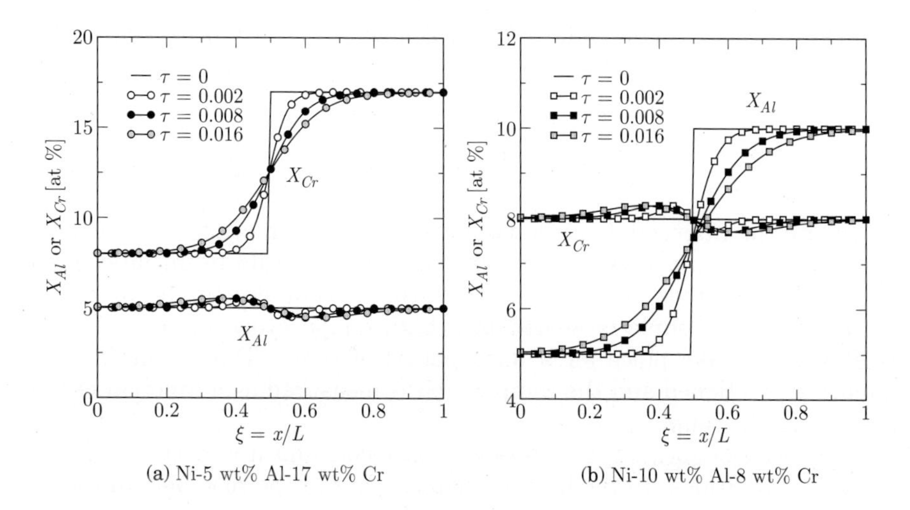

Fig. 10.15 Computed composition profiles in two diffusion couples each with one of the compositions held constant.

the couple, signifying that this species diffuses "up" its composition gradient. Clearly, this is due to the interaction with the movement of the other species. Another interesting feature of this solution is obtained if we plot the locus of all compositions over the couple on the ternary composition diagram. To obtain such a plot, the compositions X_{Al} and X_{Ni} are determined for each location ξ in Fig. 10.15 and placed at points on the ternary composition diagram. The result is shown in Fig. 10.16(a). Since the composition curves are self-similar, all times yield the same result. The compositions are not situated on a straight line due to the diffusive interactions between species.

Finally, one might wonder if there are couples for which the compositions do indeed lie on a straight line, and where diffusion does not run "uphill." Such couples do exist, corresponding to the eigenvectors of the diffusivity tensor. These are analogous to the principal directions for stress and strain, along which the tensors are diagonal. The eigenvalues of $[D']$ are $\lambda_D^{1,2} = 1.1964, 0.3763$. The corresponding eigenvectors are

$$\boldsymbol{E}_1 = \left\{ \begin{array}{c} 1.0000 \\ 0.5684 \end{array} \right\} \qquad \boldsymbol{E}_2 = \left\{ \begin{array}{c} 1.0000 \\ -1.8050 \end{array} \right\} \tag{10.81}$$

The composition profile for an alloy chosen along $\boldsymbol{E}_1$ is shown in Fig. 10.16(b), and it does indeed display a more "normal" appearance. The locus of compositions for this alloy, as well as another along $\boldsymbol{E}_2$, is also shown in Fig. 10.16(a). As can be seen, they do lie on a straight line in composition space.

This excursion into the field of multicomponent diffusion was intended to demonstrate that such problems can be readily solved, assuming that the necessary data are available. The amount of data required for even a fairly simple ternary alloy is quite large, and experimental results

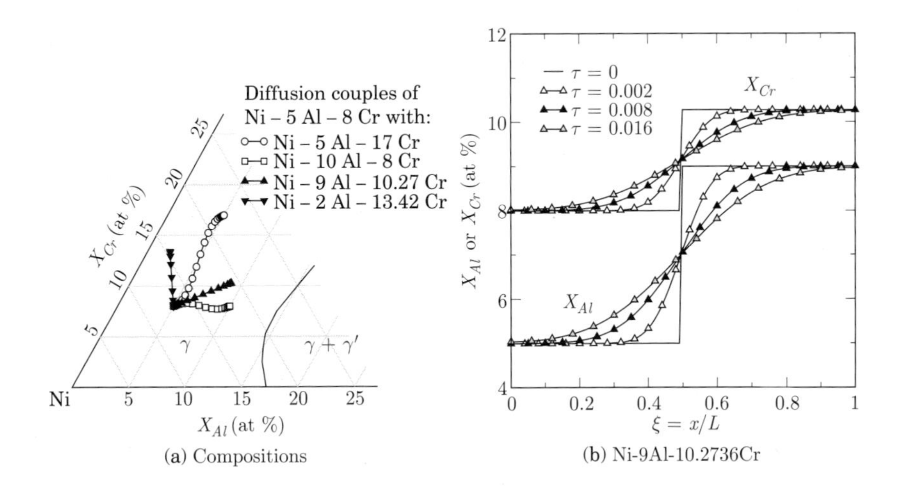

Fig. 10.16 (a) The locus of compositions attained in the various diffusion couples. (b) An alloy along an eigenvector of $[D']$.

are generally unavailable. For such cases, one resorts to thermodynamic and kinetic databases when needed.

10.5 SUMMARY

This chapter developed a variety of models for microsegregation during solidification. The basic phenomenon is such that solidification proceeds faster than diffusion is able to equilibrate the composition, particularly in the solid state. As a result, more solute is rejected into the liquid than would be found for equilibrium solidification. In many cases, non-equilibrium phases are obtained after solidification. Several models were examined for the case of limited diffusion in the solid state, all of which were essentially one-dimensional. Nonetheless, they still represent reasonable models for the compositions between secondary dendrite arms. For more complex microstructures, an averaged form was employed to simulate the coupled evolution of the composition with temperature. We then considered post-solidification heat treatment processes intended to relieve the microsegregation profile. These included homogenization, solution treatment, and heat treatment of peritectic alloys.

10.6 EXERCISES

Exercise 10.1. Segregation in Fe-C alloys.
Suzuki et al [14] found the following approximate relationship between cooling rate $\dot{T}$ and λ_2 for Fe-C alloys containing between 0.1 wt% and 0.9 wt%C:

$$\lambda_2 \approx \left(1.757 \times 10^3 \ \mu\text{m s}^{0.38} \ \text{K}^{-0.38}\right) \dot{T}^{-0.38} \tag{10.82}$$

Consider an alloy of Fe-0.1 wt%C and $\dot{T} = 1$ K s^{-1}.

(a) Use the Fe-C phase diagram to determine the approximate solidification range for this alloy.

(b) Using the following data for δ-Fe-C alloys, compute the various Fourier numbers in Eqs. (10.5)-(10.7) to show that the lever rule should apply.

$$\alpha \approx 6.1 \times 10^{-6} \text{ m}^2 \text{ s}^{-1}; \ D_\ell \approx 2 \times 10^{-8} \text{ m}^2 \text{ s}^{-1}; \ D_s \approx 6 \times 10^{-9} \text{ m}^2 \text{ s}^{-1}$$

Exercise 10.2. Boundary conditions in averaged model.
This exercise provides the details for the derivation of the balance equation for the interdendritic fluid, Eq. (10.46)

(a) The first term on the right hand side of Eq. (10.45) represents the solute flux from the outer boundary of the interdendritic liquid. Evaluate this flux in terms of the flux into the extradendritic liquid.

(b) The second term on the right hand of Eq. (10.45) represents the solute flux into the solid-liquid interface. Begin with the interfacial boundary condition, Eq. (4.152), and simplify it for the case where $\rho_s = \rho_\ell = \rho$ and $\mathbf{v}_s = 0$. Then, take the average form over the interface to evaluate the second term on the right hand side of Eq. (10.45).

(c) Substitute the results from the preceding two parts into Eq. (10.45) to obtain Eq. (10.46)

Exercise 10.3. Microsegregation in Al-3 wt%Cu.
Consider an Al-3 wt%Cu alloy that solidifies with an equiaxed dendritic microstructure.

(a) Compute and plot the $g_s - T$ curve for this alloy using both the lever rule and Gulliver-Scheil models. Indicate the expected fraction eutectic for each cases.

(b) Suppose that the local solidification time t_f is 20 minutes. Compute Fo_s, and then add $g_s - T$ curves to your graph computed using the Clyne-Kurz and Ohnaka models.

(c) Sketch the microstructure expected when t_f is 20 minutes, and estimate the time and temperature for solution treatment of this alloy.

Exercise 10.4. Microsegregation in Pb-Sn alloys.
Consider a Pb-20 wt%Sn alloy that solidifies with an equiaxed dendritic microstructure.

(a) Find the equilibrium phase diagram for this alloy, and use it to determine the appropriate value of k_0.

(b) Compute and plot $g_s - T$ curve for this alloy using both the lever rule and Gulliver-Scheil models. Indicate the expected fraction eutectic for each cases.

Exercise 10.5. Diffusion couple.
Consider a diffusion couple consisting of the following alloys: (Ni-5 at%Al-5 at%Cr) and (Ni-5 at%Al-10 at%Cr). The couple is welded together, and then heated to $1100°$C. Write an explicit finite difference program to compute the evolution of the solute profile in both halves of the diffusion couple. Plot your results in the same form as those in Fig. 10.15 and 10.16(a).

Exercise 10.6. Microsegregation in Fe-Au alloys.
Using the phase diagram and composition profiles in Fig. 10.8, sketch the microstructure for alloy X_0 at $1232°$C and $1010°$C. Indicate the various phases, and try to be accurate as their relative amounts.

10.7 REFERENCES

[1] R. M. Brick, A. W. Pense, and R. B. Gordon. *Structure and properties of engineering materials*. McGraw-Hill, New York, 4th edition, 1977.

[2] H. D. Brody and M. C. Flemings. Solute redistribution in dendritic solidification. *Trans. Metall. Soc AIME*, 236:615, 1966.

[3] Y. Z. Chen, F. Liu, G. C. Yang, N. Liu, C. L. Yang, and Y. N. Zhou. Suppression of peritectic reaction in the undercooled peritectic Fe-Ni melts. *Scripta Mat.*, 57:779–782, 2007.

[4] T. W. Clyne and W. Kurz. Solute redistribution during solidification with rapid solid state diffusion. *Metall. Trans.*, 12A:965, 1981.

[5] G. H. Gulliver. *Metallic alloys*. Griffin, 1922.

[6] S. Kobayashi. Mathematical analysis of solute redistribution during solidification based in a columnar dendrite model. *Trans. ISIJ*, 28:728–735, 1988.

[7] D. Larouche, C. Laroche, and M. Bouchard. Analysis of differential scanning calorimetric measurments performed on a binary aluminium alloy. *Acta Mat.*, 51:2161–2170, 2003.

[8] I. Ohnaka. Mathematical analysis of solute redistribution during solidification with diffusion in the solid phase. *Trans. ISIJ*, 26:1045, 1986.

[9] A. Roósz and H. E. Exner. Complete model for microsegregation during columnar dendrite growth. In T. S. Piwonka, V. Voller, and L. Katgerman, editors, *Modeling of Casting, Welding and Advanced Solidification Processes - VI*, pages 243–250. TMS, 1993.

[10] J. A. Sarreal and G. J. Abbaschian. The effect of solidification rate on microsegregation. *Metall. Trans.*, 17A:2063–2073, 1986.

[11] E. Scheil. Bemerkungen zur Schichtkristallbildung (Retrograde saturation curves). *Zeitschrift für Metallkunde*, 34:70–72, 1942.

[12] S. N. Singh, B. P. Bardes, and M. C. Flemings. Solution treatment of cast Al-4.5%Cu alloy. *Met. Trans.*, 1:1383, 1971.

[13] S. Sundarraj and V. R. Voller. The binary alloy problem in an expanding domain: the microsegregation problem. *Intl. J. Heat Mass Transfer*, 36:713–723, 1993.

[14] A. Suziki, T. Suzuki, Y. Nagaoka, and Y. Iawata. *Nippon Kingaku Gakkai Shuho*, 32: 7804, 1969.

[15] Ph. Thévoz and M. Rappaz. Solute diffusion model for equiaxed dendritic growth. *Acta Metall.*, 35:1487–1497, 1987.

[16] Ph. Thévoz and M. Rappaz. Solute diffusion model for equiaxed dendritic growth: analytical solution. *Acta Metall.*, 35:2929–2933, 1987.
[17] M. S. Thompson, J. E. Morral, and Jr. A. D. Romig. Applications of the square root diffusivity to diffusion in Ni-Al-Cr alloys. *Metall. Trans. A*, 21A:2679–2685, 1990.
[18] R. Trivedi and K. Somboonsuk. Constrained dendritic growth and spacing. *Mat. Sci. and Engrg.*, 65:65–74, 1984.
[19] V. R. Voller and S. Sundarraj. Comprehensive microsegregation model for binary alloys. In T. S. Piwonka, V. Voller, and L. Katgerman, editors, *Modeling of Casting, Welding and Advanced Solidification Processes - VI*, pages 251–258. TMS, 1993.
[20] C. Y. Wang and C. Beckermann. A unified solute diffusion model for columnar and equiaxed dendritic solidification. *Mat. Sci. and Engng.*, A171:199–211, 1993.

CHAPTER 11

MACRO- AND MICROSTRUCTURES

11.1 INTRODUCTION

The first three chapters of Part II have described how nucleation and growth of the major types of microstructures occur under simple thermal conditions, e.g., at a fixed undercooling or under a fixed thermal gradient and velocity of the isotherms (i.e., Bridgman growth conditions). We examined the competition of various microstructures or morphologies during columnar growth, e.g., stable vs. metastable phases or dendrites vs. eutectics (coupled zone). In some cases, such as for peritectics growing at low velocity, as demonstrated in Chap. 9, the competition of the two solid phases might involve not only growth, but also nucleation kinetics. The present chapter further addresses how nucleation and growth kinetics influence the formation of microstructures under more realistic thermal conditions typical of most solidification processes. We will see that, in many practical applications, the final microstructure observed after solidification results from a competition between columnar structures moving with the isotherms and grains nucleating ahead of the solidification front. This can occur for the same phase, e.g., columnar and equiaxed dendrites, or for the same morphology, e.g., columnar and equiaxed eutectic grains.

Such a competition is illustrated in Fig. 11.1 for a dendritic Al-7wt%Si alloy directionally solidified (DS) in a vertical thermal gradient. A similar grain macrostructure is shown in a 2D schematic form for an ingot solidified from the lateral and bottom sides. Both figures display several grains that nucleated at the bottom surface in contact with a water-cooled chill for the DS ingot or at the cold surface of the mold for the 2D ingot. From this *outer equiaxed zone*, *columnar* grains form as a result of growth competition. Grains with an unfavorable orientation can be overgrown either by their neighbors, or by new grains having a more favorable orientation that nucleate ahead of them. Since the thermal gradient decreases with the distance to the coldest surface(s), columnar growth may be totally arrested by grains nucleating in the liquid ahead of the columnar front, and these new grains might maintain an elongated shape

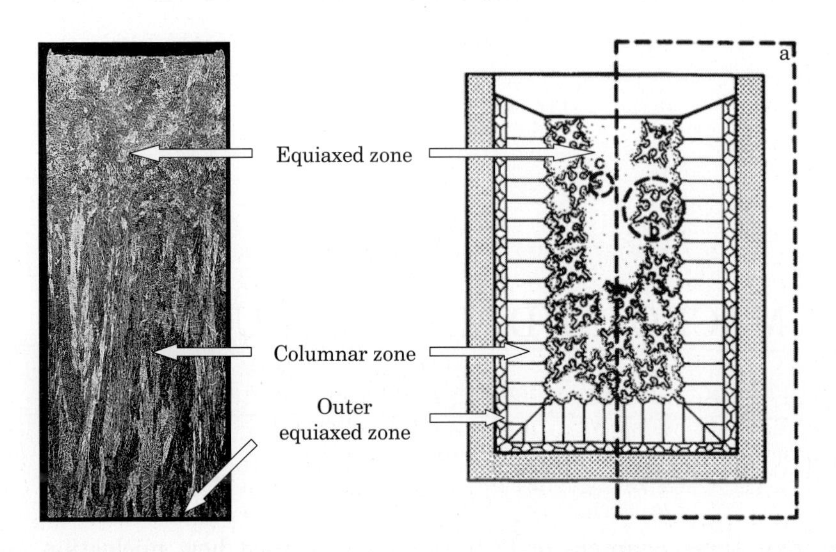

Fig. 11.1 The grain structure as observed in a directionally solidified Al-7wt%Si casting (left, after Gandin and Rappaz [8]) and schematically illustrated for a 2D casting (right, after Kurz and Fisher [12]). The outer equiaxed, columnar and (inner) equiaxed zones are indicated.

if they grow in a thermal gradient. As the thermal gradient decreases further, near the center of the 2D casting or near the top insulated surface of the directionally solidified ingot, the new grains become increasingly equiaxed, eventually forming what is called the *inner equiaxed zone* (top equiaxed zone for the DS ingot). The transition between columnar and equiaxed grains of the same phase, the so-called Columnar-to-Equiaxed Transition (CET), has been studied quite extensively. The reverse competition, from the outer equiaxed zone to columnar grains, near a chill surface, has received less attention, but is governed by essentially the same phenomena.

This chapter considers such competing growth mechanisms, with the objective of predicting the grain structure that will appear under a specific solidification practice. In order to tackle such problems, it is first necessary to understand how nucleation and growth mechanisms combine to give a final microstructure. We therefore first address the case of equiaxed grains growing in a uniform temperature field in Sect. 11.2. Grains nucleating and growing in a thermal gradient, but without considering any preferential growth direction, are treated in Sect. 11.3. The orientation-driven competition occurring among columnar grains is considered in Sect. 11.4, followed by an examination of the CET in Sect. 11.5. The last section of this chapter, Sect. 11.6, describes the coupling of the nucleation-growth models with macroscopic heat- and fluid-flow calculations. Emphasis will be placed on one particular model coupling grain nucleation and growth using Cellular Automata (CA) with Finite Element (or Finite Difference) methods for the macroscopic heat flow, the so-called CAFE approach. Such models provide a fairly simple and robust means for integrating the most important phenomena into a single model, providing a direct calculation of the grain structure.

11.2 EQUIAXED GRAINS GROWING IN A UNIFORM TEMPERATURE FIELD

The present section considers several types of equiaxed structures, illustrated in Fig. 11.2. A *globular* or *globular-dendritic* microstructure found in an inoculated Al-1wt%Cu specimen is shown in Fig. 11.2(a). This structure occurs when solutal or thermal instabilities are unable to develop, e.g., when the grain radius, R_g, is small with respect to a characteristic instability wavelength of the interface (see Sect. 8.2.3). Such situations are frequently encountered in inoculated alloys, where there is a high density of nuclei and the cooling rate is moderate. When the grain size and/or the cooling rate is large enough, morphological instabilities develop and the equiaxed grains of a primary phase become *dendritic* (Fig. 11.2b). When the thermal gradient is small, the extent of the primary arms is the same along all the primary crystallographic growth directions. For example, regular $\langle 100 \rangle$ dendrites produce a 6-branched, equiaxed star-like (or

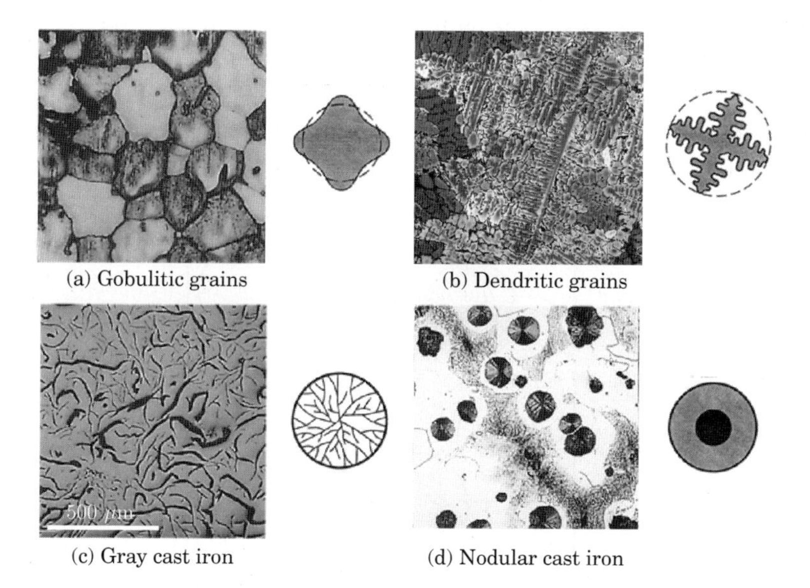

Fig. 11.2 Equiaxed grain types frequently encountered in solidification: (a) globulitic grains in Al-1wt%Cu (grain size $\approx 100\mu$m), (b) dendritic grains in an Al-7wt%Si alloy (grain size ≈ 2 mm, Courtesy of A. Dahle), (c) gray cast iron, and (d) nodular cast iron (typical graphite nodule size $\approx 50\mu$m) (Courtesy of G. Lesoult).

cruciform) grain in 3D, or a 4-branched equiaxed cross in 2D as indicated in the sketch of Fig. 11.2b.

Eutectic alloys, such as cast iron or Al-Si, can also grow with an equiaxed morphology if they are inoculated. Figure 11.2(c) shows *lamellar gray cast iron*, for which diffusion of carbon occurs between lamellae in the liquid phase. Figure 11.2(d) corresponds to *spheroidal graphite iron* (or *nodular cast iron*), in which carbon diffuses through the austenite shell surrounding the graphite nodule. The growth kinetics of these two eutectic morphologies were described in Chap. 9, and the kinetics of dendritc growth was presented in Chap. 8.

11.2.1 Nucleation and growth of equiaxed eutectic grains

Before considering the complexity of real castings, where numerous different morphologies can grow, we first develop a model for a population of equiaxed grains nucleating and growing in a small volume element, as illustrated in Fig. 11.3. The derivation formalizes the pioneering work of Oldfield [13], also developed in the review articles by Stefanescu [24] and by Rappaz [19]. The volume element is assumed to be small enough that the temperature within it can be regarded as uniform. Such an assumption is valid for an entire sample with volume V and surface area A if the Biot number $\mathrm{Bi} = h_T V/Ak \ll 1$, where h_T is the heat transfer coefficient and k the thermal conductivity. This assumption is also valid for an element inside a casting for which the maximum internal temperature difference is small in comparison to the undercooling needed to resolve the microstructure (described further below). For the sake of simplicity in the model development, we also take $\rho_s = \rho_\ell = \rho$. The evolution of the solid fraction in such a volume element is obtained by integrating the average

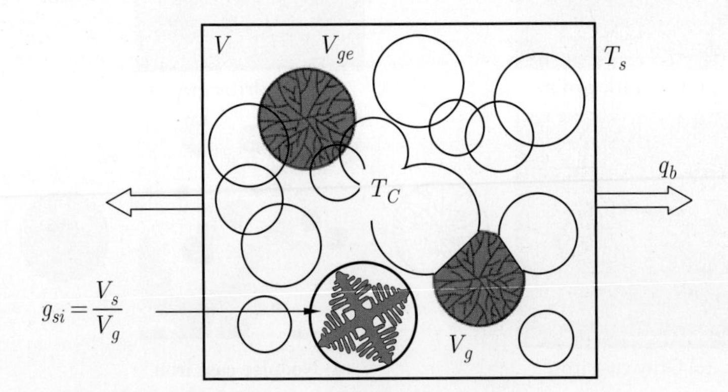

Fig. 11.3 A schematic of equiaxed grains growing in a small volume element of uniform temperature. The extended volume of a eutectic grain, V_{ge}, the actual volume, V_g, when impingement is accounted for, and the internal volume fraction of solid, g_{si}, of a dendritic grain are displayed.

heat flow equation, Eq. (4.137) (with $v = 0$ and $\dot{R}_q = 0$), over the volume element, with the result:

$$q_b \frac{A}{V} = \frac{d\langle \rho h \rangle}{dt} = \langle \rho c_p \rangle \frac{dT}{dt} - \rho L_f \frac{dg_s}{dt} \tag{11.1}$$

where q_b is the external heat flux leaving the surface A of the specimen, $\langle \rho h \rangle$ and $\langle \rho c_p \rangle$ are the average volumetric enthalpy and specific heat, respectively, and ρL_f is the volumetric latent heat of fusion.

Consider first spherical grains that nucleate randomly in the volume at a rate designated $\dot{n}_g(t)$, as described in Chap. 7. The number of new nuclei formed per unit volume between times t_n and $t_n + dt_n$ is given by $dn_g = \dot{n}_g(t_n)dt_n$. As the grains grow, they will eventually impinge on one another, which must be accounted for by the model. The *extended volume fraction of solid* of the grains $g_{se}(t)$, defined as the volume that would be occupied by the grains if there were no impingement (see Fig. 11.3), is obtained by integrating over all the grains:

$$g_{se}(t) = \int_0^t \frac{4\pi R_g^3(t_n, t)}{3} dn_g(t_n) = \int_0^t \frac{4\pi R_g^3(t_n, t)}{3} \dot{n}_g dt_n \tag{11.2}$$

where $R_g(t_n, t)$ is the radius of the grains nucleated at time t_n and observed at time t. If the growth rate $v_g(\Delta T)$ is known, one can also write

$$R_g(t_n, t) = R_g(t_n, t_n) + \int_{t_n}^t v_g(t')dt' = R_g(t_n, t_n) + \int_{t_n}^t v_g(\Delta T(t'))dt' \tag{11.3}$$

where $R_g(t_n, t_n)$ is the nucleation radius of the grains at time t_n.

A fairly simple model that accounts for the decreased growth rate due to impingement is described by the Kolmogorov-Johnson-Mehl-Avrami (KJMA) relationship. In this model the grains are assumed to be randomly distributed in space, and the increment dg_s is equal to dg_{se} multiplied by the probability of the actual interface of the grains being in contact with the liquid. This probability is given by $(1 - g_s)$, leading to, after integration

$$g_s(t) = 1 - \exp(-g_{se}(t)) \tag{11.4}$$

Differentiating Eq. (11.2) with respect to time and using Eq. (11.4) gives

$$\begin{aligned} \frac{dg_s}{dt} &= (1 - g_s)\frac{dg_{se}}{dt} \\ &= (1 - g_s)\left[\dot{n}_g(t)\frac{4\pi R_g^3(t, t)}{3} + \int_0^t 4\pi R_g^2(t_n, t) v_g(t_n, t) dn_g(t_n)\right] \end{aligned} \tag{11.5}$$

The first term in the square brackets corresponds to the increase in volume fraction associated with nucleation, whereas the second term represents the increase in solid fraction associated with growth. Since the initial nucleus size is small, the first term is negligible in comparison to the second. When the velocity of the grains, $v_g(t_n, t)$, is independent of the time t_n at which they nucleated, Eq. (11.5) can also be written in the form

$$\frac{dg_s}{dt} = \mathcal{S}_V^{s\ell} v_g(t) = \left[(1 - g_s) \int_0^t 4\pi R_g^2(t_n, t) dn_g(t_n)\right] v_g(t) \tag{11.6}$$

where $\mathcal{S}_V^{s\ell}$ is the specific surface of the solid-liquid interface, i.e., the integral of the spherical surface of all the grains multiplied by the impingement factor, $(1-g_s)$. At this stage, it is useful to introduce the second-order average grain radius, $\overline{R_g^2}(t)$, and the total density of the grains, $n_g(t)$:

$$n_g(t) = \int_0^t \dot{n}_g(t_n) dt_n \quad \text{and} \quad \overline{R_g^2}(t) = \frac{1}{n_g(t)} \int_0^t R_g^2(t_n, t) dn_g(t_n) \tag{11.7}$$

thus allowing the increase in the solid fraction to be expressed as

$$\frac{dg_s}{dt} = n_g(t) 4\pi \overline{R_g^2}(t) v_g(t)(1 - g_s(t)) \tag{11.8}$$

One can clearly see the individual contributions of nucleation (increase of $n_g(t)$), of growth $(4\pi\overline{R_g^2}(t)v(t))$ and of impingement $(1 - g_s(t))$.

Let us now adopt the athermal nucleation model developed in Chap. 7, and further assume that the growth kinetics depend only on the undercooling. The integral over time in Eq. (11.6) can then be converted into one over the temperature or undercooling:

$$\frac{dg_s}{dt} = (1 - g_s) v_g(\Delta T) \int_0^{\Delta T} 4\pi R_g^2(t_n, t) \frac{dn_g}{d(\Delta T_n)} d(\Delta T_n) \tag{11.9}$$

where

$$R_g(t_n, t) = \int_{t_n}^t v_g(\Delta T(t')) dt' \tag{11.10}$$

Combining Eq. (11.9) with the heat balance given in Eq. (11.1) gives:

$$\frac{d\langle\rho h\rangle}{dt} = q_b \frac{A}{V} = \langle\rho c_p\rangle \frac{dT}{dt} - \rho L_f \left[(1 - g_s) \int_0^{\Delta T} 4\pi R_g^2(t_n, t) \frac{dn_g}{d(\Delta T_n)} d(\Delta T_n)\right] v_g(t) \tag{11.11}$$

An illustration of the type of result that such a micro-model of equiaxed solidification of eutectic grains can provide is presented in Fig. 11.4. The cooling curve shown to the right exhibits a typical *recalescence*, a period during which the temperature of the specimen $T(t)$ increases while its average enthalpy $\langle\rho h(t)\rangle$ decreases, due to the release of the latent heat of fusion generated by the growing grains. By definition, recalescence begins at the local minimum indicated by the maximum undercooling ΔT_{max} in Fig. 11.4, and terminates at the local maximum ΔT_{min}. These extrema of the cooling curve correspond to points where $dT/dt = 0$, i.e., to the time when the last term of Eq. (11.11) is equal to the heat extraction rate per unit volume, $d\langle\rho h\rangle/dt = q_b(A/V) < 0$. In the model, before

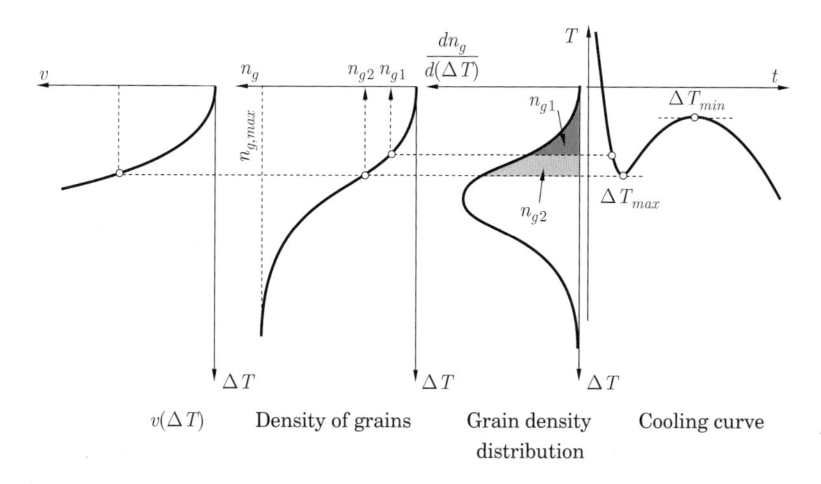

Fig. 11.4 The convolution principle for equiaxed solidification.

the start of recalescence, grains are assumed to nucleate continuously from $\Delta T = 0$ up to the current undercooling. The athermal nucleation law depicted next to the cooling curve highlights, in gray, the density of nuclei that have become active. At the onset of recalescence, ΔT_{max}, nucleation stops and the density of nuclei in the specimen is designated as n_{g2}. The growth rate at this undercooling is given by the growth kinetics law, $v(\Delta T_{max})$, shown in the curve to the left.

Figure 11.5 provides an illustrative example of such a calculation. Cooling curves were measured for gray cast iron samples each having

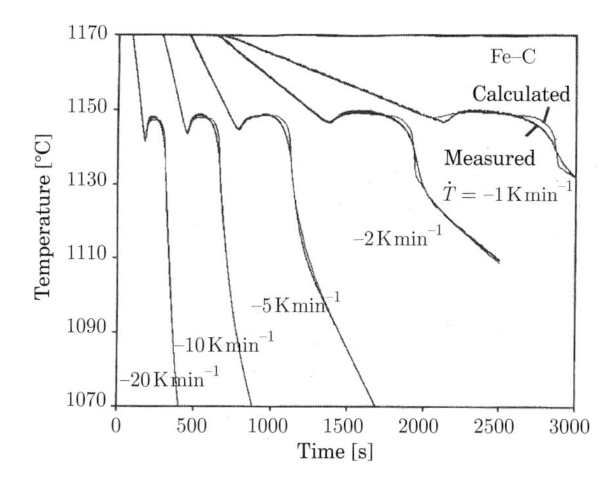

Fig. 11.5 Cooling curves measured and calculated for an Fe-C alloy of eutectic composition solidified at various cooling rates. Growth kinetics: $v_g = 0.04\,\mu\mathrm{m\,s^{-1}\,K^{-2}}\,\Delta T^2$, Gaussian athermal nucleation law with the following parameters: mean undercooling $19.5\,\mathrm{K}$, standard deviation $5.0\,\mathrm{K}$ and $n_{g,max} = 60\,\mathrm{mm^{-3}}$. (After Zou Jie. [31].)

a volume of about 1 cm^3, solidified at various cooling rates. As can be seen, increasing the cooling rate produces a deeper recalescence ($\Delta T_{max} - \Delta T_{min}$), a larger ΔT_{max} and consequently a smaller average grain size (larger grain density). Simulation results, obtained using a model similar to that of Eq. (11.11), fit the experimental data very well over the entire range of samples. The model parameters providing the best fit to the experimental data are presented in the caption of Fig. 11.5. It should be noted that fitting the experimental results by adjusting the parameters is a reasonable action if one has a well-defined foundry practice, because it is possible to calibrate the model for the process using one set of experiments, such as the ones shown here, and then use the parameters obtained from fitting the model when modeling the solidification of larger and more complex castings.

11.2.2 Transition from globular to dendritic grain morphologies

The analysis presented in the previous section in the context of eutectic grains can also be applied to the formation of globular grains of a primary phase, provided that the overlap of the solute layers surrounding the grains is neglected. This is sometimes referred to as *hard impingement*. However, as shown in Sect. 8.2.2, the spherical growth of a primary phase in a supersaturated melt is unstable with respect to morphological perturbations once the sphere exceeds a certain radius. The critical radius at which the sphere becomes unstable to disturbance of mode m is given by (see Sect. 8.2.3)

$$R^* = \left(\frac{(m+1)(m+2)}{2} + 1 \right) R_c \tag{11.12}$$

where $R_c = 2\Gamma_{s\ell}/\Delta T$ is the radius of the critical nucleus.

For a cubic crystal, we choose $m = 4$, whereupon the spherical grain remains stable until $R^* \approx 16 R_c$. Assuming that the solute layer has not yet reached the limit of the grain R_{g0}, i.e., $C_\ell(R_{g0}) = C_0$ where C_0 is the nominal composition of the alloy, the undercooling ΔT is given by

$$\Delta T = m_\ell (C_\ell(R_{g0}) - C_\ell^*) = m_\ell (C_0 - C_\ell^*) \tag{11.13}$$

As in Chap. 8, Eq. (8.35), the solute balance at the interface is made using the Zener approximation, i.e., approximating the solute layer around the grain as a linear profile from the solid up to a distance equal to R_g. This gives

$$v_g C_\ell^* (1 - k_0) = -D_\ell \frac{C_0 - C_\ell^*}{R^*} \tag{11.14}$$

Rearranging Eq. (11.14) recovers the approximate relationship $\Omega = \mathrm{Pe}_g$, where the supersaturation Ω and grain Péclet number Pe_g are defined as

$$\Omega = \frac{(C_\ell^* - C_0)}{C_\ell^*(1 - k_0)} \quad \text{and} \quad \mathrm{Pe_g} = \frac{v_g R^*}{D_\ell} \tag{11.15}$$

For small undercoolings ($C_\ell^* \cong C_0$), Equations (11.12)–(11.15) allow to recover an equation similar to Eq. (8.37) with $D_s = 0$

$$R^{*2} v_g = 32 \frac{D_\ell \Gamma_{s\ell}}{k_0 \Delta T_0} \tag{11.16}$$

The growth rate v_g also can be estimated from a simple heat balance at the scale of the grain. Assuming that the heat extraction rate per unit volume, $d\langle \rho h \rangle / dt$, is constant, one can express the heat balance for a grain in a similar form to Eq. (11.1)

$$q_b 4\pi R_{g0}^2 = \frac{d\langle \rho h \rangle}{dt} \frac{4\pi}{3} R_{g0}^3 = \langle \rho c_p \rangle \frac{4\pi}{3} R_{g0}^3 \dot{T}_0 = -4\pi R^{*2} v_g \rho L_f \tag{11.17}$$

where q_b is the (negative) external heat flux leaving the grain, $\dot{T}_0$ the cooling rate before the start of solidification, and R_{g0} the final grain size. Note that the specific heat variation during solidification has been neglected, i.e., we assume that the solidification was nearly isothermal at the onset of growth. This heat balance allows us to relate the product $R^{*2} v_g$ to process parameters, such as the heat extraction rate or cooling rate before solidification. Replacing $R^{*2} v_g$ in Eq. (11.17) by the value obtained in Eq. (11.16) allows to obtain the critical grain radius $R_{g0,c}$ characteristic of the globular-to-dendritic transition

$$R_{g0,c} = \left(-96 \frac{D_\ell \Gamma_{s\ell}}{k_0 \Delta T_0} \frac{L_f}{c_p \dot{T}_0} \right)^{1/3} \tag{11.18}$$

It is interesting to note that this expression gives the same cooling rate (or solidification time) dependence as the coarsening law for secondary dendrite arms seen in Sect. 8.4.4. Diepers and Karma [5] developed a very similar expression where the factor 96 in Eq. (11.18) was replaced by a function $A(\varepsilon)$, where ε is the anisotropy of $\gamma_{s\ell}$.

For a given alloy and cooling rate, Eq. (11.18) permits the calculation of the maximum final grain radius, $R_{g0,c}$, below which the grain remains globular over its entire growth stage. If the final grain size is larger than this value, the grain will grow first as a sphere and then destabilize once $R_g(t) > 16 R_c$. However, the destabilization might be limited if the solute layer reaches the limit of the grain, R_{g0}, soon after the onset of destabilization. Fully dendritic equiaxed structures such as the one shown in Fig. 11.2(b) will be observed only when $R_{g0} \gg R_{g0,c}$.

Figure 11.6 compares the results obtained using Eq. (11.18) with phase-field computations done by Diepers and Karma [5] for an Al-5wt%Cu alloy. In addition to the very close agreement between the results obtained from the two approaches, the calculated values, typically $R_{g0,c} = 100$ μm for a cooling rate of $-1\,\mathrm{K\,s^{-1}}$, is of a correct order of magnitude when compared to observations made on inoculated Al-alloys.

It should be pointed out that in real casting processes, the final grain size R_{g0} and the cooling rate $\dot{T}_0$ are interrelated. For any particular nucleation conditions and growth kinetics, the final grain size decreases as

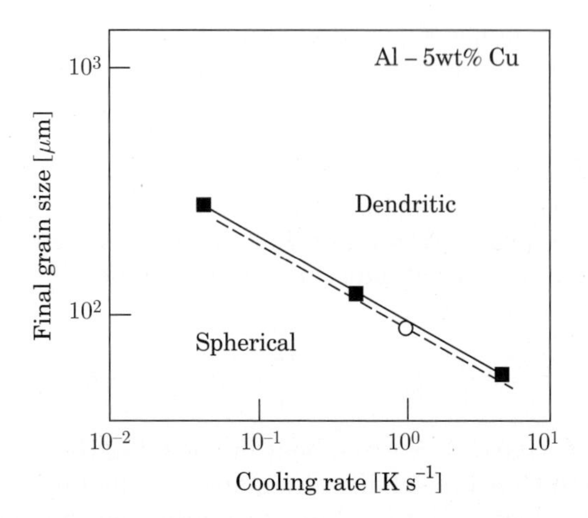

Fig. 11.6 The transition between globular and dendritic growth as a function of the cooling rate for an Al-5wt%Cu alloy. The dashed line (with an open circle) corresponds to Eq. (11.18), whereas the continuous line (with filled squares) was calculated by phase-field simulation by Diepers and Karma [5].

the cooling rate increases. In the following section, the growth analysis of an equiaxed primary phase is restricted to fully dendritic grains, i.e., to cases where $R_{g0} \gg R_{g0,c}$, a situation typically encountered in non-inoculated alloys. The case of fully globulitic grains can be treated in a very similar manner to the eutectic case, provided that the "hard" impingement of eutectic grains associated with the KJMA model is replaced by a "soft" impingement via the interaction of the solute layers. For a treatment of such mixed equiaxed structures, i.e., globular, globular-dendritic or dendritic structures in the same casting, see Beckermann et al. [3] and Combeau et al. [1].

11.2.3 Nucleation and growth of equiaxed dendritic grains

The modeling of equiaxed dendritic grains is not that different from that of eutectics, with the exception that the grains are only partially solid (see Fig. 11.3). Chapter 10 introduced the concept of a *grain envelope*, defined as a "smooth surface" going through all of the dendrite arm tips, within which a solute can be considered to have a nearly uniform composition. In cubic crystals, for which the preferred dendritic growth direction is $\langle 100 \rangle$, this envelope has an octahedral shape, bounded by (111) faces in 3D (a square in 2D, Fig. 11.7). In fact, the faces should be slightly concave, since secondary dendrite arms emitted behind a free-growing tip require a certain amount of time before they can escape the solute field of their neighbors. For the moment, we develop a simpler model wherein the grain

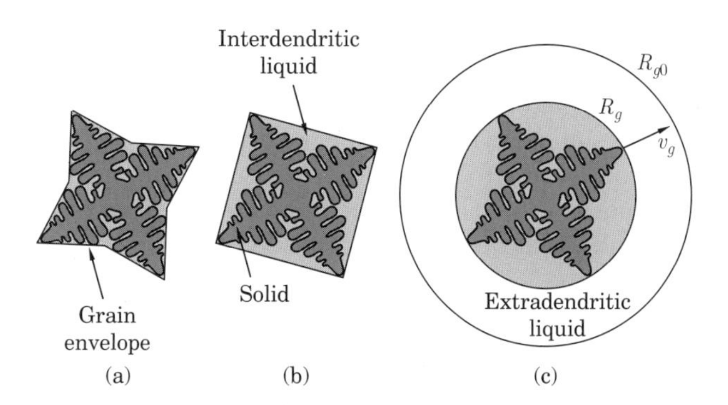

Fig. 11.7 A 2D schematic of an equiaxed dendrite surrounded by a cruciform (a), square (b) and spherical envelope (c).

envelope is approximated by a sphere, as illustrated in Fig. 11.7. We leave the more realistic shape to be addressed in Sect. 11.6.3.

A similar model was introduced in Chap. 10, where we identified the volume fractions of the solid g_s, of the interdendritic liquid g_d and of the extradendritic liquid g_e, along with the constraint $g_s + g_d + g_e = 1$. Note that the grain fraction g_g introduced for equiaxed eutectic morphologies is given in this case by $g_g = g_s + g_d$. The velocity of the envelope is equal to the grain velocity v_g. Solute conservation equations for the solid (Eq. (10.44)), the interdendritic liquid (Eq. (10.46)) and the extradendritic liquid (Eq. (10.49)) were also written. Following the developments of Rappaz and Thévoz [18], it is now assumed that the extradendritic liquid composition $\langle C \rangle_e$ remains at the nominal composition C_0, i.e., we neglect the solute content $(C_\ell - C_0)$ within the solute boundary layer surrounding the envelope. Assuming that the interdendritic liquid is well mixed, Eq. (10.49) becomes

$$\frac{d(g_e \langle C \rangle_e)}{dt} = C_0 \frac{dg_e}{dt} = C_\ell^* \frac{dg_e}{dt} - \mathcal{S}_V^{de} D_\ell \frac{C_0 - C_\ell^*}{\delta_e} \tag{11.19}$$

where $\mathcal{S}_V^{de}$ is the specific surface of the envelope, C_ℓ^* is the composition of the liquid within the envelope and δ_e is the thickness of the solute boundary layer surrounding the envelope. We further assume that the velocity of the grain envelope v_g is uniform and equal to that of the primary dendrite tips. Note that this approximation is equivalent to the presumption of a spherical envelope. Since $dg_e/dt = -\mathcal{S}_V^{de} v_g$, the result is simple:

$$\delta_e = \frac{D_\ell}{v_g} \tag{11.20}$$

Consequently, the average extent of the solute boundary layer outside of the grain envelope is neither the grain radius, nor the dendrite tip radius. Instead, it is approximately equal to the extent of the solute layer characterizing a steady planar front. The introduction of a solute layer outside the grain envelope provides an alternative to the KJMA hard impingement model: the equiaxed dendritic grains start to slow down as soon as their surrounding solute layers impinge, i.e., when $(R_g + \delta_e) \gtrsim R_{g0}$.

On the other hand, if $\langle C \rangle_e = C_0$, this also means that the sum of the average compositions of the interdendritic liquid and the average solid composition must be equal to the nominal composition

$$g_s \langle C \rangle_s + g_d C_\ell^* = (g_s + g_d) C_0 = g_g C_0 \tag{11.21}$$

Dividing by g_g and introducing the *internal volume fraction of solid* of the grains $g_{si} = g_s / g_g = g_s / (g_s + g_d)$, we obtain

$$g_{si} \langle C \rangle_s + (1 - g_{si}) C_\ell^* = C_0 \tag{11.22}$$

The physical interpretation of this equation is fairly straightforward. The solute is entirely conserved within the grain envelope, with no solute transport occurring outside. During the early part of the solidification, the dendritic grains grow at temperatures close to the liquidus temperature, for which all of the microsegregation models discussed in Chap. 10 predict essentially the same behavior. Therefore, the composition of the solid can be assumed to be uniform, i.e., $\langle C \rangle_s \approx C_s^* = k_0 C_\ell^*$, and Eq. (11.22) recovers the lever rule approximation at the scale of the grain envelope:

$$g_{si} = \frac{g_s}{g_g} = \frac{g_s}{g_s + g_d} = \Omega = \frac{C_\ell^* - C_0}{C_\ell^* (1 - k_0)} \tag{11.23}$$

Thus, the internal volume fraction of solid of an equiaxed dendritic grain is roughly proportional to the supersaturation Ω or undercooling ΔT.

Using these results in Eq. (11.2), the extended volume fraction of solid in the volume element can be written as

$$g_{se}(t) = \int_0^t \frac{4\pi R_g^3(t_n, t)}{3} g_{si}(\Omega(t)) dn_g(t_n) \tag{11.24}$$

With the help of Eq. (11.23), and under the assumption that the dendrite growth kinetics developed in Chap. 8 dictates that of the grain envelope $v_g(\Delta T)$, the procedure for coupling the grain growth to the thermal balance is the same as that for eutectic grains. By neglecting the impingement term, one can show (see Exercise 11.4) that the local heat balance in this case becomes

$$\frac{d\langle \rho h \rangle}{dt} = q_b \frac{A}{V} = \langle \rho c_p \rangle \frac{dT}{dt} - \rho L_f \frac{dg_s}{dt}$$

with

$$\frac{dg_s}{dt} = n_g(t) 4\pi \overline{R_g^2}(t) \Omega(T(t)) v_g(t) + n_g(t) \frac{4\pi}{3} \overline{R_g^3}(t) \frac{d\Omega(T(t))}{dt} \tag{11.25}$$

where the following notation has been introduced:

$$\overline{R_g^3}(t) = \frac{1}{n_g(t)} \int_0^t R_g^3(t_n, t) dn_g(t_n)$$

$$R_g(t_n, t) = \int_{t_n}^t v_g(\Delta(t'))dt'$$

$$\frac{d\Omega}{dt} = \frac{d}{dt}\left(\frac{C_\ell^* - C_0}{C_\ell^*(1 - k_0)}\right) \approx -\frac{1}{k_0 \Delta T_0}\frac{dT}{dt}$$

There are two important differences in this result, as compared to the equiaxed eutectic case. The latent heat release from the growth of the grain envelope is weighted by the internal solid fraction of the grains, $g_{si} = \Omega$, and the evolution of the internal solid fraction contributes proportionally to the cooling rate. Figure 11.8 shows the result of a calculation for an Al-7wt% Si specimen solidified at various cooling rates. The growth kinetics $v_g(\Delta T)$ in this case comes from the LGK model, i.e., Ivantsov solution for a parabolic tip and marginal stability criterion (see Sect. 8.4.2). The parameters of the Gaussian athermal nucleation law were adjusted so as to give the best agreement between the calculations and the experiments for the final grain size and maximum undercooling before recalescence.

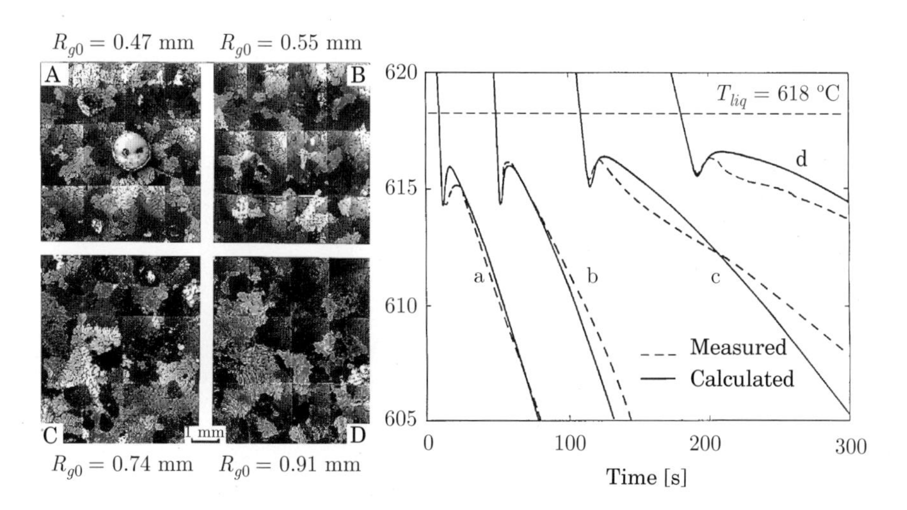

Fig. 11.8 Grain structures (left) and cooling curves for a small Al - 7 wt% Si specimen solidified at various cooling rates. The calculated cooling curves (continuous lines) were computed using a micro-model of solidification based on $g_{si} = \Omega$, the growth kinetics of this alloy and a Gaussian distribution of nuclei. The predicted average grain radii for the 4 specimens can be compared qualitatively to the micrographs. $v_g = 2.9\ \mu\text{m}\,\text{s}^{-1}\,\text{K}^{-2}\ \Delta T^2 + 1.5\ \mu\text{m}\,\text{s}^{-1}\,\text{K}^{-3}\Delta T^3$. A Gaussian athermal nucleation law was used with a mean undercooling of $3.5\,\text{K}$, standard deviation of 0.8 K and $n_{g,max} = 3 \times 10^9\ \text{m}^{-3}$. (After Thévoz [26].)

11.3 GRAINS NUCLEATING AND GROWING IN A THERMAL GRADIENT

Most standard treatments of solidification assume that a positive thermal gradient produces columnar grains, and that equiaxed grains always grow in a zero or negative thermal gradient. However, "equiaxed" grains can also nucleate and grow in a positive thermal gradient. If the temperature difference across the grain is larger than the local undercooling, the growth speed of the eutectic interface or of the dendrite tips is not uniform along the periphery, and the resulting grain has a slightly elongated shape, rather than being truly equiaxed. Figure 11.9 shows an example of such grains, revealed through an X-ray synchrotron image taken *in-situ* during the growth of "equiaxed" Al-Ni grains in a thermal gradient G. The grains extend longer primary $\langle 100 \rangle$ dendrite arms in the direction of the thermal gradient. The overall shape of one of these grains is outlined by an ellipse, and a circle has been placed at the nucleation center.

An idealized model of the configuration of such grains is illustrated in Fig. 11.10(a). For this presentation, we assume that the grain is fully solid. This is appropriate for eutectics, but a similar treatment can be developed for dendritic grains with preferential growth directions (See Gandin et al. [9]). Since it is in a temperature gradient, the warmer portion of the interface has a lower undercooling, and thus grows more slowly than its colder part.

One can perform a numerical solution for the evolution of the grain shape [15]. Here, we prefer to derive an analytical solution for the evolution

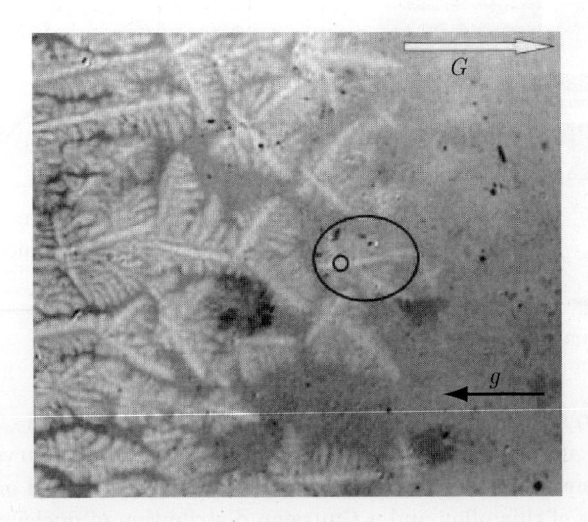

Fig. 11.9 A synchrotron X-ray radiograph of *in-situ* growth of equiaxed grains in an Al-3.5wt%Ni alloy, growing in a vertical thermal gradient G. The shape of one grain and its nucleation center are outlined with an ellipse and a circle, respectively. (Courtesy of G. Reinhart and B. Billia [20].)

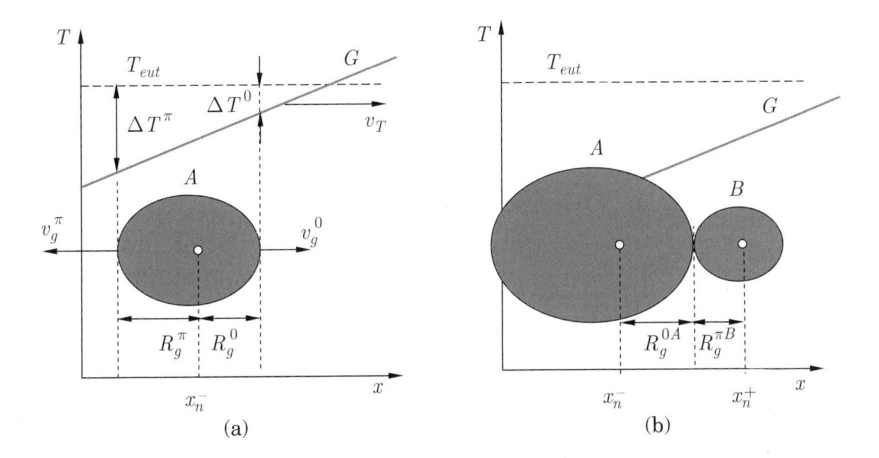

Fig. 11.10 (a) A single "equiaxed" eutectic grain growing in a thermal gradient after having nucleated at position x_n^-. (b) The impingement of this grain with another one nucleated later at position x_n^+ (b). (After Rappaz et al. [15].)

of the distances $R_g^0(t)$ and $R_g^\pi(t)$ perpendicular to the isotherms, to get an idea of the change in aspect ratio of the grain during growth. To that end, we idealize the thermal conditions by taking both the thermal gradient G, and the velocity of the isotherm v_T, to be constant

$$T(x,t) = T_{eut} - G(v_T t - x) \tag{11.26}$$

The reference frame has been defined such that the eutectic isotherm, T_{eut}, is located at $x = 0$ at time $t = 0$. We take the growth kinetics of the eutectic interface to be given by the relationship derived in Chap. 9, $v = A(\Delta T)^2$. The evolution of the two distances, $R_g^0(t)$ and $R_g^\pi(t)$, of a grain nucleating at position $x_n^- = 0$ can be expressed as

$$\begin{aligned} v_g^0 &= \frac{dR_g^0}{dt} = AG^2(v_T t - R_g^0)^2 \\ v_g^\pi &= \frac{dR_g^\pi}{dt} = AG^2(v_T t + R_g^\pi)^2 \end{aligned} \tag{11.27}$$

Integrating these equations between the time of nucleation t_n, and the current time t, and introducing the following notation:

$$\Delta T_{col} = \sqrt{v_T/A} \quad ; \quad \Delta T'_n = \frac{\Delta T_n}{\Delta T_{col}} \quad ; \quad \Delta T'_g(t) = \frac{\Delta T_g(t)}{\Delta T_{col}} = \frac{G v_T t}{\Delta T_{col}} \tag{11.28}$$

We obtain:

$$\begin{aligned} R_g^0(t) &= \frac{\Delta T_{col}}{G}\left[\Delta T'_g(t) - \frac{\tanh(\Delta T'_g(t) - \Delta T'_n) + \Delta T'_n}{1 + \Delta T'_n \tanh(\Delta T'_g(t) - \Delta T'_n)}\right] \\ R_g^\pi(t) &= -\frac{\Delta T_{col}}{G}\left[\Delta T'_g(t) - \frac{\tan(\Delta T'_g(t) - \Delta T'_n) + \Delta T'_n}{1 - \Delta T'_n \tan(\Delta T'_g(t) - \Delta T'_n)}\right] \end{aligned} \tag{11.29}$$

Here, ΔT_{col} is the undercooling of a steady-state columnar front moving at the velocity of the isotherms v_T, $\Delta T'_n$ is the nucleation undercooling normalized by the columnar undercooling, and $\Delta T'_g(t)$ is the undercooling at the nucleation center position (i.e., $x = 0$) at the current time t, normalized by the columnar undercooling. It is interesting to note that the warmest part of the grain, $R_g^0(t)$, tends toward the steady-state eutectic front position, i.e., $R_g^0(t) \rightarrow v_T t - \Delta T_C/G$. For the coldest part of the grain, the growth speed $v_g^\pi(t)$ diverges as a tangent function since the front encounters increasingly cooler regions. The corresponding undercoolings of both fronts are given by

$$\Delta T^0(t) = \Delta T_{col} \frac{\tanh(\Delta T'_g(t) - \Delta T'_n) + \Delta T'_n}{1 + \Delta T'_n \tanh(\Delta T'_g(t) - \Delta T'_n)}$$
$$\Delta T^\pi(t) = \Delta T_{col} \frac{\tan(\Delta T'_g(t) - \Delta T'_n) + \Delta T'_n}{1 - \Delta T'_n \tan(\Delta T'_g(t) - \Delta T'_n)} \tag{11.30}$$

Figure 11.11 shows the evolution of R_g^0 and $-R_g^\pi$ for typical growth conditions, together with the trajectory of the eutectic isotherm. Note that $R_g^\pi < 0$ corresponds to growth at the colder tail of the ellipse. Thus, in this simple analysis, a single "equiaxed" grain growing in a positive thermal gradient is more elongated in the colder direction. However, since the grain nucleates at an undercooling ΔT_n that is smaller than that of a steady-state columnar front ΔT_{col}, other nucleation events can also occur ahead of the growing grain. Such a situation is illustrated in Fig. 11.10(b) for two grains that nucleate along a single line parallel to G. Let us assume that the liquid melt contains a uniform density of nucleation sites n_g, all of which become active at the same nucleation undercooling ΔT_n. The

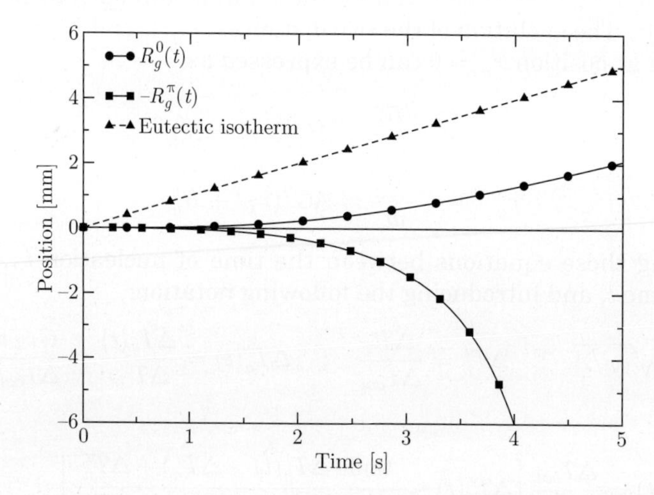

Fig. 11.11 The evolution of the extreme positions $R_g^0(t)$ and $R_g^\pi(t)$ of a eutectic grain growing in a thermal gradient. $G = 10^3\,\mathrm{K\,m^{-1}}$, $v_T = 10^{-3}\,\mathrm{m\,s^{-1}}$, $v_g = 10^{-4}[\mathrm{m\,s^{-1}\,K^{-2}}]\,\Delta T^2$, $\Delta T_n = 0\,\mathrm{K}$.

distance L separating the two nucleation centers shown in Fig. 11.10 is proportional to $n_g^{-1/3}$ and the delay in nucleation events is given by L/v_T. Therefore, the warmer portion of grain A and the colder portion of grain B will impinge at a time t_i when $R_g^{0A}(t_i) + R_g^{\pi B}(t_i) = L$. Using Eqs. (11.29) for R_g^0 and R_g^π gives the impingement condition as

$$\frac{\tanh(\Delta T'_g(t_i) - \Delta T'_n) + \Delta T'_n}{1 + \Delta T'_n \tanh(\Delta T'_g(t_i) - \Delta T'_n)} = \frac{\tan(\Delta T'_g(t_i - L/v_T) - \Delta T'_n) + \Delta T'_n}{1 - \Delta T'_n \tan(\Delta T'_g(t_i - L/v_T) - \Delta T'_n)} \tag{11.31}$$

Equation (11.31) can be solved numerically to find the time of impingement t_i, which in turn gives R_g^{0A} and $R_g^{\pi B}$. The *shape factor* or *elongation factor* S for these "equiaxed" grains can then be computed as

$$S(G, v_T, \Delta T_n, L = n_g^{-1/3}) = \frac{R_g^0(t_i)}{R_g^\pi(t_i - L/v_T)} > 1 \tag{11.32}$$

Figure 11.12 shows the elongation factor as a function of the process parameters G and v_T, for a particular set of athermal nucleation parameters ΔT_n and n_g, and a specified growth kinetics law. When S is close to one, the grains are truly equiaxed. This occurs when the thermal gradient is low or the grain density is high. On the other hand, large values of S correspond to columnar grains, since the growth in colder portion of the grain is effectively zero in that case. In fact, Eq. (11.32) can be used as a criterion to determine the columnar-to-equiaxed transition, discussed in detail in Sect. 11.5. Although the analytical expressions we have developed appear to be somewhat complicated, the model is very useful for understanding the gradual transition from columnar to elongated

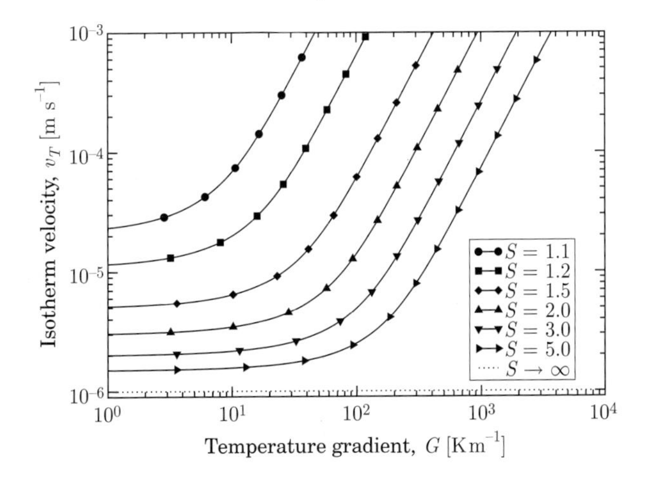

Fig. 11.12 Shape factors for "equiaxed" grains growing in a thermal gradient G with a velocity v_T of the isotherms. The nucleation model parameters are: $\Delta T_n = 0.1\,\text{K}$, $L = 10^{-3}$ m (density of grains $n_g = 10^9\,\text{m}^{-3}$). The growth kinetics are given by: $v_g = 10^{-4}\,\text{m}\,\text{s}^{-1}\,\text{K}^{-2}\,\Delta T^2$.

"equiaxed", to fully equiaxed grains in the unidirectionally solidified casting shown in Fig. 11.1. We also note that the phase field simulations by Badillo and Beckermann [2] of dendritic grains growing in a thermal gradient exhibit the same type of behavior as that shown in Fig. 11.12.

11.4 COLUMNAR GRAINS

Columnar microstructures in eutectics and dendrites are somewhat easier to model than their equiaxed counterparts, because the velocity is known. Indeed, as the undercooling at the growth front is typically just a few degrees, the velocity of the eutectic front or of the columnar dendrite tips is directly linked to that of the eutectic or liquidus isotherms, respectively, as illustrated in Fig. 11.13. The velocity is determined by the thermal conditions imposed at a scale much larger than the microstructural features that we consider here. In the case of a columnar eutectic, the lamellae or fibers grow perpendicular to the isotherms, and thus $v^* \approx v_T(T_{eut})$. For dendrites, however, the growth direction is constrained to follow preferred crystallographic directions, e.g., $\langle 100 \rangle$ for cubic metals (see Chap. 8). Under steady-state conditions, the tip velocity is therefore given by

$$v^* = \frac{v_T(T_{liq})}{\cos\phi} \tag{11.33}$$

where ϕ is the angle between the thermal gradient G and the primary dendrite trunk direction. This implies that dendrites whose preferred direction is inclined with respect to G must grow faster than their counterparts that are well aligned with G in order to maintain their relative tip positions. Since the dendrite tip undercooling increases monotonically

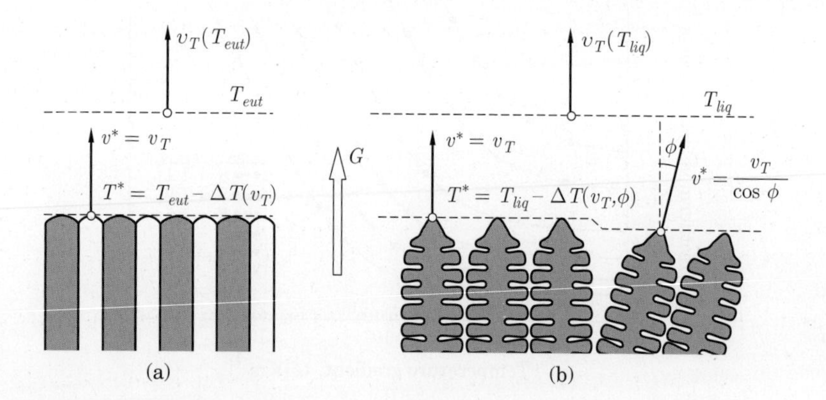

Fig. 11.13 Velocities of (a) eutectic isotherm and eutectic front, and (b) liquidus isotherm and dendrite tips. Because dendrites are constrained to grow along well-defined directions, their velocity, and thus their undercooling, depend on their angle of inclination with respect to the thermal gradient.

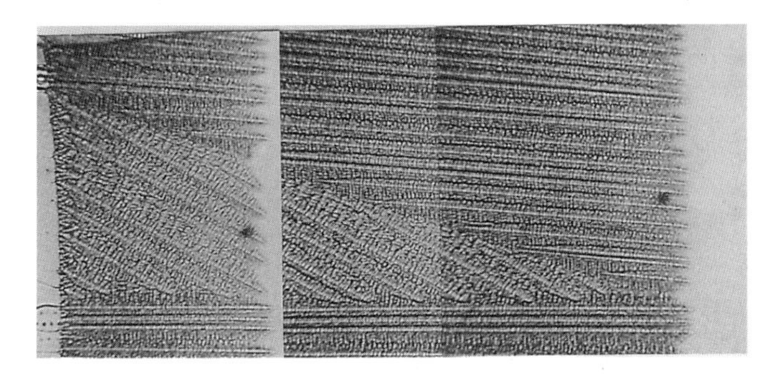

Fig. 11.14 A montage of micrographs displaying the growth of 3 grains of succinonitrile-acetone. The left panel shows the misaligned dendrite tips of the central grain slightly behind the well-aligned tips of its neighbors, and the right hand panel shows the elimination of the central grain. (After the film of Esaka et al., reprinted from Gandin and Rappaz [8].)

with velocity, inclined dendrites have a higher undercooling, and therefore their tips must lie slightly behind those of well-aligned dendrites (Fig. 11.13b).

Figure 11.14 shows a sequence of micrographs taken during the directional solidification of a succinonitrile-acetone alloy. The dendrites grow from left to right, beginning with three initial grains with different orientations. The grains at the top and bottom have a $\langle 100 \rangle$ direction that is fairly well aligned with the horizontal thermal gradient, whereas the middle grain is misoriented by about 30°. As one can see in the left panel, the dendrites in the center grain are slightly more undercooled than those of the two other grains. At the lower grain boundary, where the $\langle 100 \rangle$ dendrites trunks are *converging*, the misoriented dendrite tips hit the well-aligned dendrites slightly behind their tips and become blocked, making this grain boundary coincide with the orientation of the well-aligned dendrites. At the upper grain boundary, where the $\langle 100 \rangle$ dendrite trunks are *diverging*, a region of liquid continuously opens. The two dendrites bordering the grain boundary can extend secondary arms into the gap, which then initiate tertiary arms parallel to the primary trunks. Eventually, some of these tertiaries become new primary trunks. This complex branching mechanism produces a grain boundary at an angle that is approximately a bisector of the respective $\langle 100 \rangle$ directions in the two grains. As illustrated in Fig. 11.14, the center grain is progressively eliminated during growth.

Although the mechanisms of grain selection have been shown to be slightly more complex at low misorientation [30], it is evident from this example that growth competition among columnar dendrites results in a natural selection of grains with a small angle ϕ between their $\langle 100 \rangle$ direction and the thermal gradient G [27]. The elimination of unfavorably oriented grains as the growth proceeds is accompanied by an evolution of

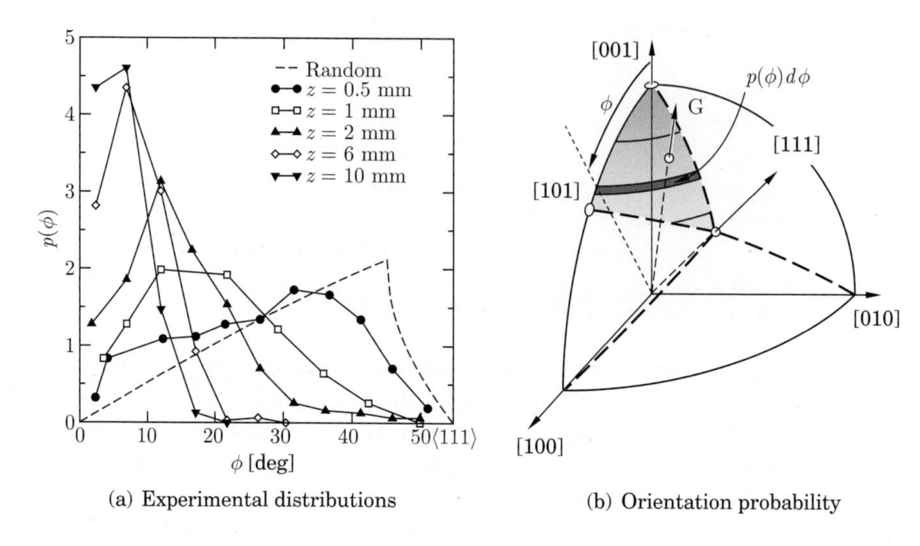

(a) Experimental distributions (b) Orientation probability

Fig. 11.15 (a) The evolution of the orientation distribution with the distance z from the chill surface in a directionally solidified Ni-based superalloy casting [7]. The distribution sharpens and shifts toward $[100]$ over a very small distance. (b) A schematic diagram displaying the probability $p(\phi)d\phi$ of finding the thermal gradient direction G in the range $[\phi, \phi + d\phi]$ to the closest $\langle 100 \rangle$ direction in a cubic symmetry crystal.

the distribution of grain orientations; from random at the surface, to columnar grains preferentially oriented at decreasing values of ϕ. The biasing of the distribution toward certain angles is called a *preferred orientation* or *texture*. One can quantify the texture development by measuring the distribution $p(\phi)$ of grains at various locations in a casting. Figure 11.15 provides an example, taken from a Ni-based superalloy casting that was directionally solidified vertically from a chill surface [7]. The left panel shows several orientation distributions, measured using Electron Back-Scattered Diffraction (EBSD) techniques at various distances from the chill. One can clearly see that as the distance from the chill increases, the distribution becomes sharper and the peak of the distribution shifts toward the $\langle 100 \rangle$ direction.

It may seem peculiar that the dashed curve indicating the random distribution of orientations in Fig. 11.15(a) does not have a constant probability for all angles ϕ. The shape of the curve can be understood with reference to Fig. 11.15(b), which shows one octant of a unit sphere in orientation space. The surface of the sphere is divided into several spherical triangles, all of which are equivalent because the crystal cannot distinguish one $\langle 100 \rangle$ direction from another. Thus, it is sufficient to discuss the light gray spherical triangle with vertices $[001]$, $[101]$ and $[111]$. The angle ϕ is measured as a polar angle from $[001]$. The probability density $p(\phi)d\phi$ of finding the thermal gradient G at an angle between ϕ and $\phi + d\phi$ from the closest $\langle 100 \rangle$ direction (in this case $[001]$) is indicated as a dark gray band of width $d\phi$ (two other ϕ-circles corresponding to lower and larger values of ϕ are also indicated). The probability density goes to zero at the $[001]$ pole because there is only one possible orientation satisfying this condition. The probability density increases with the polar angle ϕ from $[001]$, until it reaches a maximum probability at $\phi = \pi/4$. The probability density then falls off rapidly with increasing ϕ in this spherical triangle because most of the thermal gradient orientations fall into the next spherical triangle, i.e., the thermal gradient becomes closer to either $[100]$ or $[010]$. The probability density goes to zero again at $[111]$, corresponding to $\phi = \cos^{-1}(1/\sqrt{3}) = 54.7°$.

From the numerous randomly oriented grains formed at the chill surface, those with the largest misorientation with respect to the thermal gradient, corresponding to the tail of the distribution where ϕ is largest, should be eliminated first. This is clearly the case in Fig. 11.15(a). As growth proceeds, the distribution sharpens and its peak rotates toward the $[100]$ direction. One should notice that the selection process occurs rapidly, within just 10-mm range of the chill. As the misorientation between grains decreases, so does the difference in their respective velocities. The grain elimination mechanism is therefore much less effective once ϕ for all grains is less than about 10°. It is also interesting to note that since the probability of having a grain with a $\langle 100 \rangle$ direction exactly aligned with n is zero, the peak in the distribution will never reach $\phi = 0$, because there were no grains in the original distribution with that orientation. Note also that the mechanism of successive elimination of less favorably oriented grains implies that the average width of the remaining columnar grains will increase with the distance from the chill, as suggested by Fig. 11.14. The interested reader is referred to Gandin et al. [7] for further details.

The mechanism for grain selection is commonly used for the production of directionally solidified and single crystal turbine blades. Figure 11.16 shows a simple blade cast using the lost wax process described in Chap. 1. In this case, the mold was covered by lateral insulating wool and attached to a water-cooled copper chill. As the melt is poured into the

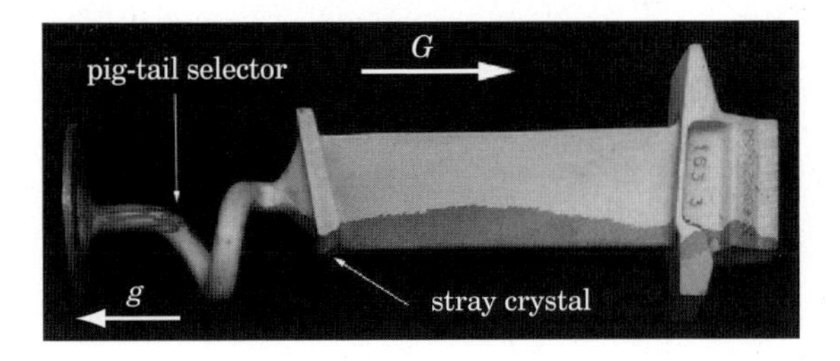

Fig. 11.16 An image of a nickel-based superalloy turbine blade produced by directional solidification. The vertical thermal gradient during the process goes from left to right in the figure. (After Rappaz et al. [17].)

mold, many grains nucleate at the chill surface and start to grow. The grain selection process described above occurs over the first few millimeters. Thus, the grains entering the helical "pig-tail" have their $\langle 100 \rangle$ directions aligned within $\approx 10^\circ$ of the vertical thermal gradient. The pig-tail eliminates most of the remaining grains, selecting those that can extend their secondary arms in the most efficient way into the first turn of the helix. After one turn in the helix, only a single grain emerges and starts to propagate by branching laterally into the entire blade. One can see that, in this particular blade, another stray grain nucleates before the arms of the main grain can reach the corner of the blade platform. If this stray grain is well-oriented with respect to the thermal gradient, it continues to grow, parallel to the initial grain, giving rise to the bicrystal blade seen here.

Let us consider further the conditions that lead to the formation of such undesirable stray crystals. The extension of a columnar grain into an open region of liquid, such as the blade platform, is a critical step in growing a dendritic single crystal. Figure 11.17 presents a montage of micrographs from a model succinonitrile-acetone system, solidified from below toward a re-entrant corner. Consider first the case where the isotherms are horizontal. When the primary trunk emerges from the narrower part below, secondary arms propagate rapidly across the surface into the undercooled liquid, since the undercooling they experience is larger than that of the dendrite tips. Before they reach the end of the platform, the undercooling may be large enough for a new grain to nucleate at the corner, which leads to a stray crystal. Two additional phenomena may enhance the formation of stray crystals at such a corner: (i) corners radiate more energy, and are therefore colder than the center (i.e., the liquidus isotherm is not flat but concave); (ii) corners promote the formation of nuclei, as discussed in Chap. 7.

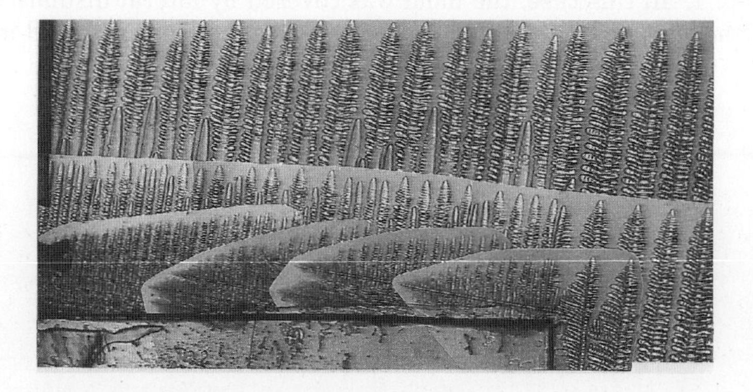

Fig. 11.17 A montage of micrographs from directional solidification of a succinonitrile-acetone alloy, showing the propagation of a columnar grain into the open space after a re-entrant corner. (After Rappaz et al. [17].)

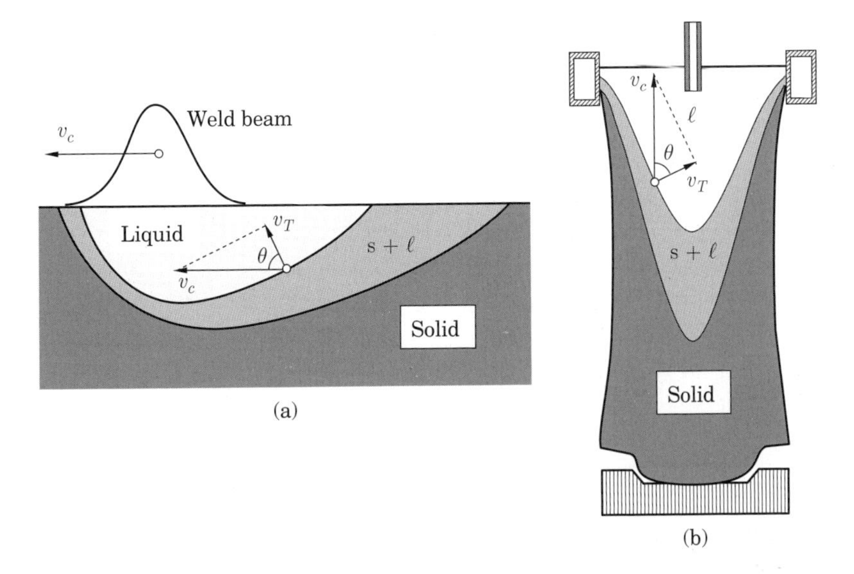

Fig. 11.18 The relationship between the continuous steady-state solidification process velocity and the isotherm velocity for two processes: (a) welding and (b) continuous casting of Al.

Thus far, all of the examples considered have been essentially one-dimensional, with an interface speed v_T equal to the process speed v_c. In more complex solidification processes, this is not the case. Figure 11.18 illustrates two such processes, welding and continuous casting, where the process speed is set by the welding speed and the withdrawal rate, respectively. In steady state operation of such processes, v_T and v_c are related by

$$v_T = v_c \cos\theta \tag{11.34}$$

where θ is the angle between the two velocities, as indicated in the figure. By substituting Eq. (11.34) into Eq. (11.33), one finds the steady growth velocity of the dendrites in such steady-state processes

$$v^* = v_c \frac{\cos\theta}{\cos\phi} \tag{11.35}$$

Equation (11.35) can be used to understand the origin of the microstructure formed during e-beam welding of single crystals. The shape of the liquid pool can be determined from a thermal calculation, giving the normal to the liquidus surface, and thus the angle θ in Eq. (11.35). The dendrite trunk variant that will be selected during growth is the one with its $\langle 100 \rangle$ direction having the lowest velocity, i.e., the smallest angle ϕ with respect to the normal or the smallest undercooling. The selection process is shown schematically in Fig. 11.19 for e-beam welding together with the microstructure observed in a cross-section perpendicular to the welding direction.

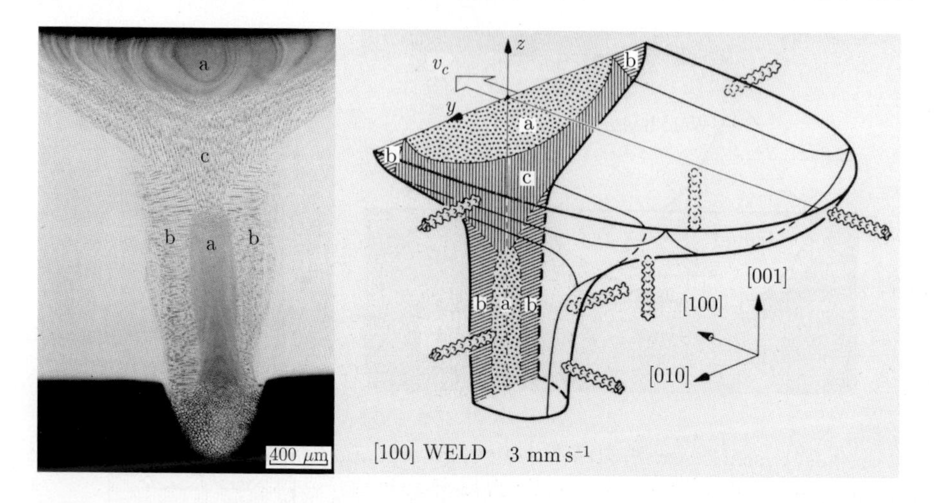

Fig. 11.19 (Left) Microstructure observed in a cross-section of an e-beam weld made in a Fe-15wt%Ni-15wt%Cr single crystal. (Right) A schematic diagram of the 3D pool shape with the various ⟨100⟩ dendrite growth variants. (After Rappaz et al. [16].)

11.5 COLUMNAR-TO-EQUIAXED TRANSITION

11.5.1 Hunt's criterion

We now consider more formally the transition from a columnar to an equiaxed microstructure, commonly abbreviated CET. The discussion is presented in the context of the schematic drawing shown in Fig. 11.20, displaying columnar grains growing in a temperature gradient G, with isotherms moving at velocity v_T. We consider the possible nucleation and growth of equiaxed crystals ahead of the columnar front, as sketched in the figure. This topic was introduced in Sect. 11.3, where we took into account the shape factor for grains nucleating and growing in a thermal gradient, and used the analysis to predict a gradual transition from fully columnar grains to truly equiaxed ones.

This section presents the derivation of another criterion, first developed by Hunt [11]. The physical basis for this criterion, with reference to Fig. 11.20, is the following: Equiaxed grains can nucleate and grow in the undercooled region ahead of columnar dendrites. If the volume fraction occupied by such grains when the columnar front arrives is large enough, they will block the progress of the columnar front. Let us make this physical argument more quantitative by adopting the simple nucleation model that all grains nucleate at a fixed undercooling ΔT_n, with a density n_g. The extended fraction of these grains, g_g, at the level of the columnar front is given by

$$g_g = n_g \frac{4\pi}{3} R_g^3(t_n, t_{col}) \qquad (11.36)$$

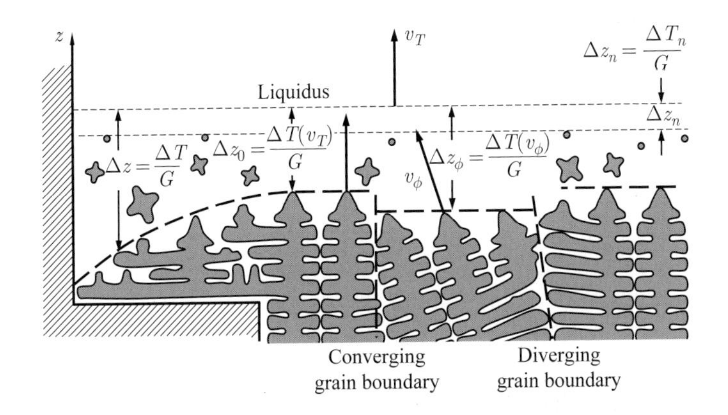

Fig. 11.20 A schematic illustration of the growth of columnar grains, showing the competition near a converging and a diverging grain boundary. We also show the extension of the left columnar grain into the open space after a re-entrant corner, and the competition with equiaxed grains that nucleate in the undercooled region ahead of the columnar dendrite front. (Redrawn from Gandin and Rappaz [8].)

where t_n is the time at which the grains nucleate, t_{col} is the time at which the columnar front arrives, and $R_g(t_n, t_{col})$ corresponds to their radius at the level of the columnar front. To form a tractable model, it is assumed that the grains do not move or settle, that their growth kinetics are given by $v_g(\Delta T) = A\Delta T^2$, and that the cooling rate is constant. One can then compute the equiaxed grain radius R_g as

$$R_g(t_n, t_{col}) = \int_{t_n}^{t_{col}} v_g(t)dt = \frac{A}{\Delta \dot{T}} \int_{\Delta T_n}^{\Delta T_{col}} \Delta T^2 d(\Delta T) = \frac{A}{3 \mid \dot{T} \mid}(\Delta T_{col}^3 - \Delta T_n^3) \tag{11.37}$$

Substituting this expression into Eq. (11.36) gives the fraction of equiaxed grains g_g just ahead of the columnar front

$$g_g = n_g \frac{4\pi}{81} \frac{A^3}{\mid \dot{T} \mid^3}(\Delta T_{col}^3 - \Delta T_n^3)^3 = n_g \frac{4\pi}{81} \frac{A^3}{(Gv_T)^3} \left(\left(\frac{v_T}{A} \right)^{3/2} - \Delta T_n^3 \right)^3 \tag{11.38}$$

Hunt defined a criterion for the CET as a particular value of g_g beyond which columnar grains are stopped and equiaxed grains take over.

To illustrate the use of the model, we consider again the model system used in Sect. 11.3. Figure 11.21 augments the data in Fig. 11.12 with the curves $g_g = 0.01$ and $g_g = 0.99$ computed using Eq. (11.38). The $G - v_T$ diagram is computed employing the same parameters as those used to calculate the shape factors in Fig. 11.12. For $G - v_T$ combinations below the curve $g_g = 0.01$ (open square symbols), the grain structure is columnar, and above the curve $g_g = 0.99$ (open circle symbols), the structure will be equiaxed. Overall, the shapes of the two new curves are very similar to those shown in Fig. 11.12 for the shape factor of grains nucleating

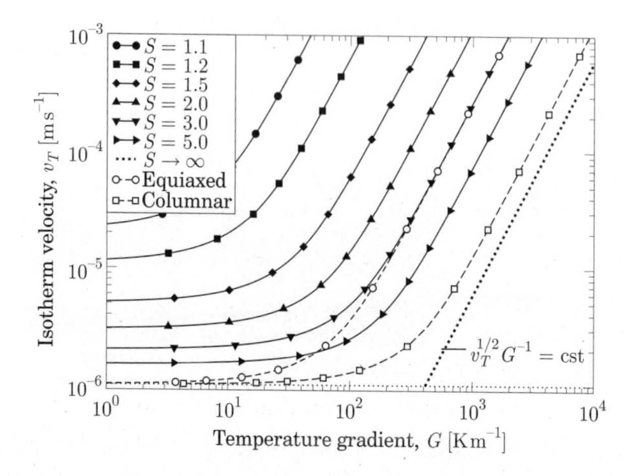

Fig. 11.21 The superposition of Hunt's criterion from Eq. (11.38) onto the shape factors for elongated grains from Sect. 11.3.

and growing in a thermal gradient. One can observe two asymptotic limits. One branch is found in the lower left-hand corner of the graph, corresponding to small values of both v_T and G. In this region, the velocity of the columnar grains is very low, and accordingly their undercooling ΔT_{col} is very low. In this region of the $G - v_T$ diagram, the transition from columnar to equiaxed growth occurs when the undercooling $\Delta T_{col}(v_T)$ is equal to the nucleation undercooling for equiaxed grains ΔT_n. In equation form, this can be expressed as

$$\Delta T_n = \Delta T_{col} = \sqrt{v_T / A} \tag{11.39}$$

Since ΔT_n has been assumed to be fixed, this branch of the curve then gives $v_T = A \Delta T_n^2$, which is a constant, corresponding to the horizontal dotted line just above $v_T = 10^{-6}\,\mathrm{m\,s^{-1}}$ in Fig. 11.21.

The other branch is found at larger thermal gradients and velocities. In this region, $\Delta T_{col} \gg \Delta T_n$. By neglecting ΔT_n on the right-hand side of Eq. (11.38) and performing a little algebra on the remaining terms, we obtain:

$$g_g \approx n_g \frac{4\pi}{81} \left(\frac{\sqrt{v_T}}{G} \right)^3 \tag{11.40}$$

Thus, in this region, curves of constant g_g correspond to $\sqrt{v_T}/G = \text{cst}$, as indicated. The transition from columnar to equiaxed (i.e., $g_g = 0.01$ to $g_g = 0.99$ in this case) occurs over a range of $v_T - G$ values that depends strongly upon the density of the grains n_g. Although it is used extensively in solidification models, Hunt's criterion does not predict a gradual transition from elongated "equiaxed" grains to truly equiaxed ones, as we found in Sect. 11.3.

Now consider a process, such as solidification from a chill, where G and v_T are both large at the beginning of solidification near the chill. This corresponds to a point toward the upper right-hand corner of Fig. 11.21, which implies that one should expect to find an initial columnar zone. As the distance from the chill increases, heat extraction is less effective, and both G and v_T decrease. This corresponds to movement to the left in Fig. 11.21, which implies that the equiaxed fraction should gradually increase with distance, until the structure becomes entirely equiaxed.

11.5.2 Microsegregation and cooling curves

In Chap. 10, we introduced several models for microsegregation and computed the fraction solid-temperature curve for each case. These models did not include any information regarding the microstructure, however. Now that models for columnar and equiaxed growth have been developed, it would be useful and interesting to consider the effect of including the nucleation and growth kinetics associated with these two morphologies. The problem is fairly complex, as there are numerous possible combinations of microstructures that can form. For example, one could have columnar primary dendrites followed by a columnar eutectic front; equiaxed dendritic grains separated by eutectics growing in columnar form; equiaxed dendrites with equiaxed eutectic grains either surrounding the primary phase grains or nucleating and growing independently; etc. We consider two of these cases here: equiaxed dendrites/eutectic nodules forming in a specimen of nearly uniform temperature ("equiaxed case"); and columnar dendrites/eutectic front growing in a positive thermal gradient ("columnar case").

Cooling curves under constant heat extraction rate associated with these two cases are shown schematically in Fig. 11.22, as plots of the average enthalpy per unit volume $\langle \rho h \rangle$ vs. temperature T. The dotted line

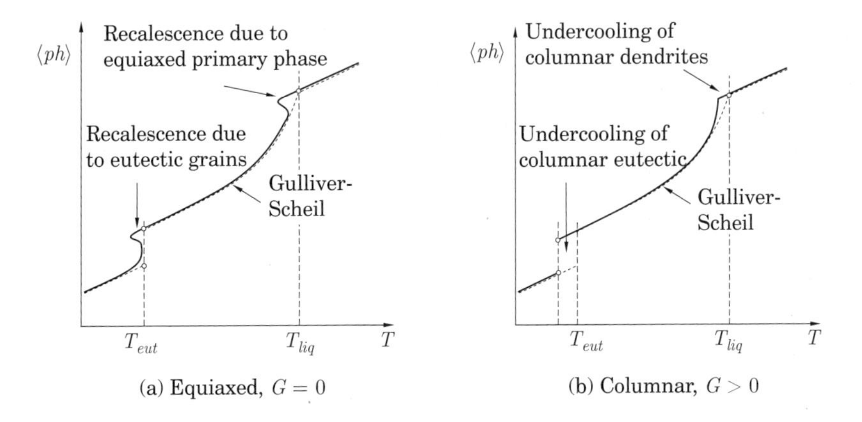

Fig. 11.22 The average enthalpy per unit volume as a function of the temperature for a hypoeutectic alloy (a) with the formation of equiaxed dendrites and eutectics in a specimen of spatially uniform temperature; (b) with columnar dendrites and eutectics growing in a positive thermal gradient. The dotted curves correspond to the Gulliver-Scheil microsegregation model.

represents the cooling curve corresponding to the Gulliver-Scheil model, included as a reference. For the equiaxed case with $G \approx 0$ (Fig. 11.22a), the average enthalpy decreases linearly with time if the heat extraction rate is constant. In that case, the expected cooling curve can be obtained by rotating the axes 90°. *Recalescence*, defined as an increase in temperature while the enthalpy decreases, is predicted near both the liquidus and eutectic temperatures. This case can be simulated using the equiaxed growth models presented earlier in this chapter by setting the volume fraction of the liquid available for eutectic growth to $(1 - g_s(T^*_{eut}))$, where $g_s(T^*_{eut})$ is the volume fraction of the primary dendrites when nucleation of equiaxed eutectic grains begins.

No recalescence is expected in the case of columnar growth. The slope of the enthalpy changes at the dendrite tip position, corresponding to the location where $T = T^* = (T_{liq} - \Delta T(v^*))$, with $\Delta T(v^*)$ corresponding to the tip undercooling associated with the velocity $v^* \approx v_T(T_{liq})$. Similarly, the columnar eutectic front induces a jump in the enthalpy at a temperature $T^* = (T_{eut} - \Delta T(v^*))$, where $\Delta T(v^*)$ is now the eutectic undercooling for $v^* \approx v_T(T_{eut})$. Although the analysis of this case seems easier at first because the velocity is known (see Sect. 11.4), solute is not necessarily conserved in the previous analysis. Two approaches can be used to address this problem for columnar structures. The first and easiest solution, adopted by many authors, is to simply introduce a Gulliver-Scheil microsegregation model *truncated* at the actual tip temperature, T^*. This procedure leads to a small enthalpy jump at T^* from the extended enthalpy of the liquid to the dashed Gulliver-Scheil curve. At low velocities, the tip undercooling is usually small, and thus this procedure gives satisfactory results despite the fact that it does not conserve solute. For rapid solidification, where the tip undercooling is no longer negligible in comparison to the solidification interval, more complex relationships for $g_s(T)$ and $\langle \rho h(T) \rangle$ have been derived based on an overall solute balance. The interested reader should see the references of Kurz and Giovanola [10] or Wang and Beckermann [28] for further details.

11.6 MICRO-MACROSCOPIC MODELS

In practice, solidification processes rarely produce the ideal solidification conditions considered thus far. As explained in the introduction of this chapter, real castings often include both columnar and equiaxed morphologies. Therefore, it is necessary to couple the microscopic models of grain structure formation, developed in the previous sections, to a simulation of the macroscopic heat and mass transfer.

11.6.1 Thermal conditions

Thermal conditions, in particular the thermal gradient G and the speed v_T of the liquidus or eutectic isotherm, are the primary factors influencing the

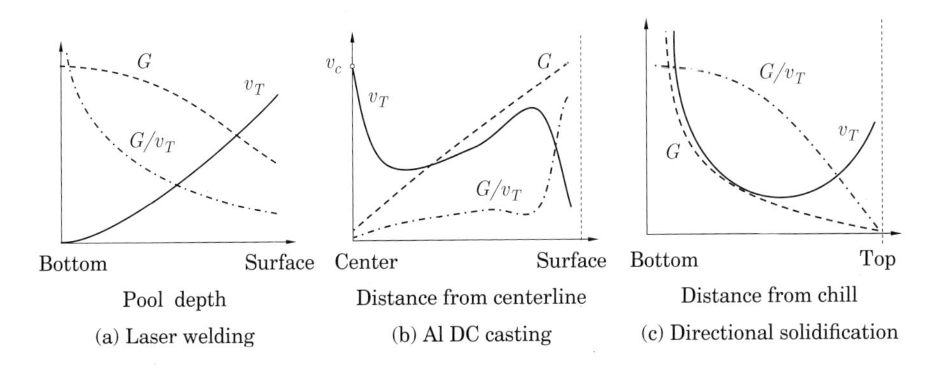

Fig. 11.23 The thermal gradient, G, the velocity of the liquidus isotherm, v_T, and the ratio G/v_T for three solidification processes. Panels (a) and (b) correspond to steady-state processes, whereas case (c) is unsteady.

formation of microstructure (and defects). Before describing more quantitative models of microstructure formation in the next sections, let us first consider qualitatively the effect of these two parameters.

Figure 11.23 depicts schematically the thermal gradient in the liquid G, the velocity of the liquidus isotherm v_T, and the ratio G/v_T for welding, DC casting and unsteady directional solidification (DS) from a chill. The evolution of G and v_T is presented as a function of some relevant position parameter for each case. It is interesting to note that in the two steady-state processes, the gradient is maximum when the velocity is minimum: at the bottom of the weld pool, corresponding to the transition between remelting and solidification; and at the surface of the continuously cast slab. In the latter case, the evolution of the liquidus isotherm velocity, deduced from Eq. (11.34), is non-monotonic and reflects the gradual change of the angle θ between the casting direction and the normal to the liquidus isotherm. This angle is zero at the center of the slab due to symmetry, leading to v_T taking on its maximum value of v_c. Closer to the surface, the transition from the direct contact with the water-cooled mold to the secondary water cooling induces a local maximum in v_T.

In the case of unsteady DS ingot, Fig. 11.23(c) shows that the situation is markedly different, at least near the water-cooled bottom surface where the thermal gradient in the liquid and the velocity decrease simultaneously. Once the superheat has been eliminated, the velocity of the liquidus isotherm might increase near the top surface as shown, while the thermal gradient continues to decrease.

The microstructures that are expected to form under various combinations of G and v_T are summarized in Fig. 11.24, based on a similar figure by Kurz and Fisher [12]. The first two curves that can be recognized readily are those separating the columnar and equiaxed regions, calculated using Hunt's CET criterion as in Fig. 11.21. We have also added in the lower right hand corner, the constitutional undercooling criterion for a stable planar front, given in Eq. (8.44). Lines of constant cooling

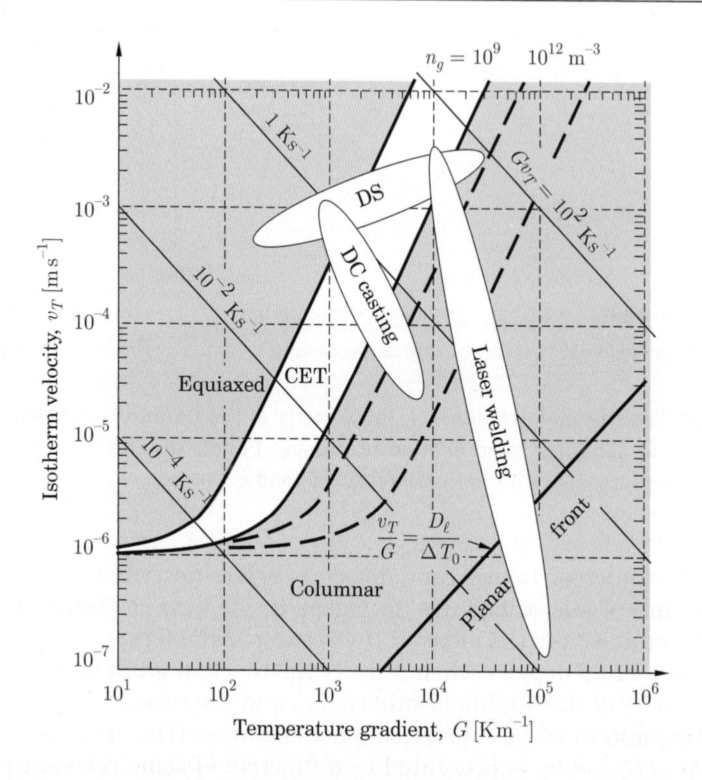

Fig. 11.24 A $G - v_T$ diagram showing the limit of constitutional undercooling (line separating the planar front and columnar region), the CET according to Hunt's criterion for two nucleant densities, and typical (G, v_T) values for the 3 processes shown in Fig. 11.23. The secondary dendrite arm spacing $\lambda_2 \sim (Gv_T)^{(-1/3)}$. $D_\ell = 3 \times 10^{-9}$ m^{2}s^{-1}, $\Delta T_0 = 100\,$K, other values for the CET are identical to those of Fig. 11.21, with $n_g = 10^9$ m^{-3} (continuous lines) and $n_g = 10^{12}$ m^{-3} (dashed lines). (Derived from Kurz and Fisher [12].)

rates given by $Gv_T = \text{Cst}$ have been included to help indicate the expected length scale of the microstructure. Recall from Sects. 8.4.3 and 8.4.4 that $\lambda_1 \sim G^{-1/2} v_T^{-1/4}$ and $\lambda_2 \sim (Gv_T)^{-1/3}$. Since the spacing of regular eutectics varies as $v_T^{-1/2}$, almost independent of the thermal gradient, the variation in eutectic spacing would be given instead by horizontal lines in the diagram. Finally, typical (G, v_T) values for the three solidification processes considered in Fig. 11.23 are indicated by white elliptical domains.

Let us consider individually the microstructure formation in the three processes shown in Figs. 11.23 and 11.24.

Laser welding

Solidification starts at the edges of the weld pool with a large thermal gradient and a very low velocity. Figure 11.24 indicates that solidification should begin with a planar front, which is indeed very often observed as a thin featureless layer at the weld periphery. Most of the weld consists of columnar dendrites that become finer as one approaches the top surface. Elongated equiaxed grains are occasionally observed near the weld centerline. The final microstructure will also depend strongly on Marangoni convection, as discussed in Chap. 1, which is capable of detaching fragments from the columnar dendrites, thus promoting a high density of "nuclei" with $\Delta T_n = 0$. We discuss this phenomenon further in Sect. 11.6.4.

DC casting

One can draw similar conclusions for this process as for laser welding, except that in DC casting G/v_T is not sufficient to produce planar front growth. Columnar dendrites form near the surface of the slab and equiaxed grains at the center. The location of the CET depends on the nucleant density and the associated undercooling. Although some dendrite fragments may detach due to melt convection, the CET in DC casting is affected much more strongly by the inoculation. This is indicated in Fig. 11.24 by including two sets of CET boundaries: solid lines corresponding to $n_g = 10^9\ \mathrm{mm}^{-3}$, and dashed lines corresponding to $n_g = 10^{12}\ \mathrm{mm}^{-3}$. Increasing n_g shifts the CET transition to the right, indicating that the slab then becomes fully equiaxed. The dendrite spacings are not drastically different through the slab, because the product Gv_T is fairly constant.

Directional solidification

Columnar dendrites are observed at the bottom of the DS casting, whereas equiaxed grains are expected to form near the top of the sample. In this case, the length scale of the microstructure becomes coarser as the solidification proceeds, because the cooling rate decreases with distance from the chill.

11.6.2 Analytical models of microstructure formation

As shown in Sect. 11.4, it is fairly straightforward to predict the progress of columnar solidification. Neglecting the small undercooling of the dendrite tips or eutectic front (Fig. 11.22), a macroscopic thermal calculation is first performed with one of the microsegregation models for $g_s(T)$ described in Chap. 10. Since the speed of the corresponding isotherm is known, the undercooling of the front, as well as other microstructural parameters such as the dendrite tip radius, the primary trunk spacing or the eutectic spacing, can be calculated using a growth kinetics model, as discussed in Chaps. 8 and 9. On the other hand, the secondary dendrite arm spacing λ_2 is independent of the solidification mode (columnar or equiaxed), because λ_2 comes from a coarsening law, i.e., it depends only on the local solidification time. Thus, under normal solidification conditions, all of the microstructural parameters in columnar solidification can be deduced using a macroscopic heat and mass transfer, followed by a post-processing calculation. However, it should be emphasized that such an approach cannot predict grain selection, transverse grain size and texture evolution occurring within the columnar zone. This can only be achieved through the use of the stochastic models, discussed in Sect. 11.6.3.

For equiaxed structures, the growth rate of the grains is not directly related to the speed of the isotherms, but rather to the local undercooling. Consider the simple model for a sample where the liquid and the solid are fixed in space (i.e., where there is no convection and no sedimentation of the grains), but the temperature is not uniform. We then need to couple the microscopic model of equiaxed solidification with the heat flow equation (Eq. (4.137) with $v = 0$ and $\dot{R}_q = 0$):

$$\frac{\partial \langle \rho h \rangle}{\partial t} = \nabla \cdot \langle k \nabla T \rangle \tag{11.41}$$

This is straightforward since the time derivative of $\langle \rho h \rangle$ is provided by the microscopic solidification model (cf. Eq. (11.11) or (11.25)). Various numerical schemes, such as the enthalpy method (Sect. 6.3.1), can be used to solve the evolution of the average enthalpy, temperature and solid fraction at each node. In addition to the prediction of a possible recalescence in the cooling curves (Fig. 11.22), such micro-macro coupling provides information about grain structure and microstructure at the scale of the casting. For further details on these methods, see for example the review of Rappaz [19].

Although this analytical micro-macro coupling can be implemented fairly easily into standard heat and mass transfer codes, it suffers from several drawbacks: (i) only equiaxed morphologies can be modeled, leading to relatively poor results being obtained when mixed equiaxed and columnar structures are present; (ii) the method is further limited to cases where the equiaxed grain size is much smaller than the temperature inhomogeneities; and (iii) the method does not provide a direct representation of the microstructure. Most of these shortcomings can be overcome by the stochastic approaches we describe next.

11.6.3 Stochastic models of microstructure formation

In the 1990's, *Cellular Automata* (CA) models were developed to incorporate the phenomena described in this chapter. This technique, initially developed in 2D, was later extended to 3D and coupled with finite element (FE) heat flow calculations, resulting in the so-called *CAFE* model presented in this section. Details of the model can be found in the series of papers by Gandin et al. [6–8; 11].

The starting point of the CAFE approach lies in resolving the various length scales governing heat flow, convection and solute diffusion. Heat flow can be modeled accurately on a fairly coarse mesh at the scale of the geometric features of a casting. Solute diffusion, on the other hand, requires a much finer mesh, typically one order of magnitude smaller than the dendrite tip radius (i.e., $< 1\ \mu$m). The CAFE approach was developed with the notion that computers are not powerful enough to resolve the details of the dendritic structure for every single grain in an entire casting. The model therefore assumes that the growth kinetics of the dendrite tips and arms, $v(\Delta T)$, is known, e.g., from one of the models given in Chap. 8. It then simulates grain growth at the scale of the dendrite arms, typically $50 - 100\ \mu$m, using the growth kinetics law, and the thermal field computed in a macroscale calculation.

The approach is shown schematically in Fig. 11.25, illustrating a portion of a FE mesh along with the underlying microstructure. At each time step, the heat flow equation given in Eq. (11.41) is solved using an enthalpy scheme (described in Sect. 6.3.1). This provides for each node i the temperature T_i^t and the enthalpy variation $\Delta\langle\rho h\rangle_i^t$ over the time step Δt. These quantities are then interpolated onto a much finer structured grid of regular square (2D) or cubic (3D) cells, shown at the bottom of Fig. 11.25. A connectivity map provides the relationship between the cell μ contained in element e_μ, and the FEM nodes of this element (right of Fig. 11.25). The interpolation of the macroscopic fields is given by

$$T_\mu^t = \sum_{i \in e_\mu} N_i(\boldsymbol{x}_\mu) T_i^t; \qquad \Delta\langle\rho h\rangle_\mu^t = \sum_{i \in e_\mu} N_i(\boldsymbol{x}_\mu) \Delta\langle\rho h\rangle_i^t \tag{11.42}$$

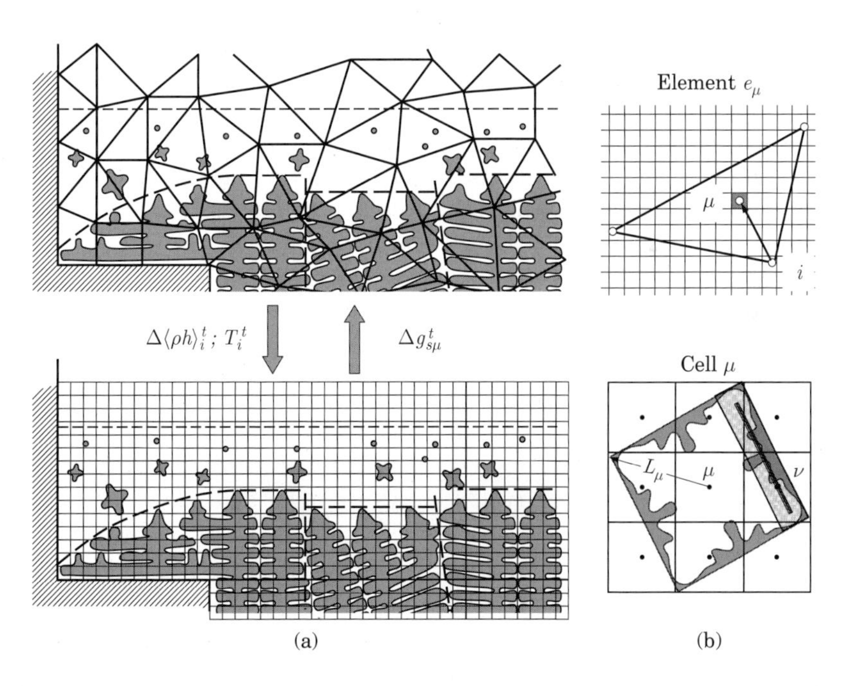

Fig. 11.25 (a) A schematic showing the coupling between finite element heat flow computations and Cellular Automata calculations for nucleation and growth. (b) Schematic representation of dendrite growth (confined to a square envelope) from nucleation cell μ belonging to an element e_μ after it has captured the first-nearest neighbor cells ν. The growth process is then repeated from the neighbor cells, also considering the extent of the dendritic front at the time of capture. (See Gandin et al. [6; 8].)

where the summation is taken over the nodal points i belonging to the element e_μ containing cell μ, N_i are the element interpolation functions, and $\boldsymbol{x}_\mu$ is the coordinate of the centroid of cell μ. Next, the nucleation and growth of the grains are calculated at the scale of the cells as described below, releasing a certain amount of latent heat $\rho L_f \Delta g_{s\mu}^t$ within the cells. A *restriction* is then made to deduce the latent heat release for each node i

$$\rho L_f \Delta g_{si}^t = \rho L_f \sum_{e_i} \sum_{\mu \in e_i} \Delta g_{s\mu}^t N_i(\boldsymbol{x}_\mu) \left(\sum_{e_i} \sum_{\mu \in e_\mu} N_i(\boldsymbol{x}_\mu) \right)^{-1} \tag{11.43}$$

where the summation is now carried out over all of the cells contained in all of the elements e_i that contain i as a nodal point. Once the variation of solid fraction at the FE nodes is known, ΔT_i^t can be computed from the relationship: $\Delta\langle\rho h\rangle_i^t = \langle\rho c_p\rangle \Delta T_i^t - \rho L_f \Delta g_{si}^t$. After the temperature has been updated for all of the nodes, the algorithm proceeds to the next time step.

Nucleation and growth modeling at the level of the CA cells is carried out as follows. At the start of the calculation, all cells are given a solidification index $I_{s\mu} = 0$, indicating that they are entirely liquid. A certain number of *nucleation cells* are chosen randomly within the liquid, based on a preset athermal nucleation law. Each nucleation cell μ_n is assigned a critical nucleation undercooling $\Delta T_{n\mu_n}$ following this distribution. A similar procedure can be performed for nucleation at the domain boundary by a random selection of cells attached to this boundary. If the macroscopic solidification calculation produces a sufficiently large undercooling $\Delta T(\boldsymbol{x}_{\mu_n}) > \Delta T_{n\mu_n}$ in nucleation cell μ_n while the cell is still entirely liquid, a new grain is generated. Its crystallographic orientation, (one angle in 2D or 3 angles in 3D) is selected randomly from a pre-defined set of orientation classes, and the solidification index $I_{s\mu_n}$ is set to an integer value identifying this class. Subsequent growth within such cells is then calculated by integrating the growth kinetics law over time. The position of the dendrite tip (see Fig. 11.25, bottom right) is given by

$$L_\mu(t) = \int_{t_n}^{t} v(\Delta T_\mu(t'))dt' \tag{11.44}$$

At some point, the local square (octahedron in 3D) associated with the envelope of the dendritic network can "capture" a neighboring cell ν if it is still liquid. The solidification index from cell μ is then propagated to this cell by setting $I_{s\nu} = I_{s\mu}$, and the growth process is re-initialized in cell ν using the local undercooling ΔT_ν. Care must be taken to ensure that the final grain orientation is not biased to the basis directions of the cellular grid. See Gandin et al. [6; 8] for details.

Finally, the variation of solid fraction within each cell is calculated according to a truncated Gulliver-Scheil microsegregation model $g_s^{GS}(T)$. The initial value of g_s in nucleation cell μ_n is set to $g_s^{GS}(T_{liq} - \Delta T_{n\mu_n})$, whereas that of a captured cell ν is set to $g_s(T_{liq} - \Delta T_\nu(t_c))$, where t_c is the time of its capture. After cell ν is captured by cell μ, its evolution of g_s follows the Gulliver-Scheil model.

Two examples of CAFE simulations are shown in Fig. 11.26. The first one is related to the 1D Al-7wt%Si casting discussed in Fig. 11.1 at the beginning of this chapter. Figure 11.26(a) shows a simulated "metallographic section" of the computed 3D grain structure. Since the growth kinetics of the dendrites is known, adjustment of the three parameters in the Gaussian nucleation law allows to reproduce the features seen in the experimental ingot fairly accurately. We find an outer equiaxed zone at the bottom of the ingot, near the chill, from which a columnar zone forms up to approximately mid-height. Elongated "equiaxed" grains begin to appear at mid-height, and truly equiaxed grains become visible near the top of the casting. The CAFE model also predicts cooling curves that are consistent with measured ones, showing a recalescence in the upper part of the casting where the gradient is nearly zero, and monotonically decreasing cooling curves with a small inflection point closer to the chill where the gradient is large.

The second example shows that the CAFE method can be applied equally well to castings of arbitrary shapes, in this case a directionally solidified nickel-based superalloy turbine blade (Fig. 11.26(b)). The thermal field was first computed. The CA algorithm was then run after introducing a random distribution of nuclei at the bottom surface. The grain

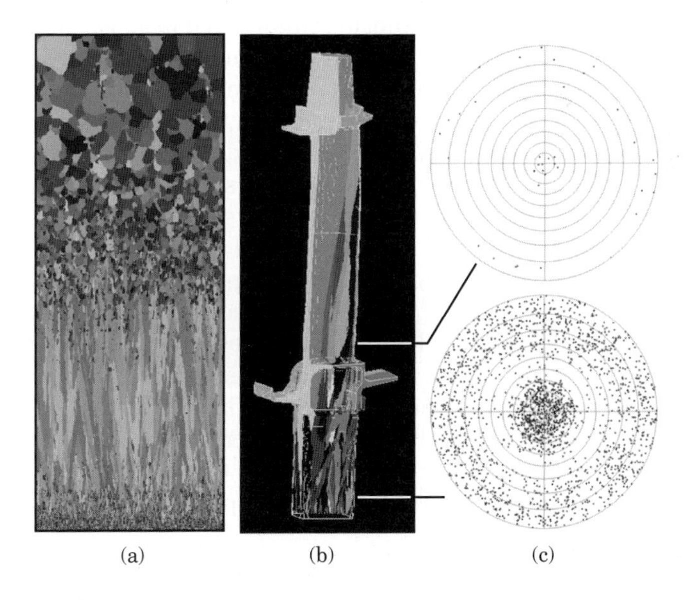

Fig. 11.26 (a) Grain structure from a 3D simulation of a directionally solidified Al-7wt%Si alloy, shown in a longitudinal cross section (to be compared with Fig. 11.1). (b) The final grain structure predicted by the CAFE approach in a DS turbine blade, and (c) the corresponding $\langle 100 \rangle$ pole figure at two heights. The vertical thermal gradient is placed at the center of the pole figures. (From Gandin et al. [6; 8].)

selection process is readily apparent, and only a few of the original grains survived at mid-height. Figure 11.26(c) shows ⟨100⟩ pole figures [1] obtained from the output of the CAFE model. If the grain orientation distribution was random, the ⟨100⟩ directions would cover the pole figure uniformly. It can be seen that, near the bottom of the blade, the most unfavorably oriented grains ($30° \leq \phi \leq 54.7°$) have started to disappear, leaving a ring that is nearly devoid of ⟨100⟩ points at the corresponding angles in the pole figure. As columnar growth proceeds, the surviving grains are those that have a ⟨100⟩ direction nearly aligned with the thermal gradient, as explained in Sect. 11.4. Above the first platform of the blade, there are only about 10 remaining columnar grains with a ⟨100⟩ direction within $10°$ from the center of the pole figure. One should notice that the other ⟨100⟩ directions of these grains are close to the equator and are still randomly oriented, since no selection occurs for the secondary dendrite arms. The evolution in grain orientation for the experimental version of this casting was already shown in Fig. 11.15. Further examples and details related to CAFE applications can be found in the series of papers by Gandin et al. [6; 8].

11.6.4 Influence of convection

Up to now, very little consideration has been given to the effect of convection on the grain structure. In particular, we have assumed that any columnar or equiaxed grains that form remain fixed in space during growth. While this is a reasonable assumption for columnar growth, it is clearly inappropriate for equiaxed structures. For example, the Al-Ni equiaxed grains observed by X-ray radiography in Fig. 11.9 actually fell on the columnar front and on each other, because the solid density is greater than that of the surrounding liquid. In some alloys, the solid is less dense than the surrounding liquid, causing equiaxed grains to float rather than to sediment as a result of buoyancy. This is the case in Al-Cu when $C_0 \gtrsim 10$ wt% [21], and also in various cast irons. Equiaxed grains can also be entrained by convection currents in the liquid. Their trajectory will not follow exactly that of the surrounding liquid, leading to macrosegregation (see Chap. 14). The interactions between fluid flow and grain structure is a very complex topic, and we provide only a brief introduction in this section.

The most obvious effect that convection can have on grain structure formation is a *modification of the thermal field*. Chapters 4 and 14 describe natural convection and its influence on the shape of the isotherms through

[1] A ⟨100⟩ pole figure is produced by placing each grain at the center of the unit sphere shown in Fig. 11.15 and extending all of its ⟨100⟩ directions that intersect the Northern hemisphere. The intersection points, 3 for an arbitrary orientation, are then projected onto the equatorial plane using a line connecting each intersection point to the South pole. In Fig. 11.26(c), the North pole is at the center of the pole figure, corresponding to the longitudinal axis of the blade, i.e., close to the thermal gradient direction. For each grain that has a ⟨100⟩ direction close to the North pole, two other ⟨100⟩ points are located near the equator given by the outer circle.

advection of heat. This will in turn affect the speed of the isotherms v_T and the thermal gradient G, especially in the liquid phase where convection is the strongest. In particular, convection tends to decrease the thermal gradient in the liquid, and as described in Fig. 11.24, this promotes equiaxed structures. In some cases, strong convection associated with metal injection can increase the thermal gradient in certain locations, which may promote the formation of columnar *twinned dendrites* in DC castings of aluminum. Twinned dendrites in Al alloys grow with $\langle 110 \rangle$ trunks split in their center by coherent (111) twin planes [22]. Secondary arms on both side of each trunk grow along $\langle 110 \rangle$ or$\langle 100 \rangle$ directions.

In the absence of convection and heterogeneous nucleation in the bulk liquid, one would expect to find fully columnar structures. When convection is present, however, it can promote the formation of nuclei by a process called *dendrite fragmentation*. When the fluid velocity v_ℓ has a component along the thermal gradient larger than the speed of the isotherms v_T, local remelting occurs, as illustrated in Fig. 11.27. Since solute-rich liquid (for $k_0 < 1$) is transported from deep in the mushy zone toward the liquidus isotherm, the average composition at a given height of the mushy zone increases (e.g., from the open circle to the filled square in Fig. 11.27), leading to partial solutal remelting of secondary arms followed by their detachment from the primary trunks. These fragments are then entrained by the flow and may grow or remelt, depending upon the thermal conditions. Near the top of a DS ingot or at the thermal center of a casting, the temperature in the remaining liquid is fairly uniform (i.e., $G_\ell \approx 0$) and equal to that of the columnar dendrite tips. Since the remaining liquid is undercooled, dendrite fragments can survive such conditions, leading to equiaxed grains in such regions. Note also that detached fragments that subsequently remelt consume latent heat from the melt, leading to an overall reduction in the superheat of the remaining liquid.

Fluid flow is not the only way to obtain fragments from a columnar zone. A sudden change in thermal conditions, e.g., induced by thermal contraction of the solid shell and the subsequent loss of thermal contact

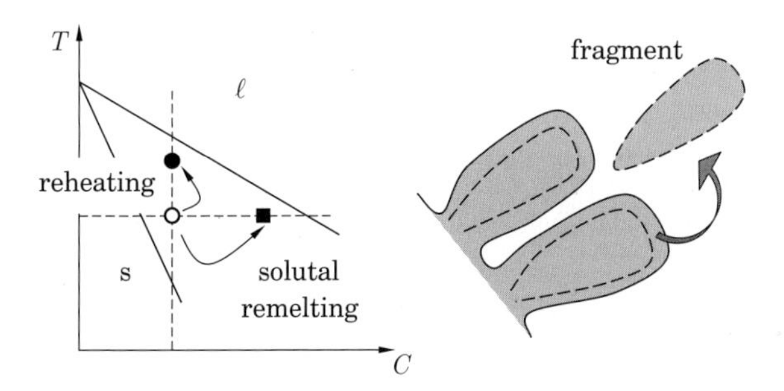

Fig. 11.27 A schematic of the remelting of a secondary dendrite arm as well as fragmentation due to reheating (open circle to filled circle) or a local change in composition (open circle to filled square).

with a cold mold, can induce remelting (Fig. 11.27, open circle to filled circle). Figure 11.28 shows such partial remelting and detachment of side arms in columnar dendrites in a succinonitrile-camphor alloy after an abrupt change in growth rate and thermal gradient during Bridgman solidification. Fragmentation by thermal effects without convection is of course much less effective in promoting a transition from columnar to equiaxed morphologies, because the fragments remain more or less in place. This effect can nevertheless be detrimental to the production of single crystal turbine blades, leading to so-called *zebra grains* that are slightly misoriented with respect to the main grain.

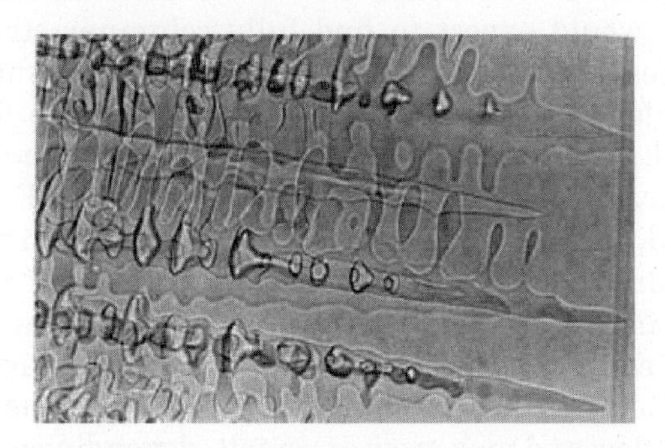

Fig. 11.28 The partial remelting and detachment of dendrites side arms in the succinonitrile-camphor alloy during a change in growth rate and thermal gradient during Bridgman solidification. (After Sato et al. [23].)

The phenomenon of fragmentation by fluid flow can be used to advantage in industrial processes for refining the grain structures and producing more homogeneous properties. In many continuous or semi-continuous casting processes, electromagnetic stirring (EMS) of the melt is applied for this purpose. Figure 11.29 shows macrostructures in two 20-mm diameter Cu-Sn continuously cast bars, one produced with and and the other without EMS. When no EMS was used, fully columnar grains grew from the surface in contact with the graphite mold, leading to undesirable centerline

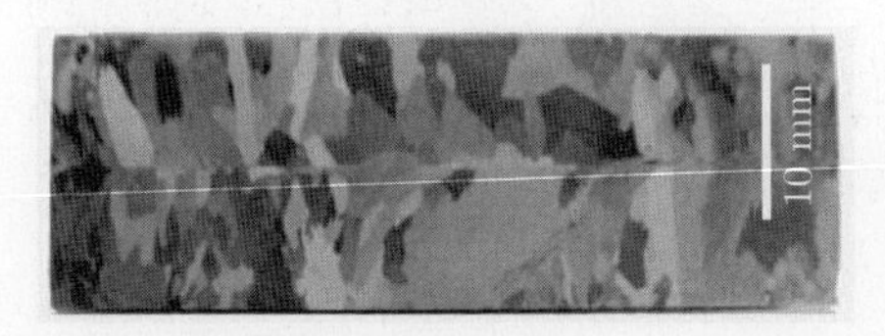

(a) Without EMS

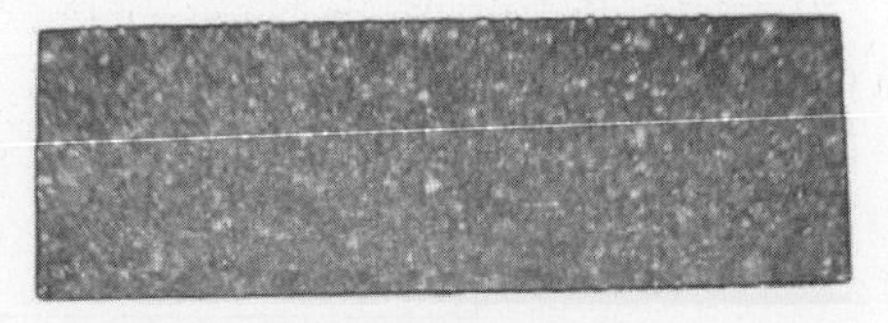

(b) With EMS

Fig. 11.29 Grain structures in continuously cast Cu-Sn base alloys (a) without and (b) with electromagnetic stirring. (Courtesy of Swissmetal Boillat.)

segregation. Applying EMS under otherwise identical conditions produced a very fine structure, with no centerline segregation.

Convection also alters the kinetics and direction of dendrite growth, as explained in Sect. 8.5.1, where it was shown that the growth kinetics depends on the growth Péclet number $v^* R/(2D_\ell)$ as well as on the fluid flow Péclet number $v_\ell R/(2D_\ell)$. For an equiaxed dendritic grain moving with a velocity $(\boldsymbol{v}_s - \boldsymbol{v}_\ell)$ relative to the liquid, we saw that the growth is enhanced on the leading edges and retarded on the trailing edges. As an example of this phenomenon, Fig. 11.30(a) shows the shape of a denser NH_4Cl crystal falling while growing in an undercooled NH_4Cl-water melt at uniform temperature. The falling crystal experiences an apparent incoming flow, causing the downward pointing primary arm to grow faster than the upward pointing one. Primary arms growing perpendicular to the relative movement have their orientation biased downward slightly by the same mechanism. This also occurs for dendrite trunks of columnar grains growing in a flow parallel to the isotherms, as discussed in Sect. 8.5.2.

In order to model the structures that result from the movement of grains during solidification, one must also take into account the momentum transfer between the solid and liquid phases. Consider again the single grain shown in Fig. 11.30(a), growing and settling in an otherwise quiescent liquid. The settling velocity v_s can be computed from the equilibrium between buoyancy and drag forces. The buoyancy force is given by $gV_s(\rho_s - \rho_\ell)$, where V_s is the volume of the solid. The drag force on the dendritic grain is complicated to express analytically, because the flow passes both through and around the grain. Beckermann and co-workers deduced the drag force for an equiaxed cruciform dendritic grain by observing the settling and growth of ammonium chloride dendritic grains in an

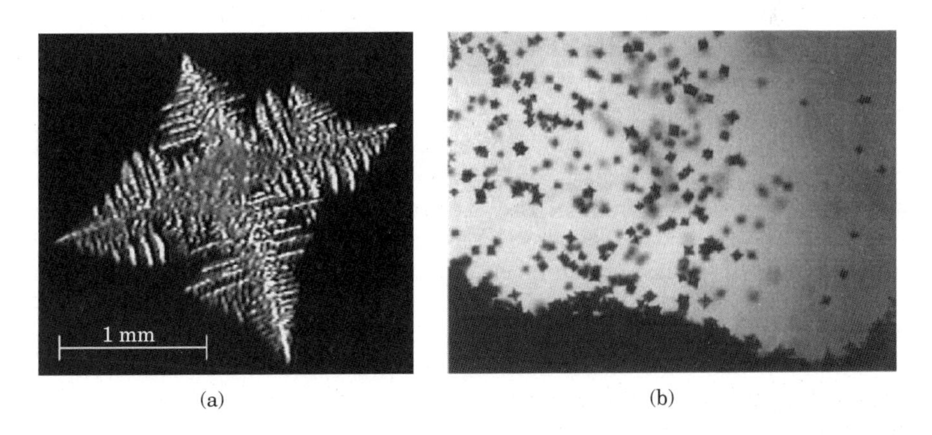

Fig. 11.30 (a) The shape of an ammonium chloride dendritic crystal growing and settling in an undercooled NH_4Cl-water melt. (After Badillo et al. [3]) (b) The movement and settling of a family of ammonium chloride crystals in a rectangular cavity cooled from all sides. (After Beckermann and Wang [4].)

undercooled bath. They expressed their results in terms of the grain envelope model shown in Fig. 11.7, which gave rise to an expression for v_s given by

$$\frac{\rho_\ell v_s^2}{2}\frac{\pi R_{eq}^2 C_D}{KS_e} - gV_s(\rho_s - \rho_\ell) = 0 \tag{11.45}$$

where R_{eq} is the equivalent radius of a sphere of volume $V_s + V_d$, and C_D is a drag coefficient given by

$$C_D = \frac{12}{\text{Re}}(1 + 0.24\text{Re}^{0.687}) \quad \text{with} \quad \text{Re} = \frac{R_{eq}v_s}{\nu_\ell} = \frac{R_{eq}v_s}{\mu_\ell/\rho_\ell} \tag{11.46}$$

The term KS_e is defined as the ratio of the Stokes settling velocity of the dendritic grain v_s and the Stokes velocity v_{eq} of the equivalent sphere:

$$KS_e = \frac{v_s}{v_{eq}} = 1.26\log_{10}(6.135\xi)\frac{2\zeta^2 + 3(1 - \tanh(\zeta)/\zeta)}{2\zeta^2(1 - \tanh(\zeta)/\zeta)} \tag{11.47}$$

where $\zeta = R_{eq}/\sqrt{K}$ and K is the permeability of the dendritic grain. Permeability of dendritic structures is discussed in more detail in Chap. 12. The final parameter ξ is a factor that measures the spheroidicity of the grain

$$\xi = \frac{4\pi R_{eq}^2}{\mathcal{S}_V^{de}} \tag{11.48}$$

where $\mathcal{S}_V^{de}$ is the effective surface of the envelope of the grain, assumed to be a cruciform in 3D. The dependence of KS_e on ξ in Eq. (11.47) reflects the non-spherical shape of the dendrite envelope, whereas the dependence on ζ represents the penetration of the flow inside the envelope. In most cases, $\zeta \gg 1$ and the last term is approximately one. This means that the flow goes around the grain as if it were fully solid, but the drag force is nonetheless modified with respect to an equivalent solid sphere due to the cruciform shape of the grain.

Settling of a population of equiaxed grains in an otherwise stagnant liquid is much more complicated than the settling of just one grain. For such cases, one must consider the exchange of momentum between both phases. The interested reader is referred to the works by Beckermann and co-workers given at the end of this chapter for further treatment of this problem. It is sufficient for our purposes to say that the expression for the exchange of momentum between the two phases is similar to that described for a single settling grain, provided that the velocity is replaced by the relative velocity $(\boldsymbol{v}_s - \boldsymbol{v}_\ell)$, and the density of the grains is accounted for.

As an example of the application of such two-phase models, Fig. 11.31 shows results from a simulation of solidification in a rectangular cavity containing a NH_4Cl-H_2O alloy, which is cooled on all faces. This simulation models the experiments of Jackson and Hunt in this system. Figure 11.31(a) shows the velocity field and the solid fraction (black region

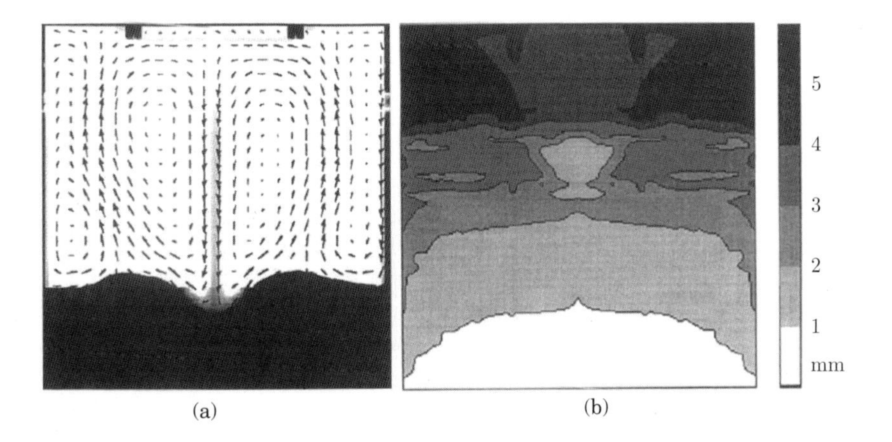

Fig. 11.31 (a) The volume fraction of solid and the velocity field in NH_4Cl-70wt%H_2O during solidification. (b) The corresponding average grain size after solidification. (After Beckermann and Wang [4].)

corresponds to $g_s > 0.03$) at one time during solidification. The buoyancy flow due to the cold vertical walls induces counter-rotating vortices in the center of the cavity. This flow is reinforced by the settling of the grains. The descending liquid at the center and periphery of the cavity creates depressions in the sedimented region where equiaxed grains accumulate, whereas a rising flow has a tendency to create mounds, as observed in experiments. The final average grain size, shown in Fig. 11.31(b), displays smaller grains located near the bottom of the cavity and larger ones above. This example is continued in Chap. 14, where it is demonstrated that one can explain the accumulation of ammonium chloride-rich grains in this region, leading to a water-lean macrosegregation, and where we also describe the adaptation of stochastic models for solidification to include grain movement, liquid flow and macrosegregation.

11.7 SUMMARY

This chapter has demonstrated how the concepts of nucleation and growth kinetics can be integrated into a macroscopic approach to solidification. This so-called micro-macro coupling is fairly straightforward for columnar structures, as their growth rate is directly related to the speed of the isotherms. Equiaxed microstructures, for which the growth rate is proportional to the local undercooling, require a combination of nucleation and growth kinetics models. The concept of extended volume fraction of solid and grain impingement was introduced for eutectics, where the diffusion layer thickness (on the order of the lamellar spacing) is much smaller than the grain size. We called this form "hard impingement." The growth of equiaxed primary dendritic phases must be handled more thoroughly due to the ability of the grains to gradually change their morphology from globulitic to dendritic as solidification proceeds. The transition between these two morphologies was analyzed. Grains that remain globular can be treated similarly to equiaxed eutectic ones, provided that the solute boundary layer surrounding the grains (comparable to the grain radius) is accounted for. This was referred to as "soft impingement." Fully dendritic structures were described using the concepts of grain envelope and internal volume fraction of solid, first introduced in Chap. 10. The internal volume fraction of solid was shown to be on the order of the supersaturation. The conditions leading to a transition from columnar to equiaxed structures were also analyzed. In cases where analytical models provide a sound basis for the understanding of each type of morphology, stochastic simulations based on cellular automata were able to combine all of them into a single model. These approaches also provided a direct representation of the grain structure and the possibility of predicting the grain selection and competition occurring during columnar growth. Finally, the effects of convection on grain structures were briefly described.

11.8 EXERCISES

Exercise 11.1. Biot number.
Write the heat balance for a representative element with volume V, losing heat at its surface of area A. Show that the criterion for the temperature to be considered uniform within the element is that the Biot number, $\mathrm{Bi} = h_T V/Ak$, is small as compared to one. h_T is the heat transfer coefficient (Eq. (6.15)) and k is the thermal conductivity.

Exercise 11.2. Cooling curve and recalescence.
Assume that equiaxed eutectic grains nucleate with a density n_g at time $t = 0$ in a small specimen of uniform temperature $T = T_{eut}$. Further assume that their interface velocity, as well as the heat extraction rate $\rho\dot{h} < 0$, is constant. Using Eq. (11.9) at an early stage of solidification, i.e., when $g_s \ll 1$, deduce the evolution of the solid fraction and of the temperature. Determine the time and temperature at which recalescence occurs. Why does one still see recalescence even though $\Delta T_n = 0$? What happens if, under the same assumptions, we set the initial temperature below the minimum of the recalescence just computed?

Exercise 11.3. Growth of globular grains.
If the grain size is smaller than the critical radius $R_{g0,c}$ given by Eq. (11.18), equiaxed grains of a primary phase grow with a globular structure. How could you adapt the equiaxed model developed for eutectic growth to describe this situation?

Exercise 11.4. Equiaxed dendritic grains.
Derive Eq. (11.25) for equiaxed dendritic growth.

Exercise 11.5. Eutectic grain growing in a thermal gradient.
Assuming a constant thermal gradient and cooling rate, show that the temperature at the warmer part of an equiaxed eutectic grain nucleated at an undercooling ΔT_n tends towards the steady-state undercooling of columnar dendrites, ΔT_{col}, in the form of a hyperbolic tangent. Demonstrate that it diverges on the coldest side as a tangent (Eq. (11.30)). What difference would you see if the grain is an equiaxed dendrite?

Exercise 11.6. Columnar dendritic growth.
Deduce the theoretical distribution of the angle ϕ characterizing the closest $\langle 100 \rangle$ direction with respect to a fixed direction n for randomly oriented grains with cubic symmetry (curve labeled "t" in Fig. 11.15).

Exercise 11.7. Columnar dendritic grain selection.
Determine the space separating the dendrite tip front between two grains in directional solidification when one grain has a $\langle 100 \rangle$ direction perfectly aligned with the thermal gradient G and the other has an angle ϕ between $\langle 100 \rangle$ and G. Discuss this relationship with respect to typical primary and secondary dendrite arm spacings (about 200 and 50 μm, respectively). Use the following data: $G = 100\,\mathrm{K\,cm^{-1}}$, $v_T = 1\,\mathrm{mm\,s^{-1}}$ and growth kinetics given by $v^* = 10^{-4}\,\mathrm{m\,s^{-1}K^{-2}}\,\Delta T^2$.

Exercise 11.8. Dendrite extension in a turbine blade platform.
Consider the experiment using an organic alloy illustrated in Fig. 11.17 and drawn schematically in Fig. 11.20. Assume that the reentrant corner has an extension L parallel to the $\langle 100 \rangle$ direction of the secondary dendrite arms. The tips of the $\langle 001 \rangle$ dendrites follow the liquidus isotherm and the temperature field is given by the equation: $T(x, y, z, t) = T_{liq} - G(v_T t - z)$, i.e., perfect Bridgman conditions of flat isotherms moving at a constant speed v_T in a uniform thermal gradient G. Assuming growth kinetics $v = A\Delta T^2$ for the dendrite tips and arms, estimate the maximum undercooling of the secondary arms at the end of the reentrant corner. What is the influence of the thermal and geometric parameters on the tendency to form a spurious grain at this corner?

Exercise 11.9. Dendritic growth in a single crystal spot weld.
Consider a single crystal with an (001) surface on which a spot weld is made. In the absence of convection, the shape of the melt pool is approximated by a hemisphere. Based on Eq. (11.33) and a minimum undercooling criterion, sketch the dendritic pattern observed at the top surface and in a (100) cross section. What happens if the top surface is a (110) surface?

Exercise 11.10. Outer equiaxed zone of a casting.
Figure 11.1 mentions an outer equiaxed zone close to the surface of the mold. Such a zone is not always observed in castings, however. Based on Hunt's criterion (Sect. 11.5), how can you explain the formation of this zone, before columnar grains take over with the formation of another equiaxed zone at the center?

11.9 REFERENCES

[1] B. Appolaire, H. Combeau, and G. Lesoult. Modeling of equiaxed growth in multicomponent alloys accounting for convection and for the globular/dendritic morphological transition. *Mater. Sci. Eng.*, 487:33–45, 2008.

[2] A. Badillo and C. Beckermann. Phase-field simulation of the columnar-to-equiaxed transition in alloy solidification. *Acta Mater.*, 54:2015, 2006.

[3] A. Badillo, D. Ceynar, and C. Beckermann. Growth of equiaxed dendritic crystals settling in an undercooled melt, Part 1: Tip kinetics. *J. Cryst. Growth.*, 309:197, 2007. Part 2: Internal solid fraction. Ibid, 309:216, 2007.

[4] C. Beckermann and C. Y. Wang. Equiaxed dendritic solidification with convection: Part III. Comparisons with NH_4Cl-H_20 experiments. *Metall. Mater. Trans.*, 27A:2784, 1996.

[5] H.-J. Diepers and A. Karma. Globular-dendritic transition in equiaxed alloy solidification. In M. Rappaz, R. Trivedi, and C. Beckermann, editors, *Solidification Processes and Microstructures: A Symposium in Honor of Prof. W. Kurz*, page 369, TMS Publ., Warrendale, 2004.

[6] Ch.-A. Gandin, J.-L. Desbiolles, M. Rappaz, and Ph. Thévoz. A three-dimensional Cellular Automaton -Finite Element model for the prediction of solidification grain structures. (See also references therein.). *Metall. Mater. Trans.*, 30A:3153, 1999.

[7] Ch.-A. Gandin, M. Rappaz, D. West, and B. L. Adams. Grain texture evolution during the columnar growth of dendritic alloys. *Metall. Trans.*, 26A:1543, 1995.

[8] Ch.-A. Gandin and M. Rappaz. A coupled Finite Element - Cellular Automaton model for the prediction of dendritic grain structures in solidification processes. *Acta Met. Mater.*, 42:2233, 1994.

[9] Ch. A. Gandin, R. J. Schaefer, and M. Rappaz. Analytical and numerical predictions of dendritic grain envelopes. *Acta Mater.*, 44:3339, 1996.

[10] B. Giovanola and W. Kurz. Modeling of microsegregation under rapid solidification conditions. *Metall. Mater. Trans. A*, 21:260, 1990.

[11] J. D. Hunt. Steady state columnar and equiaxed growth of dendrites and eutectic. *Mater. Sci. Eng*, 65:75, 1984.

[12] W. Kurz and D. J. Fisher. *Fundamentals of solidification*. Trans. Tech. Publ., Aedermansdorf, Switzerland, 4th edition, 1998.

[13] W. Oldfield. A quantitative approach to casting solidification: freezing of cast iron. *Trans. ASM*, 59:945–960, 1966.

[14] M. Rappaz and W. J. Boettinger. On dendritic solidification of multicomponent alloys with unequal liquid diffusion coefficients. *Acta Met. Mater.*, 47:3205, 1999.

[15] M. Rappaz, Ch. Charbon, and R. Sasikumar. About the shape of eutectic grains solidifying in a thermal gradient. *Acta Met. Mater.*, 42:2365, 1994.

[16] M. Rappaz, S. A. David, J. M. Vitek, and L. A. Boatner. Development of microstructures in Fe-15Ni-15Cr single-crystal electron beam welds. *Metall. Trans.*, 20A:1125, 1989. See also: *Metall. Trans.*, 21A:1767, 1990.

[17] M. Rappaz, Ch.-A. Gandin, J.-L. Desbiolles, and Ph. Thévoz. Prediction of grain structures in various solidification processes. *Metall. Mater.Trans.*, 27A:695, 1996.

[18] M. Rappaz and Ph. Thévoz. Solute diffusion model for equiaxed dendritic growth: Analytical solution. *Acta Met.*, 35:2929, 1987. See also: *Acta Met.*, 35:1487, 1987.

[19] M. Rappaz. Modelling of microstructure formation in solidification processes. *Int. Mater. Rev.*, 34:93, 1989.

[20] G. Reinhart, N. Mangelinck-Noel, H. Nguyen-Thi, T. Schenk, J. Gastaldi, B. Billia, P. Pino, J. Hartwig, and J. Baruchel. Investigation of columnar-equiaxed transition and equiaxed growth of aluminium based alloys by X-ray radiography. *Mater. Sci. Eng A*, 413-414:384, 2005.

[21] R. S. Rerko, H. C. de Groh III, and C. Beckermann. Effect of melt convection and solid transport on macrosegregation and grain structure in equiaxed Al-Cu alloys. *Mater. Sci. Eng*, A347:186, 2003.

[22] M. Salgado-Ordorica and M. Rappaz. Twinned dendrit growth in binary aluminum alloys. *Acta Mater.* 56:5708, 2008.

[23] T. Sato, W. Kurz, and K. Ikawa. Experiments on dendrite branch detachment in the succinonitrile-camphor alloy. *Trans. Japan Inst. Met.*, 28:1012, 1987.

[24] D. M. Stefanescu. Solidification and modeling of cast iron: a short history of the defining moments. *Mater. Sci. Eng A*, 413-414:322, 2005.

[25] H. Takatani, C.-A. Gandin, and M. Rappaz. EBSD characterisation and modelling of columnar dendritic grain growing in the presence of fluid flow. *Acta Mater.*, 48:675, 2000.

[26] Ph. Thévoz. *Modélisation de la solidification dendritique équiaxe.* PhD thesis, EPFL, Lausanne, Switzerland, 1988. Pdf available at http://library.epfl.ch/theses/.

[27] D. Walton and B. Chalmers. The origin of the preferred orientation in the columnar zone of ingots. *Trans. Metall. Soc. AIME*, 215:447, 1959.

[28] C. Y. Wang and C. Beckermann. Prediction of columnar to equiaxed transition during diffusion-controlled dendritic alloy solidification. *Metall. Mater. Trans.*, 25A:1, 1994.

[29] C. Y. Wang and C. Beckermann. Equiaxed dendritic solidification with convection. (See also references therein). *Metall. Mater. Trans.*, 27A:2765, 1996.

[30] Y. Z. Zhou, A. Volek, and N. R. Green. Mechanism of competitive grain growth in directional solidification of a nickel-base superalloy. *Acta Mater.*, 56:2631, 2008.

[31] J. Zou and M. Rappaz. Experiment and modeling of gray cast iron solidification. In V. R. Voller, M. S.Stachowicz, and B. G. Thomas, editors, *Materials Processing in the Computer Age*, page 335, TMS Publ., Warrendale, 1991.

Part III

Defects

Compared to forming processes, solidification of a melt can produce very complex shapes in a single operation, thus limiting the number of subsequent joining steps. However, this major advantage is offset in some cases by the formation of *defects*, which may prevent the use of castings for high integrity automotive parts, such as suspension arms, or aircraft components such as retractable landing gear. Among those defects, *porosity* and *hot tears* or *hot cracks* are the most serious. They are both induced by a lack of feeding of the mushy zone, i.e., some volume changes cannot be compensated by liquid flow and voids form. Unless closed by a hot isostatic pressing (HIP) operation, these voids can initiate premature fractures.

In the case of porosity, the volume change is associated with *solidification shrinkage*. In most materials, the liquid is less dense than the solid. The difference in density in turn requires a flow from the liquid, through the mushy zone to the regions where final solidification takes place. The pressure decreases with distance into the mush, due to viscous losses, and if the pressure loss is large enough, a pore might nucleate and grow. This type of porosity is often referred to as *shrinkage porosity* because it occurs deep in the mushy zone and is primarily induced by the density difference. Such pores become filled by gaseous elements that come out of the melt. Among those elements, hydrogen is the most important one in metallic alloys. Hydrogen becomes dissolved in the melt from the reaction between the melt and water either adsorbed at the surface of the mold or simply contained in the air, thus leading to the formation of oxides and hydrogen. Other gases are also important in some alloys. Examples include nitrogen in nickel-base superalloys and carbon monoxide in steel. If the composition of such gaseous elements in the melt is high, voids can form fairly early in the mushy zone and are referred to as *gas porosity*. Although the pore morphology and the main mechanism leading to their formation are quite different, we will see in Chap. 12 that both types of porosity are described using the same governing phenomena and equations.

In the case of hot tearing, the volume change is associated with *strains* in the partially coherent solid, which add to solidification shrinkage. These strains are induced by the contraction of the solid in a thermal gradient and are localized at the weakest parts of the mushy zone, the grain boundaries. Although the equations governing both defects are similar, the topic of hot tearing is presented in Chap. 13.

Another problem encountered in solidification processes is the inhomogeneity of compositions at large scale, known as *macrosegregation*. Indeed, such defects cannot be eliminated after solidification since they involve solute diffusion at long range and thus very long homogenization times (typically a few centuries for diffusion distance on the order of 10 cm!). Although the melt itself usually has a uniform composition, microsegregation at the scale of the microstructure induces local variations of the concentration. If the enriched (or depleted) liquid moves with respect to the solid, this leads to a final average composition that differs from the nominal one. Among the various sources of macrosegregation, we will distinguish in Chap. 14 four mechanisms: solidification shrinkage, natural or forced convection, grain sedimentation and deformation of the solid.

CHAPTER 12

POROSITY

12.1 INTRODUCTION

One of the most important types of defects that can form in cast products is porosity. There are many types of porosity (see for example the book by Campbell [1]), but the most severe is *microporosity*, the subject of this chapter. For convenience, we will use the terms porosity and microporosity interchangeably. Microporosity has a strong negative effect on mechanical properties, especially on ductility and fatigue life, because internal pores act as local stress concentrators and crack initiation sites. There are two main causes of microporosity in castings: *shrinkage porosity*, due to the volume change upon solidification combined with restricted feeding of liquid to the final solidification region; and *gas porosity*, due to the condensation of dissolved gases in the melt upon freezing, as a result of the difference in solubility of such gases in the liquid and solid phases.

The two types of porosity manifest themselves in different ways: where they form in the mushy zone, their morphologies and the extent to which they interconnect are quite distinct. This chapter is devoted to the understanding and control of both types of porosity.

Figure 12.1(a) shows a micrograph of an Al-10wt%Cu alloy that was allowed to remain in an atmosphere of N_2-10%H_2 for an extended period of time, causing it to become saturated with hydrogen. Figure 12.1(b) shows shrinkage porosity in an Al-4.5wt%Cu alloy. Although both alloys exhibit microporosity, the morphology of the pores is clearly different. As one can see in Fig. 12.1(a), the pore formed at an early stage of solidification, perhaps even in the liquid phase, due to the melt being saturated with hydrogen. The pore appears as a nearly spherical hole about 300μm in diameter, with dendrites growing around it, thus indicating that the pore existed prior to the development of the dendritic network. The pore is initiated by the presence of excess dissolved hydrogen in the melt, caused by the rejection of hydrogen by the solid. This is a typical case of "*gas porosity*". In contrast, Fig. 12.1(b) shows several very fine pores (in black), which appear in the interstices between the primary dendrites (in gray). The Al-Al_2Cu eutectic appears as white in this image, obtained using X-ray

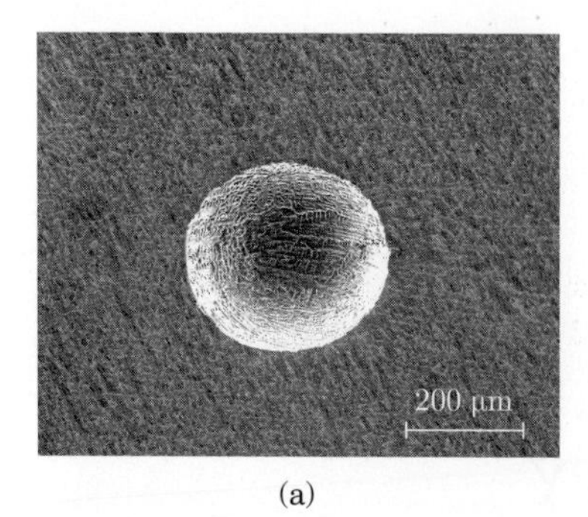

(a)

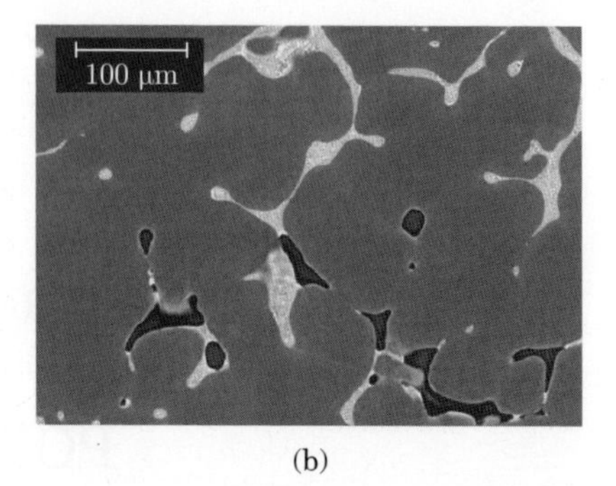

(b)

Fig. 12.1 (a) Gas porosity as observed by secondary electrons in scanning electron microscopy in an Al-10 wt%Cu alloy saturated with hydrogen in the liquid. (b) Shrinkage porosity observed by X-ray tomography in an Al-4.5 wt%Cu alloy. (Courtesy of M. Felberbaum.)

tomography. This is typical of "*shrinkage porosity*". Since the melt does not contain much dissolved gas in this case, the initial solidification occurs without the formation of bubbles as in Fig. 12.1(a). Due to its low copper content, the eutectic in this alloy forms at very high volume fraction of the primary phase (around 90%). The secondary phase, Al_2Cu, is much denser than the primary aluminum, which necessitates a significant amount of fluid flow through the dendritic network to compensate for the solidification shrinkage. Since feeding through densely packed dendrite arms is made difficult, shrinkage porosity appears as fine holes, typically a few tens of microns in size, but usually interconnected in a complex extended network, as described further below.

The formation of porosity in solidifying alloys can now be visualized *in-situ* using X-ray radiography or X-ray tomography. Lee and Hunt [11] were the first to apply this technique to visualize the formation of porosity in Al-Cu alloys, using a micro-focus X-ray source. They were able to observe the evolution of the pore density and the mean pore radius, even though the limited resolution available at that time (about 25 μm) and the beam intensity were fairly limiting factors. The availability of X-ray synchrotron radiation improved the situation dramatically. A resolution below 1 μm can be attained in several synchrotron facilities, and the beam intensity is high enough to allow *in-situ* X-ray tomography.

Figures 12.2 and 12.3 show time sequences obtained using such techniques, illustrating well the difference between gas and shrinkage porosity. The first set of images corresponds to the directional solidification of a 200 μm-thick Al-30wt%Cu alloy contained between two quartz plates. X-rays penetrated the specimen during growth, and their absorption was proportional to the atomic number of the elements. Thus, regions of low Cu concentration appear white and regions with high concentration appear dark. In this experiment, columnar dendrites grew downward at an angle

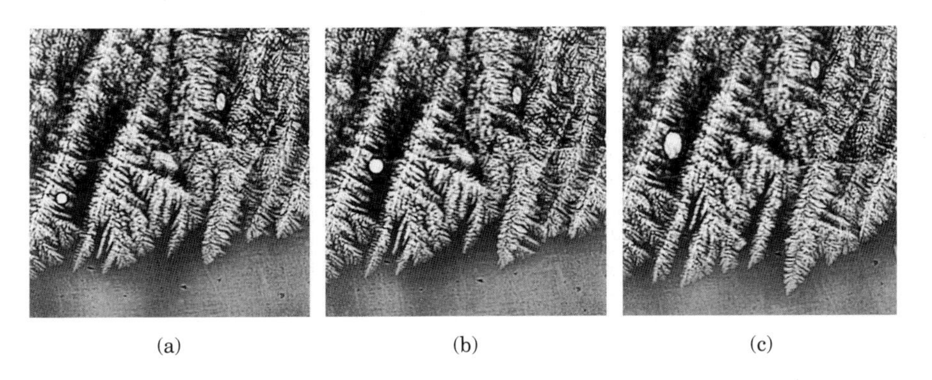

Fig. 12.2 A sequence of three X-ray radiographs showing the formation of gas porosity in a directionally solidified Al-30wt%Cu alloy, and its subsequent entrapment by the eutectic front. (Courtesy of R. Mathiesen, Sintef, Norway [12].)

of about 30° from the vertical thermal gradient. The liquid became enriched in Cu as solidification proceeded, and thus appears darker near the dendrite roots in comparison to the region ahead of the dendrite tips. The interdendritic eutectic front can be seen as a thin, nearly horizontal line at approximately mid-height of the images. Because the solidification interval is only about 10 K, the liquid can flow very easily in between the dendrites. Thus, shrinkage porosity is unlikely to form in this alloy and the small pores observed in these images are believed to be gas porosity.

Let us focus our attention on the pore located to the left in the images. At the onset of this sequence, a small pore nucleates and is trapped within the dendrite arms (small white spherical bubble). In fact, the origin of this bubble can be traced to earlier frames, revealing it to be an elongated air bubble at the surface of the mold, probably in a tiny groove, that was present prior to the start of solidification. As the dendrite front approached the air bubble, some liquid, probably induced by the density difference, caused the bubble to escape from the groove and become a spherical bubble trapped among the dendrites. It then moved toward the eutectic front in Fig. 12.2(b) under the influence of buoyancy forces. As the eutectic front passed around the pore, it became more ellipsoidal (Fig. 12.2c). The smaller pore seen in the top right corner in this sequence followed a similar evolution. We can thus conclude that gas porosity forms early in the mushy zone, is associated with a certain gas precipitation (in this case the gas was already present) and is nearly spherical because it grows in a fairly open dendritic network.

In contrast to gas porosity, shrinkage porosity forms deep in the mushy zone in fairly large solidification interval alloys. It is induced primarily by the difficulty to provide the shrinkage flow required by the final solidification of a solid phase that is denser than the interdendritic liquid, e.g., the Al_2Cu phase in Al-Cu alloys. Figure 12.3 illustrates the formation of shrinkage porosity in an Al-10wt%Cu alloy through a sequence of micrographs obtained by X-ray tomography. In this case, the specimen

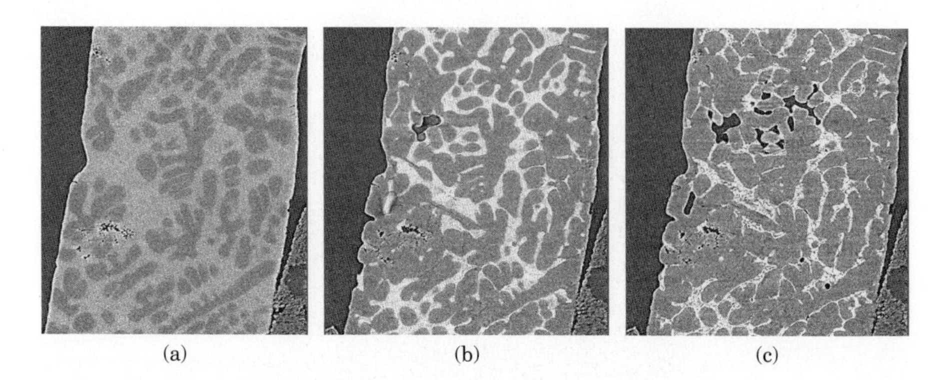

Fig. 12.3 A sequence of three X-ray tomography images from a longitudinal section, showing the formation of shrinkage porosity in a 1 mm diameter Al-10wt%Cu cylinder. (Courtesy of M. Suéry and L. Salvo, INPG, France.)

was about 1 mm in diameter and 2 to 3 mm long. The liquid was self-contained by its oxide skin during heating in an air-blowing furnace. It was then solidified by air cooling, while simultaneously being examined by X-ray tomography. Figure 12.3 shows a longitudinal section of the reconstructed 3D microstructure. As the specimen cools, one first observes the formation of equiaxed dendrites (Fig. 12.3a). Note that the contrast is opposite to that of the radiographs shown in Fig. 12.2, i.e., the dendrites are darker than the Cu-rich liquid. Some very fine porosity (black region) is already present at the early stages of cooling, probably associated with the presence of entrained oxide. Notice that the surface of the specimen has already deformed inward as a result of solidification shrinkage and thermal contraction of both the liquid and solid phases, producing a sharp cusp on the left, in a region where the dendrites are widely spaced, and a smooth S-shape in the upper right corner. The fraction of primary phase increases with time, and the lighter colored liquid is close to the eutectic composition at the time of Fig. 12.3(b). A pore with a size on the order of the secondary dendrite arm spacing subsequently can be seen to have nucleated in the upper left part of the specimen. The image in Fig. 12.3(c) suggests that new pores formed as the solidification of the eutectic phase was completed. The reconstructed 3D view of the porosity, shown in Fig. 12.4, clearly indicates that this was actually a single pore with a very complex shape, resulting from growth in the tortuous space left between the solid dendrites. This is a typical feature of shrinkage porosity.

Both types of porosity involve the nucleation and growth of pores, in one case due mainly to the evolution of dissolved gases, and in the other as a result of failure to feed solidification shrinkage. Although the two types each have different morphology, both can be described using the same set of governing equations to determine the pressure drop in the mushy zone associated with liquid feeding and the segregation/precipitation of dissolved gases. Models for the two cases are developed in Sects. 12.3 and 12.4–12.5, respectively. Before doing so, we first extend the average mass

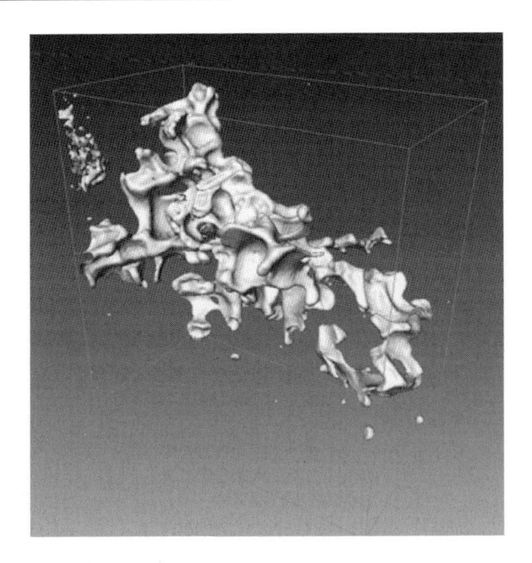

Fig. 12.4 A 3D view of the shrinkage micropore shown in Fig. 12.3 as obtained by X-ray tomography. (Courtesy of M. Suéry and L. Salvo, INPG, France.)

balance equation derived in Sect. 4.2 to the case where a compressible phase is present. Section 12.6 addresses the issue of boundary conditions to apply to the problem of liquid feeding, and Sect. 12.7 presents a few applications of the theory and models presented in this chapter.

12.2 GOVERNING EQUATIONS

Consider a fixed representative volume element (RVE) of the mushy zone within which the liquid, various solid phases and pores coexist. The average mass conservation equation for such a volume element, derived in Chap. 4, is

$$\frac{\partial \langle \rho \rangle}{\partial t} + \nabla \cdot \langle \rho \boldsymbol{v} \rangle = 0 \tag{12.1}$$

where the density $\langle \rho \rangle$ is averaged over the various phases

$$\langle \rho \rangle = \rho_\ell g_\ell + \rho_p g_p + \sum_{\nu}^{N} \rho_\nu g_\nu = \rho_\ell g_\ell + \rho_p g_p + \langle \rho \rangle_s g_s \tag{12.2}$$

The subscript p refers to the pore, and $\langle \rho \rangle_s$ is the density averaged over the N solid phases. Neglecting the density of the gaseous phase, and introducing the pore-free average density $\langle \rho_0(T) \rangle = (\rho_\ell g_\ell + \langle \rho \rangle_s g_s)/(g_\ell + g_s)$, the average density can be expressed as

$$\langle \rho \rangle = \langle \rho_0 \rangle (1 - g_p) \approx \langle \rho_0 \rangle - \rho_\ell g_p \tag{12.3}$$

since $g_p + g_s + g_\ell = 1$. Following the same procedure, the average mass flow $\langle \rho \boldsymbol{v} \rangle$ is given by

$$\langle \rho \boldsymbol{v} \rangle = \rho_\ell g_\ell \langle \boldsymbol{v} \rangle_\ell + \langle \rho \rangle_s g_s \langle \boldsymbol{v} \rangle_s \tag{12.4}$$

We have neglected the mass transport associated with the gaseous phase, and assumed that all of the solid phases have the same average velocity $\langle \boldsymbol{v} \rangle_s$ at the scale of the REV. Inserting Eqs. (12.3) and (12.4) into Eq. (12.1) yields

$$\frac{\partial \langle \rho_0 \rangle}{\partial t} - \rho_\ell \frac{\partial g_p}{\partial t} + \nabla \cdot (\rho_\ell g_\ell \langle \boldsymbol{v} \rangle_\ell) + \nabla \cdot (\langle \rho \rangle_s g_s \langle \boldsymbol{v} \rangle_s) = 0 \tag{12.5}$$

The term $g_p \partial \rho_\ell / \partial t$ has been neglected in Eq. (12.5) since it is much smaller than the others. The leading term, $\partial \langle \rho_0 \rangle / \partial t$, represents the change in density due to the combined effects of solidification shrinkage and thermal contraction of the solid and liquid phases. This change in density must be compensated by three phenomena: void growth in the liquid (second term), inward flow of the interdendritic liquid (third term) or compression of the solid phase (fourth term). If the solid phase is under a tensile load, arising for instance from external constraints, void growth and interdendritic flow must produce additional compensating volumes, which can lead to hot tears, as discussed in Chap. 13. Three cases are illustrated in Fig. 12.5: the compensation of shrinkage by interdendritic flow in (a), void formation in (b) and the additional effect of tensile strains in (c).

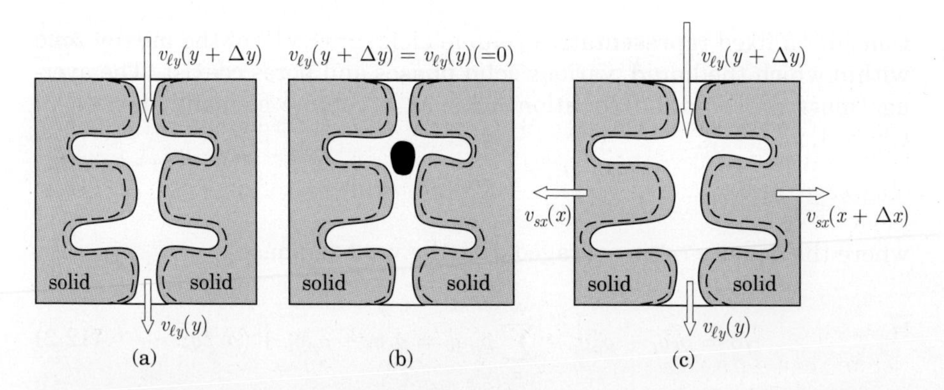

Fig. 12.5 Schematic views of microstructure and flow during directional solidification along the $y-$axis. In (a), solidification shrinkage occurs between times t (dashed interface) and $t + \Delta t$ (solid line) and is compensated by liquid flow along the $y-$axis. Notice that the incoming flow is larger than the outgoing flow. In (b), solidification shrinkage is compensated by pore growth alone. In (c), tensile strains on the dendrites perpendicular to the growth direction add to the solidification shrinkage to induce more liquid flow.

We focus on porosity for the remainder of this chapter, ignoring strains in the solid for now, to be taken up again in Chap. 13. The mass conservation equation then reduces to

$$\frac{\partial \langle \rho_0 \rangle}{\partial t} - \rho_\ell \frac{\partial g_p}{\partial t} + \nabla \cdot \left(\rho_\ell g_\ell \langle \boldsymbol{v} \rangle_\ell \right) = 0 \tag{12.6}$$

There are two unknown fields in this equation, the pore fraction $g_p(\boldsymbol{x}, t)$ and the velocity field $\langle \boldsymbol{v} \rangle_\ell (\boldsymbol{x}, t)$. We treat first the case of fluid flow in the mushy zone without pore formation, and then return to consider pore formation.

12.3 INTERDENDRITIC FLUID FLOW AND PRESSURE DROP

12.3.1 Darcy equation

When no voids form, $g_p = 0$ and Eq. (12.6) reduces to

$$\frac{\partial \langle \rho_0 \rangle}{\partial t} + \nabla \cdot \left(\rho_\ell g_\ell \langle \boldsymbol{v} \rangle_\ell \right) = 0 \tag{12.7}$$

In order to solve this equation, we treat the mushy zone as a porous medium, using the Darcy form of the momentum equation presented in detail in Chap. 4 (Eq. 4.114). This is a good approximation for our application, as the pressure variations associated with liquid flow in regions of low solid fraction are small. Darcy's Law relates the velocity to pressure drop

$$\langle \boldsymbol{v}_\ell \rangle = g_\ell \langle \boldsymbol{v} \rangle_\ell = -\frac{K}{\mu_\ell} \left(\nabla p_\ell - \rho_\ell \boldsymbol{g} \right) \tag{12.8}$$

where $\langle \boldsymbol{v}_\ell \rangle = g_\ell \langle \boldsymbol{v} \rangle_\ell$ is the superficial velocity of the liquid phase, K is the permeability of the mushy zone (assumed to be isotropic), μ_ℓ the viscosity of the interdendritic liquid and p_ℓ its pressure. Measurements of the permeability of alloys show that the data fit the Carman-Kozeny model given in Eq. (4.116)

$$K = \frac{1}{5(\mathcal{S}_V^{s\ell})^2} \frac{g_\ell^3}{(1 - g_\ell)^2} \tag{12.9}$$

Here, $\mathcal{S}_V^{s\ell} = A_{s\ell}/V_s$ is the intrinsic specific solid-liquid surface. A compilation of such data from various sources is shown in Fig. 12.6.

Combining Eqs. (12.6) and (12.8) gives

$$\nabla \cdot \left(\rho_\ell \frac{K}{\mu_\ell} \left(\nabla p_\ell - \rho_\ell \boldsymbol{g} \right) \right) + \rho_\ell \frac{\partial g_p}{\partial t} = \frac{\partial \langle \rho_0 \rangle}{\partial t} \tag{12.10}$$

Taking $g_p = 0$ gives a second-order partial differential equation for the pressure

$$\nabla \cdot \left(\rho_\ell \frac{K}{\mu_\ell} \left(\nabla p_\ell - \rho_\ell \boldsymbol{g} \right) \right) = \frac{\partial \langle \rho_0 \rangle}{\partial t} \tag{12.11}$$

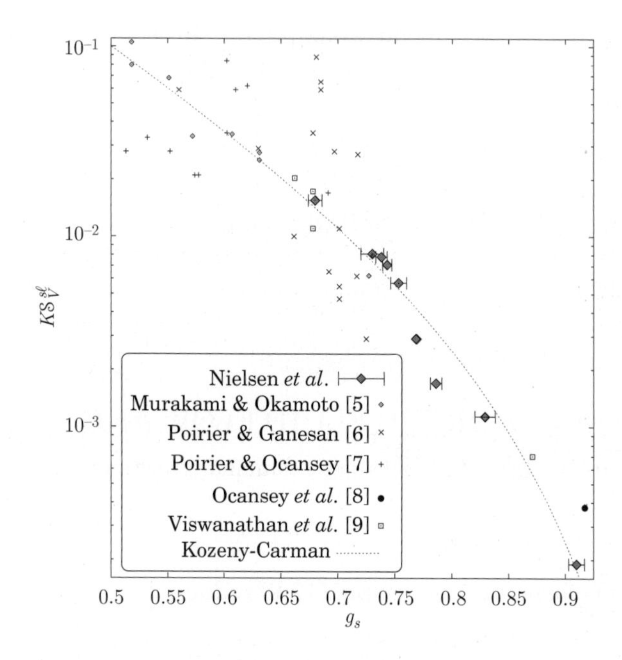

Fig. 12.6 The permeability K normalized by the square of the effective solid-liquid interface $\mathbb{S}_V^{s\ell}$, measured by a variety of authors for various volume fractions of solid g_s. The dotted curve corresponds to the Carman-Kozeny relationship. (Courtesy of Ø. Nielsen, Sintef, Norway [13].)

The solution of Eq. (12.11) requires boundary conditions at the limits of the mushy zone, i.e., where the first solid appears and the last interdendritic liquid disappears. As demonstrated later, one usually specifies the pressure at the liquidus front and the velocity of the last liquid to freeze. We first develop a simple one-dimensional model, which will lead to the so-called *Niyama criterion* for the formation of shrinkage porosity.

12.3.2 Niyama criterion

We showed in Chap. 4 that the mass balance at a solidifying interface requires that $\langle v \rangle_{\ell n} = -\beta v_n^*$, where $v_n^* = \boldsymbol{v}^* \cdot \boldsymbol{n}$ is the normal velocity of the solidification front and $\beta = (\rho_s - \rho_\ell)/\rho_\ell$ is the solidification shrinkage (Eq. (4.57)). Let us now consider one-dimensional steady-state columnar solidification, in which the temperature varies as $T(x,t) = T(x - v_T t)$, where v_T is the speed of the isotherms and x is perpendicular to the interface. For a pure substance or an alloy of eutectic composition, where the solidification front is isothermal, we obtain then $v^* = v_T$ and accordingly $\langle v \rangle_{\ell x} = -\beta v_T$. This case is illustrated on the left in Fig. 12.7. For an alloy that has a freezing range, we begin with Eq. (12.7) in 1D

$$\frac{\partial \langle \rho_0 \rangle}{\partial t} + \frac{\partial (g_\ell \rho_\ell \langle v \rangle_{\ell x})}{\partial x} = 0 \tag{12.12}$$

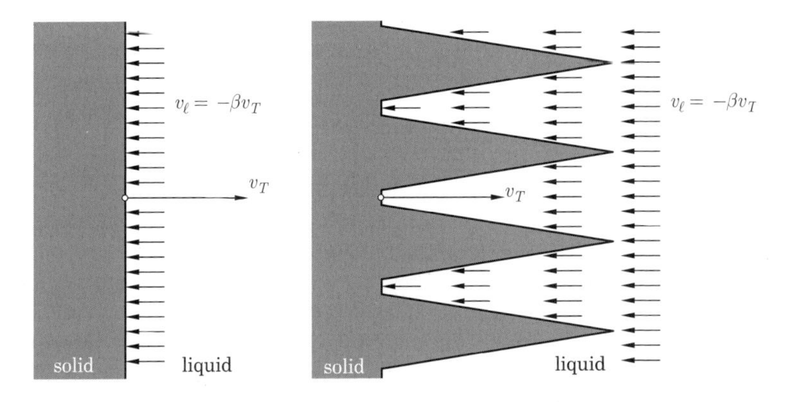

Fig. 12.7 Flow induced by shrinkage in a one-dimensional situation for two cases: (left) a planar front and (right) a schematic columnar growth in which g_s is a linear function of space.

In the steady-state reference frame attached to solid moving at velocity v_T, we have

$$\frac{\partial\langle\rho_0\rangle}{\partial t} = -v_T\frac{\partial\langle\rho_0\rangle}{\partial x} \tag{12.13}$$

Substituting this result in Eq. (12.12) gives

$$-v_T\frac{\partial\langle\rho_0\rangle}{\partial x} + \frac{\partial(g_\ell\rho_\ell\langle v\rangle_{\ell x})}{\partial x} = 0 \tag{12.14}$$

This implies that $-v_T\langle\rho_0\rangle + g_\ell\rho_\ell\langle v\rangle_{\ell x} = \mathrm{cst}$. If we now take both ρ_s and ρ_ℓ to be constant, but $\rho_s \neq \rho_\ell$, and we replace $\langle\rho_0\rangle$ by $(g_s\rho_s + g_\ell\rho_\ell)$, the constant cst can be determined to be $-\rho_s v_T$ by enforcing the boundary condition at the liquidus front: $\langle v\rangle_{\ell x}(g_\ell = 1) = -\beta v_T$. One then obtains the interesting result that

$$\langle v\rangle_{\ell x} = -\beta v_T \qquad \forall g_\ell \tag{12.15}$$

In words, the microscopic velocity of the fluid in the mushy zone is uniform under steady state conditions (Fig. 12.7, right) and only the superficial velocity $\langle v_{\ell x}\rangle = g_\ell\langle v\rangle_{\ell x}$, which is proportional to the mass flow rate, varies. The flow rate attains its maximum near the liquidus, since it has to compensate for the shrinkage of the entire mushy zone. As one moves deeper into the mushy zone, part of the flow has already been used to compensate shrinkage of the solid ahead of this location, leaving only the flow to compensate for the remaining part of the mushy zone. This is shown schematically in Fig. 12.7(right), where the arrows representing $\langle v\rangle_{\ell x}$ all have the same length, but their number decreases. Note that the velocity at the root of the dendrites when any interdendritic eutectic solidifies is still given by $-\beta v_T$. Note also that Eq. (12.15) may also be used when ρ_s is not constant, provided that a variable $\beta = (\rho_s/\rho_\ell - 1)$ is introduced. Equation (12.15) is not valid if ρ_ℓ is not constant.

After deriving this relationship, Niyama used the Darcy equation (Eq. 12.8) with $g = 0$ to write an expression in terms of the pressure. It is left as an exercise to show that integrating the resulting equation from the tip of the dendrites x_{liq} to any point $x(T)$ in the mushy zone yields

$$\Delta p_\ell = p_\ell(T_{liq}) - p_\ell(T) = \beta\mu_\ell v_T \int_{x(T)}^{x_{liq}} \frac{g_\ell(x)}{K(g_\ell(x))} dx = \frac{\beta\mu v_T}{G} \int_{g_\ell}^{1} \frac{g'_\ell}{K(g'_\ell)} \frac{dT}{dg'_\ell} dg'_\ell \tag{12.16}$$

where G is the thermal gradient. If a eutectic reaction occurs at temperature T_{eut}, the total pressure drop of the liquid across the mushy zone should be obtained by replacing the lower bound g_ℓ by the fraction of eutectic g_{eut} forming between the dendrites. The integral depends on the alloy only via the relationship $g_\ell(T)$ and the typical spacing of the microstructure entering into the permeability. This equation states that the pressure drop across the mushy zone is proportional to the solidification shrinkage β, the viscosity μ_ℓ and the ratio v_T/G, This last term can be understood by recognizing that increasing the velocity of the isotherms causes the interdendritic fluid to move faster, whereas an increase in the thermal gradient reduces the extent of the mushy zone and thus facilitates feeding. If one takes the Carman-Kozeny relationship for the permeability, Eq. (12.16) can be integrated analytically. Two special cases are considered:

- An idealized alloy in which there is no eutectic reaction and $dT/dg_\ell = \Delta T_0$ is constant, where $\Delta T_0 = (T_{liq} - T_{sol})$ is the solidification interval of the alloy. In this case, Eq. (12.16) simplifies to

$$\begin{aligned} \Delta p'_\ell &= \Delta p_\ell \frac{G}{5(S_V^{s\ell})^2 \beta\mu_\ell \Delta T_0 v_T} \\ &= \int_{g_\ell}^{1} \frac{(1-g'_\ell)^2}{g'^2_\ell} dg'_\ell = -g_\ell + g_\ell^{-1} + 2\ln g_\ell \end{aligned} \tag{12.17}$$

- An alloy following a Gulliver-Scheil microsegregation model, for which we relate g_ℓ and T using Eq. (10.23)

$$\frac{T - T_f}{T_{liq} - T_f} = g_\ell^{k_0 - 1} \tag{12.18}$$

with this relationship, $dT/dg_\ell = m_\ell C_0(k_0 - 1)g_\ell^{k_0-2} = k_0 \Delta T_0 g_\ell^{k_0-2}$ and the integration of Eq. (12.16) yields

$$\begin{aligned} \Delta p'_\ell &= \Delta p_\ell \frac{G}{5(S_V^{s\ell})^2 \beta\mu_\ell \Delta T_0 v_T} = k_0 \int_{g_\ell}^{1} \frac{(1-g'_\ell)^2}{g'^2_\ell} g'^{(k_0-2)}_\ell dg'_\ell \\ &= k_0 \frac{2 - g_\ell^{k_0-3}\left(k_0^2(1-g_\ell)^2 + k_0(1-g_\ell)(5g_\ell - 3) + 2(1 - 3g_\ell + 3g_\ell^2)\right)}{(k_0-1)(k_0-2)(k_0-3)} \end{aligned} \tag{12.19}$$

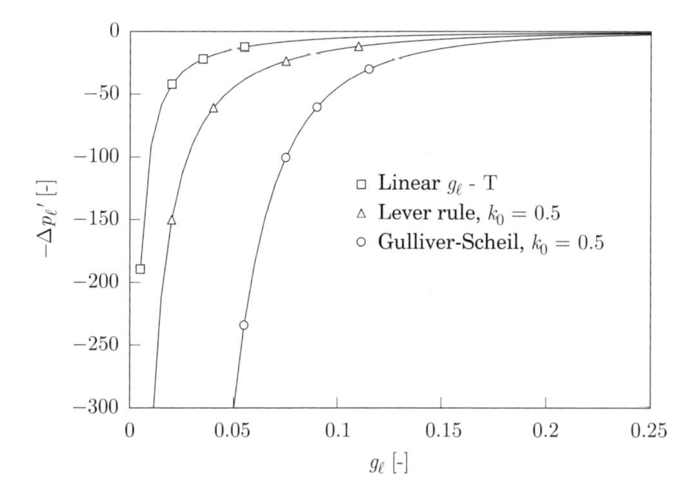

Fig. 12.8 The normalized pressure in the liquid as a function of g_ℓ for an alloy of solidification interval ΔT_0 with three microsegregation models $g_\ell(T)$. The pressure drop was normalized by the factor $5(\mathcal{S}_V^{s\ell})^2\beta\mu_\ell\Delta T_0 v_T/G$.

Figure 12.8 shows the pressure difference $-\Delta p_\ell = p_\ell(g_\ell) - p_\ell(1)$ normalized by the factor $5(\mathcal{S}_V^{s\ell})^2\beta\mu_\ell\Delta T_0 v_T/G$ for three cases: a linear $g_\ell - T$ relationship, the Gulliver-Scheil, and the lever rule with $k_0 = 0.5$ (for the lever rule case, the analytical solution is left as an exercise). As can be seen, the pressure in the liquid decreases fastest for the Gulliver-Scheil model since $g_\ell(x)$ increases most rapidly for that case. The pressure decreases significantly only over the last $10-15\%$ volume fraction of liquid. This illustrates the difficulty in accurately modeling the pressure drop in a mushy zone because it occurs over a very narrow range of solid fraction. For realistic values of the materials properties: $\mu_\ell = 10^{-3}$ Pa s, $\beta = 0.1$, $\mathcal{S}_V^{s\ell} = 6\lambda_2^{-1}$, $\lambda_2 = 50\,\mu$m, $\Delta T_0 = 100\,$K, $v_T = 1$ mm s^{-1} and $G = 10^3$ K m^{-1}, the normalization constant $5(\mathcal{S}_V^{s\ell})^2\beta\mu_\ell\Delta T_0 v_T/G = 720$ Pa. One can use Fig. 12.8 to deduce that an alloy solidifying according to the Gulliver-Scheil microsegregation model and having 10% final eutectic ($\Delta p_\ell' \approx 50$) experiences a pressure drop of about 36 kPa over the mushy zone.

Niyama noted that the isotherm speed is difficult to measure, whereas the cooling rate $\dot{T}$ and G are relatively easy. He therefore replaced v_T by $-\dot{T}/G$. The factor that scales Δp_ℓ thus becomes proportional to $G^2/\dot{T}$. Niyama then performed experiments to correlate the appearance of porosity in steel castings (Fig. 12.9) with this factor, from which he concluded that porosity occurs when:

$$\frac{G}{\sqrt{-\dot{T}}} < 1 \ \mathrm{K}^{1/2}\mathrm{min}^{1/2}\mathrm{cm}^{-1} \tag{12.20}$$

One should note that this criterion was developed specifically for shrinkage porosity in steels. Despite this fact, it is sometimes used for other alloys, even though no similar experimental correlations have been

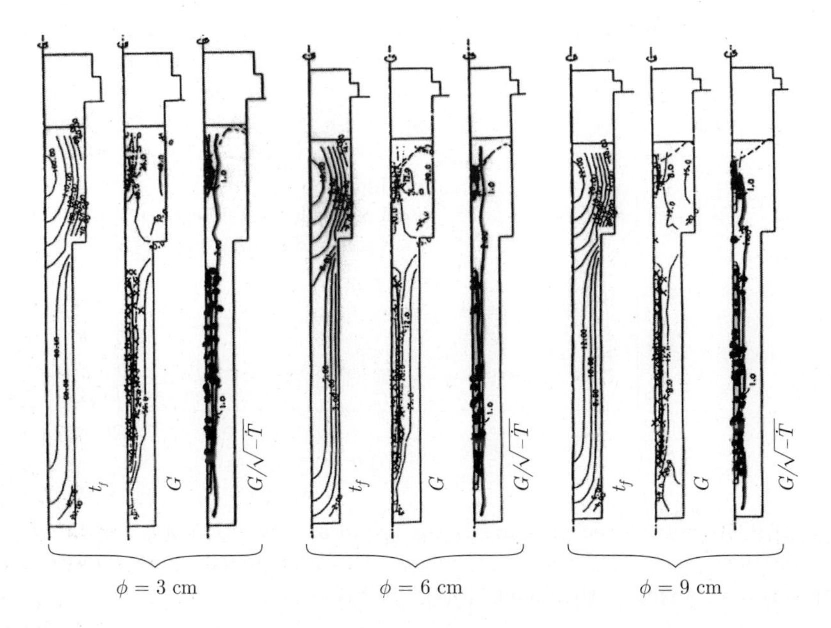

Fig. 12.9 Longitudinal sections of steel cylinders of various diameters ($\phi = 3$ (left), 6 (middle) and 9 cm (right)). For each cylinder, the location of porosity is indicated with crosses. Moreover, 3 types of iso-value maps have been drawn: the solidification time t_f (left), the thermal gradient G (middle) and the Niyama criterion $G/\sqrt{-\dot{T}}$ where $\dot{T}$ (denoted R in the original paper of Niyama), is the cooling rate. (After Niyama [14].)

done. We note that this criterion should never be applied to predict gas porosity, which forms by a completely different mechanism. When porosity is created, the pressure is relieved, as we shall see in the next section, but one then ends up with two unknown fields: the pressure in the liquid p_ℓ and the fraction of pores g_p. The relationship between the two requires a state equation to relate the amount of gas dissolved in the liquid and the pressure in the pore. This is addressed in the next section.

12.4 THERMODYNAMIC OF GASES IN SOLUTION

In order to understand gas porosity, it is important to first study the thermodynamic behavior of dissolved gases in liquids. To that end, we begin this section by describing the more familiar case of a bottle of carbonated water.

When the bottle is closed with a cap, the partial pressure p_g of CO_2 in the gas just below the cap is slightly larger than the atmospheric pressure, p_0. The gas reacts with water according to

$$CO_2 + H_2O \rightleftarrows H_2CO_3 \tag{12.21}$$

If we assume equilibrium between the species, the liquid has a composition C_{g0} of carbonic acid H_2CO_3 equal to its *solubility* composition $C^{eq}_{g\ell}(p_g, T_1)$. (Since one molecule of CO_2 gives one molecule of H_2CO_3, we hereafter use the term "gas composition" in the liquid). When one opens the bottle, the pressure of the gas falls to p_0 and the gas bubbles are observed to form in the liquid, usually at the glass wall. We conclude that the liquid is now supersaturated, so that $C_{g0} > C^{eq}_{g\ell}(p_0, T_1)$. This implies that the solubility limit of the gas in the liquid decreases as the pressure in the gas reservoir decreases. Suppose now that we pour the carbonated liquid over ice so that its temperature decreases from T_1 to $T_2 < T_1$. Experience shows that many more gas bubbles form, and we deduce from this that the solubility limit also decreases with temperature, i.e., $C^{eq}_{g\ell}(p_0, T_1) > C^{eq}_{g\ell}(p_0, T_2)$. Finally, an experiment not so frequently performed is to freeze the liquid at constant temperature and pressure. If one were to do so, bubbles would again form in the liquid. Since p_0 and T_2 were kept constant in our "thought experiment", we could conclude that the liquid became supersaturated because the solubility of the gas is lower in the solid than in the liquid. This implies that solidification increases the gas content in the liquid such that its composition exceeds the saturation limit, which remains fixed at $C^{eq}_{g\ell}(p_0, T_2)$. Summarizing these observations in equation form, we obtain

$$\frac{\partial C^{eq}_{g\ell}}{\partial p_g} > 0; \quad \frac{\partial C^{eq}_{g\ell}}{\partial T} > 0; \quad \frac{C^{eq}_{gs}}{C^{eq}_{g\ell}} = k_{0g} < 1 \tag{12.22}$$

These observations also hold for most gases that are dissolved in liquid metals.

Following this qualitative introduction, let us now be more quantitative with regard to the equilibrium of a molten metal, a solid and a gas, using the illustration in Fig. 12.10. Assume for the time being that the

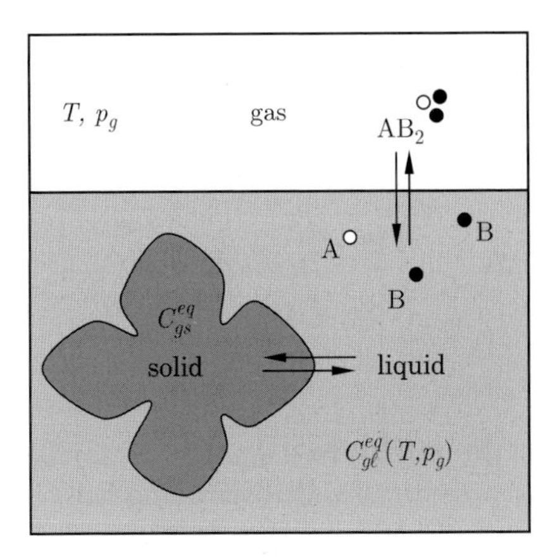

Fig. 12.10 The equilibrium between a gas, a liquid and a solid phase, with the schematic representation of a gas AB_2 and elements A and B in solution.

metal is pure, except for a few gas-forming elements that might be dissolved in ppm quantities. In order for the solid and liquid phases to coexist, the system must be at its melting point. At equilibrium, there will be an exchange between the three phases such that the chemical potentials of all the elements are equal, as discussed in Chap. 2. Let us focus first on the equilibrium between dissolved elements A and B and the liquid, for which we have

$$N_A \mathrm{A} + N_B \mathrm{B} \rightleftarrows \mathrm{A}_{N_A}\mathrm{B}_{N_B}(g) \tag{12.23}$$

where N_A and N_B are the stoichiometric coefficients for the formation of a molecule $\mathrm{A}_{N_A}\mathrm{B}_{N_B}$ in the gaseous phase. For example, one could have the reaction H + H $\rightleftarrows$ H_2 in most metals, the reaction C + O $\rightleftarrows$ CO in steel, etc. At equilibrium, the Gibbs free energy of the reaction in Eq. (12.23) is given by

$$G^m_{A_{N_A}B_{N_B}}(T,p_g) - N_A G^m_A(T,p_g) - N_B G^m_B(T,p_g) = 0 \tag{12.24}$$

The free energies of the components A and B dissolved in the liquid can be obtained from those of the elements in their "natural" state, i.e, undissolved gases, using the activity coefficients defined in Chap. 2 for a solute element dissolved in a solvent

$$\begin{aligned} G^m_A(T,p_g) &= G^{m0}_A(T,p_g) + \mathcal{R}T \ln a_{A\ell}(T,X_A) \\ G^m_B(T,p_g) &= G^{m0}_B(T,p_g) + \mathcal{R}T \ln a_{B\ell}(T,X_B) \end{aligned} \tag{12.25}$$

Note that the pressure-dependence of the chemical activities $a_{A\ell}$ and $a_{B\ell}$ has been omitted because the compressibility of the liquid is very low. Inserting Eqs. (12.25) into Eq. (12.24) gives

$$\begin{aligned} \Delta G^m_g(T,p_g) &= G^m_{A_{N_A}B_{N_B}}(T,p_g) - N_A G^{m0}_A(T,p_g) - N_B G^{m0}_B(T,p_g) \\ &= \mathcal{R}T(N_A \ln a_{A\ell} + N_B \ln a_{B\ell}) = \mathcal{R}T \ln\left((a_{A\ell})^{N_A}(a_{B\ell})^{N_B}\right) \end{aligned} \tag{12.26}$$

The term $\Delta G^m_g(T,p_g)$ is the Gibbs free energy for the formation of the gas phase at pressure p_g. A pressure dependence can be extracted by introducing the equation of state for the gas. For example, for an ideal gas, we have

$$\begin{aligned} \Delta G^m_g(T,p_g) &= \Delta G^m_g(T,p_0) + \int_{p_0}^{p_g} V^m dp = \Delta G^m_g(T,p_0) + \int_{p_0}^{p_g} \frac{\mathcal{R}T}{p} dp \\ &= \Delta G^m_g(T,p_0) + \mathcal{R}T \ln \frac{p_g}{p_0} \end{aligned} \tag{12.27}$$

where $\Delta G^m_g(T,p_0)$ is the Gibbs free energy of formation of the gaseous phase at the reference pressure p_0. Substituting Eq. (12.27) into Eq. (12.26), one finally obtains

$$\frac{p_g}{p_0} = (a_{A\ell})^{N_A}(a_{B\ell})^{N_B} \exp \frac{-\Delta G^m_g(T,p_0)}{\mathcal{R}T} \tag{12.28}$$

Recall from Chap. 2 that ΔG_g^m is defined in terms of the enthalpy and entropy of formation. We may thus write

$$\Delta G_g^m(T, p_0) = \Delta H_g^m(T, p_0) - T\Delta S_g^m(T, p_0) \tag{12.29}$$

The chemical activities of elements A and B in the liquid phase, $a_{A\ell}$ and $a_{B\ell}$, respectively, depend on the concentrations of gas-forming elements in the liquid as well as the chemical composition of the liquid. For an ideal solution, the chemical activities are simply equal to the molar compositions, $a_{A\ell} = X_{A\ell}^{eq}$ and $a_{B\ell} = X_{B\ell}^{eq}$. Chemical activity in dilute solutions is better represented by Raoult's law

$$a_{A\ell} = f_{A\ell}^o X_{A\ell}^{eq} \qquad \text{and} \qquad a_{B\ell} = f_{B\ell}^o X_{B\ell}^{eq} \tag{12.30}$$

where $f_{A\ell}^o$ and $f_{B\ell}^o$ represent the *activity coefficients* of species A and B, respectively, in the pure liquid. These coefficients are equal to 1 for an ideal solution, and are larger than 1 when the gas has a positive enthalpy of mixing, and smaller than 1 for negative mixing enthalpy.

Combining Eqs. (12.30) and (12.29), the equilibrium condition finally becomes

$$\frac{p_g}{p_0} = \left(A_g \exp\frac{-\Delta H_g^m(T, p_0)}{\mathcal{R}T}\right)(X_{A\ell}^{eq})^{N_A}(X_{B\ell}^{eq})^{N_B} \tag{12.31}$$

where

$$A_g = \left(\exp\frac{\Delta S_g^m(T, p_0)}{\mathcal{R}}\right)(f_{A\ell}^o)^{N_A}(f_{B\ell}^o)^{N_B} \tag{12.32}$$

For a diatomic gas, such as H_2, one recovers the well-known Sievert's law

$$\frac{p_{H_2}}{p_0} = A_{H_2} \exp\frac{-\Delta H_{H_2}^m}{\mathcal{R}T}(X_{H\ell}^{eq})^2 \tag{12.33}$$

or

$$X_{H\ell}^{eq} = (A_{H_2})^{-1/2} \exp\frac{\Delta H_{H_2}^m}{2\mathcal{R}T}\left(\frac{p_{H_2}}{p_0}\right)^{1/2} \tag{12.34}$$

We have omitted the variables (T, p_0) in the enthalpy of formation to reduce clutter. In words, Sievert's law states that the concentration of an element such as hydrogen dissolved in a liquid increases as $\sqrt{p_g}$. This is consistent with our observation in the simple carbonated liquid experiment. However, the temperature dependence is not so obvious: it depends on the sign of the enthalpy of mixing. If $\Delta H_g^m < 0$, the equilibrium concentration effectively decreases when T decreases. However, if $\Delta H_g^m > 0$, the concentration in the liquid increases when the temperature decreases, which is the opposite of what we observed in our carbonated water experiment.

Turning our attention now to the solid, the compositions of gas-forming elements in the solid phase can be obtained from those in the liquid using their respective partition coefficients

$$X_{As}^{eq} = k_{0A} X_{A\ell}^{eq}; \quad X_{Bs}^{eq} = k_{0B} X_{B\ell}^{eq} \tag{12.35}$$

In general, the partition coefficients k_{0A} and k_{0B} are functions of the thermodynamic variables (temperature, pressure, solute and gas concentrations). We will assume them to be constant, given by their values in the pure metal at the melting point.

In the case of an alloy, we must consider the effect of solute elements on the equilibrium between the gas, the liquid and the solid phases. We assume that the partition coefficients for gas-forming elements are identical to those of the pure metal, and it is therefore sufficient to consider only the equilibrium between the gas and the liquid. The activity coefficients in the alloy are modified from those in the pure melt according to [17]

$$f_{A\ell} = f_{A\ell}^{o} 10^{c} \tag{12.36}$$

where

$$c = \sum_{I=1}^{N_s} e_A^I C_{I\ell} + r_A^I (C_{I\ell})^2 \tag{12.37}$$

where e_A^I and r_A^I are the first- and second-order interaction coefficients between solute element I and the gas, respectively. Table 12.1 lists the values of these coefficients for various solute elements for hydrogen in aluminum alloys.

Using these data and Eqs. (12.31)–(12.37) we can compute the equilibrium solubility of H in several Al alloys as a function of the temperature. The result is plotted in Fig. 12.11. It can be seen that Si decreases the

Table 12.1 First and 2nd-order interaction coefficients for H in Al alloys. Other values for this system: A_{H_2} = 4.11 (cc_{STP}/100 g metal)$^{-2}$ and $\Delta H_{H_2}^{m}$ = 3000 Jmole^{-1}K^{-1}. (After Sigworth and Engh) [17].

Solute element I	e_H^I [wt%$^{-1}$]	r_H^I [wt%$^{-2}$]
Cerium	−0.08	0.0
Copper	0.03	−0.0004
Chromium	0.0	0.0
Iron	0.0	0.0
Magnesium	−0.01	0.0
Manganese	0.06	0.0
Nickel	0.04	0.0
Thorium	−0.006	0.0
Titanium	−0.1	0.0
Silicon	0.03	−0.0008
Tin	0.004	0.0

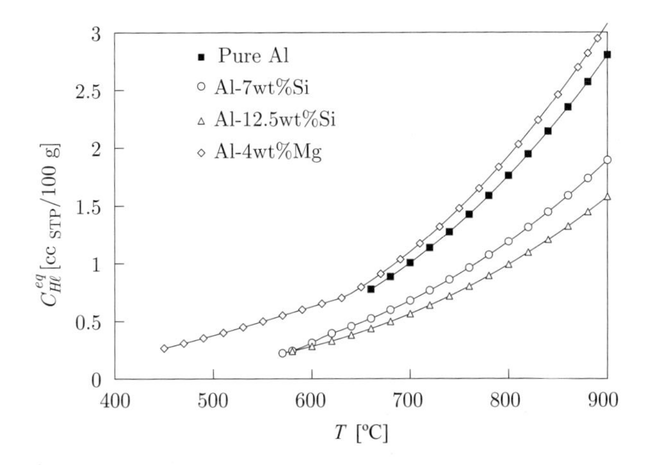

Fig. 12.11 The equilibrium composition of hydrogen in molten aluminum as a function of the temperature, showing the effect of Si and Mg additions.

solubility of hydrogen in aluminum alloys, whereas Mg increases it. Note the rather strange units "cc_{STP} per 100 g of metal" used to measure the solubility limit of hydrogen in Fig. 12.11. When these are used instead of mole fraction, the constant A_{H_2} in Eqs. (12.33) and (12.34) must have units of $(cc_{STP}/100\ g)^{-2}$! Knowing that 1 mole of gas $A_{N_A}B_{N_B}$ under standard conditions of pressure (1.013×10^5 Pa) and temperature (273.15 K) occupies 22.4 liters, the conversion between this unit into mass fraction and mole fraction is

$$\begin{aligned} \frac{1cc_{STP}}{100g} &\longrightarrow C_{I\ell} = \frac{N_I}{22.4 \times 10^3\ (\text{cm}^3/\text{mol})} \frac{\mathcal{M}_I\ (\text{g/mol})}{100\ g} \qquad \text{I} = \text{A, B} \\ \frac{1cc_{STP}}{100g} &\longrightarrow X_{I\ell} = \frac{N_I}{22.4 \times 10^3\ (\text{cm}^3/\text{mol})} \frac{\mathcal{M}_\ell\ (\text{g/mol})}{100\ g} \qquad \text{I} = \text{A, B} \qquad (12.38) \end{aligned}$$

where $\mathcal{M}_A$, $\mathcal{M}_B$ and $\mathcal{M}_\ell$, are the molecular weights of the elements composing the gas phase and the liquid. For hydrogen, $N_H = 2$ and $\mathcal{M}_H = 1$ g mol^{-1}. In this case, 1 cc_{STP}/100 g corresponds to 0.89×10^{-6}, i.e., approximately 1 ppm by weight. In aluminum, this is equivalent to 24 single atoms of hydrogen dissolved in 1 million atoms of Al. Although this does not seem like much, it is more than enough to create pores, as we shall see shortly.

The thermodynamics of gases dissolved in aluminum alloys is fairly simple, compared to other alloys, because hydrogen is the only gas that plays a significant role in the formation of porosity. Volatile solute elements such as Zn have been shown to have a negligible influence on the formation of porosity in such alloys. In nickel-base superalloys, the situation is more complicated, as both hydrogen and nitrogen can be dissolved to a significant extent in the melt. In steels, hydrogen, nitrogen and oxygen (that reacts with carbon to form CO) make the system even more complicated. In copper-based alloys, both hydrogen and oxygen can be dissolved in the melt, thus rendering the formation of water vapor possible. In brass alloys, the vapor pressure of zinc at the melting point of copper is fairly large and can contribute to the formation of porosity. For more practical information on the role of gases in the formation of porosity in various metallic alloys, the reader is referred to the book of J. Campbell [1].

The model presented above for a single gas-forming element has been extended to multiple dissolved gases in multi-component alloys by Poirier and co-workers [8] and by Couturier and Rappaz [4]. The equilibrium condition in this case is based on the *partial pressure* of each gas-forming species, since the total pressure in the gas phase is given by the sum of all the partial pressures. For the sake of simplicity, we restrict ourselves in the following section to the case where there is only one diatomic gas, such as H_2 or N_2. Equation (12.23) then simplifies to $2A \rightleftarrows A_2$, and we then refer to the "gas composition" of the liquid, $C_{g\ell}$ or $X_{g\ell}$ (instead of gas-forming element composition). Our goal is now to employ Sievert's law in order to model the nucleation and growth of pores.

12.5 NUCLEATION AND GROWTH OF PORES

During solidification of alloys, all of the conditions required for bubbles to form are present: the solubility of gas-forming elements in the liquid decreases, due to both the pressure drop through the mushy zone associated with feeding (Sect. 12.3), and to the temperature decrease in the mushy zone (see Fig. 12.11), while gas-forming elements are simultaneously and massively rejected into the liquid upon solidification because k_{0g} is much smaller than unity. Since dissolved gases diffuse rapidly, it is reasonable to use the lever rule to compute the local compositions in the solid and liquid phases.

Consider a volume element that is initially entirely liquid, having a volume V_0, gas composition C_{g0} and density $\rho_{\ell 0}$. At some time during solidification, the total volume is V. In the absence of pores, the solute balance over the volume can be written as

$$V_0 \rho_{\ell 0} C_{g0} = V \langle \rho_0 \rangle C_{g0} = V_s \rho_s C_{gs} + V_\ell \rho_\ell C_{g\ell} \tag{12.39}$$

where $V = (V_s + V_\ell)$ is the volume at a given time of solidification and $\langle \rho_0 \rangle = g_s \rho_s + g_\ell \rho_\ell$ is the average density of the pore-free material. Dividing by V and introducing the equilibrium condition $C_{gs} = k_{0g} C_{g\ell}$, one obtains

$$C_{g\ell} = \frac{\langle \rho_0 \rangle C_{g0}}{g_s \rho_s k_{0g} + g_\ell \rho_\ell} \tag{12.40}$$

Figure 12.12 presents an example of the dissolved H enrichment of the liquid during solidification computed using Eq. (12.40) for an Al-7%Si

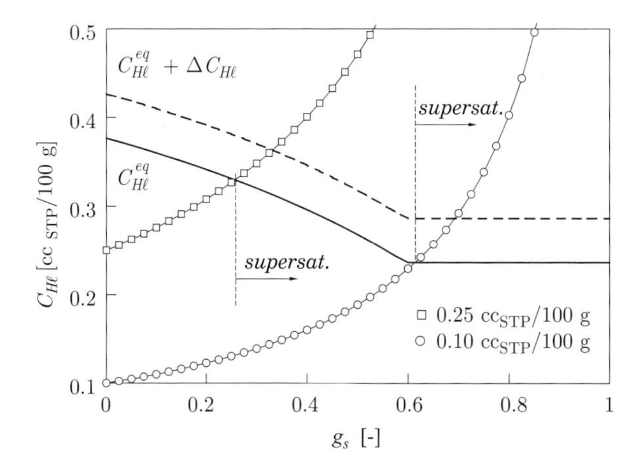

Fig. 12.12 The evolution of the hydrogen composition in the liquid, $C_{H\ell}(g_s)$, as a function of the volume fraction of solid for an Al-7wt%Si alloy, for two different initial compositions, $C_{H0} = 0.1$ and 0.25 cc_{STP}/100 g. The continuous solid curve corresponds to the equilibrium saturation composition $C_{H\ell}^{eq}(T(g_s))$ for p_0 = 1 atm. The arrows indicate the point at which liquid becomes supersaturated. The dashed curve corresponds to a supersaturation of 0.05 cc_{STP}/100 g over the equilibrium value.

alloy and two different values of C_{g0} (curves with symbols). The equilibrium hydrogen composition, $C_{H\ell}^{eq}$, shown as a function of T in Fig. 12.11, has been converted to be a function of g_s using a Gulliver-Scheil microsegregation model for this alloy (continuous line). The graph shows that for values of g_s where $C_{H\ell} > C_{H\ell}^{eq}$, the melt will be supersaturated with hydrogen, making it possible to precipitate pores. Notice that this occurs at a much lower g_s when the initial gas content is high ($C_{H0} = 0.25$ cc_{STP}/100 g). Notice that for $C_{H0} = 0.1$ cc_{STP}/100 g, the melt does not become supersaturated until eutectic solidification begins, near $g_s = 0.6$. Figure 12.12 also shows as a dashed line the points at which the melt becomes supersaturated by an amount $\Delta C_{H\ell} = 0.05$ $\text{cc}_{\text{STP}}/100$ g. This becomes important when we discuss pore nucleation and growth in the next section.

Before addressing this topic, we must first develop the relationship between the gas porosity and pressure under the assumption of local equilibrium. To that end, it is important to understand that one can consider either the material (Lagrangian) frame, or the laboratory (Eulerian) frame, both of which are sketched in Fig. 12.13. The mass balance of Eq. (12.39) is modified when porosity forms (in either reference frame)

$$V\langle\rho_0\rangle C_{g0} = (V_s\rho_s k_{0g} + V_\ell\rho_\ell)C_{g\ell}^{eq}(T, p_\ell) + \frac{V_p p_\ell}{\mathcal{R}T}\mathcal{M}_g N_g \tag{12.41}$$

where V_p is the volume of the pore and $\mathcal{M}_g$ is the molar mass of the gas in the dissolved state (e.g., for hydrogen with $n_H = 2$, one has $\mathcal{M}_H n_H = \mathcal{M}_{H_2}$). Note that the pressure in the pore has been assumed equal to that in the

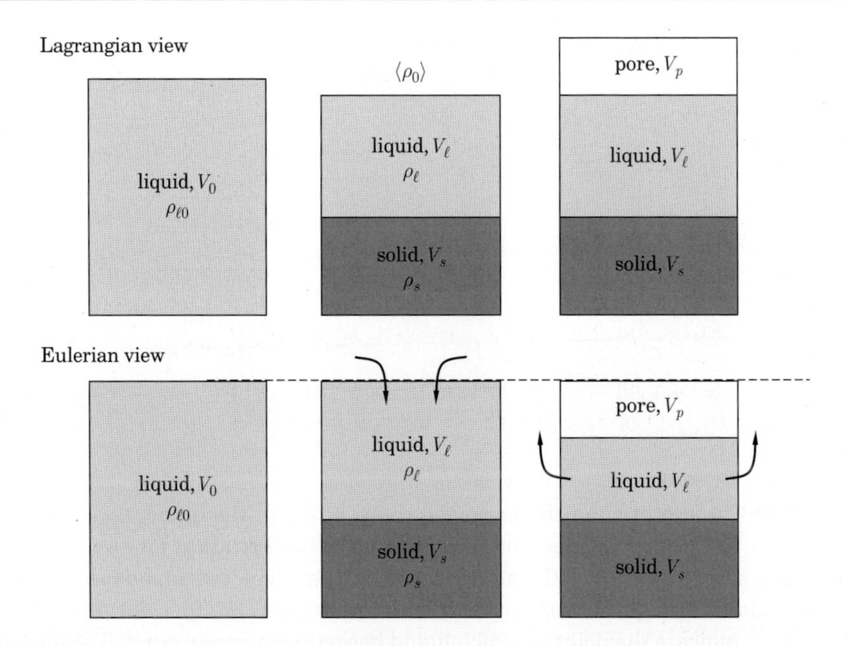

Fig. 12.13 The evolution of a closed volume in which solid and gas form from the liquid phase. In the Lagrangian view no mass is exchanged, whereas in the Eulerian view, mass enters or leaves to compensate volume changes.

liquid, i.e., $V_p p_\ell / \mathcal{R}T$ gives the number of moles of gas in the pore. The compositions in the liquid and solid phases are now given by their respective equilibrium solubility limits. In the Lagrangian frame, the volume V on the left hand side of Eq. (12.41) is still $(V_s + V_\ell)$, whereas the volume is given by $(V_s + V_\ell + V_p)$ (see Fig. 12.13, top). Dividing Eq. (12.41) by V gives

$$(1 - g_p^L)\langle\rho_0\rangle C_{g0} = (g_s^L \rho_s k_{0g} + g_\ell^L \rho_\ell) C_{g\ell}^{eq}(T, p_\ell) + \frac{g_p^L p_\ell}{\mathcal{R}T} \mathcal{M}_g N_g \qquad (12.42)$$

The superscript "L" has been added to indicate the Lagrangian representation. In practice, however, the volume fractions are almost always calculated on a fixed mesh, i.e., in an Eulerian frame (Fig. 12.13, bottom). In this frame, the volume required to compensate shrinkage can come from both the liquid, as discussed in Sect. 12.3, and porosity. Clearly, the presence of porosity implies that less liquid will flow into the volume element. In fact, liquid might even be pushed out if the volume fraction of porosity is larger than the shrinkage. It is easy to demonstrate that Eq. (12.41) becomes in the Eulerian frame

$$\langle\rho_0\rangle C_{g0} = (g_s^E \rho_s k_{0g} + (1 - g_s^E)\rho_\ell) C_{g\ell}^{eq}(T, p_\ell) + \frac{g_p^E p_\ell}{\mathcal{R}T} \mathcal{M}_g N_g \qquad (12.43)$$

Equation (12.43) provides an expression relating the volume fraction of porosity g_p to the pressure in the liquid p_ℓ. One could use this expression to replace g_p in Eq. (12.10), so as to have the pressure in the liquid as the only unknown field. The resulting equation could be solved, after specifying appropriate boundary conditions at the limits of the mushy zone (see Sect. 12.6). Unfortunately, such a simple procedure would fail because it neglects three essential contributions:

1. Pore *nucleation* requires some degree of supersaturation to overcome surface tension,
2. Pore *growth*, which usually occurs within a well-developed network of dendrites/grains, is also strongly affected by surface tension,
3. *Diffusion* of dissolved gases toward the pores must occur during growth. This implies that the composition is not uniform in the liquid.

These three topics are treated next.

12.5.1 Pore Nucleation

As we discussed in the context of condensed phases in Chap. 7, nucleation of a gas phase (bubble) in the liquid requires a certain amount of supersaturation to overcome the gas-liquid interfacial energy, i.e., the gas composition in the liquid $C_{g\ell}$ must be higher than the equilibrium solubility limit $C_{g\ell}^{eq}(p_\ell, T)$. Let us first show that the very large interfacial energy between the two phases typically $\gamma_{\ell g} \approx 10\,\gamma_{s\ell}$, implies that homogenous nucleation of gas bubbles will never occur. The pressure p_g inside a spherical bubble of radius R_p is given by

$$p_g = p_\ell + \frac{2\gamma_{\ell g}}{R_p} \qquad (12.44)$$

Taking $\gamma_{\ell g} = 1\,\mathrm{J\,m^{-2}}$ and a typical radius of a nucleating bubble R_p of a few nanometers, the pressure difference $(p_g - p_\ell)$ is on the order of several GPa! If the pressure in the gas bubble, which has to be positive, is close to atmospheric pressure, such a large pressure difference would require the liquid to have a very significant negative pressure. Although negative pressures in a liquid are perfectly legitimate, values in the GPa range are much too high. We therefore conclude that porosity must be nucleated heterogeneously, and turn our attention to the substrates that can facilitate such nucleation.

Considering the values of $\gamma_{\ell g}$ and $\gamma_{s\ell}$, it is very unlikely that the solid-liquid interface is a good location for the promotion of heterogeneous pore nucleation (see Fig. 12.14). Indeed, most metals are very well wetted by their own melt, and it would be very difficult for a gas phase to form there. Therefore one should look for foreign particles or phases that form during solidification or are already present in the melt, but are not well wetted by the liquid, i.e., for which the interfacial energy $\gamma_{f\ell} \gg \gamma_{s\ell}$. This is the case for numerous oxides in contact with liquid metals. For example, interfacial energies $\gamma_{f\ell}$ of 2.5 and 1.6 $\mathrm{J\,m^{-2}}$ have been measured for Al_2O_3 and

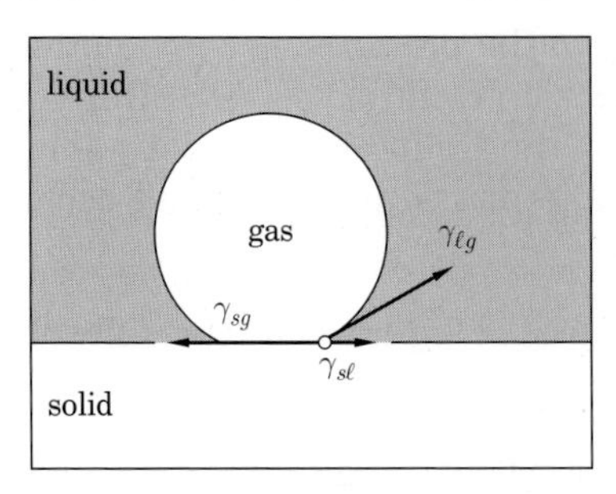

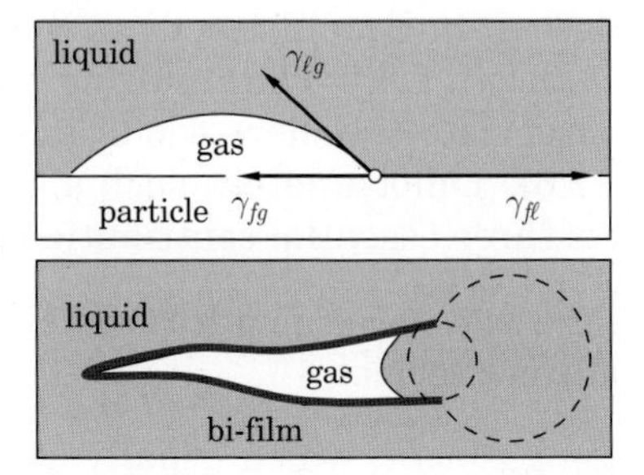

Fig. 12.14 The formation of a gas bubble at (left) the solid-liquid interface and (top right) the surface of a non-wetting particle. (bottom right) The growth of a gas pocket pre-existing in a bi-film of oxide.

ZrO_2 ceramics, respectively, in contact with liquid metals at their melting point. These values were almost the same for many different metallic melts. Although the wetting angle of a bubble forming at the surface of such a foreign particle immersed within a liquid still depends on the interfacial energy between the particle and the gas, γ_{fg}, such oxide particles are potent nucleation sites for the formation of bubbles by the mechanisms illustrated in Fig. 12.14. As discussed in Chap. 7, nucleation on these types of particles can even be athermal if a gaseous phase already exists at its surface, which can occur, for example, if the oxide comes from the surface of the melt and is entrained along with some air in the liquid during pouring. Campbell has clearly demonstrated that the presence of Al_2O_3 bi-films in aluminum alloys is the main cause of gas porosity nucleation, as shown in Fig. 12.14. Nevertheless, a small supersaturation is still required for the bubble to grow from the bi-film. In this scenario, the smallest radius of curvature associated with a hemispherical cap emerging from a bi-film might be on the order of $10-100\ \mu$m, in which case $(p_g - p_\ell) \approx 20 - 200\,$kPa. This is a much more realistic range of values for the pressures in the mushy zone (see Sect. 12.3).

As discussed in Chap. 7, the initial nucleation of pores is not well understood, and all of the existing porosity models are therefore based on a set of very simple assumptions. In many cases, it is simply assumed that a density of potent nucleation sites, having density n_p, become active at a given supersaturation $\Delta C_{g\ell}^{heter} = C_{g\ell}^{heter} - C_{g\ell}^{eq}(p_\ell, T)$. This is analogous to the athermal nucleation models discussed in Chap. 7 for a single set of characteristic nuclei. Note that supersaturation, i.e., composition departure from equilibrium, and pressure difference between the bubble and the liquid are related:

$$\begin{aligned}\Delta C_{g\ell}^{heter} &= C_{g\ell}^{heter} - C_{g\ell}^{eq}(p_\ell, T) = C_{g\ell}^{eq}(p_g, T) - C_{g\ell}^{eq}(p_\ell, T)\\ &= C_{g\ell}^{eq}\left(p_\ell + \frac{2\gamma_{\ell g}}{R_p}, T\right) - C_{g\ell}^{eq}(p_\ell, T)\end{aligned} \tag{12.45}$$

Given this relation, some models define nucleation in terms of a critical pressure difference, or a critical nucleus size. Since supersaturation is required for nucleation to occur, this clearly delays the onset of bubble nucleation, as demonstrated in Fig. 12.12, where the dashed curve corresponds to $(C_{H\ell}^{eq}(g_s(T), p_0) + \Delta C_{H\ell})$ with $\Delta C_{H\ell} = 0.05$ cc$_{\mathrm{STP}}/100$ g. The intersections of this curve with the actual hydrogen compositions of the liquid $C_{H\ell}(g_s)$ (curves with square and circle symbols) are shifted to higher values of g_s, as compared to those determined under equilibrium conditions, corresponding to the solid curve.

12.5.2 The role of curvature during growth

Once a pore nucleats, it grows. For small bubbles growing freely in the melt, where dynamic and gravitational effects can be neglected, the pore will remain spherical. Since the pore curvature is uniform, its internal pressure must also be uniform. The mass balance of gas is fairly easy to construct for this geometry. One replaces the pressure in the liquid p_ℓ in Eqs. (12.41) or (12.43) by that in the bubble, i.e., $p_g = p_\ell + 2\gamma_{\ell g}/R_p$. Further, the fraction of pores, $g_p = V_p/V$, is related to the radius R_p and to the pore density n_p according to

$$g_p = n_p \frac{4\pi}{3} R_p^3 \tag{12.46}$$

so that Eq. (12.43) becomes

$$\langle \rho_0 \rangle C_{g0} = (g_s^E \rho_s k_{0g} + (1 - g_s^E)\rho_\ell) C_{g\ell}^{eq}(T, p_g) + \frac{g_p^E p_g}{\mathcal{R}T} \mathcal{M}_g N_g \tag{12.47}$$

with

$$p_g = p_\ell + \frac{2\gamma_{\ell g}}{\mathcal{R}_p} = p_\ell + 2\gamma_{\ell g} \left(\frac{4\pi n_p}{3 g_p^E} \right)^{1/3} \tag{12.48}$$

When the pores form at high volume fraction of solid (shrinkage porosity), they are constrained to grow within the well-developed dendrite or grain network. The fact that p_g is uniform implies that the interface between the pore and the remaining liquid has the same mean curvature $\overline{\kappa}$ everywhere. The interface between the pore and the solid is defined by the shape of the dendrites or the grains when their growth was stopped due to the absence of liquid. This situation is shown schematically in Fig. 12.15 for a 2D dendritic network. Note that the Laplace-Young condition at the gas-liquid-pore triple junction must be satisfied (Fig. 12.14, left) and thus that the contact angle is not necessarily $180°$. During the growth of both the pore and the solid, the radius of curvature of the interface between the pore and the liquid decreases continuously to finally give the complex 3D pore morphology shown in Fig. 12.4. In that figure, the most curved portions of the pores, probably in contact with the last interdendritic liquid, have radii of $5 - 10$ μm, indicating a capillary overpressure of $200 - 400$ kPa. This contribution is clearly of the same order of magnitude as the pressure drop associated with liquid feeding.

At this time, it is still unclear how to relate g_p and p_ℓ for such complex pore morphologies. Future *in-situ* X-ray tomography experiments

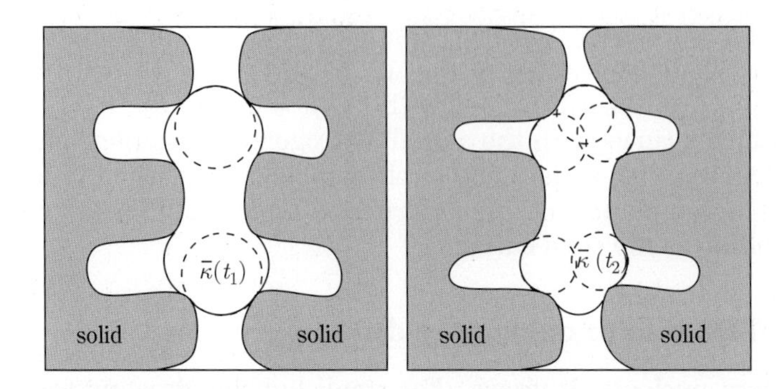

Fig. 12.15 A 2D schematic view of pore growth between dendrite arms at two solidification times, $t_1 < t_2$. The mean curvature, $\bar{\kappa}(t)$, of the pore surface in contact with the liquid is uniform at any instant and increases with time.

should shed some light on the subject. At present, one can only construct approximate models such as setting the pore radius to be a function of g_s rather than g_p in Eq. (12.44). This is a crude way to incorporate in the model the space remaining between the grains or dendrite arms. Such a simple model can be obtained from a regular arrangement of spheres (for globulitic grains) or cylinders (for dendrite arms). A simple geometry calculation gives the $R_p(g_s)$ relationship shown in Fig. 12.16, when it is normalized by the grain diameter or the secondary dendrite arm spacing, respectively.

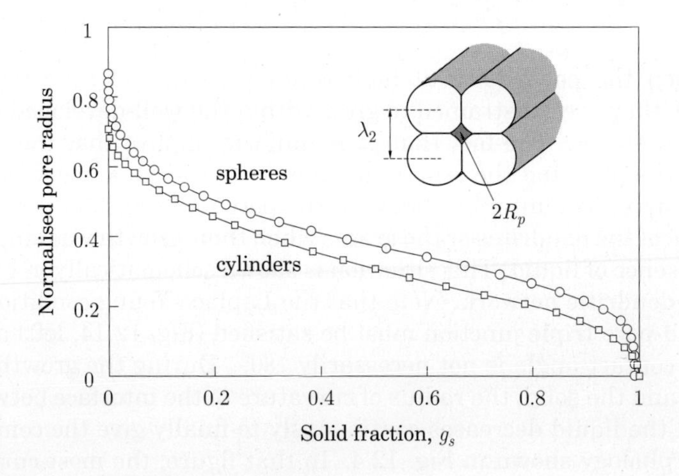

Fig. 12.16 The pore radius R_p normalized by the secondary arm spacing (dendrite morphology) or grain diameter (globular equiaxed structure) as a function of the volume fraction of solid g_s. The insert shows the geometry for the former case. (After Couturier et al. [5].)

12.5.3 Contribution of gas diffusion during growth

Although the diffusion of a gaseous element such as hydrogen in aluminum is quite fast ($D_{sH} = 5.6 \times 10^{-8}$ m^{2}s^{-1} and $D_{\ell H} = 3.2 \times 10^{-7}$ m^{2}s^{-1} at the melting point), it can contribute nonetheless to non-equilibrium effects. Indeed, the growth of a pore is controlled by the rate of diffusion of gas in the supersaturated medium to the pore surface. This effect is particularly important when the density of pores is low and the solidification time is short, i.e., if the Fourier number $D_{\ell H}t_f/L^2$ associated with the process is small (L is half of the typical distance between two pores). Although D_{sH} is not much less than $D_{\ell H}$, gas diffusion in the solid phase is nevertheless very limited because the solubility limit is very low. Thus, diffusion occurs mainly in the interdendritic liquid, as illustrated in Fig. 12.17. To construct a simple model, let us assume that we have spherical pores growing inside a fully liquid region. The density of such pores is n_p, and thus the characteristic radius of the spherical volume within which the pore can grow is given by $R_{p0} = (3/(4\pi n_p))^{1/3}$. Assuming quasi-steady state diffusion, the gas composition field in the liquid has the same form as the carbon profile in the austenite shell surrounding a graphite nodule (Eq. 9.19)

$$C_{g\ell}(r) = \frac{1}{R_{p0} - R_p}\left[C_{g0}R_{p0}\left(1 - \frac{R_p}{r}\right) - C^*_{g\ell}R_p\left(1 - \frac{R_{p0}}{r}\right)\right] \qquad (12.49)$$

where $C^*_{g\ell} = C_{g\ell}(r = R_p)$ is the gas composition of the liquid in contact with the pore, equal to the solubility limit $C^{eq}_{g\ell}(T, p_g)$ associated with the pressure in the pore p_g, and $C_{g0} = C_{g\ell}(r = R_{p0})$ is the gas composition at the limit of the domain. Since this latter boundary condition does not satisfy a zero-flux balance for a closed volume, C_{g0} has to be calculated from an overall mass balance of gas over the entire domain (pore plus liquid).

Under the present assumptions, the flux of gas species at the pore-liquid interface controls the growth rate of the pore. For diatomic gas such as hydrogen, one has

$$4\pi R_p^2 D^{\ell}_{H}\rho_\ell\left(\frac{dC_{H\ell}}{dr}\right)_{r=R_p} = \frac{N_H \mathcal{M}_H}{\mathcal{R}T}\frac{d(V_p p_p)}{dt} \qquad (12.50)$$

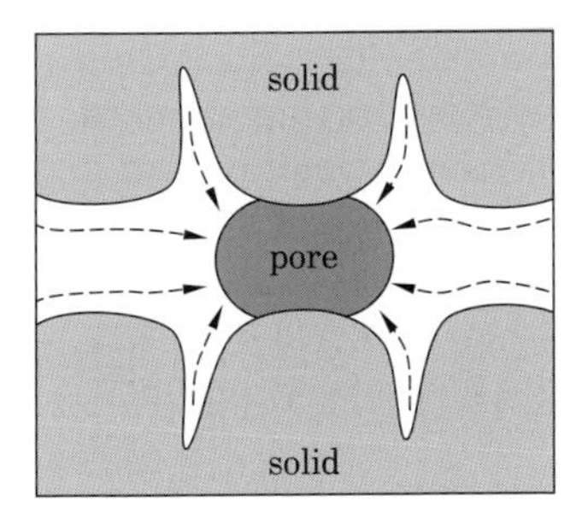

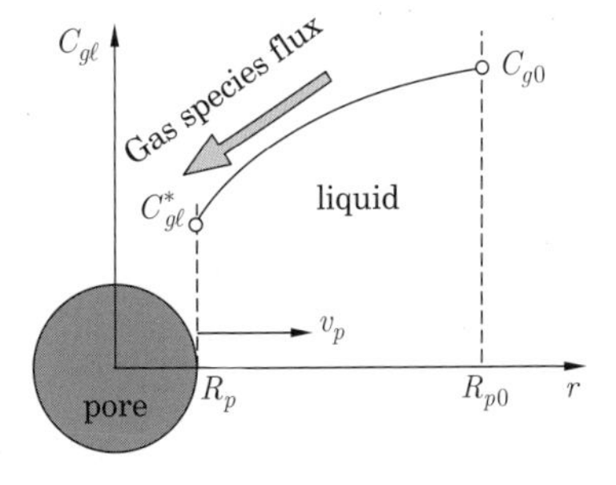

Fig. 12.17 Schematic views of (left) gas diffusion in between dendrite arms, and (right) composition profile in an ideal spherical geometry.

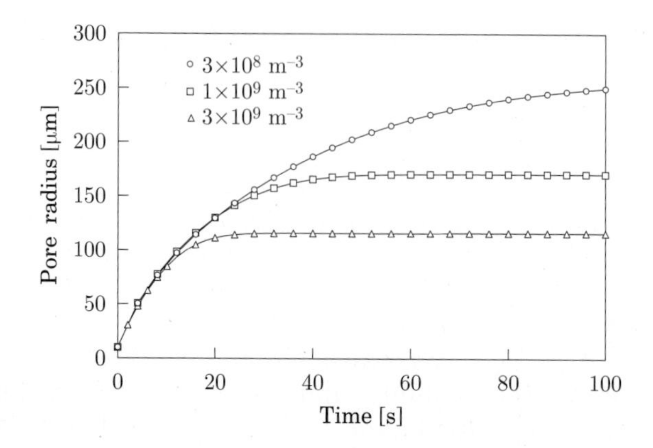

Fig. 12.18 Temporal evolution of the radius of the pore for three different pore densities. The other parameters are: $D_{\ell H} = 10^{-8}\,\mathrm{m^2 s^{-1}}$, $C_{H0} = 0.8$ and $C^{eq}_{H\ell}(p_\ell = 10^5\mathrm{Pa}) = 0.3$ cc$_{\mathrm{STP}}$/100 g, pore nucleation radius $R_{pn} = 10\,\mu$m, $\gamma_{\ell g} = 0.9\,\mathrm{J\,m^{-2}}$ and $T = 933\,$K.

where the factor $n_H \mathcal{M}_H / RT = \mathcal{M}_{H_2}/\mathcal{R}T$ has been introduced to match the units. Inserting the composition profile given by Eq. (12.49) and considering that $p_p = p_\ell + 2\gamma_{\ell g}/R_p$, Eq. (12.50) can be solved for the radius of the pore as a function of time. Figure 12.18 shows the evolution of the pore radius for three different pore densities n_p. The various parameters are given in the figure caption. As one would expect, the final pore radius increases with the size R_{p0} of the domain within which it can grow. More importantly, the time needed to reach equilibrium also increases with R_{p0}. The scaling techniques described in Chap. 4 give, as a first approximation, that this time is given by R^2_{p0}/D^ℓ_H.

12.5.4 Summary of the coupling between pressure and pore fraction

In summary, porosity formation can be modeled by solving the combined mass conservation – Darcy equation (Eq. 12.10). We showed in this section that the relationship between the two unknown fields, $p_\ell(\boldsymbol{x}, t)$ and $g_p(\boldsymbol{x}, t)$, is obtained from a local mass balance of the gas-forming element, including non-equilibrium effects associated with the nucleation and growth of pores. We identified gas and shrinkage porosity at the beginning of this chapter as distinct entities, however we have shown that they differ only by the relative importance of the various contributions entering into the same set of equations. The formation of gas porosity at an early stage of solidification gives rise to a very small pressure drop in the mushy zone (typically a few kPa if no pores form). The supersaturation of the melt necessary to overcome the curvature contribution (typically 20 kPa for a pore radius of 100 μm) is primarily due to an increasing amount of gas segregated into the liquid. In the case of shrinkage porosity, which forms much deeper in the mushy zone of large solidification interval alloys, the pressure drop is much more important (typically a few 100 kPa if no pores form). The curvature contribution is also considerable in this case, as pores have to nucleate and grow within a well-developed dendritic network. A certain amount of gas remains in the pores as nature abhors vacuum!

In general approaches for the modeling of porosity formation, the same set of governing equations can therefore be used to relate g_p and p_ℓ, in the mass balance-Darcy equation, regardless of the type of porosity. Equation (12.10) is parabolic, which requires appropriate initial and boundary conditions. In order for the Darcy equation to have meaning, the boundary conditions must be applied at the periphery of the solution domain, which corresponds to the boundaries of the mushy zone. This topic is developed in the next section.

12.6 BOUNDARY CONDITIONS

The appropriate boundary conditions required for the modeling of porosity formation depend on the location of the mushy zone within a casting. Two important cases are shown in Fig. 12.19.

Close to a mold wall, where some of liquid still exists, the interdendritic liquid velocity must be zero, since there is no source of liquid. At such a boundary, one has

$$\langle v_n \rangle = g_\ell \langle v \rangle_n = -\frac{K}{\mu_\ell}\left(\frac{\partial p_\ell}{\partial n} - \rho_\ell g_n\right) = 0 \qquad (12.51)$$

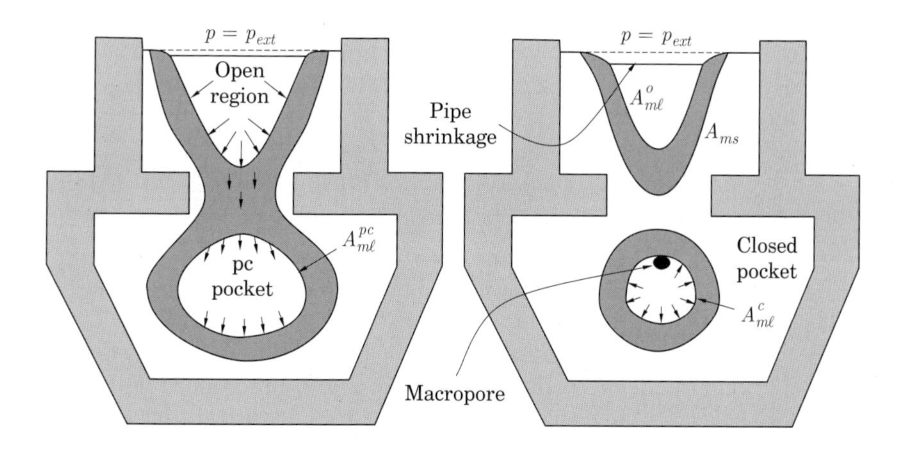

Fig. 12.19 (left) A schematic view of an alloy solidifying in a mold and exhibiting a semi-open region of liquid connected via the mushy zone to an open pocket. (right) As solidification proceeds, the semi-open region becomes closed. A_{ms} is the boundary separating the mushy zone and the fully solid region, and $A_{m\ell}$ separates the fully liquid and mushy regions.

At a boundary $g_s = 1$, such as that denoted A_{ms} in Fig. 12.19, two cases should be distinguished. If there is no eutectic reaction, the pressure drop calculation should normally be performed up to $g_s = 1$. However, at this limit, the permeability K and thus the pressure drop calculation diverge (see Fig. 12.8). The calculation has no physical meaning once the primary phase starts to coalesce, i.e., once the continuous liquid films transform into isolated liquid pockets at approximately $g_s = 0.95$ (see Chap. 13). As it is no longer possible to feed the remaining liquid pockets, A_{ms} is defined as the isovalue $g_s = 0.95$, and the boundary condition that should be applied at such a boundary is still given by Eq. (12.51).

However, if some eutectic is formed before coalescence, a new condition needs to be applied at T_{eut}. Using the sketch in Fig. 12.7, one has

$$\langle v_n \rangle = g_{eut} \langle v \rangle_n = -g_{eut} \beta v_{T,eut} = -\frac{K_{eut}}{\mu_\ell} \left(\frac{\partial p_\ell}{\partial n} - \rho_\ell g_n \right) \qquad (12.52)$$

where g_{eut} is the volume fraction of the remaining liquid at the eutectic front and $K_{eut} = K(g_{eut})$. The velocity of the eutectic isotherm $v_{T,eut}$ is obtained from a separate thermal field calculation. Equation (12.52) is applied at the eutectic interface, which is available directly when using a front-tracking method. The boundary condition can be applied approximately in fixed grid approaches, for example using a penalty method. In the latter approach, the eutectic reaction can be spread over a small solidification interval, and thus over a few mesh points. As soon as $g_\ell < g_{eut}$, the permeability is limited to K_{eut} and the calculation of the pressure drop is continued until $g_s = 1$.

For a boundary of the mushy zone in contact with the liquid, such as at the surface $A_{m\ell}$ corresponding to the liquidus, a pressure boundary condition can be imposed if the pressure in the liquid is known. One such case is that of an *open region of liquid* in contact with external air (Fig. 12.19). In this case, one obtains

$$p_\ell(\boldsymbol{x}) = p_{ext} + \rho_\ell g (z_{ext} - z) \qquad \text{for } \boldsymbol{x} \in A^o_{m\ell} \qquad (12.53)$$

where z_{ext} is the height of the top free surface where a pressure p_{ext} is applied, and z is that of the point considered on the liquidus isotherm surface $A^o_{m\ell}$.

For a *closed-liquid pocket*, i.e., a pocket totally surrounded by the solid and/or the mold, no feeding is possible and a macropore will form at the highest point (i.e, lowest pressure) of $A^c_{m\ell}$. A cavitation pressure condition $p_\ell = p_c$ must be applied. Note that the value of p_c should normally follow the same gas model as that established for microporosity, but with a much greater radius of curvature.

When the liquid pocket is *partially closed* (pc), i.e., connected to an open region of liquid through the mushy zone, an *integral boundary condition* is applied before cavitation occurs. A pressure $p_{\ell z_0}$ must apply at a height z_0 of this pc liquid pocket, and the pressure at the other points of this region is given by

$$p_\ell(\boldsymbol{x}) = p_{\ell z_0} + \rho_\ell g(z_0 - z) \tag{12.54}$$

The pressure $p_{\ell z_0}$ is in fact unknown, but since any flow that enters the pc pocket must also exit, its value must produce a velocity field at $A^{pc}_{m\ell}$ satisfying the overall mass balance equation

$$\int_{A^{pc}_{m\ell}} \rho_\ell \boldsymbol{v}_\ell \cdot \boldsymbol{n} dS = 0 \tag{12.55}$$

If $p_{\ell z_0}$, which is part of the problem, falls below p_c, then a pressure condition corresponding to that of a closed pocket must be applied and a macropore forms.

For a surface of the mushy zone in contact with the environment, the height of the surface itself can decrease, up to some volume fraction of solid, provided that grains are allowed to move. This is called "mass feeding". In such a case, the pressure at this surface can be imposed as p_{ext}. When the grains are hindered from moving, as occurs with columnar structures or densely packed equiaxed grains, the velocity at the free surface is set to zero, as we did for a mold wall. A more realistic condition would be to model the suction of interdendritic liquid at such a location with the formation of liquid menisci of negative curvature.

A test problem illustrating these boundary conditions is presented in Fig. 12.20. We consider a one-dimensional "casting" containing two liquid

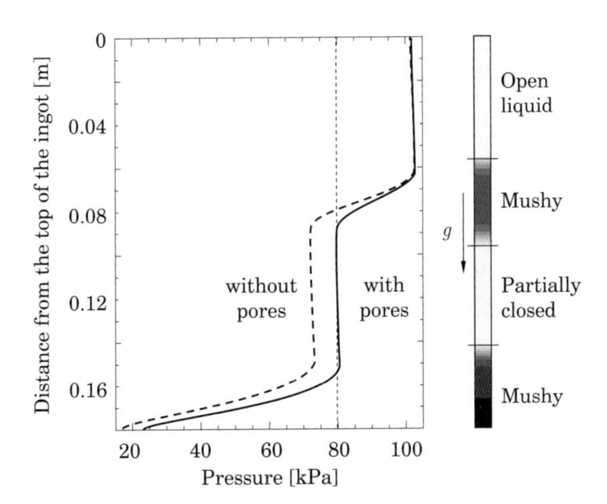

Fig. 12.20 The pressure drop in a one-dimensional model with two liquid pockets, one open to the ambient pressure and the other partially closed. The grey level to the right indicates the volume fraction of solid, where white represents the liquid. The dashed curve gives the pressure without any cavitation pressure p_c set in the partially closed liquid pocket, i.e., without pores, whereas $p_c = 80\,\text{kPa}$ for the continuous curve. A macropore forms in the latter case. (After Péquet et al. [15].)

pockets. The pocket on the top (represented in white to the right of the diagram) is open to the atmosphere and the pressure at its top surface is thus fixed to 101 kPa. Gravity causes p_ℓ to increase slightly with distance from the free surface. A first mushy region (in grey) is present in the middle of the casting as a result of the thermal conditions imposed on the lateral sides. It is not fully solid, and thus allows some feeding to the second, partially closed (pc) liquid pocket in the lower half of the casting. Finally, a mushy region (dark grey) is present at the bottom. The pressure in the pc liquid pocket is calculated by applying the integral boundary conditions (Eqs. (12.54) and (12.55)). If no porosity is allowed to form in the entire casting, a pressure drop of slightly more than 30 kPa occurs over the first mushy zone, induced by the flow necessary to feed this region itself and the last mushy zone on the bottom. The pressure profile for this case is given by the dashed curve in Fig. 12.20. If we set a cavitation pressure, beyond which a macropore forms in a liquid pocket, then the pressure profile given by the solid curve is obtained. Its minimum value is equal to the prescribed cavitation pressure (80 kPa) and is found at the highest portion of the pc pocket. The pressure then increases again slightly due to gravity, before a large pressure drop occurs in the lowest mushy zone.

It should be noted that once the pressure field in the mushy zone has been calculated, the velocity there is known, and one can thus deduce the overall liquid flow at the mushy zone-liquid boundaries. This flow causes the upper surface of an open region of liquid to fall, thereby producing *pipe shrinkage*. For a closed pocket, the flow integral gives the increase in volume of the macropore. This approach makes it possible to determine the distribution of various types of shrinkage: pipe shrinkage at upper free surfaces, macropores in closed cavities and microporosity.

12.7 APPLICATION OF THE CONCEPTS

The purpose of this section is to present a few well-selected examples that best illustrate the application of some of the concepts introduced in this chapter. As already explained in Sect. 12.5.4, the same concepts can be used for gas or shrinkage porosity. However, depending on the characteristics of the alloy (solidification interval, density), the amount of dissolved gases and the solidification process parameters, the relative importance of the various contributions (pressure drop, gas concentration of the liquid, curvature contribution) will vary from one case to another, leading to an implicit classification of gas and shrinkage porosity. More details concerning porosity modeling can be found in review articles by Stefanescu [18] and Lee et al. [10], whereas practical information regarding porosity formation and control is given by Campbell [1].

We begin with an example that illustrates the relationship between boundary conditions and the resulting microporosity. Figure 12.21 shows

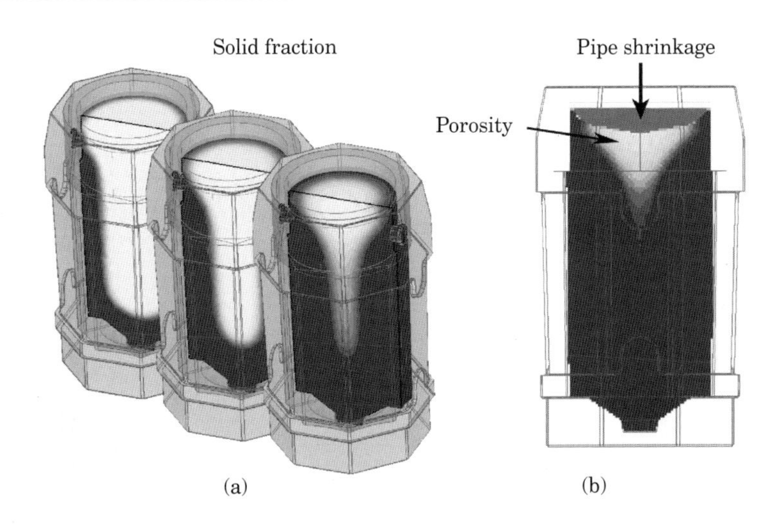

Fig. 12.21 (a) Three stages of solidification of a 200 t, 5 m high, tool steel ingot, showing the evolution of the volume fraction of solid ($g_s = 0$, white; $g_s = 1$, dark). (b) The final pipe shrinkage appearing at the top surface (in light gray) and the increased microporosity developing just below. The black region has no microporosity, whereas the white region corresponds to a maximum pore fraction equal to an unfed solidification shrinkage. (Courtesy of ESI group.)

porosity formation underneath the top surface of a large unrisered ingot. Three times during solidification are shown on the left. The gray level corresponds to the volume fraction of solid (white is the liquid). Due to the nearly axisymmetric thermal conditions, the remaining liquid has a "finger"-shape in the center of the ingot. The top surface of the casting moves down as feeding takes place during freezing. The final pipe shrinkage is shown in light gray to the right. Since feeding of the region that was last to solidify was made difficult in the simulation, an increased amount of porosity appears just below this surface.

An example demonstrating the influence of composition on the computed pressure drop and microporosity formation in an aluminum alloy is shown in Fig. 12.22 [4]. In this axisymmetric gravity casting of an Al-0.6wt%Mg-0.7wt%Si-0.1wt%Fe-2.3wt%Cu alloy, solidification was calculated using a Gulliver-Scheil microsegregation model and a uniform heat transfer coefficient on all boundaries. The three figures to the left show the computed pressure in the liquid, whereas those on the right give the corresponding microporosity fraction. The three cases (a)–(c) show results for different models used for the effect of solute composition on microporosity, the hydrogen composition being fixed. Case (a) assumes that the composition has no influence on the hydrogen solubility. In case (b), the solubility limit is calculated using the nominal composition C_{I0} of the alloy in Eq. (12.37), whereas in (c) the evolving compositions $C_{I\ell}(T(t))$ are utilized in Eq. (12.37) to compute the equilibrium hydrogen solubility limit.

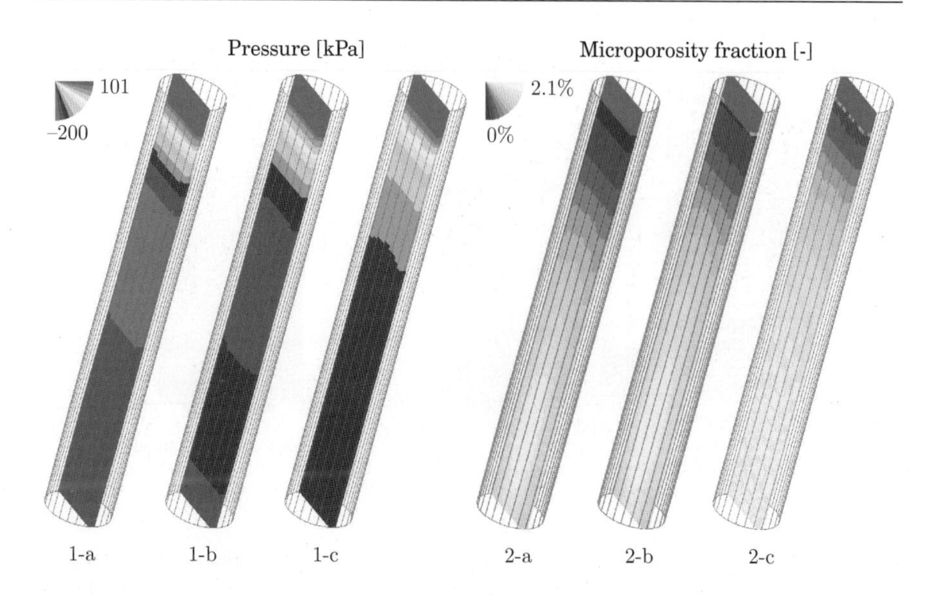

Fig. 12.22 (1) The pressure in the liquid and (2) the pore fraction in a cylinder of Al-2.3wt%Cu-0.7wt%Si-0.1wt%Fe-0.6wt%Mg calculated for 3 hydrogen solubility models: (a) no influence of alloy composition; (b) influence calculated using the nominal composition C_{I0}; (c) influence calculated using $C_{I\ell}(T)$. $C_{H0} = 0.15$ $\mathrm{cc_{STP}}/100\,\mathrm{g}$. (After Couturier and Rappaz [4].)

Table 12.1 shows that Cu and Si decrease the solubility limit of hydrogen in liquid aluminum, whereas Mg increases it. In this alloy, the Mg content is lower than that of Cu, and, thus one would expect that, for a fixed nominal hydrogen content, the final porosity fraction would be greater if the activity of the solute elements is taken into account. This effect can be observed in Fig. 12.22(b), and is magnified in (c) when the enrichment of the liquid in Cu during solidification is accounted for. It should be noted that although these calculations were done for a fixed initial content of hydrogen, this is probably difficult to achieve in practice, because Mg tends to enhance the oxidation reaction between Al and water vapor, thus increasing the nominal hydrogen content of the melt.

Lee and co-workers [20] used similar concepts, combined with a Cellular Automaton (CA), to model microporosity formation in experimental wedge castings of various Al-Cu-Si alloys (Fig. 12.23). Small specimens were cut for X-ray tomography observations at various distances from the chill surface, thus representing several cooling rates. The three views at the top of Fig. 12.23 correspond to experimental measurements at a cooling rate of -2.9 $\mathrm{Ks^{-1}}$ for three different alloys, Al-4wt% Cu, Al-7wt% Si and W319 (3.57wt%Cu, 7.63wt% Si plus minor elements). Simulation results for nominally identical conditions appear in the bottom row. The initial hydrogen content in the experiments varied between 0.17 and 0.25 $\mathrm{cc_{STP}}$/100 g, whereas a fixed value of 0.24 $\mathrm{cc_{STP}}$/100 g was used in the simulations. The CA calculation was not coupled to a mesoscopic pressure drop

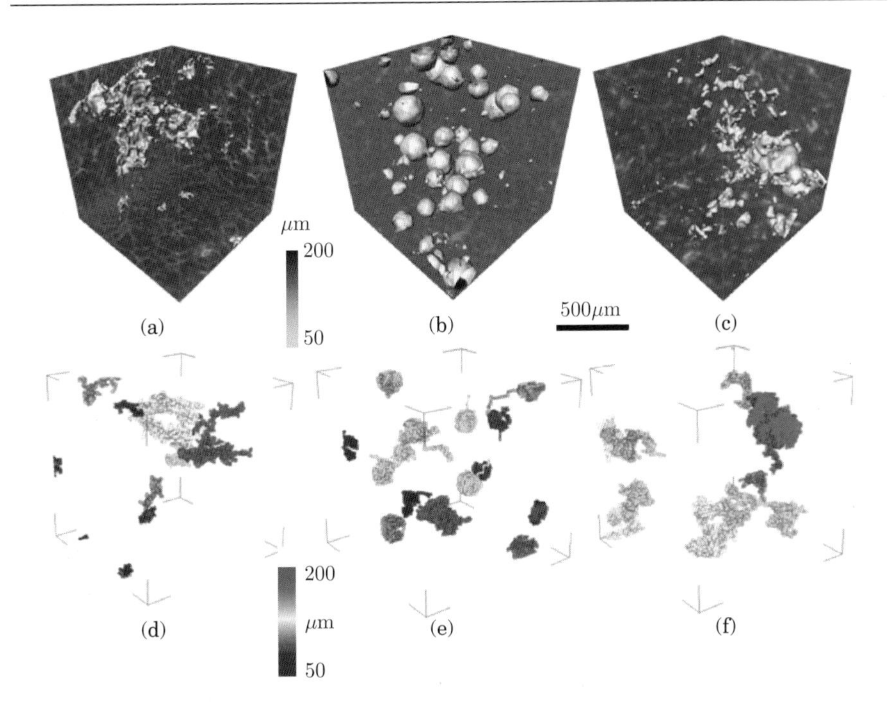

Fig. 12.23 Comparison of (top) experimentally observed and (bottom) simulated pore formation in binary and ternary alloys solidified at $\dot{T} = -2.9$ K s^{-1}: (a), (b) and (c) are 3D micro-tomography views of the pore morphology in Al-4wt%Cu, Al-7wt%Si and W319 alloys, respectively. (d), (e) and (f) are results from 3D simulations of pore formation in the same alloys. The domain size is 1.2 mm^3 and the gray level in the simulated pores represents the equivalent pore diameter. Nominal hydrogen content: 0.24 cc$_{STP}$/100 g. (Courtesy of J.S. Wang and P.D. Lee [20].)

calculation, and the simulation results therefore reflect only the effects of composition on the hydrogen solubility and on the space available for pore growth. This may have led to the fraction of porosity being underestimated. In the model, the pores were allowed to grow in the constraining network of solid according to CA-based criteria that consider neighboring cells with the largest fraction of liquid, leaving the maximum space available for pore growth (maximum radius of curvature) and exhibiting the maximum hydrogen composition gradient, i.e., the fastest growth rate. When comparing the results for the three alloys, both the X-ray tomography and the CA simulations show that the Al-Si alloy exhibits spherical pores since the amount of liquid is nearly 50% at the eutectic temperature. In the other two Cu-bearing alloys, the pores have a much more intricate or tortuous shape, especially in the binary Al-Cu alloy, which has the smallest amount of eutectic. The fraction of porosity is also lower in this alloy since the curvature overpressure is the highest (minimum space available for growth) and the alloy is free of Si, a solute element reducing the solubility of hydrogen (see Table 12.1).

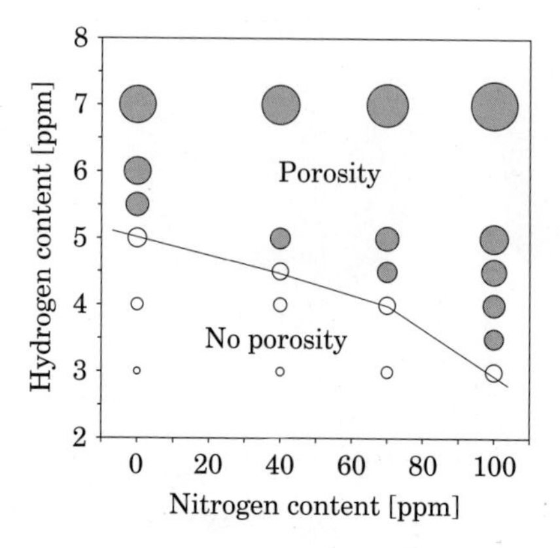

Fig. 12.24 The effect of gas content on porosity formation in AISI 8620 steel castings. The solid line separates castings expected to have porosity from those without it. Composition of the alloy (in wt%): 0.186 C, 0.83 Mn, 0.28 Si, 0.56 Cr, 0.55 Ni, 0.16 Mo, 0.015 P and 0.015 S. (From Sung et al. [19].)

Figure 12.24 shows the conditions that should lead to porosity formation in an AISI 8620 steel casting. The model for this result was developed by Sung et al. [19] The simulations track the partitioning of the alloying elements during solidification, as in the case of aluminum alloys, but here the evolution of two gases, hydrogen and nitrogen, has to be considered. In low carbon steels, one must also take into account the peritectic reaction where the primary ferrite dendrites can transform into austenite. Figure 12.24 displays the predictions for porosity formation during solidification of a 25.4-mm thick plate castings in a ceramic-shell mold. The circle size in the figure is proportional to the sum of the equilibrium partial pressures of the two gases, $p_g = (p_{H_2} + p_{N_2})$, corresponding to their actual atomic fractions $X_{H\ell}^{eq}$ and $X_{N\ell}^{eq}$ in the interdendritic liquid. When the pressure p_g exceeds the actual pressure p_ℓ in the liquid, there is a chance for pore nucleation if the curvature contribution is neglected. In Fig. 12.24, the liquid pressure is represented by circles of nearly the same diameter connected by the continuous line. Above this line, the liquid is supersaturated in gas whereas below it, supersaturation has not yet been reached. As expected, the trend to form pores increases when the content of either gas is raised, but the alloy is much more sensitive to hydrogen content than to nitrogen. One can use this type of calculation to examine the effect of alloying elements such as Ti. When no Ti is present, as little as 3 ppm of hydrogen leads to porosity provided that the melt initially contains 100 ppm of nitrogen. By adding Ti to "tie up" nitrogen in the form of TiN precipitates, as much as 4.5 ppm of hydrogen in the melt can be tolerated.

In our final example, we examine the influence of limited gas diffusion in the liquid. Figure 12.25 presents a comparison of measured

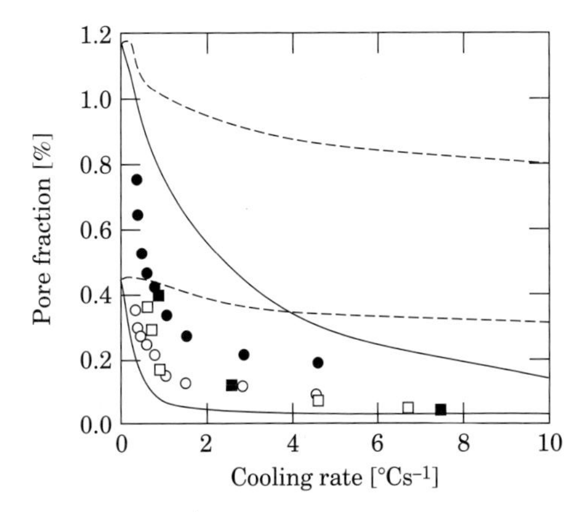

Fig. 12.25 The volume fraction of porosity in 1D Al-7%Si-0.37%Mg castings as a function of the local cooling rate for two initial hydrogen concentrations: 0.13 (open symbols) and 0.26 (filled symbols) $cc_{STP}/100$ g. The circles and squares correspond to two sets of experimental data. The continuous and dashed curves were calculated with and without consideration of the limited hydrogen diffusion, respectively. $D_{\ell H} = 3.8 \times 10^{-6} \exp(-2315/T)$ m^{2}s^{-1}. (After Carlson et al. [2].)

and simulated porosity fractions in aluminum alloys solidified at various cooling rates. The experimental data were obtained for two hydrogen contents in A356 (Al-7wt%Si-1wt%Mg) and compiled by Carlson et al. [2]. The calculated amount of porosity was studied as a function of various parameters: the hydrogen solubility in the silicon phase, the density of pores, the cooling conditions, etc. Carslon et al. found that these factors were insufficient to explain the observed rapid decrease of g_p with cooling rate. They then investigated the effect of a finite diffusion rate of hydrogen (Sect. 12.5.3). The two flat dashed curves were calculated assuming infinitely fast diffusion for the two nominal hydrogen contents C_H^0, whereas the two continuous curves considered a finite-rate diffusion. In order to model the growth of pores in a constrained network of solid, the authors weighted the effective liquid-pore interfacial area entering the diffusion equation by a factor $(1 - g_s)^3$ with respect to that of spherical pores. These results demonstrated that the diffusion of hydrogen can be important during the formation of porosity, especially when the density of pores is low (or the distance separating two pores is large). Since the solid remains supersaturated, the porosity level can evolve after solidification, for example during heat treatment.

12.8 SUMMARY

This chapter has demonstrated that very small amounts (in the ppm range) of dissolved gases in a metallic melt can lead to the formation of porosity.

Several factors encourage porosity formation: (i) The solubility limit in the liquid (and in the solid) decreases as the temperature decreases; (ii) The local gas content in the liquid in the mushy zone increases due to segregation: (iii) Solidification shrinkage and thermal contraction tend to draw the liquid into the mushy zone, inducing a pressure drop that can be substantial in alloys with a large solidification interval. It has been shown that this last contribution is minimized when the (G/v_T) ratio is as large as possible. The only factor that opposes the formation of porosity is the supersaturation required for its nucleation and growth. Working with a clean melt, free of oxides and foreign particles that are not wetted by the liquid metal, is the best way to substantially decrease the density of heterogeneous nucleation sites from which pores may develop. Further care must be taken not to contaminate the melt during casting. The absence of available nucleation sites also increases the characteristic distance over which gases must diffuse to reach the pores. As a result, their growth is reduced as the alloy becomes increasingly supersaturated in gas. For a fixed (G/v_T) ratio, more gas can be kept in solution and the pore fraction reduced in a fine microstructure. Indeed, we have seen that the pressure inside the pore exceeds the local pressure in the liquid by an amount given by $2\gamma_{\ell g}R_p^{-1}$, where the radius of curvature of the pores R_p is dictated, in most cases, by the remaining liquid spaces in between the dendrite arms.

Most of the phenomena described in this chapter have been implemented in various solidification modeling codes. A few examples illustrating how pipe shrinkage, macro- and microporosity can be predicted using such codes were given in the last section. Although quantitative comparisons with experiments are difficult due to the complexity of the problem and the lack of accurate physical data (nucleation parameters, gas content, etc), such models contribute to the identification of the key parameters that favor porosity formation and thus play a significant part in the casting optimization.

12.9 EXERCISES

Exercise 12.1. Pore fraction measurement.
The fraction of pores is an entity fairly difficult to measure. One of the techniques is to use densimetry. How could one combine densimetry and HIP (Hot Isostatic Pressure) to determine the pore fraction? Establish a procedure to determine precisely the fraction of pores with these combined techniques.

Exercise 12.2. Reduced pressure casting.
A common test used in the aluminium industry to assess the amount of hydrogen in a melt is to cast it under reduced pressure. Suppose that when the metal is cast under atmospheric pressure (1 bar), the fraction of pores is g_p. With the help of the equations developed in this chapter, explain

why g_p increases substantially when the metal is cast under a pressure of 80 mbar.

Exercise 12.3. Blistering.
When a component having subsurface porosity, i.e., porosity located just below the surface, is heated up for example for a homogenization treatment, it might develop blisters at the surface of the casting, thus making it uneven. Can you find an explanation?

Exercise 12.4. Pressure losses.
Estimate for a typical sand casting the orders of magnitude of: the metallostatic pressure, the dynamic contribution to the pressure in the melt due to natural convection, the pressure drop in the mushy zone, and the contribution due to curvature to the internal pressure in the pores. Compare these values with atmospheric pressure. Which is the most important phenomenon for porosity formation?

Exercise 12.5. Carbonated beverages.
When you shake a bottle of sparkling water (or beer, or champagne!) and you open it immediately afterward, there is usually a burst of bubbles. Why? If you do not open the bottle after shaking and leave it for a while, the burst of bubbles does not occur when you open it. Why?

Exercise 12.6. Mentos experiment.
A famous experiment consists of dropping a Mentos candy into a bottle of sparkling water, beer, champagne or better Coca Cola. A jet of fluid suddenly erupts from the bottle. Explain why.

Exercise 12.7. Cereal expansion.
During the extrusion of cereals, water is mixed with the flour. The mixture is then sheared in a double-screw helix, during which the temperature may exceed 100°C, and vapor bubbles can form. As a result, the product expands transversely up to 10 times the cross section of the die after it exits the extrusion machine. The amount of expansion is proportional to the amount of water present in the extruded product. However, past a certain water content, the transverse expansion starts to decrease. Can you find an explanation based on the region where nucleation of the vapor bubbles occurs? As the vapor bubbles expand very rapidly and up to very large volume fractions, do you see a major difference in the equation governing their growth, compared to what has been presented here for porosity in metals?

Exercise 12.8. Porosity in Al.
An aluminum slab of $10 \times 2 \times 0.5$ m^3 has a hydrogen content of 0.15 cc$_{\mathrm{STP}}$/ 100 g of metal. Compute the total volume of pores if equilibrium conditions

until the end of solidification are assumed. What is the corresponding fraction of porosity? Perform this calculation for two alloys, Al-7wt%Si and Al-4wt%Mg, using the solubilities given in Fig. 12.11.

Exercise 12.9. Niyama pressure drop.
Using the Niyama approach, derive an expression for the pressure drop in a binary alloy following the lever rule microsegregation model. As the analytical solution is fairly complicated, make the integration over g_ℓ for the particular case $k_0 = 0.5$. You should obtain the following solution:

$$\Delta p'_\ell = 4k_0 \int_{g_\ell}^{1} \left(\frac{1 - g'_\ell}{g'_\ell(1 + g'_\ell)} \right)^2 dg'_\ell$$
$$= 4k_0 \left(-3 + 4\ln 2 + \frac{1}{g_\ell} + \frac{4}{1 + g_\ell} + 4\ln\frac{g_\ell}{1 + g_\ell} \right) \tag{12.56}$$

where the normalisation factor is the same as that used for the Gulliver-Scheil solution (Eq. (12.19)). Compare this result with that shown in Fig. 12.8.

12.10 REFERENCES

[1] J. Campbell. *Castings*. Butterworth-Heinemann, 2nd edition, Oxford, 2003.

[2] K. D. Carlson, Z. Lin, and C. Beckermann. Modeling the effect of finite-rate hydrogen diffusion on porosity formation in aluminum alloys. *Met. Mater. Trans. B*, 38:541-55, 2007.

[3] P. C. Carman. Fluid flow through granular beds. *Trans. Inst. Chem. Engng*, 15:150, 1937.

[4] G. Couturier and M. Rappaz. Effect of volatile elements on porosity formation in solidifying alloys. *Mod. Sim. Mater. Sc. Engng.*, 14:25371, 2006.

[5] G. Couturier and M. Rappaz. Modeling of porosity formation in multicomponent alloys in the presence of several dissolved gases and volatile solute elements. In Y. T. Zhu and Qigui Wang, editors, *Simulation of Aluminum Shape Casting Processing*, page 143, Warrendale, PA, USA, 2006. TMS Publ.

[6] H. P. G. Darcy. *Les fontaines publiques de la ville de Dijon*. Victor-Dalmont, Paris, 1856.

[7] L. S. Darken, R. W. Gurry, and M. B. Bever. *Physical chemistry of metals*. Mc Graw Hill, New York, 1953.

[8] S. D. Felicelli, D. R. Poirier, and P. K. Sung. A model for prediction of pressure and redistribution of gas- forming elements in multicomponent casting alloys. *Met. Mater. Trans. B*, 31:1283-92, 2000.

[9] K. Kubo and R. D. Pehlke. Mathematical modeling of porosity formation in solidification. *Met. Trans. B*, 16:359-66, 1985.

[10] P. D. Lee, A. Chirazi, and D. See. Modeling microporosity in aluminum-silicon alloys: A review. *J. Light Metals*, 1:15-30, 2001.

[11] P. D. Lee and J. D. Hunt. Hydrogen porosity in directional solidified aluminium-copper alloys: In-situ observation. *Acta Mater.*, 45:4155-69, 1997.

[12] R. Mathiesen and L. Arnberg. The real-time, high-resolution X-ray video microscopy of solidification in aluminum alloys. *J. Metals*, 59:20-26, 2007.

[13] Ø. Nielsen, L. Arnberg, A. Mo, and H. Thevik. Experimental determination of mushy zone permeability in aluminum-copper alloys with equiaxed microstructures. *Met. Mater. Trans. A*, 30A:2455, 1999.

[14] E. Niyama, T. Uchida, M. Morikawa, and S. Saito. A method of shrinkage prediction and its application to steel casting practice. *AFS Int. Cast Met. J.*, 9:52-63, 1982.

[15] C. Péquet, M. Gremaud, and M. Rappaz. Modeling of microporosity, macroporosity, and pipe-shrinkage formation during solidification of alloys using a mushy-zone refinement method: Applications to aluminium alloys. *Met. Mater. Trans. A*, 33:2095-106, 2002.

[16] D. R. Poirier, K. Yeum, and A. L. Maples. A thermodynamic prediction for microporosity formation in aluminium-rich Al-Cu alloys. *Metall. Trans. A*, 18:1979-87, 1987.

[17] G. K. Sigworth and T. A. Engh. Chemical and kinetic factors related to hydrogen removal from aluminum. *Met. Trans. B*, 13:447, 1982.

[18] D. M. Stefanescu. Computer simulation of shrinkage related defects in metal castings: A review. *Int. J. Cast Metals Res.*, 18:129-43, 2005.

[19] P. K. Sung, D. R. Poirier, and S. D. Felicelli. Continuum model for predicting microporosity in steel castings. *Model. Simul. Mater. Sci. Eng.*, 10:551-568, 2002.

[20] J. S. Wang and P. D. Lee. Simulating tortuous 3D morphology of microporosity formed during solidification of Al-Si-Cu alloys. *Intl. J. Cast Metals Res.*, 2007.

CHAPTER 13

DEFORMATION DURING SOLIDIFICATION AND HOT TEARING

13.1 INTRODUCTION

We saw in the previous chapter that porosity can be caused by the inability to feed solidification shrinkage due to pressure losses through the mushy zone (shrinkage porosity), or by the segregation of too much dissolved gas in the liquid (gas porosity). Although the same equations are used to describe the two cases, they differ in both the mechanism and possible preventive measures for pore formation. Gas porosity can be eliminated by an appropriate melt treatment, such as degassing, which removes excess dissolved gas prior to casting, or by preventing reactions between the melt and oxygen or water which lead to excess gas in the melt. However, the alloy's tendency to form shrinkage porosity is a more intrinsic characteristic, caused by the formation of a dense phase during late-stage solidification. Consequently, it is harder to avoid.

One way to eliminate shrinkage porosity is to *compress* the mushy zone during freezing. Indeed, we show in Chap. 14 that the partially coherent solid of a mushy zone can be compacted or expanded like a sponge. Putting this concept in the context of Eq. (12.5), in regions of a casting that are under compression, i.e., where $\nabla \cdot \boldsymbol{v}_s < 0$, the deformation of the solid can counterbalance the density increase and thus maintain $g_p = 0$, even when there is no feeding, i.e., for $\boldsymbol{v}_\ell = 0$. This is the idea behind the squeeze casting process, in which an extra compressive stress is applied at certain locations at prescribed times.

However, in many solidification processes such as shape casting, continuous casting or welding, some locations are subjected to tension rather than to compression during freezing. In this case, the deformation of the solid skeleton adds to the solidification shrinkage, thereby inducing even more liquid suction in the mushy zone (see Fig. 12.5). A lack of feeding of mushy regions under tension will result in what is called *hot*

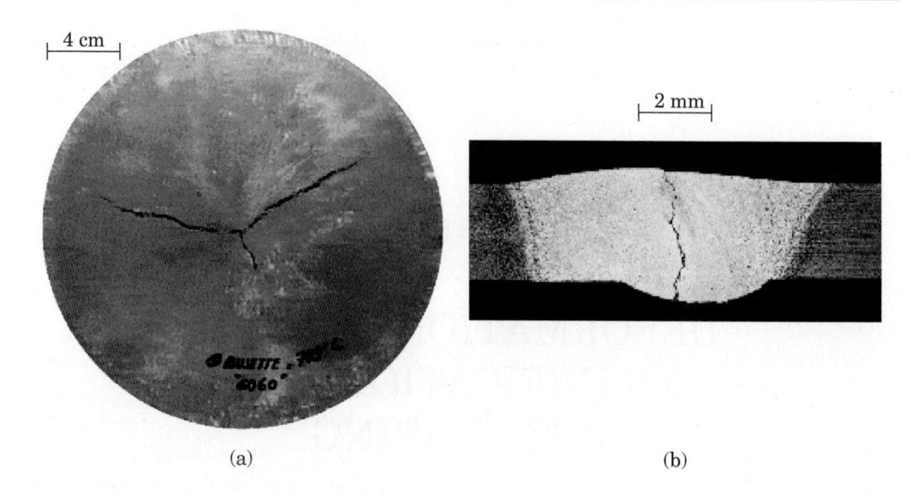

Fig. 13.1 Hot cracks observed (a) at the center of a DC cast and (b) at the weld centerline of an aluminum alloy.

tearing or *hot cracking*, a defect that is similar to shrinkage porosity, but that also requires tensile stresses. Whereas microporosity is fairly uniformly distributed within the mushy zone, hot tears are localized at grain boundaries. Figure 13.1(a) and (b) show, respectively, hot tears formed at the center of a DC-cast Al billet, and along the weld centerline of an aluminum sheet. During both processes, the tears occur in regions where the thermal strains are the largest.

Feeding has already been treated in the chapter on porosity. Therefore, the present chapter focuses on how and where tensile deformation arises during solidification. We show in Sect. 13.2 that non-uniform cooling of the solid induces non-uniform thermal contractions, deformations and stresses. Near the solidus temperature, the elastic limit of the solid is low and viscoplastic deformation of the solid normally occurs at constant volume, i.e., with $\nabla \cdot \boldsymbol{v}_s = 0$. We demonstrate in Sect. 13.3 that a two-phase region with a certain amount of remaining liquid has a much more complicated constitutive response, and includes the possibility of viscoplastic deformation where $\nabla \cdot \boldsymbol{v}_s \neq 0$.

The last two sections of the chapter are dedicated to characteristics, observations and measurements of hot tearing (Sect. 13.4), and to hot cracking models (Sect. 13.5).

13.2 THERMOMECHANICS OF CASTINGS

13.2.1 Origins of thermal stresses

In this section, we explore, through the use of simple models, the origin of deformation and stress during solidification. Dimensional changes are required to accommodate the volume change upon solidification, and more importantly, the thermal contraction of the solid upon cooling to room temperature. These dimensional changes produce stresses if one or more of the following conditions are met: non-uniform temperature, non-uniform mechanical properties, constraints induced by the mold. We illustrate this concept now, using a simple 1-D model of a thin, hot coating at temperature T_{sol} attached to a cold substrate at temperature T_0, and examine the stress state of the assembled pair when equilibrium is established at T_0. One might encounter such a situation in plasma spraying of a thin (and hot) coating over a component.

In order to keep the analysis simple, we consider both the coating and the substrate to be thin enough that they are both in a state of plane stress. We first address the case where the substrate is constrained at its ends, as illustrated in Fig. 13.2 (top). The assembly is assumed to be free to deform in the direction perpendicular to the page, and thus the only non-zero stress is σ_{xx}. It is conceptually useful to separate the process into two steps. In the first step, we imagine that the coating is detached from the substrate and cooled to T_0 without constraint. This would be accompanied by thermal shrinkage ΔL_x, and the associated thermal strain ε_{xx}^{th} is given by

$$\varepsilon_{xx}^{th} = \frac{\Delta L_x}{L_x} = \alpha_{Ts}(T_0 - T_{sol}) < 0 \tag{13.1}$$

where L_x is the nominal length of the substrate and coating and $\alpha_{Ts} = \beta_{Ts}/3$ is the linear thermal expansion coefficient of the coating. In the second step, we re-attach the cold coating to the undeformed substrate, which requires the layer to be mechanically stretched until its length is once again L_x. The *total strain* of the coating is then zero, since its length

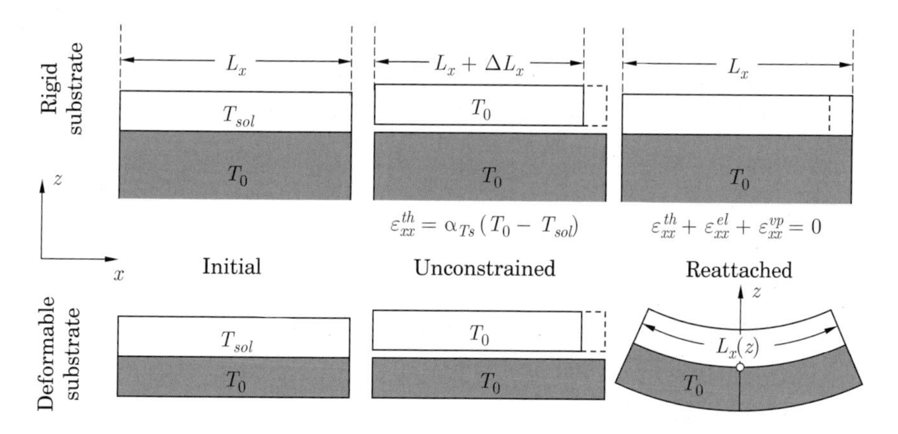

Fig. 13.2 A simple block model of thermal contraction for the case of a thin hot solid cooling over a (top) large undeformable and (bottom) deformable substrate.

remains unchanged from its original value, which implies that the mechanical strain balances the thermal strain,

$$\varepsilon_{xx} = \varepsilon_{xx}^{th} + \varepsilon_{xx}^{el} + \varepsilon_{xx}^{vp} = 0 \tag{13.2}$$

where the superscripts "el" and "vp" refer to elastic and viscoplastic components of the mechanical strain, respectively.

If the deformation is entirely elastic, then $\varepsilon_{xx}^{vp} = 0$ and the elastic stress is obtained using Eq. (13.1) and Hooke's law

$$\sigma_{xx} = E\alpha_{Ts}(T_{sol} - T_0) > 0 \tag{13.3}$$

where E is the elastic modulus of the coating. The case of an elastic-viscoplastic solid is treated in Exercise 13.1.

Let us now consider the case where the substrate is also deformable, and bending is permitted in the x-z plane, as illustrated in Fig. 13.2 (bottom). In this case, when the cooled coating is re-attached to the substrate, the coating remains under tension at the interface, whereas the substrate is under compression. This causes bending of the pair. To provide a more quantitative example, suppose that the coating and the substrate each have the same thickness, e, the same elastic modulus, E, and that the deformation is entirely elastic. The length of any chord, $L'_x(z)$, is shown in Ex. 13.2 to be

$$L'_x(z) = L_x\left[1 - \alpha_{Ts}(T_{sol} - T_0)\left(\frac{1}{2} + \frac{3z}{4e}\right)\right] \quad -e \leq z \leq e \tag{13.4}$$

where z is the distance measured from the centerline. The chord length at the interface position ($z = 0$) is reduced by an amount equal to half of the free thermal contraction of the coating, and the radius of curvature at the interface, R, is given by

$$\frac{1}{R} = -\frac{1}{L_x}\frac{dL'_x(z)}{dz} \Rightarrow R = \frac{4e}{3\alpha_{Ts}(T_{sol} - T_0)} \tag{13.5}$$

Using a typical value for aluminum, $\alpha_{Ts} = 2 \times 10^{-5}$ K^{-1}, and a temperature difference of 500 K, the radius of curvature $R = 133\ e$. The stress in the substrate is given by

$$\sigma_{xx}^{sub} = E\frac{L'_x(z) - L_x}{L_x} = -E\alpha_{Ts}(T_{sol} - T_0)\left(\frac{1}{2} + \frac{3z}{4e}\right) \quad -e \leq z \leq 0 \tag{13.6}$$

The length for the free coating at T_0 is not L_x, but rather $L_x(1 - \alpha_{Ts}(T_{sol} - T_0))$, and the stress in the coating therefore becomes

$$\begin{aligned}\sigma_{xx}^{coat} &= E\frac{L'_x(z) - L_x(1 - \alpha_{Ts}(T_{sol} - T_0))}{L_x} \\ &= E\alpha_{Ts}(T_{sol} - T_0)\left(\frac{1}{2} - \frac{3z}{4e}\right) \quad 0 \leq z \leq e\end{aligned} \tag{13.7}$$

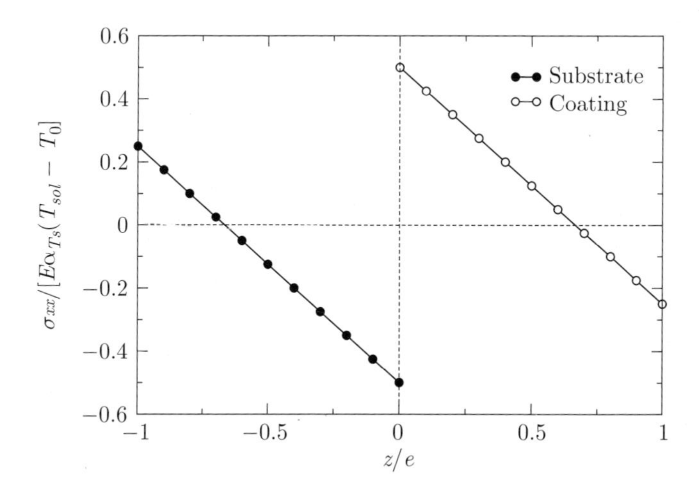

Fig. 13.3 Stress in a coating-substrate of equal thickness and elastic modulus as a function of the normalized thickness, z/e. The stress has been normalized by the factor $E\alpha_{Ts}(T_{sol} - T_0)$.

Thus, we see that the stresses in the coating and substrate at the interface are tensile and compressive, respectively, but as one moves towards the free surfaces, the sign of the stress changes in both entities (see Fig. 13.3).

Although simple, this "block model" serves to illustrate two important features of thermal deformation: (i) A component that is not constrained by external forces, e.g., by a rigid mold or welding clamps, bends towards the hottest region (Fig. 13.2); (ii) The interfacial region at the interior of the curvature tends to be under tension, whereas the external part is subjected to compression. Such simple models are very helpful to better understand more complex solidification processes, such as those presented in Sect. 13.2.3.

13.2.2 General formalism for a fully solid material

In Sect. 4.3.2, we demonstrated that the total mechanical strain ε_{ij} can be decomposed into elastic and viscoplastic contributions, ε_{ij}^{el} and ε_{ij}^{vp}, respectively. In order to model thermal stresses, as in the simple thermoelastic block model described in the previous section, we also include the thermal strain, ε_{ij}^{th}, induced by thermal contractions. In certain materials, one must also consider an additional strain contribution, ε_{ij}^{tr}, induced by solid state transformations, e.g., the volume change of approximately 4% associated with the austenite-martensite transformation in steels. The total strain thus becomes the sum of four contributions:

$$\varepsilon_{ij} = \varepsilon_{ij}^{el} + \varepsilon_{ij}^{vp} + \varepsilon_{ij}^{th} + \varepsilon_{ij}^{tr} \tag{13.8}$$

If the material is isotropic, the thermal and transformation strain tensors each reduce to a single scalar material parameter, such that

$$\varepsilon_{ij}^{th}(\boldsymbol{x},t) = \alpha_{Ts}(T(\boldsymbol{x},t) - T_{ref})\delta_{ij}; \qquad \varepsilon_{ij}^{tr}(\boldsymbol{x},t) = \frac{\beta_{tr}}{3} g_{tr}(\boldsymbol{x},t)\delta_{ij} \tag{13.9}$$

where T_{ref} is a reference temperature at which the strain is defined to be zero. Notice that in one dimension, ε_{xx}^{th} reduces to Eq. (13.1). In the expression for ε_{ij}^{tr}, g_{tr} is the volume fraction of solid that has transformed into the new phase, and β_{tr} is the volumetric expansion coefficient associated with the transformation. In general, when computing thermal stresses, the temperature field is obtained first in a separate calculation. Computation of g_{tr} requires an additional microscopic model of transformation. In most cases, such models are obtained from Time-Temperature-Transformation (TTT) diagrams (cf. Jacot et al. [15] for more details). Since the thermal and transformational strains depend only on temperature (and possibly on the cooling rate), they become "loads" in the momentum balance equations that are subsequently solved for the displacement field, $u_i(\boldsymbol{x},t)$. Appropriate constitutive models are supplied in order to relate the stress to the elastic and viscoplastic strains (see Chap. 4). Although these equations are generally nonlinear, many commercial codes are available for their solution.

In most solidification problems, it is not necessary to consider the heat generated by deformation in the energy equation. The thermal and stress-deformation fields are nevertheless coupled in several ways. One form of coupling derives from the temperature dependence of the mechanical and physical properties of the material. The shape of the isotherms, through the magnitude and orientation of the thermal gradient, affects the deformation and stresses, as we saw in the simple block model. Deformation can also affect the thermal field by altering the boundary conditions. For example, air gap formation decreases the heat transfer coefficient between the mold and casting by as much as an order of magnitude. On the other hand, increased contact pressure in other regions can raise the local heat transfer coefficient by an almost as large a factor.

13.2.3 Examples

Hot tearing is a major problem in continuous casting processes as well as in welding, as illustrated in Fig. 13.1. Figure 13.4 demonstrates the mechanism of stress/strain build-up for both processes. In Direct Chill (DC) casting of aluminum, the melt is initially poured onto a bottom block inserted inside the open mold. The thermal gradient is therefore vertical, and according to the insight gained from the simple block model of Sect. 13.2.1, the solidified ingot has a tendency to curve upwards, and thus to lift off from the bottom block. This so-called "butt curl" can be as large as several centimeters at the edges of a 2 m wide slab, and can thereby strongly affect the thermal exchange with the bottom block. The width of the ingot in this bottom region is close to that of the rectangular

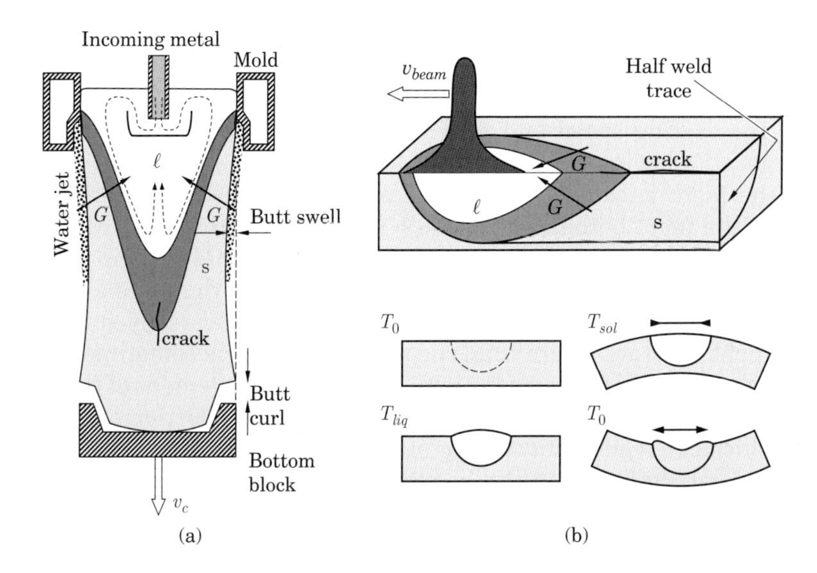

Fig. 13.4 A schematic of hot tear formation (a) in a DC casting of aluminum and (b) during e-beam or laser welding. The shapes of the transverse cross-sections before and after welding are shown at the bottom of the figure.

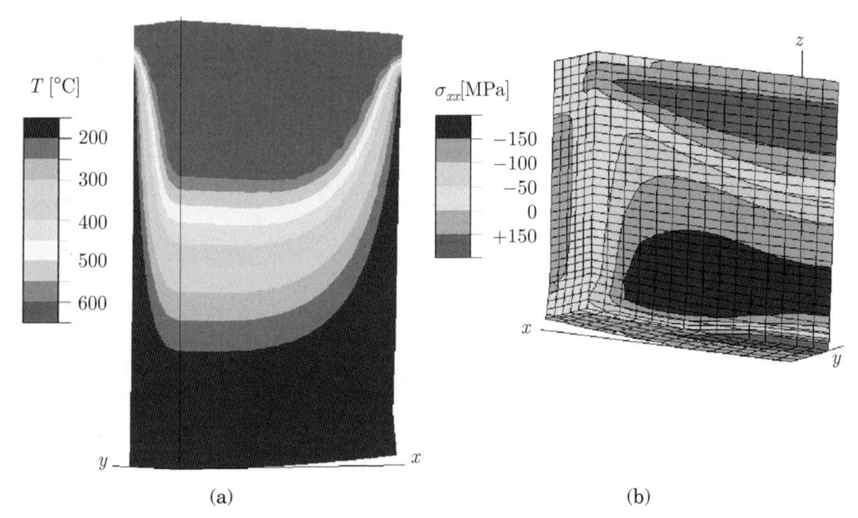

Fig. 13.5 The steady-state temperature profile (a) and the σ_{xx} component of the stress tensor (b) as calculated for a DC casting of an AA1201 alloy at 60 mm/min. (After Drezet and Rappaz [9].)

mold, decreased by the thermal contraction of the solid ($\alpha_{Ts}(T_{sol} - T_0) \approx 1.2\%$ for aluminum). As the ingot is withdrawn, a steady-state regime is established, for which the thermal gradient on both sides of the rolling faces of the ingot becomes inclined due to lateral cooling from water jets. Figure 13.5 presents a calculated temperature profile for a DC cast ingot.

The angle of the temperature gradient with respect to the vertical axis increases with the depth of the liquid sump, the latter being a function of the ingot Péclet number, $v_c L_x / \alpha_T$, where α_T is the thermal diffusivity, v_c the casting speed and L_x a characteristic transverse dimension of the ingot. Following the argument of Sect. 13.2.1, the lateral faces pull in as the ingot is extracted from the mold. As a result, the steady-state width of the ingot becomes about 10% smaller than the inner width of the mold under typical DC casting conditions. One should keep in mind that this value measures the horizontal displacement of the faces and not the actual, local deformation of the solid. The fact that the steady-state portion of the ingot contracts more than the start-up region is sometimes referred to as "butt swell" of the ingot bottom. The inward movement of the two lateral faces also induces tensile stresses in the central portion of the mushy zone, which can lead to hot tearing in susceptible alloys. Figure 13.5 shows the temperature field and the σ_{xx} component of the stress tensor during DC casting of a nearly pure aluminum alloy.

It should be noted that the rolling faces of a DC-cast aluminum ingot do not pull in uniformly because their edges are mechanically reinforced by the solidification of the short sides of the ingot. In order to compensate for this non-uniformity, manufacturers commonly use convex molds, whose shapes are optimized to produce the flattest possible ingots for subsequent rolling operations. Designing such a mold is no simple task, since the final shape of the ingot depends on the inclination of the thermal gradient on each of its sides and therefore on the sump depth. Thus, the cross-sectional shape varies with casting speed, in addition to cooling conditions, alloy properties, etc. Figure 13.6 shows the cross section profile of DC-cast AA5182 ingots (essentially an Al-4.5%Mg alloy) when cast in such a mold

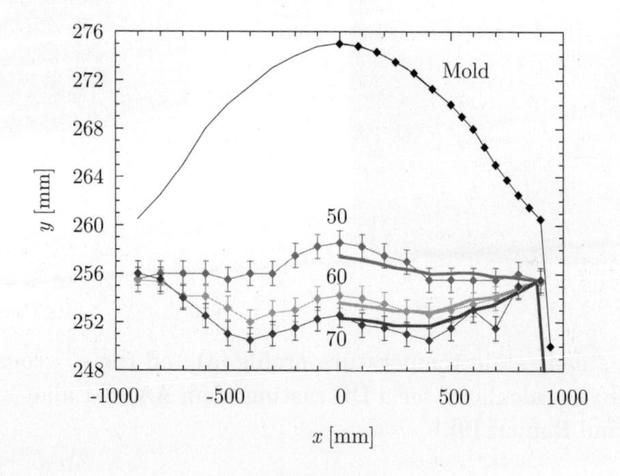

Fig. 13.6 Convex shape of a DC casting mold, and final cross section of DC-cast ingots obtained at steady state in this mold when casting at speeds $v_c = 50, 60$ and 70 mm/min. The profiles shown with thin lines and symbols correspond to measured values, whereas the thick lines represent simulation results. (After Drezet and Rappaz [9].)

at various casting speeds. The internal cross section of the mold itself is given by the most convex curve (note that the $x-$ and $y-$scales are not the same). The other curves correspond to measured (symbols) and calculated (heavy lines) profiles of ingots cast at various speeds. As expected, the larger the velocity, the deeper is the sump depth and the more inclined is the lateral thermal gradient. This induces a larger pull-in of the faces at the center, while the contraction of the ingot edges is almost independent of the casting speed.

Thermal stresses in welding can be explained in a very similar way to those during continuous casting (Fig. 13.4b). As the weld heat source moves, the material located ahead of it melts, and solidification takes place behind it. For the sake of simplicity, let us again use the simple block model from Sect. 13.2.1, assuming that only the weld trace (i.e., the material that melts and re-solidifies) heats and cools during welding, whereas the base material remains cold. The bottom right part of Fig. 13.4 illustrates the sequence of steps in the deformation of the weld. Ahead of the heat source, the plate is flat at a temperature T_0 and stress-free. When the weld trace is heated up to the solidus temperature T_{sol}, it wants to expand. This causes the plate to bend in the direction opposite that of the thermal gradient, and the weld trace section will be under local compression. Once the weld trace melts (marked as T_{liq} in Fig. 13.4), the stresses there are relieved and the plate goes back to being flat, provided that the deformation of the base material was elastic. The expansion in the solid state and upon melting causes the volume of the weld pool to be slightly larger than the initial volume of the cold material. Behind the moving heat source, solidification occurs without stress. Note that the volume of the weld trace returns to its original size, but the surface is no longer flat since solidification started at the edges. During cooling, the weld trace tries to shrink, but because it is constrained, it is subjected to tension. As discussed earlier, in such situations, the plate curves in the direction of the thermal gradient and the weld centerline, now under tension, is the most sensitive to hot cracking. Naturally, the stress state of the weld depends on the boundary conditions imposed on the plate. A strongly clamped weldment experiences higher

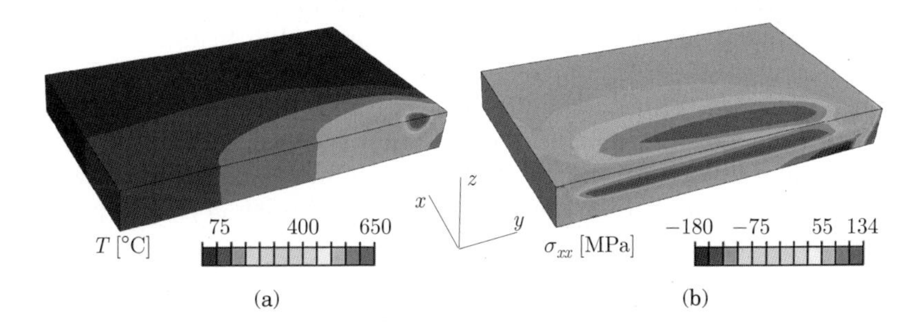

Fig. 13.7 (a) The temperature field, and (b) the transverse component parallel to the welded surface component of the stress tensor, σ_{xx}, during laser welding of an aluminum alloy. (After Drezet et al. [8].)

stresses and less deformation. As an example, Fig. 13.7 shows the temperature field and the σ_{xx} component of the stress tensor during laser welding of an aluminum alloy.

Finally, we note that thermomechanical aspects are also important for shape castings, especially in permanent molds, since they affect the heat transfer coefficient during solidification as well as the final shape dimensions and the residual stress state. When close to sharp re-entrant corners, they can also induce hot tearing (see Exercise 13.4).

13.3 DEFORMATION OF THE MUSHY ZONE

In the previous section, we examined the development of stresses and strains in fully solid materials. We now shift our attention to deformation of the mushy zone. The behavior of the semi-solid is quite different from both that of the solid, which we treated as an elastic-visco-plastic material, and of the liquid, in which state most metals can be modeled as Newtonian fluids. The two-phase mixture has several characteristics making its rheology significantly more complicated than that of its constituent solid and liquid phases:

1. *The stress state is inhomogeneous.* The solid and liquid phases bear unequal loads, and the phases can separate during deformation.

2. *The microstructure affects the response.* The morphology of the solid phase has a significant effect on the mechanical behavior. A network of dendrites can become interlocked at relatively low volume fraction of solid, whereas a microstructure consisting of globular grains can be easily deformed up to relatively high volume fractions of solid.

3. *During deformation, the microstructure changes, and its evolution may depend on the deformation itself.* As the temperature decreases, the solid fraction increases, and at a certain point, which depends on the microstructure (see the previous item), the structure rigidifies. Further, even if the temperature is maintained constant, the microstructure coarsens over time and solid bridges may form between neighboring particles. This will, in turn, affect the mechanical response. Finally, the presence of stress during solidification can affect the morphology of the solid. A familiar example of this last phenomenon is found in home-made ice cream, where the shear applied during freezing alters the morphology of the ice crystals from dendritic to globulitic.

These phenomena make it difficult to measure consistent rheological properties, which has led to the development of a wide variety of testing devices and procedures, as discussed in the following section. Prior to this, however, it is useful to present a qualitative description to provide a context for the results obtained in such tests. A common apparatus for measuring rheological properties places the material in a narrow gap,

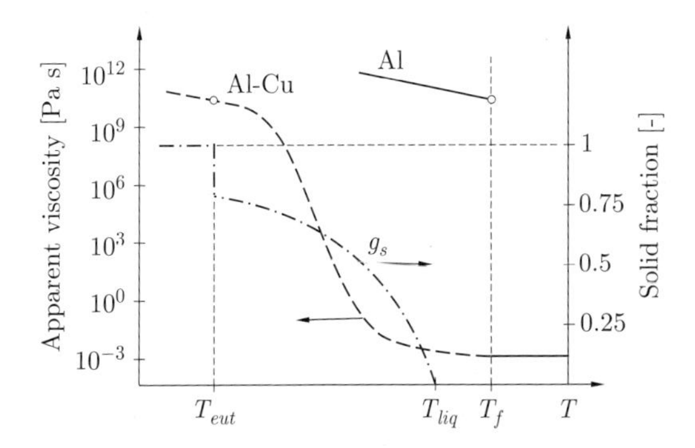

Fig. 13.8 A semi-quantitative illustration of the development of the apparent viscosity of a metal (in this case pure aluminum and an Al-Cu alloy) during solidification.

e.g., between parallel plates. Shear is applied, and measurements of the relative speed and force on the plates are readily converted into shear stress, σ_{xy}, and strain rate, $\dot{\varepsilon}_{xy}$.[1] One then defines the "apparent viscosity", η, by the relation

$$\sigma_{xy} = \eta(T, g_s, \mathcal{S}_V, \dot{\varepsilon}_{eq})\dot{\varepsilon}_{xy} \tag{13.10}$$

where we have noted that the apparent viscosity (also called the non-Newtonian viscosity) may depend on the temperature, T, solid fraction, g_s, microstructure (as represented by the specific surface area $\mathcal{S}_V^{s\ell}$), and the magnitude of the strain rate, $\dot{\varepsilon}_{eq}$ (see Chap. 4), in turn defined as

$$\dot{\varepsilon}_{eq} = \frac{\sqrt{2}}{3}\sqrt{(\dot{\varepsilon}_{xx} - \dot{\varepsilon}_{yy})^2 + (\dot{\varepsilon}_{xx} - \dot{\varepsilon}_{zz})^2 + (\dot{\varepsilon}_{yy} - \dot{\varepsilon}_{zz})^2 + 6\dot{\varepsilon}_{xy}^2 + 6\dot{\varepsilon}_{xz}^2 + 6\dot{\varepsilon}_{yz}^2} \tag{13.11}$$

When discussing the various tests in the next section, each of which may produce more complex stress states than simple shear, it is most convenient to use the von Mises or equivalent stress, defined in Eq. (4.88) and repeated here to facilitate the presentation

$$\sigma_{eq} = \frac{1}{\sqrt{2}}\left[(\sigma_{xx} - \sigma_{yy})^2 + (\sigma_{yy} - \sigma_{zz})^2 + (\sigma_{zz} - \sigma_{xx})^2 + 6\sigma_{xy}^2 + 6\sigma_{yz}^2 + 6\sigma_{zx}^2\right]^{1/2} \tag{13.12}$$

Rewriting Eq. (13.10) in terms of σ_{eq} and $\dot{\varepsilon}_{eq}$, we have

$$\sigma_{eq} = 3\eta(T, g_s, \mathcal{S}_V, \dot{\varepsilon}_{eq})\dot{\varepsilon}_{eq} \tag{13.13}$$

A schematic plot of η vs. T at a fixed shear rate is shown in Fig. 13.8. For pure aluminum, there is a sharp transition at the melting point between the Newtonian viscosity in the liquid and the non-Newtonian viscosity in the solid, in this case computed from Eq. (13.13) using a measured

[1]The strain rate tensor, $\dot{\boldsymbol{\varepsilon}}$, defined as twice the rate of the deformation tensor, $\boldsymbol{D}$, is more commonly used in this context.

value of the flow stress of about 3 MPa at a strain rate of 10^{-4} s^{-1}. For an alloy, the response is more complicated. We sketch the general trend of the apparent viscosity, which at first rises slowly with solid fraction, corresponding to a microstructure consisting of isolated particles surrounded by liquid, and then begins to increase sharply as the particles start to interact. The exact solid fraction and temperature where this occurs vary with the microstructure, as discussed further in the following sections. Eventually, the fully solid alloy shows a similar solid state behavior to that of pure aluminum.

Finally, we should point out that the non-Newtonian viscosity is a measure of an aggregate property of the semi-solid mixture, i.e., it is an average property over a volume element that includes both phases. The actual shear rate in the solid (10^{-5} to 10^{-2} s^{-1}) is much smaller than that in the liquid (1 to 10^{3} s^{-1}). These shear rates are considerably larger than those associated with the flow in the mushy zone related to feeding solidification shrinkage.

13.3.1 Rheological measurements on semi-solid alloys

As noted above, the rheology of semi-solid materials is a very complex phenomenon, from both experimental and theoretical perspectives. The material response depends strongly on the microstructure in a way that cannot be captured easily. In particular, it is not sufficient to use a single parameter such as temperature or solid fraction, because the microstructure evolves with time through processes such as coarsening, coalescence and fragmentation. For this reason, experimental measurements performed at any specific solid fraction upon solidification from the liquid provide very different results compared to measurements carried out at the same value of g_s, but upon reheating of a fully solid specimen. In addition, the tests themselves modify the microstructure. For example, solid bridges established among the various grains can be broken during the initial steps of a shearing test. The apparent viscosity of the semi-solid often displays a maximum, followed by a decrease to a steady value as grain bridges are broken by shearing. This is an example of a time-dependent response, called *thixotropy*. The same mechanism accounts for the observation that mushy alloys are generally *shear thinning*, i.e., the viscosity decreases with an increased shear rate. Numerous rheological tests have been devised, and we discuss only a few of them in this section. We refer the interested reader to the review by Eskin et al. [10], for a good overview of the various tests, with a particular focus on Al alloys.

Shear tests on semi-solid alloys

Shear tests have been carried out during solidification and reheating using Couette, planar or tubular viscometers. These and several other examples of shearing devices are illustrated in Fig. 13.9. For lower volume fractions of solid, a blade-type viscometer has also been used to measure the

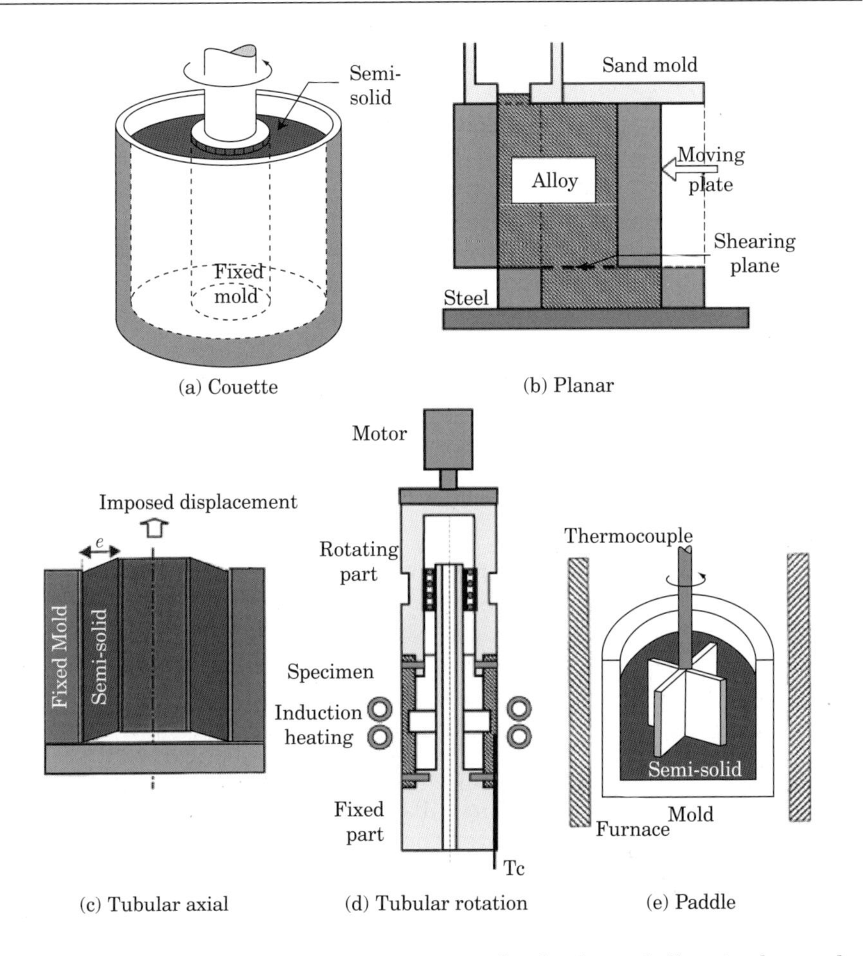

Fig. 13.9 Major shear devices used to measure the rheology of alloys in the mushy state: (a) Couette viscometer, (b) planar shearing device (Sumitomo et al. [28]), (c) tubular with axial shear (Ludwig et al. [17]), (d) tubular with azimuthal shear (Grasso et al. [13]), and (e) blade viscometer (Dahle and Arnberg [5].)

temperature or solid fraction, where a significant torque first appears on the axis holding the blade. A typical result from a tubular axial shear test on a grain-refined Al-2wt%Cu alloy is shown in Fig. 13.10. At a fixed volume fraction of solid, $g_s = 0.99$ (Fig. 13.10a), there is an initial increase in σ_{eq} with equivalent strain ε_{eq} up to about 10%–20% strain, after which the material flows at essentially constant stress, which is typical of a viscous material. This stress, called the *saturation stress*, σ_{sat}, increases with the strain rate. The relation between σ_{sat} and $\dot{\varepsilon}_{eq}$ is often written in the form of a power law

$$\sigma_{sat} \propto \dot{\varepsilon}_{eq}^{m} \tag{13.14}$$

where m is called the *strain rate sensitivity*. When $m < 1$, the material displays a shear thinning behavior, and the case $m = 1$ corresponds to a

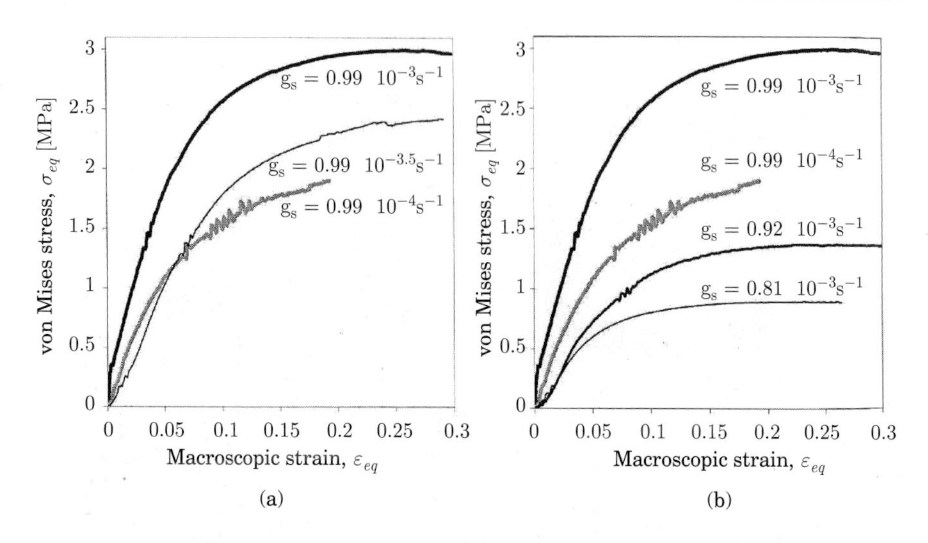

Fig. 13.10 (a) Stress-strain curves for an Al-2%Cu alloy solidified up to $g_s = 0.99$, measured with an axial tubular shear device at various shear rates. (b) The effect of the volume fraction of solid on the deformation behavior. (After O. Ludwig et al. [17].)

Newtonian fluid. Analysis of the data in Fig. 13.10(a) demonstrates that for $g_s = 0.99$, $m = 0.22$, and the data in Fig. 13.10(b) show that m increases as the solid fraction decreases ($m = 0.33$ for $g_s = 0.81$). One would expect that $m \to 1$ as $g_s \to 0$ to recover the Newtonian behavior of the liquid phase.

Figure 13.10(b) also shows that the saturation stress ranges between 1 and 3 MPa as g_s varies between 0.8 and 1. At lower volume fractions of solid, σ_{sat} is significantly lower, typically 5 to 30 kPa for $g_s = 0.3 - 0.5$ [28]. The observed behavior is sometimes non-monotonic: after the stress reaches a maximum, it may decrease before saturating at a lower value. The peak stress corresponds to the initial stress required to separate grains linked by solid bridges, thus making them more globular, and leading to a final state of lower apparent viscosity. As this domain of g_s is less relevant to hot tearing, we do not discuss such measurements further. We note, however, that this behavior is important for numerous processes, e.g., in thixocasting, where an alloy is molded in the semi-solid state.

As noted earlier, the stress-strain response of an alloy also depends on its microstructure, in addition to g_s and $\dot{\varepsilon}$. For example, the shear resistance of directionally solidified specimens measured with the rotational shear device Fig. 13.9(d) is larger when the stress is applied in a plane perpendicular, as opposed to parallel, to the grain boundaries. Both resistances are larger than that of the same alloy when it has a fine equiaxed microstructure (see Grasso [13]). Grain boundaries and the solid bridges that cross them are responsible for such behavior, as discussed in more detail in Sect. 13.3.2.

Tensile tests on semi-solid alloys
Some of the devices commonly used to measure the tensile properties of mushy zones during solidification are shown in Fig. 13.11. In the simplest apparatus, shown in Fig. 13.11(a), a specimen is attached to the jaws of a standard tensile testing machine and the central portion is completely

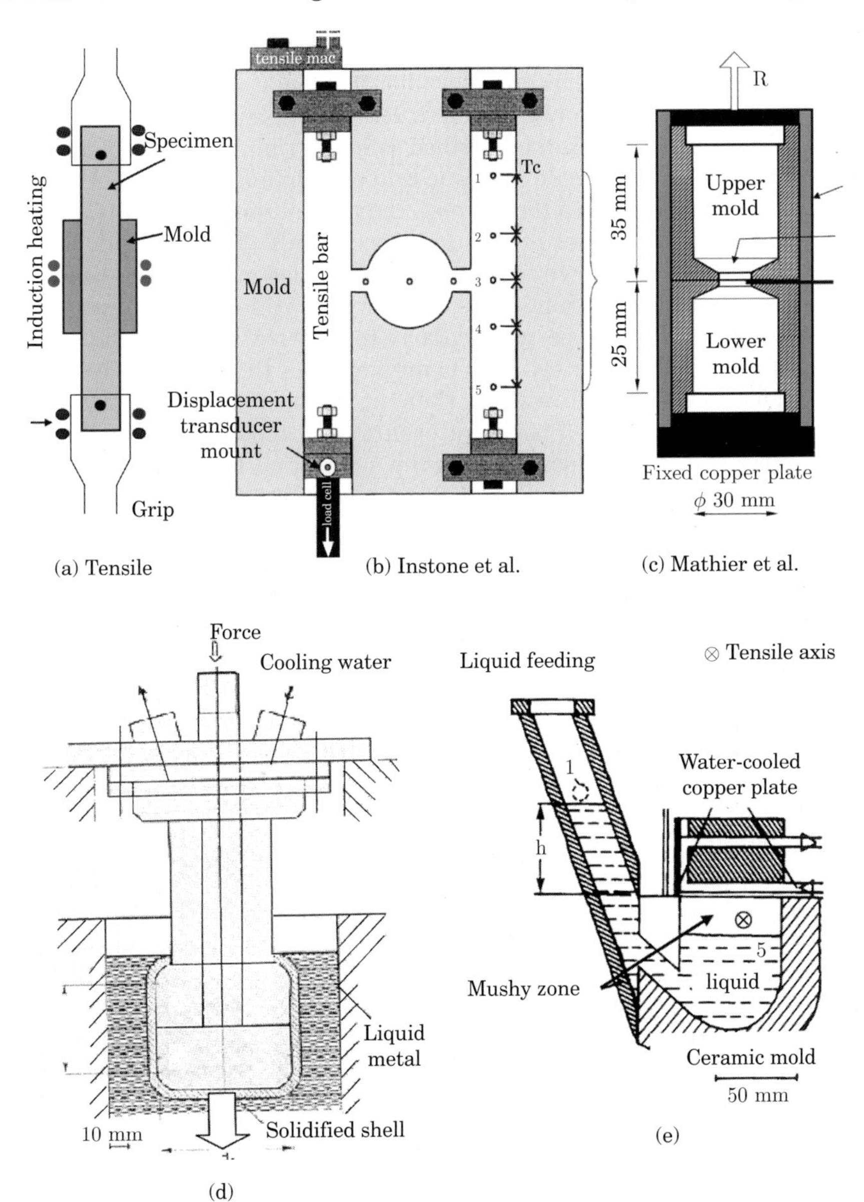

Fig. 13.11 Common devices used to perform tensile tests on mushy zones: (a) simple tensile test, (b) device of Instone et al. [14], (c) device of Mathier et al. [18], (d) device of Ackermann et al. [1], and (e) device of Ohm and Engler [21].

remelted, then the stress-strain curve is directly measured during re-solidification. Two bars are cast simultaneously in Instone's device, shown in Fig. 13.11(b), which can also be used for hot tearing tests. One bar is used to measure the stress-strain curve, and the other is instrumented with thermocouples to monitor the temperature. A hot spot is created at the center in order to localize the deformation. In Mathier's device, shown in Fig. 13.11(c), an inoculated alloy is directionally solidified in a two-part mold. The temperature gradient is made fairly small, so as to have a more or less uniform volume fraction of solid in the restriction of the mold at mid-height. At a prescribed solid fraction, the upper part of the mold is raised by the tensile machine and the force is recorded. Although this procedure resembles a tensile test, the mushy zone is actually sheared from the re-entrant corner of the upper mold half. The device developed by Ackermann et al., shown in Fig. 13.11(d), is also more like a shear testing apparatus and is in some ways similar to Mathier's. However, in this case, the load is applied perpendicularly to the thermal gradient. A two-piece water-cooled copper cylinder is immersed in the liquid bath. A solid crust starts to form at the surface, then, at a preset time, the bottom part is pushed away from its upper counterpart. Columnar dendrites that had grown from the surface located near the separation line between the two copper mold halves are therefore sheared transverse to their growth direction. Since shear is initiated at the dendrite roots, this test is probably the one that most closely resembles a real hot tearing situation. Finally, in the device of Ohm and Engler shown in Fig. 13.11(e), an alloy is solidified from an upper surface with a bottom feeding. Once a predetermined solid fraction is reached, the upper cooling plate is removed and the stress is applied perpendicularly to the solidification direction by a tensile testing machine.

Figure 13.12 shows stress-strain curves measured by Vernède [33] on an Al-2%Cu alloy, using a device similar to that of Ohm and Engler [21]. The stress-strain curves on the left were measured using various strain rates and $g_s = 0.92$. The maximum stress, σ_{max}, is in the range of 200-400 kPa, with a weak dependence on the strain rate. Notice that

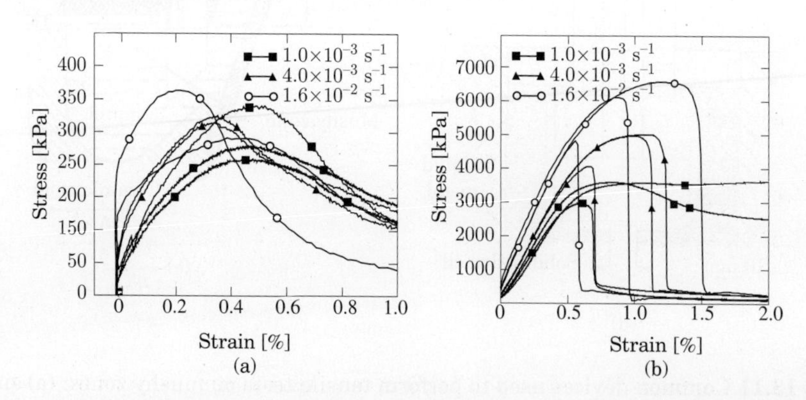

Fig. 13.12 Stress-strain curve measured with a device similar to that of Engler et al. for Al-2%Cu at (a) $g_s = 0.92$ and (b) $g_s = 0.99$ at various strain rates. (After Vernède [33].)

the maximum tensile strength of the alloy is considerably smaller than the shear strength of the same alloy shown in Fig. 13.10, and the overall shape of the curve is quite different as well. In these tests, Vernède found that after reaching a maximum, the stress decreases smoothly to reach a nearly flat plateau at around 50–100 kPa. It seems that the microstructure, after an initial breakage of the bonds between the grains, is able to re-establish a weaker resistance, probably by some liquid flow compensating the movement and rupture of the solid skeleton. On the other hand, the situation for $g_s = 0.99$, shown in Fig. 13.12(b), is markedly different. The maximum resistance of the mushy zone (3 to 7 MPa) is comparable to that during shear (Fig. 13.10) and is strongly dependent upon the strain rate. At the two highest strain rates, rupture of the specimen is very abrupt whereas, at the lowest value (10^{-3} s^{-1}), the ductility is much larger and the structure does not show any significant loss of resistance.

The transition in behavior between $g_s = 0.9$ and $g_s = 1$ is examined further in Fig. 13.13, in which we can see the maximum stress prior to rupture for an inoculated Al-1%Cu alloy as a function of g_s. These results were obtained by Mathier et al. [18] using the device shown in Fig. 13.11(c). Ludwig et al. [17] found similar results, displayed in Fig. 13.13(b), in inoculated Al-2%Cu and Al-4%Cu alloys, using the tensile device from Fig. 13.11(a). One can see that there is a fairly sharp transition at $g_s \approx 0.9$–0.95. Below this range, the mushy zone is weak whereas above, it resembles more closely that of a solid. Novikov [10] reported ultimate tensile strength (UTS) values for several Al-Cu alloys as functions of temperature and Cu content, as shown in Fig. 13.14. Although these tests exhibit a similar behavior, it should be pointed out that the measurements in this case were carried out on remelted samples. Thus, the microstructure is not necessarily the same as that of the samples shown in Fig. 13.13, for which measurements were performed during solidification.

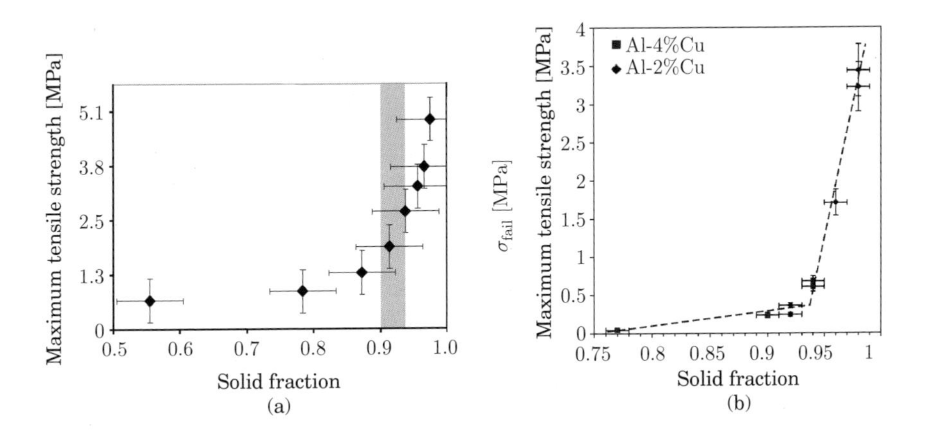

Fig. 13.13 The maximum tensile strength measured in (a) an Al-1%Cu alloy using the tensile device of Mathier et al. and (b) Al-2%Cu and Al-4%Cu alloys using a tensile device. All alloys were inoculated and the measurements were performed upon solidification. (After Mathier et al. [18] and Ludwig et al. [17].)

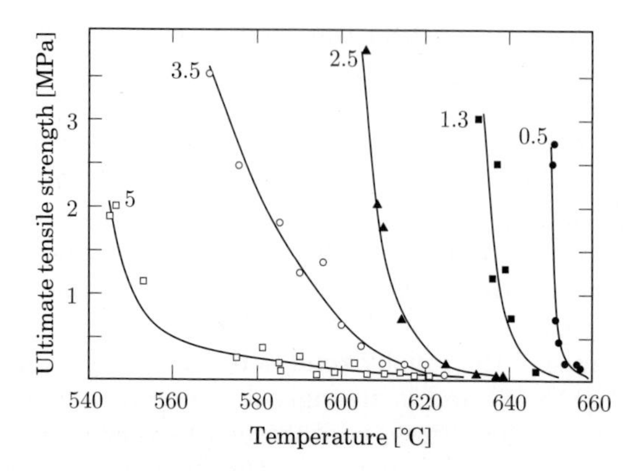

Fig. 13.14 Ultimate tensile strength of several Al-Cu alloys of various compositions indicated in wt%, measured using a loading rate of 10 N/s. (Redrawn from Novikov, see ref. of Eskin et al. [10].)

In summary, the strength of the mushy zone, both in tension and shear, increases with g_s and with $\dot{\varepsilon}$, at least at high volume fractions of solid. This type of behavior is correlated with the morphology of the microstructure in Sect. 13.3.2.

Ductility of mushy zones

If the maximum strength increases with g_s, then what about the ductility, defined as the maximum strain before cracking? Most semi-solids exhibit a *Brittle Temperature Range* (BTR), as illustrated in Fig. 13.15 for an inoculated Al-2wt%Cu alloy. The maximum strain at fracture observed in the experiments shown in Fig. 13.12 is reported as a function of g_s. At low temperature ($g_s \approx 1$), the strength of the alloy is high (several MPa) as seen in the preceding section. The ductility is also high. At higher temperature, when $g_s < 0.9$, the strength of the alloy is low... but its ductility in this case is also fairly high! The reason for this behavior lies in the mechanisms introduced in the previous section. Although solid bridges at such temperatures are easily broken, the liquid also can flow readily to fill the empty spaces. Therefore, the mushy zone can adapt its morphology to the deformation without opening a crack. Between these two limits, the mushy zone exhibits a brittle behavior, especially near the value of g_s where a sharp increase of σ_{max} is observed ($g_s \simeq 0.95$, see Fig. 13.13). This is because the remaining liquid is present in the form of continuous films, instead of being in the form of isolated droplets in the solid at higher g_s. These films make the material brittle. However, unlike the case where $g_s < 0.9$, an opening of the mushy zone along such films cannot be compensated by liquid flow because the permeability is too low. The ductility measured around $g_s = 0.95$ is nearly zero, regardless of the strain rate. Similar data have been obtained on remelted aluminum alloys by Novikov [10].

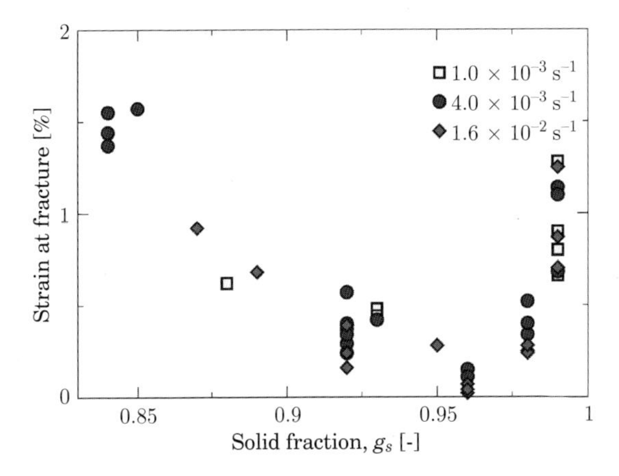

Fig. 13.15 The ductility of an inoculated Al-2%Cu alloy at various strain rates, as measured with a tensile testing device similar to that of Ohm and Engler (see Fig. 13.11e). (After Vernède [33].)

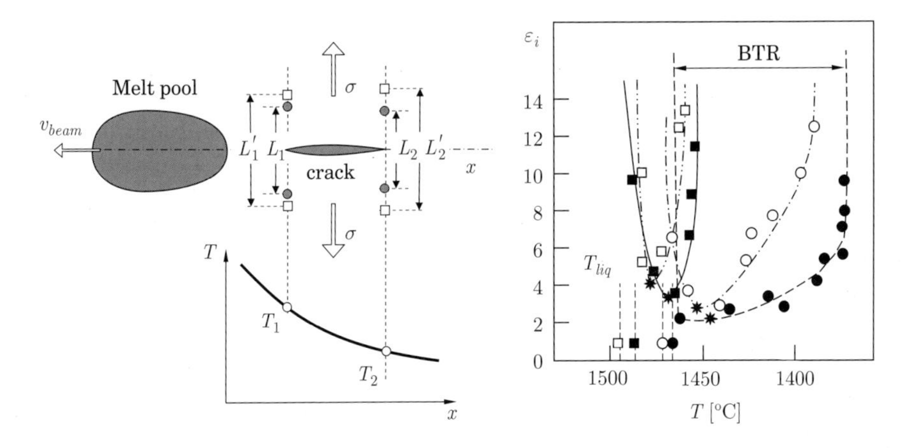

Fig. 13.16 Deformation at the upstream (1) and downstream (2) crack tip position during e-beam welding of steel for 4 concentrations: 0.56%C (filled circles), 0.41%C (open circles), 0.32%C (filled squares) and 0.21%C (open squares). The deformation $\varepsilon_i = (L'_i - L_i)/L_i$ at the crack initiation is indicated with a star. The other values are for both ends of the crack. Furthermore, the liquidus of each alloy is given with a vertical dashed line. (Redrawn from Matsuda et al. [19].)

Another process in which loads are applied during solidification is welding. Tensile stresses build up behind the weld pool, making the weld centerline particularly sensitive to hot cracking (see Sect. 13.2.3). Matsuda et al. [19] estimated the hot ductility of various carbon steels by applying a stress perpendicular to the welding direction, while simultaneously using an optical microscope to observe crack formation along the weld centerline, as well as the lateral displacement of various surface defects (inclusions, pits, etc.). The experimental setup is shown schematically in Fig. 13.16.

The deformation at both crack tips could be estimated and related to the temperature field. The strains ε_1 and ε_2 measured at the crack tips also showed a BTR. At some level of deformation, a crack initiates behind the liquidus temperature (indicated by star symbols in Fig. 13.16). The crack propagates as the two parts of the welds are pulled apart, but it stops at a position close to the liquid pool because at low enough g_s, the crack can be filled with interdendritic liquid. On the other side of the crack, when the fraction of solid reaches a value close to unity, the material is sufficiently ductile for the crack to stop there as well. As the carbon concentration increases (as does the solidification interval), the BTR shifts with the respective liquidus (also indicated in Fig. 13.16) and widens. The reduction of the ductility and the strength build-up of the mushy zone as the solid fraction increases is discussed further in the next section.

13.3.2 Coherency

The term "*coherency*" is commonly used in discussions of hot tearing, but unfortunately not always with the same meaning by all authors! The coherency point can be defined as the volume fraction of solid, or the temperature, at which a given alloy starts to develop mechanical resistance. But the questions are: in which mechanical test and for which morphology? In order to better understand the complexity of defining a unique coherency point, it is necessary to study the evolution of the mushy zone over the entire solidification range.

Let us begin by considering a single dendritic grain. Figure 13.17 shows a sequence of six micrographs taken during growth of succinonitrile-acetone columnar dendrites, redrawn from the original observations for clarity. The specimen was grown between two glass plates which were drawn at constant velocity through a fixed temperature gradient, using a device similar to that described in Chap. 8. Optical micrographs were taken at various times, of which $t = 0$ represents the time at which the dendrite tips passed under the microscope. Observing the microstructure

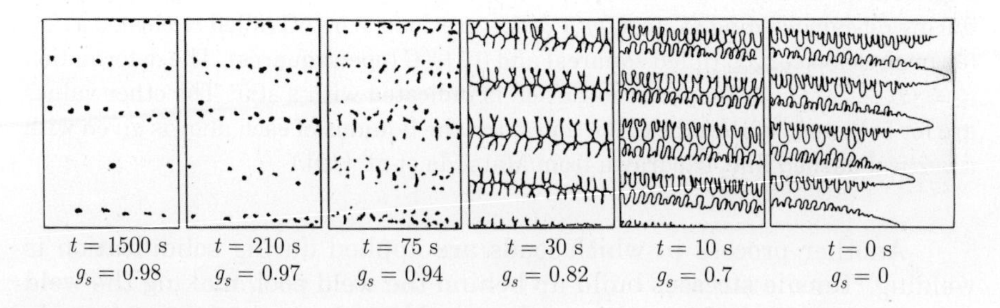

Fig. 13.17 Growth and coalescence of dendrite arms within a single grain of succinonitrile-acetone. The microstructure has been redrawn from the original micrographs to highlight the solid-liquid interface. The volume fractions of solid are indicated below each image.

deeper within the mushy zone, i.e., with higher g_s, one first notices that the dendrite arms become coarser and thicker. At $t = 30$ s, the dendritic network fills nearly the entire frame, but g_s is still only 0.82. Note that, at this point, the columnar dendrites remain separated from their neighbors by continuous liquid films. Between $t = 30$ s ($g_s = 0.82$) and $t = 75$ s ($g_s = 0.94$), a dramatic change in topology occurs. The dendrite arms *coalesce* or *bridge*, transforming continuous liquid films into isolated liquid droplets. Coalescence or bridging, not to be confused with coarsening, is defined as the replacement of two solid-liquid interfaces by one solid bridge. This *intragranular coalescence* or *intragranular bridging* does not have to surmount any grain boundary energy barrier, and thus occurs at a fairly low volume fraction of solid (~ 0.9), as soon as two dendrite arms touch. The temperature at which this takes place is designated T_b^{intra}. Below $g_s = 0.82$, dendrite trunks do not pose much resistance to tensile stresses applied perpendicular to their growth direction, whereas above $g_s = 0.94$, the mushy zone behaves as a coherent solid, slightly weakened by the presence of the liquid droplets.

Next, we consider the simplest possible polycrystalline specimen, consisting of two columnar dendritic grains growing with a symmetric tilt boundary between them (Fig. 13.18a). One can see again the same intragranular coalescence occurring at around T_b^{intra} within each grain. Below this temperature, the grains are already close to being fully solid, with a few isolated liquid droplets. On the other hand, dendrite arms located on both sides of the future grain boundary must overcome a grain boundary energy $\gamma_{gb}(\theta)$, which depends on the relative misorientation between the two grains. For values of θ small enough, γ_{gb} is less than two times $\gamma_{s\ell}$, and thus there would be an decrease in energy upon coalescence (Fig. 13.18(c)).

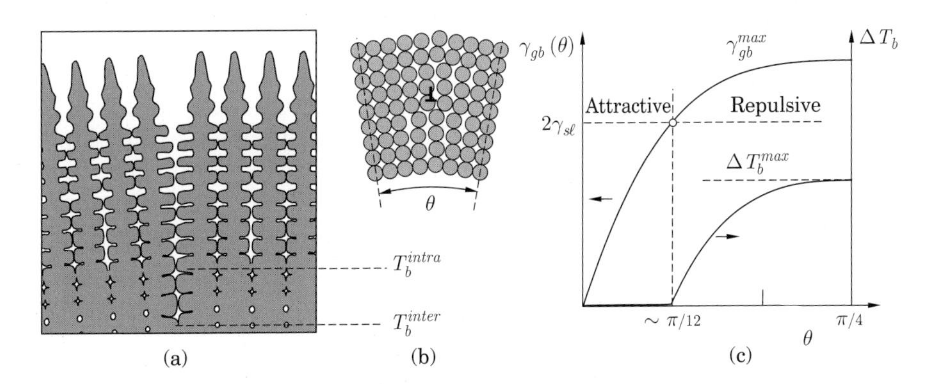

Fig. 13.18 (a) A schematic diagram of two columnar dendritic grains in a symmetric tilt boundary configuration. (b) The configuration of the atoms located at the grain boundary after solidification. (c) The surface energy of a tilt boundary, $\gamma_{gb}(\theta)$, as a function of the misorientation angle θ. The bridging undercooling ΔT_b is shown on the right hand scale. The transition between an attractive and repulsive boundary occurs at around $\pi/12$.

Such a boundary is thus "attractive" and coalescence occurs around T_b^{intra}. However, for larger misorientations, we have $\gamma_{gb}(\theta) > 2\gamma_{s\ell}$, and the grain boundary is said to be "repulsive." It has been shown, in this case, that intergranular bridging occurs at a lower temperature $T_b^{inter} < T_b^{intra}$. For a pure material, where T_b^{intra} is equal to T_f, coalescence at a repulsive grain boundary occurs at an undercooling $\Delta T_b(\theta)$ given by

$$\Delta T_b = \frac{\gamma_{gb}(\theta) - 2\gamma_{s\ell}}{\Delta S_f \delta}, \quad \text{for } \gamma_{gb}(\theta) > 2\gamma_{s\ell} \tag{13.15}$$

ΔS_f is the volumetric entropy of fusion and δ is the thickness of the diffuse solid-liquid interface. Although the value of δ is not very well defined, it must be on the order of a few nanometers. Thus, ΔT_b can have a value of several tens of degrees. Figure 13.18 shows the tilt grain boundary energy as given by the Read-Shockley relation, where $\gamma_{gb}(\theta)$ increases for $0 < \theta <\sim \pi/12$ before reaching a maximum value $\gamma_{gb}^{max} > 2\gamma_{s\ell}$. The intersection of $\gamma_{gb}(\theta)$ and $2\gamma_{s\ell}$ separates attractive and repulsive boundaries. The maximum value γ_{gb}^{max} defines the minimum temperature T_b^{min} (or maximum undercooling ΔT_b^{max}) below which all intergranular liquid films in a polycrystalline specimen are solid.

In alloys, one also has to account for diffusion of solute. In particular, for repulsive boundaries, back-diffusion occurs in the solid while the liquid film is still present. This decreases the final composition in the liquid if $k_0 < 1$, thus increasing the temperature at which coalescence occurs. [25].

In summary, for a polycrystalline columnar specimen in which all of the grain boundaries are nearly parallel to the thermal gradient, all of the dendrites are surrounded by liquid films for $T > T_b^{intra}$. Accordingly, there is little mechanical resistance in the direction perpendicular to the thermal gradient, i.e., in the direction along which thermal strains develop. However, any lateral separation of the dendrites can be compensated by interdendritic liquid flow, since the permeability is high. Therefore, the range is characterized by high ductility and low strength. On the other hand, for $T < T_b^{min}$, all of the boundaries are solid, and the specimen exhibits both high ductility and high strength. For $T_b^{min} < T < T_b^{intra}$, films of liquid remain at some of the grain boundaries, thereby weakening the material. Any opening of such grain boundaries by thermal strains cannot be compensated easily by liquid flow, since the permeability of the mushy zone is low in this range ($\sim 0.94 < g_s < 1$). Therefore, this range typically corresponds to low ductility and an intermediate strength of the mushy zone. The material is very sensitive to hot cracking, due to the remaining liquid films acting as a brittle phase and concentrating the thermal strains.

Equiaxed grain structures behave somewhat differently from the columnar structures that we have discussed up until now. The arrangement of grain boundaries is random, thus causing the topology of any remaining liquid films to be very intricate, especially in 3-D. The gradual transition of a continuous liquid with isolated grains or clusters to a continuous solid network with isolated droplets or films of liquid is called

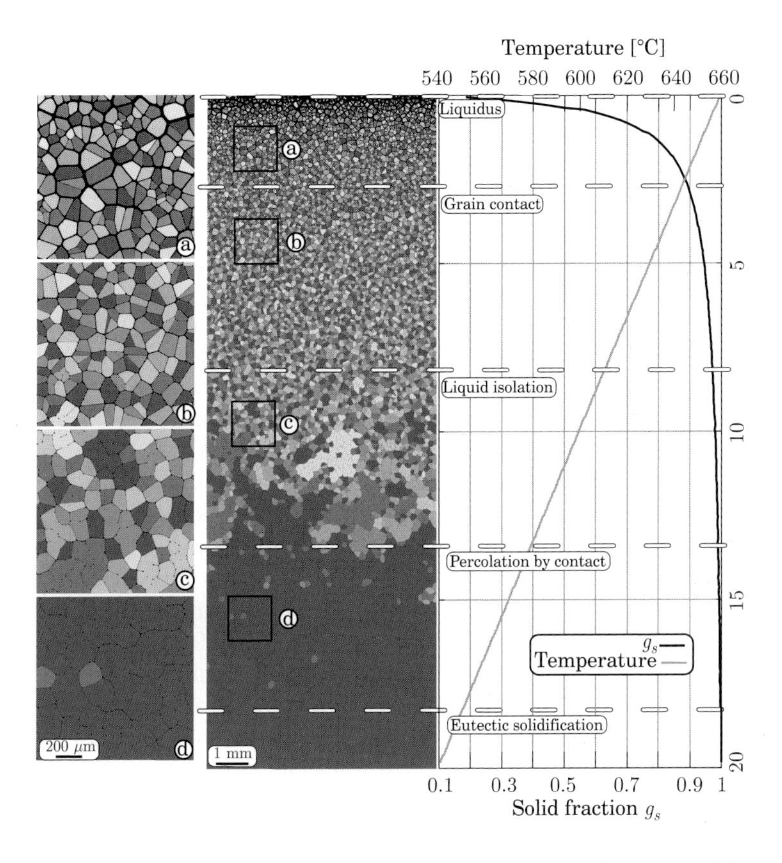

Fig. 13.19 The evolution of a population of globular equiaxed grains solidifying in a fixed thermal gradient, $G = 60$ K cm^{-1}, and cooling rate, $\dot{T} = -1$ K s^{-1}, for an Al-1wt%Cu alloy. The temperature profile $T(z)$ is shown to the right together with the evolution $g_s(z)$ calculated in horizontal sections. Each cluster of grains (including individual grains) is represented by a particular grey level in the central grain structure, whereas the liquid is shown in black. Enlargements of the grain structure at four locations are displayed to the left. (After Vernède et al. [32; 33].)

"*percolation*," illustrated in Fig. 13.19 for a 2-D population of globular equiaxed grains solidifying in a fixed thermal gradient at constant cooling rate. The results shown in the figure were obtained using a *granular approach*, in which the solidification of each grain from its corresponding nucleation center was calculated using a microsegregation model. The cumulative solid fraction $g_s(z)$ in sections perpendicular to the vertical thermal gradient is displayed to the right, together with the temperature profile $T(z)$. The nucleation centers were randomly distributed in the domain, and each grain has a random orientation. When the solid-liquid interfaces of two neighboring globular grains fall within a predefined interaction distance, coalescence is calculated according to the model given in

Fig. 13.18, taking into account the mutual misorientation of the grains. For further details of the model, see Vernède et al. [32; 33].

Close to the liquidus (zone labeled (a)), all of the grains are surrounded by a liquid film, highlighted in black. The transition to region (b) called "grain contact" is characterized by a maximum specific solid-liquid area. Near $g_s = 0.9$, the increase of the perimeter of the grains in contact with the liquid due to solidification is counterbalanced by grain boundaries that become solid. The beginning of region (c) is identified as having the first isolated liquid boundaries, i.e., boundaries that are no longer connected to the liquid channel network. At this stage, fairly large clusters of grains shaded with the same grey level can be distinguished. Finally, the grains form a continuous solid network across the transverse dimension of the specimen in zone (d) (percolation). For this dilute Al-Cu alloy, percolation occurs above the eutectic temperature. The last two transitions (liquid isolation and percolation) occur between $g_s = 0.97$ and $g_s = 1$.

Although such a 2-D granular approach cannot represent the full complexity of 3-D grain structures (e.g., both the liquid and the solid can be continuous in 3-D, whereas this is not possible in 2-D), it does provide valuable insight. For instance, the values of g_s characterizing the various transitions of the mushy zone seen in Fig. 13.19 are intrinsic to the problem. They can be obtained from a single simulation performed over a very large number of grains or from an *average* over many simulations with small numbers of grains. The *fluctuations* of these values, however, decrease with increasing domain size, and the exponent of this decrease can be deduced from percolation theory. Therefore, although the topology of equiaxed grain structures is complex, it appears that transitions are well characterized. Coherency by percolation seems to occur above 0.97 in 2-D, and at a somewhat lower value in 3-D.

Finally, it should be pointed out that some authors define coherency as the point where a measurable torque appears on the paddle of the device shown in Fig. 13.9(e) when stirring a mixture of equiaxed grains forming in a uniform temperature liquid. In this case, coherency measures the "packing" of equiaxed grains: for dendritic morphologies, this occurs for $g_s \simeq 0.2$, whereas it is delayed to about $g_s \simeq 0.5$ for globular grains. The latter value, roughly equal to the density of regular spheres packed in a cubic network, corresponds to a situation where globular grains are unable to pass each other easily. However, this definition is not readily applied to the study of hot tearing, where specimens are subjected to tensile stresses.

13.3.3 Two-phase approach

Now that we have described qualitatively how the mushy zone evolves as $g_s \to 1$, and how this influences its rheological behavior, it is time to consider the semi-solid. As we are mainly interested in hot tearing, we will focus our attention on the behavior at high volume fractions of solid.

Using the development from Chap. 4, the quasi-static momentum equation averaged over the solid phase becomes (see Eq. (4.103))

$$\nabla \cdot (g_s \langle \boldsymbol{\sigma} \rangle_s) + g_s \rho_s \boldsymbol{g} = \boldsymbol{\Gamma}_s^{\sigma *} \tag{13.16}$$

Here, $\langle \boldsymbol{\sigma} \rangle_s = - \langle p \rangle_s \boldsymbol{I} + \langle \boldsymbol{\tau} \rangle_s$ is the average stress tensor and $\boldsymbol{g}$ is the gravity vector. We have neglected both transient and inertial terms in writing Eq. (13.16). For the momentum balance equation in the liquid phase, the Darcy term is much larger than the other terms involving $\langle \boldsymbol{v} \rangle_\ell$ at high g_s, and thus Eq. (4.105) becomes

$$-\nabla \cdot (g_\ell \langle p \rangle_\ell) + g_\ell \rho_\ell \boldsymbol{g} = -\boldsymbol{\Gamma}_\ell^{\sigma *} \tag{13.17}$$

Following the development of Chap. 4 and introducing Darcy's equation for the flow in the liquid phase, we have

$$-g_\ell \nabla \langle p \rangle_\ell + g_\ell \rho_\ell \boldsymbol{g} = -\boldsymbol{\Gamma}_\ell^{\sigma *} + \langle p \rangle_\ell \nabla g_\ell = \frac{g_\ell^2 \mu_\ell}{K} \left(\langle \boldsymbol{v} \rangle_\ell - \langle \boldsymbol{v} \rangle_s \right) \tag{13.18}$$

Here, the average velocity of the liquid $\langle v_\ell \rangle$ in Eq. (4.114) is replaced by $g_\ell \langle \boldsymbol{v} \rangle_\ell^r = g_\ell (\langle \boldsymbol{v} \rangle_\ell - \langle \boldsymbol{v} \rangle_s)$, where the superscript "r" refers to the velocity of the fluid relative to the solid. Combining Eq. (13.18) with mass conservation from Eq. (12.5), and assuming that $g_p = 0$, i.e., $\langle \rho_0 \rangle = \langle \rho \rangle$, gives

$$\frac{\partial \langle \rho \rangle}{\partial t} + \nabla \cdot (\langle \rho \rangle \langle \boldsymbol{v} \rangle_s) - \nabla \cdot \left[\frac{\rho_\ell K}{\mu_\ell} (\nabla \langle p \rangle_\ell - \rho_\ell \boldsymbol{g}) \right] = 0 \tag{13.19}$$

Note that, in order to obtain this form, the contribution $\langle \rho \boldsymbol{v} \rangle = (\rho_s g_s \langle \boldsymbol{v} \rangle_s + \rho_\ell g_\ell \langle \boldsymbol{v} \rangle_\ell)$ was replaced by $(\langle \rho \rangle \langle \boldsymbol{v} \rangle_s + \rho_\ell g_\ell (\langle \boldsymbol{v} \rangle_\ell - \langle \boldsymbol{v} \rangle_s) = (\langle \rho \rangle \langle \boldsymbol{v} \rangle_s + \rho_\ell g_\ell \langle \boldsymbol{v} \rangle_\ell^r)$.

We then neglect the contribution of the surface energy in Eq. (4.108) so that $\boldsymbol{\Gamma}_s^{\sigma *} = \boldsymbol{\Gamma}_\ell^{\sigma *}$. The average momentum equation for the solid is then obtained by combining Eqs. (13.16) and (13.18):

$$\nabla \cdot (g_s \langle \boldsymbol{\sigma} \rangle_s) + g_s \rho_s \boldsymbol{g} = \langle p \rangle_\ell \nabla g_\ell - \frac{g_\ell^2 \mu_\ell}{K} \left(\langle \boldsymbol{v} \rangle_\ell - \langle \boldsymbol{v} \rangle_s \right) \tag{13.20}$$

By adding Eqs. (13.18) and (13.20), we are able to recover the static equilibrium of the solid-liquid mixture:

$$\nabla \cdot \langle \boldsymbol{\sigma} \rangle + \langle \rho \rangle \boldsymbol{g} = \nabla \cdot (g_s \langle \boldsymbol{\sigma} \rangle_s - g_\ell \langle p \rangle_\ell \mathbf{I}) + (g_s \rho_s + g_\ell \rho_\ell) \boldsymbol{g} = 0 \tag{13.21}$$

where $\langle \boldsymbol{\sigma} \rangle$ is the average stress acting on the mixture. Although this quantity is effectively the response that can be measured on a mushy zone, the viscoplastic behavior of the solid phase itself is not influenced by the pressure in the liquid phase. Therefore, we introduce the effective stress $\langle \hat{\boldsymbol{\sigma}} \rangle$

$$\langle \hat{\boldsymbol{\sigma}} \rangle = \langle \boldsymbol{\sigma} \rangle + \langle p \rangle_\ell \mathbf{I} \tag{13.22}$$

for which a constitutive law will have to be given. Using this definition in Eq. (13.21), we obtain finally

$$\nabla \cdot \langle \hat{\boldsymbol{\sigma}} \rangle + \langle \rho \rangle \boldsymbol{g} = \nabla \langle p \rangle_\ell \tag{13.23}$$

It follows from this development that the momentum equations for the liquid (Eq. 13.18) and for the mushy solid (Eq. 13.23) are *coupled*. The velocity of the solid appears in Darcy's equation while the pressure gradient in the liquid enters into the momentum equation of the mushy solid. However, the velocity of the solid deep in the mushy zone is very small. The pressure in the liquid is also much smaller than the stresses in the solid, as already noted in the introduction of this section. Neglecting this direct coupling, one then has for the solid

$$\nabla \cdot \langle \hat{\boldsymbol{\sigma}} \rangle + \langle \rho \rangle \boldsymbol{g} = 0 \tag{13.24}$$

and for the liquid

$$-\nabla \langle p \rangle_\ell + \rho_\ell \boldsymbol{g} = \frac{g_\ell \mu_\ell}{K} \langle \boldsymbol{v} \rangle_\ell = \frac{\mu_\ell}{K} \langle \boldsymbol{v}_\ell \rangle \tag{13.25}$$

This approximate treatment has the great advantage of *decoupling* the equations for the solid and liquid phases. This means that the momentum balance for the mushy solid, Eq. (13.24), is that of a porous medium instead of a mushy zone filled with liquid, and thus can be solved without including the presence of the liquid phase to determine the deformation and the velocity field of the solid. Since the solid is porous, its rheological behavior should account for the fact that $\nabla \cdot \langle \boldsymbol{v} \rangle_s \neq 0$. Once the velocity field of the solid $\langle \boldsymbol{v} \rangle_s$ is known, this gives an additional contribution in the average mass conservation, established in Chap. 12, Eq. (12.5). When combined with Eq. (13.25), one is able to calculate the pressure in the liquid, but now taking into account the compression/expansion of the mushy solid. If $g_p = 0$, i.e., there is no porosity, we obtain the pressure in the liquid. If, on the other hand, $g_p \neq 0$, the pressure field in the liquid and the amount of porosity must be coupled through a microscopic model of gas segregation-precipitation, as explained in detail in Chap. 12.

An accurate solution to this problem requires a good rheological model for the solid skeleton. At low volume fractions of liquid, the solid can be considered to be homogeneous, but simply weakened by the presence of isolated liquid pockets/localized films. In this case, the rheological behavior can be described by, for instance, a modified form of Norton's law (see Drezet and Eggeler [7]):

$$\dot{\varepsilon}_{eq}^{vp} = A \left(\frac{\hat{\sigma}_{eq}}{g_s} \right)^n \exp\left(-\frac{Q}{RT} \right) \tag{13.26}$$

where $\dot{\varepsilon}_{eq}^{vp}$ is the equivalent strain rate defined in Eq. (13.11) and $\hat{\sigma}_{eq}$ is the equivalent von Mises stress defined in Eq. (13.12). Further, A is a material parameter, n is the stress-sensitivity coefficient and Q is the activation energy, all of which depend on the alloy. Note that we have dropped the $\langle \cdot \rangle$ symbols for the sake of clarity. This relation can be modified to account for the fraction of liquid that is effectively weakening the solid. For example, Van Haaften et al. [30] have replaced g_s by $(1 - g_{\ell gb})$, where $g_{\ell gb}$ is the fraction of wetted grain boundaries.

The theory of the mechanics of porous media is used to describe the response of the mushy zone to complex multi-axial stress states at lower solid fraction, when the liquid and the solid phases are both interconnected. The essential physics that must be captured is that the solid and liquid phases are able to rearrange under loading in such a way that the volume depends on pressure, even though both individual phases are incompressible. A sponge saturated with water is a useful analogy. When compressed, the sponge expresses water into the surrounding environment, and the observed yield stress for the sponge is thus much lower than that of the solid making up the sponge skeleton itself. Similarly, the sponge can take up liquid when it is under hydrostatic tension.

The pressure dependence of the mechanical response of the mushy material can be incorporated by modeling it as a material whose yield stress depends on the hydrostatic stress. Numerous models have been proposed with this attribute, however we discuss only two of the simplest ones here. In the Gurson model, the yield stress of the mushy zone σ_Y^{mz}, is written

$$\sigma_Y^{mz} = \sqrt{I_1 \hat{\sigma}_{eq}^2 + I_2 \hat{p}_s^2} \tag{13.27}$$

where $I_1(g_s, \dot{\varepsilon}_{eq}^p, ...)$ and $I_2(g_s, \dot{\varepsilon}_{eq}^p, ...)$ are material properties, and the effective pressure, $\hat{p}_s$, is defined in the usual way as

$$\hat{p}_s = -\frac{1}{3}\mathrm{tr}(\hat{\boldsymbol{\sigma}}) \tag{13.28}$$

This model is symmetric with respect to compression and tension. The Gurson yield surface is an ellipse in the $\hat{p}_s - \hat{\sigma}_{eq}$ space, as illustrated in Fig. 13.20(a), where we show that σ_Y^{mz} increases with g_s. Notice that for $g_s \to 1$, we recover the incompressible viscoplastic response, which in this model implies that $I_2 \to 0$.

Three types of mechanical tests that can be used to determine I_1 and I_2 are also shown in the figure. Point "E" represents an *eudometric* or *drained compression test*, where the liquid is allowed to flow out of the domain (through filters) under hydrostatic load. Point "S" is a pure shear test, in which $\hat{p}_s = 0$. Point "U" represents a uniaxial tensile test that can

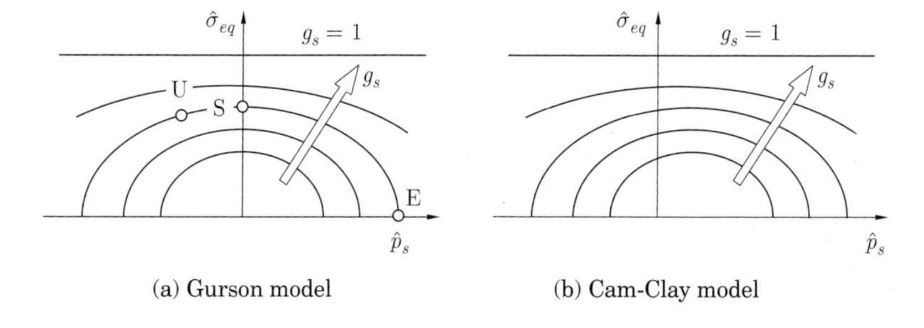

Fig. 13.20 The yield surface of a porous medium according to (a) the Gurson and (b) Cam-Clay models, for several values of the volume fraction of solid, g_s. The dots labeled "U", "S" and "E" correspond to various mechanical tests (see text).

be decomposed into non-zero shear and hydrostatic components. Where would a uniaxial compression test appear on this figure?

The Gurson model cannot reproduce the behavior that most porous media are more difficult to compress than to expand. The Cam-Clay model is one of the simplest models that does capture this phenomenon. The yield surface in this case is still an ellipse in the $\hat{p}_s - \hat{\sigma}_{eq}$ space, expanding with g_s, but its center is shifted toward positive values of $\hat{p}_s$. Note that the same suite of tests described for the Gurson model can be carried out to define the parameters of the Cam-Clay model.

13.4 HOT TEARING

As demonstrated in the first part of this chapter, thermal stresses and strains during casting and welding develop mainly in the fully solid part. These deformations are transmitted to the mushy zone and tend to localize at the weakest regions, i.e., at the continuous liquid films remaining between the grains. Some of these films are still present even at low temperatures for the reasons mentioned in Sect. 13.3.2, and the low permeability of the mushy zone in such regions prevents feeding of any grain boundary opening. This leads to hot tearing. The next section reviews the main characteristics of such defects, after which we proceed to describe hot tearing tests and measures of hot tearing susceptibility for various alloys.

13.4.1 Characteristics of hot tears

Figure 13.1 illustrated hot tears formed in two industrial processes: DC casting and welding. We later describe several laboratory test experiments that can produce such defects under well-controlled conditions, but for the moment, we simply present some examples from such tests to illustrate the phenomena and characteristics of hot tearing.

If the strains develop at an early stage of solidification, the opening can be filled by intergranular liquid, as demonstrated in Fig. 13.21(a). This "healed" hot tear exhibits a negative macrosegregation for $k_0 < 1$. This phenomenon is discussed further in Chap. 14.

A longitudinal section of an Al-4.5%Cu casting solidified in a ring mold test is shown in Fig. 13.21(b). The first solid forms on a central water-cooled tube, and tensile stresses are induced because the solid sticks on the tube, which leads to cracks forming along grain boundaries, nearly perpendicular to the tube axis. Figure 13.22 displays an SEM image of one of these cracks. Since a continuous liquid film remained present at the time the grain boundary opened, it is not surprising to observe a fairly smooth surface with only a few spikes. The smooth part of the surface has a wavy appearance, with a characteristic spacing of bumps corresponding to the "secondary" dendrite arm spacing at the end of solidification. These bumps are thus understood to be secondary dendrite arm ends that were

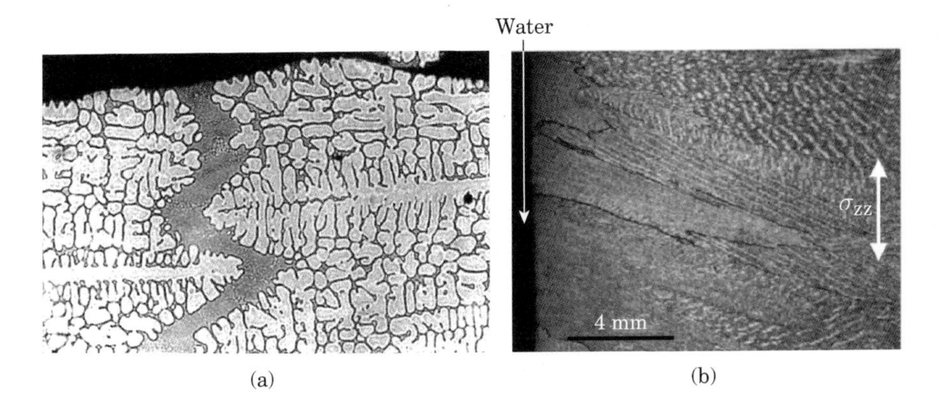

Fig. 13.21 (a) An optical micrograph of a "healed" or filled hot tear in an Al-10%Cu specimen. (From Spittle and Cushway [27]); (b) A hot crack observed in a cylindrical Al-4.5wt%Cu ingot cooled from the center (ring mold test).

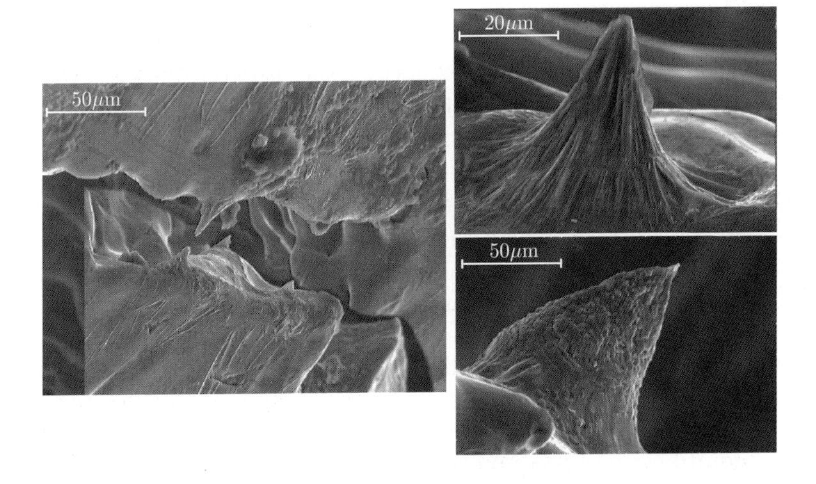

Fig. 13.22 SEM micrograph (left) showing the two sides of the hot tear present in Fig. 13.21(b), with spikes facing each other on the upper and bottom lips of the crack. Enlarged SEM views of a spike associated with the last interdendritic liquid solidification (top right) and of a solid spike deformed during the formation of the crack (bottom right). (After Farup et al. [11].)

pulled apart by the tensile strain. The spikes have a dimension at their roots that is also on the order of the secondary dendrite arm spacing. Some of them correspond to liquid menisci that formed and solidified during the opening of the grain boundary (Fig. 13.22, top right). Others show a torn surface, indicating that they were already solid bridges before formation of the crack (Fig. 13.22, bottom right).

Deeper insight into hot tear formation has been gained by *in-situ* observations on organic alloys. Using a device similar to that developed for

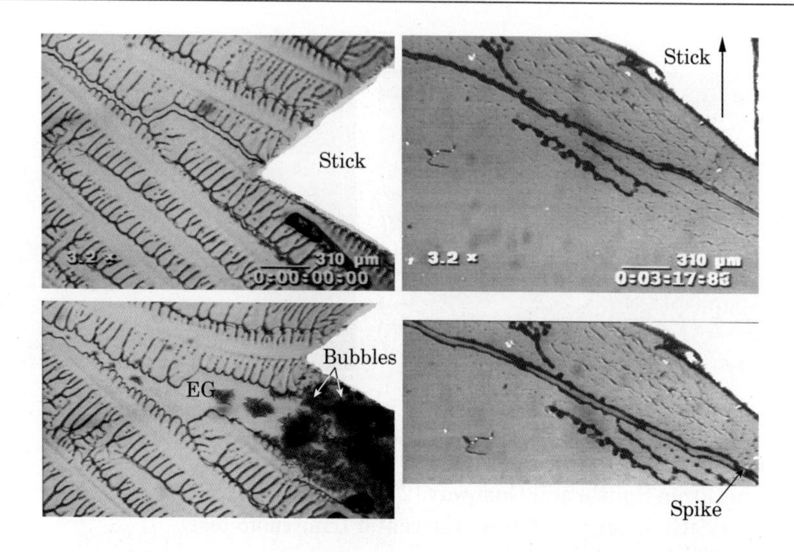

Fig. 13.23 Two sequences from hot tearing experiments in a succinonitrile-acetone alloy. (left) Deformation at low solid fraction produces a heated hot tear. (right) Deformation at high solid fraction produces a void between the solid segments. The temperature gradient is horizontal and toward the left in both cases. (After Farup et al. [11].)

the observation of directional solidification of succinonitrile-acetone alloys, Farup et al. [11] inserted a pulling stick to produce strains, as shown in Fig. 13.23. The stick could be placed close to a grain boundary during growth, and then pulled transverse to the thermal gradient in order to produce mechanical deformation that mimics thermal strains. If straining was applied early during solidification (Fig. 13.23 (left)), interdendritic liquid could flow into the opening, thus producing a healed hot tear. This liquid is leaner in solute than the liquid it replaces. On the other hand, thermal equilibrium is reached very rapidly causing the new liquid to be highly undercooled. This leads to the formation of equiaxed grains that nucleate almost instantaneously (dark equiaxed dendritic grains labeled "EG" in Fig. 13.23 (left)). If the pulling rate is high, the depression in the liquid can even induce a few pores within the filling liquid (dark bubbles in Fig. 13.23 (left)). Note that at this time, the dendrite arms within the grains have not yet coalesced, i.e., $T > T_b^{intra}$.

Pulling the stick at a temperature $T < T_b^{intra}$ leads to a different type of behavior, as shown in Fig. 13.23 (right). The dendritic network in this case is much more compact, and the secondary arms within each grain are well coalesced. However, along the boundary that separates the two grains (dark line nearly parallel to the edge of the white pulling stick), they have not yet bridged. When the upper grain is pulled with the stick in the direction of the arrow, the crack forms. Although not clearly seen in the figure, some liquid was still present along this boundary. The crack nucleated at the extremities of the pulled region and the liquid formed

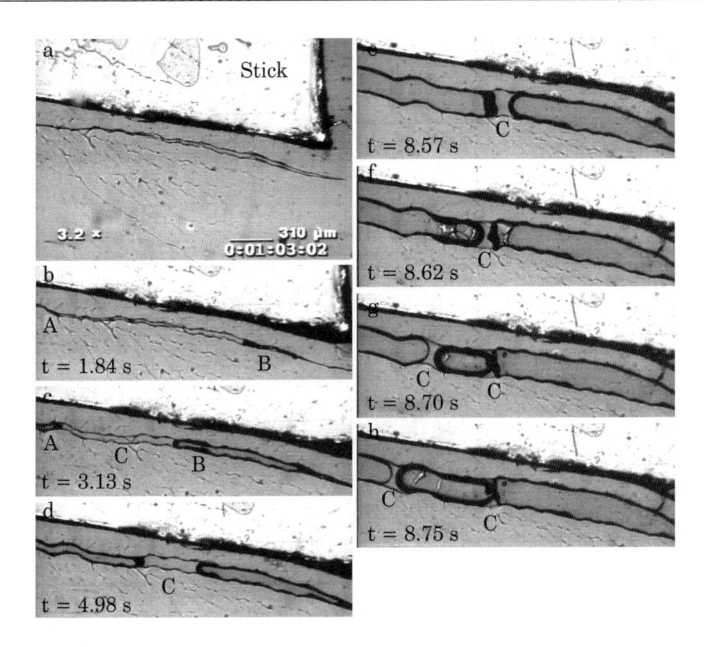

Fig. 13.24 A sequence showing the formation of a hot-crack at a grain boundary during the directional solidification of a succinonitrile-acetone alloy, induced by the slow transverse pulling of the upper grain with a mylar stick. (After Farup et al. [11].)

a meniscus that was then arrested at a position where a solid spike was already present (labeled in the figure).

The sequence for a similar experiment shown in Fig. 13.24 was taken at a slightly lower volume fraction of solid than in Fig. 13.23 (right). The grain boundary located near the pulling stick could not be fed from the left (i.e., from the region closer to the dendrite tips), but it contained somewhat more liquid at the time of pulling than in the previous sequence. Pores nucleated at the boundary, first at point B, then at point A, and the remaining liquid accumulated at point C, once again forming a meniscus that stopped at a small solid asperity. At $t = 8.57s$, the liquid meniscus becomes so stretched across the crack that mechanical equilibrium cannot be maintained at the solid-liquid-bubble junction. Within a few tenths of a second, the meniscus breaks down, leaving a solid spike on each side of the crack. The remaining liquid travels very quickly toward the upper part of the crack, where it can establish a new mechanical equilibrium. These *in-situ* observations of spike formation are consistent with the appearance of the crack surfaces in metallic systems after solidification seen in Fig. 13.22.

These results clearly show that grain boundaries play a key role in hot tear formation, in addition to the thermomechanical aspects discussed earlier. This is further illustrated in Fig. 13.25, which shows four different

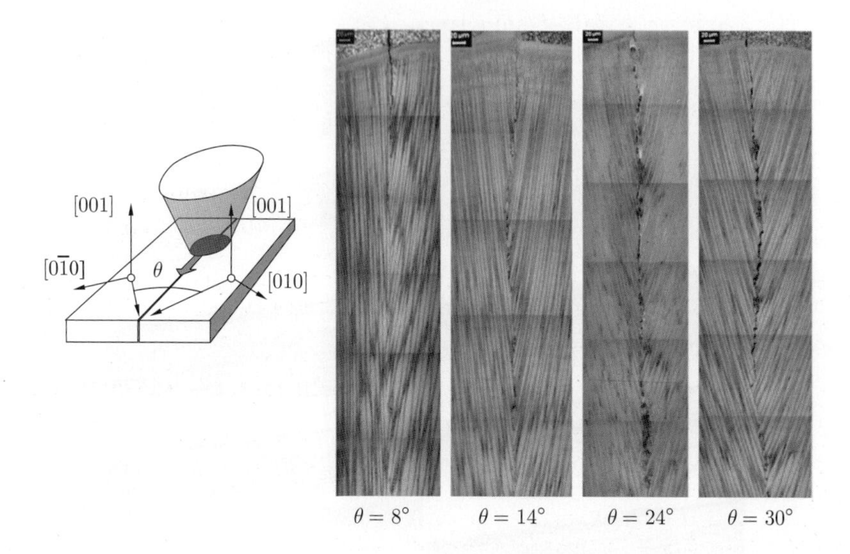

Fig. 13.25 The formation of hot cracks in laser welding of bi-crystals. The welding configuration of the two superalloy single crystals is shown to the left. Micrographs of the top surface of the weld trace are displayed for several increasing values of the misorientation, θ, to the right. The welding direction goes from top to bottom. (After Wang et al. [34].)

laser welds, each made using two misoriented MC2 nickel-based superalloy single crystals. The crystal pairs were oriented such that their upper surfaces corresponded to (001) planes, but their [100] directions were purposely misoriented by a prescribed angle θ. Thus, each bi-crystal weld created a pure tilt boundary similar to those described in Sect. 13.3.2. The misorientation can be clearly identified from the $\langle 100 \rangle$ dendrites growing in a plane parallel to the top surface. The welding parameters were identical in each experiment. It is interesting to note that no crack formed for $\theta < 14°$, whereas the weld cracked when $\theta > 24°$. The first case corresponds to an attractive grain boundary, i.e.,$\gamma_{gb} < 2\gamma_{s\ell}$, for which solid bridges form at $T \sim T_b^{intra}$. For the high angle boundaries, on the other hand, the difficulty of bridging maintains a liquid film at the boundary between the two crystals down to lower temperatures and renders the weld much more sensitive to hot cracking.

In summary, hot cracks have the following characteristics:

- They form under the combined effects of thermal strains and a lack of feeding. They develop in regions subject to tensile stresses (e.g., hot spots, weld trace centerlines, centers of billets, etc.).

- They are mostly intergranular. The cracks probably follow a path along repulsive grain boundaries, i.e., boundaries for which $\gamma_{gb} > 2\gamma_{s\ell}$ and liquid films persist down to lower temperatures.

- They exhibit a fairly smooth surface with bumps and occasional spikes. The bumps correspond to dendrite arms that had not yet coalesced with neighbors at the time when the crack started to form, whereas the spikes are due to the solidification of last liquid menisci or the stretching of solid bridges.

From these observations, we obtain a better understanding of why alloys that are the most sensitive to hot cracks have the following characteristics:

- They reach (tensile) coherency at high volume fractions of solid. This is the case for dilute alloys, for which the primary phase typically represents 95% or more of the microstructure.
- They have a large thermal expansion coefficient, i.e., thermal strains are large.
- They have high strength.

13.4.2 Hot tearing tests and hot tear sensitivity

Tests of the mushy zone rheology, especially in tension, provide a certain measure of the Hot Cracking Sensitivity (HCS) of alloys. For example, the HCS increases with the width of the brittle temperature range (BTR), since the mushy zone does not offer a significant resistance to deformation over a large temperature range. Although several of the devices shown in Fig. 13.11 (Instone, Ackermann, Ohm and Engler) have been used to assess the HCS of alloys, they are fairly complicated to implement and control. In this section, we review several simpler tests for measuring the HCS of alloys.

The basic idea of all HCS tests is to induce tensile stresses in a region of a casting that is still mushy. The Tatur test (Fig. 13.26(a)) comprises a casting with several bars of varying lengths linking an outer ring to an inner cylindrical feeder. The overall contraction of the bars is restrained by the outer ring, which induces tensile thermal strains in the bars during solidification. The thermal strains reach their maxima at hot spots found near the centers of the bars. The amount of contraction is proportional to the bar length, so the longer bars might crack while the smaller ones remain sound. Figure 13.27(a) summarizes the measurements of Rosenberg et al. [26], for which such a device was used to investigate the HCS for several Al-Cu alloys. The vertical axis represents the length of the bars, with a symbol indicating their soundness (cracked or not), and the horizontal axis gives the alloy composition. The authors superimposed the phase diagram on the plot. The line separating cracked (black squares) and sound (open circles) bars exhibits a "V-curve" with a minimum near the maximum solubility limit of Cu in Al, $C_0 \simeq 5$ wt%. One might thus conclude that the maximum HCS can be correlated with the solidification interval of the alloy. However, we see later that this notion is too simplistic, as it does not account for microsegregation, grain morphology and grain boundary phenomena.

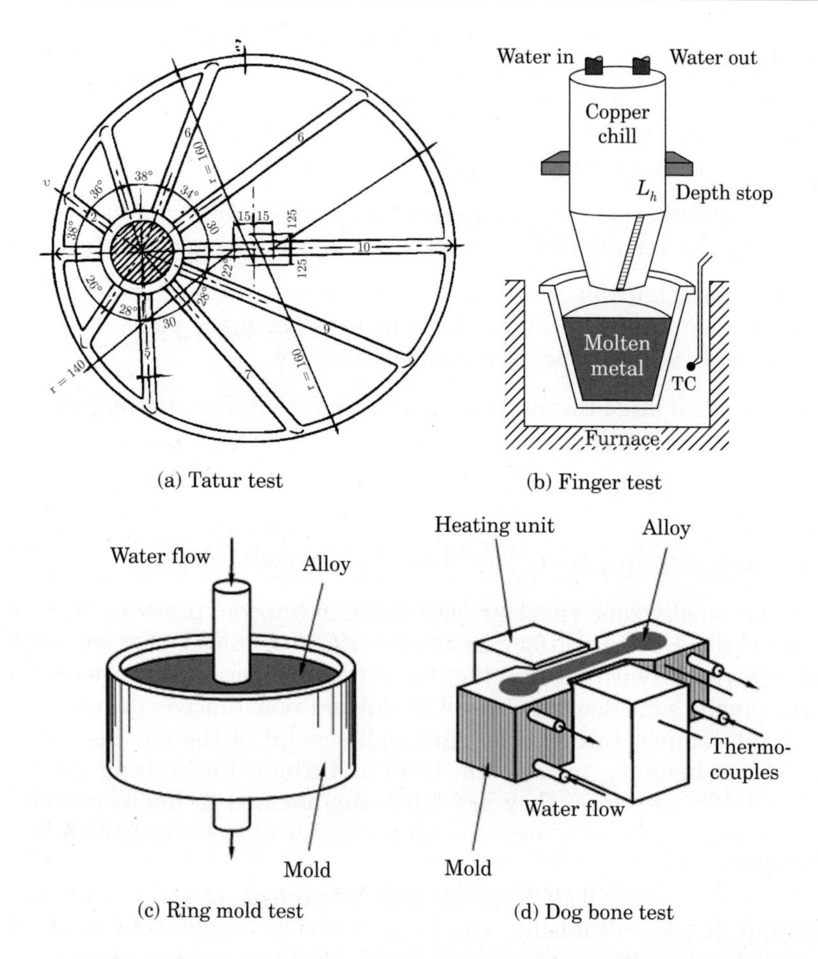

Fig. 13.26 Hot tearing test devices: (a) Tatur test, (b) finger test, (c) ring mold test and (d) dog bone test.

Figure 13.26(b) shows the apparatus for the so-called "finger test," designed by Warrington and McCarthy [35], in which a water-cooled conical copper mold, with a thin line of an insulating ceramic of width L_h deposited along a generating line, is dipped into the molten alloy. The ceramic locally reduces the heat transfer coefficient, and thus produces a hot spot of the same width in the crust solidifying around the cone. In such a configuration, the thermal stress is mainly tangential, i.e., $\sigma_{\theta\theta}$ is the largest component of the stress tensor. As the deformation is localized at the hot spot, it is amplified by a factor $2\pi R(z)/L_h$, where $R(z)$ is the radius of the cone at height z. The height z to which a crack propagates thus measures the *resistance* of the alloy to hot tearing, i.e., the inverse of a HCS. Figure 13.27(b) shows the HCS of a binary Al-Cu alloy measured with such

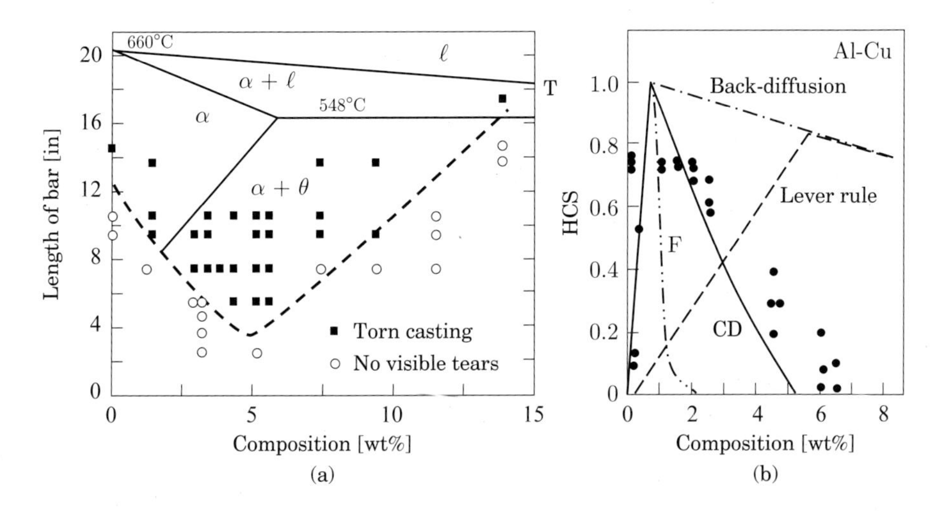

Fig. 13.27 The hot cracking sensitivity of Al-Cu alloys as a function of the composition: (a) Filled squares and open circles indicate whether the cast bar of a particular length in a Tatur test was cracked or had no visible tears, respectively. (After Rosenberg et al. [26]). (b) HCS (arbitrary units) measured using the finger test of Warrington and McCartney. The various curves and lines are from theoretical models discussed in Sect. 13.5. (After Campbell [3].)

a device as a function of the composition. The form is called a "Λ-curve" and is the reverse of the "V-curve" shown in Fig. 13.27(a). Notice that the maximum in the Λ-curve occurs at a much lower Cu content (around 0.7 wt%Cu) than the minimum of the V-curve in Fig. 13.27(a). The various theoretical models indicated by curves and lines superimposed on the experimental data in Fig. 13.27(b) are discussed in the next section.

A similar device to the finger test is the ring mold (Fig. 13.26(c)), in which an alloy contained in a cylindrical mold is solidified starting from a water-cooled tube running through the center. As the alloy solidifies on the tube, two types of stresses are generated: hoop stresses $\sigma_{\theta\theta}$ (as in the finger test) and vertical stresses σ_{zz} due to the solid crust sticking on the tube. As the radius of the solid layer increases, the thermal stresses decrease and at some point, both the axial and horizontal cracks stop. The HCS of the alloy is defined as the length of the crack. The cracks shown in Fig. 13.21(a) were obtained with such a device. Finally, the last device for HCS measurements of alloys is the so-called "dog bone test," presented in Fig. 13.26(d). In this test, the contraction of a cast bar is restrained by two cylindrical extremities, while a hot spot is maintained at the center of the bar by a heating device or thicker insulation. The density of cracks can be determined either after solidification by optical inspection, or during solidification by measuring the electrical resistivity of the casting (the higher the resistivity, the larger the density of transverse cracks).

13.5 HOT TEARING CRITERIA AND MODELS

The treatment of hot tearing described in the preceding section represents the state of the art circa 1990. The various tests and physical explanations provide a semi-quantitative basis for understanding HCS and the role of alloy characteristics in hot tearing. Since then, and along with the rise of computational codes capable of predicting microstructures, came the desire to include more sophisticated and quantitative approaches for predicting hot tearing during solidification. These models are reviewed in the present section, in more or less chronological order, which also coincides with increasing complexity.

Solidification interval criteria
The simplest hot tearing models define the HCS solely as a function of the alloy composition, regardless of solidification conditions (i.e., temperature gradient, cooling rate, etc.) or of microstructural considerations (grain size, grain morphology, etc.). In this view, a pure material or an alloy with a eutectic or azeotrope composition should have no mushy zone and therefore present no risk of hot cracking. For other compositions, the hot cracking tendency should increase with the extent of the mushy zone, and therefore with the solidification interval. This approach leads to diagrams such as those in Figs. 13.27(a) and (b). Notice that if one uses the equilibrium solidification interval computed via the lever rule, the maximum HCS is expected to occur at the maximum solubility composition of the alloy. While this seems to be the case in the measurements by Rosenberg et al. [26] using the Tatur test, more recent investigations do not corroborate these results (see Fig. 13.27(b)).

A first simple correction to such models is to consider back-diffusion. For example, in the Gulliver-Scheil model, the maximum solidification interval should correspond to C_0 approaching (but not equal to) zero. If one chooses the Brody-Flemings or one of its derivative microsegregation models (Chap. 10), the maximum in the solidification interval can be made to coincide with the maximum HCS (dashed-dotted curve labeled "back-diffusion" in Fig. 13.27(b)). However, this shows nothing more than the fact that hot tearing occurs deep in the mushy zone of alloys with large solidification intervals, when feeding is difficult and coherency at the grain boundary must occur through coalescence of the primary phase. Notice that such a model is not capable of matching the entire dataset, especially the sharp decrease with composition of the HCS past the maximum (see Fig. 13.27(b)).

Mixed models
Several improvements can be made to the simple model that considers only the solidification interval of the alloy. For example, Feurer took into account the feeding ability of the mushy zone and solidification shrinkage [12]. Such approaches typically neglect mechanical strains and are

therefore closer to the criteria for porosity formation described in Chap. 12. Such a model, shown as the curve labeled "F" in Fig. 13.27(b), corresponds to a hydrostatic tension model based on a pressure drop calculation (see Chap. 12 and the book of Campbell [3]). The maximum HCS is well predicted with such models, since it is associated with the solidification interval of the alloy, and thus with microsegregation. However, the predicted HCS decreases too rapidly with composition past the maximum, indicating that such models do not properly capture hot tearing mechanisms.

Clyne and Davies [4] developed a criterion based on the time spent by the mushy zone in the critical zone where continuous liquid films exist, but with a permeability too low to allow feeding. Their criterion can be expressed as

$$\mathrm{HCS} = \frac{t_V}{t_R} = \frac{t_{g_s=0.99} - t_{g_s=0.9}}{t_{g_s=0.9} - t_{g_s=0.4}} \tag{13.29}$$

where t_V and t_R correspond to the time spent by the mushy zone in the vulnerable region (estimated to be between $g_s = 0.9$ and $g_s = 0.99$) and the time where feeding by the movement of the liquid (and solid) can occur (estimated to be between $g_s = 0.4$ and $g_s = 0.9$), respectively. These times can be determined from a purely thermal calculation. As can be seen in Fig. 13.27(b) (continuous curve labeled "CD"), this model predicts a rapid decrease of the HCS past the maximum, as observed in the finger test measurements. However, it does not reproduce the right trends when applied to processes such as DC casting (Suyitno et al. [29]) or when using a constant cooling rate instead of a constant heat extraction rate (Rappaz et al. [24]).

Ohm et al. [21] considered the maximum resistance that a liquid film can pose to tensile stresses, corresponding to the nucleation of a bubble in a liquid film of thickness h. Assuming that the liquid wets its solid perfectly, one has:

$$\sigma_{max} = \frac{2\gamma_{\ell g}}{h} \tag{13.30}$$

where $\gamma_{\ell g}$ is the interfacial energy between the liquid and the gas. Strain rate dependence can also be introduced by calculating viscous flow in the liquid film induced by the opening of the grain boundary. Such models make it possible to recover the dependence of the maximum strength on g_s, since the liquid film thickness decreases and the strength increases for $g_s \to 1$. Lahaie and Bouchard [16] extended this concept by considering a hexagonal arrangement of grains, the flow necessary to accommodate the grain re-arrangement and the interlocking of the grains. However, these models fail to predict the BTR as they do not take into account feeding at higher temperatures.

It is also worth mentioning that several criteria have been developed based on the strain rate, or accumulated strain, or maximum strength of the mushy zone, sometimes compared to other quantities, such as the cooling rate. For example, the difference between the strain at rupture $\varepsilon_{max}(T)$ and the thermal contraction ε^{th} was considered by Novikov (see the review of Eskin et al. [10]). A "reserve" of plasticity ε_{res}, which is equivalent to HCS^{-1}, is then defined as

$$\varepsilon_{res} = \frac{1}{\Delta T_{BTR}} \int_{T_{coh}}^{T_{sol}} (\varepsilon_{max}(T) - \varepsilon^{th}) dT \qquad (13.31)$$

where ΔT_{BTR}, T_{coh} and T_{sol} are the temperature interval of the BTR, the coherency temperature and the temperature where $g_s = 1$, respectively. Other authors have defined an HCS as the inverse of the maximum resistance in traction, σ_{max}^{-1}, of the mushy zone, the strain rate $\dot{\varepsilon}$ normalized by the cooling rate $\dot{T}$, or using a modified Griffith fracture criterion.

Overall, these models include one or two of the multiple aspects of hot tear formation: the solidification interval, feeding, strain or strain rate, ductility induced at high g_s by coalescence and at lower g_s by feeding, nucleation of a bubble in a liquid film, etc. In the next section, we describe two-phase models, which are certainly a much better starting point for the modeling of hot tearing, even though they still neglect certain phenomena.

Two-phase models

Following Niyama's approach to modeling porosity formation (Chap. 12), Rappaz, Drezet and Gremaud [24] derived the first 2-phase model for hot tearing (usually referred to as the RDG criterion). The model, illustrated in Fig. 13.28, assumes steady-state solidification conditions, $\rho_s = \mathrm{cst} \neq \rho_\ell = \mathrm{cst}$ and the absence of pore formation ($g_p = 0$). The average liquid velocity $\langle v \rangle_\ell$ is assumed to be solely in the x-direction, parallel to the

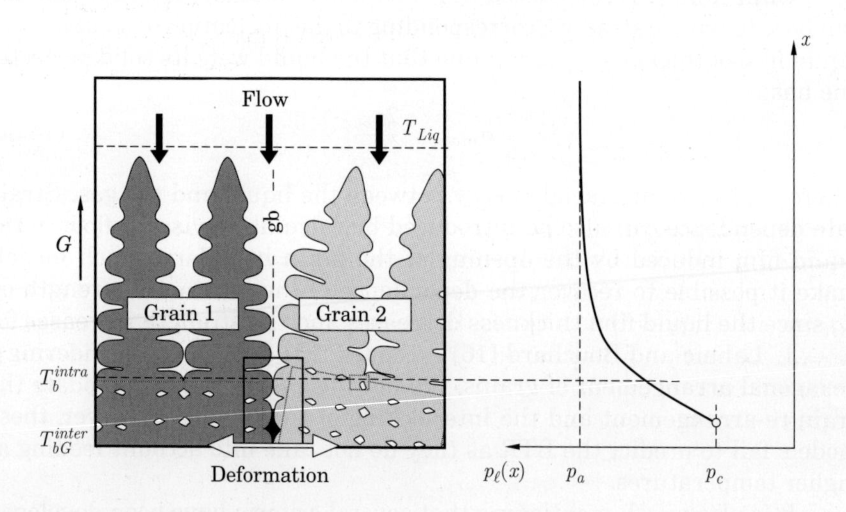

Fig. 13.28 (left) The directional solidification of columnar dendrites with thermal strains perpendicular to the thermal gradient. (right) Pressure drop within the liquid induced by solidification shrinkage and thermal contraction. (After Rappaz et al. [24].)

thermal gradient G, whereas the velocity of the solid $\langle v \rangle_s$ is assumed to be solely in the transverse y-direction. Under these assumptions, the steady-state mass conservation equation (Eq. (12.5) with $g_p = 0$) can be written as

$$-\beta v_T \frac{\partial g_s}{\partial x} + \frac{\partial (g_\ell v_{\ell x})}{\partial x} + (1+\beta)\frac{\partial (g_s v_{sy})}{\partial y} = 0 \qquad (13.32)$$

As previously defined, $\beta = (\rho_s/\rho_\ell - 1)$ corresponds to the shrinkage factor and v_T to the velocity of the isotherms. Note that the intrinsic average symbol, $\langle \cdot \rangle$, has been omitted, i.e., $\langle v \rangle_{\ell x} = v_{\ell x}$ and $\langle v \rangle_{sy} = v_{sy}$. If we further assume that g_s depends only on x, the last term of Eq. (13.32) is equal to $(1+\beta) g_s \dot{\varepsilon}_{syy}(x)$, where $\dot{\varepsilon}_{syy}(x) = \partial v_{sy}/\partial y$ is the strain rate in the solid, which may vary with x. As in the case of the development of Niyama's criterion, Eq. (13.32) can be integrated once appropriate boundary conditions are specified. At the tips of the dendrites, the liquid flow must compensate both the shrinkage and the overall solid deformation. Using this condition, one can show that the velocity in the liquid at any point x of the mushy zone is given by

$$g_\ell v_{\ell x} = -(1+\beta)\dot{E} - g_\ell \beta v_T \qquad (13.33)$$

where $\dot{E}(x)$ is the cumulative average deformation rate of the solid, defined as

$$\dot{E}(x) = \int_0^x g_s \dot{\varepsilon}_{syy}(x) dx \qquad (13.34)$$

The position $x = 0$ corresponds to the roots of the dendrites where coalescence has occurred. For $\dot{\varepsilon}_{syy} = 0$, one recovers the result obtained in Chap. 12, that $v_{\ell x} = -\beta v_T$ everywhere in the mushy zone. Returning to the more general case and using Darcy's equation for $v_{\ell x}$ allows us to compute the pressure drop across the mushy zone

$$\Delta p_{\ell,max} = (1+\beta)\mu_\ell \int_0^L \frac{\dot{E}(x)}{K} dx + v_T \beta \mu_\ell \int_0^L \frac{g_\ell}{K} dx \qquad (13.35)$$

Here, L is the total extent of the mushy zone, i.e., from the tips of the dendrites to the roots where bridging or coalescence occurs.

The last term in Eq. (13.35) is the same as the right hand side of Eq. (12.16): it is due to solidification shrinkage and can be transformed into an integral over the temperature field. Doing the same for the first term, which is associated with strain-induced liquid flow, and using the Carman-Kozeny relation for the permeability K, we have

$$\Delta p_{\ell,max} = \Delta p_{\ell,max}^{\varepsilon_s} + \Delta p_{\ell,max}^{\beta} = \frac{5(S_V^{s\ell})^2(1+\beta)\mu}{G} \int_{g_{\ell,b}}^1 \frac{\dot{E}(g_\ell)(1-g_\ell)^2}{g_\ell^3} \frac{dT}{dg_\ell} dg_\ell$$

$$+ \frac{5(S_V^{s\ell})^2 v_T \beta \mu}{G} \int_{g_{\ell,b}}^1 \frac{(1-g_\ell)^2}{g_\ell^2} \frac{dT}{dg_\ell} dg_\ell \qquad (13.36)$$

where, $S_V^{s\ell}$ is the specific solid-liquid interfacial area. The fraction of liquid $g_{\ell,b} = (1 - g_{s,b})$ in the lower bound of the integrals corresponds to the value of g_ℓ at coalescence. For the continuous liquid films remaining within a grain, the value of $g_{s,b}$ is given by $g_s(T_b^{intra})$, whereas for continuous liquid films at a grain boundary, we have $g_{s,b} = g_s(T_b^{inter})$. As for many other models, this important parameter of the RDG model is not precisely defined: it lies between 0.9 and 0.96, as discussed in Sect. 13.3.2. If one takes a linear model for microsegregation, i.e., $dT/dg_\ell = \Delta T_0$, and also assumes that $\dot{\varepsilon}_{syy}$ is uniform across the mushy zone, the first integral of Eq. 13.36 can be found analytically (see Exercise 13.7), to obtain

$$\begin{aligned}\Delta p_{\ell,max}^{\varepsilon_s}{}' &= \Delta p_{\ell,max}^{\varepsilon_s} \frac{G^2}{5(S_V^{s\ell})^2(1+\beta)\mu\Delta T_0^2 \dot{\varepsilon}_{syy}} \\ &= -\frac{g_{\ell,b}^2}{4} + 2g_{\ell,b} - 3\ln g_{\ell,b} - \frac{2}{g_{\ell,b}} + \frac{1}{4g_{\ell,b}^2}\end{aligned} \tag{13.37}$$

The normalized pressure $\Delta p_{\ell,max}^{\varepsilon_s}{}'$ is plotted in Fig. 13.29 as a function of $g_{\ell.b}$. Retaining the assumption of constant $\dot{\varepsilon}_{syy}$ over the mushy zone, Eq. (13.36) can also be developed using the Gulliver-Scheil or lever rule approximations. The resulting analytical expressions are too complicated to be given here. The pressure drop $\Delta p_{\ell,max}^{\varepsilon_s}{}'$ calculated using the Gulliver-Scheil approximation is also given in Fig. 13.29 for the case $k_0 = 0.5$. Comparing this result with the shrinkage-induced pressure drop for a similar case (Fig. 12.8), one can see that the choice of a microsegregation model has a much larger effect for deformation-induced pressure drop. This is due to the fact that the cumulative strain rate $\dot{E}$ is also averaged with $g_s(T(x))$ in Eq. (13.34).

The RDG model can be used to develop a criterion for the appearance of hot tears as follows: if the pressure in the liquid at the roots of the

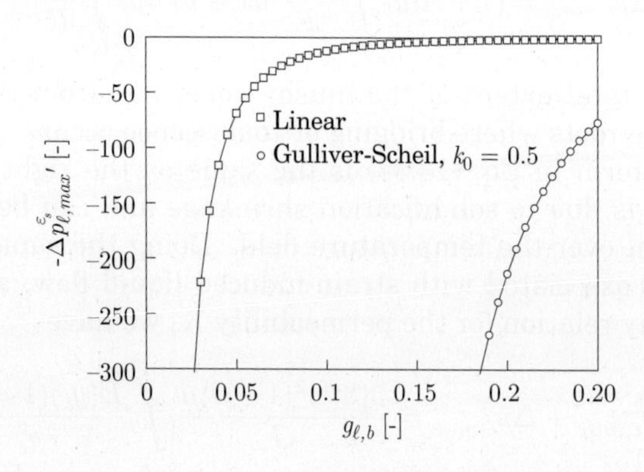

Fig. 13.29 The pressure drop normalized by $5(S_V^{s\ell})^2(1+\beta)\mu\Delta T_0^2\dot{\varepsilon}_{syy}G^{-2}$, $\Delta p_{\ell,max}^{\varepsilon_s}{}'$, due to thermal strains according to the RDG model and two different microsegregation models.

dendrites falls below a certain "cavitation" pressure p_c, a crack will form. This criterion can be employed in two ways:

- For a given alloy ($\Delta T_0, k_0, \ldots$), microstructure ($\mathcal{S}_V$), and arbitrary pressure drop (e.g., $\Delta p_c = p_a - p_c$), Eq. (13.36) or its analytical form for a specific microsegregation model (e.g., Eq. (13.37)) makes it possible to deduce the maximum strain rate $\dot{\varepsilon}_{s,max}$ beyond which a crack will form. In other words, $\dot{\varepsilon}_{s,max}^{-1}$ is an estimate of the alloy HCS. The HCS is in this case proportional to ΔT_0^2, and a sharper decrease is thus introduced when increasing the solute content beyond the maximum solidification interval $\Delta T_{0,max}$ (see Fig. 13.27). This criterion provides a much better match to the experimental observations and the physical mechanisms of hot tearing than the model of Clyne and Davies. It also has the advantage that it can be used for both constant heat extraction rate and constant cooling rate.

- Since the criterion includes a strain-rate dependence, it can be used in real solidification processes to calculate the liquid pressure distribution. First, a thermomechanical calculation is performed using the usual single phase approach, i.e., the two-phase nature of the mushy zone is not considered and the plastic deformation occurs at constant volume, i.e., $\nabla \cdot \boldsymbol{v}_s = \mathrm{tr}(\dot{\varepsilon}_s^{vp}) = 0$ even when $g_\ell \neq 0$. This calculation implies that the viscoplastic strain rate perpendicular to the thermal gradient $\dot{\varepsilon}_{\perp G}^{vp}$ is equal to $-\dot{\varepsilon}_{\parallel G}^{vp} \neq 0$. The component $\dot{\varepsilon}_{\perp G}^{vp}$, which is the most likely contributor to the opening of the grain boundaries, can be added to solidification shrinkage for the calculation of the non-steady state pressure drop in the liquid. This is done with the help of the mass conservation equation (Eq. (12.5) with $g_p = 0$).

Suyitno et al. [29] compared the predictions of eight different hot cracking criteria in simulations of DC casting of aluminum alloys, and found that the RDG model best reproduced the experimental trends. Nevertheless, this model is limited by two main approximations: first, the rheology of the mushy zone is not properly accounted for, and the extraction of $\dot{\varepsilon}_{\perp G}$ is therefore a fairly poor way of representing the compression/expansion behavior of the mushy solid; secondly, the localization of strains and feeding at grain boundaries is not considered and the lower bound $g_{\ell,b}$ of the integrals is selected arbitrarily.

The first shortcoming has been addressed by M'Hamdi et al. [20], who coupled the momentum equation for the solid phase (Eq. 13.20) to that for the liquid flow (Eq. 13.19). The velocity field in the solid phase $\langle \boldsymbol{v} \rangle_s$ could thereby be related to the pressure in the liquid $\langle p \rangle_\ell$. In this model, the solid phase is considered to be compressible, and it uses the rheological model of Ludwig et al. [17], which is similar to the Gurson model described in Sect. 13.3.3. Since $\mathrm{tr}(\dot{\varepsilon}_s^{vp}) \neq 0$ in the mushy zone, the volume change of the solid skeleton has a direct influence on the liquid pressure. Mathier et al. [18] decoupled the equations for $\langle \boldsymbol{v} \rangle_s$ and $\langle p \rangle_\ell$ by noting that the pressure drop in the liquid is much smaller than the stresses in the solid. It

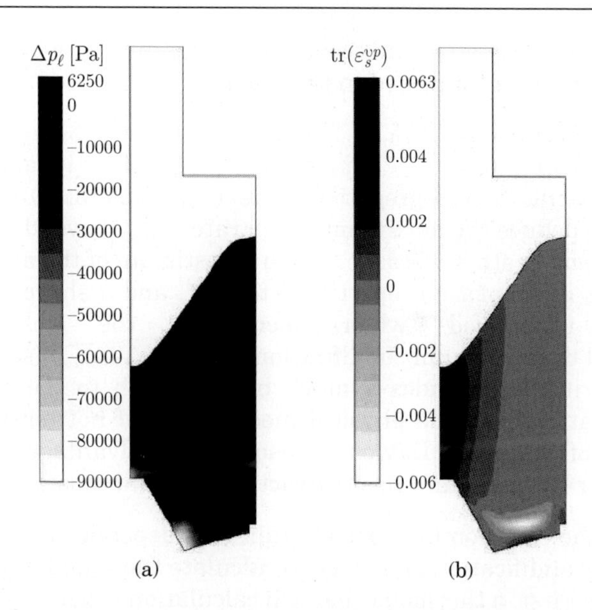

Fig. 13.30 Two-phase modeling of DC casting for an AA6060 axisymmetric billet: (a) The pressure drop in the liquid at $g_s = 0.98$; (b) The volumetric plastic strain $\mathrm{tr}(\varepsilon_s^{vp})$ cumulated over the entire solidification time. (After M'Hamdi et al. [20].)

thereby became possible to calculate the deformation of the solid with the help of Eq. (13.24), neglecting the pressure in the liquid. The pressure drop in the liquid induced by solidification shrinkage and solid deformations was subsequently calculated using Eq. (13.19) in which $\langle v \rangle_s$ is already known.

As an example, Fig. 13.30 presents a simulation using this two-phase approach for the start-up phase of a DC-cast AA6060 alloy (Al-0.45wt%Si-0.45wt%Mg). The contour plot in (a) shows the liquid pressure drop when $g_s = 0.98$, while that in (b) represents the volumetric viscoplastic strain, $\mathrm{tr}(\varepsilon_s^{vp})$, accumulated over the entire solidification time. Both plots indicate that the part of the DC-cast ingot that is most sensitive to hot cracking is the axis of the billet, just above the bottom block, which is also consistent with observations.

The second limitation of the RDG model, the neglect of strain and feeding localization at the grain boundaries, must be tackled using "granular models." An example of such an approach to modeling the percolation of an ensemble of grains was presented in Sect. 13.3.2. Since each grain and grain boundary are considered individually, the feeding through the network of liquid films can be calculated by taking the flow in each liquid channel to follow a simple law of viscous flow. Along each channel, the flow decreases due to solidification shrinkage and possible movement of the grains. At any triple junction, the total liquid flux must sum to zero. Further details can be found in the work of Vernède et al. [31–33]. Figure 13.31 displays a small volume element of uniform temperature

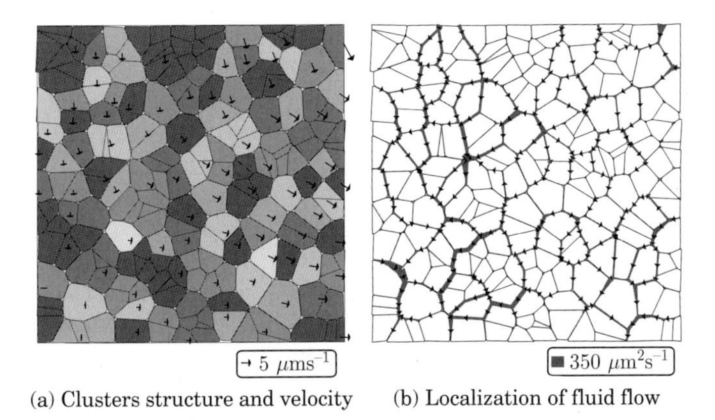

(a) Clusters structure and velocity (b) Localization of fluid flow

Fig. 13.31 Granular modeling of a small 2-D volume element of uniform temperature with $g_s = 0.975$, subjected to horizontal straining. The grains are assumed to be rigid, and their movements are indicated by arrows in (a). The flow induced by the grain movement at the onset of straining with a deformation rate of 10^{-3} s^{-1} of the vertical boundaries of the domain is shown in (b). The width of the boundary is scaled to be proportional to the flow rate, using the scale given below the picture. (After Vernède et al. [31; 33].)

containing approximately 200 grains, modeled using such a method. At $g_s = 0.975$, the grains at the left and right boundaries are translated horizontally to the left and right, respectively. The strain rate, computed as the difference in velocity of the two boundaries divided by their separation, is 10^{-3} s^{-1}. Although the globular equiaxed grains remain separated by liquid films, their movement is influenced by the ability of the liquid to flow between them and by the pressure gradients created in the liquid channels. The liquid flow in each grain boundary is represented in Fig. 13.31(b) with a line of which the width is proportional to the flux. The velocity of each (rigid) grain is indicated with small arrows. This work has been extended to include also deformation of the individual grains [23].

13.6 SUMMARY

This chapter began by presenting the necessary concepts for understanding the development of stresses and deformation in castings during solidification, first for a fully solid domain, then extended to the mushy zone. At low volume fractions of solid, the solid-liquid mixture can be treated as a viscous fluid, whereas at high g_s, its rheology closely resembles that of a viscoplastic medium. After introducing the notion of coherency, it was shown that the mushy zone has a much more complex behavior when such a point is reached. Two-phase approaches as well as rheological behavior typical of porous media must be used in this case.

We then described the most important phenomena associated with hot tear formation. It was shown that hot tears form predominantly in alloys that have large solidification intervals due to both a lack of feeding and the development of thermal strains in the mushy zone. Tears are located at intergranular boundaries where continuous films of liquid still exist, whereas the primary phase within grains already forms a fully coherent solid. The remaining liquid films tend to localize the strains and render the mushy zone extremely brittle. In a given casting, hot tears are localized in regions where the solid skeleton is under expansion. Two-phase models are capable of providing reasonable predictions of the regions of castings most susceptible to hot tearing formation, but not the direct influence of grain structures. For example, it is well-known that a fine grain size reduces the HCS of an alloy, since strains are more evenly distributed among many more grain boundaries. Such effects can only be accounted for by granular models. These are limited to small volume elements but can be coupled to two-phase approaches.

13.7 EXERCISES

Exercise 13.1. Stress and strain in a thin coating deposited over a rigid substrate.
Consider the coating in Fig. 13.2(top) in which the coating has an ideal elasto-viscoplastic behavior (cf. Fig. 4.10) and the substrate is rigid. Demonstrate that the final stress and strain state in the coating is:

$$\sigma_{xx} = E\alpha_{Ts}(T_{sol} - T_0) \qquad \text{and } \varepsilon_{xx}^{vp} = 0 \qquad \text{if } -\varepsilon_{xx}^{th} < \frac{\sigma_Y}{E}$$

or

$$\varepsilon_{xx}^{vp} = \alpha_{Ts}(T_{sol} - T_0) - \frac{\sigma_Y}{E} \qquad \text{and } \sigma_{xx} = \sigma_Y \qquad \text{if } -\varepsilon_{xx}^{th} > \frac{\sigma_Y}{E} \tag{13.38}$$

Exercise 13.2. Stress and strain in a thin coating deposited over a deformable substrate.
Considering only the component $\sigma_{xx}(z)$ of the stress tensor for the situation shown in Fig. 13.2(bottom), write down the two equilibrium conditions: (i) Sum of forces equal to zero; (ii) Sum of moments equal to zero. Assuming elastic deformation in both the substrate and the coating, the total strain of the assembly $\varepsilon_{xx}(z)$ is a linear function of z. Using Hooke's law, derive the shape of the assembly, given in Eqs. (13.4) and (13.5).

Exercise 13.3. Stresses in galvanized steel sheets.
In order to protect steel sheets against corrosion, they are continuously coated on both sides with a thin layer of Zn-0.2wt%Al by immersion in a liquid bath. Suppose that the temperatures of the sheet and coating are uniform over the thickness during the entire process. Calculate the stress and strain at room temperature in both materials. Use the following

thermal expansion data: α_{Ts}(Zn) = 35×10^{-6} K^{-1}, α_{Ts}(Fe) = 12×10^{-6} K^{-1}. Treat the deformation of Fe as purely elastic, and that of the zinc coating as ideally elastic-plastic. Perform the calculation with E(Fe) = 200 GPa, $E(\mathrm{Zn}) = 90$ GPa, σ_Y(Zn) = 28 MPa, a sheet thickness of 0.8 mm, a coating thickness of 20μm, and T_{sol} = 410°C.

Exercise 13.4. Hot cracking near a sharp re-entrant corner.
Consider in 2-D the sharp re-entrant corner of a shape casting solidifying in a steel mold. Based on a simple block-type model similar to that in Fig. 13.2, explain why such a corner is more sensitive to hot cracking than the lateral face away from the corner.

Exercise 13.5. Thermal contraction in DC casting.
Estimate the thermal contraction rate at the surface and in the middle of a DC-cast Al ingot, given that: α_{Ts}(Al) = 20×10^{-6} K^{-1}, $v_c = 1$ mm s^{-1}, the thermal gradient G equals 5 and 100 K cm^{-1} at the center and surface, respectively, close to the solidus temperature. Why is the hot cracking tendency the highest at the center?

Exercise 13.6. Thermo-mechanics of laser welding.
Consider a point source moving at constant velocity v_b. The steady-state temperature field is given by the Rosenthal solution (see Chap. 1):

$$T(x, y, z) = T_0 + \frac{\beta I}{2\pi k r} \exp\left(-\frac{v_b(r+z)}{2\alpha}\right) \tag{13.39}$$

where I is the power of the heat source, β is the absorption coefficient, k is the thermal conductivity and α is the thermal diffusivity. The radius, r, measured from the point of incidence of the beam on the surface is given by $(x^2 + y^2 + z^2)^{1/2}$, and z measures the distance relative to the point of incidence in the welding direction. Calculate the thermal gradient at various points of the solidus surface T_{sol} and the solidification speed. From that, deduce the cooling rate at the solidus and the thermal contraction rate $\dot{\varepsilon}_{ii}^{th}$. What are the possible means of decreasing the hot cracking sensitivity of the welds? Application: $I = 500$ W, $k = 100$ W m^{-1}K^{-1}, $\alpha = 5 \times 10^{-5}$ m^{2}s^{-1}, $\beta = 0.2$, $v_b = 0.1$ m s^{-1}, $\beta_T = 60 \times 10^{-6}$ K^{-1}, $T_{sol} = 500$°C.

Exercise 13.7. Pressure drop due to thermal strains.
Using the RDG model (Eq. (13.36)), demonstrate that the normalized pressure drop across the mushy zone due to a uniform strain rate perpendicular to the thermal gradient is given by Eq. (13.37), provided that g_s is a linear function of temperature.

13.8 REFERENCES

[1] P. Ackermann, W. Kurz, and W. Heinemann. In-situ testing of solidifying aluminium and Al-Mg shells. *Mater. Sci. Eng.*, 75:79, 1985.

[2] J. C. Borland. Generalised theory of super-solidus cracking in welds. *Br. Weld. J.*, 7:508, 1960.
[3] J. Campbell. *Castings*. Butterworth-Heinemann, Oxford, 2nd edition, 2003.
[4] T. W. Clyne and G. J. Davies. The influence of composition on solidification cracking suceptibility in binary alloy systems. *J. Brit. Foundry*, 74:65, 1981.
[5] A. H. Dahle and L. Arnberg. Development of strength in solidifying aluminium alloys. *Acta Mater.*, 45:547, 1997.
[6] C. H. Dickhaus, L. Ohm, and S. Engler. Mechanical properties of solidifying shells of aluminium alloys. *Trans. AFS*, 118:677, 1993.
[7] J.-M. Drezet and G. Eggeler. High apparent creep activation-energies in mushy zone microstructures. *Scripta Met. Mater.*, 31:757, 1994.
[8] J.-M. Drezet, V. Mathier, and D. Alléhaux. FEM modelling of laser beam welding of aluminium alloys with special attention to hot cracking in transient regimes. In H. Cerjak et al., editors, *Mathematical Modelling of Weld Phenomena 8*, page 137, Graz, Austria, 2007. Verlag Techn. Univ. Graz.
[9] J.-M. Drezet and M. Rappaz. Modelling of ingot distortions during direct chill casting of aluminium alloys. *Met. Mater. Trans. A*, 37A:3214, 1996.
[10] D. G. Eskin, Suyitno, and L. Katgerman. Mechanical properties in the semi-solid state and hot tearing of aluminium alloys. *Prog. Mater. Sci.*, 49:629, 2004.
[11] I. Farup, J.-M. Drezet, and M. Rappaz. In-situ observation of hot tearing formation in succinonitrile-acetone. *Acta Mater.*, 49:1261, 2001.
[12] U. Feurer. Mathematisches Modell der Warmrissneigung von binaeren Aluminiumlegierungen. *Giesserei Forsch.*, 2:75, 1976.
[13] P.-D. Grasso, J.-M. Drezet, J.-D. Wagnière, and M. Rappaz. Shear behaviour of a semi-solid Al-4.5 wtpct Cu alloy in relation with its microstructure. In M. G. Chu, D. A. Granger, and Q. Han, editors, *Solidification of Aluminum Alloys*, pages 399–408. TMS-AIME, 2004.
[14] S. Instone, D. StJohn, and J. Grandfield. New apparatus for characterising tensile strength development and hot cracking in the mushy zone. *Int. J. Cast Met. Res.*, 12: 441, 2000.
[15] A. Jacot, M. Swierkosz, J. Rappaz, M. Rappaz, and D. Mari. Modelling of electromagnetic heating, cooling and phase transformations during surface hardening of steels. *J. Physique IV C1*, 6:203, 1996.
[16] D. J. Lahaie and M. Bouchard. Physical modeling of the deformation mechanisms of semisolid bodies and a mechanical criterion for hot tearing. *Met. Mater. Trans.*, 32B: 697, 2001.
[17] O. Ludwig, J.-M. Drezet, C. L. Martin, and M. Suéry. Rheological behavior of Al-Cu alloys during solidification. *Met. Mater. Trans.*, 36A:1525, 2005.
[18] V. Mathier, P.-D. Grasso, and M. Rappaz. A new tensile test for aluminum alloys in the mushy state: experimental method and numerical modeling. *Met. Mater. Trans. A*, 39A: 1399, 2008.
[19] F. Matsuda, H. Nakagawa, S. Katayama, and Y. Arata. Weld metal cracking and improvement of 25%Cr-20%Ni (AISI 310S) fully austenitic stainless steel. *Trans. Japan. Weld. Soc.*, 13:115, 1982.
[20] M. M'Hamdi, A. Mo, and H. G. Fjaer. TearSim: a two-phase model addressing hot tearing formation during aluminum direct chill casting. *Met. Mater. Trans.*, 37A:3069, 2006.
[21] L. Ohm and S. Engler. Festigkeiteigenschaften erstarrender Randschalen aus Aluminiumlegierungen. *Giessereiforschung*, 42:149, 1990.
[22] A. B. Phillion, S. L. Cockcroft, and P. D. Lee. X-ray micro-tomographic observations of hot tear damage in an Al-Mg commercial alloy. *Scripta Mater.*, 55:489, 2006.
[23] A. B. Phillion, S. L. Cockcroft, and P. D. Lee. A three-phase simulation of the effect of microstructural features on semi-solid tensile deformation. *Acta Mater.*, 56(16):4328–4338, 2008.
[24] M. Rappaz, M. Gremaud, and J.-M. Drezet. A new hot-tearing criterion. *Met. Mater. Trans.*, 30A:449, 1999.

[25] M. Rappaz, A. Jacot, and W. Boettinger. Last stage solidification of alloys: a theoretical study of dendrite arm and grain coalescence. *Met. Mater. Trans.*, 34A:467, 2003.

[26] R. A. Rosenberg, M. C. Flemings, and H. F. Taylor. Nonferrous binary alloys hot tearing. *Trans. AFS*, 68:518, 1960.

[27] J. A. Spittle and A. A. Cushway. Influences of superheat and grain structure on hot-tearing susceptibilities of Al-Cu castings. *Met. Technol.*, 10:6, 1983.

[28] T. Sumitomo, D. H. StJohn, and T. Steinberg. The shear behaviour of partially solidified Al-Si-Cu alloys. *Mater. Sci. Eng.*, A289:18, 2000.

[29] Suyitno, W. H. Kool, and L. Katgerman. Hot tearing criteria evaluation for direct-chill casting of an Al-4.5 pct Cu alloy. *Met. Mater. Trans.*, 36A:1537, 2005.

[30] W. M. van Haaften, W. H. Kool, and L. Katgerman. Tensile behaviour of DC-cast AA5182 in solid and semi-solid state. In K. Ehrke and W. Schneider, editors, *Continuous Casting*, page 237. Wiley-VCH Verlag GmbH, 2000.

[31] S. Vernède, J. A. Dantzig, and M. Rappaz. A mesoscale granular model for the mechanical behavior of alloys during solidification. *Acta Mat.*, 57:1554–1569, 2009.

[32] S. Vernède, P. Jarry, and M. Rappaz. A granular model of equiaxed mushy zones: Formation of a coherent solid and localization of feeding. *Acta Mater.*, 54:4023, 2006.

[33] S. Vernède. *A granular model of solidification as applied to hot tearing*. PhD Thesis No 3795, EPFL, Station 12, 1015 Lausanne, Switzerland, 2007.

[34] N. Wang, S. Mokadem, M. Rappaz, and W. Kurz. Solidification cracking of superalloy single- and bi-crystals. *Acta Mater.*, 52:3173, 2004.

[35] D. Warrington and D. G. McCartney. Development of a new hot-cracking test for aluminium alloys. *Cast Metals*, 2:134, 1989.

CHAPTER 14

MACROSEGREGATION

14.1 INTRODUCTION

Solute composition inhomogeneities at the macroscopic scale of a casting – a phenomenon commonly known as *macrosegregation* – are undesirable for several reasons. First, such inhomogeneities cannot be removed readily by heat treatment, as the time required for solid state diffusion to occur at this scale is prohibitive. For instance, for $D_s \approx 10^{-12}$ m^{2}s^{-1} and $L = 1$ m, the diffusion time is in the range of 10^{12} s or 32,000 years! Second, macrosegregation results in spatial variations of the mechanical properties after heat treatment due to the concomitant spatial variations in the amount, nature and size of the precipitates. Third, macrosegregation can lead to the formation of gross compositional defects such as freckles, also called segregated chimneys, which are described later on in this chapter. Finally, there is a cost associated with macrosegregation for certain expensive alloys. Consider gold alloys, for instance: A rating of 18 carats requires an absolute *minimum* of 75 wt% Au everywhere in the material. Thus, if the composition is not uniform, certain regions will have even higher gold content, thereby adding to the overall cost. It is therefore important to understand, model and control macrosegregation, which is the subject of this chapter.

It was seen in Chap. 5 that the directional solidification of an alloy with $k_0 < 1$, at constant velocity with a planar front, consists of three stages: the initial transient during which the composition in the solid C_s is lower than the nominal composition C_0; the steady state regime where $C_s = C_0$; and the final transient where $C_s > C_0$. This implies segregation at the scale of the whole specimen, i.e., macrosegregation. Macrosegregation can occur even during the steady-state regime of planar front growth if there is a certain amount of convection. This case, briefly presented in Sect. 14.2, will enhance our understanding of what takes place in the more general case of dendritic solidification.

Indeed, in most cases encountered during the production of alloys or shape castings – as demonstrated in Chap. 8 – the thermal conditions are such that the planar front is unstable, causing dendrites or cells to form.

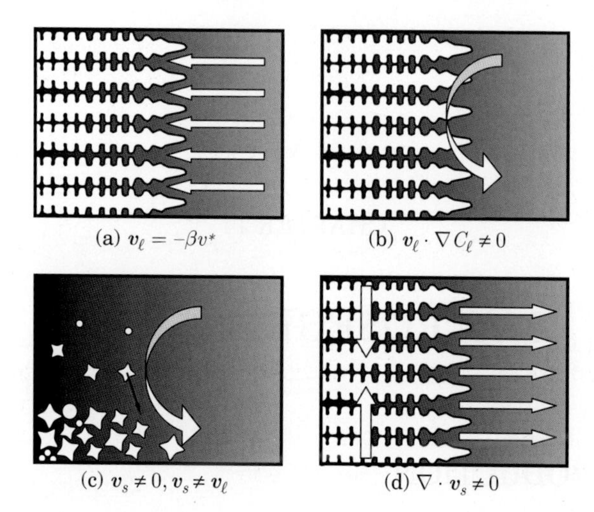

Fig. 14.1 Various types of macrosegregation induced by (a) solidification shrinkage, (b) fluid flow, (c) grain movement and (d) deformation of the solid.

This induces *microsegregation* at the scale of the secondary dendrite arm spacing, as discussed in Chap. 10. Despite this localization of segregation at the micron scale, macrosegregation can nevertheless occur by several mechanisms, as illustrated in Fig. 14.1. In all of them, macrosegregation takes its origin in the *relative movement of one phase with respect to the other*. Indeed, as the solid and liquid phases do not accept the same amount of solute, microsegregation will give rise to macrosegregation when one phase moves relative to the other. The four main causes of macrosegregation illustrated in Fig. 14.1, and treated in the following sections, are:

(a) *Macrosegregation associated with solidification shrinkage.* In this case, the suction of the liquid induced by the moving interface is opposite to the movement of the isotherms, leading to what is known as *inverse segregation* (Sect. 14.4).

(b) *Macrosegregation associated with natural or forced convection.* If the velocity of the liquid is parallel to the composition isopleths, i.e., perpendicular to the solute gradient, no macrosegregation occurs. However, macrosegregation is induced when the flow enters or exits the mushy zone, i.e., when $v_\ell \cdot \nabla C_\ell \neq 0$ (Sect. 14.5).

(c) *Macrosegregation associated with grain movement.* As long as the grains move with the same velocity as the liquid, there is no macrosegregation. However, equiaxed grains have a tendency to sediment, or, in certain cases, to float, giving rise to macrosegregation (Sect. 14.6.1).

(d) *Macrosegregation associated with deformation of the mushy solid.* In this case, the overall velocity of the solid is small but $\nabla \cdot v_s \neq 0$. In other words, the mushy solid behaves like a sponge causing the liquid to be expelled (when in compression) or sucked in (when in traction) (Sect. 14.6.2).

Macrosegregation at a given location in a casting is said to be *negative* when the average composition measured over a representative volume element is lower than the nominal value. It is *positive* in the opposite case. But which composition are we talking about? Before detailing each source of macrosegregation in Sects. 14.4 to 14.6, it is worthwhile to re-emphasize in Sect. 14.3 what has been presented in Chaps. 3 and 4 concerning the composition field and the general equations governing macrosegregation for dendritic solidification. Before doing so, we present some general aspects of macrosegregation during planar front solidification. It should be emphasized that, unless stated otherwise, we will always assume that $k_0 < 1$ and $m_\ell < 0$ throughout this chapter.

14.2 MACROSEGREGATION DURING PLANAR FRONT SOLIDIFICATION

14.2.1 Thermal convection in a pure material

Before describing convection and macrosegregation induced during directional solidification of alloys, let us first briefly recall the basics of natural convection in pure substances already seen in Chaps. 4 and 6. Using the Boussinesq approximation, i.e., assuming $\beta_T \Delta T_c \ll 1$, where ΔT_c is a characteristic temperature difference, and taking all other properties to be constant, the governing equations controlling heat and mass transport in a fully liquid region are given by (see Sect. 4.6)

$$\nabla \cdot \boldsymbol{v}_\ell = 0 \tag{14.1}$$

$$\frac{\partial \boldsymbol{v}_\ell}{\partial t} + \boldsymbol{v}_\ell \cdot \nabla \boldsymbol{v}_\ell - \nu_\ell \nabla^2 \boldsymbol{v}_\ell = \frac{\nabla p}{\rho_{\ell 0}} - \boldsymbol{g}\beta_T(T - T_0) \tag{14.2}$$

$$\frac{\partial T}{\partial t} + \boldsymbol{v}_\ell \cdot \nabla T - \alpha_0 \nabla^2 T = 0 \tag{14.3}$$

where β_T is the volumetric thermal expansion coefficient ($= V^{-1}dV/dT$), $\boldsymbol{g}$ is the gravity vector and T_0 is the reference temperature at which $\rho_\ell = \rho_{\ell 0}$. The hydrostatic contribution $\rho_{\ell 0}\boldsymbol{g}$ is included in the pressure, as described in Sect. 4.6 (Eq. (4.166)). In order to understand the mechanism by which flow is induced by thermal gradients, we take the curl of Eq. (14.2) and use the vector identity $\nabla \times (\nabla p) \equiv 0$ to obtain

$$\frac{\partial \boldsymbol{w}_\ell}{\partial t} + (\boldsymbol{v}_\ell \cdot \nabla)\boldsymbol{w}_\ell - (\boldsymbol{w}_\ell \cdot \nabla)\boldsymbol{v}_\ell - \nu_\ell \nabla^2 \boldsymbol{w}_\ell = -\beta_T \nabla T \times \boldsymbol{g} = \frac{1}{\rho_{\ell 0}} \nabla \rho_\ell \times \boldsymbol{g} \tag{14.4}$$

where $\boldsymbol{w}_\ell = \nabla \times \boldsymbol{v}_\ell$ is the *vorticity* vector. If the fluid simply rotates with angular velocity Ω about a point in the $x-y$ plane, i.e., $v_{\ell\theta}(r) = r\Omega$, the vorticity vector component is parallel to the z-axis and $w_{\ell z} > 0$ corresponds to a counterclockwise rotation. If $w_{\ell z} < 0$, on the other hand, the rotation is clockwise. Consider a fluid initially at rest, i.e., $\boldsymbol{v}_\ell = \boldsymbol{w}_\ell = 0$. Equation (14.4) then resumes to:

$$\frac{\partial \boldsymbol{w}_\ell}{\partial t} = -\beta_T \nabla T \times \boldsymbol{g} = \frac{1}{\rho_{\ell 0}} \nabla \rho_\ell \times \boldsymbol{g} \tag{14.5}$$

This clearly indicates that a temperature gradient ∇T (or density gradient $\nabla \rho_\ell = -\rho_{\ell 0} \beta_T \nabla T$) in a gravitational field can induce a rotation of the fluid. Using this equation, one can then distinguish several cases:

- $\nabla \rho_\ell \times \boldsymbol{g} \neq 0$
  Such a situation is illustrated in Fig. 14.2. A square cavity has thermally insulated bottom and top walls, its right boundary is maintained at a high temperature, and its left boundary is maintained at a low temperature. The gravity vector points vertically down. As expected, solving Eqs. (14.1)-(14.3) for this configuration gives a counterclockwise velocity field and isotherms distorted in the sense of the fluid rotation.

- $\nabla \rho_\ell$ is parallel to $\boldsymbol{g}$, i.e., $-\rho_{\ell 0} \beta_T G_z g_z > 0$ and $\nabla \rho_\ell \times \boldsymbol{g} = 0$
  Such a situation would occur if, for instance, the cold boundary were at the bottom and the hot boudary at the top, with insulated vertical sides. Since the colder/denser liquid is situated below the warmer/less dense liquid, no movement would be induced and the configuration is *stable* or *stagnant*.

- $\nabla \rho_\ell$ is opposite of $\boldsymbol{g}$, i.e., $-\rho_{\ell 0} \beta_T G_z g_z < 0$ and $\nabla \rho_\ell \times \boldsymbol{g} = 0$
  In this case, the hot boundary is at the bottom while the cold boundary is at the top. Since the warmer/less dense liquid has a tendency to

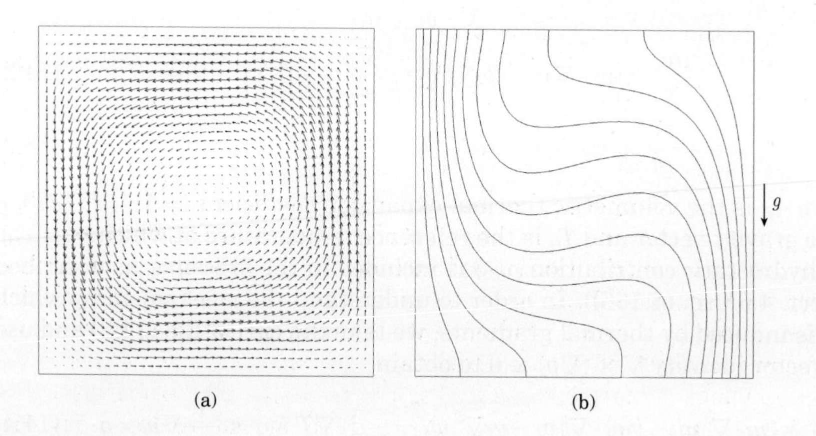

Fig. 14.2 (a) The steady state velocity field and (b) isotherms in a square cavity with a cold boundary to the left, a hot boundary to the right and two horizontal thermally insulated boundaries (Grashof number Gr = 10^5). (Courtesy of J.-L. Desbiolles.)

rise, such a situation would be *unstable*. Beyond a critical Rayleigh number, in this geometry $\mathrm{Ra} = g\beta_T \Delta T L^3/(\nu_\ell \alpha) > 1708$, convection occurs. If the upper boundary is replaced by a free surface, the critical Rayleigh number for convective instability decreases to approximately 1100.

With these considerations in mind, it is now possible to describe the more interesting situation of planar front solidification of alloys.

14.2.2 Convection during directional solidification of a binary alloy

For alloys, in addition to Eqs. (14.1)-(14.3) for the liquid phase, solute transport must also be included,

$$\frac{\partial C_\ell}{\partial t} + \boldsymbol{v}_\ell \cdot \nabla C_\ell - D_\ell \nabla^2 C_\ell = 0 \tag{14.6}$$

For the moment, we focus on the composition of the liquid during directional solidification with a planar front, taken to be horizontal, and thus perpendicular to $\boldsymbol{g}$. We consider solidification both from below and from above. The thermal gradient $\boldsymbol{G}$ is taken as externally imposed and may be either parallel ($\boldsymbol{G} \cdot \boldsymbol{g} = G_z g_z = Gg > 0$) or anti-parallel to gravity ($\boldsymbol{G} \cdot \boldsymbol{g} = G_z g_z = -Gg < 0$). These two cases correspond to unstable and stable thermal configurations, respectively if $\beta_T > 0$. Considering now the contribution of the solutal field, we will sketch the density field in the liquid, which is related to the temperature and composition by the constitutive model given in Eq. (4.52)

$$\rho_\ell = \rho_{\ell 0}(1 - \beta_T(T - T_0) - \beta_C(C_\ell - C_{\ell 0})) \tag{14.7}$$

where β_C is the solutal expansion coefficients of the liquid phase, $\beta_C = V^{-1}\partial V/\partial C_\ell = -\rho^{-1}\partial \rho/\partial C_\ell$. Computing the dot product of $\boldsymbol{g}$ with the gradient of Eq. (14.7) gives

$$\nabla \rho_\ell \cdot \boldsymbol{g} = \rho_{\ell 0} g_z \left(-\beta_T \frac{\partial T}{\partial z} - \beta_C \frac{\partial C_\ell}{\partial z} \right) = -\rho_{\ell 0} \beta_T G_z g_z \left(1 + \frac{\beta_C G_{Cz}}{\beta_T G_z} \right) \tag{14.8}$$

where $G_{Cz} = \partial C_\ell / \partial z$ is the vertical solutal gradient. When compared to the case of the pure fluid discussed in Sect. 14.2.1, one can see that the stability of the liquid during solidification of a binary alloy depends on the sign of the term inside parentheses, i.e., on the relative magnitude of the thermal and solutal gradient terms, $\beta_T G_z$ and $\beta_C G_{Cz}$. It was demonstrated in Chap. 8 that, in the absence of convection, planar front growth is stable provided that $m_\ell G_{Cz} < G_z$. Let us assume that we are just at the limit of stability, $m_\ell G_{Cz} = G_z$. Equation (14.8) becomes in that case

$$\nabla \rho_\ell \cdot \boldsymbol{g} = -\rho_{\ell 0} \beta_T G_z g_z \left(1 + \frac{\beta_C}{m_\ell \beta_T} \right) = -\rho_{\ell 0} \beta_T G_z g_z \, (1 + B) \tag{14.9}$$

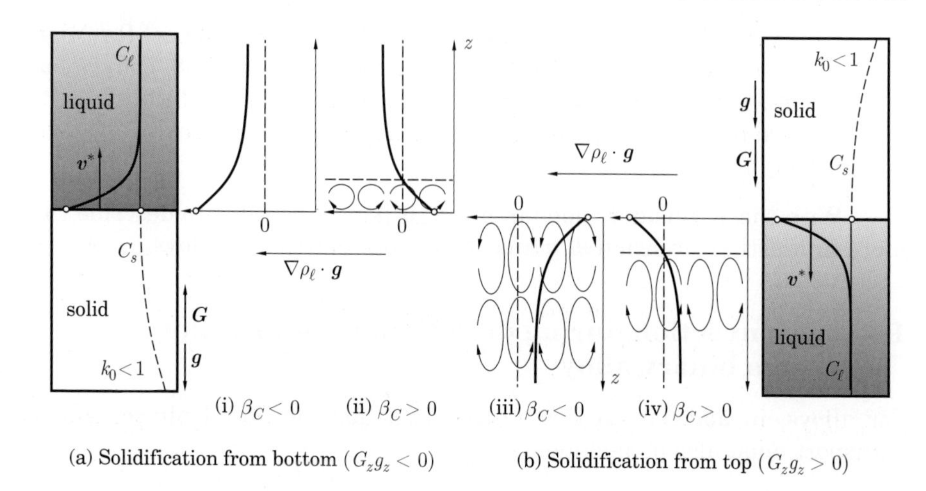

Fig. 14.3 A schematic illustration of planar front solidification of a binary alloy during the steady state phase of directional solidification. The composition profile in the liquid and in the solid are displayed with solid and dashed lines, respectively. In cases (i) and (ii), solidification proceeds from below, i.e., anti-parallel to gravity. Cases (iii) and (iv) correspond to solidification from above. The associated buoyancy force $\nabla\rho_\ell \cdot \boldsymbol{g}$ is shown for the four cases. In the regions where $\nabla\rho_\ell \cdot \boldsymbol{g} < 0$, the liquid has a tendency to be unstable.

The coefficient $B = \beta_C/(m_\ell \beta_T)$ is a property of the alloy alone, not of the geometry nor the processing conditions. Figure 14.3 shows the various possibilities that can be encountered during vertical solidification of a binary alloy with $k_0 < 1$ and $m_\ell < 0$. Figure 14.3(a) presents the solute profile in the solid derived in Chap. 5 for $v_\ell = 0$, during the initial transient, up to a point where steady state is reached. Recall that, at steady state, the composition at the interface in the liquid C_ℓ^* is equal to C_0/k_0, where C_0 is the nominal composition, and the solute boundary layer thickness δ is equal to D_ℓ/v_T. The composition profile at steady state in the liquid is also shown in Fig. 14.3(a). When considering this steady state situation and $\beta_T > 0$, four situations can be distinguished depending on the value of B and the direction of growth:

- $\beta_C < 0 \Rightarrow B > 0$, and $G_z g_z < 0$ (Fig. 14.3(i)).
  This situation corresponds, for example, to the growth of Al-Cu against gravity, where denser Cu is rejected into the liquid and thus $\nabla\rho_\ell \cdot \boldsymbol{g} > 0$ everywhere, including in the solute boundary layer. Since B is positive, the solute gradient reinforces the stability of the liquid layer at the solid-liquid interface. No macrosegregation is expected.

- $\beta_C > 0 \Rightarrow B < 0$, and $G_z g_z < 0$ (Fig. 14.3(ii)).
  In this case, the rejected solute is less dense (e.g., such as in Cu-Sn or Fe-C) and $B < 0$. (Note that the same situation might occur with

a denser solute element and $k_0 > 1$). If $B > -1$, the configuration remains stable even in the boundary layer. However, if $B < -1$, such as for the case represented schematically in Fig. 14.3(ii), a convective instability might develop in the region close to the interface where $\nabla \rho_\ell \cdot \boldsymbol{g} < 0$.

- $\beta_C < 0 \Rightarrow B > 0$ and $G_z g_z > 0$ (Fig. 14.3(iii)).
  Since the solidification occurs in the direction of gravity, the hotter liquid finds itself in the lower position and thus displays a thermal instability that is reinforced in the solute boundary layer since the rejected solute is denser than the solvent.

- $\beta_C > 0$, i.e., $B < 0$ and $G_z g_z > 0$ (Fig. 14.3(iv)).
  Although this situation is thermally unstable, it could be stabilized close to the interface if $B < -1$.

Certainly, the cases described above are merely indicative of when convection, and thus macrosegregation, can be expected. They were obtained under the very restrictive assumptions of a planar front growing at the limit of constitutional undercooling in a well-defined vertical thermal gradient. Once convection starts, the thermal and solutal fields evolve according to Eqs. 14.3 and 14.6. The flow can then become rather complex, as illustrated in Fig. 14.4, which shows plumes of a Sn-rich liquid moving upward during the planar front solidification of a Cu-Sn alloy, for which $\beta_C > 0$ and $G_z g_z < 0$.

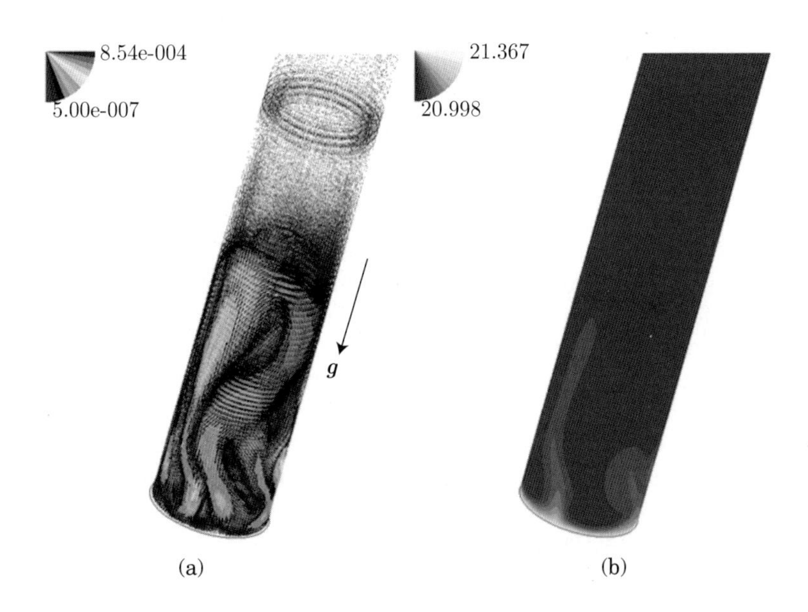

Fig. 14.4 Sn-rich plumes of liquid during vertically upward directional solidification in a fixed thermal gradient at low speed for a Cu-Sn alloy in a 6 mm OD/4 mm ID tube. (a) Velocity field and (b) Sn composition. (From F. Kohler [22].)

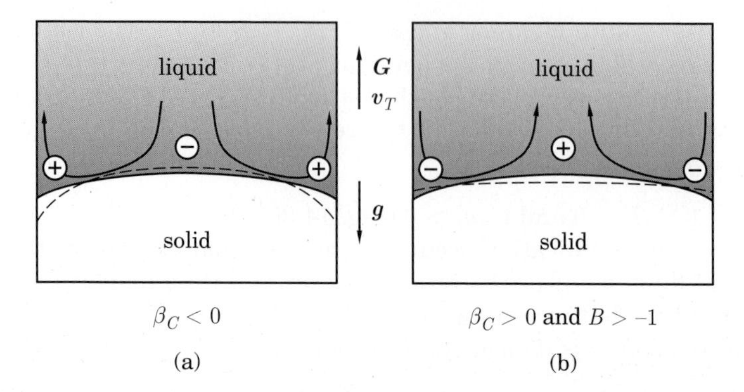

Fig. 14.5 Convection and segregation for a constitutionally stable front growing anti-parallel to gravity in the presence of a slight radial thermal gradient: (a) $\beta_C < 0$, (b) $\beta_C > 0$ but $B > -1$. The solid line represents the boundary of an isothermal solid-liquid front, whereas the dashed line displays the non-isothermal front when convection, indicated by the arrows, is accounted for. The $\oplus$ and $\ominus$ signs indicate regions of positive and negative segregation respectively.

Although certain configurations might at first appear to be stable in a vertical thermal gradient, some convection may still occur when a radial thermal gradient exists (Fig. 14.5). Consider the stable arrangement in Fig. 14.3(b) where $\beta_C < 0$, i.e., $B > 0$ and $G_z g_z < 0$. If the thermal gradient at the solid-liquid interface has a small radial component $G_r = \partial T/\partial r > 0$, then an isothermal solidification front, corresponding to a fixed composition for a binary alloy, will be slightly convex as shown in Fig. 14.5(a) (solid line). Since $\beta_C < 0$, the solute-rich liquid is no longer stable because $\nabla \rho_\ell \times \boldsymbol{g} \neq 0$ except at the center of the specimen. The enriched liquid then starts moving toward the mold wall, as indicated by the arrows. This tends to enrich the outer regions, producing positive segregation there, whereas the center of the specimen becomes depleted (negative segregation). Since $k_0 < 1$ and $m_\ell < 0$, the associated liquidus $T_{liq}(C_\ell(r))$ becomes a function of the radius, and is thus lower at the outer regions and larger at the center. This effect tends to increase in turn the curvature of the solid-liquid front (dashed line), thereby inducing even more convection and segregation. The same reasoning can be applied to the case of an initially concave solid-liquid interface (i.e., $G_r < 0$), with the same result. The interface becomes more concave when the composition field is considered. Thus, this situation becomes potentially unstable in the sense that a slightly concave/convex isothermal front has a tendency to display a more significant curvature due to macrosegregation.

Consider next the convex solid-liquid interface, but now with $\beta_C > 0$, i.e., $B < 0$, while the other conditions are unchanged. For $B > -1$, $\nabla \rho_\ell \cdot \boldsymbol{g}$ remains positive, as the situation continues to be stagnant in a vertical thermal gradient. Surprisingly, for a small component G_r, the interface tends to become increasingly flat (Fig. 14.5b). Indeed, following the same line of reasoning as before, the flow ahead of the front would be in the opposite sense, since the solute is less dense than the solvent. Thus, with a convex isothermal front, negative segregation occurs near the wall, whereas positive segregation is observed at the center. This behavior has a tendency to increase T_{liq} near the wall and to decrease it near the center, i.e., the originally convex isotherm becomes flattened, as indicated by the dashed line.

Now that we have described qualitatively the basic phenomena of macrosegregation in the case of a planar front – a rather rare situation, encountered almost exclusively in single crystal growth – we turn our attention to macrosegregation associated with the more common case of dendritic solidification. The next section presents some general concepts and basic equations, and the subsequent sections focus on specific causes of macrosegregation.

14.3 COMPOSITION FIELD AND GOVERNING EQUATIONS

For dendritic solidification, it is no longer practical to track the solid-liquid front because the microstructure is too fine. The average equations introduced in Chap. 4 are therefore employed, however certain other aspects also need to be specified. As in our previous models, we assume complete mixing in the liquid phase within a small representative volume element. The average mass composition $\langle C \rangle_M$ is defined as

$$\langle C \rangle_M = \int_0^{f_s} C_s df'_s + C_\ell f_\ell \qquad (14.10)$$

where f_s and f_ℓ are the mass fractions of the solid and liquid phases and C_s, C_ℓ are the associated compositions. The index "M" has been added to indicate that the average is taken over the mass fractions of the phases. Note that the composition in the solid phase can be "frozen", as in the Gulliver-Scheil approximation, or be a function of time if back-diffusion takes place (see Chap. 10). For sufficiently fast diffusion in the solid, C_s is uniform and Eq. (14.10) reduces to the lever rule

$$\langle C \rangle_M = C_s f_s + C_\ell f_\ell \qquad (14.11)$$

In a casting, the total mass of the solvent and solute remain constant as solidification proceeds (i.e., there is no evaporation) even if solidification shrinkage causes the overall volume to change. This can be written

$$\int_M \langle C \rangle_M(\boldsymbol{x}, t) dM = C_0 M_0 \qquad \forall t \qquad (14.12)$$

where M_0 and C_0 are the initial mass and solute compositions of the liquid poured into the mold, respectively. In the absence of macrosegregation, the average composition $\langle C \rangle_M$ is independent of the location in the casting. Since $M = M_0$, we would then have $\langle C \rangle_M = C_0$ at every point in the casting. Thus, macrosegregation clearly refers to *compositions* (mass or atomic fractions) and not to volumetric concentrations. However, when deriving the average conservation equations in Chap. 4, a volume-averaged formulation was used. Based on a fixed volume element, the average *volumetric concentrations* $\langle \rho C \rangle$ are defined naturally as

$$\langle \rho C \rangle = \int_0^{g_s} \rho_s C_s dg_s' + g_\ell \rho_\ell C_\ell \tag{14.13}$$

where g_s and g_ℓ refer to the volume fractions of solid and liquid, respectively. It should be clear then that the volumetric concentrations vary during solidification, regardless of whether macrosegregation takes place or not. Indeed, if $\rho_s > \rho_\ell$, the volume of the solid is smaller than that of the initially poured liquid, and the volumetric concentration thus becomes higher.

Therefore, as the derivation of the macrosegregation is performed in a volumetric framework and macrosegregation refers to compositions, it is necessary to be able to convert between $\langle \rho C \rangle$ and $\langle C \rangle_M$. This is straightforward, as one has

$$\langle C \rangle_M = \frac{\int_0^{M_s} C_s dM + M_\ell C_\ell}{M_s + M_\ell} = \frac{\int_0^{V_s} \rho_s C_s dV + \rho_\ell V_\ell C_\ell}{\rho_s V_s + \rho_\ell V_\ell} \tag{14.14}$$

or, after division by $(V_s + V_\ell)$,

$$\langle C \rangle_M = \frac{\int_0^{g_s} \rho_s C_s dg_s' + \rho_\ell g_\ell C_\ell}{\rho_s g_s + \rho_\ell g_\ell} = \frac{\langle \rho C \rangle}{\langle \rho \rangle} \quad \Rightarrow \quad \langle \rho C \rangle = \langle \rho \rangle \langle C \rangle_M \tag{14.15}$$

Macrosegregation is governed by the average solute conservation equation derived in Eq. (4.155). Neglecting the source term and considering only one solute species, so that we may drop the index used there, we have

$$\frac{\partial \langle \rho C \rangle}{\partial t} + \nabla \cdot \langle \rho C \boldsymbol{v} \rangle = \nabla \cdot \langle \rho D \nabla C \rangle \tag{14.16}$$

Since macrosegregation is frequently a result of buoyancy-driven flows, let us take this as our "reference" case and examine the relative importance of the advective and diffusive terms in Eq. (14.16). It was demonstrated in Sect. 4.6 that the characteristic velocity for buoyancy-driven flow is $v_c = \sqrt{g\beta_T \Delta T L} \approx 0.025$ m/s. If we now consider macrosegregation at the scale of the dendrite arms (say λ_1) caused by such a flow, the important dimensionless group is the solutal Péclet number, given by

$$\mathrm{Pe}_C = \frac{v_c \lambda_1}{D_\ell} \approx \frac{(0.025 \text{ m/s})(10^{-4} \text{ m})}{10^{-9} \text{ m}^2/\text{s}} = 2.5 \times 10^3 \gg 1 \tag{14.17}$$

This implies that the diffusive term is negligible in comparison with the convective term and it is therefore omitted for the remainder of the discussion. Equation (14.16) then becomes

$$\frac{\partial \langle \rho C \rangle}{\partial t} + \nabla \cdot \langle \rho C \boldsymbol{v} \rangle = 0 \tag{14.18}$$

The removal of the diffusive term, however, creates a problem with regard to satisfying the boundary conditions, since the highest spatial derivative of C in Eq. (14.18) is now only first order. In practice, one usually artificially *increases* the diffusivity while keeping Pe_C large relative to unity, in order to stabilize the solution and control the size of the solute boundary layer.

Using the definitions of the average quantities defined in Chap. 4, and assuming complete mixing in the liquid phase, Eq. (14.18) becomes

$$\frac{\partial}{\partial t}\left(\int_0^{g_s} \rho_s C_s dg_s' + g_\ell \rho_\ell C_\ell\right) + \nabla \cdot \left(\boldsymbol{v}_s \int_0^{g_s} \rho_s C_s dg_s' + g_\ell \rho_\ell C_\ell \boldsymbol{v}_\ell\right) = 0 \tag{14.19}$$

If the lever rule approximation is used, Eq. (14.18) becomes

$$\frac{\partial \left(g_s \rho_s C_s + g_\ell \rho_\ell C_\ell\right)}{\partial t} + \nabla \cdot \left(g_s \rho_s C_s \boldsymbol{v}_s + g_\ell \rho_\ell C_\ell \boldsymbol{v}_\ell\right) = 0 \tag{14.20}$$

Inserting the expressions from Eq. (14.15) into Eq. (14.18) and dividing by $\langle \rho C \rangle$, we have

$$\frac{1}{\langle C \rangle_M} \frac{\partial \langle C \rangle_M}{\partial t} = -\frac{1}{\langle \rho \rangle} \frac{\partial \langle \rho \rangle}{\partial t} - \frac{\nabla \cdot \langle \rho C \boldsymbol{v} \rangle}{\langle \rho C \rangle} \tag{14.21}$$

Recall the average mass conservation equation (Eq. (4.63)):

$$\frac{\partial \langle \rho \rangle}{\partial t} + \nabla \cdot \langle \rho \boldsymbol{v} \rangle = \frac{\partial \left(g_s \rho_s + g_\ell \rho_\ell\right)}{\partial t} + \nabla \cdot \left(g_s \rho_s \boldsymbol{v}_s + g_\ell \rho_\ell \boldsymbol{v}_\ell\right) = 0 \tag{14.22}$$

Extracting the time derivative of the average density from this equation and inserting the result into Eq. (14.21), gives

$$\frac{1}{\langle C \rangle_M} \frac{\partial \langle C \rangle_M}{\partial t} = \frac{\nabla \cdot \langle \rho \boldsymbol{v} \rangle}{\langle \rho \rangle} - \frac{\nabla \cdot \langle \rho C \boldsymbol{v} \rangle}{\langle \rho C \rangle} \tag{14.23}$$

It can be seen from either Eq. (14.21) or (14.23) that no macrosegregation will appear, i.e., $\langle C \rangle_M$ will be constant, if the right hand side is zero. Thus, the right hand side contains all of the sources of macrosegregation: solidification shrinkage (variation of $\langle \rho \rangle$), convection in the liquid induced by solidification shrinkage, natural or forced convection (term $\boldsymbol{v}_\ell$ in the $\langle . \rangle$), movement and deformation of the solid (term $\boldsymbol{v}_s$ in the $\langle . \rangle$). We first explore the case of macrosegregation induced solely by solidification shrinkage, then consider the other possible sources of macrosegregation.

14.4 MACROSEGREGATION INDUCED BY SOLIDIFICATION SHRINKAGE

Flemings and Nereo [12] were the first to model macrosegregation induced by interdendritic liquid flow. Their result, known as Flemings' macrosegregation criterion, was obtained for 1-D solidification under the following additional assumptions: (i) the density of the solid and liquid phases are constant, but not equal; (ii) the solid is fixed; and (iii) microsegregation occurs according to a Gulliver-Scheil model, i.e., no diffusion in the solid and complete mixing in the liquid (see Chap. 10). Taking the solidification direction along the x-axis, the solute and mass conservation balances, Eqs. (14.19) and (14.22), become

$$\rho_s \frac{\partial}{\partial t} \int_0^{g_s(t)} C_s(g_s') dg_s' + \rho_\ell \frac{\partial (g_\ell C_\ell)}{\partial t} + \rho_\ell \frac{\partial (g_\ell C_\ell v_{\ell x})}{\partial x} = 0 \qquad (14.24)$$

$$(\rho_\ell - \rho_s) \frac{\partial g_\ell}{\partial t} + \rho_\ell \frac{\partial (g_\ell v_{\ell x})}{\partial x} = 0 \qquad (14.25)$$

The assumption of the Gulliver-Scheil model implies that the integral in Eq. (14.24) is independent of time, and its time derivative is therefore simply equal to $-\rho_s k_0 C_\ell \partial g_\ell / \partial t$. Expanding the last term of Eq. (14.24) using the chain rule, then combining it with Eq. (14.25) and dividing by ρ_ℓ gives

$$C_\ell \frac{\rho_s}{\rho_\ell}(1-k_0)\frac{\partial g_\ell}{\partial t} = -g_\ell \left(1 + v_{\ell x} \frac{\partial C_\ell / \partial x}{\partial C_\ell / \partial t}\right) \frac{\partial C_\ell}{\partial t} \qquad (14.26)$$

Since we are in the mushy zone and local equilibrium is assumed to apply at the solid-liquid interface, the composition of the liquid for a binary alloy is directly related to the temperature. One therefore has $\partial C_\ell / \partial x = G/m_\ell$ and $\partial C_\ell / \partial t = \dot{T}/m_\ell$, where G is the thermal gradient and $\dot{T}$ is the cooling rate. Inserting these expressions into Eq. (14.26) and dividing the result by $g_\ell C_\ell$ gives

$$\frac{dg_\ell}{g_\ell} = -\frac{\rho_\ell}{\rho_s(1-k_0)}\left(1 - \frac{v_{\ell x}}{v_T}\right)\frac{dC_\ell}{C_\ell}$$

or

$$\frac{dg_\ell}{g_\ell} = -\frac{\rho_\ell}{\rho_s(1-k_0)}\left(1 - \frac{\mathbf{v}_\ell \cdot \nabla T}{\mathbf{v}_T \cdot \nabla T}\right)\frac{dC_\ell}{C_\ell} \qquad (14.27)$$

or

$$\frac{d(\ln g_\ell)}{d(\ln C_\ell)} = -\frac{\rho_\ell}{\rho_s(1-k_0)}\left(1 - \frac{\mathbf{v}_\ell \cdot \nabla T}{\mathbf{v}_T \cdot \nabla T}\right) \qquad (14.28)$$

where $v_T = -G/\dot{T}$ is the speed of the isotherms, assumed to be along the x-direction. Note that the ratio of the densities $\rho_\ell/\rho_s = (1+\beta)^{-1}$ when β is defined as we have done in this text, $\beta = (\rho_s/\rho_\ell - 1)$. Flemings designated the shrinkage factor β' as $(1 - \rho_\ell/\rho_s)$, in which case $\rho_\ell/\rho_s = (1-\beta')$. To first order, β and β' are equal since $(1+\epsilon)^{-1} \simeq 1 - \epsilon$. A similar

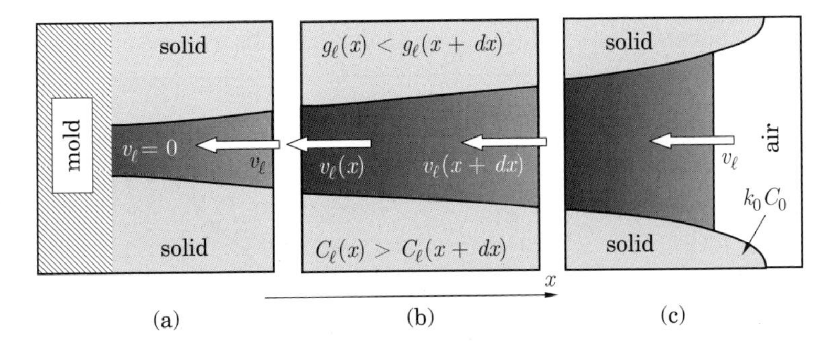

Fig. 14.6 A schematic illustration of shrinkage-induced macrosegregation (a) at the onset of solidification, (b) during steady state and (c) at the end of solidification.

expression can be obtained under the assumption of lever rule solidification (see Exercise 14.4).

Equation (14.27) can be used to obtain a better understanding of macrosegregation associated with different times during solidification, as illustrated schematically in Fig. 14.6. There are three distinct regions: initial solidification, steady state portion, and final transient. These three regions are discussed separately in the following sections.

14.4.1 Initial solidification at the mold surface

At the solid-mold surface (Fig. 14.6a), the velocity of the liquid is zero. In this case, integration of Eq. (14.27) gives:

$$g_\ell = \left(\frac{C_\ell}{C_0} \right)^{\rho_\ell/(\rho_s(k_0-1))} \tag{14.29}$$

Using Eq. (14.29), one can show (see Exercise 14.6) that the final composition of the solid at the mold surface is given by

$$\frac{\langle C \rangle_M (g_s = 1)}{C_0} = k_0 \left(\frac{\rho_s}{\rho_\ell}(k_0 - 1) + 1 \right)^{-1} \tag{14.30}$$

The macrosegregation predicted by Eq. (14.30) is plotted as a function of the solidification shrinkage coefficient $\beta = (\rho_s/\rho_\ell - 1)$ in Fig. 14.7 for three different values of the partition coefficient k_0. It is clear that for any real alloy, for which $k_0 \neq 0$ and $\beta \neq 0$, the final composition of the solid $\langle C \rangle_M$ at the mold wall is not equal to C_0. In fact, when $k_0 < 1$, $\langle C \rangle_M$ at the mold wall is *larger* than the nominal composition. This can be understood by considering a thin volume element at the surface of the mold (Fig. 14.6a). The first solid to form has a composition $k_0 C_0$. As a certain amount of shrinkage occurs, liquid needs to flow in to compensate for this change

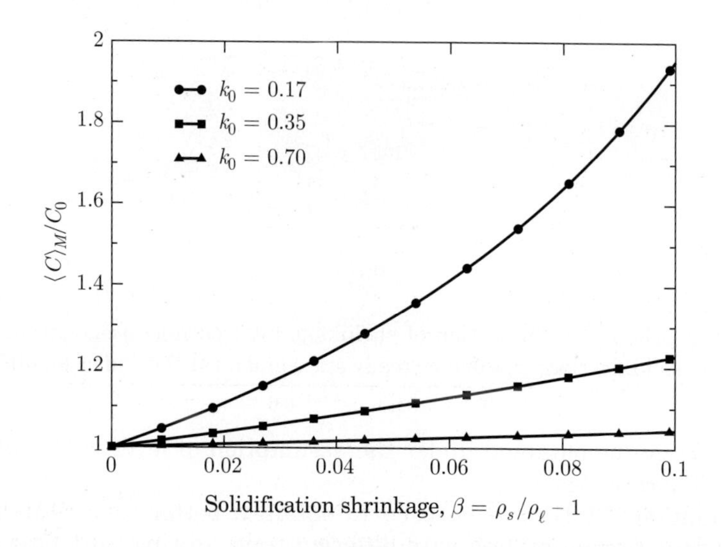

Fig. 14.7 The final composition of the solid $\langle C \rangle_M(g_s = 1)$ at the surface of the mold normalized by the nominal composition C_0 as a function of the solidification shrinkage coefficient $\beta = (\rho_s/\rho_\ell - 1)$ for three values of the partition coefficient k_0.

in volume. However, as solidification progresses from the surface of the mold toward the bulk of the casting, the liquid that flows into this volume element has a composition C_ℓ that is increasingly larger than C_0. This gives rise to a larger final solid composition compared to the nominal composition.

It was demonstrated in Chap. 5 that the analysis of plane-front solidification of an alloy with $k_0 < 1$ and $\beta = 0$ predicts an initial transient in which $C_s < C_0$. The length of the transient is dictated by the distance required to establish a steady-state solute profile ahead of the moving planar front. The present analysis for dendritic solidification, with $\beta \neq 0$, predicts the opposite, i.e., that the composition at the surface is *greater* than C_0, thus giving rise to the term *inverse segregation*. For dendritic solidification, the extent of the region of the casting close to the mold for which $\langle C \rangle_M$ is larger than the nominal composition C_0 typically corresponds to the length of the mushy zone when g_s at the mold wall is equal to 1, or in other words to $\Delta T_0/G$ where G is the average thermal gradient near the mold wall and ΔT_0 is the solidification interval. Beyond this distance, steady state is reached. This topic is discussed next.

14.4.2 Steady state

It was shown in Sect. 12.3.2 that $v_{\ell x} = -\beta v_T$ everywhere in the mushy zone under conditions of steady-state one-dimensional freezing. Inserting

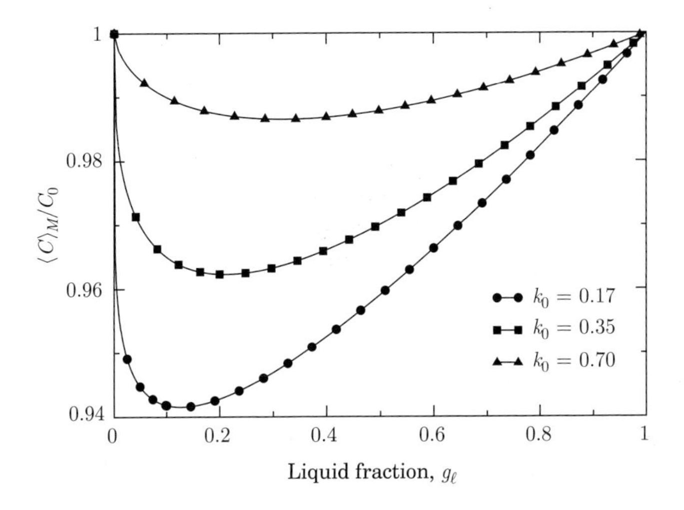

Fig. 14.8 The mean composition $\langle C \rangle_M$ normalized by the nominal composition C_0 during steady-state, directional solidification, plotted as a function of the liquid fraction g_ℓ for $\beta = 0.1$, and the Gulliver-Scheil approximation, for three values of the partition coefficient.

this relation into Eq. (14.27), recovers the Gulliver-Scheil equation,

$$\frac{dg_\ell}{g_\ell} = -\frac{1}{(1-k_0)}\frac{dC_\ell}{C_\ell} \quad \text{or} \quad g_\ell = \left(\frac{C_\ell}{C_0}\right)^{1/(k_0-1)} \tag{14.31}$$

By substituting Eq. (14.31) into Eq. (14.21), it can be shown (see Exercise 14.3) that the mean composition varies with g_ℓ as

$$\langle C \rangle_M = C_0 \frac{g_\ell^{k_0}(\rho_\ell - \rho_s) + \rho_s}{\rho_s(1-g_\ell) + \rho_\ell g_\ell} \quad \text{or} \quad \frac{\langle \rho C \rangle}{C_0} = g_\ell^{k_0}(\rho_\ell - \rho_s) + \rho_s \tag{14.32}$$

This result is plotted in Fig. 14.8, indicating a slight decrease in the mean composition within the mushy zone as the fraction of liquid decreases, after which the nominal composition C_0 is recovered in the fully solid state. In other words, there is no macrosegregation in this region. However, everywhere in the mushy zone, the mean composition $\langle C \rangle_M \neq C_0$.

14.4.3 Final transient

Thus far, we have shown that solidification shrinkage produces an excess of solute at the external surface of the mold and gives rise to a steady-state composition equal to C_0 after the initial transient. As a result, the end of solidification must be accompanied by a solute depletion, as shown schematically in Fig. 14.6(c). When the tips of the dendrites reach the top liquid surface, they solidify with a composition close to kC_0. At that instant, the top liquid surface has to move inside the mushy zone to compensate for the density change induced by the solidification shrinkage. Therefore, the last solid composition at the free surface must be k_0C_0 (instead of C_0), if one neglects the tip undercooling. Moreover, the distance over which the final solid composition $\langle C \rangle_M$ is smaller than C_0 is dictated by the extent of the shrinkage pipe, i.e., the region near the top surface where porosity can be observed in between the dendrites.

Experimental evidence for the phenomena we have described is given in Fig. 14.9, which shows the final measured composition profile in a directionally solidified Al-3.4wt%Cu ingot, as well as a numerical solution of the average 1-D heat, solute and mass conservation equations described in Sect. 14.5.2. The composition profile shows a slight increase in Cu content near the mold, and a sharp drop at the end of solidification. The plateau is slightly higher than the nominal composition due to several factors not accounted for in the simplified analysis given above, but incorporated in the numerical model. These phenomena include non-steady state conditions, temperature- and solute-dependence of ρ_s and ρ_ℓ, as well as a nonlinear phase diagram.

In certain cases where thermal contact with the mold is lost due to thermal strains, local reheating can occur before the alloy becomes fully solid at the surface. This produces *exudation* at the surface, i.e., the formation of an uneven surface due to the bleeding-out of a solute-rich liquid. Such a situation is common in DC-casting of aluminum alloys (cf. Thevik et al. [33]).

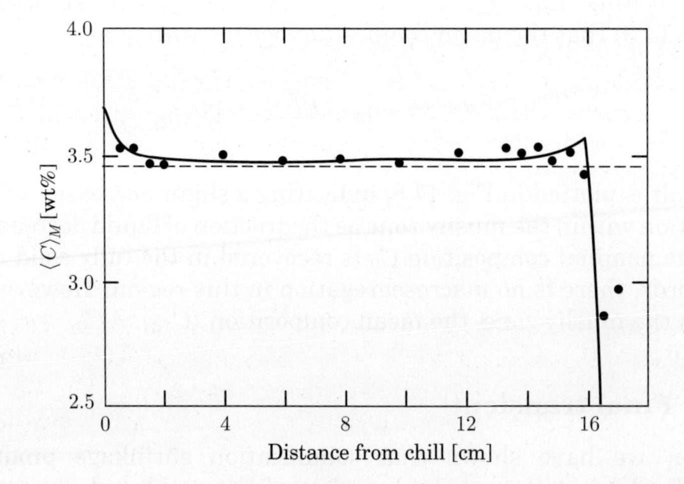

Fig. 14.9 The final average composition $\langle C \rangle_M$ in a directionally solidified Al-3.4wt%Cu alloy: the filled circles represent experimental points and the solid line is the result of a simulation. (After Rousset et al. [29])

14.5 MACROSEGRAGATION INDUCED BY FLUID FLOW

The present section considers macrosegregation induced by more complex flow patterns in the fluid, retaining the assumption (for now) that $\boldsymbol{v}_s = 0$. In one dimension, the fluid flow is limited to that induced by solidification shrinkage. In higher dimensions, this is no longer the case. Fluid flow can enter and exit the mushy zone while still ensuring an overall mass balance. We begin with a simple analysis based on Flemings' criterion, and then develop a more general formulation.

14.5.1 Analysis based on Flemings' criterion

Starting from Eq. (14.23), multiplying by $\langle \rho C \rangle$ and setting $\boldsymbol{v}_s = 0$, it is easy to demonstrate that

$$\langle \rho \rangle \frac{\partial \langle C \rangle_M}{\partial t} = (\langle C \rangle_M - C_\ell)\, \nabla \cdot (\rho_\ell g_\ell \boldsymbol{v}_\ell) - \rho_\ell g_\ell \boldsymbol{v}_\ell \cdot \nabla C_\ell \tag{14.33}$$

This equation is the basis for computing macrosegregation induced solely by fluid flow. Since solidification shrinkage has already been discussed, let us assume that $\rho_s = \rho_\ell$ in order to focus on this new phenomenon. Thus, $\langle \rho \rangle$ is constant and $\nabla \cdot (\rho_\ell g_\ell \boldsymbol{v}_\ell) = 0$ by Eq. (14.22), so the macrosegregation equation simplifies to

$$\frac{\partial \langle \rho C \rangle}{\partial t} = -\rho_\ell g_\ell \boldsymbol{v}_\ell \cdot \nabla C_\ell \tag{14.34}$$

The variation of $\langle \rho C \rangle$ occurs, in this case, directly from $\boldsymbol{v}_\ell \cdot \nabla C_\ell$. Thus, as mentioned at the beginning of this chapter, macrosegregation does not take place if fluid flow is parallel to the composition isopleths, i.e., if $\boldsymbol{v}_\ell \cdot \nabla C_\ell = 0$.

In a binary alloy, the gradient of the liquid composition in the mushy zone can be directly linked to the temperature gradient via the liquidus curve, giving

$$\frac{\partial \langle \rho C \rangle}{\partial t} = -m_\ell \rho_\ell g_\ell \boldsymbol{v}_\ell \cdot \nabla T = -m_\ell \rho_\ell g_\ell G v_{\ell\perp} \tag{14.35}$$

where the component of the fluid velocity perpendicular to the isotherms in the direction of the thermal gradient, $v_{\ell\perp} = \boldsymbol{v}_\ell \cdot \nabla T / G$ has been introduced. Therefore, when $v_{\ell\perp} = 0$, no macrosegregation occurs. For $k_0 < 1$ and $m_\ell < 0$, the average volumetric concentration $\langle \rho C \rangle$ or composition $\langle C \rangle_M$ increases (positive segregation) when $v_{\ell\perp} > 0$, i.e., where a liquid flow leaves the mushy zone. Conversely, where the flow enters the mushy zone, $v_{\ell\perp} < 0$, negative segregation occurs.

If solidification shrinkage is now added to this simple line of reasoning, Flemings' analysis tells us that the no-macrosegregation case under

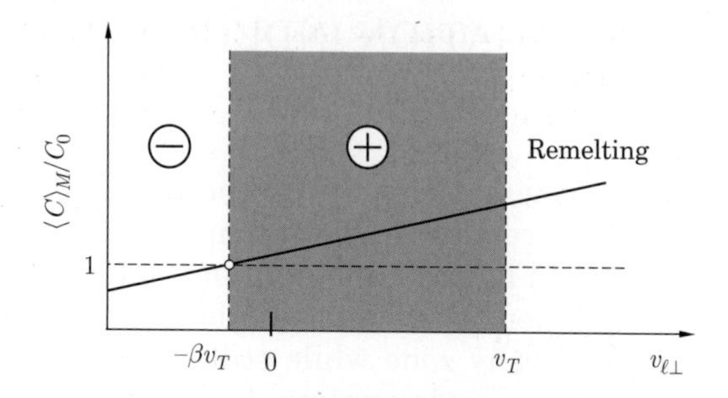

Fig. 14.10 The average composition, normalized by the nominal composition, as a function of the flow velocity in a multi-dimensional situation.

steady state conditions occurs when $v_{\ell\perp} = -\beta v_T$ instead of $v_{\ell\perp} = 0$ for $\beta = 0$. Fig. 14.10 summarizes the following cases:

- For $v_{\ell\perp} = -\beta v_T$, there is no macrosegregation, $\langle C\rangle_M = C_0$.
- For $v_{\ell\perp} < -\beta v_T$, negative segregation occurs, $\langle C\rangle_M < C_0$.
- For $v_{\ell\perp} > -\beta v_T$, positive segregation occurs, $\langle C\rangle_M > C_0$.

In the last case, an interesting situation occurs when the velocity of the liquid $v_{\ell\perp}$ exceeds the speed of the isotherms v_T. Using Eq. (14.28), it can be seen that the expression $d(\ln g_\ell)/d(\ln C_\ell)$ changes sign when $v_{\ell\perp} > -\beta v_T$. For $k_0 < 1$, the fraction of liquid increases with the liquid composition. This is equivalent to a local *remelting* of the alloy, and such a situation is depicted schematically in Fig. 14.11. Suppose that a columnar mushy zone grows at steady state with a velocity v_T. If a small parcel of liquid deep in the mushy zone, where C_ℓ is large, moves toward the dendrite tips because $v_{\ell\perp} > v_T$, the average composition in the liquid at this location increases. On the other hand, its temperature remains almost unchanged since $\alpha \gg D_\ell$. In order to maintain local equilibrium at the solid-liquid interface, i.e., to keep $C_\ell = C_{\ell 2}$, remelting of the

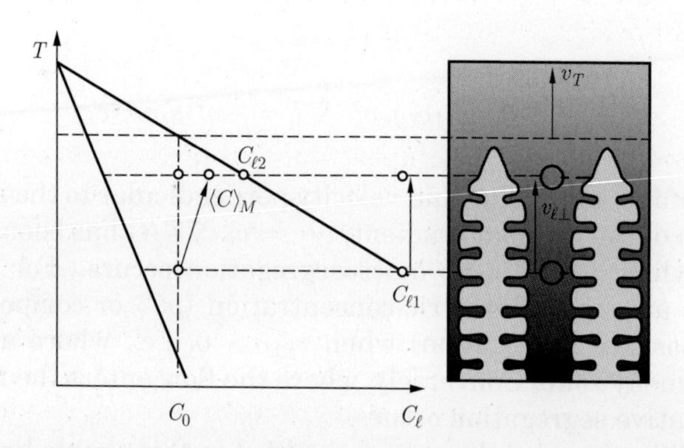

Fig. 14.11 Local remelting of a mushy zone when $v_{\ell\perp} > v_T$.

solute-lean solid must occur. From the standpoint of the average composition $\langle C \rangle_M$, increasing the liquid composition increases $\langle C \rangle_M$ at the highest location. If the temperature remains fixed, this clearly decreases the solid fraction, i.e., remelting occurs. This phenomenon is responsible for *freckle formation* and *dendrite fragmentation*, a topic that is discussed further in Sect. 14.5.3.

14.5.2 General approach

In the one-dimensional model of shrinkage-induced macrosegregation, the velocity field could be determined from a simple mass balance. Combining this result with solute transport allowed us to deduce Flemings' criterion. In general, however, the liquid velocity that dictates macrosegregation has to be obtained by solving the momentum balance equation. For the sake of simplicity, we first develop the governing equations for the case where $\beta = 0$. The average form of the balance equations for the liquid phase were developed in Chap. 4, and are repeated here for convenience

$$\nabla \cdot \langle \boldsymbol{v} \rangle = 0 \tag{14.36}$$

$$\rho_\ell \frac{\partial \langle \boldsymbol{v} \rangle}{\partial t} + \frac{\rho_\ell}{g_\ell}(\langle \boldsymbol{v} \rangle \cdot \nabla)\langle \boldsymbol{v} \rangle - \mu_\ell \nabla^2 \langle \boldsymbol{v} \rangle + \frac{\mu_\ell g_\ell}{K}\langle \boldsymbol{v} \rangle = \rho_\ell g_\ell \boldsymbol{g} - g_\ell \nabla p_\ell \tag{14.37}$$

$$\rho \frac{\partial \langle h \rangle}{\partial t} + \rho_\ell c_{p\ell} \langle \boldsymbol{v} \rangle \cdot \nabla T - \nabla \cdot (\langle k \rangle \nabla T) = 0 \tag{14.38}$$

$$\rho \frac{\partial \langle C \rangle}{\partial t} + \rho_\ell \langle \boldsymbol{v} \rangle \cdot \nabla C_\ell - \nabla \cdot \langle \rho D \nabla C \rangle = 0 \tag{14.39}$$

All of the terms that appear in these equations have been defined in Chap. 4. In particular, the average enthalpy is given by (see Eq. (4.138))

$$\rho \langle h \rangle = \rho_s g_s h_s + \rho_\ell g_\ell h_\ell = \rho \int_{T_{ref}}^{T} \langle c_p(T') \rangle \mathrm{d}T' + \rho L_f g_\ell \tag{14.40}$$

If we further assume that the compositions follow the lever rule, the average concentration $\langle \rho C \rangle = \rho \langle C \rangle$ is given by

$$\rho \langle C \rangle = \rho(g_s C_s + g_\ell C_\ell)) = \rho C_\ell (g_\ell + k_0 g_s) \tag{14.41}$$

In order to model solute- or temperature-driven buoyant convection, the buoyancy term that appears in the momentum equation for the liquid phase has to be developed according to Eq. (14.7). Forced convection is introduced via boundary conditions, whereas electromagnetic stirring of the liquid would be included as a body force $\boldsymbol{j}_{ind} \times \boldsymbol{B}$ in the momentum equation. Here, $\boldsymbol{j}_{ind}$ is the induced current in the metal and $\boldsymbol{B}$ is the induction field, both obtained from a solution of Maxwell's equations.

A total of 10 unknown fields appear in these conservation equations: $(v_{\ell x}, v_{\ell y}, v_{\ell z})$, p_ℓ, $\langle h \rangle$, T, g_s, $\langle C \rangle$, C_s and C_ℓ, all functions of x and t. Equations (14.36)-(14.39) provide a total of six relations between the variables, and consequently, four additional constitutive relations are needed. The definitions of the average enthalpy (Eq. (14.40)) and composition (Eq. (14.41)) provide two of these, and another relation, already used in defining the average composition, is given by the phase diagram: $C_s = k_0 C_\ell$. As in our earlier analysis of inverse segregation, the last condition can be provided by either setting $g_\ell = 1$, in which case C_ℓ is free, or by setting $T = T_{liq}(C_\ell)$ for $1 > g_\ell > 0$. The extension to multi-component alloys is left as an exercise. However, it should be re-emphasized that the relationship $\langle h \rangle(T, g_\ell)$ for a binary alloy depends on the fluid flow history via the evolution of the local average composition $\langle C \rangle$.

The numerical "tricks" that have been developed to solve Eqs. (14.38)-(14.39) will not be discussed in detail. However, it is worth mentioning one of them. The equations are usually solved in a sequential fashion at each time step, rather than as a set of fully coupled equations. The new fields available from the solution of one equation, e.g., the velocity and pressure obtained by solving the momentum and mass balance equations, are used in the solution of the heat and solute transport. Some authors then iterate in order to find a consistent solution for all field variables within each time step, while others perform a single iteration per time step. The latter requires Δt to be kept small in order for the error control to be maintained.

In order to demonstrate the application of these modeling techniques, let us consider experiments performed by Hebditch and Hunt [19], where two different Sn-Pb alloys were solidified in the apparatus shown schematically in Fig. 14.12. All of the walls of the $10 \times 6 \times 1.3$ cm^3 box were insulated, with the exception of the left side, which was cooled at a controlled rate in order to effect solidification while inducing buoyant convection. A Sn-5 wt% Pb alloy that rejects lead upon solidification, and a Pb-48 wt% Sn alloy that rejects tin upon solidification were solidified. In one alloy, the solutal effects enhance the thermal buoyancy, and in the other the two effects compete with each other.

Compositions measured in the solidified block along the indicated horizontal lines are shown in Fig. 14.13, along with the results of calculations carried out by Ahmad et al. [1] using the methods described

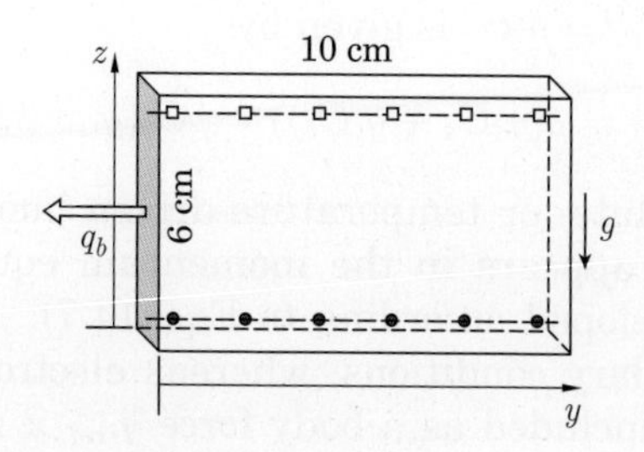

Fig. 14.12 A schematic of the box used to measure macrosegregation in Sn-Pb alloys. All faces were insulated except for the one on the left, which was cooled. Compositions were measured after solidification at the points indicated along horizontal lines near the top and bottom.

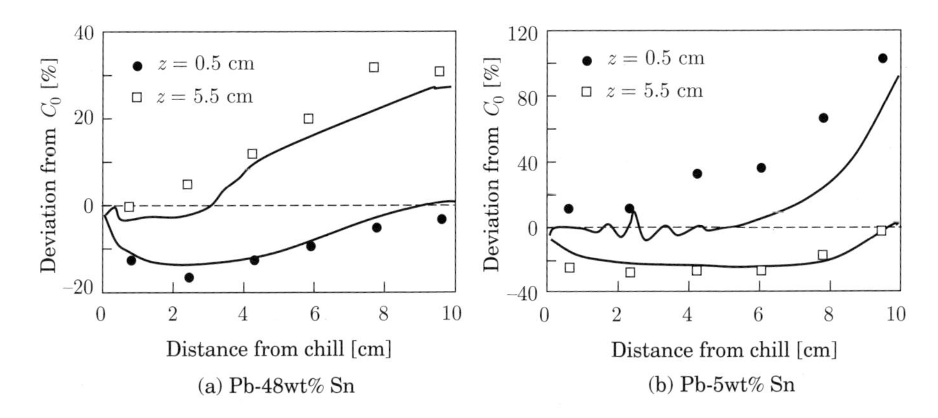

Fig. 14.13 The relative variation in mass fraction, $\langle C \rangle_M / C_0 - 1$, at the end of solidification for two Sn-Pb alloys as a function of the distance from the cold wall. The symbols correspond to measured compositions, whereas the solid lines represent computational results by Ahmad et al. [1].

in this section. The calculations match the measured data reasonably well, and the results can be understood as follows. In the absence of any solutal effects, thermal buoyancy would produce a counterclockwise (ccw) flow in the cavity, as described in Sect. 14.2. This thermal ccw convection is reinforced in the case of the Sn-5wt% Pb alloy, since the lead rejected by the growth of tin dendrites is denser than the mean composition ($\beta_C < 0$). In the case of the Pb-48wt% Sn alloy, however, the rejected tin makes the liquid less dense, thus inducing a clockwise (cw) convection. As can be seen in Fig. 14.13, solutal convection seems to dominate in both cases, giving rise to positive segregation at the bottom-right corner of the cavity for the Sn-5wt% Pb alloy and at the top-right corner for its Pb-48wt% Sn counterpart. In the case of the Sn-5wt% Pb alloy, the flow field traveling toward the bottom-right corner of the cavity is so strong that it induces veins of high solute content, called *freckles*, as shown in Fig. 14.14. Freckles are discussed in more detail in the next section.

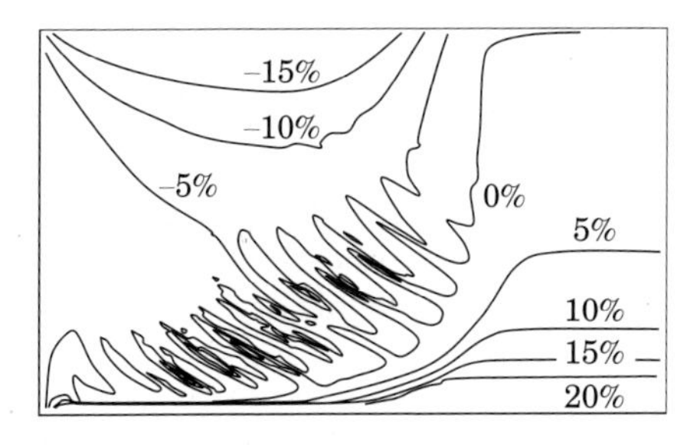

Fig. 14.14 The contours of the relative composition variation, $\langle C \rangle_M / C_0 - 1$, for an Sn-5wt%Pb alloy. (Redrawn from Ahmad et al. [1])

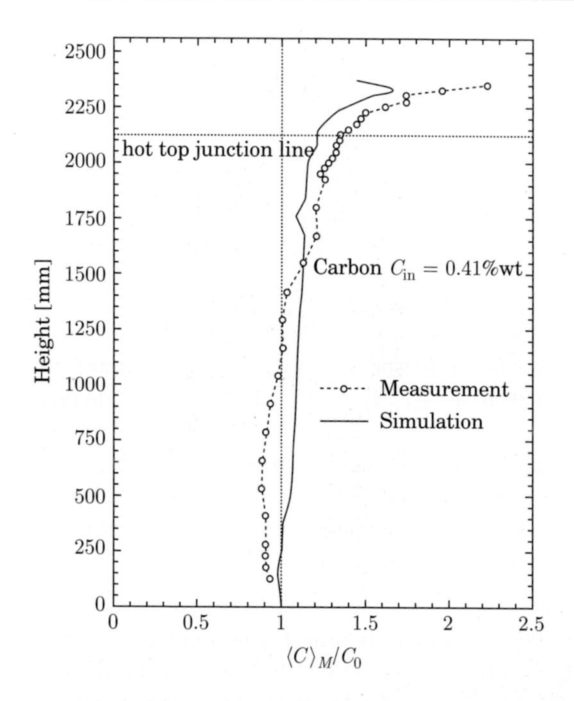

Fig. 14.15 The average calculated and measured carbon composition profile at the end of solidification in a large 2-D steel casting, as a function of the height. (Reprinted from Gu and Beckermann [18].)

Figure 14.15 displays the results of an extension of the binary alloy model, presented here to a multi-component alloy. Both solidification shrinkage and natural convection have been accounted for in this model, which also includes a dynamic calculation of the solidification path. The calculation was performed in 2-D representing a slice from a large steel ingot. The computed carbon composition profile is compared with measurements along the height of the ingot, and we see once again a reasonable agreement between the calculations and the experimental results.

One of the challenges faced when modeling fluid flow-driven macrosegregation is that the size of the domain is generally large, but at the same time the computational resolution must be high. As can be seen from Eq. (14.34), macrosegregation in a volume of liquid with an initially uniform composition C_0 occurs in the region where both $\boldsymbol{v}_\ell \neq 0$ and $\nabla C_\ell \neq 0$. Except for shrinkage-induced macrosegregation, which takes place everywhere in the mushy zone, these conditions are met in a zone near the liquidus. Indeed, because the permeability of the mushy zone decreases rapidly as g_ℓ decreases, liquid flow does not penetrate deeply between the dendrites. On the other hand, the initial value of ∇C_ℓ is zero in the liquid outside of the mushy zone. Consequently, in order to model macrosegregation accurately, a fine mesh is required close to the liquidus isotherm. Of course, the liquidus moves during solidification, so in order to ensure

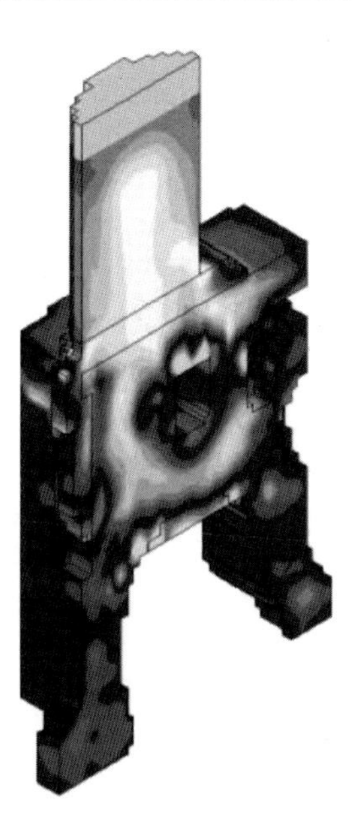

Fig. 14.16 A carbon composition map for a large 3-D steel casting. (Reprinted Schneider and Beckermann [31].)

adequate resolution, one must have either a fine mesh everywhere in the casting, or implement of a multi-grid [20] or adaptive meshing techniques [15].

The task becomes further complicated when dealing with multi-component alloys. One way in which this aspect of the problem can be simplified is to specify the solidification path through the phase diagram *a priori*. In this case, the relationship between enthalpy, composition, fraction of liquid and temperature is calculated as a single time for any given location. The average heat flow equation then allows the computation of the solid fraction, which, in turn, can be used to determine the convection and solute species transport. Figure 14.16 shows an example of such a calculation for a large steel casting. While such a procedure allows the identification of positive and negative segregation zones in a casting, it cannot predict local remelting and therefore freckle formation, a topic that is treated next.

14.5.3 Freckle formation

Macrosegregation not only renders the composition inhomogeneous at the scale of the entire product, but it can also induce segregation at the mesoscopic scale through the formation of *freckles*. This important defect in castings typically arises when the velocity of the liquid in the mushy zone $v_{\ell\perp}$ exceeds that of the isotherm v_T, as explained in Sect. 14.5.1. If local remelting of the mushy zone occurs, the associated local increase in permeability allows the liquid to flow more easily, thus enhancing remelting and macrosegregation. In such remelted zones, the liquid composition is of course higher (Fig. 14.10). Such positive segregation zones were created in the cavity shown in Fig. 14.14 for the Sn-5wt% Pb alloy by the outgoing ccw flow induced by the initially horizontal density gradient, i.e., $\nabla \rho_\ell \times \boldsymbol{g}$ pointing out of the page.

The condition that most often leads to freckle formation appears when $\nabla \rho$ is anti-parallel to gravity (i.e., $\nabla \rho \times \boldsymbol{g} = 0$ and $\nabla \rho \cdot \boldsymbol{g} < 0$). This situation, which can give rise to Rayleigh-Bénard instabilities in a pure liquid, now occurs directly within the mushy zone. An example of freckle formation in a directionally solidified Ni-based superalloy ingot is presented in Fig. 14.17. There are two major solute elements in such alloys that induce a vertical density gradient: lighter Al rejected into the liquid ($k_0^{Al} < 1$) and heavier W removed from the liquid ($k_0^{W} > 1$). Both factors contribute to an interdendritic liquid that becomes less dense as solidification progresses. In such a situation, the fluid would like to rise, thereby leaving the mushy zone. At the same time, however, a certain amount of fluid must enter for the mass to be conserved. This leads to localized jets of solute rising from *chimneys* in the mushy zone and commonly called *plumes*. The phenomenon is somewhat similar to a wood fire, where air enters the periphery between the wood logs and combustion products rise from the center.

Steube and Hellawell [32] used the transparent NH_4Cl-water system to visualize freckle formation as shown in Fig. 14.18. When solidified from below, NH_4Cl columnar dendrites grow upward, rejecting less dense water. This case corresponds to $\beta_C > 0$ and $G_z g_z < 0$, as sketched in Fig. 14.3(ii).

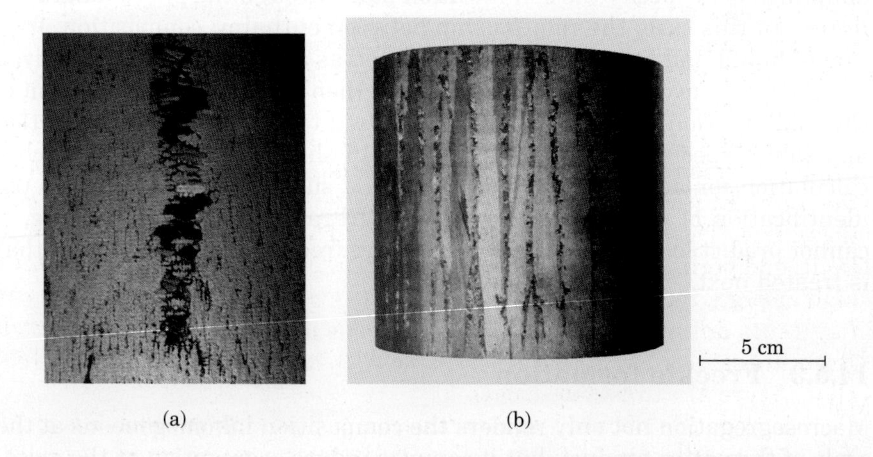

Fig. 14.17 (a) An enlargement of a freckle in a metallographic cross section and (b) freckles at the surface of a cylinder, of a MAR-M200 superalloy. (From Giamei and Kear [14].)

Fig. 14.18 Freckle formation in the NH_4Cl-water system. The top left panel shows a plume of liquid, visible due to the variation of the refraction index with composition. In the top right panel, the entrainment of some fragments by the plume is displayed using dark field illumination. The bottom left panel shows the interior of the chimney with the accumulation of fragments whereas the bottom right panel presents the actual grain structure of a freckle in a nickel-based superalloy. (Sequence taken from the movie by R.S. Steube and A. Hellawell [32].)

A water-rich plume is clearly visible in the top-left image of Fig. 14.18. Smaller, less well-defined plumes of liquid are also visible on the sides of the larger one. Interestingly, the dendrite front around the central plume is slightly higher, an indication of negative segregation (i.e., of a higher liquidus temperature for $k_0 < 1$).

The upward flow in the mushy zone can partially remelt and entrain dendrite arms in the bulk liquid. A few dendrite fragments are visible in the top-right panel. At some point, these fragments can grow and/or leave the plume. They subsequently sediment, partly on the columnar dendrite front, and perhaps even fall into the chimney. If the density of fragments is high enough, this can lead to a columnar-to-equiaxed transition. Equiaxed fragments falling into the chimney are responsible of the equiaxed grain structure seen after solidification within a chimney of a nickel-based superalloy (figure at the bottom-right).

How does a freckle form? In the case of a pure liquid with $\nabla\rho \cdot \boldsymbol{g} < 0$, it was demonstrated in Sect. 14.2 that Rayleigh-Benard vortices develop when the Rayleigh number, $\mathrm{Ra} = g\Delta\rho L^3/(\rho_{\ell 0}\nu_\ell \alpha)$, exceeds a critical value. Note that the thermal and solutal contributions have been grouped into $\Delta\rho = -\rho_{\ell 0}(\beta_T \Delta T + \beta_C \Delta C)$. To describe freckle formation, Worster [35] and then Beckermann et al. [2] introduced a "mushy zone" Rayleigh number defined as

$$\mathrm{Ra}^{\mathrm{mz}}(h) = \frac{g\Delta\rho(h)\overline{K}(h)h}{\rho_{\ell 0}\nu_\ell \alpha} = \frac{g\Delta\rho(h)\overline{K}(h)h}{\mu_\ell \alpha} \tag{14.42}$$

where $\Delta\rho(h)$ is the density variation observed over the depth h in the mushy zone measured from the dendrite tips, and $\rho_{\ell 0}$ is the density at the dendrite tip position. $\overline{K}(h) = K(\overline{g_s}(h))$ represents the average permeability over the depth h, where $\overline{g_s}(h)$ is the average volume fraction over the depth h. It is interesting to note that, as we penetrate deeper into the mushy zone, $\Delta\rho(h)$ increases at the same time as $\overline{K}(h)$ decreases. This leads to $\mathrm{Ra}^{\mathrm{mz}}(h)$ going through a maximum at a critical depth h_{crit}, which corresponds to the location where the driving forces for, and the resistance to, convection are balanced.

Figure 14.19 shows the mushy zone Rayleigh number $\mathrm{Ra}^{\mathrm{mz}}(g_s)$ for a model binary alloy close to the Ni-Al system, as a function of the volume fraction of solid g_s, for three sets of processing parameters (G, v_T). As can be seen, the maximum value of $\mathrm{Ra}^{\mathrm{mz}}(g_s)$ is reached at a fairly low volume fraction of solid, typically around 0.2. Below this value, the driving force for convection is not large enough, whereas above $g_s = 0.2$, the average permeability is too low. This indicates that freckles should occur close to the tip of the dendrites (3 or 4 mm behind the tips for the present conditions). In order for a freckle to form, however, a critical Rayleigh number must be reached. It has been found from both numerical simulations and experiments that freckles form when the maximum value of $\mathrm{Ra}^{\mathrm{mz}}$ exceeds 0.25 (horizontal dashed line in Fig. 14.19).

It should be pointed out that this value is for columnar dendrites growing exactly parallel to the gradient. If the casting or the dendrites are

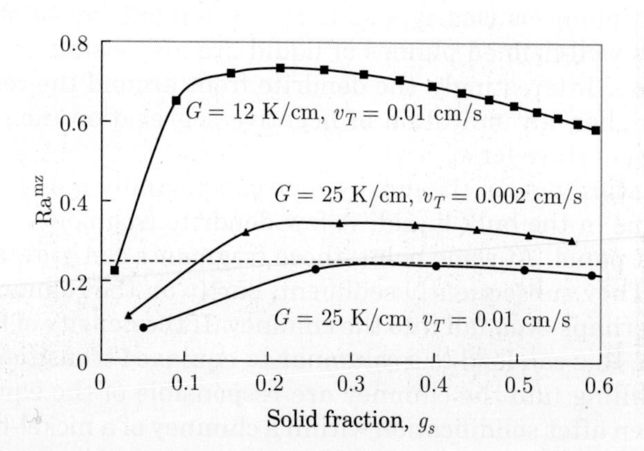

Fig. 14.19 The mushy zone Rayleigh number for a model alloy similar to Ni-6wt%Al. Parameters used: $\beta_T = 4 \times 10^{-5}$ K^{-1}, $\beta_C = 1 \times 10^{-2}$ wt%$^{-1}$, $m_\ell = -5\,\mathrm{K\,wt\%}^{-1}$, $k_0 = 0.5$, $\alpha\nu_\ell = 5 \times 10^{-12}$ m^{4}s^{-2}, $g_s(T)$ calculated with the Gulliver-Scheil equation, and permeability given by Carman-Kozeny: $\overline{K}(\overline{g}_s, \lambda_1) = 6 \times 10^{-4}\lambda_1^2(1-\overline{g}_s)^3(\overline{g}_s)^{-2}$, primary dendrite arm λ_1 given by $\lambda_1 = AG^{-1/2}v_T^{-1/4}$ with $A = 2.5 \times 10^{-3}$ K$^{1/2}$m$^{3/4}$s$^{-1/4}$ (see Chap. 8).

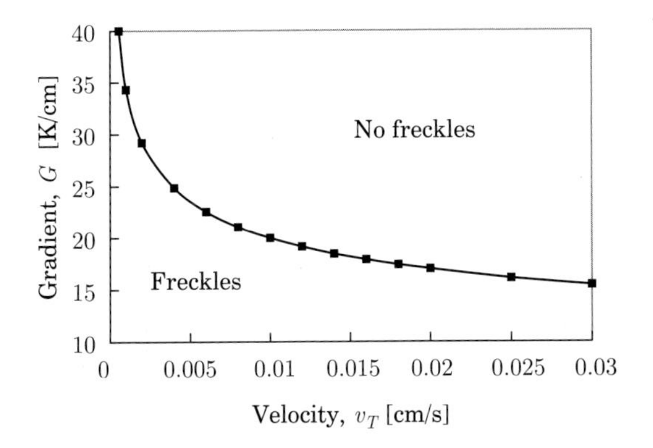

Fig. 14.20 The processing window of freckles during the directional solidification of the model alloy from Fig. 14.19.

inclined with respect to gravity, the critical Rayleigh number decreases. Fig. 14.20 presents a processing map for the alloy shown in Fig. 14.19 derived using this idea. A series of calculations was made in which the thermal gradient G for which $\mathrm{Ra}_{\max}^{\mathrm{mz}} = 0.25$ was calculated for different values of the isotherm velocity v_T. The set of points $G(v_T)$ obtained in this way divides the processing regions into an area where no freckles are expected and where they form.

Figure 14.21 displays the results from some numerical simulations of freckle formation in 2-D and 3-D. One of the problems encountered in

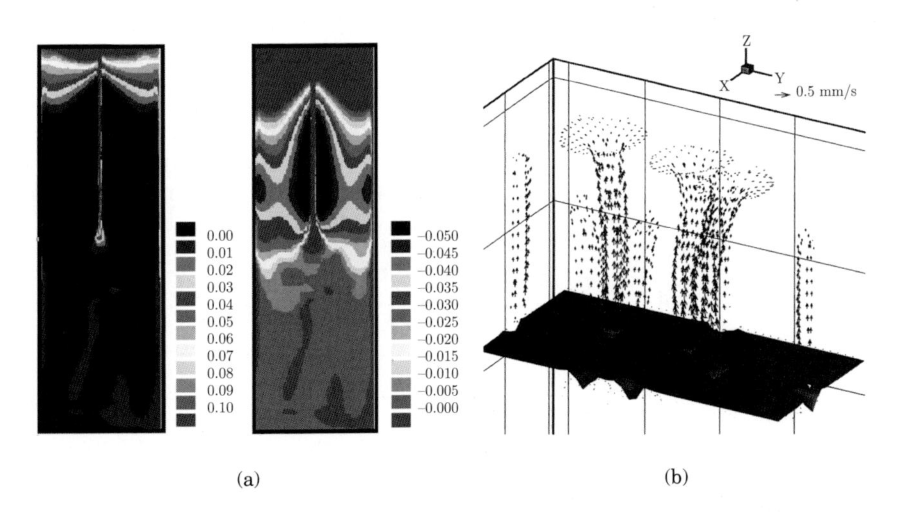

Fig. 14.21 (a) 2-D FVM simulation of a freckle formation in an Ni-Al alloy. The left and right panels show positive and negative segregation index maps ($\langle C \rangle_M / C_0 - 1$), respectively. (From H. Combeau et al. [6]). (b) 3-D FEM simulation of a freckle formation in a small rectangular cavity containing a Ni-6wt%Al-5wt%W-6wt%Ta superalloy, showing the fluid flow pattern. (From S.D. Felicelli et al. [10].)

the simulation of freckle formation is the mesh size. Although the tendency toward freckling is well reproduced by the simulations, the chimney is always confined to a very narrow region, perhaps as small as one grid spacing. The accuracy of the fluid flow calculation within the channel is obviously poor. However, the overall flow field is controlled to a larger degree by the ability of the mushy zone to bring fresh liquid into the channel, i.e., by the permeability of the mush in the neighborhood of the freckle, thereby partially offsetting this problem.

We conclude this section with two final remarks about freckle formation. Freckles initiate in regions where the permeability is the largest. As was seen in Chap. 8, the primary spacing is not highly constrained, and may exhibit large fluctuations. Freckles are most likely to start at the location of the largest primary spacing. This also explains why freckles are very often observed at the surface of a casting, where dendrites can be seen to diverge from it. The microstructure there is thus naturally more permeable due to the branching mechanisms that are necessary to fill the space.

In large ingots, where the interdendritic liquid is less dense than the initial liquid (e.g., Fe-C), cooling from the side and the bottom will induce U-shaped isotherms. The flow moving upward will have a tendency to take the easiest path, i.e., move in the direction of the liquidus from the point of initiation. This leads to so-called "A-type" segregate channels, which are simply inclined freckles.

14.6 MACROSEGREGATION INDUCED BY SOLID MOVEMENT

In this section, we consider macrosegregation produced by solid deformation and the movement of solid grains relative to the liquid, e.g., the settling of equiaxed grains to the bottom of a casting (Figs. 14.1(c) and (d)). If Eqs. (14.18) and (14.22) are combined, we obtain, after some manipulation (see Nicolli et al. [26]),

$$\begin{aligned}\frac{\mathrm{D}^s\langle C\rangle_M}{\mathrm{D}t} &= \frac{\partial\langle C\rangle_M}{\partial t} + \boldsymbol{v}_s\cdot\nabla\langle C\rangle_M \\ &= (C_\ell - \langle C\rangle_M)\left[\frac{1}{\langle\rho\rangle}\frac{\mathrm{D}^s\langle\rho\rangle}{\mathrm{D}t} + \nabla\cdot\boldsymbol{v}_s\right] - \frac{\rho_\ell}{\langle\rho\rangle}g_\ell(\boldsymbol{v}_\ell - \boldsymbol{v}_s)\cdot\nabla C_\ell\end{aligned} \tag{14.43}$$

The shorthand notation $\mathrm{D}^s\langle\cdot\rangle/\mathrm{D}t$ indicates that the advective velocity is $\boldsymbol{v}_s$. We emphasize that this is not the material derivative, the meaning of which is somewhat ambiguous in the context of a mixture comprised of phases moving at different speeds. The variation in density can be further expanded, taking ρ_ℓ to be constant and $\langle\rho\rangle \approx \rho_s$, with the result

$$\frac{1}{\langle\rho\rangle}\frac{\mathrm{D}^s\langle\rho\rangle}{\mathrm{D}t} = \frac{\langle\rho\rangle_s - \rho_\ell}{\langle\rho\rangle}\frac{\mathrm{D}^s g_s}{Dt} + \frac{g_s}{\langle\rho\rangle}\frac{\mathrm{D}^s\langle\rho\rangle_s}{Dt} \approx \beta\frac{\mathrm{D}^s g_s}{Dt} - g_s\beta_{Ts}\frac{\mathrm{D}^s T}{Dt} \tag{14.44}$$

Inserting Eq. (14.44) into Eq. (14.43) and rearranging terms finally gives

$$\frac{\mathrm{D}^s\langle C\rangle_M}{\mathrm{D}t} \approx (C_\ell - \langle C\rangle_M)\left[\beta\frac{\mathrm{D}^s g_s}{\mathrm{D}t} - g_s\beta_{Ts}\frac{\mathrm{D}^s T}{\mathrm{D}t} + \nabla\cdot\boldsymbol{v}_s\right] - \frac{\rho_\ell}{\langle\rho\rangle}g_\ell(\boldsymbol{v}_\ell - \boldsymbol{v}_s)\cdot\nabla C_\ell \qquad (14.45)$$

Equation (14.45) isolates the various contributions to macrosegregation to the right hand side. When the average composition $\langle C\rangle_M$ is different from the liquid composition C_ℓ, solidification shrinkage (first term in the square bracket), thermal contraction of the solid (second term) and deformation of the solid skeleton (third term) each lead to a variation of the average composition. The final term in Eq. (14.45) represents macrosegregation due to the relative movement of the solid and liquid phases. The contributions of solidification shrinkage and liquid movement with a fixed solid have already been reviewed (Sects. 14.4 and 14.5). With the help of Eq. (14.45), it becomes possible to explore the two remaining causes of macrosegregation illustrated in Fig. 14.1, namely grain movement and solid deformation.

14.6.1 Macrosegregation induced by grain movement

In order to see the influence of grain movement on macrosegregation, let us consider the case of fairly low volume fractions of solid under which conditions equiaxed grains can move almost freely and are completely surrounded by the liquid. Again, to focus on the new cause of macrosegregation, we take ρ_s and ρ_ℓ to be constant and equal. Under this assumption, the first two terms in the square bracket of Eq. (14.45) vanish. We now develop two cases, corresponding to the two remaining terms on the right hand side of Eq. (14.45).

Consider first the isothermal solidification of a small volume element, consisting of globulitic grains dispersed in a liquid of uniform composition (Fig. 14.22(a) with the solute layer in the liquid $\delta_{C\ell}$ assumed to be much larger than the grain size). This means that $\nabla C_\ell \approx 0$ and that the last

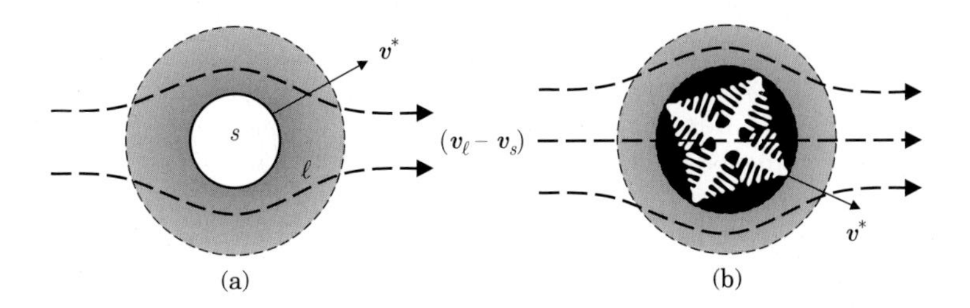

Fig. 14.22 The solute profile associated with the solidification of equiaxed (a) globular and (b) dendritic grains. The relative flow of liquid $(\boldsymbol{v}_\ell - \boldsymbol{v}_s)$ is represented with bold dashed lines: in the case of dendritic growth, the flow can either go around the dendrite (*curved flow lines*) or through it (*straight flow line*).

term of Eq. (14.45) can be neglected. In this case, macrosegregation can be induced only from the contribution $(C_\ell - \langle C \rangle_M)\nabla \cdot \boldsymbol{v}_s$. If all of the grains were to have the same velocity, then $\nabla \cdot \boldsymbol{v}_s = 0$, and no macrosegregation would occur. However, if the grains do not all have the same velocity, they tend to be compacted in regions where $\nabla \cdot \boldsymbol{v}_s < 0$, and depleted where $\nabla \cdot \boldsymbol{v}_s > 0$. Since $C_\ell > \langle C \rangle_M$ for $k_0 < 1$, such movement is accompanied by macrosegregation.

For dendritic grains (Fig. 14.22(b)), the situation is more complex due to the presence of a solute-rich interdendritic liquid between the dendrite arms. If one neglects the solute layer surrounding the grains macrosegregation depends on whether the flow of liquid passes through or around the grains during their movement. If the grains are compact, i.e., if they present a high internal volume fraction of solid, the flow will tend to go around them, in which case the interdendritic liquid contained within the grain envelope will be transported with the solid. Since the composition of the extragranular liquid C_ℓ is equal to C_0 and the mean composition of the grain $\langle C \rangle_M$ is also equal to C_0, macrosegregation will not be induced by their movement regardless of their velocity, even if $\nabla \cdot \boldsymbol{v}_s \neq 0$.

If we now include the solute boundary layer around globulitic or compact dendritic grains, and the flow going through open dendritic grains, then $\nabla C_\ell \neq 0$ at the same time as $\boldsymbol{v}_s \neq \boldsymbol{v}_\ell$. The contribution of the last term of Eq. (14.45) thus adds to that associated with the term $(C_\ell - \langle C \rangle_M)\nabla \cdot \boldsymbol{v}_s$ to produce macrosegregation.

Beckermann and co-workers [34] have studied extensively this difficult problem of segregation induced by grain movement [4]. They considered the various phenomena occurring during the growth and movement of grains using the two-phase micro-macroscopic approach discussed in Chap. 11. For globulitic grains, this requires the specification of the interaction of the flow with the solute boundary layer $\delta_{C\ell}$ surrounding the grains. For dendritic structures, an additional parameter (referred to as κ_ν in their papers) is introduced to describe the relative amounts of the flow going around and through the grains. The reader is referred to their publications for more detailed information on this matter. Once these microscopic aspects of solidification are described and the velocity fields $\boldsymbol{v}_s$ and $\boldsymbol{v}_\ell$ are computed from the average momentum equation, Equation (14.45) provides the basis for solving the governing equations for the evolution of the average composition $\langle C \rangle_M$.

Figure 14.23 depicts the macrosegregation associated with grain movement in a small cavity containing a binary Al-4wt%Cu alloy. As solidification proceeds from the left boundary, the cold, Cu-rich liquid tends to rotate counterclockwise as shown in Fig. 14.23(a) where the velocity field in the liquid $\boldsymbol{v}_\ell$ is displayed along with contours of constant solid fraction (various grey levels). The isotherms shown in Fig. 14.23(c) are distorted from vertical by this flow, and the effect is reinforced, in this case, by entrainment of the grains as they partially remelt and absorb latent heat near the bottom of the cavity, contributing further to the cooling of this region. For $C_0 = 4\,\text{wt\%Cu}$, the solid grains are denser than the liquid,

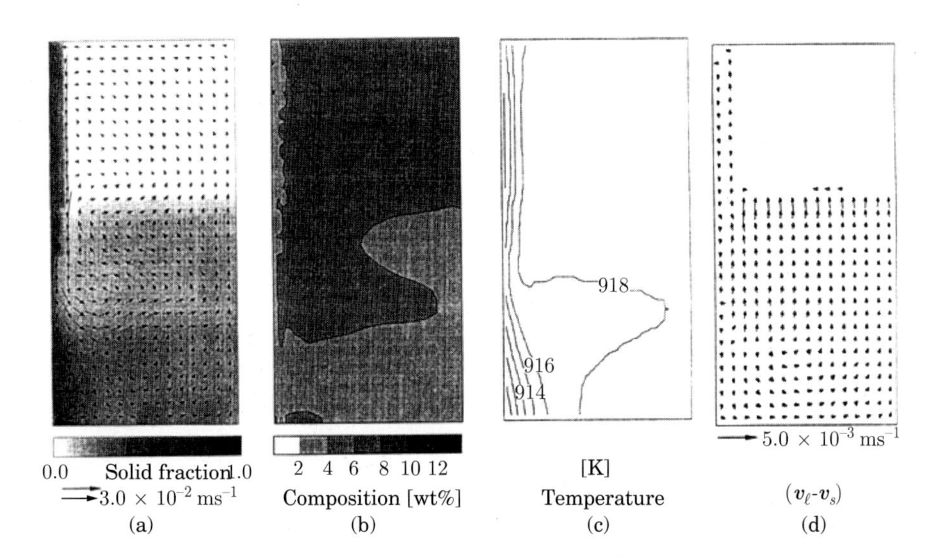

Fig. 14.23 A simulation of macrosegregation induced by grain movement and sedimentation in Al-4wt% Cu: (a) Volume fraction of solid and velocity of the liquid, (b) average composition of the solid-liquid mixture, (c) temperature, and (d) relative velocity $(\boldsymbol{v}_\ell - \boldsymbol{v}_s)$. (After Wang and Beckermann [34].)

despite their lower Cu content, and tend to fall more rapidly than the liquid. As they sediment to the bottom of the cavity, negative segregation is produced there (Fig. 14.23(b)). Recall that, in the absence of grain movement, positive segregation occurs at this location (see Fig. 14.14 for the Sn-Pb alloy). In the present case, the segregation is due to both the relative velocity $(\boldsymbol{v}_\ell - \boldsymbol{v}_s)$ shown in 14.23(d) and the term $\nabla \cdot \boldsymbol{v}_s$ in Eq. (14.45).

The calculation of macrosegregation associated with grain transport has been coupled with a Cellular Automata (CA) by Guillemot et al. [16] As an example, these authors simulated the solidification of a Pb-48wt% Sn alloy, from the left boundary of the apparatus shown in Fig. 14.12. Since convection is primarily governed by solutal gradients in this case, the tin-rich liquid has a tendency to rotate clockwise, inducing a positive segregation at the top-right corner of the cavity. However, equiaxed lead-rich grains formed during solidification tend to sediment close to the bottom-left boundary, thus slightly reinforcing the negative segregation of tin in this region. This initiates an instability in the flow and the associated segregation map, leading to the formation of freckles pointing in the upper-right direction as seen in Fig. 14.24.

14.6.2 Macrosegregation induced by solid deformation

Consider now a fairly dense mushy zone, e.g., $g_s \gtrsim 0.5$. Equation (14.45) can still be employed, but we now must focus on solid contraction and

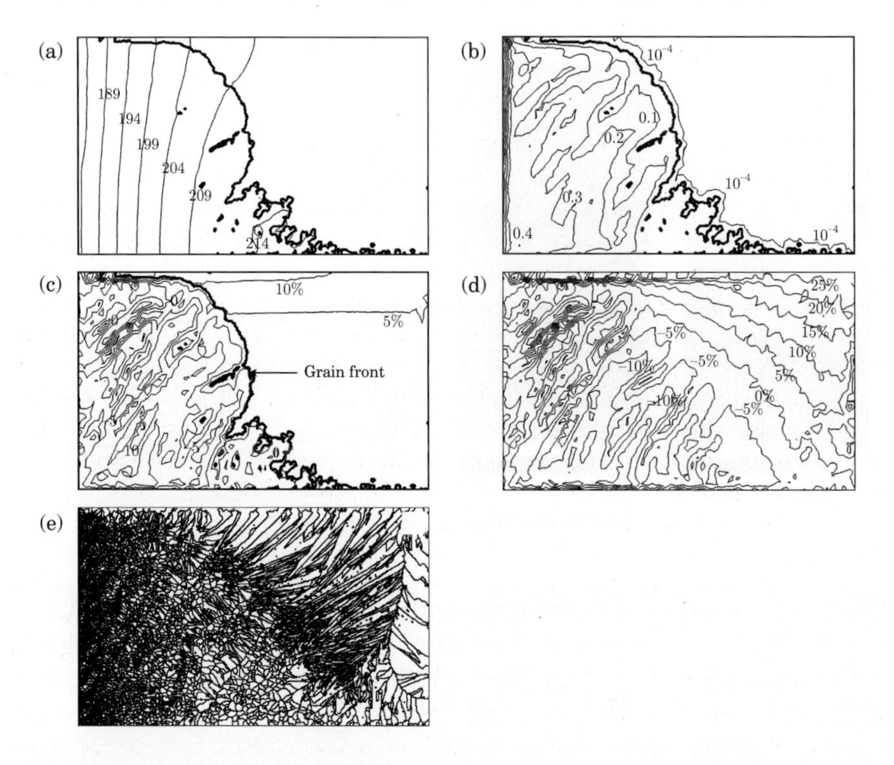

Fig. 14.24 Coupled Cellular Automata (CA) and Finite Element (FE) simulations of grain structures and macrosegregation in a Pb-48wt%Sn alloy: (a) temperature field, (b) solid fraction, and (c) relative composition $(\langle C \rangle_M - C_0)/C_0$ after 100 s of solidification; (d) final relative segregation map, and (e) grain structure. (After Guillemot et al. [16])

deformation. We first note that $\nabla \cdot \boldsymbol{v}_s$ includes both the viscoplastic deformation and the thermal contraction of the solid, i.e.,

$$\nabla \cdot \boldsymbol{v}_s = \mathrm{tr}(\dot{\varepsilon}_s^{vp}) + \mathrm{tr}(\dot{\varepsilon}_s^{th}) = \mathrm{tr}(\dot{\varepsilon}_s^{vp}) + \beta_{Ts} \frac{\mathrm{D}^s T}{\mathrm{D}t} \tag{14.46}$$

Inserting Eq. (14.46) into Eq. (14.45) gives

$$\frac{\mathrm{D}^s \langle C \rangle_M}{\mathrm{D}t} \approx (C_\ell - \langle C \rangle_M) \left[\beta \frac{\mathrm{D}^s g_s}{\mathrm{D}t} + g_\ell \beta_{Ts} \frac{\mathrm{D}^s T}{\mathrm{D}t} + \mathrm{tr}(\dot{\varepsilon}_s^{vp}) \right] - \frac{\rho_\ell}{\langle \rho \rangle} g_\ell (\boldsymbol{v}_\ell - \boldsymbol{v}_s) \cdot \nabla C_\ell \tag{14.47}$$

A similar expression has been derived by Lesoult, et al. [23] using a fully Lagrangian approach. Equation (14.47) can be understood by considering the mushy zone to be a saturated sponge, consisting of two incompressible phases: the solid skeleton and the liquid in the pores. Due to their very different rheological properties, deformation proceeds by a compression/expansion of the solid skeleton along with expulsion/suction of the liquid.

Let us first consider equiaxed solidification in a small volume element in which the temperature and composition in the liquid are uniform. Under such conditions, Eq. (14.47) indicates that compression of the mushy solid, $\mathrm{tr}(\dot{\varepsilon}_s^{vp}) < 0$, leads to negative segregation since $C_\ell > \langle C \rangle_M$. Indeed, compaction of equiaxed grains formed at an early stage of solidification up to a fully dense solid results in a solid of average composition $\langle C \rangle_M$ closer to $k_0 C_0$ than C_0. The opposite behavior, i.e., positive segregation, occurs for expansion under otherwise equivalent conditions.

We next consider a healed hot tear, in which a grain boundary opened by thermal strains perpendicular to the thermal gradient is filled by liquid flowing in the direction opposite to the thermal gradient (Fig. 13.21). The last term in Eq. (14.47) can no longer be neglected since $\boldsymbol{v}_s \cdot \nabla C_\ell = 0$ but $\boldsymbol{v}_\ell \cdot \nabla C_\ell \neq 0$. Because feeding mainly occurs opposite to the thermal gradient, the expansion term $\mathrm{tr}(\dot{\varepsilon}_s^{vp}) > 0$ in Eq. (14.47) adds to the shrinkage contribution, thus inducing more flow. Examining Fig. 14.10, it can be deduced that tensile strain in the solid now gives rise to a negative segregation. Comparing this case with the previous one, we see that the same deformation mode of the solid, in this case expansion, leads to a positive segregation for $\nabla C_\ell = 0$ and to a negative segregation in the case of hot tearing when the liquid flow is parallel to ∇C_ℓ. This clearly demonstrates that deformation-induced macrosegregation depends on the velocity field in both the solid and liquid phases.

A good example of deformation-induced segregation is found in thixoforming, a process where a solid billet is reheated up to $g_s \approx 0.5$ and then formed into a desired shape by a forging operation. A typical alloy used in this process is Al-7wt%Si, for which $g_s = 0.5$ gives a primary solid phase of Al surrounded by a liquid of near-eutectic composition. The billet, which can still be handled as a solid, is then placed in the shot sleeve of a molding machine, as shown schematically in Fig. 14.25(a). Rapidly forcing the charge through a small orifice into a mold induces very high strain rates (typically 10^3 s^{-1} or higher), which tends to separate the grains, and typically brings the average viscosity of the slurry down to a few Pa s (similar to honey). This *shear-thinning* behavior, where the viscosity decreases with the strain rate, gives rise to localized bands of high shear rate where the liquid is squeezed out of the semi-solid mixture. This produces positive segregation near the tube wall and in veins inside the billet, as shown in Fig. 14.25.

The main difficulty in modeling deformation-induced segregation is the determination of the strains in the mushy solid. Indeed, when the velocity field in the solid $\boldsymbol{v}_s$ in Eq. (14.47) is known, the other unknowns that have already been encountered in convection-induced segregation (Sect. 14.5.2) can be computed. We first show an example in which $\boldsymbol{v}_s$ is estimated *a priori* for the continuous casting of steel (see Fig. 1.7). This

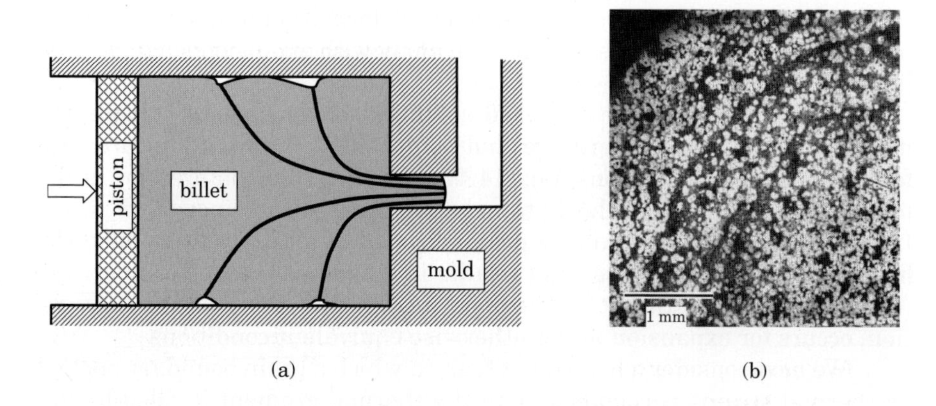

Fig. 14.25 (a) A schematic illustration of the thixoforming process, and (b) the veins of eutectic composition at the exit die during thixoforming of an Al-Mg-Si alloy, viewed in a cross section. (L. Orgéas et al. [27].)

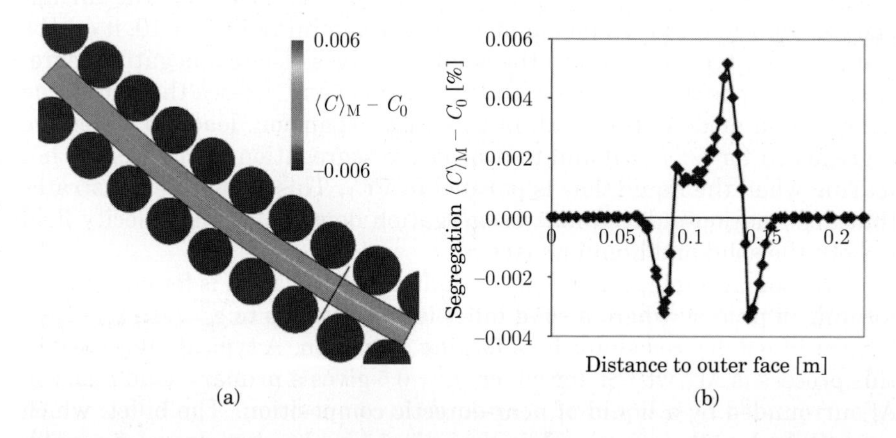

Fig. 14.26 (a) The computational domain showing contours of the deviation of the carbon composition from its nominal value for Fe-0.18wt%C. A transverse composition profile is shown in (b). (V.D. Fachinotti et al. [9].)

process takes advantage of the high-temperature strength and ductility of steel, as well as of its low thermal conductivity and large volumetric specific and latent heat of fusion, to cast at high speeds the steel from vertical to horizontal direction while its core is still liquid, as illustrated in Fig. 14.26(a). The liquid pool is usually more than 10 m deep.

During this process, deformation occurs through bending and unbending at the transitions between curved and linear sections, as well as by bulging of the shell between adjacent rolls due to the large metallostatic pressure head (close to 1 MPa). Bulging, shown schematically in Fig. 14.27(a), creates *centerline segregation* by alternate squeezing and expanding the shell. Figure 14.28(a) shows the carbon composition profile measured in a cross section of a continuously cast slab. The profile exhibits

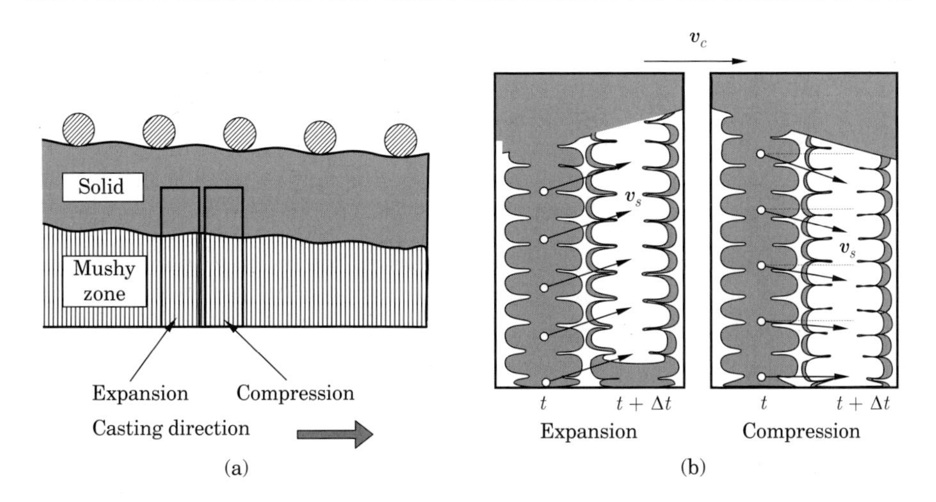

Fig. 14.27 (a) A schematic illustration of bulging between rolls in continuous casting of steel; (b) expansion/compression stages experienced by the mushy zone. (After T. Kajitani et al. [21].)

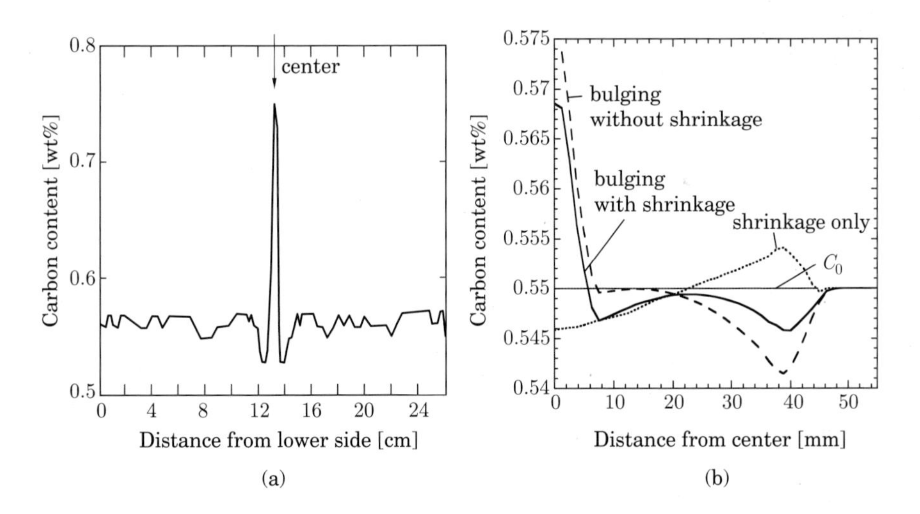

Fig. 14.28 (a) The carbon composition measured within a steel ingot as a function of the distance to the lower surface. (Redrawn from Miyazawa and Schwerdtfeger [25]). (b) The carbon composition calculated when various contributions to macrosegregation are considered. The maximum amount of bulging is in this case 0.1 mm. (From Kajitani et al. [21].)

a strong positive segregation at the center of the slab (with approximately 35% more carbon than the nominal composition). On each side of this positively segregated region, two zones of negative segregation can also be observed.

Centerline segregation can be understood with the aid of Fig. 14.27. In each section of the casting machine, the mushy zone expands slightly after each roll before being compressed by the next one. If no solidification were to take place in between, each section of the casting machine would be "self-fed", i.e., mass would be conserved within such a section, the compression stage providing mass to the expansion. However, in addition to shrinkage, solidification also induces an asymmetry between the expansion and compression stages. The compression occurs at a slightly higher volume fraction of solid than the expansion, as shown schematically in Fig. 14.27. The columnar dendrites are assumed to simply translate parallel to the shell between t and $t + \Delta t$ during expansion (white dendrite), while they also grow (grey dendrite at $t + \Delta t$). During this step, $\nabla \cdot \boldsymbol{v}_s = 0$ according to this simple model. When compression occurs, solid dendrites have filled the space opened during expansion and their velocity at the center is thus horizontal. Assuming that the dendrites just above the solid shell move with it, the field $\boldsymbol{v}_s$ is not uniform and deformation occurs. For example, a simple model is obtained by linear interpolation of $\boldsymbol{v}_s$.

When using this simple velocity field, shown schematically in Fig. 14.27(b), and calculating the extent of bulging beforehand by way of a standard code, the solution to the macrosegregation problem is described by Eqs. (14.38)-(14.39), modified to account for the contribution of $\boldsymbol{v}_s$. Further details can be found in the work of Kajitani et al. [21]. Figure 14.28(b) shows the calculated macrosegregation profile across a steel slab, and Fig. 14.29 presents the corresponding relative velocity field, $(\boldsymbol{v}_\ell - \boldsymbol{v}_s)$. The effect of the compression-expansion cycles is clearly visible in this last figure, whereas the segregation profile reveals the positive centerline segregation peak in between the two negatively segregated regions (Fig. 14.28(b)).

The carbon composition profile shown in Fig. 14.26(b) was calculated using a rheological model for the solid phase identical to the one described

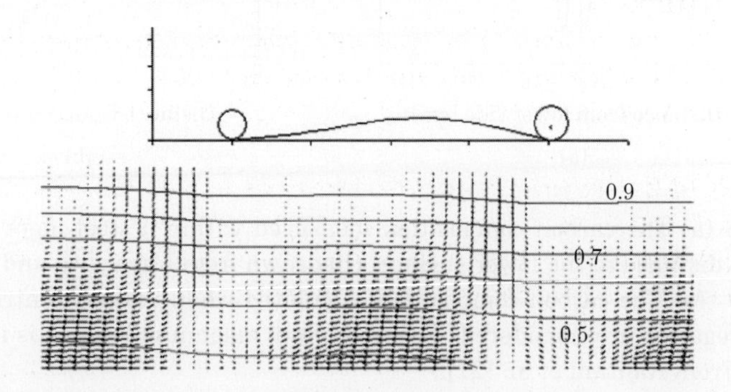

Fig. 14.29 The relative velocity field $(\boldsymbol{v}_\ell - \boldsymbol{v}_s)$ associated with the calculated bulging (shown amplified at the top) occurring between two successive rolls during the continuous casting of steel. The continuous lines show isovolume fractions of solid. (From Kajitani et al. [21].)

in more detail in Chap. 13 (Sect. 13.3.3). This model can predict the deformations of the solid phase in the mushy zone, i.e., when $g_s < 1$ and $\nabla \boldsymbol{v}_s \neq 0$. Details of the model can be found in the publication by Fachinotti et al. [9] The predicted profile is not symmetric, due to the curvature of the slab and gravity.

14.7 SUMMARY

This chapter has demonstrated that solutal inhomogeneities at the macroscopic scale of a casting arise from the relative movements of the solid and liquid phases. Several sources of macrosegregation have been discussed: (i) solidification shrinkage; (ii) natural or forced convection; (iii) grain movement and (iv) deformation of the solid. The general equations governing macrosegregation could be derived from the average mass and solute balances. We demonstrated that, when the solid is fixed and solidification shrinkage is neglected, liquid flowing in the direction opposite to the thermal gradient gives rise to negative segregation (for $k_0 < 1$) and enhanced solidification. Conversely, liquid flowing in the same direction as the thermal gradient induces positive segregation and, in some cases, there may be even local remelting. When the solid is not fixed, macrosegregation resulting from grain movement or deformation of the solid skeleton depends on the velocity field in both phases, $\boldsymbol{v}_\ell$ and $\boldsymbol{v}_s$, as well as on the solute gradient in the liquid, ∇C_ℓ.

14.8 EXERCISES

Exercise 14.1. Macrosegregation during planar front growth with $k_0 > 1$.
Analyze the convection-segregation encountered during directional solidification of a binary alloy with $k_0 > 1$, growing with a planar front, for each of the four situations shown in Fig. 14.3: $\beta_C \gtrless 0$ and $G_z g_z \gtrless 0$.

Exercise 14.2. Average composition over a two-phase domain.
A specimen is held for a long period of time in the semi-solid state, so that the composition of the solid and liquid phases can be considered uniform. After quenching, the microstructure consists of the primary solid phase, called "s", and of the quenched liquid, called "ℓ". An EDX measurement of the composition is then made randomly at the surface of the specimen. This implies that the chance of the EDX measurement to hit the solid is given by the volume fraction of solid g_s. Show that the average of many composition measurements is given by:

$$\langle C \rangle = g_s C_s + g_\ell C_\ell = \frac{f_s C_s \rho_\ell + f_\ell C_\ell \rho_s}{f_s \rho_\ell + f_\ell \rho_s} \neq \langle C \rangle_M$$

Exercise 14.3. Mean weight composition in Flemings' criterion.
Using the Gulliver-Scheil microsegregation model and the steady state relationship between liquid fraction and liquid concentration (Eq. 14.31), show that the mean composition $\langle C \rangle_M$ during solidification is given by Eq. (14.32):

$$\langle C \rangle_M = C_0 \frac{g_\ell^{k_0}(\rho_\ell - \rho_s) + \rho_s}{\rho_s(1 - g_\ell) + \rho_\ell g_\ell} \tag{14.48}$$

Start from Eq. (14.21), replace $v_{\ell x}$ by $-\beta v_T$ and integrate the equation knowing that $v_T \partial/\partial x = \partial/\partial t$ under steady state conditions. The mean volumetric concentration $\langle \rho C \rangle$ has to be integrated first using the Gulliver-Scheil relation.

Exercise 14.4. Flemings' criterion with lever rule.
Assuming lever rule for microsegregation, show that Flemings' criterion for macrosegregation (Eqs. (14.19) and (14.22)) becomes:

$$\frac{dg_\ell}{g_\ell} = -\frac{\rho_\ell}{\rho_s(1 - k_0)} \left(1 + k_0 \frac{\rho_s g_s}{\rho_\ell g_\ell} - \frac{\mathbf{v}_\ell \cdot \nabla T}{\mathbf{v}_T \cdot \nabla T} \right) \frac{dC_\ell}{C_\ell}$$

Exercise 14.5. Flemings' criterion with lever rule: steady state.
Under 1D steady state conditions, for which $\mathbf{v}_\ell \cdot \nabla T = -\beta \mathbf{v}_T \cdot \nabla T$, show that Flemings' criterion derived in the previous exercise recovers the lever rule:

$$C_\ell = \frac{C_0}{g_\ell(1 - k_0) + k_0} \tag{14.49}$$

Start with Eq. (14.21), then show that the mean weight composition $\langle C \rangle_M$ is given by

$$\frac{\langle C \rangle_M}{C_0} = \frac{\rho_\ell g_\ell + k_0 \rho_s (1 - g_\ell)}{(\rho_s(1 - g_\ell) + \rho_\ell g_\ell)(g_\ell(1 - k_0) + k_0)} \tag{14.50}$$

or:

$$\frac{\langle \rho C \rangle}{C_0} = \frac{\rho_\ell g_\ell + k_0 \rho_s (1 - g_\ell)}{g_\ell(1 - k_0) + k_0} \tag{14.51}$$

Compare and discuss this result, illustrated in Fig. 14.30, with that obtained using the Gulliver-Scheil model shown in Fig. 14.8.

Exercise 14.6. Composition at the mold surface.
When using the Gulliver-Scheil model, Flemings' criterion for macrosegregation for $\mathbf{v}_\ell = 0$ gives Eq. (14.29) for the liquid composition evolution at the surface of the mould. Using this relationship, show that the final composition of the solid is given by Eq. (14.30), i.e.,

$$\frac{\langle C \rangle_M}{C_0} = k_0 \left(\frac{\rho_s}{\rho_\ell}(k_0 - 1) + 1 \right)^{-1} \tag{14.52}$$

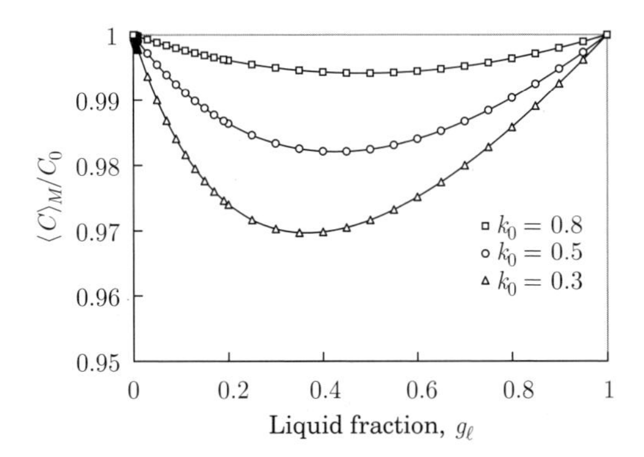

Fig. 14.30 Mean weight composition $\langle C \rangle_M$ normalized with the nominal composition C_0 during steady state directional solidification as a function of the fraction of liquid g_ℓ for the lever rule approximation and three values of the partition coefficient.

Exercise 14.7. Fluid flow-induced macrosegregation in a ternary alloy.

Using the same assumptions made in Sect. 14.5.2 to derive the governing equations for macrosegregation in a binary alloy ($\rho_\ell = \rho_s$, $\boldsymbol{v}_s = 0$, lever rule for all solute species), extend the equations of macrosegregation to a simple ternary alloy. Point out the differences between the ternary and binary alloy cases.

Exercise 14.8. Grain movement- and deformation-induced macrosegregation.

Using the definition of the derivative $\mathrm{D}^s\langle\cdot\rangle/\mathrm{D}t = \partial\langle\cdot\rangle/\partial t + \boldsymbol{v}_s \cdot \langle\cdot\rangle$ defined in Sect. 14.6, derive Eq. (14.43) from the average mass balance (Eq. (14.22)) and the average solute balance (Eq. (14.21)). Under the assumption $\rho_\ell = \text{cst}$, derive Eq. (14.44). Then combine this result with Eq. (14.21) to finally obtain the various contributions to macrosegregation given in Eq. (14.45).

Exercise 14.9. Compression of a mushy zone.

Sketch the macrosegregation profile in the two following situations, pointing out the differences between them:

- The solidification of a domain with a planar front up to a given volume fraction of solid g_s,
- The solidification of the same domain with a dendritic morphology up to the same volume fraction of solid, followed by a compression of the mushy zone into a fully dense solid with the liquid outside.

Assume the temperature of the domain to be uniform in both cases.

14.9 REFERENCES

[1] N. Ahmad, H. Combeau, J.-L. Desbiolles, T. Jalanti, G. Lesoult, M. Rappaz, and C. Stomp. Numerical simulation of macrosegregation: A comparison between finite volume method and finite element method predictions and a confrontation with experiments. *Metall. Mater. Trans.*, 29A:617-630, 1997.

[2] C. Beckermann, J. P. Gu, and W. J. Boettinger. Development of a freckle predictor via Rayleigh number method for single-crystal nickel-base superalloy castings. *Metall. Mater. Trans*, 31A:2545-2557, 2000.

[3] C. Beckermann and C. Y. Wang. Multi-phase/-scale modeling of transport phenomena in alloy solidification. In C. L. Tien, editor, *Annual Review of Heat Transfer VI*, volume 6, pages 115-198. Begell House, 1995.

[4] C. Beckermann. Modeling of macrosegregation: Applications and future needs. *Intl. Mater. Rev.*, 34:243-261, 2002.

[5] M. Bellet and V. D. Fachinotti. A two-phase two-dimensional finite element thermomechanics and macrosegregation model of the mushy zone. Application to continuous casting. In Ch-A. Gandin and M. Bellet, editors, *Modeling of Casting, Welding and Advanced Solidification Processes - XI*, pages 169-176. TMS-AIME, 2006.

[6] H. Combeau, B. Appolaire, and G. Lesoult. Recent progress in understanding and prediction of macro and mesosegregations. In B. G. Thomas and C. Beckermann, editors, *Modeling of Casting, Welding and Advanced Solidification Processes - VIII*, pages 245-256. TMS-AIME, 1998.

[7] H. Combeau and G. Lesoult. Simulation of freckle formation and related segregation during directional solidification of metallic alloys. In T. S. Piwonka et al., editor, *Modeling of Casting, Welding and Advanced Solidification Processes - VI*, pages 201-208. TMS-AIME, 1993.

[8] H. Combeau, M. Zaloznik, B. Rabia, S. Charmond, S. Hans, and P. E. Richy. Prediction of the macrosegregation in steel ingots: influence of the motion and the growth of the equiaxed grains. In P. D. Lee, A. Mitchell, J. P. Bellot, and A. Jardy, editors, *Liquid Metal Processing and Casting*, page 127, 2007.

[9] V. D. Fachinotti, S. Le Corre, N. Triolet, M. Bobadilla, and M. Bellet. Two-phase thermomechanical and macrosegregation modelling of binary alloys solidification with emphasis on the secondary cooling stage of steel slab continuous casting. *Intl. J. Num. Meth. Eng.*, 67:1341-1384, 2006.

[10] S. D. Felicelli, D. R. Poirier, and J. C. Heinrich. Modeling of freckle formation in three dimensions during solidification of multicomponent alloys. *Metall. Mater. Trans.*, 29B: 847-855, 1998.

[11] M. C. Flemings, R. Mehrabian, and G. E. Nereo. Macrosegregation: Part II. *Trans. AIME*, 242:41-49, 1968.

[12] M. C. Flemings and G. E. Nereo. Macrosegregation: Part I. *Trans. AIME*, 239:1449-1461, 1967.

[13] M. C. Flemings and G. E. Nereo. Macrosegregation: Part III. *Trans. AIME*, 242:50-55, 1968.

[14] A. F. Giamei and B. H. Kear. On the nature of freckles in nickel base superalloys. *Metall. Trans.*, 1:2185-2192, 1970.

[15] S. Gouttebroze, W. Liu, H. Combeau, and M. Bellet. 2D/3D simulation of macrosegregation: A comparison between codes on a small cavity and on a large ingot. In Ch-A. Gandin and M. Bellet, editors, *Modeling of Casting, Welding and Advanced Solidification Processes - XI*, pages 227-234. TMS-AIME, 2006.

[16] G. Guillemot, Ch-A. Gandin, and H. Combeau. Modeling of macrosegregation and solidification grain structures with a coupled cellular automaton - finite element model. *ISIJ Intern.*, 46:880-895, 2006.

[17] J. P. Gu, C. Beckermann, and A. F. Giamei. Motion and remelting of dendrite fragments during directional solidification of a nickel-base superalloy. *Metall. Mater. Trans.*, 28A (7):1533-1542, 1997.

[18] J. P. Gu and C. Beckermann. Simulation of convection and macrosegregation in a large steel ingot. *Metall. Mater. Trans.*, 30A(5):1357-1366, 1999.

[19] D. J. Hebditch and J. D. Hunt. Observations of ingot macrosegregation on model systems. *Metall. Trans.*, 5(7):1557-1564, 1974.

[20] T. Kaempfer and M. Rappaz. Modeling of macrosegregation during solidification processes using an adaptive domain decomposition method. *Modeling Simul. Mater. Sci. Eng.*, 11:575-597, 2003.

[21] T. Kajitani, J.-M. Drezet, and M. Rappaz. Numerical simulation of deformation-induced segregation in continuous casting of steel. *Met. Mater. Trans.*, 32A:1479-1488, 2001.

[22] F. Kohler. *Peritectic solidification of Cu-Sn: Microstructure competition at low speed.* PhD thesis No 4037, EPFL, Station 12, 1015 Lausanne, Switzerland, 2008.

[23] G. Lesoult, Ch-A. Gandin, and N. T. Niane. Segregation during solidification with spongy deformation of the mushy zone. *Acta Mater.*, 51:5263-83, 2003.

[24] S. Liu and A. Hellawell. Experiments with constrained chimney-plume flows in the system ammonium chloride-water: Comparison with the unconstrained case. *J. Fluid Mech.*, 388:21-48, 1999.

[25] K. Miyazawa and K. Schwerdtfeger. Macrosegregation in continuously cast steel slabs: Preliminary theoretical investigation on the effect of steady state bulging. *Arch. Eisenhüttenwes.*, 52:415-422, 1981.

[26] L. C. Nicolli, A. Mo, and M. M'Hamdi. Modeling of macrosegregation caused by volumetric deformation in a coherent mushy zone. *Metall. Mater. Trans.*, 36A:433-442, 2005.

[27] L. Orgéas, J.-P. Gabathuler, T. Imwinkelried, C. Paradies, and M. Rappaz. Modelling of semi-solid processing using a modified temperature-dependent power-law model. *Modelling Simul. Mater. Sci. Eng.*, 11:533-574, 2003.

[28] S. D. Ridder, S. Kou, and R. Mehrabian. Effect of fluid flow on macrosegregation in axi-symmetric ingots. *Metall. Trans.*, 12B:435-447, 1981.

[29] P. Rousset, M. Rappaz, and B. Hannart. Modeling of inverse segregation and porosity formation in directionally solidified aluminum-alloys. *Metall. Mater. Trans.*, 26A(9): 2349-2358, 1995.

[30] J. R. Sarazin and A. Hellawell. Channel formation in Pb-Sn, Pb-Sb and Pb-Sn-Sb alloy ingots and comparison with the system NH_4Cl-H_2O. *Metall. Trans.*, 19A, 1988.

[31] M. C. Schneider and C. Beckermann. Formation of macrosegregation by multicomponent thermosolutal convection during solidification of steel. *Metall. Mater. Trans.*, 26A:2373-2388, 1995.

[32] R. S. Steube and A. Hellawell. The use of a transparent aqueous analogue to demonstrate the development of segregation channels during alloy solidification, vertically upwards. *Intl. Video J. Engng Res.*, 3:1-16, 1993.

[33] H. J. Thevik, A. Mo, and T. Rusten. A mathematical model for surface segregation in aluminum direct chill casting. *Metall. Mater. Trans.*, 30B:135-142, 1999.

[34] C. Y. Wang and C. Beckermann. Equiaxed dendritic solidification with convection: 2. Numerical simulations for an Al-4wt% Cu alloy. *Metall. Mater. Trans.*, 27A(12):2765-2783, 1996.

[35] M. G. Worster. Instabilities of the liquid and mushy regions during solidification of alloys. *J. Fluid Mech.*, 237:649-669, 1992.

INDEX